DATABASE SYSTEMS

Design, Implementation, and Management

12e

Carlos Coronel | Steven Morris

CENGAGE
Learning

Australia • Brazil • Mexico • Singapore • United Kingdom • United States

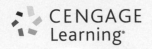
CENGAGE
Learning®

Database Systems: Design, Implementation, and Management, 12th Edition

Carlos Coronel and Steven Morris

Vice President, General Manager: Science, Math & Quantitative Business: Balraj S. Kalsi

Product Director: Mike Schenk

Sr. Product Team Manager: Joe Sabatino

Content Development Manager: Jennifer King

Content Developer: Ted Knight

Product Assistant: Adele Scholtz

Marketing Director: Michele McTighe

Content Project Manager: Nadia Saloom

Media Developer: Chris Valentine

Manufacturing Planner: Ron Montgomery

Marketing Communications Manager: Dan Murphy

Production Service: Cenveo Publisher Services

Senior Art Director: Michelle Kunkler

Cover and Internal Designer: Tippy McIntosh

Cover Art Credit: agsandrew/iStock/ Getty Images Plus/Getty Images

Internal Design Image: silver tiger/ Shutterstock

Intellectual Property
    Analyst: Christina Ciaramella
    Project Manager: Kathryn Kucharek

For product information and technology assistance, contact us at **Cengage Learning Customer & Sales Support, 1-800-354-9706**

For permission to use material from this text or product, submit all requests online at **www.cengage.com/permissions** Further permissions questions can be emailed to **permissionrequest@cengage.com**

Screenshots for this book were created using Microsoft Access® and Visio® and were used with permission from Microsoft. Microsoft and the Office logo are either registered trademarks or trademarks of Microsoft Corporation in the United States and/or other countries.

Oracle is a registered trademark, and Oracle12c and MySQL are trademarks of Oracle Corporation.

iPhone, iPad, and iPod are registered trademarks of Apple Inc.

Library of Congress Control Number: 2015955694

Student Edition ISBN: 978-1-305-62748-2

Loose Leaf Edition ISBN: 978-1-305-86679-9

Cengage Learning
20 Channel Center Street
Boston, MA 02210
USA

Cengage Learning is a leading provider of customized learning solutions with employees residing in nearly 40 different countries and sales in more than 125 countries around the world. Find your local representative at **www.cengage.com.**

Cengage Learning products are represented in Canada by Nelson Education, Ltd.

To learn more about Cengage Learning Solutions, visit **www.cengage.com**

Purchase any of our products at your local college store or at our preferred online store **www.cengagebrain.com**

Printed in the United States of America
Print Number: 02   Print Year: 2017

Dedication

To the treasures in my life: To Victoria, for 26 wonderful years. Thank you for your unending support, for being my angel, my sweetie, and most importantly, my best friend. To Carlos Anthony, who is an awesome older brother to all. Thank you for your words of wisdom, hard-working attitude, and for giving us reasons to be happy. You are still young; your best times are still to come. To Gabriela Victoria, who is the image of brilliance, beauty, and faithfulness. Thank you for being the sunshine in my cloudy days. Your future is bright and endless. To Christian Javier, who is smarter than of all of us. Thank you for being the youthful reminder of life's simple beauties. Keep challenging yourself to new highs. To my parents, Sarah and Carlos, thank you for your sacrifice and example. To all of you, you are all my inspiration. "TQTATA."

Carlos Coronel

To Pamela, from high school sweetheart through 26 years of marriage, you are the beautiful love of my life who has supported, encouraged, and inspired me. More than anyone else, you are responsible for whatever successes I have achieved. To my son, Alexander Logan, your depth of character is without measure. You are my pride and joy. To my daughter, Lauren Elizabeth, your beauty and intensity take my breath away. You are my heart and soul. Thank you all for the sacrifices you have made that enabled me to pursue this dream. I love you so much more than I can express. To my mother, Florence Maryann, and to the memory of my father, Alton Lamar, together they instilled in me the desire to learn and the passion to achieve. To my mother-in-law, Connie Duke, and to the memory of my father-in-law, Wayne Duke, they taught me to find joy in all things. To all of you, with all my love, I dedicate this book.

Steven Morris

For Peter

To longtime colleague and friend, Peter Rob: Your drive and dedication to your students started this book. Your depth of knowledge, attention to detail, and pursuit of excellence made it succeed. Your patience and guidance continue to light our path. It is our sincere hope that, as we move forward, we can continue to live up to your standard. Enjoy your retirement, my friend; you have surely earned it.

Carlos Coronel and Steven Morris

Brief Contents

The following appendixes are included on the Instructor and Student Companion Sites at *www.cengagebrain.com*.

Appendix A1: Designing Databases with Visio Professional 2010: A Tutorial

Appendix A2: Designing Databases with Visio 2013: A Tutorial

Appendix B: The University Lab: Conceptual Design

Appendix C: The University Lab: Conceptual Design Verification, Logical Design, and Implementation

Appendix D: Converting an ER Model into a Database Structure

Appendix E: Comparison of ER Model Notations

Appendix F: Client/Server Systems

Appendix G: Object-Oriented Databases

Appendix H: Unified Modeling Language (UML)

Appendix I: Databases in Electronic Commerce

Appendix J: Web Database Development with ColdFusion

Appendix K: The Hierarchical Database Model

Appendix L: The Network Database Model

Appendix M: MS Access Tutorial

Appendix N: Creating a New Database Using Oracle 12c

Appendix O: Data Warehouse Implementation Factors

Contents

Part 3: **Advanced Design and Implementation** 245

Chapter 7: Introduction to Structured Query Language (SQL) **246**

Chapter 8: Advanced SQL **340**

The following appendixes are included on the Instructor and Student Companion Sites at *www.cengagebrain.com*.

It is our great pleasure to present the twelfth edition of *Database Systems*. We are grateful and humbled that so many of our colleagues around the world have chosen this text to support their classes. We wrote the first edition of this book because we wanted to explain the complexity of database systems in a language that was easy for students to understand. Over the years, we have maintained this emphasis on reaching out to students to explain complex concepts in a practical, approachable manner. This book has been successful through eleven editions because the authors, editors, and the publisher paid attention to the impact of technology and to adopter questions and suggestions. We believe that this twelfth edition successfully reflects the same attention to such factors.

In many respects, rewriting a book is more difficult than writing it the first time. If the book is successful, as this one is, a major concern is that the updates, inserts, and deletions will adversely affect writing style and continuity of coverage. The combination of superb reviewers and editors, plus a wealth of feedback from adopters and students of the previous editions, helped make this new edition the best yet.

Changes to The Twelfth Edition

In this twelfth edition, we added some new features and reorganized some coverage to provide a better flow of material. Aside from enhancing the already strong coverage of database design, we made other improvements in the topical coverage. In particular, the continued growth of Big Data and NoSQL technologies have challenged the status quo in the database industry. Therefore, we created an entire new chapter, Big Data Analytics and NoSQL, to help students grasp the key aspects of these complex new technologies and challenges. The twelfth edition also presents a major step forward in the integration of digital content with the text through online, automatically graded exercises to improve student outcomes. Here are a few of the highlights of changes in the twelfth edition:

- New coverage of Big Data challenges beyond the traditional 3 Vs

- Expanded coverage of Hadoop, the Hadoop Distributed File System (HDFS), and MapReduce

- Updated coverage of cloud data services and their impact on DBAs

- Expanded coverage of NoSQL databases, including key-value databases, document databases, column-oriented database, and graph databases

- New coverage of the emerging NewSQL technologies

- Improved coverage of data visualization

- Added coverage of new sequence and identity capabilities in Oracle and SQL Server

- Complete redesign of the look and feel of the text and layout to improve readability and visual appeal

- Embedded key term definitions within the text

This twelfth edition continues to provide a solid and practical foundation for the design, implementation, and management of database systems. This foundation is built on the notion that, while databases are very practical, their successful creation depends on understanding the important concepts that define them. It's not easy to come up with the proper mix of theory and practice, but the previously mentioned feedback suggests that we largely succeeded in our quest to maintain the proper balance.

The Approach: A Continued Emphasis On Design

As the title suggests, ***Database Systems: Design, Implementation, and Management*** covers three broad aspects of database systems. However, for several important reasons, special attention is given to database design.

- The availability of excellent database software enables people with little experience to create databases and database applications. Unfortunately, the "create without design" approach usually paves the road to a number of database disasters. In our experience, many database system failures are traceable to poor design and cannot be solved with the help of even the best programmers and managers. Nor is better DBMS software likely to overcome problems created or magnified by poor design. Even the best bricklayers and carpenters can't create a good building from a bad blueprint.

- Most vexing problems of database system management seem to be triggered by poorly designed databases. It hardly seems worthwhile to use scarce resources to develop excellent database management skills merely to use them on crises induced by poorly designed databases.

- Design provides an excellent means of communication. Clients are more likely to get what they need when database system design is approached carefully and thoughtfully. In fact, clients may discover how their organizations really function once a good database design is completed.

- Familiarity with database design techniques promotes understanding of current database technologies. For example, because data warehouses derive much of their data from operational databases, data warehouse concepts, structures, and procedures make more sense when the operational database's structure and implementation are understood.

Because the practical aspects of database design are stressed, we have covered design concepts and procedures in detail, making sure that the numerous end-of-chapter problems and cases are sufficiently challenging so students can develop real and useful design skills. We also make sure that students understand the potential and actual conflicts between database design elegance, information requirements, and transaction processing speed. For example, it makes little sense to design databases that meet design elegance standards while they fail to meet end-user information requirements. Therefore, we explore the use of carefully defined trade-offs to ensure that the databases meet end-user requirements while conforming to high design standards.

Topical Coverage

The Systems View

The book's title begins with ***Database Systems***. Therefore, we examine the database and design concepts covered in Chapters 1–6 as part of a larger whole by placing them within the systems analysis framework of Chapter 9. Database designers who fail to understand that the database is part of a larger system are likely to overlook important design requirements. In fact, Chapter 9, Database Design, provides the map for the advanced database design developed in Appendixes B and C. Within the larger systems framework, we can also explore issues such as transaction management and concurrency control (Chapter 10), distributed database management systems (Chapter 12), business intelligence and data warehouses (Chapter 13), database connectivity and web technologies (Chapter 15), and database administration and security (Chapter 16).

PART 1

Database Concepts

1 Database Systems

2 Data Models

Chapter 9

Database Design

In this chapter, you will learn:
- That a sound database design is the foundation for a successful information system, and that the database design must reflect the information system of which the database is a part
- That successful information systems are developed within a framework known as the Systems Development Life Cycle (SDLC)
- That within the information system, the most successful databases are subject to frequent evaluation and revision within a framework known as the Database Life Cycle (DBLC)
- How to conduct evaluation and revision within the SDLC and DBLC frameworks
- About database design strategies: top-down versus bottom-up design and centralized versus decentralized design

Preview

Databases are a part of a larger picture called an information system. Database designs that fail to recognize this fact are not likely to be successful. Database designers must recognize that the database is a critical means to an end rather than an end in itself. Managers want the database to serve their management needs, but too many databases seem to force managers to alter their routines to fit the database requirements.

Information systems don't just happen; they are the product of a carefully staged development process. Systems analysis is used to determine the need for an information system and to establish its limits. Within systems analysis, the actual information system is created through a process known as systems development.

The creation and evolution of information systems follows an iterative pattern called the Systems Development Life Cycle (SDLC), which is a continuous process of creation, maintenance, enhancement, and replacement of the information system. A similar cycle applies to databases: the database is created, maintained, enhanced, and eventually replaced. The Database Life Cycle (DBLC) is carefully traced in this chapter, and is shown in the context of the larger Systems Development Life Cycle.

At the end of the chapter, you will be introduced to some classical approaches to database design: top-down versus bottom-up and centralized versus decentralized.

Data Files Available on cengagebrain.com

Note

Because it is purely conceptual, this chapter does not reference any data files.

Database Design

The first item in the book's subtitle is *Design*, and our examination of database design is comprehensive. For example, Chapters 1 and 2 examine the development and future of databases and data models, and illustrate the need for design. Chapter 3 examines the details of the relational database model; Chapter 4 provides extensive, in-depth, and practical database design coverage; and Chapter 5 explores advanced database design topics. Chapter 6 is devoted to critical normalization issues that affect database efficiency and effectiveness. Chapter 9 examines database design within the systems framework and maps the activities required to successfully design and implement the complex, real-world database developed in Appendixes B and C. Appendix A, Designing Databases with Visio Professional: A Tutorial, provides a good introductory tutorial for the use of a database design tool.

Because database design is affected by real-world transactions, the way data is distributed, and ever-increasing information requirements, we examine major database features that must be supported in current-generation databases and models. For example, Chapter 10, Transaction Management and Concurrency Control, focuses on the characteristics of database transactions and how they affect database integrity and consistency. Chapter 11, Database Performance Tuning and Query Optimization, illustrates the need for query efficiency in a world that routinely generates and uses terabyte-size databases and tables with millions of records. Chapter 12, Distributed Database Management Systems, focuses on data distribution, replication, and allocation. In Chapter 13, Business Intelligence and Data Warehouses, we explore the characteristics of databases that are used in decision support and online analytical processing. Chapter 14, Big Data Analytics and NoSQL, explores the challenges of designing nonrelational databases to use vast global stores of unstructured data. Chapter 15, Database Connectivity and Web Technologies, covers the basic database connectivity issues in a web-based data world, development of web-based database front ends, and emerging cloud-based services.

Implementation

The second portion of the subtitle is *Implementation*. We use Structured Query Language (SQL) in Chapters 7 and 8 to show how relational databases are implemented and managed. Appendix M, Microsoft Access Tutorial, provides a quick but comprehensive guide to implementing an MS Access database. Appendixes B and C demonstrate the design of a database that was fully implemented; these appendixes illustrate a wide range of implementation issues. We had to deal with conflicting design goals: design elegance, information requirements, and operational speed. Therefore, we carefully audited the initial design in Appendix B to check its ability to meet end-user needs and establish appropriate implementation protocols. The result of this audit yielded the final design developed in Appendix C. While relational databases are still the appropriate database technology to use in the vast majority of situations, Big Data issues have created an environment in which special

PART 3

Advanced Design and Implementation

7 Introduction to Structured Query Language (SQL)

8 Advanced SQL

9 Database Design

requirements can call for the use of new, nonrelational technologies. Chapter 14, Big Data Analytics and NoSQL, describes the types of data that are appropriate for these new technologies and the array of options available in these special cases. The special issues encountered in an Internet database environment are addressed in Chapter 15, Database Connectivity and Web Technologies, and in Appendix J, Web Database Development with ColdFusion.

Management

The final portion of the subtitle is *Management*. We deal with database management issues in Chapter 10, Transaction Management and Concurrency Control; Chapter 12, Distributed Database Management Systems; and Chapter 16, Database Administration and Security. Chapter 11, Database Performance Tuning and Query Optimization, is a valuable resource that illustrates how a DBMS manages data retrieval. In addition, Appendix N, Creating a New Database Using Oracle 12c, walks you through the process of setting up a new database.

PART 6
Database Administration

16 Database Administration and Security

Teaching Database: A Matter of Focus

Given the wealth of detailed coverage, instructors can "mix and match" chapters to produce the desired coverage. Depending on where database courses fit into the curriculum, instructors may choose to emphasize database design or database management. (See Figure 1.)

The hands-on nature of database design lends itself particularly well to class projects in which students use instructor-selected software to prototype a system that they design for the end user. Several end-of-chapter problems are sufficiently complex to serve as projects, or an instructor may work with local businesses to give students hands-on experience. Note that some elements of the database design track are also found in the database management track, because it is difficult to manage database technologies that are not well understood.

The options shown in Figure 1 serve only as a starting point. Naturally, instructors will tailor their coverage based on their specific course requirements. For example, an instructor may decide to make Appendix I an outside reading assignment and make Appendix A a self-taught tutorial, and then use that time to cover client/server systems or object-oriented databases. The latter choice would serve as a gateway to UML coverage.

FIGURE 1

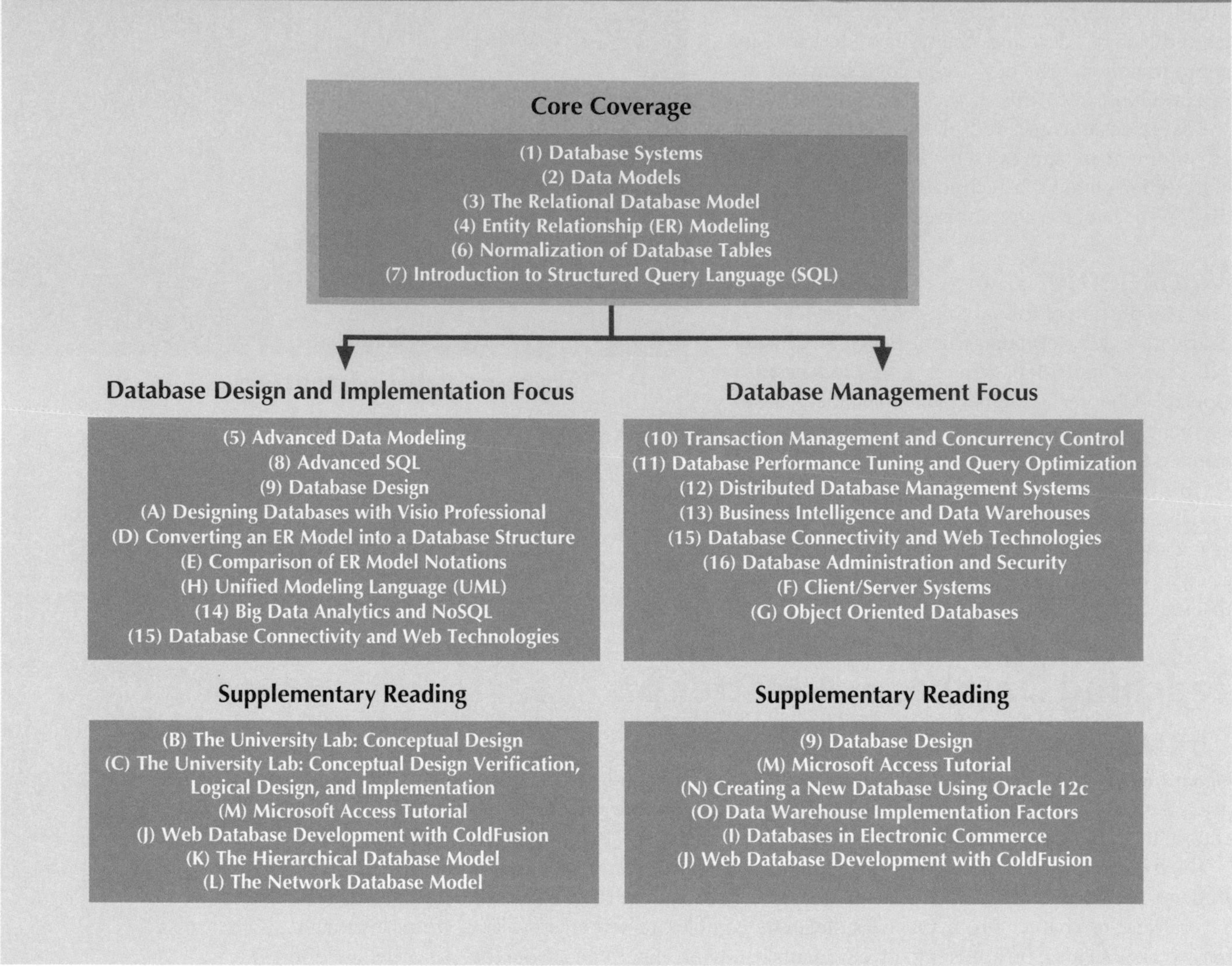

Core Coverage

(1) Database Systems
(2) Data Models
(3) The Relational Database Model
(4) Entity Relationship (ER) Modeling
(6) Normalization of Database Tables
(7) Introduction to Structured Query Language (SQL)

Database Design and Implementation Focus

(5) Advanced Data Modeling
(8) Advanced SQL
(9) Database Design
(A) Designing Databases with Visio Professional
(D) Converting an ER Model into a Database Structure
(E) Comparison of ER Model Notations
(H) Unified Modeling Language (UML)
(14) Big Data Analytics and NoSQL
(15) Database Connectivity and Web Technologies

Database Management Focus

(10) Transaction Management and Concurrency Control
(11) Database Performance Tuning and Query Optimization
(12) Distributed Database Management Systems
(13) Business Intelligence and Data Warehouses
(15) Database Connectivity and Web Technologies
(16) Database Administration and Security
(F) Client/Server Systems
(G) Object Oriented Databases

Supplementary Reading

(B) The University Lab: Conceptual Design
(C) The University Lab: Conceptual Design Verification, Logical Design, and Implementation
(M) Microsoft Access Tutorial
(J) Web Database Development with ColdFusion
(K) The Hierarchical Database Model
(L) The Network Database Model

Supplementary Reading

(9) Database Design
(M) Microsoft Access Tutorial
(N) Creating a New Database Using Oracle 12c
(O) Data Warehouse Implementation Factors
(I) Databases in Electronic Commerce
(J) Web Database Development with ColdFusion

Online Content boxes draw attention to material at *www.cengagebrain.com* for this text and provide ideas for incorporating this content into the course.

Online Content

All of the databases used to illustrate the material in this chapter (see the Data Files list at the beginning of the chapter) are available at *www.cengagebrain.com*. The database names match the database names shown in the figures.

Notes highlights important facts about the concepts introduced in the chapter.

Note

A null is no value at all. It does *not* mean a zero or a space. A null is created when you press the Enter key or the Tab key to move to the next entry without making an entry of any kind. Pressing the Spacebar creates a blank (or a space).

A variety of **four-color figures**, including ER models and implementations, tables, and illustrations, clearly illustrate difficult concepts.

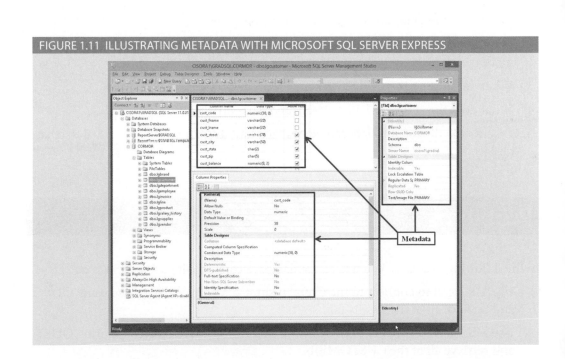

FIGURE 1.11 ILLUSTRATING METADATA WITH MICROSOFT SQL SERVER EXPRESS

Summary

- An information system is designed to help transform data into information and to manage both data and information. Thus, the database is a very important part of the information system. Systems analysis is the process that establishes the need for an information system and its extent. Systems development is the process of creating an information system.

A robust **Summary** at the end of each chapter ties together the major concepts and serves as a quick review for students.

Key Terms

bottom-up design	Database Life Cycle (DBLC)	module coupling
boundaries	database role	physical design
centralized design	decentralized design	scope
clustered tables	description of operations	systems analysis
cohesivity	differential backup	systems development
computer-aided software engineering (CASE)	full backup	Systems Development Life Cycle (SDLC)
	information system	
conceptual design	logical design	top-down design
database development	minimal data rule	transaction log backup
database fragment	module	virtualization

An alphabetic list of **Key Terms** summarizes important terms.

Review Questions

1. What is an information system? What is its purpose?

2. How do systems analysis and systems development fit into a discussion about information systems?

3. What does the acronym SDLC mean, and what does an SDLC portray?

4. What does the acronym DBLC mean, and what does a DBLC portray?

5. Discuss the distinction between centralized and decentralized conceptual database design.

Review Questions challenge students to apply the skills learned in each chapter.

Problems

In the following exercises, you will set up database connectivity using MS Excel.

1. Use MS Excel to connect to the Ch02_InsureCo MS Access database using ODBC, and retrieve all of the AGENTs.

2. Use MS Excel to connect to the Ch02_InsureCo MS Access database using ODBC, and retrieve all of the CUSTOMERs.

Problems become progressively more complex as students draw on the lessons learned from the completion of preceding problems.

MindTap® for Database Systems 12e

MindTap® combines learning tools—such as readings, multimedia, activities, and assessments—into a singular learning path that guides students through the course. You'll find a full ebook as well as a robust set of auto-gradable homework problems. Multiple-choice homework questions developed from the end-of-chapter review questions confirm students' understanding of core concepts and key terms. Higher-level assignments enable students to practice database design concepts in an automated environment, and chapter quizzes help prepare students for exams. Students will also benefit from the chapter-opening videos created by the authors, as well as study tools such as crossword puzzles and key-term flashcards.

MindTap® is designed to be fully integrated with any Learning Management System and can be used as a stand-alone product or in conjunction with a print textbook.

Appendixes

Fifteen online appendixes provide additional material on a variety of important areas, such as using Microsoft® Visio® and Microsoft® Access®, ER model notations, UML, object-oriented databases, databases and electronic commerce, and Adobe® ColdFusion®.

Database, SQL Script, and ColdFusion Files

The online materials for this book include all of the database structures and table contents used in the text. For students using Oracle®, MySQL, and Microsoft SQL Server™, SQL scripts are included to help students create and load all tables used in the SQL chapters (7 and 8). In addition, all ColdFusion scripts used to develop the web interfaces in Appendix J are included.

Instructor Resources

Database Systems: Design, Implementation, and Management, Twelfth Edition, includes teaching tools to support instructors in the classroom. The ancillary material that accompanies the textbook is listed below. They are available on the web at *www.cengagebrain.com.*

Instructor's Manual

The authors have created this manual to help instructors make their classes informative and interesting. Because the authors tackle so many problems in depth, instructors will find the *Instructor's Manual* especially useful. The details of the design solution process are shown in the *Instructor's Manual,* as well as notes about alternative approaches that may be used to solve a particular problem.

SQL Script Files for Instructors

The authors have provided teacher's SQL script files to allow instructors to cut and paste the SQL code into the SQL windows. (Scripts are provided for Oracle, MySQL, and MS SQL Server.) The SQL scripts, which have all been tested by Cengage Learning, are a major convenience for instructors. You won't have to type in the SQL commands, and the use of the scripts eliminates typographical errors that are sometimes difficult to trace.

ColdFusion Files for Instructors

The ColdFusion web development solutions are provided. Instructors have access to a menu-driven system that allows teachers to show the code as well as its execution.

Databases

For many chapters, Microsoft® Access® instructor databases are available that include features not found in the student databases. For example, the databases that accompany Chapters 7 and 8 include many of the queries that produce the problem solutions. Other Access databases, such as the ones that accompany Chapters 3, 4, 5, and 6, include implementations of the design problem solutions to allow instructors to illustrate the effect of design decisions. In addition, instructors have access to all the script files for Oracle, MySQL, and MS SQL Server so that all the databases and their tables can be converted easily and precisely.

Cengage Learning Testing Powered by Cognero

A flexible, online system that allows you to:

- Author, edit, and manage test bank content from multiple Cengage Learning solutions

- Create multiple test versions in an instant

- Deliver tests from your LMS, your classroom, or wherever you want

Start right away!
Cengage Learning Testing Powered by Cognero works on any operating system or browser.

- No special installs or downloads needed

- Create tests from school, home, the coffee shop—anywhere with Internet access

What will you find?

- Simplicity at every step. A desktop-inspired interface features drop-down menus and familiar, intuitive tools that take you through content creation and management with ease.

- Full-featured test generator. Create ideal assessments with your choice of 15 question types (including true/false, multiple-choice, opinion scale/Likert, and essay). Multi-language support, an equation editor, and unlimited metadata help ensure your tests are complete and compliant.

- Cross-compatible capability. Import and export content into other systems.

PowerPoint® Presentations

Microsoft PowerPoint slides are included for each chapter. Instructors can use the slides in a variety of ways—for example, as teaching aids during classroom presentations or as printed handouts for classroom distribution. Instructors can modify these slides or include slides of their own for additional topics introduced to the class.

Figure Files

Figure files for solutions are presented in the *Instructor's Manual* to allow instructors to create their own presentations. Instructors can also manipulate these files to meet their particular needs.

Acknowledgments

Regardless of how many editions of this book are published, they will always rest on the solid foundation created by the first edition. We remain convinced that our work has become successful because that first edition was guided by Frank Ruggirello, a former Wadsworth senior editor and publisher. Aside from guiding the book's development, Frank also managed to solicit the great Peter Keen's evaluation (thankfully favorable) and subsequently convinced Peter Keen to write the foreword for the first edition. Although we sometimes found Frank to be an especially demanding taskmaster, we also found him to be a superb professional and a fine friend. We suspect Frank will still see his fingerprints all over our current work. Many thanks.

A difficult task in rewriting a book is deciding what new approaches, topical coverage, and changes to depth of coverage are appropriate for a product that has successfully weathered the test of the marketplace. The comments and suggestions made by the book's adopters, students, and reviewers play a major role in deciding what coverage is desirable and how that coverage is to be treated.

Some adopters became extraordinary reviewers, providing incredibly detailed and well-reasoned critiques even as they praised the book's coverage and style. Dr. David Hatherly, a superb database professional who is a senior lecturer in the School of Information Technology, Charles Sturt University–Mitchell, Bathhurst, Australia, made sure that we knew precisely what issues led to his critiques. Even better for us, he provided the suggestions that made it much easier for us to improve the topical coverage in earlier editions. All of his help was given freely and without prompting on our part. His efforts are much appreciated, and our thanks are heartfelt.

We also owe a debt of gratitude to Professor Emil T. Cipolla, who teaches at St. Mary College. Professor Cipolla's wealth of IBM experience turned out to be a valuable resource when we tackled the embedded SQL coverage in Chapter 8.

Every technical book receives careful scrutiny by several groups of reviewers selected by the publisher. We were fortunate to face the scrutiny of reviewers who were superbly qualified to offer their critiques, comments, and suggestions—many of which strengthened this edition. While holding them blameless for any remaining shortcomings, we owe these reviewers many thanks for their contributions:

Mubarak Banisaklher, Bethune Cookman University

David Bell, Pacific Union College

Yurii Boreisha, Minnesota State University, Moorhead

Laurie Crawford, Franklin University

Mel Goetting, Shawnee State University

Jeff Guan, University of Louisville

William Hochstettler, Franklin University

Laurene Hutchinson, Louisiana State University, Baton Rouge

Nitin Kale, University of Southern California, Los Angeles

Gerald Karush, Southern New Hampshire University

Michael Kelly, Community College of Rhode Island

Timothy Koets, Grand Rapids Community College

Klara Nelson, The University of Tampa

Chiso Okafor, Roxbury Community College

Brandon Olson, The College of St. Scholastica

James Reneau, Shawnee State University

Julio Rivera, University of Alabama at Birmingham

Ruth Robins, University of Houston, Downtown

Samuel Sambasivam, Azusa Pacific University

Paul Seibert, North Greenville University

Ronghua Shan, Dakota State University

Andrew Smith, Marian University

Antonis Stylianou, University of North Carolina, Charlotte

Brian West, University of Louisiana at Lafayette

Nathan White, McKendree University

In some respects, writing books resembles building construction: When 90 percent of the work seems done, 90 percent of the work remains to be done. Fortunately for us, we had a great team on our side.

- We are deeply indebted to Deb Kaufmann for her help and guidance. Deb has been everything we could have hoped for in a development editor and more. Deb has been our editor for almost all the editions of this book, and the quality of her work shows in the attention to detail and the cohesiveness and writing style of the material in this book.

- After writing so many books and twelve editions of *this* book, we know just how difficult it can be to transform the authors' work into an attractive product. The production team, both at Cengage Learning (Nadia Saloom) and Cenveo Publisher Services (Saravanakumar Dharman), have done an excellent job.

- We also owe Jennifer King and Ted Knight, our Content Developers, special thanks for their ability to guide this book to a successful conclusion.

We also thank our students for their comments and suggestions. They are the reason for writing this book in the first place. One comment stands out in particular: "I majored in systems for four years, and I finally discovered why when I took your course." And one of our favorite comments by a former student was triggered by a question about the challenges created by a real-world information systems job: "Doc, it's just like class, only easier. You really prepared me well. Thanks!"

Special thanks go to a very unique and charismatic gentleman. For over 20 years, Peter Rob has been the driving force behind the creation and evolution of this book. This book originated as a product of his drive and dedication to excellence. For over 22 years, he was the voice of *Database Systems* and the driving force behind its advancement. We wish him peace in his retirement, time with his loved ones, and luck on his many projects.

Last, and certainly not least, we thank our families for their solid support at home. They graciously accepted the fact that during more than a year's worth of rewriting, there would be no free weekends, rare free nights, and even rarer free days. We owe you much, and the dedications we wrote are but a small reflection of the important space you occupy in our hearts.

Carlos Coronel and Steven Morris

PART 1
Database Concepts

Chapter 1

Database Systems

In this chapter, you will learn:

- The difference between data and information
- What a database is, the various types of databases, and why they are valuable assets for decision making
- The importance of database design
- How modern databases evolved from file systems
- About flaws in file system data management
- The main components of the database system
- The main functions of a database management system (DBMS)

Preview

Organizations use data to keep track of their day-to-day operations. Such data is used to generate information, which in turn is the basis for good decisions. Data is likely to be managed most efficiently when it is stored in a database. Databases are involved in almost all facets and activities of our daily lives: from school, to work, to medical care, government, nonprofit organizations, and houses of worship. In this chapter, you will learn what a database is, what it does, and why it yields better results than other data management methods. You will also learn about various types of databases and why database design is so important.

Databases evolved from computer file systems. Although file system data management is now largely outmoded, understanding the characteristics of file systems is important because file systems are the source of serious data management limitations. In this chapter, you will also learn how the database system approach helps eliminate most of the shortcomings of file system data management.

Data Files and Available Formats

	MS Access	Oracle	MS SQL	My SQL		MS Access	Oracle	MS SQL	My SQL
CH01_Text	✓	✓	✓	✓	CH01_Problems	✓	✓	✓	✓
CH01_Design_Example	✓	✓	✓	✓					

Data Files Available on cengagebrain.com

1-1 **Why Databases?**

So, why do we need databases? In today's world, data is ubiquitous (abundant, global, every-where) and pervasive (unescapable, prevalent, persistent). From birth to death, we generate and consume data. The trail of data starts with the birth certificate and continues all the way to a death certificate (and beyond!). In between, each individual produces and consumes enormous amounts of data. As you will see in this book, databases are the best way to store and manage data. Databases make data persistent and shareable in a secure way. As you look at Figure 1.1, can you identify some of the data generated by your own daily activities?

FIGURE 1.1 THE PERVASIVE NATURE OF DATABASES

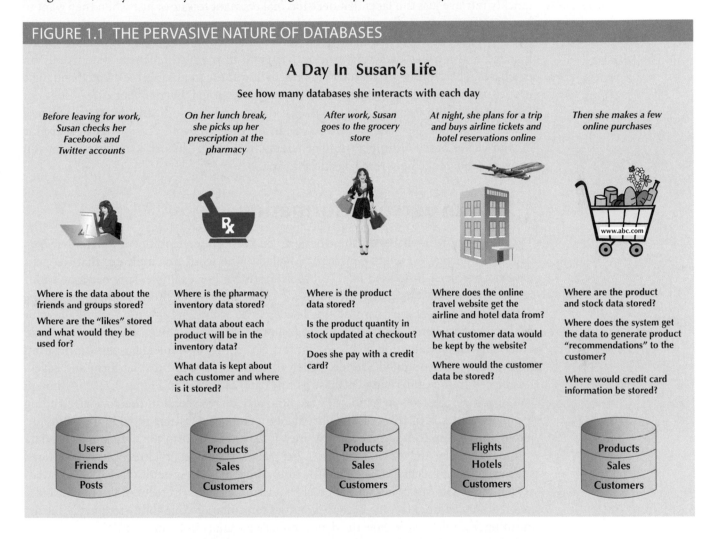

Data is not only ubiquitous and pervasive, it is essential for organizations to survive and prosper. Imagine trying to operate a business without knowing who your customers are, what products you are selling, who is working for you, who owes you money, and to whom you owe money. All businesses have to keep this type of data and much more. Just as important, they must have that data available to decision makers when necessary. It can be argued that the ultimate purpose of all business information systems is to help businesses use information as an organizational resource. At the heart of all of these systems are the collection, storage, aggregation, manipulation, dissemination, and man-agement of data.

Depending on the type of information system and the characteristics of the busi-ness, this data could vary from a few megabytes on just one or two topics to terabytes covering hundreds of topics within the business's internal and external environment.

Telecommunications companies, such as Sprint and AT&T, are known to have systems that keep data on trillions of phone calls, with new data being added to the system at speeds up to 70,000 calls per second! Not only do these companies have to store and manage immense collections of data, they have to be able to find any given fact in that data quickly. Consider the case of Internet search staple Google. While Google is reluctant to disclose many details about its data storage specifications, it is estimated that the company responds to over 91 million searches per day across a collection of data that is several terabytes in size. Impressively, the results of these searches are available almost instantly.

How can these businesses process this much data? How can they store it all, and then quickly retrieve just the facts that decision makers want to know, just when they want to know it? The answer is that they use databases. Databases, as explained in detail throughout this book, are specialized structures that allow computer-based systems to store, manage, and retrieve data very quickly. Virtually all modern business systems rely on databases. Therefore, a good understanding of how these structures are created and their proper use is vital for any information systems professional. Even if your career does not take you down the amazing path of database design and development, databases will be a key component of the systems that you use. In any case, you will probably make decisions in your career based on information generated from data. Thus, it is important that you know the difference between data and information.

1-2 Data versus Information

To understand what drives database design, you must understand the difference between data and information. **Data** consists of raw facts. The word *raw* indicates that the facts have not yet been processed to reveal their meaning. For example, suppose that a university tracks data on faculty members for reporting to accrediting bodies. To get the data for each faculty member into the database, you would provide a screen to allow for convenient data entry, complete with drop-down lists, combo boxes, option buttons, and other data-entry validation controls. Figure 1.2(a) shows a simple data-entry form from a software package named Sedona. When the data is entered into the form and saved, it is placed in the underlying database as raw data, as shown in Figure 1.2(b). Although you now have the facts in hand, they are not particularly useful in this format. Reading through hundreds of rows of data for faculty members does not provide much insight into the overall makeup of the faculty. Therefore, you transform the raw data into a data summary like the one shown in Figure 1.2(c). Now you can get quick answers to questions such as "What percentage of the faculty in the Information Systems (INFS) department are adjuncts?" In this case, you can quickly determine that 20 percent of the INFS faculty members are adjunct faculty. Because graphics can enhance your ability to quickly extract meaning from data, you show the data summary pie chart in Figure 1.2(d).

Information is the result of processing raw data to reveal its meaning. Data processing can be as simple as organizing data to reveal patterns or as complex as making forecasts or drawing inferences using statistical modeling. To reveal meaning, information requires *context*. For example, an average temperature reading of 105 degrees does not mean much unless you also know its context: Is this reading in degrees Fahrenheit or Celsius? Is this a machine temperature, a body temperature, or an outside air temperature? Information can be used as the foundation for decision making. For example, the data summary for the faculty can provide accrediting bodies with insights that are useful in determining whether to renew accreditation for the university.

Keep in mind that raw data must be properly *formatted* for storage, processing, and presentation. For example, dates might be stored in Julian calendar formats within the database, but displayed in a variety of formats, such as day-month-year or month/day/year, for

data
Raw facts, or facts that have not yet been processed to reveal their meaning to the end user.

information
The result of processing raw data to reveal its meaning. Information consists of transformed data and facilitates decision making.

FIGURE 1.2 TRANSFORMING RAW DATA INTO INFORMATION

a) Data entry screen

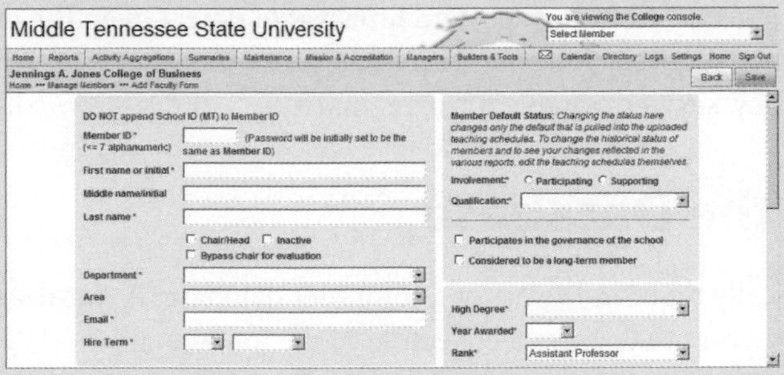

b) Raw data

Id	LastName	MidName	FirstName	DeptCode	Office	Email	Rank	HireYear	Degree
1	Washington	A.	George	MGMT	N135	gwashington@mtsu.edu	Professor	2001	Ph.D.
2	Adams		John	FIN	N313	jadams@mtsu.edu	Professor	1984	Ph.D.
3	Jefferson	L.	Thomas	ECON		tjefferson@mtsu.edu	Instructor	2002	M.B.A.
4	Madison	D.	James	FIN	N236	jmadison@mtsu.edu	Associate Professor	1994	Ph.D.
5	Monroe	N.	James	ACCT	N411	jmonroe@mtsu.edu	Assistant Professor	1995	Ph.D.
6	Adams	Q.	John	ACCT	N418	jqadams@mtsu.edu	Associate Professor	1989	Ph.D.
7	Jackson	C.	Andrew	ECON	N303	ajackson@mtsu.edu	Associate Professor	1999	Ph.D.
8	Van Buren	T.	Martin	FIN	N306	mvanburen@mtsu.edu	Professor	1989	Ph.D.
9	Harrison	R.	William	MKTG	N118	wharrison@mtsu.edu	Professor	1994	Ph.D.
10	Tyler	M.	John	MGMT		Jtyler@mtsu.edu	Assistant Professor	2000	Ed.D.
11	Polk		Cheryl	MKTG	N143	cpolk@mtsu.edu	Associate Professor	2002	Ph.D.
12	Taylor	G.	Zachery	ACCT	N415	ztaylor@mtsu.edu	Associate Professor	1996	Ph.D.
13	Fillmore		Millard	JCB	N219	mfillmore@mtsu.edu	Professor	1992	Ph.D.
14	Pierce	A.	Franklin	MKTG	N359	pfranklin@mtsu.edu	Instructor	2005	M.B.A.
15	Buchanan	T.	James	MGMT	N146	jbuchanan@mtsu.edu	Associate Professor	1996	D.B.A.
16	Lincoln	W.	Larry	MGMT	N150	llincoln@mtsu.edu	Associate Professor	1996	Ph.D.
18	Johnson		Andrew	ISYS	N360	ajohnson@mtsu.edu	Professor	1987	Ph.D.
19	Grant		Katie	MKTG	N120	kgrant@mtsu.edu	Assistant Professor	1989	D.B.A.
20	Rutherford		Hayes	ACCT	N408	hrutherford@mtsu.edu	Professor	1992	Ph.D.
21	Grafield	T.	Denise	ACCT		dgarfield@mtsu.edu	Assistant Professor	2018	Ph.D.
22	Arthur		Emily	ACCT	N413	earthur@mtsu.edu	Associate Professor	2003	J.D.
23	Cleveland	G.	Robert	ACCT	N401	rcleveland@mtsu.edu	Associate Professor	1997	Ph.D.
24	Harrison	X	Patricia	BULA	N406	pharrison@mtsu.edu	Associate Professor	2001	J.D.
25	McKinley	B.	Priscilla	ISYS	N363	pmckinley@mtsu.edu	Adjunct	1994	M.S.
26	Roosevelt	F.	Hillary	MGMT	N104	hroosevelt@mtsu.edu	Associate Professor	2002	Ph.D.
27	Wilson		Laura	BCEN	N448	lwilson@mtsu.edu	Professor	1992	Ph.D.
28	Harding		Warren	MKTG	N114	wharding@mtsu.edu	Professor	1984	Ed.D.
29	Coolidge		Calvin	ECON	N316	ccoolidge@mtsu.edu	Professor	1975	Ph.D.
30	Hoover		Lisa	MGMT		lhoover@mtsu.edu	Adjunct	1978	M.B.A.
31	Truman		Betty	ACCT	N416	btruman@mtsu.edu	Professor	1971	Ed.D.
32	Johnson		Robert	BCEN	N240	rjohnsonr@mtsu.edu	Professor	2001	Ph.D.

c) Information in summary format

Rank	COUNT	%/INFS	TOT/COL	%/COL. TOT.	%/COL. FAC.
Adjunct	5	20.00%	23	21.74%	3.27%
Assistant Professor	2	8.00%	28	7.14%	1.31%
Associate Professor	9	36.00%	37	24.32%	5.88%
Instructor	2	8.00%	18	11.11%	1.31%
Professor	7	28.00%	47	14.89%	4.58%

d) Information in graphical format

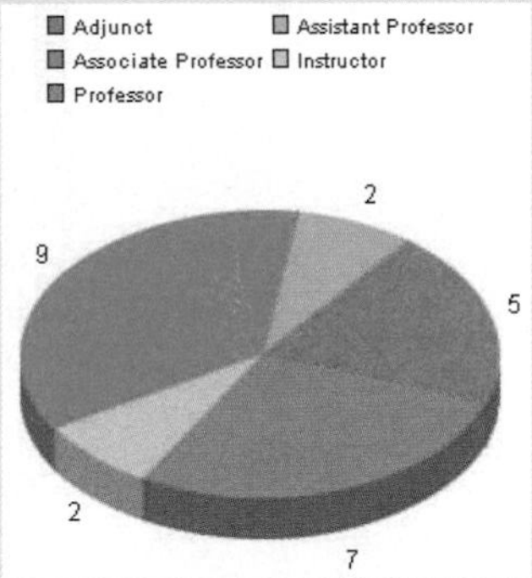

different purposes. Respondents' yes/no responses might need to be converted to a Y/N or 0/1 format for data storage. More complex formatting is required when working with complex data types, such as sounds, videos, or images.

In this "information age," production of accurate, relevant, and timely information is the key to good decision making. In turn, good decision making is the key to business survival in a global market. We are now said to be entering the "knowledge age."[1]

Data is the foundation of information, which is the bedrock of **knowledge**—that is, the body of information and facts about a specific subject. Knowledge implies familiarity, awareness, and understanding of information as it applies to an environment. A key characteristic of knowledge is that "new" knowledge can be derived from "old" knowledge.

Let's summarize some key points:

- Data constitutes the building blocks of information.

- Information is produced by processing data.

- Information is used to reveal the meaning of data.

- Accurate, relevant, and timely information is the key to good decision making.

- Good decision making is the key to organizational survival in a global environment.

knowledge
The body of information and facts about a specific subject. Knowledge implies familiarity, awareness, and understanding of information as it applies to an environment. A key characteristic is that new knowledge can be derived from old knowledge.

[1] Peter Drucker coined the phrase "knowledge worker" in 1959 in his book *Landmarks of Tomorrow*. In 1994, Esther Dyson, George Keyworth, and Dr. Alvin Toffler introduced the concept of the "knowledge age."

Timely and useful information requires accurate data. Such data must be properly generated and stored in a format that is easy to access and process. In addition, like any basic resource, the data environment must be managed carefully. **Data management** is a discipline that focuses on the proper generation, storage, and retrieval of data. Given the crucial role that data plays, it should not surprise you that data management is a core activity for any business, government agency, service organization, or charity.

1-3 **Introducing the Database**

Efficient data management typically requires the use of a computer database. A **database** is a shared, integrated computer structure that stores a collection of the following:

- End-user data—that is, raw facts of interest to the end user

- **Metadata**, or data about data, through which the end-user data is integrated and managed

The metadata describes the data characteristics and the set of relationships that links the data found within the database. For example, the metadata component stores information such as the name of each data element, the type of values (numeric, dates, or text) stored on each data element, and whether the data element can be left empty. The metadata provides information that complements and expands the value and use of the data. In short, metadata presents a more complete picture of the data in the database. Given the characteristics of metadata, you might hear a database described as a "collection of *self-describing* data."

A **database management system (DBMS)** is a collection of programs that manages the database structure and controls access to the data stored in the database. In a sense, a database resembles a very well-organized electronic filing cabinet in which powerful software (the DBMS) helps manage the cabinet's contents.

1-3a Role and Advantages of the DBMS

The DBMS serves as the intermediary between the user and the database. The database structure itself is stored as a collection of files, and the only way to access the data in those files is through the DBMS. Figure 1.3 emphasizes the point that the DBMS presents the end user (or application program) with a single, integrated view of the data in the database. The DBMS receives all application requests and translates them into the complex operations required to fulfill those requests. The DBMS hides much of the database's internal complexity from the application programs and users. The application program might be written by a programmer using a programming language, such as Visual Basic. NET, Java, or C#, or it might be created through a DBMS utility program.

Having a DBMS between the end user's applications and the database offers some important advantages. First, the DBMS enables the data in the database *to be shared* among multiple applications or users. Second, the DBMS *integrates* the many different users' views of the data into a single all-encompassing data repository.

Because data is the crucial raw material from which information is derived, you must have a good method to manage such data. As you will discover in this book, the DBMS helps make data management more efficient and effective. In particular, a DBMS provides these advantages:

- *Improved data sharing.* The DBMS helps create an environment in which end users have better access to more and better-managed data. Such access makes it possible for end users to respond quickly to changes in their environment.

data management
A process that focuses on data collection, storage, and retrieval. Common data management functions include addition, deletion, modification, and listing.

database
A shared, integrated computer structure that houses a collection of related data. A database contains two types of data: end-user data (raw facts) and metadata.

metadata
Data about data; that is, data about data characteristics and relationships. See also *data dictionary*.

database management system (DBMS)
The collection of programs that manages the database structure and controls access to the data stored in the database.

FIGURE 1.3 THE DBMS MANAGES THE INTERACTION BETWEEN THE END USER AND THE DATABASE

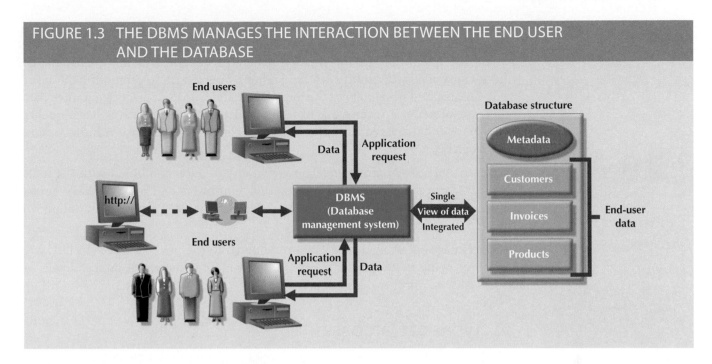

- *Improved data security.* The more users access the data, the greater the risks of data security breaches. Corporations invest considerable amounts of time, effort, and money to ensure that corporate data is used properly. A DBMS provides a framework for better enforcement of data privacy and security policies.

- *Better data integration.* Wider access to well-managed data promotes an integrated view of the organization's operations and a clearer view of the big picture. It becomes much easier to see how actions in one segment of the company affect other segments.

- *Minimized data inconsistency.* **Data inconsistency** exists when different versions of the same data appears in different places. For example, data inconsistency exists when a company's sales department stores a sales representative's name as Bill Brown and the company's personnel department stores that same person's name as William G. Brown, or when the company's regional sales office shows the price of a product as $45.95 and its national sales office shows the same product's price as $43.95. The probability of data inconsistency is greatly reduced in a properly designed database.

- *Improved data access.* The DBMS makes it possible to produce quick answers to ad hoc queries. From a database perspective, a **query** is a specific request issued to the DBMS for data manipulation—for example, to read or update the data. Simply put, a query is a question, and an **ad hoc query** is a spur-of-the-moment question. The DBMS sends back an answer (called the **query result set**) to the application. For example, when dealing with large amounts of sales data, end users might want quick answers to questions (ad hoc queries). Some examples include the following:

  - What was the dollar volume of sales by product during the past six months?

  - What is the sales bonus figure for each of our salespeople during the past three months?

  - How many of our customers have credit balances of $3,000 or more?

data inconsistency
A condition in which different versions of the same data yield different (inconsistent) results.

query
A question or task asked by an end user of a database in the form of SQL code. A specific request for data manipulation issued by the end user or the application to the DBMS.

ad hoc query
A "spur-of-the-moment" question.

query result set
The collection of data rows returned by a query.

- *Improved decision making.* Better-managed data and improved data access make it possible to generate better-quality information, on which better decisions are based. The quality of the information generated depends on the quality of the underlying data. **Data quality** is a comprehensive approach to promoting the accuracy, validity, and timeliness of the data. While the DBMS does not guarantee data quality, it provides a framework to facilitate data quality initiatives. Data quality concepts will be covered in more detail in Chapter 16, Database Administration and Security.

- *Increased end-user productivity.* The availability of data, combined with the tools that transform data into usable information, empowers end users to make quick, informed decisions that can make the difference between success and failure in the global economy.

The advantages of using a DBMS are not limited to the few just listed. In fact, you will discover many more advantages as you learn more about the technical details of databases and their proper design.

1-3b Types of Databases

A DBMS can be used to build many different types of databases. Each database stores a particular collection of data and is used for a specific purpose. Over the years, as technology and innovative uses of databases have evolved, different methods have been used to classify databases. For example, databases can be classified by the number of users supported, where the data is located, the type of data stored, the intended data usage, and the degree to which the data is structured.

The number of users determines whether the database is classified as single user or multiuser. A **single-user database** supports only one user at a time. In other words, if user A is using the database, users B and C must wait until user A is done. A single-user database that runs on a personal computer is called a **desktop database**. In contrast, a **multiuser database** supports multiple users at the same time. When the multiuser database supports a relatively small number of users (usually fewer than 50) or a specific department within an organization, it is called a **workgroup database**. When the database is used by the entire organization and supports many users (more than 50, usually hundreds) across many departments, the database is known as an **enterprise database**.

Location might also be used to classify the database. For example, a database that supports data located at a single site is called a **centralized database**. A database that supports data distributed across several different sites is called a **distributed database**. The extent to which a database can be distributed and the way in which such distribution is managed are addressed in detail in Chapter 12, Distributed Database Management Systems.

Both centralized and decentralized (distributed) databases require a well-defined infrastructure (hardware, operating systems, network technologies, etc.) to implement and operate the database. Typically, the infrastructure is owned and maintained by the organization that creates and operates the database. But in recent years, the use of cloud databases has been growing in popularity. A **cloud database** is a database that is created and maintained using cloud data services, such as Microsoft Azure or Amazon AWS. These services, provided by third-party vendors, provide defined performance measures (data storage capacity, required throughput, and availability) for the database, but do not necessarily specify the underlying infrastructure to implement it. The data owner does not have to know, or be concerned about, what hardware and software is being used to support their database. The performance capabilities can be renegotiated with the

data quality
A comprehensive approach to ensuring the accuracy, validity, and timeliness of data.

single-user database
A database that supports only one user at a time.

desktop database
A single-user database that runs on a personal computer.

multiuser database
A database that supports multiple concurrent users.

workgroup database
A multiuser database that usually supports fewer than 50 users or is used for a specific department in an organization.

enterprise database
The overall company data representation, which provides support for present and expected future needs.

centralized database
A database located at a single site.

distributed database
A logically related database that is stored in two or more physically independent sites.

cloud database
A database that is created and maintained using cloud services, such as Microsoft Azure or Amazon AWS.

general-purpose database
A database that contains a wide variety of data used in multiple disciplines.

cloud provider as the business demands on the database change. For example, during the 2012 presidential election in the United States, the Obama campaign used a cloud database hosted on infrastructure capabilities purchased from Amazon. The campaign did not have to buy, install, configure, or maintain any hardware, operating systems, or network devices. It simply purchased storage and processing capacity for its data and applications. As the demands on the database increased, additional processing and storage capabilities could be purchased as needed.

In some contexts, such as research environments, a popular way of classifying databases is according to the type of data stored in them. Using this criterion, databases are grouped into two categories: general-purpose and discipline-specific databases. **General-purpose databases** contain a wide variety of data used in multiple disciplines—for example, a census database that contains general demographic data and the LexisNexis and ProQuest databases that contain newspaper, magazine, and journal articles for a variety of topics. **Discipline-specific databases** contain data focused on specific subject areas. The data in this type of database is used mainly for academic or research purposes within a small set of disciplines. Examples of discipline-specific databases include financial data stored in databases such as CompuStat or CRSP (Center for Research in Security Prices), geographic information system (GIS) databases that store geospatial and other related data, and medical databases that store confidential medical history data.

The most popular way of classifying databases today, however, is based on how they will be used and on the time sensitivity of the information gathered from them. For example, transactions such as product or service sales, payments, and supply purchases reflect critical day-to-day operations. Such transactions must be recorded accurately and immediately. A database that is designed primarily to support a company's day-to-day operations is classified as an **operational database**, also known as an **online transaction processing (OLTP) database**, **transactional database**, or **production database**. In contrast, an **analytical database** focuses primarily on storing historical data and business metrics used exclusively for tactical or strategic decision making. Such analysis typically requires extensive "data massaging" (data manipulation) to produce information on which to base pricing decisions, sales forecasts, market strategies, and so on. Analytical databases allow the end user to perform advanced analysis of business data using sophisticated tools.

Typically, analytical databases comprise two main components: a data warehouse and an online analytical processing front end. The **data warehouse** is a specialized database that stores data in a format optimized for decision support. The data warehouse contains historical data obtained from the operational databases as well as data from other external sources. **Online analytical processing (OLAP)** is a set of tools that work together to provide an advanced data analysis environment for retrieving, processing, and modeling data from the data warehouse. In recent times, this area of database application has grown in importance and usage, to the point that it has evolved into its own discipline: business intelligence. The term **business intelligence** describes a comprehensive approach to capture and process business data with the purpose of generating information to support business decision making. Chapter 13, Business Intelligence and Data Warehouses, covers this topic in detail.

Databases can also be classified to reflect the degree to which the data is structured. **Unstructured data** is data that exists in its original (raw) state—that is, in the format in which it was collected. Therefore, unstructured data exists in a format that does not lend itself to the processing that yields information. **Structured data** is the result of formatting unstructured data to facilitate storage, use, and the generation of information. You apply structure (format) based on the type of processing that you intend to perform

discipline-specific database
A database that contains data focused on specific subject areas.

operational database
A database designed primarily to support a company's day-to-day operations. Also known as a *transactional database*, *OLTP database*, or *production database*.

online transaction processing (OLTP) database
See operational database.

transactional database
See operational database.

production database
See operational database.

analytical database
A database focused primarily on storing historical data and business metrics used for tactical or strategic decision making.

data warehouse
A specialized database that stores historical and aggregated data in a format optimized for decision support.

online analytical processing (OLAP)
A set of tools that provide advanced data analysis for retrieving, processing, and modeling data from the data warehouse.

business intelligence
A set of tools and processes used to capture, collect, integrate, store, and analyze data to support business decision making.

unstructured data
Data that exists in its original, raw state; that is, in the format in which it was collected.

on the data. Some data might not be ready (unstructured) for some types of processing, but they might be ready (structured) for other types of processing. For example, the data value 37890 might refer to a zip code, a sales value, or a product code. If this value represents a zip code or a product code and is stored as text, you cannot perform mathematical computations with it. On the other hand, if this value represents a sales transaction, it must be formatted as numeric.

To further illustrate the concept of structure, imagine a stack of printed paper invoices. If you want to merely store these invoices as images for future retrieval and display, you can scan them and save them in a graphic format. On the other hand, if you want to derive information such as monthly totals and average sales, such graphic storage would not be useful. Instead, you could store the invoice data in a (structured) spreadsheet format so that you can perform the requisite computations. Actually, most data you encounter is best classified as semistructured. **Semistructured data** has already been processed to some extent. For example, if you look at a typical webpage, the data is presented in a prearranged format to convey some information. The database types mentioned thus far focus on the storage and management of highly structured data. However, corporations are not limited to the use of structured data. They also use semistructured and unstructured data. Just think of the valuable information that can be found on company emails, memos, and documents such as procedures, rules, and webpages. Unstructured and semistructured data storage and management needs are being addressed through a new generation of databases known as XML databases. **Extensible Markup Language (XML)** is a special language used to represent and manipulate data elements in a textual format. An **XML database** supports the storage and management of semistructured XML data.

Table 1.1 compares the features of several well-known database management systems.

structured data
Data that has been formatted to facilitate storage, use, and information generation.

semistructured data
Data that has already been processed to some extent.

Extensible Markup Language (XML)
A metalanguage used to represent and manipulate data elements. Unlike other markup languages, XML permits the manipulation of a document's data elements.

TABLE 1.1

TYPES OF DATABASES

PRODUCT	NUMBER OF USERS			DATA LOCATION		DATA USAGE		XML
	SINGLE USER	MULTIUSER		CENTRALIZED	DISTRIBUTED	OPERATIONAL	ANALYTICAL	
		WORKGROUP	ENTERPRISE					
MS Access	X	X		X		X		
MS SQL Server	X[3]	X	X	X	X	X	X	X
IBM DB2	X[3]	X	X	X	X	X	X	X
MySQL	X	X	X	X	X	X	X	X
Oracle RDBMS	X[3]	X	X	X	X	X	X	X

With the emergence of the World Wide Web and Internet-based technologies as the basis for the new "social media" generation, great amounts of data are being stored and analyzed. **Social media** refers to web and mobile technologies that enable "anywhere, anytime, always on" human interactions. Websites such as Google, Facebook, Twitter, and LinkedIn capture vast amounts of data about end users and consumers. This data grows exponentially and requires the use of specialized database systems. For example, as of 2015, over 500 million tweets were posted every day on Twitter, and that number continues to grow. As a result, the MySQL database Twitter was using to store user content was frequently overloaded by demand.[2,3] Facebook faces

XML database
A database system that stores and manages semistructured XML data.

social media
Web and mobile technologies that enable "anywhere, anytime, always on" human interactions.

[2] Vendor offers single-user/personal DBMS version.
[3] www.internetlivestats.com/twitter-statistics/

similar challenges. With over 500 terabytes of data coming in each day, it stores over 100 petabytes of data in a single data storage file system. From this data, its database scans over 200 terabytes of data each hour to process user actions, including status updates, picture requests, and billions of "Like" actions.[4] Over the past few years, this new breed of specialized database has grown in sophistication and widespread usage. Currently, this new type of database is known as a NoSQL database. The term **NoSQL** (Not only SQL) is generally used to describe a new generation of database management systems that is not based on the traditional relational database model. NoSQL databases are designed to handle the unprecedented volume of data, variety of data types and structures, and velocity of data operations that are characteristic of these new business requirements. You will learn more about this type of system in Chapter 2, Data Models.

This section briefly mentioned the many different types of databases. As you learned earlier, a database is a computer structure that houses and manages end-user data. One of the first tasks of a database professional is to ensure that end-user data is properly structured to derive valid and timely information. For this, good database design is essential.

1-4 Why Database Design is Important

A problem that has evolved with the use of personal productivity tools such as spreadsheets and desktop database programs is that users typically lack proper data-modeling and database design skills. People naturally have a "narrow" view of the data in their environment. For example, consider a student's class schedule. The schedule probably contains the student's identification number and name, class code, class description, class credit hours, class instructor name, class meeting days and times, and class room number. In the mind of the student, these various data items compose a single unit. If a student organization wanted to keep a record of the schedules of its members, an end user might make a spreadsheet to store the schedule information. Even if the student makes a foray into the realm of desktop databases, he or she is likely to create a structure composed of a single table that mimics his or her view of the schedule data. As you will learn in the coming chapters, translating this type of narrow view of data into a single two-dimensional table structure is a poor database design choice.

Database design refers to the activities that focus on the design of the database structure that will be used to store and manage end-user data. A database that meets all user requirements does not just happen; its structure must be designed carefully. In fact, database design is such a crucial aspect of working with databases that most of this book is dedicated to the development of good database design techniques. Even a good DBMS will perform poorly with a badly designed database.

Data is one of an organization's most valuable assets. Data on customers, employees, orders, and receipts is all vital to the existence of a company. Tracking key growth and performance indicators are also vital to strategic and tactical plans to ensure future success; therefore, an organization's data must not be handled lightly or carelessly. Thorough planning to ensure that data is properly used and leveraged to give the company the most benefit is just as important as proper financial planning to ensure that the company gets the best use from its financial resources.

NoSQL
A new generation of database management systems that is not based on the traditional relational database model.

database design
The process that yields the description of the database structure and determines the database components. The second phase of the Database Life Cycle.

---

[4] Josh Constine, "How big is Facebook's data? 2.5 billion pieces of content and 500+ terabytes of data ingested every day," *Tech Crunch,* August 22, 2012, *http://techcrunch.com/2012/08/22/how-big-is-facebooks-data-2-5-billion-pieces-of-content-and-500-terabytes-ingested-every-day/*

Because current-generation DBMSs are easy to use, an unfortunate side effect is that many computer-savvy business users gain a false sense of confidence in their ability to build a functional database. These users can effectively navigate the creation of database objects, but without the proper understanding of database design, they tend to produce flawed, overly simplified structures that prevent the system from correctly storing data that corresponds to business realities, which produces incomplete or erroneous results when the data is retrieved. Consider the data shown in Figure 1.4, which illustrates the efforts of an organization to keep records about its employees and their skills. Some employees have not passed a certification test in any skill, while others have been certified in several skills. Some certified skills are shared by several employees, while other skills have no employees that hold those certifications.

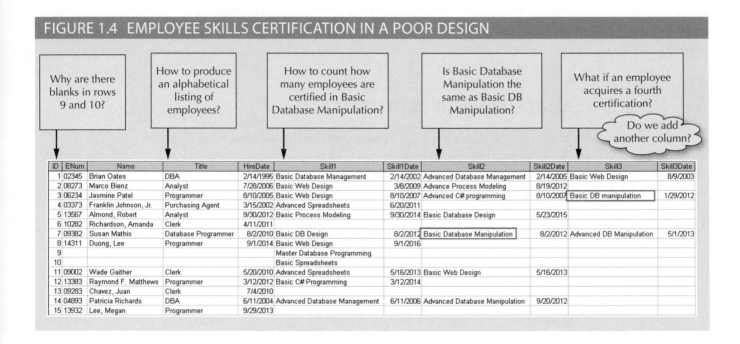

FIGURE 1.4 EMPLOYEE SKILLS CERTIFICATION IN A POOR DESIGN

Based on this storage of the data, notice the following problems:

- It would be difficult, if not impossible, to produce an alphabetical listing of employees based on their last names.

- To determine how many employees are certified in Basic Database Manipulation, you would need a program that counts the number of those certifications recorded in Skill1 and places it in a variable. Then the count of those certifications in Skill2 could be calculated and added to the variable. Finally, the count of those certifications in Skill3 could be calculated and added to the variable to produce the total.

- If you redundantly store the name of a skill with each employee who is certified in that skill, you run the risk of spelling the name differently for different employees. For example, the skill *Basic Database Manipulation* is also entered as *Basic DB Manipulation* for at least one employee in Figure 1.4, which makes it difficult to get an accurate count of employees who have the certification.

- The structure of the database will have to be changed by adding more columns to the table when an employee is certified in a fourth skill. It will have to be modified again if an employee is certified in a fifth skill.

Contrast this poor design with that shown in Figure 1.5, where the design has been improved by decomposing the data into three related tables. These tables contain all of the same data that was represented in Figure 1.4, but the tables are structured so that you can easily manipulate the data to view it in different ways and answer simple questions.

FIGURE 1.5 EMPLOYEE SKILL CERTIFICATIONS IN A GOOD DESIGN

Database name: Ch01_Text

Table name: EMPLOYEE

Employee_ID	Employee_FName	Employee_LName	Employee_HireDate	Employee_Title
02345	Johnny	Jones	2/14/1995	DBA
03373	Franklin	Johnson	3/15/2002	Purchasing Agent
04893	Patricia	Richards	6/11/2004	DBA
06234	Jasmine	Patel	8/10/2005	Programmer
08273	Marco	Bienz	7/28/2006	Analyst
09002	Ben	Joiner	5/20/2010	Clerk
09283	Juan	Chavez	7/4/2010	Clerk
09382	Jessica	Johnson	8/2/2010	Database Programmer
10282	Amanda	Richardson	4/11/2011	Clerk
13383	Raymond	Matthews	3/12/2012	Programmer
13567	Robert	Almond	9/30/2012	Analyst
13932	Megan	Lee	9/29/2013	Programmer
14311	Lee	Duong	9/1/2014	Programmer

Table name: CERTIFIED

Employee_ID	Skill_ID	Certified_Date
02345	100	2/14/2002
02345	110	8/9/2003
02345	180	2/14/2005
03373	120	6/20/2011
04893	180	6/11/2006
04893	220	9/20/2012
06234	110	8/10/2007
06234	200	8/10/2007
06234	210	1/29/2012
08273	110	3/8/2009
08273	190	8/19/2012
09002	110	5/16/2013
09002	120	5/16/2013
09382	140	8/2/2012
09382	210	8/2/2012
09382	220	5/1/2013
13383	170	3/12/2014
13567	130	9/30/2014
13567	140	5/23/2015
14311	110	9/1/2016

Table name: SKILL

Skill_ID	Skill_Name	Skill_Description
100	Basic Database Management	Create and manage database user accounts.
110	Basic Web Design	Create and maintain HTML and CSS documents.
120	Advanced Spreadsheets	Use of advanced functions, user-defined functions, and macroing.
130	Basic Process Modeling	Create core business process models using standard libraries.
140	Basic Database Design	Create simple data models.
150	Master Database Programming	Create integrated trigger and procedure packages for a distributed environment.
160	Basic Spreadsheets	Create single tab worksheets with basic formulas
170	Basic C# Programming	Create single-tier data aware modules.
180	Advanced Database Management	Manage Database Server Clusters.
190	Advance Process Modeling	Evaluate and Redesign cross-functional internal and external business processes.
200	Advanced C# Programming	Create multi-tier applications using multi-threading
210	Basic Database Manipulation	Create simple data retrieval and manipulation statements in SQL.
220	Advanced Database Manipulation	Use of advanced data manipulation methods for multi-table inserts, set operations, and correlated subqueries.

With the improved structure in Figure 1.5, you can use simple commands in a standard data manipulation language to do the following:

- Produce an alphabetical listing of employees by last name:

  SELECT * FROM EMPLOYEE ORDER BY EMPLOYEE_LNAME;

- Determine how many employees are certified in Basic Database Manipulation:

  SELECT Count(*)
  FROM SKILL JOIN CERTIFIED ON SKILL.SKILL_ID = CERTIFIED.SKILL_ID
  WHERE SKILL_NAME = 'Basic Database Manipulation';

You will learn more about these commands in Chapter 7, Introduction to Structured Query Language.

Note that because each skill name is stored only once, the names cannot be spelled or abbreviated differently for different employees. Also, the additional certification of an employee with a fourth or fifth skill does not require changes to the structure of the tables.

Proper database design requires the designer to identify precisely the database's expected use. Designing a transactional database emphasizes accurate and consistent data and operational speed. Designing a data warehouse database emphasizes the use of historical and aggregated data. Designing a database to be used in a centralized, single-user environment requires a different approach from that used in the design of a distributed, multiuser database. This book emphasizes the design of transactional, centralized, single-user, and multiuser databases. Chapters 12 and 13 also examine critical issues confronting the designer of distributed and data warehouse databases.

Designing appropriate data repositories of integrated information using the two-dimensional table structures found in most databases is a process of decomposition. The integrated data must be decomposed properly into its constituent parts, with each part stored in its own table. Further, the relationships between these tables must be carefully considered and implemented so the integrated view of the data can be recreated later as information for the end user. A well-designed database facilitates data management and generates accurate and valuable information. A poorly designed database is likely to become a breeding ground for difficult-to-trace errors that may lead to poor decision making—and poor decision making can lead to the failure of an organization. Database design is simply too important to be left to luck. That's why college students study database design, why organizations of all types and sizes send personnel to database design seminars, and why database design consultants often make an excellent living.

1-5 **Evolution of File System Data Processing**

Understanding what a database is, what it does, and the proper way to use it can be clarified by considering what a database is not. A brief explanation of the evolution of file system data processing can be helpful in understanding the data access limitations that databases attempt to overcome. Understanding these limitations is relevant to database designers and developers because database technologies do not make these problems magically disappear—database technologies simply make it easier to create solutions that avoid these problems. Creating database designs that avoid the pitfalls of earlier systems requires that the designer understand these problems and how to avoid them; otherwise, the database technologies are no better (and are potentially even worse!) than the technologies and techniques they have replaced.

1-5a Manual File Systems

To be successful, an organization must develop systems for handling core business tasks. Historically, such systems were often manual, paper-and-pencil systems. The papers within these systems were organized to facilitate the expected use of the data. Typically, this was accomplished through a system of file folders and filing cabinets. As long as a collection of data was relatively small and an organization's business users had few reporting requirements, the manual system served its role well as a data repository. However, as organizations grew and as reporting requirements became more complex, keeping track of data in a manual file system became more difficult. Therefore, companies looked to computer technology for help.

1-5b Computerized File Systems

Generating reports from manual file systems was slow and cumbersome. In fact, some business managers faced government-imposed reporting requirements that led to weeks of intensive effort each quarter, even when a well-designed manual system was used. Therefore, a **data processing (DP) specialist** was hired to create a computer-based system that would track data and produce required reports.

Initially, the computer files within the file system were similar to the manual files. A simple example of a customer data file for a small insurance company is shown in Figure 1.6. (You will discover later that the file structure shown in Figure 1.6, although typically found in early file systems, is unsatisfactory for a database.)

data processing (DP) specialist
The person responsible for developing and managing a computerized file processing system.

FIGURE 1.6 CONTENTS OF THE CUSTOMER FILE

Database name: Ch01_Text

C_NAME	C_PHONE	C_ADDRESS	C_ZIP	A_NAME	A_PHONE	TP	AMT	REN
Alfred A. Ramas	615-844-2573	218 Fork Rd., Babs, TN	36123	Leah F. Hahn	615-882-1244	T1	100.00	05-Apr-2016
Leona K. Dunne	713-894-1238	Box 12A, Fox, KY	25246	Alex B. Alby	713-228-1249	T1	250.00	16-Jun-2016
Kathy W. Smith	615-894-2285	125 Oak Ln, Babs, TN	36123	Leah F. Hahn	615-882-2144	S2	150.00	29-Jan-2017
Paul F. Olowski	615-894-2180	217 Lee Ln., Babs, TN	36123	Leah F. Hahn	615-882-1244	S1	300.00	14-Oct-2016
Myron Orlando	615-222-1672	Box 111, New, TN	36155	Alex B. Alby	713-228-1249	T1	100.00	28-Dec-2016
Amy B. O'Brian	713-442-3381	387 Troll Dr., Fox, KY	25246	John T. Okon	615-123-5589	T2	850.00	22-Sep-2016
James G. Brown	615-297-1228	21 Tye Rd., Nash, TN	37118	Leah F. Hahn	615-882-1244	S1	120.00	25-Mar-2017
George Williams	615-290-2556	155 Maple, Nash, TN	37119	John T. Okon	615-123-5589	S1	250.00	17-Jul-2016
Anne G. Farriss	713-382-7185	2119 Elm, Crew, KY	25432	Alex B. Alby	713-228-1249	T2	100.00	03-Dec-2016
Olette K. Smith	615-297-3809	2782 Main, Nash, TN	37118	John T. Okon	615-123-5589	S2	500.00	14-Mar-2017

C_NAME = Customer name
C_PHONE = Customer phone
C_ADDRESS = Customer address
C_ZIP = Customer zip code

A_NAME = Agent name
A_PHONE = Agent phone
TP = Insurance type
AMT = Insurance policy amount, in thousands of $
REN = Insurance renewal date

The description of computer files requires a specialized vocabulary. Every discipline develops its own terminology to enable its practitioners to communicate clearly. The basic file vocabulary shown in Table 1.2 will help you to understand subsequent discussions more easily.

TABLE 1.2

BASIC FILE TERMINOLOGY

TERM	DEFINITION
Data	Raw facts, such as a telephone number, a birth date, a customer name, and a year-to-date (YTD) sales value. Data has little meaning unless it has been organized in some logical manner.
Field	A character or group of characters (alphabetic or numeric) that has a specific meaning. A field is used to define and store data.
Record	A logically connected set of one or more fields that describes a person, place, or thing. For example, the fields that constitute a record for a customer might consist of the customer's name, address, phone number, date of birth, credit limit, and unpaid balance.
File	A collection of related records. For example, a file might contain data about the students currently enrolled at Gigantic University.

Using the proper file terminology in Table 1.2, you can identify the file components shown in Figure 1.6. The CUSTOMER file contains 10 records. Each record is composed of 9 fields: C_NAME, C_PHONE, C_ADDRESS, C_ZIP, A_NAME, A_PHONE, TP, AMT, and REN. The 10 records are stored in a named file. Because the file in Figure 1.6 contains customer data for the insurance company, its filename is CUSTOMER.

When business users wanted data from the computerized file, they sent requests for the data to the DP specialist. For each request, the DP specialist had to create programs to retrieve the data from the file, manipulate it in whatever manner the user had requested, and present it as a printed report. If a request was for a report that had been run previously, the DP specialist could rerun the existing program and provide the printed results to the user. As other business users saw the new and innovative ways in which customer data was being reported, they wanted to be able to view their data in similar fashions. This generated more requests for the DP specialist to create more computerized files of other business data, which in turn meant that more data management programs had to be created, which led to even more requests for reports. For example, the sales department at the insurance company created a file named SALES, which helped track daily sales efforts. The sales department's success was so obvious that the personnel department manager demanded access to the DP specialist to automate payroll processing and other personnel functions. Consequently, the DP specialist was asked to create the AGENT file shown in Figure 1.7. The data in the AGENT file was used to write checks, keep track of taxes paid, and summarize insurance coverage, among other tasks.

FIGURE 1.7 CONTENTS OF THE AGENT FILE

Database name: Ch01_Text

A_NAME	A_PHONE	A_ADDRESS	ZIP	HIRED	YTD_PAY	YTD_FIT	YTD_FICA	YTD_SLS	DEP
Alex B. Alby	713-228-1249	123 Toll, Nash, TN	37119	01-Nov-2000	26566.24	6641.56	2125.30	132737.75	3
Leah F. Hahn	615-882-1244	334 Main, Fox, KY	25246	23-May-1986	32213.78	8053.44	2577.10	138967.35	0
John T. Okon	615-123-5589	452 Elm, New, TN	36155	15-Jun-2005	23198.29	5799.57	1855.86	127093.45	2

A_NAME	= Agent name	YTD_PAY	= Year-to-date pay
A_PHONE	= Agent phone	YTD_FIT	= Year-to-date federal income tax paid
A_ADDRESS	= Agent address	YTD_FICA	= Year-to-date Social Security taxes paid
ZIP	= Agent zip code	YTD_SLS	= Year-to-date sales
HIRED	= Agent date of hire	DEP	= Number of dependents

As more and more computerized files were developed, the problems with this type of file system became apparent. While these problems are explored in detail in the next section, the problems basically centered on having many data files that contained related—often overlapping—data with no means of controlling or managing the data consistently across all of the files. As shown in Figure 1.8, each file in the system used its own application program to store, retrieve, and modify data. Also, each file was owned by the individual or the department that commissioned its creation.

The advent of computer files to store company data was significant; it not only established a landmark in the use of computer technologies, it also represented a huge step forward in a business's ability to process data. Previously, users had direct, hands-on access to all of the business data. But they didn't have the tools to convert that data into the information they needed. The creation of computerized file systems gave them improved tools for manipulating the company data that allowed them to create new

FIGURE 1.8 A SIMPLE FILE SYSTEM

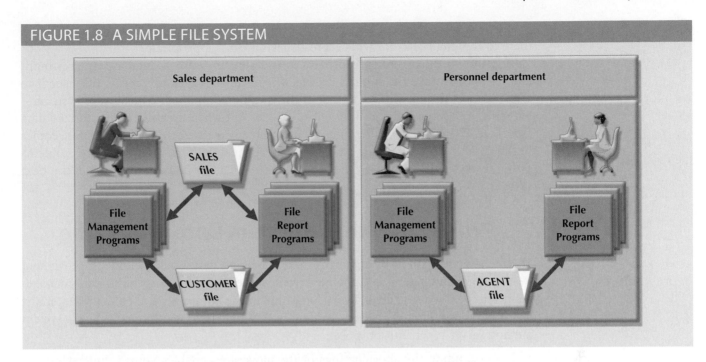

information. However, it had the additional effect of introducing a schism between the end users and their data. The desire to close the gap between the end users and the data influenced the development of many types of computer technologies, system designs, and uses (and misuses) of many technologies and techniques. However, such developments also created a split between the ways DP specialists and end users viewed the data.

- From the DP specialist's perspective, the computer files within the file system were created to be similar to the manual files. Data management programs were created to add to, update, and delete data from the file.

- From the end user's perspective, the systems separated the users from the data. As the users' competitive environment pushed them to make more and more decisions in less time, users became frustrated by the delay between conceiving of a new way to create information from the data and the point when the DP specialist actually created the programs to generate that information.

1-5c File System Redux: Modern End-User Productivity Tools

The users' desire for direct, hands-on access to data helped to fuel the adoption of personal computers for business use. Although not directly related to file system evolution, the ubiquitous use of personal productivity tools can introduce the same problems as the old file systems.

Personal computer spreadsheet programs such as Microsoft Excel are widely used by business users, and they allow the user to enter data in a series of rows and columns so the data can be manipulated using a wide range of functions. The popularity of spreadsheet applications has enabled users to conduct sophisticated data analysis that has greatly enhanced their ability to understand the data and make better decisions. Unfortunately, as in the old adage "When the only tool you have is a hammer, every problem looks like a nail," users have become so adept at working with spreadsheets that they tend to use them to complete tasks for which spreadsheets are not appropriate.

A common misuse of spreadsheets is as a substitute for a database. Interestingly, end users often take the limited data to which they have direct access and place it in a spreadsheet format similar to that of the traditional, manual data storage systems— which is precisely what the early DP specialists did when creating computerized data files. Due to the large number of users with spreadsheets, each making separate copies of the data, the resulting "file system" of spreadsheets suffers from the same problems as the file systems created by the early DP specialists, which are outlined in the next section.

1-6 Problems with File System Data Processing

The file system method of organizing and managing data was a definite improvement over the manual system, and the file system served a useful purpose in data management for over two decades—a very long time in the computer era. Nonetheless, many problems and limitations became evident in this approach. A critique of the file system method serves two major purposes:

- Understanding the shortcomings of the file system enables you to understand the development of modern databases.

- Many of the problems are not unique to file systems. Failure to understand such problems is likely to lead to their duplication in a database environment, even though database technology makes it easy to avoid them.

The following problems associated with file systems, whether created by DP specialists or through a series of spreadsheets, severely challenge the types of information that can be created from the data as well as the accuracy of the information:

- *Lengthy development times.* The first and most glaring problem with the file system approach is that even the simplest data-retrieval task requires extensive programming. With the older file systems, programmers had to specify what must be done and how to do it. As you will learn in upcoming chapters, modern databases use a nonprocedural data manipulation language that allows the user to specify what must be done without specifying how.

- *Difficulty of getting quick answers.* The need to write programs to produce even the simplest reports makes ad hoc queries impossible. Harried DP specialists who worked with mature file systems often received numerous requests for new reports. They were often forced to say that the report will be ready "next week" or even "next month." If you need the information now, getting it next week or next month will not serve your information needs.

- *Complex system administration.* System administration becomes more difficult as the number of files in the system expands. Even a simple file system with a few files requires creating and maintaining several file management programs. Each file must have its own file management programs that allow the user to add, modify, and delete records; to list the file contents; and to generate reports. Because ad hoc queries are not possible, the file reporting programs can multiply quickly. The problem is compounded by the fact that each department in the organization "owns" its data by creating its own files.

- *Lack of security and limited data sharing.* Another fault of a file system data repository is a lack of security and limited data sharing. Data sharing and security are closely related. Sharing data among multiple geographically dispersed users introduces a lot of security risks. In terms of spreadsheet data, while many spreadsheet programs provide rudimentary security options, they are not always used, and even when they are, they are insufficient for robust data sharing among users. In terms of creating data management and reporting programs, security and data-sharing features are difficult to program and consequently are often omitted from a file system environment. Such features include effective password protection, the ability to lock out parts of files or parts of the system itself, and other measures designed to safeguard data confidentiality. Even when an attempt is made to improve system and data security, the security devices tend to be limited in scope and effectiveness.

- *Extensive programming.* Making changes to an existing file structure can be difficult in a file system environment. For example, changing just one field in the original CUSTOMER file would require a program that:

  1. Reads a record from the original file.

  2. Transforms the original data to conform to the new structure's storage requirements.

  3. Writes the transformed data into the new file structure.

  4. Repeats the preceding steps for each record in the original file.

In fact, any change to a file structure, no matter how minor, forces modifications in all of the programs that use the data in that file. Modifications are likely to produce errors (bugs), and additional time is spent using a debugging process to find those errors. Those limitations, in turn, lead to problems of structural and data dependence.

1-6a Structural and Data Dependence

A file system exhibits **structural dependence**, which means that access to a file is dependent on its structure. For example, adding a customer date-of-birth field to the CUSTOMER file shown in Figure 1.6 would require the four steps described in the previous section. Given this change, none of the previous programs will work with the new CUSTOMER file structure. Therefore, all of the file system programs must be modified to conform to the new file structure. In short, because the file system application programs are affected by changes in the file structure, they exhibit structural dependence. Conversely, **structural independence** exists when you can change the file structure without affecting the application's ability to access the data.

Even changes in the characteristics of data, such as changing a field from integer to decimal, require changes in all the programs that access the file. Because all data access programs are subject to change when any of the file's data storage characteristics change (that is, changing the data type), the file system is said to exhibit **data dependence**. Conversely, **data independence** exists when you can change the data storage characteristics without affecting the program's ability to access the data.

The practical significance of data dependence is the difference between the **logical data format** (how the human being views the data) and the **physical data format** (how the computer must work with the data). Any program that accesses a file system's file must tell the computer not only what to do but how to do it. Consequently, each program must contain lines that specify the opening of a specific file type, its record specification, and its field definitions. Data dependence makes the file system extremely cumbersome from the point of view of a programmer and database manager.

structural dependence
A data characteristic in which a change in the database schema affects data access, thus requiring changes in all access programs.

structural independence
A data characteristic in which changes in the database schema do not affect data access.

data dependence
A data condition in which data representation and manipulation are dependent on the physical data storage characteristics.

data independence
A condition in which data access is unaffected by changes in the physical data storage characteristics.

logical data format
The way a person views data within the context of a problem domain.

physical data format
The way a computer "sees" (stores) data.

1-6b Data Redundancy

The file system's structure makes it difficult to combine data from multiple sources, and its lack of security renders the file system vulnerable to security breaches. The organizational structure promotes the storage of the same basic data in different locations. (Database professionals use the term **islands of information** for such scattered data locations.) The dispersion of data is exacerbated by the use of spreadsheets to store data. In a file system, the entire sales department would share access to the SALES data file through the data management and reporting programs created by the DP specialist. With the use of spreadsheets, each member of the sales department can create his or her own copy of the sales data. Because data stored in different locations will probably not be updated consistently, the islands of information often contain different versions of the same data. For example, in Figures 1.6 and 1.7, the agent names and phone numbers occur in both the CUSTOMER and the AGENT files. You only need one correct copy of the agent names and phone numbers. Having them occur in more than one place produces data redundancy. **Data redundancy** exists when the same data is stored unnecessarily at different places.

Uncontrolled data redundancy sets the stage for the following:

- *Poor data security*. Having multiple copies of data increases the chances for a copy of the data to be susceptible to unauthorized access. Chapter 16, Database Administration and Security, explores the issues and techniques associated with securing data.

- *Data inconsistency*. Data inconsistency exists when different and conflicting versions of the same data appear in different places. For example, suppose you change an agent's phone number in the AGENT file. If you forget to make the corresponding change in the CUSTOMER file, the files contain different data for the same agent. Reports will yield inconsistent results that depend on which version of the data is used.

- *Data-entry errors*. Data-entry errors are more likely to occur when complex entries (such as 10-digit phone numbers) are made in several different files or recur frequently in one or more files. In fact, the CUSTOMER file shown in Figure 1.6 contains just such an entry error: the third record in the CUSTOMER file has transposed digits in the agent's phone number (615-882-2144 rather than 615-882-1244).

- *Data integrity problems*. It is possible to enter a nonexistent sales agent's name and phone number into the CUSTOMER file, but customers are not likely to be impressed if the insurance agency supplies the name and phone number of an agent who does not exist. Should the personnel manager allow a nonexistent agent to accrue bonuses and benefits? In fact, a data-entry error such as an incorrectly spelled name or an incorrect phone number yields the same kind of data integrity problems.

islands of information
In the old file system environment, pools of independent, often duplicated, and inconsistent data created and managed by different departments.

data redundancy
Exists when the same data is stored unnecessarily at different places.

data integrity
In a relational database, a condition in which the data in the database complies with all entity and referential integrity constraints.

Note

Data that displays data inconsistency is also referred to as data that lacks data integrity. **Data integrity** is defined as the condition in which all of the data in the database is consistent with the real-world events and conditions. In other words, data integrity means that:

- Data is *accurate*—there are no data inconsistencies.

- Data is *verifiable*—the data will always yield consistent results.

1-6c Data Anomalies

The dictionary defines *anomaly* as "an abnormality." Ideally, a field value change should be made in only a single place. Data redundancy, however, fosters an abnormal condition by forcing field value changes in many different locations. Look at the CUSTOMER file in Figure 1.6. If agent Leah F. Hahn decides to get married and move, the agent name, address, and phone number are likely to change. Instead of making these changes in a single file (AGENT), you must also make the change each time that agent's name and phone number occur in the CUSTOMER file. You could be faced with the prospect of making hundreds of corrections, one for each of the customers served by that agent! The same problem occurs when an agent decides to quit. Each customer served by that agent must be assigned a new agent. Any change in any field value must be correctly made in many places to maintain data integrity. A **data anomaly** develops when not all of the required changes in the redundant data are made successfully. The data anomalies found in Figure 1.6 are commonly defined as follows:

- *Update anomalies.* If agent Leah F. Hahn has a new phone number, it must be entered in each of the CUSTOMER file records in which Ms. Hahn's phone number is shown. In this case, only four changes must be made. In a large file system, such a change might occur in hundreds or even thousands of records. Clearly, the potential for data inconsistencies is great.

- *Insertion anomalies.* If only the CUSTOMER file existed and you needed to add a new agent, you would also add a dummy customer data entry to reflect the new agent's addition. Again, the potential for creating data inconsistencies would be great.

- *Deletion anomalies.* If you delete the customers Amy B. O'Brian, George Williams, and Olette K. Smith, you will also delete John T. Okon's agent data. Clearly, this is not desirable.

On a positive note, however, this book will help you develop the skills needed to design and model a successful database that avoids the problems listed in this section.

1-7 Database Systems

The problems inherent in file systems make using a database system very desirable. Unlike the file system, with its many separate and unrelated files, the database system consists of logically related data stored in a single logical data repository. (The "logical" label reflects the fact that the data repository appears to be a single unit to the end user, even though data might be physically distributed among multiple storage facilities and locations.) Because the database's data repository is a single logical unit, the database represents a major change in the way end-user data is stored, accessed, and managed. The database's DBMS, shown in Figure 1.9, provides numerous advantages over file system management, shown in Figure 1.8, by making it possible to eliminate most of the file system's data inconsistency, data anomaly, data dependence, and structural dependence problems. Better yet, the current generation of DBMS software stores not only the data structures, but also the relationships between those structures and the access paths to those structures—all in a central location. The current generation of DBMS software also takes care of defining, storing, and managing all required access paths to those components.

data anomaly
A data abnormality in which inconsistent changes have been made to a database. For example, an employee moves, but the address change is not corrected in all files in the database.

FIGURE 1.9 CONTRASTING DATABASE AND FILE SYSTEMS

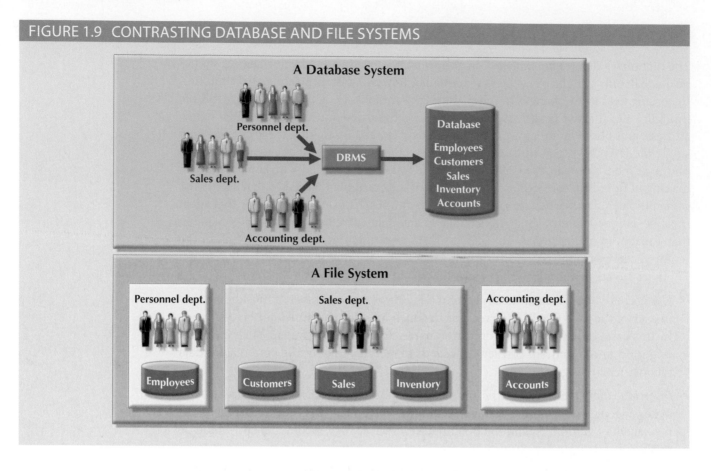

Remember that the DBMS is just one of several crucial components of a database system. The DBMS may even be referred to as the database system's heart. However, just as it takes more than a heart to make a human being function, it takes more than a DBMS to make a database system function. In the sections that follow, you'll learn what a database system is, what its components are, and how the DBMS fits into the picture.

1-7a The Database System Environment

The term **database system** refers to an organization of components that define and regulate the collection, storage, management, and use of data within a database environment. From a general management point of view, the database system is composed of the five major parts shown in Figure 1.10: hardware, software, people, procedures, and data.

Let's take a closer look at the five components shown in Figure 1.10:

- *Hardware.* Hardware refers to all of the system's physical devices, including computers (PCs, tablets, workstations, servers, and supercomputers), storage devices, printers, network devices (hubs, switches, routers, fiber optics), and other devices (automated teller machines, ID readers, and so on).

- *Software.* Although the most readily identified software is the DBMS itself, three types of software are needed to make the database system function fully: operating system software, DBMS software, and application programs and utilities.

  - *Operating system software* manages all hardware components and makes it possible for all other software to run on the computers. Examples of operating system software include Microsoft Windows, Linux, Mac OS, UNIX, and MVS.

database system
An organization of components that defines and regulates the collection, storage, management, and use of data in a database environment.

FIGURE 1.10 THE DATABASE SYSTEM ENVIRONMENT

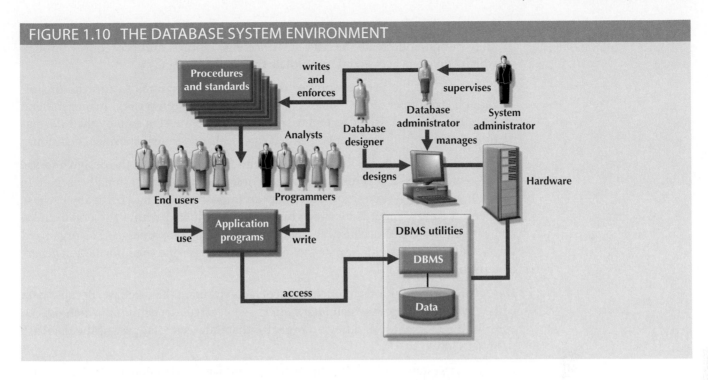

- *DBMS software* manages the database within the database system. Some examples of DBMS software include Microsoft's SQL Server, Oracle Corporation's Oracle, Oracle's MySQL, and IBM's DB2.

- *Application programs and utility software* are used to access and manipulate data in the DBMS and to manage the computer environment in which data access and manipulation take place. Application programs are most commonly used to access data within the database to generate reports, tabulations, and other information to facilitate decision making. Utilities are the software tools used to help manage the database system's computer components. For example, all of the major DBMS vendors now provide graphical user interfaces (GUIs) to help create database structures, control database access, and monitor database operations.

- *People.* This component includes all users of the database system. On the basis of primary job functions, five types of users can be identified in a database system: system administrators, database administrators, database designers, system analysts and programmers, and end users. Each user type, described next, performs both unique and complementary functions.

  - *System administrators* oversee the database system's general operations.

  - *Database administrators*, also known as DBAs, manage the DBMS and ensure that the database is functioning properly. The DBA's role is sufficiently import-ant to warrant a detailed exploration in Chapter 16, Database Administration and Security.

  - *Database designers* design the database structure. They are, in effect, the database architects. If the database design is poor, even the best application programmers and the most dedicated DBAs cannot produce a useful database environment. Because organizations strive to optimize their data resources, the database designer's job description has expanded to cover new dimensions and growing responsibilities.

- *System analysts and programmers* design and implement the application programs. They design and create the data-entry screens, reports, and procedures through which end users access and manipulate the database's data.

- *End users* are the people who use the application programs to run the organization's daily operations. For example, sales clerks, supervisors, managers, and directors are all classified as end users. High-level end users employ the information obtained from the database to make tactical and strategic business decisions.

- *Procedures.* Procedures are the instructions and rules that govern the design and use of the database system. Procedures are a critical, although occasionally forgotten, component of the system. Procedures play an important role in a company because they enforce the standards by which business is conducted within the organization and with customers. Procedures also help to ensure that companies have an organized way to monitor and audit the data that enter the database and the information generated from those data.

- *Data.* The word *data* covers the collection of facts stored in the database. Because data is the raw material from which *information* is generated, determining which data to enter into the database and how to organize that data is a vital part of the database designer's job.

A database system adds a new dimension to an organization's management structure. The complexity of this managerial structure depends on the organization's size, its functions, and its corporate culture. Therefore, database systems can be created and managed at different levels of complexity and with varying adherence to precise standards. For example, compare a local convenience store system with a national insurance claims system. The convenience store system may be managed by two people, the hardware used is probably a single computer, the procedures are probably simple, and the data volume tends to be low. The national insurance claims system is likely to have at least one systems administrator, several full-time DBAs, and many designers and programmers; the hardware probably includes several servers at multiple locations throughout the United States; the procedures are likely to be numerous, complex, and rigorous; and the data volume tends to be high.

In addition to the different levels of database system complexity, managers must also take another important fact into account: database solutions must be cost-effective as well as tactically and strategically effective. Producing a million-dollar solution to a thousand-dollar problem is hardly an example of good database system selection or of good database design and management. Finally, the database technology already in use is likely to affect the selection of a database system.

1-7b DBMS Functions

A DBMS performs several important functions that guarantee the integrity and consistency of the data in the database. Most of those functions are transparent to end users, and most can be achieved only through the use of a DBMS. They include data dictionary management, data storage management, data transformation and presentation, security management, multiuser access control, backup and recovery management, data integrity management, database access languages and application programming interfaces, and database communication interfaces. Each of these functions is explained as follows:

- *Data dictionary management.* The DBMS stores definitions of the data elements and their relationships (metadata) in a data dictionary. In turn, all programs that

FIGURE 1.11 ILLUSTRATING METADATA WITH MICROSOFT SQL SERVER EXPRESS

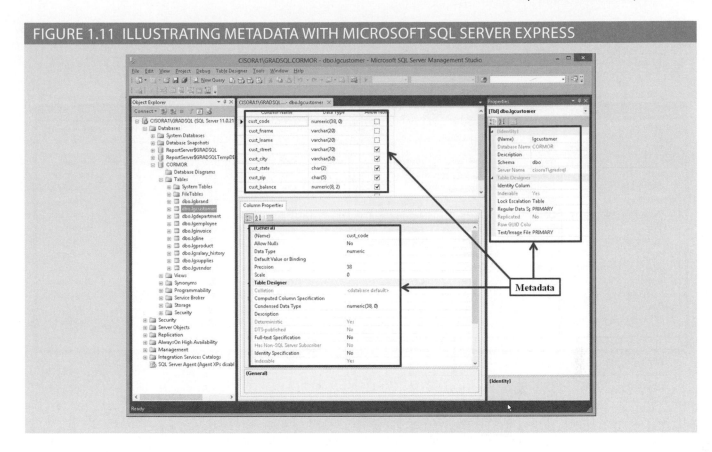

access the data in the database work through the DBMS. The DBMS uses the **data dictionary** to look up the required data component structures and relationships, thus relieving you from having to code such complex relationships in each program. Additionally, any changes made in a database structure are automatically recorded in the data dictionary, thereby freeing you from having to modify all of the programs that access the changed structure. In other words, the DBMS provides data abstraction, and it removes structural and data dependence from the system. For example, Figure 1.11 shows how Microsoft SQL Server Express presents the data definition for the CUSTOMER table.

- *Data storage management.* The DBMS creates and manages the complex structures required for data storage, thus relieving you from the difficult task of defining and programming the physical data characteristics. A modern DBMS provides storage not only for the data but for related data-entry forms or screen definitions, report definitions, data validation rules, procedural code, structures to handle video and picture formats, and so on. Data storage management is also important for database performance tuning. **Performance tuning** relates to the activities that make the database perform more efficiently in terms of storage and access speed. Although the user sees the database as a single data storage unit, the DBMS actually stores the database in multiple physical data files. (See Figure 1.12.) Such data files may even be stored on different storage media. Therefore, the DBMS doesn't have to wait for one disk request to finish before the next one starts. In other words, the DBMS can fulfill database requests concurrently. Data storage management and performance tuning issues are addressed in Chapter 11, Database Performance Tuning and Query Optimization.

data dictionary
A DBMS component that stores metadata—data about data. The data dictionary contains data definitions as well as data characteristics and relationships. May also include data that is external to the DBMS.

performance tuning
Activities that make a database perform more efficiently in terms of storage and access speed.

FIGURE 1.12 ILLUSTRATING DATA STORAGE MANAGEMENT WITH ORACLE

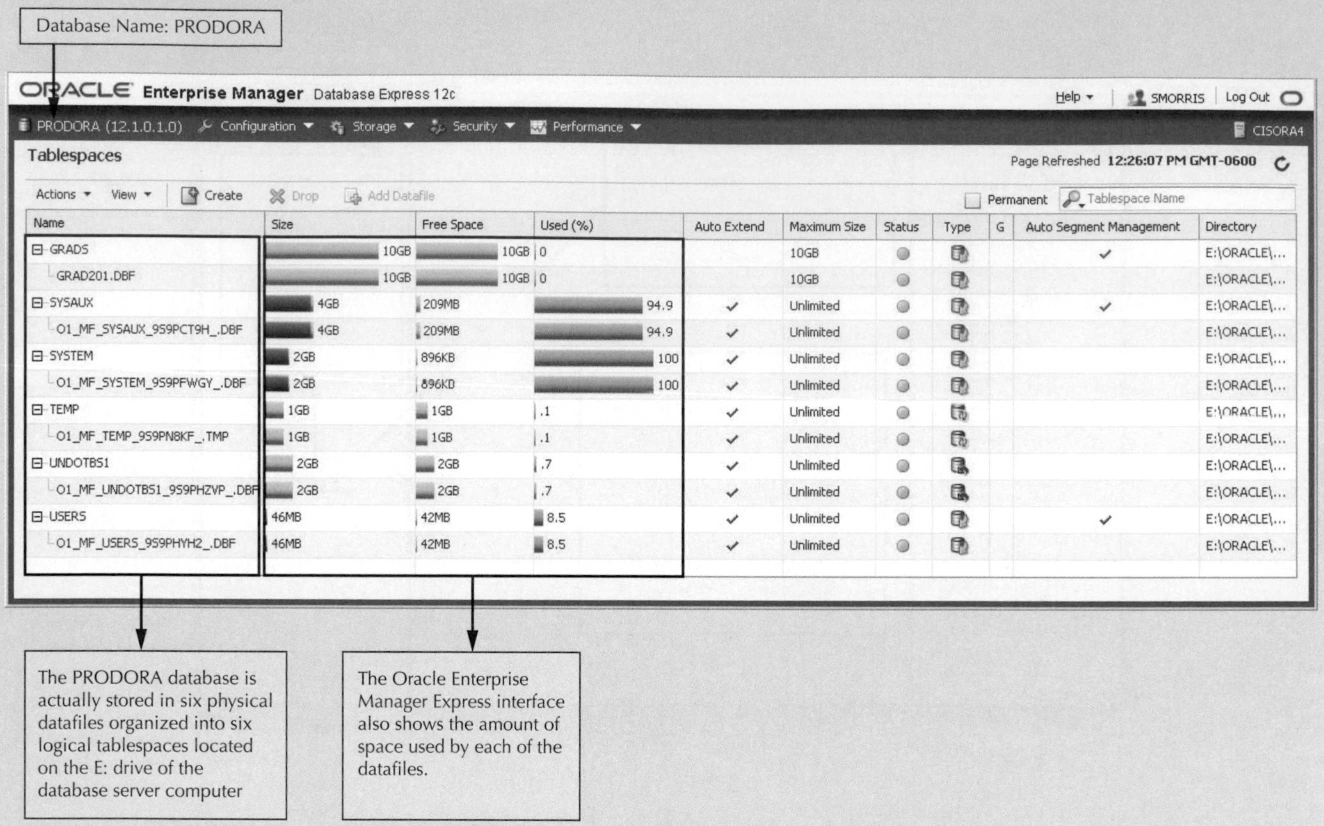

Database Name: PRODORA

The PRODORA database is actually stored in six physical datafiles organized into six logical tablespaces located on the E: drive of the database server computer

The Oracle Enterprise Manager Express interface also shows the amount of space used by each of the datafiles.

The Oracle Enterprise Manager Express GUI shows the data storage management characteristics for the PRODORA database.

- *Data transformation and presentation.* The DBMS transforms entered data to conform to required data structures. The DBMS relieves you of the chore of distinguishing between the logical data format and the physical data format. That is, the DBMS formats the physically retrieved data to make it conform to the user's logical expectations. For example, imagine an enterprise database used by a multinational company. An end user in England would expect to enter the date July 11, 2017, as "11/07/2017." In contrast, the same date would be entered in the United States as "07/11/2017." Regardless of the data presentation format, the DBMS must manage the date in the proper format for each country.

- *Security management.* The DBMS creates a security system that enforces user security and data privacy. Security rules determine which users can access the database, which data items each user can access, and which data operations (read, add, delete, or modify) the user can perform. This is especially important in multiuser database systems. Chapter 16, Database Administration and Security, examines data security and privacy issues in greater detail. All database users may be authenticated to the DBMS through a username and password or through biometric authentication such as a fingerprint scan. The DBMS uses this information to assign access privileges to various database components such as queries and reports.

- *Multiuser access control.* To provide data integrity and data consistency, the DBMS uses sophisticated algorithms to ensure that multiple users can access the database concurrently without compromising its integrity. Chapter 10, Transaction Management and Concurrency Control, covers the details of multiuser access control.

- *Backup and recovery management.* The DBMS provides backup and data recovery to ensure data safety and integrity. Current DBMS systems provide special utilities that allow the DBA to perform routine and special backup and restore procedures. Recovery management deals with the recovery of the database after a failure, such as a bad sector in the disk or a power failure. Such capability is critical to preserving the database's integrity. Chapter 16 covers backup and recovery issues.

- *Data integrity management.* The DBMS promotes and enforces integrity rules, thus minimizing data redundancy and maximizing data consistency. The data relationships stored in the data dictionary are used to enforce data integrity. Ensuring data integrity is especially important in transaction-oriented database systems. Data integrity and transaction management issues are addressed in Chapter 7, Introduction to Structured Query Language (SQL), and Chapter 10.

- *Database access languages and application programming interfaces.* The DBMS provides data access through a query language. A **query language** is a nonprocedural language—one that lets the user specify what must be done without having to specify how. **Structured Query Language (SQL)** is the de facto query language and data access standard supported by the majority of DBMS vendors. Chapter 7, Introduction to Structure Query Language (SQL), and Chapter 8, Advanced SQL, address the use of SQL. The DBMS also provides application programming interfaces to procedural languages such as COBOL, C, Java, Visual Basic.NET, and C#. In addition, the DBMS provides administrative utilities used by the DBA and the database designer to create, implement, monitor, and maintain the database.

- *Database communication interfaces.* A current-generation DBMS accepts end-user requests via multiple, different network environments. For example, the DBMS might provide access to the database via the Internet through the use of web browsers such as Mozilla Firefox, Google Chrome, or Microsoft Internet Explorer. In this environment, communications can be accomplished in several ways:

  - End users can generate answers to queries by filling in screen forms through their preferred web browser.

  - The DBMS can automatically publish predefined reports on a website.

  - The DBMS can connect to third-party systems to distribute information via email or other productivity applications.

Database communication interfaces are examined in greater detail in Chapter 12, Distributed Database Management Systems; in Chapter 15, Database Connectivity and Web Technologies; and in Appendix I, Databases in Electronic Commerce. (Appendixes are available at *www.cengagebrain.com*.)

Note

Why a Spreadsheet Is Not a Database

While a spreadsheet allows for the manipulation of data in a tabular format, it does not support even the most basic database functionality such as support for self-documentation through metadata, enforcement of data types or domains to ensure consistency of data within a column, defined relationships among tables, or constraints to ensure consistency of data across related tables. Most users lack the necessary training to recognize the limitations of spreadsheets for these types of tasks.

query language
A nonprocedural language that is used by a DBMS to manipulate its data. An example of a query language is SQL.

Structured Query Language (SQL)
A powerful and flexible relational database language composed of commands that enable users to create database and table structures, perform various types of data manipulation and data administration, and query the database to extract useful information.

1-7c Managing the Database System: A Shift in Focus

The introduction of a database system over the file system provides a framework in which strict procedures and standards can be enforced. Consequently, the role of the human component changes from an emphasis on programming (in the file system) to a focus on the broader aspects of managing the organization's data resources and on the administration of the complex database software itself.

The database system makes it possible to tackle far more sophisticated uses of the data resources, as long as the database is designed to make use of that power. The kinds of data structures created within the database and the extent of the relationships among them play a powerful role in determining the effectiveness of the database system.

Although the database system yields considerable advantages over previous data management approaches, database systems do carry significant disadvantages:

- *Increased costs.* Database systems require sophisticated hardware and software and highly skilled personnel. The cost of maintaining the hardware, software, and personnel required to operate and manage a database system can be substantial. Training, licensing, and regulation compliance costs are often overlooked when database systems are implemented.

- *Management complexity.* Database systems interface with many different technologies and have a significant impact on a company's resources and culture. The changes introduced by the adoption of a database system must be properly managed to ensure that they help advance the company's objectives. Because database systems hold crucial company data that are accessed from multiple sources, security issues must be assessed constantly.

- *Maintaining currency.* To maximize the efficiency of the database system, you must keep your system current. Therefore, you must perform frequent updates and apply the latest patches and security measures to all components. Because database technology advances rapidly, personnel training costs tend to be significant.

- *Vendor dependence.* Given the heavy investment in technology and personnel training, companies might be reluctant to change database vendors. As a consequence, vendors are less likely to offer pricing point advantages to existing customers, and those customers might be limited in their choice of database system components.

- *Frequent upgrade/replacement cycles.* DBMS vendors frequently upgrade their products by adding new functionality. Such new features often come bundled in new upgrade versions of the software. Some of these versions require hardware upgrades. Not only do the upgrades themselves cost money, it also costs money to train database users and administrators to properly use and manage the new features.

Now that you know what a database and DBMS are, and why they are necessary, you are ready to begin developing your career as a database professional.

1-8 Preparing for Your Database Professional Career

In this chapter, you were introduced to the concepts of data, information, databases, and DBMSs. You also learned that, regardless of what type of database you use (OLTP, OLAP, or NoSQL), or what type of database environment you are working in (e.g., Oracle, Microsoft, IBM, or Hadoop), the success of a database system greatly depends on how well the database structure is designed.

Throughout this book, you will learn the building blocks that lay the foundation for your career as a database professional. Understanding these building blocks and developing the skills to use them effectively will prepare you to work with databases at many different levels within an organization. A small sample of such career opportunities is shown in Table 1.3.

TABLE 1.3

DATABASE CAREER OPPORTUNITIES

JOB TITLE	DESCRIPTION	SAMPLE SKILLS REQUIRED
Database Developer	Create and maintain database-based applications	Programming, database fundamentals, SQL
Database Designer	Design and maintain databases	Systems design, database design, SQL
Database Administrator	Manage and maintain DBMS and databases	Database fundamentals, SQL, vendor courses
Database Analyst	Develop databases for decision support reporting	SQL, query optimization, data warehouses
Database Architect	Design and implementation of database environments (conceptual, logical, and physical)	DBMS fundamentals, data modeling, SQL, hardware knowledge, etc.
Database Consultant	Help companies leverage database technologies to improve business processes and achieve specific goals	Database fundamentals, data modeling, database design, SQL, DBMS, hardware, vendor-specific technologies, etc.
Database Security Officer	Implement security policies for data administration	DBMS fundamentals, database administration, SQL, data security technologies, etc.
Cloud Computing Data Architect	Design and implement the infrastructure for next-generation cloud database systems	Internet technologies, cloud storage technologies, data security, performance tuning, large databases, etc.

As you also learned in this chapter, database technologies are constantly evolving to address new challenges such as large databases, semistructured and unstructured data, increasing processing speed, and lowering costs. While database technologies can change quickly, the fundamental concepts and skills do not. It is our goal that after you learn the database essentials in this book, you will be ready to apply your knowledge and skills to work with traditional OLTP and OLAP systems as well as cutting-edge, complex database technologies such as the following:

- *Very Large Databases (VLDB).* Many vendors are addressing the need for databases that support large amounts of data, usually in the petabyte range. (A petabyte is more than 1,000 terabytes.) VLDB vendors include Oracle Exadata, IBM's Netezza, HP's Vertica, and Teradata. VLDB are now being overtaken in market interest by Big Data databases.

- *Big Data databases.* Products such as Cassandra (Facebook) and BigTable (Google) are using "columnar-database" technologies to support the needs of database applications that manage large amounts of "nontabular" data. See more about this topic in Chapter 2.

- *In-memory databases*. Most major database vendors also offer some type of in-memory database support to address the need for faster database processing. In-memory databases store most of their data in primary memory (RAM) rather than in slower secondary storage (hard disks). In-memory databases include IBM's solidDB and Oracle's TimesTen.

- *Cloud databases*. Companies can now use cloud database services to quickly add database systems to their environment while simultaneously lowering the total cost of ownership of a new DBMS. A cloud database offers all the advantages of a local DBMS, but instead of residing within your organization's network infrastructure, it resides on the Internet. See more about this topic in Chapter 15.

We address some of these topics in this book, but not all—no single book can cover the entire realm of database technologies. This book's primary focus is to help you learn database fundamentals, develop your database design skills, and master your SQL skills so you will have a head start in becoming a successful database professional. However, you first must learn about the tools at your disposal. In the next chapter, you will learn different approaches to data management and how these approaches influence your designs.

Summary

- Data consists of raw facts. Information is the result of processing data to reveal its meaning. Accurate, relevant, and timely information is the key to good decision making, and good decision making is the key to organizational survival in a global environment.

- Data is usually stored in a database. To implement a database and to manage its contents, you need a database management system (DBMS). The DBMS serves as the intermediary between the user and the database. The database contains the data you have collected and "data about data," known as metadata.

- Database design defines the database structure. A well-designed database facilitates data management and generates accurate and valuable information. A poorly designed database can lead to poor decision making, and poor decision making can lead to the failure of an organization.

- Databases can be classified according to the number of users supported, where the data is located, the type of data stored, the intended data usage, and the degree to which the data is structured.

- Databases evolved from manual and then computerized file systems. In a file system, data is stored in independent files, each requiring its own data management programs. Although this method of data management is largely outmoded, understanding its characteristics makes database design easier to comprehend.

- Some limitations of file system data management are that it requires extensive programming, system administration can be complex and difficult, making changes to existing structures is difficult, and security features are likely to be inadequate. Also, independent files tend to contain redundant data, leading to problems of structural and data dependence.

- Database management systems were developed to address the file system's inherent weaknesses. Rather than depositing data in independent files, a DBMS presents the database to the end user as a single data repository. This arrangement promotes data sharing, thus eliminating the potential problem of islands of information. In addition, the DBMS enforces data integrity, eliminates redundancy, and promotes data security.

- Knowledge of database technologies leads to many career opportunities in the ever-expanding IT industry. There is a variety of specialization within the database arena for a wide range of skills and expertise.

Key Terms

ad hoc query

analytical database

business intelligence

centralized database

cloud database

data

data anomaly

data dependence

data dictionary

data inconsistency

data independence

data integrity

data management

data processing (DP) specialist

data quality

data redundancy

data warehouse

database

database design

database management system (DBMS)

database system

desktop database

discipline-specific database

distributed database

enterprise database

Extensible Markup Language (XML)

field

file

general-purpose database

information

islands of information

knowledge

logical data format

metadata

multiuser database

NoSQL

online analytical processing (OLAP)

online transaction processing (OLTP) database

operational database

performance tuning

physical data format

production database

query

query language

query result set

record

semistructured data

single-user database

social media

structural dependence

structural independence

structured data

Structured Query Language (SQL)

transactional database

unstructured data

workgroup database

XML database

Online Content

Flashcards and crossword puzzles for key term practice are available at *www.cengage brain.com*.

Review Questions

1. Define each of the following terms:

   a. data

   b. field

   c. record

   d. file

2. What is data redundancy, and which characteristics of the file system can lead to it?

3. What is data independence, and why is it lacking in file systems?

4. What is a DBMS, and what are its functions?

5. What is structural independence, and why is it important?

6. Explain the differences among data, information, and a database.

7. What is the role of a DBMS, and what are its advantages? What are its disadvantages?

8. List and describe the different types of databases.

9. What are the main components of a database system?

10. What is metadata?

11. Explain why database design is important.

12. What are the potential costs of implementing a database system?

13. Use examples to compare and contrast unstructured and structured data. Which type is more prevalent in a typical business environment?

14. What are some basic database functions that a spreadsheet cannot perform?

15. What common problems does a collection of spreadsheets created by end users share with the typical file system?

16. Explain the significance of the loss of direct, hands-on access to business data that end users experienced with the advent of computerized data repositories.

17. Explain why the cost of ownership may be lower with a cloud database than with a traditional, company database.

Problems

Online Content

The file structures you see in this problem set are simulated in a Microsoft Access database named Ch01_ Problems, which is available at *www.cengagebrain.com*.

FIGURE P1.1 THE FILE STRUCTURE FOR PROBLEMS 1–4

PROJECT_CODE	PROJECT_MANAGER	MANAGER_PHONE	MANAGER_ADDRESS	PROJECT_BID_PRICE
21-5Z	Holly B. Parker	904-338-3416	3334 Lee Rd., Gainesville, FL 37123	16833460.00
25-2D	Jane D. Grant	615-898-9909	218 Clark Blvd., Nashville, TN 36362	12500000.00
25-5A	George F. Dorts	615-227-1245	124 River Dr., Franklin, TN 29185	32512420.00
25-9T	Holly B. Parker	904-338-3416	3334 Lee Rd., Gainesville, FL 37123	21563234.00
27-4Q	George F. Dorts	615-227-1245	124 River Dr., Franklin, TN 29185	10314545.00
29-2D	Holly B. Parker	904-338-3416	3334 Lee Rd., Gainesville, FL 37123	25559999.00
31-7P	William K. Moor	904-445-2719	216 Morton Rd., Stetson, FL 30155	56850000.00

Given the file structure shown in Figure P1.1, answer Problems 1–4.

1. How many records does the file contain? How many fields are there per record?

2. What problem would you encounter if you wanted to produce a listing by city? How would you solve this problem by altering the file structure?

3. If you wanted to produce a listing of the file contents by last name, area code, city, state, or zip code, how would you alter the file structure?

4. What data redundancies do you detect? How could those redundancies lead to anomalies?

FIGURE P1.5   THE FILE STRUCTURE FOR PROBLEMS 5–8

PROJ_NUM	PROJ_NAME	EMP_NUM	EMP_NAME	JOB_CODE	JOB_CHG_HOUR	PROJ_HOURS	EMP_PHONE
1	Hurricane	101	John D. Newson	EE	85.00	13.3	653-234-3245
1	Hurricane	105	David F. Schwann	CT	60.00	16.2	653-234-1123
1	Hurricane	110	Anne R. Ramoras	CT	60.00	14.3	615-233-5568
2	Coast	101	John D. Newson	EE	85.00	19.8	653-234-3254
2	Coast	108	June H. Sattlemeir	EE	85.00	17.5	905-554-7812
3	Satellite	110	Anne R. Ramoras	CT	62.00	11.6	615-233-5568
3	Satellite	105	David F. Schwann	CT	26.00	23.4	653-234-1123
3	Satelite	123	Mary D. Chen	EE	85.00	19.1	615-233-5432
3	Satellite	112	Allecia R. Smith	BE	85.00	20.7	615-678-6879

5. Identify and discuss the serious data redundancy problems exhibited by the file structure shown in Figure P1.5.

6. Looking at the EMP_NAME and EMP_PHONE contents in Figure P1.5, what change(s) would you recommend?

7. Identify the various data sources in the file you examined in Problem 5.

8. Given your answer to Problem 7, what new files should you create to help eliminate the data redundancies found in the file shown in Figure P1.5?

FIGURE P1.9   THE FILE STRUCTURE FOR PROBLEMS 9–10

BUILDING_CODE	ROOM_CODE	TEACHER_LNAME	TEACHER_FNAME	TEACHER_INITIAL	DAYS_TIME
KOM	204E	Williston	Horace	G	MWF 8:00-8:50
KOM	123	Cordoza	Maria	L	MWF 8:00-8:50
LDB	504	Patroski	Donald	J	TTh 1:00-2:15
KOM	34	Hawkins	Anne	W	MWF 10:00-10:50
JKP	225B	Risell	James		TTh 9:00-10:15
LDB	301	Robertson	Jeanette	P	TTh 9:00-10:15
KOM	204E	Cordoza	Maria	I	MWF 9:00-9:50
LDB	504	Williston	Horace	G	TTh 1:00-2:15
KOM	34	Cordoza	Maria	L	MWF 11:00-11:50
LDB	504	Patroski	Donald	J	MWF 2:00-2:50

9. Identify and discuss the serious data redundancy problems exhibited by the file structure shown in Figure P1.9. (The file is meant to be used as a teacher class assignment schedule. One of the many problems with data redundancy is the likely occurrence of data inconsistencies—two different initials have been entered for the teacher named Maria Cordoza.)

10. Given the file structure shown in Figure P1.9, what problem(s) might you encounter if building KOM were deleted?

11. Using your school's student information system, print your class schedule. The schedule probably would contain the student identification number, student name, class code, class name, class credit hours, class instructor name, the class meeting days and times, and the class room number. Use Figure P1.11 as a template to complete the following actions.

FIGURE P1.11 STUDENT SCHEDULE DATA FORMAT

STU_ID	STU_NAME	CLASS_CODE	CLASS_NAME	CLASS_CREDHRS	INSTR_NAME	CLASS_DAYS	CLASS_TIMES	ROOM

a. Create a spreadsheet using the template shown in Figure P1.11 and enter your current class schedule.

b. Enter the class schedule of two of your classmates into the same spreadsheet.

c. Discuss the redundancies and anomalies caused by this design.

Chapter 2

Data Models

In this chapter, you will learn:
- About data modeling and why data models are important
- About the basic data-modeling building blocks
- What business rules are and how they influence database design
- How the major data models evolved
- About emerging alternative data models and the needs they fulfill
- How data models can be classified by their level of abstraction

Preview

This chapter examines data modeling. Data modeling is the first step in the database design journey, serving as a bridge between real-world objects and the computer database.

One of the most vexing problems of database design is that designers, programmers, and end users see data in different ways. Consequently, different views of the same data can lead to database designs that do not reflect an organization's actual operation, thus failing to meet end-user needs and data efficiency requirements. To avoid such failures, database designers must obtain a precise description of the data's nature and many uses within the organization. Communication among database designers, programmers, and end users should be frequent and clear. Data modeling clarifies such communication by reducing the complexities of database design to more easily understood abstractions that define entities, relations, and data transformations.

First, you will learn some basic data-modeling concepts and how current data models developed from earlier models. Tracing the development of those database models will help you understand the database design and implementation issues that are addressed in the rest of this book. In chronological order, you will be introduced to the hierarchical and network models, the relational model, and the entity relationship (ER) model. You will also learn about the use of the entity relationship diagram (ERD) as a data-modeling tool and the different notations used for ER diagrams. Next, you will be introduced to the object-oriented model and the object/relational model. Then, you will learn about the emerging NoSQL data model and how it is being used to fulfill the current need to manage very large social media data sets efficiently and effectively. Finally, you will learn how various degrees of data abstraction help reconcile varying views of the same data.

Data Files and Available Formats

	MS Access	Oracle	MS SQL	My SQL		MS Access	Oracle	MS SQL	My SQL
CH02_InsureCo	✓	✓	✓	✓	CH02_DealCo	✓	✓	✓	✓
					CH02_TinyCollege	✓	✓	✓	✓

Data Files Available on cengagebrain.com

2-1 Data Modeling and Data Models

data modeling
The process of creating a specific data model for a determined problem domain.

data model
A representation, usually graphic, of a complex "real-world" data structure. Data models are used in the database design phase of the Database Life Cycle.

Database design focuses on how the database structure will be used to store and manage end-user data. **Data modeling**, the first step in designing a database, refers to the process of creating a specific data model for a determined problem domain. (A *problem domain* is a clearly defined area within the real-world environment, with a well-defined scope and boundaries that will be systematically addressed.) A **data model** is a relatively simple representation, usually graphical, of more complex real-world data structures. In general terms, a *model* is an abstraction of a more complex real-world object or event. A model's main function is to help you understand the complexities of the real-world environment. Within the database environment, a data model represents data structures and their characteristics, relations, constraints, transformations, and other constructs with the purpose of supporting a specific problem domain.

Note

The terms *data model* and *database model* are often used interchangeably. In this book, the term *database model* is used to refer to the implementation of a *data model* in a specific database system.

Data modeling is an iterative, progressive process. You start with a simple understanding of the problem domain, and as your understanding increases, so does the level of detail of the data model. When done properly, the final data model effectively is a "blueprint" with all the instructions to build a database that will meet all end-user requirements. This blueprint is narrative and graphical in nature, meaning that it contains both text descriptions in plain, unambiguous language and clear, useful diagrams depicting the main data elements.

Note

An implementation-ready data model should contain at least the following components:

- A description of the data structure that will store the end-user data
- A set of enforceable rules to guarantee the integrity of the data
- A data manipulation methodology to support the real-world data transformations

Traditionally, database designers relied on good judgment to help them develop a good data model. Unfortunately, good judgment is often in the eye of the beholder, and it often develops after much trial and error. For example, if each student in this class has to create a data model for a video store, it is very likely that each will come up with a different model. Which one would be correct? The simple answer is "the one that meets all the end-user requirements," and there may be more than one correct solution! Fortunately, database designers make use of existing data-modeling constructs and powerful database design tools that substantially diminish the potential for errors in database modeling. In the following sections, you will learn how existing data models are used to represent real-world data and how the different degrees of data abstraction facilitate data modeling.

2-2 **The Importance of Data Models**

Data models can facilitate interaction among the designer, the applications programmer, and the end user. A well-developed data model can even foster improved understanding of the organization for which the database design is developed. In short, data models are a communication tool. This important aspect of data modeling was summed up neatly by a client whose reaction was as follows: "I created this business, I worked with this business for years, and this is the first time I've really understood how all the pieces really fit together."

The importance of data modeling cannot be overstated. Data constitutes the most basic information employed by a system. Applications are created to manage data and to help transform data into information, but data is viewed in different ways by different people. For example, contrast the view of a company manager with that of a company clerk. Although both work for the same company, the manager is more likely to have an enterprise-wide view of company data than the clerk.

Even different managers view data differently. For example, a company president is likely to take a universal view of the data because he or she must be able to tie the company's divisions to a common (database) vision. A purchasing manager in the same company is likely to have a more restricted view of the data, as is the company's inventory manager. In effect, each department manager works with a subset of the company's data. The inventory manager is more concerned about inventory levels, while the purchasing manager is more concerned about the cost of items and about relationships with the suppliers of those items.

Applications programmers have yet another view of data, being more concerned with data location, formatting, and specific reporting requirements. Basically, applications programmers translate company policies and procedures from a variety of sources into appropriate interfaces, reports, and query screens.

The different users and producers of data and information often reflect the fable of the blind people and the elephant: the blind person who felt the elephant's trunk had quite a different view from the one who felt the elephant's leg or tail. A view of the whole elephant is needed. Similarly, a house is not a random collection of rooms; to build a house, a person should first have the overall view that is provided by blueprints. Likewise, a sound data environment requires an overall database blueprint based on an appropriate data model.

When a good database blueprint is available, it does not matter that an applications programmer's view of the data is different from that of the manager or the end user. Conversely, when a good database blueprint is not available, problems are likely to ensue. For instance, an inventory management program and an order entry system may use conflicting product-numbering schemes, thereby costing the company thousands or even millions of dollars.

Keep in mind that a house blueprint is an abstraction; you cannot live in the blueprint. Similarly, the data model is an abstraction; you cannot draw the required data out of the data model. Just as you are not likely to build a good house without a blueprint, you are equally unlikely to create a good database without first creating an appropriate data model.

2-3 **Data Model Basic Building Blocks**

The basic building blocks of all data models are entities, attributes, relationships, and constraints. An **entity** is a person, place, thing, or event about which data will be collected

entity
A person, place, thing, concept, or event for which data can be stored. See also *attribute*.

and stored. An entity represents a particular type of object in the real world, which means an entity is "distinguishable"—that is, each entity occurrence is unique and distinct. For example, a CUSTOMER entity would have many distinguishable customer occurrences, such as John Smith, Pedro Dinamita, and Tom Strickland. Entities may be physical objects, such as customers or products, but entities may also be abstractions, such as flight routes or musical concerts.

An **attribute** is a characteristic of an entity. For example, a CUSTOMER entity would be described by attributes such as customer last name, customer first name, customer phone number, customer address, and customer credit limit. Attributes are the equivalent of fields in file systems.

A **relationship** describes an association among entities. For example, a relationship exists between customers and agents that can be described as follows: an agent can serve many customers, and each customer may be served by one agent. Data models use three types of relationships: one-to-many, many-to-many, and one-to-one. Database designers usually use the shorthand notations 1:M or 1..*, M:N or *..*, and 1:1 or 1..1, respectively. (Although the M:N notation is a standard label for the many-to-many relationship, the label M:M may also be used.) The following examples illustrate the distinctions among the three relationships.

- **One-to-many (1:M or 1..*) relationship**. A painter creates many different paintings, but each is painted by only one painter. Thus, the painter (the "one") is related to the paintings (the "many"). Therefore, database designers label the relationship "PAINTER paints PAINTING" as 1:M. Note that entity names are often capitalized as a convention, so they are easily identified. Similarly, a customer (the "one") may generate many invoices, but each invoice (the "many") is generated by only a single customer. The "CUSTOMER generates INVOICE" relationship would also be labeled 1:M.

- **Many-to-many (M:N or *..*) relationship**. An employee may learn many job skills, and each job skill may be learned by many employees. Database designers label the relationship "EMPLOYEE learns SKILL" as M:N. Similarly, a student can take many classes and each class can be taken by many students, thus yielding the M:N label for the relationship expressed by "STUDENT takes CLASS."

- **One-to-one (1:1 or 1..1) relationship**. A retail company's management structure may require that each of its stores be managed by a single employee. In turn, each store manager, who is an employee, manages only a single store. Therefore, the relationship "EMPLOYEE manages STORE" is labeled 1:1.

The preceding discussion identified each relationship in both directions; that is, relationships are bidirectional:

- *One* CUSTOMER can generate *many* INVOICEs.
- Each of the *many* INVOICEs is generated by only *one* CUSTOMER.

A **constraint** is a restriction placed on the data. Constraints are important because they help to ensure data integrity. Constraints are normally expressed in the form of rules:

- An employee's salary must have values that are between 6,000 and 350,000.
- A student's GPA must be between 0.00 and 4.00.
- Each class must have one and only one teacher.

How do you properly identify entities, attributes, relationships, and constraints? The first step is to clearly identify the business rules for the problem domain you are modeling.

attribute
A characteristic of an entity or object. An attribute has a name and a data type.

relationship
An association between entities.

one-to-many (1:M or 1..*) relationship
Associations among two or more entities that are used by data models. In a 1:M relationship, one entity instance is associated with many instances of the related entity.

many-to-many (M:N or *..*) relationship
Association among two or more entities in which one occurrence of an entity is associated with many occurrences of a related entity and one occurrence of the related entity is associated with many occurrences of the first entity.

one-to-one (1:1 or 1..1) relationship
Associations among two or more entities that are used by data models. In a 1:1 relationship, one entity instance is associated with only one instance of the related entity.

constraint
A restriction placed on data, usually expressed in the form of rules. For example, "A student's GPA must be between 0.00 and 4.00." Constraints are important because they help to ensure data integrity.

2-4 **Business Rules**

When database designers go about selecting or determining the entities, attributes, and relationships that will be used to build a data model, they might start by gaining a thorough understanding of what types of data exist in an organization, how the data is used, and in what time frames it is used. But such data and information do not, by themselves, yield the required understanding of the total business. From a database point of view, the collection of data becomes meaningful only when it reflects properly defined *business rules*. A **business rule** is a brief, precise, and unambiguous description of a policy, procedure, or principle within a specific organization. In a sense, business rules are misnamed: they apply to *any* organization, large or small—a business, a government unit, a religious group, or a research laboratory—that stores and uses data to generate information.

Business rules derived from a detailed description of an organization's operations help to create and enforce actions within that organization's environment. Business rules must be rendered in writing and updated to reflect any change in the organization's operational environment.

Properly written business rules are used to define entities, attributes, relationships, and constraints. Any time you see relationship statements such as "an agent can serve many customers, and each customer can be served by only one agent," business rules are at work. You will see the application of business rules throughout this book, especially in the chapters devoted to data modeling and database design.

To be effective, business rules must be easy to understand and widely disseminated to ensure that every person in the organization shares a common interpretation of the rules. Business rules describe, in simple language, the main and distinguishing characteristics of the data *as viewed by the company*. Examples of business rules are as follows:

- A customer may generate many invoices.

- An invoice is generated by only one customer.

- A training session cannot be scheduled for fewer than 10 employees or for more than 30 employees.

Note that those business rules establish entities, relationships, and constraints. For example, the first two business rules establish two entities (CUSTOMER and INVOICE) and a 1:M relationship between those two entities. The third business rule establishes a constraint (no fewer than 10 people and no more than 30 people), two entities (EMPLOYEE and TRAINING), and also implies a relationship between EMPLOYEE and TRAINING.

2-4a Discovering Business Rules

The main sources of business rules are company managers, policy makers, department managers, and written documentation such as a company's procedures, standards, and operations manuals. A faster and more direct source of business rules is direct interviews with end users. Unfortunately, because perceptions differ, end users are sometimes a less reliable source when it comes to specifying business rules. For example, a maintenance department mechanic might believe that any mechanic can initiate a maintenance procedure, when actually only mechanics with inspection authorization can perform such a task. Such a distinction might seem trivial, but it can have major legal consequences. Although end users are crucial contributors to the development of business rules, *it pays to verify end-user perceptions*. Too often, interviews with several people who perform the

business rule
A description of a policy, procedure, or principle within an organization. For example, a pilot cannot be on duty for more than 10 hours during a 24-hour period, or a professor may teach up to four classes during a semester.

same job yield very different perceptions of what the job components are. While such a discovery may point to "management problems," that general diagnosis does not help the database designer. The database designer's job is to reconcile such differences and verify the results of the reconciliation to ensure that the business rules are appropriate and accurate.

The process of identifying and documenting business rules is essential to database design for several reasons:

- It helps to standardize the company's view of data.
- It can be a communication tool between users and designers.
- It allows the designer to understand the nature, role, and scope of the data.
- It allows the designer to understand business processes.
- It allows the designer to develop appropriate relationship participation rules and constraints and to create an accurate data model.

Of course, not all business rules can be modeled. For example, a business rule that specifies "no pilot can fly more than 10 hours within any 24-hour period" cannot be modeled in the database model directly. However, such a business rule can be represented and enforced by application software.

2-4b Translating Business Rules into Data Model Components

Business rules set the stage for the proper identification of entities, attributes, relationships, and constraints. In the real world, names are used to identify objects. If the business environment wants to keep track of the objects, there will be specific business rules for the objects. As a general rule, a noun in a business rule will translate into an entity in the model, and a verb (active or passive) that associates the nouns will translate into a relationship among the entities. For example, the business rule "a customer may generate many invoices" contains two nouns (*customer* and *invoices*) and a verb (*generate*) that associates the nouns. From this business rule, you could deduce the following:

- *Customer* and *invoice* are objects of interest for the environment and should be represented by their respective entities.
- There is a *generate* relationship between customer and invoice.

To properly identify the type of relationship, you should consider that relationships are bidirectional; that is, they go both ways. For example, the business rule "a customer may generate many invoices" is complemented by the business rule "an invoice is generated by only one customer." In that case, the relationship is one-to-many (1:M). Customer is the "1" side, and invoice is the "many" side.

As a general rule, to properly identify the relationship type, you should ask two questions:

- How many instances of B are related to one instance of A?
- How many instances of A are related to one instance of B?

For example, you can assess the relationship between student and class by asking two questions:

- In how many classes can one student enroll? Answer: many classes.
- How many students can enroll in one class? Answer: many students.

Therefore, the relationship between student and class is many-to-many (M:N). You will have many opportunities to determine the relationships between entities as you proceed through this book, and soon the process will become second nature.

2-4c Naming Conventions

During the translation of business rules to data model components, you identify entities, attributes, relationships, and constraints. This identification process includes naming the object in a way that makes it unique and distinguishable from other objects in the problem domain. Therefore, it is important to pay special attention to how you name the objects you are discovering.

Entity names should be descriptive of the objects in the business environment and use terminology that is familiar to the users. An attribute name should also be descriptive of the data represented by that attribute. It is also a good practice to prefix the name of an attribute with the name or abbreviation of the entity in which it occurs. For example, in the CUSTOMER entity, the customer's credit limit may be called CUS_CREDIT_LIMIT. The CUS indicates that the attribute is descriptive of the CUSTOMER entity, while CREDIT_LIMIT makes it easy to recognize the data that will be contained in the attribute. This will become increasingly important in later chapters when you learn about the need to use common attributes to specify relationships between entities. The use of a proper naming convention will improve the data model's ability to facilitate communication among the designer, application programmer, and the end user. In fact, a proper naming convention can go a long way toward making your model self-documenting.

2-5 The Evolution of Data Models

The quest for better data management has led to several models that attempt to resolve the previous model's critical shortcomings and to provide solutions to ever-evolving data management needs. These models represent schools of thought as to what a database is, what it should do, the types of structures that it should employ, and the technology that would be used to implement these structures. Perhaps confusingly, these models are called data models, as are the graphical data models discussed earlier in this chapter. This section gives an overview of the major data models in roughly chronological order. You will discover that many of the "new" database concepts and structures bear a remarkable resemblance to some of the "old" data model concepts and structures. Table 2.1 traces the evolution of the major data models.

2-5a Hierarchical and Network Models

The **hierarchical model** was developed in the 1960s to manage large amounts of data for complex manufacturing projects, such as the Apollo rocket that landed on the moon in 1969. The model's basic logical structure is represented by an upside-down tree. The hierarchical structure contains levels, or segments. A **segment** is the equivalent of a file system's record type. Within the hierarchy, a higher layer is perceived as the parent of the segment directly beneath it, which is called the child. The hierarchical model depicts a set of one-to-many (1:M) relationships between a parent and its children segments. (Each parent can have many children, but each child has only one parent.)

The **network model** was created to represent complex data relationships more effectively than the hierarchical model, to improve database performance, and to impose a database standard. In the network model, the user perceives the network database as a

hierarchical model
An early database model whose basic concepts and characteristics formed the basis for subsequent database development. This model is based on an upside-down tree structure in which each record is called a segment. The top record is the root segment. Each segment has a 1:M relationship to the segment directly below it.

segment
In the hierarchical data model, the equivalent of a file system's record type.

TABLE 2.1

EVOLUTION OF MAJOR DATA MODELS

GENERATION	TIME	DATA MODEL	EXAMPLES	COMMENTS
First	1960s–1970s	File system	VMS/VSAM	Used mainly on IBM mainframe systems Managed records, not relationships
Second	1970s	Hierarchical and network	IMS, ADABAS, IDS-II	Early database systems Navigational access
Third	Mid-1970s	Relational	DB2 Oracle MS SQL Server MySQL	Conceptual simplicity Entity relationship (ER) modeling and support for relational data modeling
Fourth	Mid-1980s	Object-oriented Object/relational (O/R)	Versant Objectivity/DB DB2 UDB Oracle 12c	Object/relational supports object data types Star Schema support for data warehousing Web databases become common
Fifth	Mid-1990s	XML Hybrid DBMS	dbXML Tamino DB2 UDB Oracle 12c MS SQL Server	Unstructured data support O/R model supports XML documents Hybrid DBMS adds object front end to relational databases Support large databases (terabyte size)
Emerging Models: NoSQL	Early 2000s to present	Key-value store Column store	SimpleDB (Amazon) BigTable (Google) Cassandra (Apache) MongoDB Riak	Distributed, highly scalable High performance, fault tolerant Very large storage (petabytes) Suited for sparse data Proprietary application programming interface (API)

network model
An early data model that represented data as a collection of record types in 1:M relationships.

schema
A logical grouping of database objects, such as tables, indexes, views, and queries, that are related to each other.

subschema
The portion of the database that interacts with application programs.

data manipulation language (DML)
The set of commands that allows an end user to manipulate the data in the database, such as SELECT, INSERT, UPDATE, DELETE, COMMIT, and ROLLBACK.

collection of records in 1:M relationships. However, unlike the hierarchical model, the network model allows a record to have more than one parent. While the network database model is generally not used today, the definitions of standard database *concepts* that emerged with the network model are still used by modern data models:

- The **schema** is the conceptual organization of the entire database as viewed by the database administrator.

- The **subschema** defines the portion of the database "seen" by the application programs that actually produce the desired information from the data within the database.

- A **data manipulation language** (DML) defines the environment in which data can be managed and is used to work with the data in the database.

- A schema **data definition language** (DDL) enables the database administrator to define the schema components.

As information needs grew and more sophisticated databases and applications were required, the network model became too cumbersome. The lack of ad hoc query capability put heavy pressure on programmers to generate the code required to produce even the simplest reports. Although the existing databases provided limited data independence, any structural change in the database could still produce havoc in all application programs that drew data from the database. Because of the disadvantages of the hierarchical and network models, they were largely replaced by the relational data model in the 1980s.

2-5b The Relational Model

The **relational model** was introduced in 1970 by E. F. Codd of IBM in his landmark paper "A Relational Model of Data for Large Shared Databanks" (*Communications of the ACM*, June 1970, pp. 377–387). The relational model represented a major breakthrough for both users and designers. To use an analogy, the relational model produced an "automatic transmission" database to replace the "standard transmission" databases that preceded it. Its conceptual simplicity set the stage for a genuine database revolution.

Note

The relational database model presented in this chapter is an introduction and an overview. A more detailed discussion is in Chapter 3, The Relational Database Model. In fact, the relational model is so important that it will serve as the basis for discussions in most of the remaining chapters.

The relational model's foundation is a mathematical concept known as a relation. To avoid the complexity of abstract mathematical theory, you can think of a **relation** (sometimes called a **table**) as a two-dimensional structure composed of intersecting rows and columns. Each row in a relation is called a **tuple**. Each column represents an attribute. The relational model also describes a precise set of data manipulation constructs based on advanced mathematical concepts.

In 1970, Codd's work was considered ingenious but impractical. The relational model's conceptual simplicity was bought at the expense of computer overhead; computers at that time lacked the power to implement the relational model. Fortunately, computer power grew exponentially, as did operating system efficiency. Better yet, the cost of computers diminished rapidly as their power grew. Today, even PCs, which cost a fraction of what their mainframe ancestors cost, can run sophisticated relational database software such as Oracle, DB2, Microsoft SQL Server, MySQL, and other mainframe relational software.

The relational data model is implemented through a very sophisticated **relational database management system (RDBMS)**. The RDBMS performs the same basic functions provided by the hierarchical and network DBMS systems, in addition to a host of other functions that make the relational data model easier to understand and implement (as outlined in Chapter 1, in the DBMS Functions section).

Arguably the most important advantage of the RDBMS is its ability to hide the complexities of the relational model from the user. The RDBMS manages all of the physical details, while the user sees the relational database as a collection of tables in which data is stored. The user can manipulate and query the data in a way that seems intuitive and logical.

Tables are related to each other through the sharing of a common attribute (a value in a column). For example, the CUSTOMER table in Figure 2.1 might contain a sales agent's number that is also contained in the AGENT table.

The common link between the CUSTOMER and AGENT tables enables you to match the customer to his or her sales agent, even though the customer data is stored in one table and the sales representative data is stored in another table. For example, you can easily determine that customer Dunne's agent is Alex Alby because for customer Dunne, the CUSTOMER table's AGENT_CODE is 501, which matches the AGENT table's

---

data definition language (DDL)
The language that allows a database administrator to define the database structure, schema, and subschema.

relational model
Developed by E. F. Codd of IBM in 1970, the relational model is based on mathematical set theory and represents data as independent relations. Each relation (table) is conceptually represented as a two-dimensional structure of intersecting rows and columns. The relations are related to each other through the sharing of common entity characteristics (values in columns).

table (relation)
A logical construct perceived to be a two-dimensional structure composed of intersecting rows (entities) and columns (attributes) that represents an entity set in the relational model.

tuple
In the relational model, a table row.

relational database management system (RDBMS)
A collection of programs that manages a relational database. The RDBMS software translates a user's logical requests (queries) into commands that physically locate and retrieve the requested data.

FIGURE 2.1 LINKING RELATIONAL TABLES

Table name: AGENT (first six attributes)   Database name: Ch02_InsureCo

AGENT_CODE	AGENT_LNAME	AGENT_FNAME	AGENT_INITIAL	AGENT_AREACODE	AGENT_PHONE
501	Alby	Alex	B	713	228-1249
502	Hahn	Leah	F	615	882-1244
503	Okon	John	T	615	123-5589

Link through AGENT_CODE

Table name: CUSTOMER

CUS_CODE	CUS_LNAME	CUS_FNAME	CUS_INITIAL	CUS_AREACODE	CUS_PHONE	CUS_INSURE_TYPE	CUS_INSURE_AMT	CUS_RENEW_DATE	AGENT_CODE
10010	Ramas	Alfred	A	615	844-2573	T1	100.00	05-Apr-2016	502
10011	Dunne	Leona	K	713	894-1230	T1	250.00	16-Jun-2016	501
10012	Smith	Kathy	W	615	894-2285	S2	150.00	29-Jan-2017	502
10013	Olowski	Paul	F	615	894-2180	S1	300.00	14-Oct-2016	502
10014	Orlando	Myron		615	222-1672	T1	100.00	28-Dec-2017	501
10015	O'Brian	Amy	B	713	442-3381	T2	850.00	22-Sep-2016	503
10016	Brown	James	G	615	297-1228	S1	120.00	25-Mar-2017	502
10017	Williams	George		615	290-2556	S1	250.00	17-Jul-2016	503
10018	Farriss	Anne	G	713	382-7185	T2	100.00	03-Dec-2016	501
10019	Smith	Olette	K	615	297-3809	S2	500.00	14-Mar-2017	503

Online Content

This chapter's databases are available at *www.cengagebrain.com*. For example, the contents of the AGENT and CUSTOMER tables shown in Figure 2.1 are in the database named Ch02_InsureCo.

AGENT_CODE for Alex Alby. Although the tables are independent of one another, you can easily associate the data between tables. The relational model provides a minimum level of controlled redundancy to eliminate most of the redundancies commonly found in file systems.

The relationship type (1:1, 1:M, or M:N) is often shown in a relational schema, an example of which is shown in Figure 2.2. A **relational diagram** is a representation of the relational database's entities, the attributes within those entities, and the relationships between those entities.

In Figure 2.2, the relational diagram shows the connecting fields (in this case, AGENT_CODE) and the relationship type (1:M). Microsoft Access, the database software application used to generate Figure 2.2, employs the infinity symbol (∞) to indicate the "many" side. In this example, the CUSTOMER represents the "many" side because an AGENT can have many CUSTOMERs. The AGENT represents the "1" side because each CUSTOMER has only one AGENT.

A relational table stores a collection of related entities. In this respect, the relational database table resembles a file, but there is a crucial difference between a table and a file:

relational diagram
A graphical representation of a relational database's entities, the attributes within those entities, and the relationships among the entities.

FIGURE 2.2 A RELATIONAL DIAGRAM

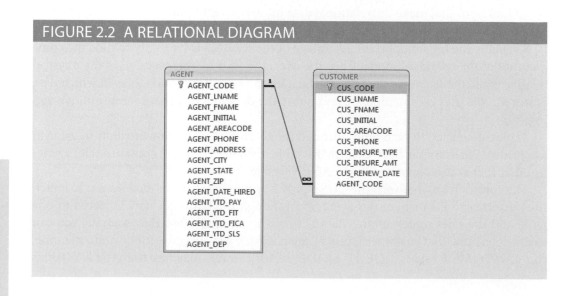

a table yields complete data and structural independence because it is a purely logical structure. How the data is physically stored in the database is of no concern to the user or the designer; the perception is what counts. This property of the relational data model, which is explored in depth in the next chapter, became the source of a real database revolution.

Another reason for the relational data model's rise to dominance is its powerful and flexible query language. Most relational database software uses Structured Query Language (SQL), which allows the user to specify what must be done without specifying how. The RDBMS uses SQL to translate user queries into instructions for retrieving the requested data. SQL makes it possible to retrieve data with far less effort than any other database or file environment.

From an end-user perspective, any SQL-based relational database application involves three parts: a user interface, a set of tables stored in the database, and the SQL "engine." Each of these parts is explained as follows:

- *The end-user interface.* Basically, the interface allows the end user to interact with the data (by automatically generating SQL code). Each interface is a product of the software vendor's idea of meaningful interaction with the data. You can also design your own customized interface with the help of application generators that are now standard fare in the database software arena.

- *A collection of tables stored in the database.* In a relational database, all data is perceived to be stored in tables. The tables simply "present" the data to the end user in a way that is easy to understand. Each table is independent. Rows in different tables are related by common values in common attributes.

- *SQL engine.* Largely hidden from the end user, the SQL engine executes all queries, or data requests. Keep in mind that the SQL engine is part of the DBMS software. The end user uses SQL to create table structures and to perform data access and table maintenance. The SQL engine processes all user requests—largely behind the scenes and without the end user's knowledge. Hence, SQL is said to be a declarative language that tells what must be done but not how. (You will learn more about the SQL engine in Chapter 11, Database Performance Tuning and Query Optimization.)

Because the RDBMS performs some tasks behind the scenes, it is not necessary to focus on the physical aspects of the database. Instead, the following chapters concentrate on the logical portion of the relational database and its design. Furthermore, SQL is covered in detail in Chapter 7, Introduction to Structured Query Language (SQL), and in Chapter 8, Advanced SQL.

2-5c The Entity Relationship Model

The conceptual simplicity of relational database technology triggered the demand for RDBMSs. In turn, the rapidly increasing requirements for transaction and information created the need for more complex database implementation structures, thus creating the need for more effective database design tools. (Building a skyscraper requires more detailed design activities than building a doghouse, for example.)

Complex design activities require conceptual simplicity to yield successful results. Although the relational model was a vast improvement over the hierarchical and network models, it still lacked the features that would make it an effective database *design* tool. Because it is easier to examine structures graphically than to describe them in text, database designers prefer to use a graphical tool in which entities and their relationships are pictured. Thus, the **entity relationship (ER) model,** or **ERM**, has become a widely accepted standard for data modeling.

Peter Chen first introduced the ER data model in 1976; the graphical representation of entities and their relationships in a database structure quickly became popular because it *complemented* the relational data model concepts. The relational data model and ERM combined to provide the foundation for tightly structured database design. ER models are normally represented in an **entity relationship diagram (ERD)**, which uses graphical representations to model database components. You will learn how to use ERDs to design databases in Chapter 4, Entity Relationship (ER) Modeling.

The ER model is based on the following components:

- *Entity*. Earlier in this chapter, an entity was defined as anything about which data will be collected and stored. An entity is represented in the ERD by a rectangle, also known as an entity box. The name of the entity, a noun, is written in the center of the rectangle. The entity name is generally written in capital letters and in singular form: PAINTER rather than PAINTERS, and EMPLOYEE rather than EMPLOYEES. Usually, when applying the ERD to the relational model, an entity is mapped to a relational table. Each row in the relational table is known as an **entity instance** or **entity occurrence** in the ER model. A collection of like entities is known as an **entity set**. For example, you can think of the AGENT file in Figure 2.1 as a collection of three agents (*entities*) in the AGENT *entity* set. Technically speaking, the ERD depicts entity sets. Unfortunately, ERD designers use the word *entity* as a substitute for *entity set*, and this book will conform to that established practice when discussing any ERD and its components.

- Each entity consists of a set of *attributes* that describes particular characteristics of the entity. For example, the entity EMPLOYEE will have attributes such as a Social Security number, a last name, and a first name. (Chapter 4 explains how attributes are included in the ERD.)

- *Relationships*. Relationships describe associations among data. Most relationships describe associations between two entities. When the basic data model components were introduced, three types of data relationships were illustrated: one-to-many (1:M), many-to-many (M:N), and one-to-one (1:1). The ER model uses the term **connectivity** to label the relationship types. The name of the relationship is usually an active or passive verb. For example, a PAINTER *paints* many PAINTINGs, an EMPLOYEE *learns* many SKILLs, and an EMPLOYEE *manages* a STORE.

Figure 2.3 shows the different types of relationships using three ER notations: the original **Chen notation**, the **Crow's Foot notation**, and the newer **class diagram notation**, which is part of the Unified Modeling Language (UML).

The left side of the ER diagram shows the Chen notation, based on Peter Chen's landmark paper. In this notation, the connectivities are written next to each entity box. Relationships are represented by a diamond connected to the related entities through a relationship line. The relationship name is written inside the diamond.

The middle of Figure 2.3 illustrates the Crow's Foot notation. The name *Crow's Foot* is derived from the three-pronged symbol used to represent the "many" side of the relationship. As you examine the basic Crow's Foot ERD in Figure 2.3, note that the connectivities are represented by symbols. For example, the "1" is represented by a short line segment, and the "M" is represented by the three-pronged "crow's foot." In this example, the relationship name is written above the relationship line.

---

entity relationship (ER) model (ERM)
A data model that describes relationships (1:1, 1:M, and M:N) among entities at the conceptual level with the help of ER diagrams. The model was developed by Peter Chen.

entity relationship diagram (ERD)
A diagram that depicts an entity relationship model's entities, attributes, and relations.

entity instance (entity occurrence)
A row in a relational table.

entity set
A collection of like entities.

connectivity
The type of relationship between entities. Classifications include 1:1, 1:M, and M:N.

Chen notation
See *entity relationship (ER) model.*

Crow's Foot notation
A representation of the entity relationship diagram that uses a three-pronged symbol to represent the "many" sides of the relationship.

class diagram notation
The set of symbols used in the creation of class diagrams.

FIGURE 2.3   THE ER MODEL NOTATIONS

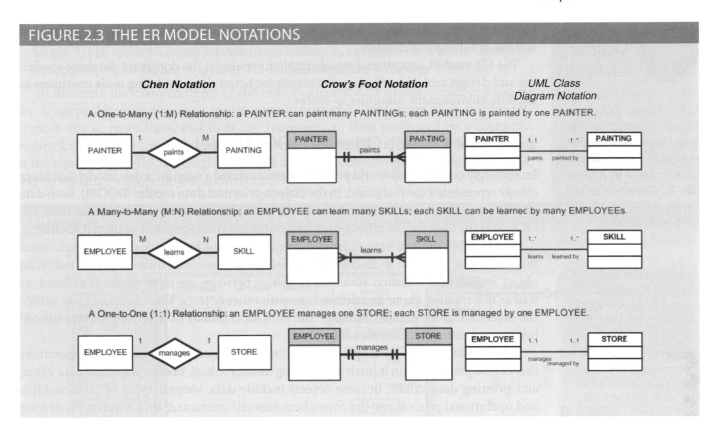

The right side of Figure 2.3 shows the UML notation (also known as the UML class notation). Note that the connectivities are represented by lines with symbols (1..1, 1..*). Also, the UML notation uses names in both sides of the relationship. For example, to read the relationship between PAINTER and PAINTING, note the following:

- A PAINTER "paints" one to many PAINTINGs, as indicated by the 1..* symbol.

- A PAINTING is "painted by" one and only one PAINTER, as indicated by the 1..1 symbol.

Note

Many-to-many (M:N) relationships exist at a conceptual level, and you should know how to recognize them. However, you will learn in Chapter 3 that M:N relationships are not appropriate in a relational model. For that reason, Microsoft Visio does not support the M:N relationship directly. Therefore, to illustrate the existence of an M:N relationship using Visio, you have to change the line style of the connector (see Appendix A, Designing Databases with Visio Professional: A Tutorial, at *www.cengagebrain.com*).

Online Content

Aside from the Chen, Crow's Foot, and UML notations, there are other ER model notations. For a summary of ER model notation symbols, see Appendix E, Comparison of ER Model Notations, at *www.cengagebrain.com*.

In Figure 2.3, entities and relationships are shown in a horizontal format, but they may also be oriented vertically. The entity location and the order in which the entities are presented are immaterial; just remember to read a 1:M relationship from the "1" side to the "M" side.

The Crow's Foot notation is used as the design standard in this book. However, the Chen notation is used to illustrate some of the ER modeling concepts whenever necessary. Most data modeling tools let you select the Crow's Foot or UML class diagram notation.

TABLE 2.2

ADVANTAGES AND DISADVANTAGES OF VARIOUS DATABASE MODELS

DATA MODEL	DATA INDEPENDENCE	STRUCTURAL INDEPENDENCE	ADVANTAGES	DISADVANTAGES
Hierarchical	Yes	No	1. It promotes data sharing. 2. Parent/child relationship promotes conceptual simplicity. 3. Database security is provided and enforced by DBMS. 4. Parent/child relationship promotes data integrity. 5. It is efficient with 1:M relationships.	1. Complex implementation requires knowledge of physical data storage characteristics. 2. Navigational system yields complex application development, management, and use; requires knowledge of hierarchical path. 3. Changes in structure require changes in all application programs. 4. There are implementation limitations (no multiparent or M:N relationships). 5. There is no data definition or data manipulation language in the DBMS. 6. There is a lack of standards.
Network	Yes	No	1. Conceptual simplicity is at least equal to that of the hierarchical model. 2. It handles more relationship types, such as M:N and multiparent. 3. Data access is more flexible than in hierarchical and file system models. 4. Data owner/member relationship promotes data integrity. 5. There is conformance to standards. 6. It includes data definition language (DDL) and data manipulation language (DML) in DBMS.	1. System complexity limits efficiency—still a navigational system. 2. Navigational system yields complex implementation, application development, and management. 3. Structural changes require changes in all application programs.
Relational	Yes	Yes	1. Structural independence is promoted by the use of independent tables. Changes in a table's structure do not affect data access or application programs. 2. Tabular view substantially improves conceptual simplicity, thereby promoting easier database design, implementation, management, and use. 3. Ad hoc query capability is based on SQL. 4. Powerful RDBMS isolates the end user from physical-level details and improves implementation and management simplicity.	1. The RDBMS requires substantial hardware and system software overhead. 2. Conceptual simplicity gives relatively untrained people the tools to use a good system poorly, and if unchecked, it may produce the same data anomalies found in file systems. 3. It may promote islands of information problems as individuals and departments can easily develop their own applications.
Entity relationship	Yes	Yes	1. Visual modeling yields exceptional conceptual simplicity. 2. Visual representation makes it an effective communication tool. 3. It is integrated with the dominant relational model.	1. There is limited constraint representation. 2. There is limited relationship representation. 3. There is no data manipulation language. 4. Loss of information content occurs when attributes are removed from entities to avoid crowded displays. (This limitation has been addressed in subsequent graphical versions.)
Object-oriented	Yes	Yes	1. Semantic content is added. 2. Visual representation includes semantic content. 3. Inheritance promotes data integrity.	1. Slow development of standards caused vendors to supply their own enhancements, thus eliminating a widely accepted standard. 2. It is a complex navigational system. 3. There is a steep learning curve. 4. High system overhead slows transactions.
NoSQL	Yes	Yes	1. High scalability, availability, and fault tolerance are provided. 2. It uses low-cost commodity hardware. 3. It supports Big Data. 4. Key-value model improves storage efficiency.	1. Complex programming is required. 2. There is no relationship support—only by application code. 3. There is no transaction integrity support. 4. In terms of data consistency, it provides an eventually consistent model.

TABLE 2.3

DATA MODEL BASIC TERMINOLOGY COMPARISON

REAL WORLD	EXAMPLE	FILE PROCESSING	HIERARCHICAL MODEL	NETWORK MODEL	RELATIONAL MODEL	ER MODEL	OO MODEL
A group of vendors	Vendor file cabinet	File	Segment type	Record type	Table	Entity set	Class
A single vendor	Global supplies	Record	Segment occurrence	Current record	Row (tuple)	Entity occurrence	Object instance
The contact name	Johnny Ventura	Field	Segment field	Record field	Table attribute	Entity attribute	Object attribute
The vendor identifier	G12987	Index	Sequence field	Record key	Key	Entity identifier	Object identifier

Note: For additional information about the terms used in this table, consult the corresponding chapters and online appendixes that accompany this book. For example, if you want to know more about the OO model, refer to Appendix G, Object-Oriented Databases.

exist without the basic conceptual framework created by the designer. Designing a usable database follows the same basic process. That is, a database designer starts with an abstract view of the overall data environment and adds details as the design comes closer to implementation. Using levels of abstraction can also be very helpful in integrating multiple (and sometimes conflicting) views of data at different levels of an organization.

In the early 1970s, the **American National Standards Institute (ANSI)** Standards Planning and Requirements Committee (SPARC) defined a framework for data modeling based on degrees of data abstraction. The resulting ANSI/SPARC architecture defines three levels of data abstraction: external, conceptual, and internal. You can use this framework to better understand database models, as shown in Figure 2.7. In the figure, the ANSI/SPARC framework has been expanded with the addition of a *physical* model to explicitly address physical-level implementation details of the internal model.

FIGURE 2.7 DATA ABSTRACTION LEVELS

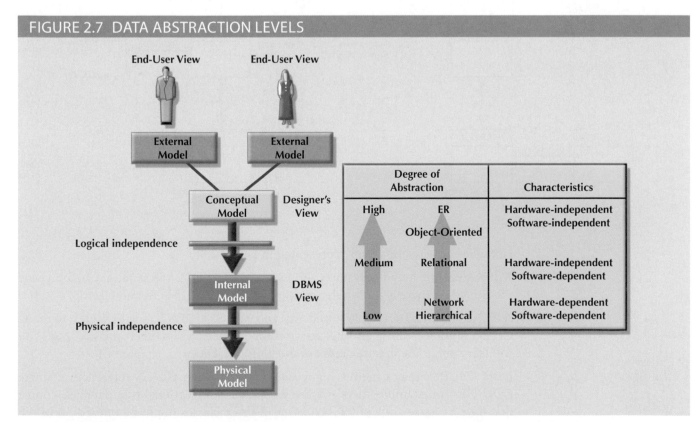

2-6a The External Model

The **external model** is the end users' view of the data environment. The term *end users* refers to people who use the application programs to manipulate the data and generate information. End users usually operate in an environment in which an application has a specific business unit focus. Companies are generally divided into several business units, such as sales, finance, and marketing. Each business unit is subject to specific constraints and requirements, and each one uses a subset of the overall data in the organization. Therefore, end users within those business units view their data subsets as separate from or external to other units within the organization.

Because data is being modeled, ER diagrams will be used to represent the external views. A specific representation of an external view is known as an **external schema**. To illustrate the external model's view, examine the data environment of Tiny College.

Figure 2.8 presents the external schemas for two Tiny College business units: student registration and class scheduling. Each external schema includes the appropriate entities, relationships, processes, and constraints imposed by the business unit. Also note that *although the application views are isolated from each other, each view shares a common entity with the other view*. For example, the registration and scheduling external schemas share the entities CLASS and COURSE.

external model
The application programmer's view of the data environment. Given its business focus, an external model works with a data subset of the global database schema.

external schema
The specific representation of an external view; the end user's view of the data environment.

FIGURE 2.8 EXTERNAL MODELS FOR TINY COLLEGE

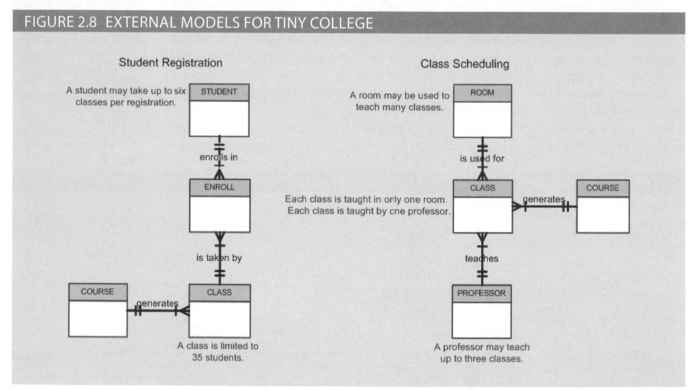

Note the entity relationships represented in Figure 2.8:

- A PROFESSOR may teach many CLASSes, and each CLASS is taught by only one PROFESSOR; there is a 1:M relationship between PROFESSOR and CLASS.

- A CLASS may ENROLL many students, and each STUDENT may ENROLL in many CLASSes, thus creating an M:N relationship between STUDENT and CLASS. (You will learn about the precise nature of the ENROLL entity in Chapter 4.)

- Each COURSE may generate many CLASSes, but each CLASS references a single COURSE. For example, there may be several classes (sections) of a database course that have a course code of CIS-420. One of those classes might be offered on MWF from 8:00 a.m. to 8:50 a.m., another might be offered on MWF from 1:00 p.m. to 1:50 p.m., while a third might be offered on Thursdays from 6:00 p.m. to 8:40 p.m. Yet, all three classes have the course code CIS-420.

- Finally, a CLASS requires one ROOM, but a ROOM may be scheduled for many CLASSes. That is, each classroom may be used for several classes: one at 9:00 a.m., one at 11:00 a.m., and one at 1:00 p.m., for example. In other words, there is a 1:M relationship between ROOM and CLASS.

The use of external views that represent subsets of the database has some important advantages:

- It is easy to identify specific data required to support each business unit's operations.

- It makes the designer's job easy by providing feedback about the model's adequacy. Specifically, the model can be checked to ensure that it supports all processes as defined by their external models, as well as all operational requirements and constraints.

- It helps to ensure *security* constraints in the database design. Damaging an entire database is more difficult when each business unit works with only a subset of data.

- It makes application program development much simpler.

2-6b The Conceptual Model

The **conceptual model** represents a global view of the entire database by the entire organization. That is, the conceptual model integrates all external views (entities, relationships, constraints, and processes) into a single global view of the data in the enterprise, as shown in Figure 2.9. Also known as a **conceptual schema**, it is the basis for the identification and high-level description of the main data objects (avoiding any database model-specific details).

The most widely used conceptual model is the ER model. Remember that the ER model is illustrated with the help of the ERD, which is effectively the basic database blueprint. The ERD is used to graphically *represent* the conceptual schema.

The conceptual model yields some important advantages. First, it provides a bird's-eye (macro level) view of the data environment that is relatively easy to understand. For example, you can get a summary of Tiny College's data environment by examining the conceptual model in Figure 2.9.

FIGURE 2.9 CONCEPTUAL MODEL FOR TINY COLLEGE

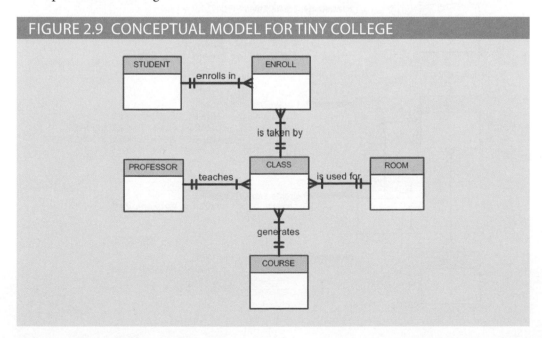

conceptual model
The output of the conceptual design process. The conceptual model provides a global view of an entire database and describes the main data objects, avoiding details.

conceptual schema
A representation of the conceptual model, usually expressed graphically. See also *conceptual model*.

Second, the conceptual model is independent of both software and hardware. **Software independence** means that the model does not depend on the DBMS software used to implement the model. **Hardware independence** means that the model does not depend on the hardware used in the implementation of the model. Therefore, changes in either the hardware or the DBMS software will have no effect on the database design at the conceptual level. Generally, the term **logical design** refers to the task of creating a conceptual data model that could be implemented in any DBMS.

2-6c The Internal Model

Once a specific DBMS has been selected, the internal model maps the conceptual model to the DBMS. The **internal model** is the representation of the database as "seen" by the DBMS. In other words, the internal model requires the designer to match the conceptual model's characteristics and constraints to those of the selected implementation model. An **internal schema** depicts a specific representation of an internal model, using the database constructs supported by the chosen database.

Because this book focuses on the relational model, a relational database was chosen to implement the internal model. Therefore, the internal schema should map the conceptual model to the relational model constructs. In particular, the entities in the conceptual model are mapped to tables in the relational model. Likewise, because a relational database has been selected, the internal schema is expressed using SQL, the standard language for relational databases. In the case of the conceptual model for Tiny College depicted in Figure 2.9, the internal model was implemented by creating the tables PROFESSOR, COURSE, CLASS, STUDENT, ENROLL, and ROOM. A simplified version of the internal model for Tiny College is shown in Figure 2.10.

The development of a detailed internal model is especially important to database designers who work with hierarchical or network models because those models require

software independence
A property of any model or application that does not depend on the software used to implement it.

hardware independence
A condition in which a model does not depend on the hardware used in the model's implementation. Therefore, changes in the hardware will have no effect on the database design at the conceptual level.

logical design
A stage in the design phase that matches the conceptual design to the requirements of the selected DBMS and is therefore software-dependent. Logical design is used to translate the conceptual design into the internal model for a selected database management system, such as DB2, SQL Server, Oracle, IMS, Informix, Access, or Ingress.

internal model
In database modeling, a level of data abstraction that adapts the conceptual model to a specific DBMS model for implementation. The internal model is the representation of a database as "seen" by the DBMS. In other words, the internal model requires a designer to match the conceptual model's characteristics and constraints to those of the selected implementation model.

internal schema
A representation of an internal model using the database constructs supported by the chosen database.

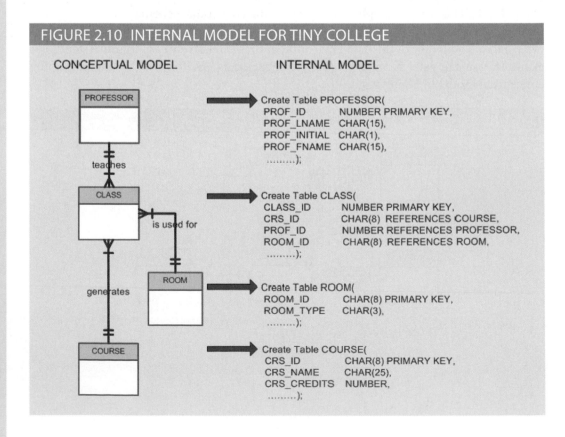

FIGURE 2.10 INTERNAL MODEL FOR TINY COLLEGE

precise specification of data storage location and data access paths. In contrast, the relational model requires less detail in its internal model because most RDBMSs handle data access path definition *transparently*; that is, the designer need not be aware of the data access path details. Nevertheless, even relational database software usually requires specifications of data storage locations, especially in a mainframe environment. For example, DB2 requires that you specify the data storage group, the location of the database within the storage group, and the location of the tables within the database.

Because the internal model depends on specific database software, it is said to be software dependent. Therefore, a change in the DBMS software requires that the internal model be changed to fit the characteristics and requirements of the implementation database model. When you can change the internal model without affecting the conceptual model, you have **logical independence**. However, the internal model is still hardware independent because it is unaffected by the type of computer on which the software is installed. Therefore, a change in storage devices or even a change in operating systems will not affect the internal model.

2-6d The Physical Model

The **physical model** operates at the lowest level of abstraction, describing the way data is saved on storage media such as magnetic, solid state, or optical media. The physical model requires the definition of both the physical storage devices and the (physical) access methods required to reach the data within those storage devices, making it both software and hardware dependent. The storage structures used are dependent on the software (the DBMS and the operating system) and on the type of storage devices the computer can handle. The precision required in the physical model's definition demands that database designers have a detailed knowledge of the hardware and software used to implement the database design.

Early data models forced the database designer to take the details of the physical model's data storage requirements into account. However, the now dominant relational model is aimed largely at the logical level rather than the physical level; therefore, it does not require the physical-level details common to its predecessors.

Although the relational model does not require the designer to be concerned about the data's physical storage characteristics, the *implementation* of a relational model may require physical-level fine-tuning for increased performance. Fine-tuning is especially important when very large databases are installed in a mainframe environment, yet even such performance fine-tuning at the physical level does not require knowledge of physical data storage characteristics.

As noted earlier, the physical model is dependent on the DBMS, methods of accessing files, and types of hardware storage devices supported by the operating system. When you can change the physical model without affecting the internal model, you have **physical independence**. Therefore, a change in storage devices or methods and even a change in operating system will not affect the internal model.

The levels of data abstraction are summarized in Table 2.4.

logical independence
A condition in which the internal model can be changed without affecting the conceptual model. (The internal model is hardware-independent because it is unaffected by the computer on which the software is installed. Therefore, a change in storage devices or operating systems will not affect the internal model.)

physical model
A model in which physical characteristics such as location, path, and format are described for the data. The physical model is both hardware- and software-dependent. See also *physical design*.

physical independence
A condition in which the physical model can be changed without affecting the internal model.

TABLE 2.4

LEVELS OF DATA ABSTRACTION

MODEL	DEGREE OF ABSTRACTION	FOCUS	INDEPENDENT OF
External	High	End-user views	Hardware and software
Conceptual		Global view of data (database model independent)	Hardware and software
Internal		Specific database model	Hardware
Physical	Low	Storage and access methods	Neither hardware nor software

Summary

- A data model is an abstraction of a complex real-world data environment. Database designers use data models to communicate with programmers and end users. The basic data-modeling components are entities, attributes, relationships, and constraints. Business rules are used to identify and define the basic modeling components within a specific real-world environment.

- The hierarchical and network data models were early models that are no longer used, but some of the concepts are found in current data models.

- The relational model is the current database implementation standard. In the relational model, the end user perceives the data as being stored in tables. Tables are related to each other by means of common values in common attributes. The entity relationship (ER) model is a popular graphical tool for data modeling that complements the relational model. The ER model allows database designers to visually present different views of the data—as seen by database designers, programmers, and end users—and to integrate the data into a common framework.

- The object-oriented data model (OODM) uses objects as the basic modeling structure. Like the relational model's entity, an object is described by its factual content. Unlike an entity, however, the object also includes information about relationships between the facts, as well as relationships with other objects, thus giving its data more meaning.

- The relational model has adopted many object-oriented (OO) extensions to become the extended relational data model (ERDM). Object/relational database management systems (O/R DBMS) were developed to implement the ERDM. At this point, the OODM is largely used in specialized engineering and scientific applications, while the ERDM is primarily geared to business applications.

- Emerging Big Data technologies such as Hadoop, MapReduce, and NoSQL provide distributed, fault-tolerant, and cost-efficient support for Big Data analytics. NoSQL databases are a new generation of databases that do not use the relational model and are geared to support the very specific needs of Big Data organizations. NoSQL databases offer distributed data stores that provide high scalability, availability, and fault tolerance by sacrificing data consistency and shifting the burden of maintaining relationships and data integrity to the program code.

- Data-modeling requirements are a function of different data views (global versus local) and the level of data abstraction. The American National Standards Institute Standards Planning and Requirements Committee (ANSI/SPARC) describes three levels of data abstraction: external, conceptual, and internal. The fourth and lowest level of data abstraction, called the physical level, is concerned exclusively with physical storage methods.

Key Terms

3 Vs	entity relationship diagram (ERD)	object/relational database management system (O/R DBMS)
American National Standards Institute (ANSI)	entity set	object-oriented data model (OODM)
attribute	eventual consistency	object-oriented database management system (OODBMS)
Big Data	extended relational data model (ERDM)	one-to-many (1:M or 1..*) relationship
business rule	Extensible Markup Language (XML)	one-to-one (1:1 or 1..1) relationship
Chen notation	external model	physical independence
class	external schema	physical model
class diagram	Hadoop	relation
class diagram notation	Hadoop Distributed File System (HDFS)	relational database management system (RDBMS)
class hierarchy	hardware independence	relational diagram
client node	hierarchical model	relational model
conceptual model	inheritance	relationship
conceptual schema	internal model	schema
connectivity	internal schema	segment
constraint	key-value	semantic data model
Crow's Foot notation	logical design	software independence
data definition language (DDL)	logical independence	sparse data
data manipulation language (DML)	MapReduce	subschema
data model	many-to-many (M:N or *..*) relationship	table
data modeling	method	tuple
data node	name node	Unified Modeling Language (UML)
entity	network model	
entity instance	NoSQL	
entity occurrence	object	
entity relationship (ER) model (ERM)		

Online Content

Flashcards and crossword puzzles for key term practice are available at *www.cengagebrain.com*.

Review Questions

1. Discuss the importance of data models.
2. What is a business rule, and what is its purpose in data modeling?
3. How do you translate business rules into data model components?
4. Describe the basic features of the relational data model and discuss their importance to the end user and the designer.

5. Explain how the entity relationship (ER) model helped produce a more structured relational database design environment.

6. Consider the scenario described by the statement "A customer can make many payments, but each payment is made by only one customer." Use this scenario as the basis for an entity relationship diagram (ERD) representation.

7. Why is an object said to have greater semantic content than an entity?

8. What is the difference between an object and a class in the object-oriented data model (OODM)?

9. How would you model Question 6 with an OODM? (Use Figure 2.4 as your guide.)

10. What is an ERDM, and what role does it play in the modern (production) database environment?

11. What is a relationship, and what three types of relationships exist?

12. Give an example of each of the three types of relationships.

13. What is a table, and what role does it play in the relational model?

14. What is a relational diagram? Give an example.

15. What is connectivity? (Use a Crow's Foot ERD to illustrate connectivity.)

16. Describe the Big Data phenomenon.

17. What does the term *3 Vs* refer to?

18. What is Hadoop and what are its basic components?

19. What is sparse data? Give an example.

20. Define and describe the basic characteristics of a NoSQL database.

21. Using the example of a medical clinic with patients and tests, provide a simple representation of how to model this example using the relational model and how it would be represented using the key-value data modeling technique.

22. What is logical independence?

23. What is physical independence?

Problems

Use the contents of Figure 2.1 to work Problems 1–3.

1. Write the business rule(s) that govern the relationship between AGENT and CUSTOMER.

2. Given the business rule(s) you wrote in Problem 1, create the basic Crow's Foot ERD.

3. Using the ERD you drew in Problem 2, create the equivalent object representation and UML class diagram. (Use Figure 2.4 as your guide.)

Using Figure P2.4 as your guide, work Problems 4–5. The DealCo relational diagram shows the initial entities and attributes for the DealCo stores, which are located in two regions of the country.

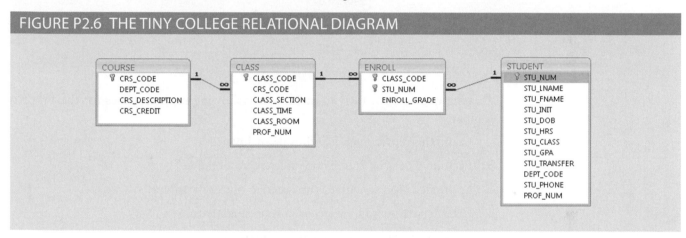

FIGURE P2.4 THE DEALCO RELATIONAL DIAGRAM

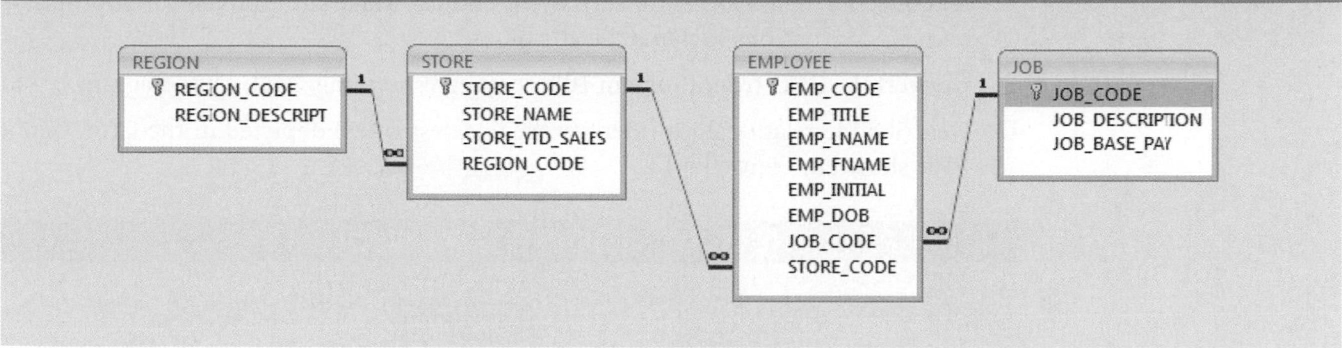

4. Identify each relationship type and write all of the business rules.

5. Create the basic Crow's Foot ERD for DealCo.

Using Figure P2.6 as your guide, work Problems 6–8. The Tiny College relational diagram shows the initial entities and attributes for the college.

FIGURE P2.6 THE TINY COLLEGE RELATIONAL DIAGRAM

6. Identify each relationship type and write all of the business rules.

7. Create the basic Crow's Foot ERD for Tiny College.

8. Create the UML class diagram that reflects the entities and relationships you identified in the relational diagram.

9. Typically, a hospital patient receives medications that have been ordered by a particular doctor. Because the patient often receives several medications per day, there is a 1:M relationship between PATIENT and ORDER. Similarly, each order can include several medications, creating a 1:M relationship between ORDER and MEDICATION.

   a. Identify the business rules for PATIENT, ORDER, and MEDICATION.

   b. Create a Crow's Foot ERD that depicts a relational database model to capture these business rules.

10. United Broke Artists (UBA) is a broker for not-so-famous artists. UBA maintains a small database to track painters, paintings, and galleries. A painting is created by a particular artist and then exhibited in a particular gallery. A gallery can exhibit many paintings, but each painting can be exhibited in only one gallery. Similarly, a painting is created by a single painter, but each painter can create many paintings. Using PAINTER, PAINTING, and GALLERY, in terms of a relational database:

    a. What tables would you create, and what would the table components be?

    b. How might the (independent) tables be related to one another?

11. Using the ERD from Problem 10, create the relational schema. (Create an appropriate collection of attributes for each of the entities. Make sure you use the appropriate naming conventions to name the attributes.)

12. Convert the ERD from Problem 10 into a corresponding UML class diagram.

13. Describe the relationships (identify the business rules) depicted in the Crow's Foot ERD shown in Figure P2.13.

FIGURE P2.13 THE CROW'S FOOT ERD FOR PROBLEM 13

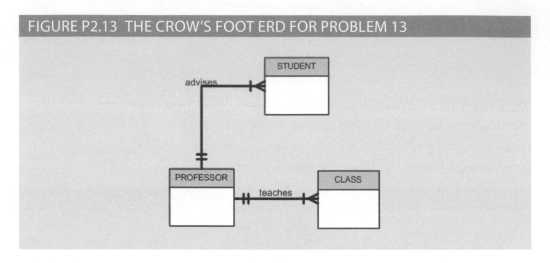

14. Create a Crow's Foot ERD to include the following business rules for the ProdCo company:

   a. Each sales representative writes many invoices.

   b. Each invoice is written by one sales representative.

   c. Each sales representative is assigned to one department.

   d. Each department has many sales representatives.

   e. Each customer can generate many invoices.

   f. Each invoice is generated by one customer.

15. Write the business rules that are reflected in the ERD shown in Figure P2.15. (Note that the ERD reflects some simplifying assumptions. For example, each book is written by only one author. Also, remember that the ERD is always read from the "1" to the "M" side, regardless of the orientation of the ERD components.)

FIGURE P2.15 THE CROW'S FOOT ERD FOR PROBLEM 15

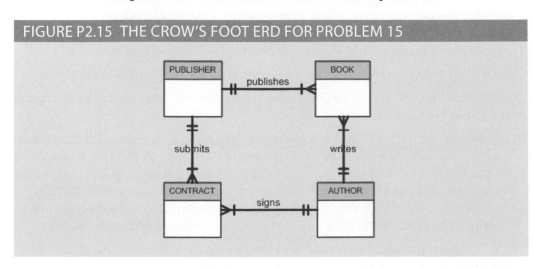

16. Create a Crow's Foot ERD for each of the following descriptions. (Note that the word *many* merely means *more than one* in the database modeling environment.)

   a. Each of the MegaCo Corporation's divisions is composed of many departments. Each department has many employees assigned to it, but each employee works for only one department. Each department is managed by one employee, and each of those managers can manage only one department at a time.

   b. During some period of time, a customer can download many ebooks from BooksOnline. Each of the ebooks can be downloaded by many customers during that period of time.

   c. An airliner can be assigned to fly many flights, but each flight is flown by only one airliner.

   d. The KwikTite Corporation operates many factories. Each factory is located in a region, and each region can be "home" to many of KwikTite's factories. Each factory has many employees, but each employee is employed by only one factory.

   e. An employee may have earned many degrees, and each degree may have been earned by many employees.

17. Write the business rules that are reflected in the ERD shown in Figure P2.17.

FIGURE P2.17 THE CROW'S FOOT ERD FOR PROBLEM 17

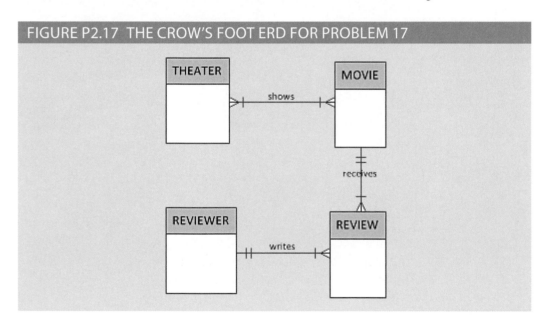

PART 2
Design Concepts

Chapter 3

The Relational Database Model

In this chapter, you will learn:

• That the relational database model offers a logical view of data
• About the relational model's basic component: relations
• That relations are logical constructs composed of rows (tuples) and columns (attributes)
• That relations are implemented as tables in a relational DBMS
• About relational database operators, the data dictionary, and the system catalog
• How data redundancy is handled in the relational database model
• Why indexing is important

Preview

In this chapter, you will learn about the relational model's logical structure and more about how ERDs (entity relationship diagrams) can be used to design a relational database. You will also learn how the relational database's basic data components fit into a logical construct known as a table, and how tables within a database can be related to one another.

After learning about tables, their components, and their relationships, you will be introduced to basic table design concepts and the characteristics of well-designed and poorly designed tables. These concepts will become your gateway to the next few chapters.

Data Files and Available Formats

	MS Access	Oracle	MS SQL	My SQL		MS Access	Oracle	MS SQL	My SQL
CH03_CollegeTry	✓	✓	✓	✓	CH03_AviaCo	✓	✓	✓	✓
CH03_CollegeTry2	✓	✓	✓	✓	CH03_BeneCo	✓	✓	✓	✓
CH03_InsureCo	✓	✓	✓	✓	CH03_CollegeQue	✓	✓	✓	✓
CH03_Museum	✓	✓	✓	✓	CH03_NoComp	✓	✓	✓	✓
CH03_SaleCo	✓	✓	✓	✓	CH03_StoreCo	✓	✓	✓	✓
CH03_TinyCollege	✓	✓	✓	✓	CH03_Theater	✓	✓	✓	✓
CH03_Relational_DB	✓	✓	✓	✓	CH03_TransCo	✓	✓	✓	✓
					CH03_VendingCo	✓	✓	✓	✓

Data Files Available on cengagebrain.com

Note

The relational model, introduced by E. F. Codd in 1970, is based on predicate logic and set theory. **Predicate logic**, used extensively in mathematics, provides a framework in which an assertion (statement of fact) can be verified as either true or false. For example, suppose that a student with a student ID of 12345678 is named Melissa Sanduski. This assertion can easily be demonstrated to be true or false. **Set theory** is a mathematical science that deals with sets, or groups of things, and is used as the basis for data manipulation in the relational model. For example, assume that set A contains three numbers: 16, 24, and 77. This set is represented as A(16, 24, 77). Furthermore, set B contains four numbers: 44, 77, 90, and 11, and so is represented as B(44, 77, 90, 11). Given this information, you can conclude that the intersection of A and B yields a result set with a single number, 77. This result can be expressed as $A \cap B = 77$. In other words, A and B share a common value, 77.

Based on these concepts, the relational model has three well-defined components:

1. A logical data structure represented by relations (see Sections 3-1, 3-2, and 3-5)

2. A set of integrity rules to enforce that the data is consistent and remains consistent over time (see Sections 3-3, 3-6, 3-7, and 3-8)

3. A set of operations that defines how data is manipulated (see Section 3-4)

3-1 A Logical View of Data

In Chapter 1, Database Systems, you learned that a database stores and manages both data and metadata. You also learned that the DBMS manages and controls access to the data and the database structure. Such an arrangement—placing the DBMS between the application and the database—eliminates most of the file system's inherent limitations. The result of such flexibility, however, is a far more complex physical structure. In fact, the database structures required by both the hierarchical and network database models often become complicated enough to diminish efficient database design. The relational data model changed all of that by allowing the designer to focus on the logical representation of the data and its relationships, rather than on the physical storage details. To use an automotive analogy, the relational database uses an automatic transmission to relieve you of the need to manipulate clutch pedals and gearshifts. In short, the relational model enables you to view data *logically* rather than *physically*.

The practical significance of taking the logical view is that it serves as a reminder of the simple file concept of data storage. Although the use of a table, quite unlike that of a file, has the advantages of structural and data independence, a table does resemble a file from a conceptual point of view. Because you can think of related records as being stored in independent tables, the relational database model is much easier to understand than the hierarchical and network models. Logical simplicity tends to yield simple and effective database design methodologies.

Because the table plays such a prominent role in the relational model, it deserves a closer look. Therefore, our discussion begins by exploring the details of table structure and contents.

3-1a Tables and Their Characteristics

The logical view of the relational database is facilitated by the creation of data relationships based on a logical construct known as a relation. Because a relation is a mathematical construct, end users find it much easier to think of a relation as a table. A *table* is perceived as a two-dimensional structure composed of rows and columns. A table is also

predicate logic
Used extensively in mathematics to provide a framework in which an assertion (statement of fact) can be verified as either true or false.

set theory
A part of mathematical science that deals with sets, or groups of things, and is used as the basis for data manipulation in the relational model.

called a *relation* because the relational model's creator, E. F. Codd, used the two terms as synonyms. You can think of a table as a *persistent* representation of a logical relation— that is, a relation whose contents can be permanently saved for future use. As far as the table's user is concerned, a table contains *a group of related entity occurrences*—that is, an entity set. For example, a STUDENT table contains a collection of entity occurrences, each representing a student. For that reason, the terms *entity set* and *table* are often used interchangeably.

Note

The word *relation*, also known as a *dataset* in Microsoft Access, is based on the mathematical set theory from which Codd derived his model. Because the relational model uses attribute values to establish relationships among tables, many database users incorrectly assume that the term *relation* refers to such relationships. Many then incorrectly conclude that only the relational model permits the use of relationships.

You will discover that the table view of data makes it easy to spot and define entity relationships, thereby greatly simplifying the task of database design. The characteristics of a relational table are summarized in Table 3.1.

TABLE 3.1

CHARACTERISTICS OF A RELATIONAL TABLE
1	A table is perceived as a two-dimensional structure composed of rows and columns.
2	Each table row (**tuple**) represents a single entity occurrence within the entity set.
3	Each table column represents an attribute, and each column has a distinct name.
4	Each intersection of a row and column represents a single data value.
5	All values in a column must conform to the same data format.
6	Each column has a specific range of values known as the attribute **domain**.
7	The order of the rows and columns is immaterial to the DBMS.
8	Each table must have an attribute or combination of attributes that uniquely identifies each row.

The database table shown in Figure 3.1 illustrates the characteristics listed in Table 3.1.

Note

Relational database terminology is very precise. Unfortunately, file system terminology sometimes creeps into the database environment. Thus, rows are sometimes referred to as *records*, and columns are sometimes labeled as *fields*. Occasionally, tables are labeled *files*. Technically speaking, this substitution of terms is not always appropriate. The database table is a logical concept rather than a physical concept, and the terms *file*, *record*, and *field* describe physical concepts. Nevertheless, as long as you recognize that the table is actually a logical concept rather than a physical construct, you may think of table rows as records and of table columns as fields. In fact, many database software vendors still use this familiar file system terminology.

FIGURE 3.1 STUDENT TABLE ATTRIBUTE VALUES

Table name: STUDENT   Database name: Ch03_TinyCollege

STU_NUM	STU_LNAME	STU_FNAME	STU_INIT	STU_DOB	STU_HRS	STU_CLASS	STU_GPA	STU_TRANSFER	DEPT_CODE	STU_PHONE	PROF_NUM
321452	Bowser	William	C	12-Feb-1985	42	So	2.84	No	BIOL	2134	205
324257	Smithson	Anne	K	15-Nov-1991	81	Jr	3.27	Yes	CIS	2256	222
324258	Brewer	Juliette		23-Aug-1979	36	So	2.26	Yes	ACCT	2256	228
324269	Oblonski	Walter	H	16-Sep-1986	66	Jr	3.09	No	CIS	2114	222
324273	Smith	John	D	30-Dec-1968	102	Sr	2.11	Yes	ENGL	2231	199
324274	Katinga	Raphael	P	21-Oct-1989	114	Sr	3.15	No	ACCT	2267	228
324291	Robertson	Gerald	T	08-Apr-1983	120	Sr	3.87	No	EDU	2267	311
324299	Smith	John	B	30-Nov-1996	15	Fr	2.92	No	ACCT	2315	230

STU_NUM	= Student number
STU_LNAME	= Student last name
STU_FNAME	= Student first name
STU_INIT	= Student middle initial
STU_DOB	= Student date of birth
STU_HRS	= Credit hours earned
STU_CLASS	= Student classification
STU_GPA	= Grade point average
STU_TRANSFER	= Student transferred from another institution
DEPT_CODE	= Department code
STU_PHONE	= 4-digit campus phone extension
PROF_NUM	= Number of the professor who is the student's advisor

Using the STUDENT table shown in Figure 3.1, you can draw the following conclusions corresponding to the points in Table 3.1:

1. The STUDENT table is perceived to be a two-dimensional structure composed of 8 rows (tuples) and 12 columns (attributes).

2. Each row in the STUDENT table describes a single entity occurrence within the entity set. (The entity set is represented by the STUDENT table.) For example, row 4 in Figure 3.1 describes a student named Walter H. Oblonski. Given the table contents, the STUDENT entity set includes eight distinct entities (rows), or students.

3. Each column represents an attribute, and each column has a distinct name.

4. All of the values in a column match the attribute's characteristics. For example, the grade point average (STU_GPA) column contains only STU_GPA entries for each of the table rows. Data must be classified according to its format and function. Although various DBMSs can support different data types, most support at least the following:

    a. *Numeric.* You can use numeric data to perform meaningful arithmetic procedures. For example, in Figure 3.1, STU_HRS and STU_GPA are numeric attributes.

    b. *Character.* Character data, also known as text data or string data, can contain any character or symbol not intended for mathematical manipulation. In Figure 3.1, STU_CLASS and STU_PHONE are examples of character attributes.

    c. *Date.* Date attributes contain calendar dates stored in a special format known as the Julian date format. In Figure 3.1, STU_DOB is a date attribute.

    d. *Logical.* Logical data can only have true or false (yes or no) values. In Figure 3.1, the STU_TRANSFER attribute uses a logical data format.

5. The column's range of permissible values is known as its **domain**. Because the STU_GPA values are limited to the range 0–4, inclusive, the domain is [0,4].

6. The order of rows and columns is immaterial to the user.

Online Content

All of the databases used to illustrate the material in this chapter (see the Data Files list at the beginning of the chapter) are available at *www.cengagebrain.com*. The database names match the database names shown in the figures.

tuple
In the relational model, a table row.

domain
In data modeling, the construct used to organize and describe an attribute's set of possible values.

7. Each table must have a primary key. In general terms, the **primary key (PK)** is an attribute or combination of attributes that uniquely identifies any given row. In this case, STU_NUM (the student number) is the primary key. Using the data in Figure 3.1, observe that a student's last name (STU_LNAME) would not be a good primary key because several students have the last name of Smith. Even the combination of the last name and first name (STU_FNAME) would not be an appropriate primary key because more than one student is named John Smith.

3-2 **Keys**

In the relational model, keys are important because they are used to ensure that each row in a table is uniquely identifiable. They are also used to establish relationships among tables and to ensure the integrity of the data. A **key** consists of one or more attributes that determine other attributes. For example, an invoice number identifies all of the invoice attributes, such as the invoice date and the customer name.

One type of key, the primary key, has already been introduced. Given the structure of the STUDENT table shown in Figure 3.1, defining and describing the primary key seem simple enough. However, because the primary key plays such an important role in the relational environment, you will examine the primary key's properties more carefully. In this section, you also will become acquainted with superkeys, candidate keys, and secondary keys.

3-2a Dependencies

The role of a key is based on the concept of determination. **Determination** is the state in which knowing the value of one attribute makes it possible to determine the value of another. The idea of determination is not unique to the database environment. You are familiar with the formula revenue − cost = profit. This is a form of determination, because if you are given the *revenue* and the *cost*, you can determine the *profit*. Given *profit* and *revenue*, you can determine the *cost*. Given any two values, you can determine the third. Determination in a database environment, however, is not normally based on a formula but on the relationships among the attributes.

If you consider what the attributes of the STUDENT table in Figure 3.1 actually represent, you will see a relationship among the attributes. If you are given a value for STU_NUM, then you can determine the value for STU_LNAME because one and only one value of STU_LNAME is associated with any given value of STU_NUM. A specific terminology and notation is used to describe relationships based on determination. The relationship is called **functional dependence**, which means that the value of one or more attributes determines the value of one or more other attributes. The standard notation for representing the relationship between STU_NUM and STU_LNAME is as follows:

STU_NUM → STU_LNAME

In this functional dependency, the attribute whose value determines another is called the **determinant** or the key. The attribute whose value is determined by the other attribute is called the **dependent**. Using this terminology, it would be correct to say that STU_NUM is the determinant and STU_LNAME is the dependent. STU_NUM functionally determines STU_LNAME, and STU_LNAME is functionally dependent on STU_NUM. As stated earlier, functional dependence can involve a determinant that comprises more than one attribute and multiple dependent attributes. Refer to the STUDENT table for the following example:

---

primary key (PK)
In the relational model, an identifier composed of one or more attributes that uniquely identifies a row. Also, a candidate key selected as a unique entity identifier. See also *key*.

key
One or more attributes that determine other attributes. See also *superkey, candidate key, primary key (PK), secondary key,* and *foreign key.*

determination
The role of a key. In the context of a database table, the statement "A determines B" indicates that knowing the value of attribute A means that the value of attribute B can be looked up.

functional dependence
Within a relation R, an attribute B is functionally dependent on an attribute A if and only if a given value of attribute A determines exactly one value of attribute B. The relationship "B is dependent on A" is equivalent to "A determines B," and is written as A → B.

determinant
Any attribute in a specific row whose value directly determines other values in that row. See also *Boyce-Codd normal form (BCNF).*

dependent
An attribute whose value is determined by another attribute.

STU_NUM → (STU_LNAME, STU_FNAME, STU_GPA)

and

(STU_FNAME, STU_LNAME, STU_INIT, STU_PHONE) → (STU_DOB, STU_HRS, STU_GPA)

Determinants made of more than one attribute require special consideration. It is possible to have a functional dependency in which the determinant contains attributes that are not necessary for the relationship. Consider the following two functional dependencies:

STU_NUM → STU_GPA

(STU_NUM, STU_LNAME) → STU_GPA

In the second functional dependency, the determinant includes STU_LNAME, but this attribute is not necessary for the relationship. The functional dependency is valid because given a pair of values for STU_NUM and STU_LNAME, only one value would occur for STU_GPA. A more specific term, **full functional dependence**, is used to refer to functional dependencies in which the entire collection of attributes in the determinant is necessary for the relationship. Therefore, the dependency shown in the preceding example is a functional dependency, but not a full functional dependency.

3-2b Types of Keys

Recall that a key is an attribute or group of attributes that can determine the values of other attributes. Therefore, keys are determinants in functional dependencies. Several different types of keys are used in the relational model, and you need to be familiar with them.

A **composite key** is a key that is composed of more than one attribute. An attribute that is a part of a key is called a **key attribute**. For example,

STU_NUM → STU_GPA

(STU_LNAME, STU_FNAME, STU_INIT, STU_PHONE) → STU_HRS

In the first functional dependency, STU_NUM is an example of a key composed of only one key attribute. In the second functional dependency, (STU_LNAME, STU_FNAME, STU_INIT, STU_PHONE) is a composite key composed of four key attributes.

A **superkey** is a key that can uniquely identify any row in the table. In other words, a superkey functionally determines every attribute in the row. In the STUDENT table, STU_NUM is a superkey, as are the composite keys (STU_NUM, STU_LNAME), (STU_NUM, STU_LNAME, STU_INIT), and (STU_LNAME, STU_FNAME, STU_INIT, STU_PHONE). In fact, because STU_NUM alone is a superkey, any composite key that has STU_NUM as a key attribute will also be a superkey. Be careful, however, because not all keys are superkeys. For example, Gigantic State University determines its student classification based on hours completed, as shown in Table 3.2.

Therefore, you can write STU_HRS → STU_CLASS.

However, the specific number of hours is not dependent on the classification. It is quite possible to find a junior with 62 completed hours or one with 84 completed hours. In other words, the classification (STU_CLASS) does not determine one and only one value for completed hours (STU_HRS).

full functional dependence
A condition in which an attribute is functionally dependent on a composite key but not on any subset of the key.

composite key
A multiple-attribute key.

key attributes
The attributes that form a primary key. See also *prime attribute*.

superkey
An attribute or attributes that uniquely identify each entity in a table. See *key*.

TABLE 3.2

STUDENT CLASSIFICATION	
HOURS COMPLETED	**CLASSIFICATION**
Less than 30	Fr
30–59	So
60–89	Jr
90 or more	Sr

One specific type of superkey is called a candidate key. A **candidate key** is a minimal superkey—that is, a superkey without any unnecessary attributes. A candidate key is based on a full functional dependency. For example, STU_NUM would be a candidate key, as would (STU_LNAME, STU_FNAME, STU_INIT, STU_PHONE). On the other hand, (STU_NUM, STU_LNAME) is a superkey, but it is not a candidate key because STU_LNAME could be removed and the key would still be a superkey. A table can have many different candidate keys. If the STUDENT table also included the students' Social Security numbers as STU_SSN, then it would appear to be a candidate key. Candidate keys are called *candidates* because they are the eligible options from which the designer will choose when selecting the primary key. The primary key is the candidate key chosen to be the primary means by which the rows of the table are uniquely identified.

Entity integrity is the condition in which each row (entity instance) in the table has its own unique identity. To ensure entity integrity, the primary key has two requirements: (1) all of the values in the primary key must be unique, and (2) no key attribute in the primary key can contain a null.

Note

A null is no value at all. It does *not* mean a zero or a space. A null is created when you press the Enter key or the Tab key to move to the next entry without making an entry of any kind. Pressing the Spacebar creates a blank (or a space).

Null values are problematic in the relational model. A **null** is the absence of any data value, and it is never allowed in any part of the primary key. From a theoretical perspective, it can be argued that a table that contains a null is not properly a relational table at all. From a practical perspective, however, some nulls cannot be reasonably avoided. For example, not all students have a middle initial. As a general rule, nulls should be avoided as much as reasonably possible. In fact, an abundance of nulls is often a sign of a poor design. Also, nulls should be avoided in the database because their meaning is not always identifiable. For example, a null could represent any of the following:

- An unknown attribute value
- A known, but missing, attribute value
- A "not applicable" condition

Depending on the sophistication of the application development software, nulls can create problems when functions such as COUNT, AVERAGE, and SUM are used. In addition, nulls can create logical problems when relational tables are linked.

In addition to its role in providing a unique identity to each row in the table, the primary key may play an additional role in the controlled redundancy that allows the

candidate key
A minimal superkey; that is, a key that does not contain a subset of attributes that is itself a superkey. See *key*.

entity integrity
The property of a relational table that guarantees each entity has a unique value in a primary key and that the key has no null values.

null
The absence of an attribute value. Note that a null is not a blank.

relational model to work. Recall from Chapter 2 that a hallmark of the relational model is that relationships between tables are implemented through common attributes as a form of controlled redundancy. For example, Figure 3.2 shows PRODUCT and VENDOR tables that are linked through a common attribute, VEND_CODE. VEND_CODE is referred to as a foreign key in the PRODUCT table. A **foreign key (FK)** is the primary key of one table that has been placed into another table to create a common attribute. In Figure 3.2, the primary key of VENDOR, VEND_CODE, was placed in the PRODUCT table; therefore, VEND_CODE is a foreign key in PRODUCT. One advantage of using a proper naming convention for table attributes is that you can identify foreign keys more easily. For example, because the STUDENT table in Figure 3.1 used a proper naming convention, you can identify two foreign keys in the table (DEPT_CODE and PROF_NUM) that imply the existence of two other tables in the database (DEPARTMENT and PROFESSOR) related to STUDENT.

FIGURE 3.2 AN EXAMPLE OF A SIMPLE RELATIONAL DATABASE

Table name: PRODUCT   **Database name: Ch03_SaleCo**
Primary key: PROD_CODE
Foreign key: VEND_CODE

PROD_CODE	PROD_DESCRIPT	PROD_PRICE	PROD_ON_HAND	VEND_CODE
001278-AB	Claw hammer	12.95	23	232
123-21UUY	Houselite chain saw, 16-in. bar	189.99	4	235
QER-34256	Sledge hammer, 16-lb. head	18.63	6	231
SRE-657UG	Rat-tail file	2.99	15	232
ZZX/3245Q	Steel tape, 12-ft. length	6.79	8	235

link

VEND_CODE	VEND_CONTACT	VEND_AREACODE	VEND_PHONE
230	Shelly K. Smithson	608	555-1234
231	James Johnson	615	123-4536
232	Annelise Crystall	608	224-2134
233	Candice Wallace	904	342-6567
234	Arthur Jones	615	123-3324
235	Henry Ortozo	615	899-3425

Table name: VENDOR
Primary key: VEND_CODE
Foreign key: none

Just as the primary key has a role in ensuring the integrity of the database, so does the foreign key. Foreign keys are used to ensure **referential integrity**, the condition in which every reference to an entity instance by another entity instance is valid. In other words, every foreign key entry must either be null or a valid value in the primary key of the related table. Note that the PRODUCT table has referential integrity because every entry in VEND_CODE in the PRODUCT table is either null or a valid value in VEND_CODE in the VENDOR table. Every vendor referred to by a row in the PRODUCT table is a valid vendor.

Finally, a **secondary key** is defined as a key that is used strictly for data retrieval purposes. Suppose that customer data is stored in a CUSTOMER table in which the customer number is the primary key. Do you think that most customers will remember their numbers? Data retrieval for a customer is easier when the customer's last name and phone number are used. In that case, the primary key is the customer number; the secondary key is the combination of the customer's last name and phone number. Keep in mind that a secondary key does not necessarily yield a unique outcome. For example, a customer's last name and home telephone number could easily yield several matches in which one family lives together and shares a phone line. A less efficient secondary key would be the combination of the last name and zip code; this could yield dozens of matches, which could then be combed for a specific match.

foreign key (FK)
An attribute or attributes in one table whose values must match the primary key in another table or whose values must be null. See *key*.

referential integrity
A condition by which a dependent table's foreign key must have either a null entry or a matching entry in the related table.

secondary key
A key used strictly for data retrieval purposes. For example, customers are not likely to know their customer number (primary key), but the combination of last name, first name, middle initial, and telephone number will probably match the appropriate table row. See also *key*.

A secondary key's effectiveness in narrowing down a search depends on how restrictive the key is. For instance, although the secondary key CUS_CITY is legitimate from a database point of view, the attribute values *New York* or *Sydney* are not likely to produce a usable return unless you want to examine millions of possible matches. (Of course, CUS_CITY is a better secondary key than CUS_COUNTRY.)

Table 3.3 summarizes the various relational database table keys.

TABLE 3.3

RELATIONAL DATABASE KEYS

KEY TYPE	DEFINITION
Superkey	An attribute or combination of attributes that uniquely identifies each row in a table
Candidate key	A minimal (irreducible) superkey; a superkey that does not contain a subset of attributes that is itself a superkey
Primary key	A candidate key selected to uniquely identify all other attribute values in any given row; cannot contain null entries
Foreign key	An attribute or combination of attributes in one table whose values must either match the primary key in another table or be null
Secondary key	An attribute or combination of attributes used strictly for data retrieval purposes

3-3 Integrity Rules

Relational database integrity rules are very important to good database design. RDBMSs enforce integrity rules automatically, but it is much safer to make sure your application design conforms to the entity and referential integrity rules mentioned in this chapter. Those rules are summarized in Table 3.4.

TABLE 3.4

INTEGRITY RULES

ENTITY INTEGRITY	DESCRIPTION
Requirement	All primary key entries are unique, and no part of a primary key may be null.
Purpose	Each row will have a unique identity, and foreign key values can properly reference primary key values.
Example	No invoice can have a duplicate number, nor can it be null; in short, all invoices are uniquely identified by their invoice number.
REFERENTIAL INTEGRITY	**DESCRIPTION**
Requirement	A foreign key may have either a null entry, as long as it is not a part of its table's primary key, or an entry that matches the primary key value in a table to which it is related; (every non-null foreign key value *must* reference an *existing* primary key value).
Purpose	It is possible for an attribute *not* to have a corresponding value, but it will be impossible to have an invalid entry; the enforcement of the referential integrity rule makes it impossible to delete a row in one table whose primary key has mandatory matching foreign key values in another table.
Example	A customer might not yet have an assigned sales representative (number), but it will be impossible to have an invalid sales representative (number).

The integrity rules summarized in Table 3.4 are illustrated in Figure 3.3.

FIGURE 3.3 AN ILLUSTRATION OF INTEGRITY RULES

Table name: **CUSTOMER**   Database name: **Ch03_InsureCo**
Primary key: **CUS_CODE**
Foreign key: **AGENT_CODE**

CUS_CODE	CUS_LNAME	CUS_FNAME	CUS_INITIAL	CUS_RENEW_DATE	AGENT_CODE
10010	Ramas	Alfred	A	05-Apr-2016	502
10011	Dunne	Leona	K	16-Jun-2016	501
10012	Smith	Kathy	W	29-Jan-2017	502
10013	Olowski	Paul	F	14-Oct-2016	
10014	Orlando	Myron		28-Dec-2016	501
10015	O'Brian	Amy	B	22-Sep-2016	503
10016	Brown	James	G	25-Mar-2017	502
10017	Williams	George		17-Jul-2016	503
10018	Farriss	Anne	G	03-Dec-2016	501
10019	Smith	Olette	K	14-Mar-2017	503

Table name: **AGENT** (only five selected fields are shown)
Primary key: **AGENT_CODE**
Foreign key: none

AGENT_CODE	AGENT_AREACODE	AGENT_PHONE	AGENT_LNAME	AGENT_YTD_SLS
501	713	228-1249	Alby	132735.75
502	615	882-1244	Hahn	138967.35
503	615	123-5589	Okon	127093.45

Note the following features of Figure 3.3.

- *Entity integrity*. The CUSTOMER table's primary key is CUS_CODE. The CUSTOMER primary key column has no null entries, and all entries are unique. Similarly, the AGENT table's primary key is AGENT_CODE, and this primary key column is also free of null entries.

- *Referential integrity*. The CUSTOMER table contains a foreign key, AGENT_CODE, that links entries in the CUSTOMER table to the AGENT table. The CUS_CODE row identified by the (primary key) number 10013 contains a null entry in its AGENT_CODE foreign key because Paul F. Olowski does not yet have a sales representative assigned to him. The remaining AGENT_CODE entries in the CUSTOMER table all match the AGENT_CODE entries in the AGENT table.

To avoid nulls, some designers use special codes, known as **flags**, to indicate the absence of some value. Using Figure 3.3 as an example, the code –99 could be used as the AGENT_CODE entry in the fourth row of the CUSTOMER table to indicate that customer Paul Olowski does not yet have an agent assigned to him. If such a flag is used, the AGENT table must contain a dummy row with an AGENT_CODE value of –99. Thus, the AGENT table's first record might contain the values shown in Table 3.5.

> **flags**
> Special codes implemented by designers to trigger a required response, alert end users to specified conditions, or encode values. Flags may be used to prevent nulls by bringing attention to the absence of a value in a table.

TABLE 3.5

A DUMMY VARIABLE VALUE USED AS A FLAG

AGENT_CODE	AGENT_AREACODE	AGENT_PHONE	AGENT_LNAME	AGENT_YTD_SLS
–99	000	000–0000	None	$0.00

Chapter 4, Entity Relationship (ER) Modeling, discusses several ways to handle nulls.

Other integrity rules that can be enforced in the relational model are the NOT NULL and UNIQUE constraints. The NOT NULL constraint can be placed on a column to ensure that every row in the table has a value for that column. The UNIQUE constraint is a restriction placed on a column to ensure that no duplicate values exist for that column.

3-4 Relational Algebra

The data in relational tables is of limited value unless the data can be manipulated to generate useful information. This section describes the basic data manipulation capabilities of the relational model. **Relational algebra** defines the theoretical way of manipulating table contents using relational operators. In Chapter 7, Introduction to Structured Query Language (SQL), and Chapter 8, Advanced SQL, you will learn how SQL commands can be used to accomplish relational algebra operations.

Note

The degree of relational completeness can be defined by the extent to which relational algebra is supported. To be considered minimally relational, the DBMS must support the key relational operators SELECT, PROJECT, and JOIN.

3-4a Formal Definitions and Terminology

Recall that the relational model is actually based on mathematical principles, and manipulating the data in the database can be described in mathematical terms. The good news is that, as database professionals, we do not have to write mathematical formulas to work with our data. Data is manipulated by database developers and programmers using powerful languages like SQL that hide the underlying math. However, understanding the underlying principles can give you a good feeling for the types of operations that can be performed, and it can help you to understand how to write your queries more efficiently and effectively.

One advantage of using formal mathematical representations of operations is that mathematical statements are unambiguous. These statements are very specific, and they require that database designers be specific in the language used to explain them. As previously explained, it is common to use the terms *relation* and *table* interchangeably. However, since the mathematical terms need to be precise, we will use the more specific term relation when discussing the formal definitions of the various relational algebra operators.

Before considering the specific relational algebra operators, it is necessary to formalize our understanding of a table.

One important aspect of using the specific term *relation* is that it acknowledges the distinction between the relation and the relation variable, or *relvar*, for short. A relation is the data that we see in our tables. A **relvar** is a variable that holds a relation. For example, imagine you were writing a program and created a variable named *qty* for holding integer data. The variable *qty* is not an integer itself; it is a container for holding integers. Similarly, when you create a table, the table structure holds the table data. The structure is properly called a relvar, and the data in the structure would be a relation. The relvar is a container (variable) for holding relation data, not the relation itself. The data in the table is a relation.

A relvar has two parts: the heading and the body. The relvar heading contains the names of the attributes, while the relvar body contains the relation. To conveniently maintain this distinction in formulas, an unspecified relation is often assigned a lower-case letter (e.g., "r"), while the relvar is assigned an uppercase letter (e.g., "R"). We could then say that *r* is a relation of type *R*, or *r(R)*.

relational algebra
A set of mathematical principles that form the basis for manipulating relational table contents; the eight main functions are SELECT, PROJECT, JOIN, INTERSECT, UNION, DIFFERENCE, PRODUCT, and DIVIDE.

relvar
Short for relation variable, a variable that holds a relation. A relvar is a container (variable) for holding relation data, not the relation itself.

3-4b Relational Set Operators

The relational operators have the property of **closure**; that is, the use of relational algebra operators on existing relations (tables) produces new relations. Numerous operators have been defined. Some operators are fundamental, while others are convenient but can be derived using the fundamental operators. In this section, the focus will be on the SELECT (or RESTRICT), PROJECT, UNION, INTERSECT, DIFFERENCE, PRODUCT, JOIN, and DIVIDE operators.

Select (Restrict) **SELECT**, also known as **RESTRICT**, is referred to as a unary operator because it only uses one table as input. It yields values for all rows found in the table that satisfy a given condition. SELECT can be used to list all of the rows, or it can yield only rows that match a specified criterion. In other words, SELECT yields a horizontal subset of a table. SELECT will not limit the attributes returned so all attributes of the table will be included in the result. The effect of a SELECT operation is shown in Figure 3.4.

FIGURE 3.4 SELECT

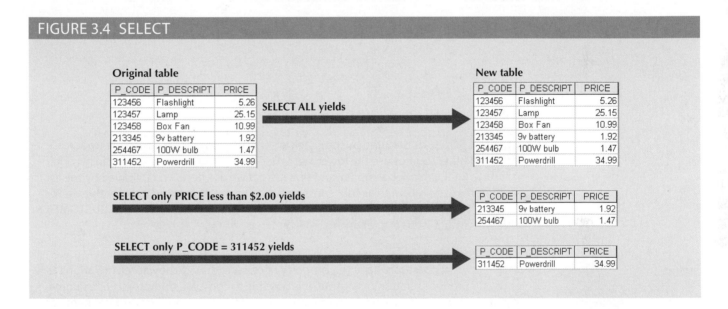

Note

Formally, SELECT is denoted by the lowercase Greek letter sigma (σ). Sigma is followed by the condition to be evaluated (called a predicate) as a subscript, and then the relation is listed in parentheses. For example, to SELECT all of the rows in the CUSTOMER table that have the value '10010' in the CUS_CODE attribute, you would write the following:

$$\sigma_{cus\_code\,=\,10010}\ (customer)$$

Project **PROJECT** yields all values for selected attributes. It is also a unary operator, accepting only one table as input. PROJECT will return only the attributes requested, in the order in which they are requested. In other words, PROJECT yields a vertical subset of a table. PROJECT will not limit the rows returned so all rows of the specified attributes will be included in the result. The effect of a PROJECT operation is shown in Figure 3.5.

closure
A property of relational operators that permits the use of relational algebra operators on existing tables (relations) to produce new relations.

SELECT
In relational algebra, an operator used to select a subset of rows. Also known as *RESTRICT*.

RESTRICT
See *SELECT*.

PROJECT
In relational algebra, an operator used to select a subset of columns.

FIGURE 3.5 PROJECT

Original table

P_CODE	P_DESCRIPT	PRICE
123456	Flashlight	5.26
123457	Lamp	25.15
123458	Box Fan	10.99
213345	9v battery	1.92
254467	100W bulb	1.47
311452	Powerdrill	34.99

PROJECT PRICE yields →

New table

PRICE
5.26
25.15
10.99
1.92
1.47
34.99

PROJECT P_DESCRIPT and PRICE yields →

P_DESCRIPT	PRICE
Flashlight	5.26
Lamp	25.15
Box Fan	10.99
9v battery	1.92
100W bulb	1.47
Powerdrill	34.99

PROJECT P_CODE and PRICE yields →

P_CODE	PRICE
123456	5.26
123457	25.15
123458	10.99
213345	1.92
254467	1.47
311452	34.99

Note

Formally, PROJECT is denoted by the Greek letter pi (π). Some sources use the uppercase letter, and other sources use the lowercase letter. Codd used the lowercase π in his original article on the relational model, and that is what we use here. Pi is followed by the list of attributes to be returned as subscripts, and then the relation listed in parentheses. For example, to PROJECT the CUS_FNAME and CUS_LNAME attributes in the CUSTOMER table, you would write the following:

$$\pi_{cus\_fname,\ cus\_lname} (customer)$$

Since relational operators have the property of closure, that is, they accept relations as input and produce relations as output, it is possible to combine operators. For example, you can combine the two previous operators to find the customer first and last name of the customer with customer code 10010:

$$\pi_{cus\_fname,\ cus\_lname} (\sigma_{cus\_code\ =\ 10010} (customer))$$

UNION
In relational algebra, an operator used to merge (append) two tables into a new table, dropping the duplicate rows. The tables must be *union-compatible*.

union-compatible
Two or more tables that have the same number of columns and the corresponding columns have compatible domains.

Union UNION combines all rows from two tables, excluding duplicate rows. To be used in the UNION, the tables must have the same attribute characteristics; in other words, the columns and domains must be compatible. When two or more tables share the same number of columns, and when their corresponding columns share the same or compatible domains, they are said to be **union-compatible**. The effect of a UNION operation is shown in Figure 3.6.

FIGURE 3.6 UNION

P_CODE	P_DESCRIPT	PRICE
123456	Flashlight	5.26
123457	Lamp	25.15
123458	Box Fan	10.99
213345	9v battery	1.92
254467	100W bulb	1.47
311452	Powerdrill	34.99

UNION

P_CODE	P_DESCRIPT	PRICE
345678	Microwave	160.00
345679	Dishwasher	500.00
123458	Box Fan	10.99

yields

P_CODE	P_DESCRIPT	PRICE
123456	Flashlight	5.26
123457	Lamp	25.15
123458	Box Fan	10.99
213345	9v battery	1.92
254467	100W bulb	1.47
311452	Powerdrill	34.99
345678	Microwave	160
345679	Dishwasher	500

Note

UNION is denoted by the symbol ∪. If the relations SUPPLIER and VENDOR are union-compatible, then a UNION between them would be denoted as follows:

supplier ∪ vendor

It is rather unusual to find two relations that are union-compatible in a database. Typically, PROJECT operators are applied to relations to produce results that are union-compatible. For example, assume the SUPPLIER and VENDOR tables are not union-compatible. If you wish to produce a listing of all vendor and supplier names, then you can PROJECT the names from each table and then perform a UNION with them.

$\pi_{supplier\_name}$ (supplier) ∪ π_{vendor_name} (vendor)

Intersect **INTERSECT** yields only the rows that appear in both tables. As with UNION, the tables must be union-compatible to yield valid results. For example, you cannot use INTERSECT if one of the attributes is numeric and one is character-based. For the rows to be considered the same in both tables and appear in the result of the INTERSECT, the entire rows must be exact duplicates. The effect of an INTERSECT operation is shown in Figure 3.7.

FIGURE 3.7 INTERSECT

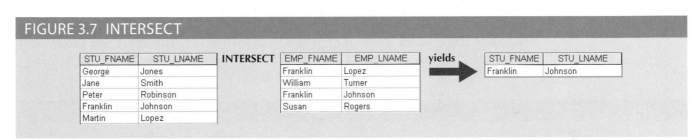

Note

INTERSECT is denoted by the symbol ∩. If the relations SUPPLIER and VENDOR are union-compatible, then an INTERSECT between them would be denoted as follows:

supplier ∩ vendor

Just as with the UNION operator, it is unusual to find two relations that are union-compatible in a database, so PROJECT operators are applied to relations to produce results that can be manipulated with an INTERSECT operator. For example, again assume the SUPPLIER and VENDOR tables are not union-compatible. If you wish to produce a listing of any vendor and supplier names that are the same in both tables, then you can PROJECT the names from each table and then perform an INTERSECT with them.

$\pi_{supplier\_name}$ (supplier) ∩ π_{vendor_name} (vendor)

INTERSECT
In relational algebra, an operator used to yield only the rows that are common to two union-compatible tables.

DIFFERENCE
In relational algebra, an operator used to yield all rows from one table that are not found in another union-compatible table.

Difference **DIFFERENCE** yields all rows in one table that are not found in the other table; that is, it subtracts one table from the other. As with UNION, the tables must be union-compatible to yield valid results. The effect of a DIFFERENCE operation is shown in Figure 3.8. However, note that subtracting the first table from the second table is not the same as subtracting the second table from the first table.

FIGURE 3.8   DIFFERENCE

STU_FNAME	STU_LNAME
George	Jones
Jane	Smith
Peter	Robinson
Franklin	Johnson
Martin	Lopez

DIFFERENCE

EMP_FNAME	EMP_LNAME
Franklin	Lopez
William	Turner
Franklin	Johnson
Susan	Rogers

yields

STU_FNAME	STU_LNAME
George	Jones
Jane	Smith
Peter	Robinson
Martin	Lopez

Note

DIFFERENCE is denoted by the minus symbol −. If the relations SUPPLIER and VENDOR are union-compatible, then an DIFFERENCE of SUPPLIER minus VENDOR would be written as follows:

supplier − vendor

Assuming the SUPPLIER and VENDOR tables are not union-compatible, producing a list of any supplier names that do not appear as vendor names, then you can use a DIFFERENCE operator.

$$\pi_{supplier\_name}(supplier) - \pi_{vendor\_name}(vendor)$$

Product PRODUCT yields all possible pairs of rows from two tables—also known as the Cartesian product. Therefore, if one table has 6 rows and the other table has 3 rows, the PRODUCT yields a list composed of 6 × 3 = 18rows. The effect of a PRODUCT operation is shown in Figure 3.9.

FIGURE 3.9   PRODUCT

P_CODE	P_DESCRIPT	PRICE
123456	Flashlight	5.26
123457	Lamp	25.15
123458	Box Fan	10.99
213345	9v battery	1.92
254467	100W bulb	1.47
311452	Powerdrill	34.99

PRODUCT

STORE	AISLE	SHELF
23	W	5
24	K	9
25	Z	6

yields

P_CODE	P_DESCRIPT	PRICE	STORE	AISLE	SHELF
123456	Flashlight	5.26	23	W	5
123456	Flashlight	5.26	24	K	9
123456	Flashlight	5.26	25	Z	6
123457	Lamp	25.15	23	W	5
123457	Lamp	25.15	24	K	9
123457	Lamp	25.15	25	Z	6
123458	Box Fan	10.99	23	W	5
123458	Box Fan	10.99	24	K	9
123458	Box Fan	10.99	25	Z	6
213345	9v battery	1.92	23	W	5
213345	9v battery	1.92	24	K	9
213345	9v battery	1.92	25	Z	6
311452	Powerdrill	34.99	23	W	5
311452	Powerdrill	34.99	24	K	9
311452	Powerdrill	34.99	25	Z	6
254467	100W bulb	1.47	23	W	5
254467	100W bulb	1.47	24	K	9
254467	100W bulb	1.47	25	Z	6

Note

PRODUCT
In relational algebra, an operator used to yield all possible pairs of rows from two tables. Also known as the Cartesian product.

PRODUCT is denoted by the multiplication symbol ×. The PRODUCT of the CUSTOMER and AGENT relations would be written as follows:

customer × agent

A Cartesian product produces a set of sequences in which every member of one set is paired with every member of another set. In terms of relations, this means that every tuple in one relation is paired with every tuple in the second relation.

Join **JOIN** allows information to be intelligently combined from two or more tables. JOIN is the real power behind the relational database, allowing the use of independent tables linked by common attributes. The CUSTOMER and AGENT tables shown in Figure 3.10 will be used to illustrate several types of joins.

FIGURE 3.10 TWO TABLES THAT WILL BE USED IN JOIN ILLUSTRATIONS

Table name: CUSTOMER

CUS_CODE	CUS_LNAME	CUS_ZIP	AGENT_CODE
1132445	Walker	32145	231
1217782	Adares	32145	125
1312243	Rakowski	34129	167
1321242	Rodriguez	37134	125
1542311	Smithson	37134	421
1657399	Vanloo	32145	231

Table name: AGENT

AGENT_CODE	AGENT_PHONE
125	6152439887
167	6153426778
231	6152431124
333	9041234445

A **natural join** links tables by selecting only the rows with common values in their common attribute(s). A natural join is the result of a three-stage process:

1. First, a PRODUCT of the tables is created, yielding the results shown in Figure 3.11.

FIGURE 3.11 NATURAL JOIN, STEP 1: PRODUCT

CUS_CODE	CUS_LNAME	CUS_ZIP	CUSTOMER.AGENT_CODE	AGENT.AGENT_CODE	AGENT_PHONE
1132445	Walker	32145	231	125	6152439887
1132445	Walker	32145	231	167	6153426778
1132445	Walker	32145	231	231	6152431124
1132445	Walker	32145	231	333	9041234445
1217782	Adares	32145	125	125	6152439887
1217782	Adares	32145	125	167	6153426778
1217782	Adares	32145	125	231	6152431124
1217782	Adares	32145	125	333	9041234445
1312243	Rakowski	34129	167	125	6152439887
1312243	Rakowski	34129	167	167	6153426778
1312243	Rakowski	34129	167	231	6152431124
1312243	Rakowski	34129	167	333	9041234445
1321242	Rodriguez	37134	125	125	6152439887
1321242	Rodriguez	37134	125	167	6153426778
1321242	Rodriguez	37134	125	231	6152431124
1321242	Rodriguez	37134	125	333	9041234445
1542311	Smithson	37134	421	125	6152439887
1542311	Smithson	37134	421	167	6153426778
1542311	Smithson	37134	421	231	6152431124
1542311	Smithson	37134	421	333	9041234445
1657399	Vanloo	32145	231	125	6152439887
1657399	Vanloo	32145	231	167	6153426778
1657399	Vanloo	32145	231	231	6152431124
1657399	Vanloo	32145	231	333	9041234445

JOIN
In relational algebra, a type of operator used to yield rows from two tables based on criteria. There are many types of joins, such as natural join, theta join, equijoin, and outer join.

natural join
A relational operation that yields a new table composed of only the rows with common values in their common attribute(s).

join columns
Columns that are used in the criteria of join operations. The join columns generally share similar values.

2. Second, a SELECT is performed on the output of Step 1 to yield only the rows for which the AGENT_CODE values are equal. The common columns are referred to as the **join columns**. Step 2 yields the results shown in Figure 3.12.

FIGURE 3.12 NATURAL JOIN, STEP 2: SELECT

CUS_CODE	CUS_LNAME	CUS_ZIP	CUSTOMER.AGENT_CODE	AGENT.AGENT_CODE	AGENT_PHONE
1217782	Adares	32145	125	125	6152439887
1321242	Rodriguez	37134	125	125	6152439887
1312243	Rakowski	34129	167	167	6153426778
1132445	Walker	32145	231	231	6152431124
1657399	Vanloo	32145	231	231	6152431124

3. A PROJECT is performed on the results of Step 2 to yield a single copy of each attribute, thereby eliminating duplicate columns. Step 3 yields the output shown in Figure 3.13.

FIGURE 3.13 NATURAL JOIN, STEP 3: PROJECT

CUS_CODE	CUS_LNAME	CUS_ZIP	AGENT_CODE	AGENT_PHONE
1217782	Adares	32145	125	6152439887
1321242	Rodriguez	37134	125	6152439887
1312243	Rakowski	34129	167	6153426778
1132445	Walker	32145	231	6152431124
1657399	Vanloo	32145	231	6152431124

The final outcome of a natural join yields a table that does not include unmatched pairs and provides only the copies of the matches.

Note a few crucial features of the natural join operation:

- If no match is made between the table rows, the new table does not include the unmatched row. In that case, neither AGENT_CODE 421 nor the customer whose last name is Smithson is included. Smithson's AGENT_CODE 421 does not match any entry in the AGENT table.

- The column on which the join was made—that is, AGENT_CODE—occurs only once in the new table.

- If the same AGENT_CODE were to occur several times in the AGENT table, a customer would be listed for each match. For example, if the AGENT_CODE 167 occurred three times in the AGENT table, the customer named Rakowski would also occur three times in the resulting table because Rakowski is associated with AGENT_CODE 167. (Of course, a good AGENT table cannot yield such a result because it would contain unique primary key values.)

equijoin
A join operator that links tables based on an equality condition that compares specified columns of the tables.

Note

Natural join is normally just referred to as JOIN in formal treatments. JOIN is denoted by the symbol ⋈. The JOIN of the CUSTOMER and AGENT relations would be written as follows:

customer ⋈ agent

Notice that the JOIN of two relations returns all of the attributes of both relations, except only one copy of the common attribute is returned. Formally, this is described as a UNION of the relvar headings. Therefore, the JOIN of the relations (c ⋈ a) includes the UNION of the relvars (C ∪ A). Also note that, as described above, JOIN is not a fundamental relational algebra operator. It can be derived from other operators as follows:

$$\pi_{cus\_code, cus\_lname, cus\_fname, cus\_initial, cus\_renew\_date, agent\_code, agent\_areacode, agent\_phone, agent\_lname, agent\_ytd\_sls}$$

$$(\sigma_{customer.agent\_code = agent.agent\_code} (customer \times agent))$$

Another form of join, known as an **equijoin**, links tables on the basis of an equality condition that compares specified columns of each table. The outcome of the equijoin does not eliminate duplicate columns, and the condition or criterion used to join the tables must be explicitly defined. In fact, the result of an equijoin looks just like the outcome shown in Figure 3.12 for Step 2 of a natural join. The equijoin takes its name from the equality comparison operator (=) used in the condition. If any other comparison operator is used, the join is called a **theta join**.

Note

In formal terms, theta join is considered an extension of natural join. Theta join is denoted by adding a theta subscript after the JOIN symbol: $\bowtie_\theta$. Equijoin is then a special type of theta join.

Each of the preceding joins is often classified as an inner join. An **inner join** only returns matched records from the tables that are being joined. In an **outer join**, the matched pairs would be retained, and any unmatched values in the other table would be left null. It is an easy mistake to think that an outer join is the opposite of an inner join. However, it is more accurate to think of an outer join as an "inner join plus." The outer join still returns all of the matched records that the inner join returns, plus it returns the unmatched records from one of the tables. More specifically, if an outer join is produced for tables CUSTOMER and AGENT, two scenarios are possible:

- A **left outer join** yields all of the rows in the CUSTOMER table, including those that do not have a matching value in the AGENT table. An example of such a join is shown in Figure 3.14.

FIGURE 3.14 LEFT OUTER JOIN

CUS_CODE	CUS_LNAME	CUS_ZIP	CUSTOMER.AGENT_CODE	AGENT.AGENT_CODE	AGENT_PHONE
1217782	Adares	32145	125	125	6152439887
1321242	Rodriguez	37134	125	125	6152439887
1312243	Rakowski	34129	167	167	6153426778
1132445	Walker	32145	231	231	6152431124
1657399	Vanloo	32145	231	231	6152431124
1542311	Smithson	37134	421		

- A **right outer join** yields all of the rows in the AGENT table, including those that do not have matching values in the CUSTOMER table. An example of such a join is shown in Figure 3.15.

FIGURE 3.15 RIGHT OUTER JOIN

CUS_CODE	CUS_LNAME	CUS_ZIP	CUSTOMER.AGENT_CODE	AGENT.AGENT_CODE	AGENT_PHONE
1217782	Adares	32145	125	125	6152439887
1321242	Rodriguez	37134	125	125	6152439887
1312243	Rakowski	34129	167	167	6153426778
1132445	Walker	32145	231	231	6152431124
1657399	Vanloo	32145	231	231	6152431124
				333	9041234445

Outer joins are especially useful when you are trying to determine what values in related tables cause referential integrity problems. Such problems are created when foreign key values do not match the primary key values in the related table(s). In fact, if you are asked to convert large spreadsheets or other "nondatabase" data into relational database

theta join
A join operator that links tables using an inequality comparison operator (<, >, <=, >=) in the join condition.

inner join
A join operation in which only rows that meet a given criterion are selected. The join criterion can be an equality condition (natural join or equijoin) or an inequality condition (theta join). The inner join is the most commonly used type of join. Contrast with *outer join*.

outer join
A relational algebra join operation that produces a table in which all unmatched pairs are retained; unmatched values in the related table are left null. Contrast with *inner join*. See also *left outer join* and *right outer join*.

left outer join
In a pair of tables to be joined, a join that yields all the rows in the left table, including those that have no matching values in the other table. For example, a left outer join of CUSTOMER with AGENT will yield all of the CUSTOMER rows, including the ones that do not have a matching AGENT row. See also *outer join* and *right outer join*.

right outer join
In a pair of tables to be joined, a join that yields all of the rows in the right table, including the ones with no matching values in the other table. For example, a right outer join of CUSTOMER with AGENT will yield all of the AGENT rows, including the ones that do not have a matching CUSTOMER row. See also *left outer join* and *outer join*.

tables, you will discover that the outer joins save you vast amounts of time and uncounted headaches when you encounter referential integrity errors after the conversions.

You may wonder why the outer joins are labeled "left" and "right." The labels refer to the order in which the tables are listed in the SQL command. Chapter 8 explores such joins in more detail.

Note

Outer join is also an extension of JOIN. Outer joins are the application of JOIN, DIFFERENCE, UNION, and PRODUCT. A JOIN returns the matched tuples, DIFFERENCE finds the tuples in one table that have values in the common attribute that do not appear in the common attribute of the other relation, these unmatched tuples are combined with NULL values through a PRODUCT, and then a UNION combines these results into a single relation. Clearly, a defined outer join is a great simplification! Left and right outer joins are denoted by the symbols ⋈ and ⋉, respectively.

Divide The **DIVIDE** operator is used to answer questions about one set of data being associated with all values of data in another set of data. The DIVIDE operation uses one 2-column table (Table 1) as the dividend and one single-column table (Table 2) as the divisor. For example, Figure 3.16 shows a list of customers and the products purchased in Table 1 on the left. Table 2 in the center contains a set of products that are of interest to the users. A DIVIDE operation can be used to determine which customers, if any, purchased every product shown in Table 2. In the figure, the dividend contains the P_CODE and CUS_CODE columns. The divisor contains the P_CODE column. The tables must have a common column—in this case, the P_CODE column. The output of the DIVIDE operation on the right is a single column that contains all values from the second column of the dividend (CUS_CODE) that are associated with every row in the divisor. Using the example shown in Figure 3.16, note the following:

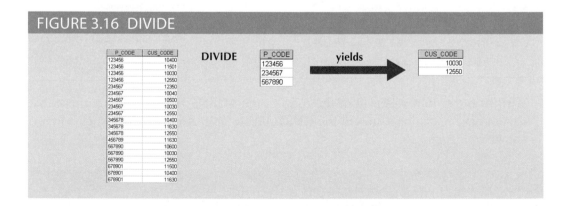

FIGURE 3.16 DIVIDE

- Table 1 is "divided" by Table 2 to produce Table 3. Tables 1 and 2 both contain the P_CODE column but do not share the CUS_CODE column.

- To be included in the resulting Table 3, a value in the unshared column (CUS_CODE) must be associated with every value in Table 2.

- The only customers associated with all of products 123456, 234567, and 567890 are customers 10030 and 12550.

DIVIDE
In relational algebra, an operator that answers queries about one set of data being associated with all values of data in another set of data.

Note

The DIVIDE operator is denoted by the division symbol ÷. Given two relations, R and S, the DIVISION of them would be written: r ÷ s.

3-5 The Data Dictionary and the System Catalog

The **data dictionary** provides a detailed description of all tables in the database created by the user and designer. Thus, the data dictionary contains at least all of the attribute names and characteristics for each table in the system. In short, the data dictionary contains metadata—data about data. Using the small database presented in Figure 3.3, you might picture its data dictionary as shown in Table 3.6.

Note

The data dictionary in Table 3.6 is an example of the *human* view of the entities, attributes, and relationships. The purpose of this data dictionary is to ensure that all members of database design and implementation teams use the same table and attribute names and characteristics. The DBMS's internally stored data dictionary contains additional information about relationship types, entity and referential integrity checks and enforcement, and index types and components. This additional information is generated during the database implementation stage.

The data dictionary is sometimes described as "the database designer's database" because it records the design decisions about tables and their structures.

Like the data dictionary, the system catalog contains metadata. The **system catalog** can be described as a detailed system data dictionary that describes all objects within the database, including data about table names, table's creator and creation date, number of columns in each table, data type corresponding to each column, index filenames, index creators, authorized users, and access privileges. Because the system catalog contains all required data dictionary information, the terms *system catalog* and *data dictionary* are often used interchangeably. In fact, current relational database software generally provides only a system catalog, from which the designer's data dictionary information may be derived. The system catalog is actually a system-created database whose tables store the user/designer-created database characteristics and contents. Therefore, the system catalog tables can be queried just like any user/designer-created table.

In effect, the system catalog automatically produces database documentation. As new tables are added to the database, that documentation also allows the RDBMS to check for and eliminate homonyms and synonyms. In general terms, **homonyms** are similar-sounding words with different meanings, such as *boar* and *bore*, or a word with different meanings, such as *fair* (which means "just" in some contexts and "festival" in others). In a database context, the word *homonym* indicates the use of the same name to label different attributes. For example, you might use C_NAME to label a customer name attribute in a CUSTOMER table and use C_NAME to label a consultant name attribute in a CONSULTANT table. To lessen confusion, you should avoid database homonyms; the data dictionary is very useful in this regard.

data dictionary
A DBMS component that stores metadata—data about data. Thus, the data dictionary contains the data definition as well as their characteristics and relationships. A data dictionary may also include data that are external to the DBMS. Also known as an *information resource dictionary*. See also *active data dictionary*, *metadata*, and *passive data dictionary*.

system catalog
A detailed system data dictionary that describes all objects in a database.

homonym
The use of the same name to label different attributes. Homonyms generally should be avoided. Some relational software automatically checks for homonyms and either alerts the user to their existence or automatically makes the appropriate adjustments. See also *synonym*.

TABLE 3.6

A SAMPLE DATA DICTIONARY

TABLE NAME	ATTRIBUTE NAME	CONTENTS	TYPE	FORMAT	RANGE	REQUIRED	PK OR FK	FK REFERENCED TABLE
CUSTOMER	CUS_CODE	Customer account code	CHAR(5)	99999	10000–99999	Y	PK	
	CUS_LNAME	Customer last name	VARCHAR(20)	Xxxxxxxx		Y		
	CUS_FNAME	Customer first name	VARCHAR(20)	Xxxxxxxx		Y		
	CUS_INITIAL	Customer initial	CHAR(1)	X				
	CUS_RENEW_DATE	Customer insurance renewal date	DATE	dd-mmm-yyyy				
	AGENT_CODE	Agent code	CHAR(3)	999			FK	AGENT
AGENT	AGENT_CODE	Agent code	CHAR(3)	999		Y	PK	
	AGENT_AREACODE	Agent area code	CHAR(3)	999		Y		
	AGENT_PHONE	Agent telephone number	CHAR(8)	999–9999		Y		
	AGENT_LNAME	Agent last name	VARCHAR(20)	Xxxxxxxx		Y		
	AGENT_YTD_SLS	Agent year-to-date sales	NUMBER(9,2)	9,999,999.99				

FK	= Foreign key
PK	= Primary key
CHAR	= Fixed character length data (1 – 255 characters)
VARCHAR	= Variable character length data (1 – 2,000 characters)
NUMBER	= Numeric data. NUMBER (9,2) is used to specify numbers with up to nine digits, including two digits to the right of the decimal place. Some RDBMS permit the use of a MONEY or CURRENCY data type.

Note

Telephone area codes are always composed of digits 0–9, but because area codes are not used arithmetically, they are most efficiently stored as character data. Also, the area codes are always composed of three digits. Therefore, the area code data type is defined as CHAR(3). On the other hand, names do not conform to a standard length. Therefore, the customer first names are defined as VARCHAR(20), indicating that up to 20 characters may be used to store the names. Character data are shown as left-aligned.

In a database context, a **synonym** is the opposite of a homonym, and indicates the use of different names to describe the same attribute. For example, *car* and *auto* refer to the same object. Synonyms must be avoided whenever possible.

3-6 Relationships within the Relational Database

You already know that relationships are classified as one-to-one (1:1), one-to-many (1:M), and many-to-many (M:N or M:M). This section explores those relationships further to help you apply them properly when you start developing database designs. This section focuses on the following points:

- The 1:M relationship is the relational modeling ideal. Therefore, this relationship type should be the norm in any relational database design.

- The 1:1 relationship should be rare in any relational database design.

- M:N relationships cannot be implemented as such in the relational model. Later in this section, you will see how any M:N relationship can be changed into two 1:M relationships.

3-6a The 1:M Relationship

The 1:M relationship is the norm for relational databases. To see how such a relationship is modeled and implemented, consider the PAINTER and PAINTING example shown in Figure 3.17.

FIGURE 3.17 THE 1:M RELATIONSHIP BETWEEN PAINTER AND PAINTING

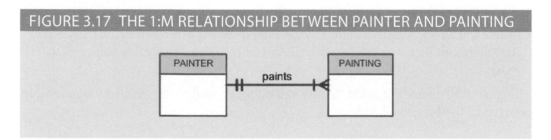

Compare the data model in Figure 3.17 with its implementation in Figure 3.18.

As you examine the PAINTER and PAINTING table contents in Figure 3.18, note the following features:

- Each painting was created by one and only one painter, but each painter could have created many paintings. Note that painter 123 (Georgette P. Ross) has three works stored in the PAINTING table.

- There is only one row in the PAINTER table for any given row in the PAINTING table, but there may be many rows in the PAINTING table for any given row in the PAINTER table.

Note

The one-to-many (1:M) relationship is easily implemented in the relational model by putting the *primary key of the "1" side in the table of the "many" side as a foreign key*.

synonym
The use of different names to identify the same object, such as an entity, an attribute, or a relationship; synonyms should generally be avoided. See also *homonym*.

FIGURE 3.18 THE IMPLEMENTED 1:M RELATIONSHIP BETWEEN PAINTER AND PAINTING

Table name: PAINTER
Primary key: PAINTER_NUM
Foreign key: none

Database name: Ch03_Museum

PAINTER_NUM	PAINTER_LNAME	PAINTER_FNAME	PAINTER_INITIAL
123	Ross	Georgette	P
126	Itero	Julio	G

Table name: PAINTING
Primary key: PAINTING_NUM
Foreign key: PAINTER_NUM

PAINTING_NUM	PAINTING_TITLE	PAINTER_NUM
1338	Dawn Thunder	123
1339	Vanilla Roses To Nowhere	123
1340	Tired Flounders	126
1341	Hasty Exit	123
1342	Plastic Paradise	126

The 1:M relationship is found in any database environment. Students in a typical college or university will discover that each COURSE can generate many CLASSes but that each CLASS refers to only one COURSE. For example, an Accounting II course might yield two classes: one offered on Monday, Wednesday, and Friday (MWF) from 10:00 a.m. to 10:50 a.m., and one offered on Thursday (Th) from 6:00 p.m. to 8:40 p.m. Therefore, the 1:M relationship between COURSE and CLASS might be described this way:

- Each COURSE can have many CLASSes, but each CLASS references only one COURSE.

- There will be only one row in the COURSE table for any given row in the CLASS table, but there can be many rows in the CLASS table for any given row in the COURSE table.

Figure 3.19 maps the ERM (entity relationship model) for the 1:M relationship between COURSE and CLASS.

FIGURE 3.19 THE 1:M RELATIONSHIP BETWEEN COURSE AND CLASS

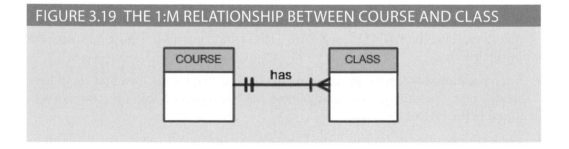

The 1:M relationship between COURSE and CLASS is further illustrated in Figure 3.20.

Using Figure 3.20, take a minute to review some important terminology. Note that CLASS_CODE in the CLASS table uniquely identifies each row. Therefore, CLASS_CODE has been chosen to be the primary key. However, the combination CRS_CODE and CLASS_SECTION will also uniquely identify each row in the class table. In other

FIGURE 3.20 THE IMPLEMENTED 1:M RELATIONSHIP BETWEEN COURSE AND CLASS

Table name: COURSE
Primary key: CRS_CODE   Database name: Ch03_TinyCollege
Foreign key: none

CRS_CODE	DEPT_CODE	CRS_DESCRIPTION	CRS_CREDIT
ACCT-211	ACCT	Accounting I	3
ACCT-212	ACCT	Accounting II	3
CIS-220	CIS	Intro. to Microcomputing	3
CIS-420	CIS	Database Design and Implementation	4
QM-261	CIS	Intro. to Statistics	3
QM-362	CIS	Statistical Applications	4

Table name: CLASS
Primary key: CLASS_CODE
Foreign key: CRS_CODE

CLASS_CODE	CRS_CODE	CLASS_SECTION	CLASS_TIME	CLASS_ROOM	PROF_NUM
10012	ACCT-211	1	MWF 8:00-8:50 a.m.	BUS311	105
10013	ACCT-211	2	MWF 9:00-9:50 a.m.	BUS200	105
10014	ACCT-211	3	TTh 2:30-3:45 p.m.	BUS252	342
10015	ACCT-212	1	MWF 10:00-10:50 a.m.	BUS311	301
10016	ACCT-212	2	Th 6:00-8:40 p.m.	BUS252	301
10017	CIS-220	1	MWF 9:00-9:50 a.m.	KLR209	228
10018	CIS-220	2	MWF 9:00-9:50 a.m.	KLR211	114
10019	CIS-220	3	MWF 10:00-10:50 a.m.	KLR209	228
10020	CIS-420	1	W 6:00-8:40 p.m.	KLR209	162
10021	QM-261	1	MWF 8:00-8:50 a.m.	KLR200	114
10022	QM-261	2	TTh 1:00-2:15 p.m.	KLR200	114
10023	QM-362	1	MWF 11:00-11:50 a.m.	KLR200	162
10024	QM-362	2	TTh 2:30-3:45 p.m.	KLR200	162

words, the *composite key* composed of CRS_CODE and CLASS_SECTION is a candidate key. Any *candidate key* must have the not-null and unique constraints enforced. (You will see how this is done when you learn SQL in Chapter 7.)

For example, note in Figure 3.18 that the PAINTER table's primary key, PAINTER_NUM, is included in the PAINTING table as a foreign key. Similarly, in Figure 3.20, the COURSE table's primary key, CRS_CODE, is included in the CLASS table as a foreign key.

3-6b The 1:1 Relationship

As the 1:1 label implies, one entity in a 1:1 relationship can be related to only one other entity, and vice versa. For example, one department chair—a professor—can chair only one department, and one department can have only one department chair. The entities PROFESSOR and DEPARTMENT thus exhibit a 1:1 relationship. (You might argue that not all professors chair a department and professors cannot be *required* to chair a department. That is, the relationship between the two entities is optional. However, at this stage of the discussion, you should focus your attention on the basic 1:1 relationship. Optional relationships will be addressed in Chapter 4.) The basic 1:1 relationship is modeled in Figure 3.21, and its implementation is shown in Figure 3.22.

FIGURE 3.21 THE 1:1 RELATIONSHIP BETWEEN PROFESSOR AND DEPARTMENT

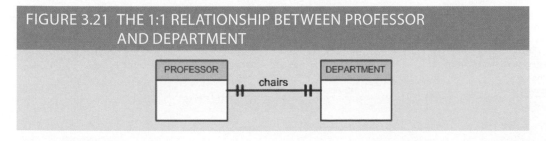

As you examine the tables in Figure 3.22, note several important features:

- Each professor is a Tiny College employee. Therefore, the professor identification is through the EMP_NUM. (However, note that not all employees are professors—there's another optional relationship.)

- The 1:1 "PROFESSOR chairs DEPARTMENT" relationship is implemented by having the EMP_NUM foreign key in the DEPARTMENT table. Note that the 1:1 relationship is treated as a special case of the 1:M relationship in which the "many" side is restricted to a single occurrence. In this case, DEPARTMENT contains the EMP_NUM as a foreign key to indicate that it is the *department* that has a chair.

FIGURE 3.22 THE IMPLEMENTED 1:1 RELATIONSHIP BETWEEN PROFESSOR AND DEPARTMENT

Table name: PROFESSOR
Primary key: EMP_NUM
Foreign key: DEPT_CODE

Database name: Ch03_TinyCollege

EMP_NUM	DEPT_CODE	PROF_OFFICE	PROF_EXTENSION	PROF_HIGH_DEGREE
103	HIST	DRE 156	6783	Ph.D.
104	ENG	DRE 102	5561	MA
105	ACCT	KLR 229D	8665	Ph.D.
106	MKT/MGT	KLR 126	3899	Ph.D.
110	BIOL	AAK 160	3412	Ph.D.
114	ACCT	KLR 211	4436	Ph.D.
155	MATH	AAK 201	4440	Ph.D.
160	ENG	DRE 102	2248	Ph.D.
162	CIS	KLR 203E	2359	Ph.D.
191	MKT/MGT	KLR 409B	4016	DBA
195	PSYCH	AAK 297	3550	Ph.D.
209	CIS	KLR 333	3421	Ph.D.
228	CIS	KLR 300	3000	Ph.D.
297	MATH	AAK 194	1145	Ph.D.
299	ECON/FIN	KLR 284	2851	Ph.D.
301	ACCT	KLR 244	4683	Ph.D.
335	ENG	DRE 208	2000	Ph.D.
342	SOC	BBG 208	5514	Ph.D.
387	BIOL	AAK 230	8665	Ph.D.
401	HIST	DRE 156	6783	MA
425	ECON/FIN	KLR 284	2851	MBA
435	ART	BBG 185	2278	Ph.D.

The 1:M DEPARTMENT employs PROFESSOR relationship is implemented through the placement of the DEPT_CODE foreign key in the PROFESSOR table.

The 1:1 PROFESSOR chairs DEPARTMENT relationship is implemented through the placement of the EMP_NUM foreign key in the DEPARTMENT table.

Table name: DEPARTMENT
Primary key: DEPT_CODE
Foreign key: EMP_NUM

DEPT_CODE	DEPT_NAME	SCHOOL_CODE	EMP_NUM	DEPT_ADDRESS	DEPT_EXTENSION
ACCT	Accounting	BUS	114	KLR 211, Box 52	3119
ART	Fine Arts	A&SCI	435	BBG 185, Box 128	2278
BIOL	Biology	A&SCI	387	AAK 230, Box 415	4117
CIS	Computer Info. Systems	BUS	209	KLR 333, Box 56	3245
ECON/FIN	Economics/Finance	BUS	299	KLR 284, Box 63	3126
ENG	English	A&SCI	160	DRE 102, Box 223	1004
HIST	History	A&SCI	103	DRE 156, Box 284	1867
MATH	Mathematics	A&SCI	297	AAK 194, Box 422	4234
MKT/MGT	Marketing/Management	BUS	106	KLR 126, Box 55	3342
PSYCH	Psychology	A&SCI	195	AAK 297, Box 438	4110
SOC	Sociology	A&SCI	342	BBG 208, Box 132	2008

- Also note that the PROFESSOR table contains the DEPT_CODE foreign key to implement the 1:M "DEPARTMENT employs PROFESSOR" relationship. This is a good example of how two entities can participate in two (or even more) relationships simultaneously.

The preceding "PROFESSOR chairs DEPARTMENT" example illustrates a proper 1:1 relationship. *In fact, the use of a 1:1 relationship ensures that two entity sets are not placed in the same table when they should not be.* However, the existence of a 1:1 relationship sometimes means that the entity components were not defined properly. It could indicate that the two entities actually belong in the same table!

Although 1:1 relationships should be rare, certain conditions absolutely require their use. In Chapter 5, Advanced Data Modeling, you will explore a concept called a generalization hierarchy, which is a powerful tool for improving database designs under specific conditions to avoid a proliferation of nulls. One characteristic of generalization hierarchies is that they are implemented as 1:1 relationships.

3-6c The M:N Relationship

A many-to-many (M:N) relationship is not supported directly in the relational environment. However, M:N relationships can be implemented by creating a new entity in 1:M relationships with the original entities.

To explore the many-to-many relationship, consider a typical college environment. The ER model in Figure 3.23 shows this M:N relationship.

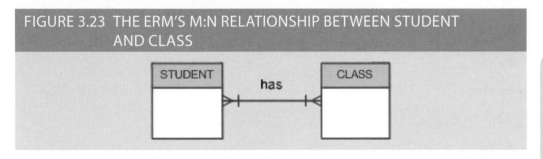

FIGURE 3.23 THE ERM'S M:N RELATIONSHIP BETWEEN STUDENT AND CLASS

Note the features of the ERM in Figure 3.23.

- Each CLASS can have many STUDENTs, and each STUDENT can take many CLASSes.

- There can be many rows in the CLASS table for any given row in the STUDENT table, and there can be many rows in the STUDENT table for any given row in the CLASS table.

To examine the M:N relationship more closely, imagine a small college with two students, each of whom takes three classes. Table 3.7 shows the enrollment data for the two students.

Given such a data relationship and the sample data in Table 3.7, you could wrongly assume that you could implement this M:N relationship simply by adding a foreign key in the "many" side of the relationship that points to the primary key of the related table, as shown in Figure 3.24.

However, the M:N relationship should *not* be implemented as shown in Figure 3.24 for two good reasons:

- The tables create many redundancies. For example, note that the STU_NUM values occur many times in the STUDENT table. In a real-world situation, additional student attributes such as address, classification, major, and home phone would also be contained in the STUDENT table, and each of those attribute values would be repeated in each of the records shown here. Similarly, the CLASS table contains much

Online Content

If you open the Ch03_TinyCollege database at *www.cengagebrain.com*, you will see that the STUDENT and CLASS entities still use PROF_NUM as their foreign key. PROF_NUM and EMP_NUM are labels for the same attribute, which is an example of the use of synonyms—that is, different names for the same attribute. These synonyms will be eliminated in future chapters as the Tiny College database continues to be improved.

Online Content

If you look at the Ch03_AviaCo database at *www.cengagebrain.com*, you will see the implementation of the 1:1 PILOT to EMPLOYEE relationship. This relationship is based on a generalization hierarchy, which you will learn about in Chapter 5.

TABLE 3.7

SAMPLE STUDENT ENROLLMENT DATA

STUDENT'S LAST NAME	SELECTED CLASSES
Bowser	Accounting 1, ACCT-211, code 10014 Intro to Microcomputing, CIS-220, code 10018 Intro to Statistics, QM-261, code 10021
Smithson	Accounting 1, ACCT-211, code 10014 Intro to Microcomputing, CIS-220, code 10018 Intro to Statistics, QM-261, code 10021

FIGURE 3.24 THE WRONG IMPLEMENTATION OF THE M:N RELATIONSHIP BETWEEN STUDENT AND CLASS

Table name: STUDENT
Primary key: STU_NUM
Foreign key: none

Database name: Ch03_CollegeTry

STU_NUM	STU_LNAME	CLASS_CODE
321452	Bowser	10014
321452	Bowser	10018
321452	Bowser	10021
324257	Smithson	10014
324257	Smithson	10018
324257	Smithson	10021

Table name: CLASS
Primary key: CLASS_CODE
Foreign key: STU_NUM

CLASS_CODE	STU_NUM	CRS_CODE	CLASS_SECTION	CLASS_TIME	CLASS_ROOM	PROF_NUM
10014	321452	ACCT-211	3	TTh 2:30-3:45 p.m.	BUS252	342
10014	324257	ACCT-211	3	TTh 2:30-3:45 p.m.	BUS252	342
10018	321452	CIS-220	2	MWF 9:00-9:50 a.m.	KLR211	114
10018	324257	CIS-220	2	MWF 9:00-9:50 a.m.	KLR211	114
10021	321452	QM-261	1	MWF 8:00-8:50 a.m.	KLR200	114
10021	324257	QM-261	1	MWF 8:00-8:50 a.m.	KLR200	114

duplication: each student taking the class generates a CLASS record. The problem would be even worse if the CLASS table included such attributes as credit hours and course description. Those redundancies lead to the anomalies discussed in Chapter 1.

- Given the structure and contents of the two tables, the relational operations become very complex and are likely to lead to system efficiency errors and output errors.

composite entity
An entity designed to transform an M:N relationship into two 1:M relationships. The composite entity's primary key comprises at least the primary keys of the entities that it connects. Also known as a *bridge entity or associative entity*. See also *linking table*.

bridge entity
See *composite entity*.

associative entity
See *composite entity*.

Fortunately, the problems inherent in the many-to-many relationship can easily be avoided by creating a **composite entity** (also referred to as a **bridge entity** or an **associative entity**). Because such a table is used to link the tables that were originally related in an M:N relationship, the composite entity structure includes—as foreign keys—*at least* the primary keys of the tables that are to be linked. The database designer has two main options when defining a composite table's primary key: use the combination of those foreign keys or create a new primary key.

Remember that each entity in the ERM is represented by a table. Therefore, you can create the composite ENROLL table shown in Figure 3.25 to link the tables CLASS and STUDENT. In this example, the ENROLL table's primary key is the combination of its foreign keys CLASS_CODE and STU_NUM. However, the designer could have decided to create a single-attribute new primary key such as ENROLL_LINE, using a different line value to identify each ENROLL table row uniquely. (Microsoft Access users might use the Autonumber data type to generate such line values automatically.)

FIGURE 3.25 CONVERTING THE M:N RELATIONSHIP INTO TWO 1:M RELATIONSHIPS

Table name: STUDENT
Primary key: STU_NUM
Foreign key: none

Database name: Ch03_CollegeTry2

STU_NUM	STU_LNAME
321452	Bowser
324257	Smithson

Table name: ENROLL
Primary key: CLASS_CODE + STU_NUM
Foreign key: CLASS_CODE, STU_NUM

CLASS_CODE	STU_NUM	ENROLL_GRADE
10014	321452	C
10014	324257	B
10018	321452	A
10018	324257	B
10021	321452	C
10021	324257	C

Table name: CLASS
Primary key: CLASS_CODE
Foreign key: CRS_CODE

CLASS_CODE	CRS_CODE	CLASS_SECTION	CLASS_TIME	CLASS_ROOM	PROF_NUM
10014	ACCT-211	3	TTh 2:30-3:45 p.m.	BUS252	342
10018	CIS-220	2	MWF 9:00-9:50 a.m.	KLR211	114
10021	QM-261	1	MWF 8:00-8:50 a.m.	KLR200	114

Because the ENROLL table in Figure 3.25 links two tables, STUDENT and CLASS, it is also called a **linking table**. In other words, a linking table is the implementation of a composite entity.

Note

In addition to the linking attributes, the composite ENROLL table can also contain such relevant attributes as the grade earned in the course. In fact, a composite table can contain any number of attributes that the designer wants to track. Keep in mind that the composite entity, *although implemented as an actual table*, is *conceptually* a logical entity that was created as a means to an end: to eliminate the potential for multiple redundancies in the original M:N relationship.

The ENROLL table shown in Figure 3.25 yields the required M:N to 1:M conversion. Observe that the composite entity represented by the ENROLL table must contain at least the primary keys of the CLASS and STUDENT tables (CLASS_CODE and STU_NUM, respectively) for which it serves as a connector. Also note that the STUDENT and CLASS tables now contain only one row per entity. The ENROLL table contains multiple occurrences of the foreign key values, but those controlled redundancies are incapable of producing anomalies as long as referential integrity is enforced. Additional attributes may be assigned as needed. In this case, ENROLL_GRADE is selected to satisfy a reporting requirement. Also note that ENROLL_GRADE is fully dependent on the composite primary key. Naturally, the conversion is reflected in the ERM, too. The revised relationship is shown in Figure 3.26.

As you examine Figure 3.26, note that the composite entity named ENROLL represents the linking table between STUDENT and CLASS.

linking table
In the relational model, a table that implements an M:M relationship. See also *composite entity*.

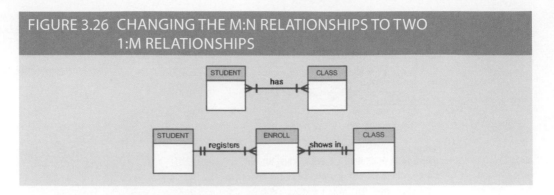

FIGURE 3.26 CHANGING THE M:N RELATIONSHIPS TO TWO 1:M RELATIONSHIPS

The 1:M relationship between COURSE and CLASS was first illustrated in Figure 3.19 and Figure 3.20. You can increase the amount of available information even as you control the database's redundancies. Thus, Figure 3.27 shows the expanded ERM, including the 1:M relationship between COURSE and CLASS shown in Figure 3.19. Note that the model can handle multiple sections of a CLASS while controlling redundancies by making sure that all of the COURSE data common to each CLASS are kept in the COURSE table.

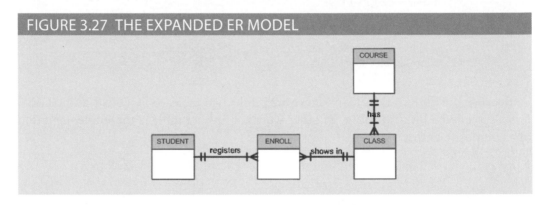

FIGURE 3.27 THE EXPANDED ER MODEL

The relational diagram that corresponds to the ERM in Figure 3.27 is shown in Figure 3.28.

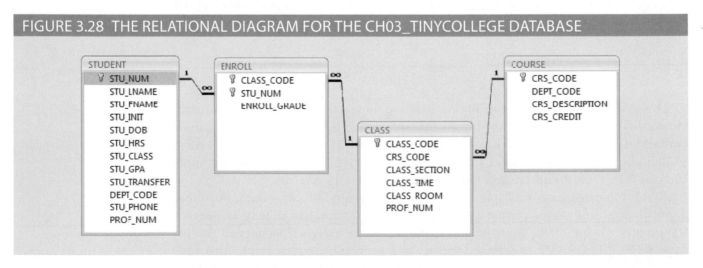

FIGURE 3.28 THE RELATIONAL DIAGRAM FOR THE CH03_TINYCOLLEGE DATABASE

The ERM will be examined in greater detail in Chapter 4 to show you how it is used to design more complex databases. The ERM will also be used as the basis for developing and implementing a realistic database design of a university computer lab in Appendixes B and C. These appendixes are available at *www.cengagebrain.com*.

3-7 **Data Redundancy Revisited**

In Chapter 1, you learned that data redundancy leads to data anomalies, which can destroy the effectiveness of the database. You also learned that the relational database makes it possible to control data redundancies by using common attributes that are shared by tables, called foreign keys.

The proper use of foreign keys is crucial to controlling data redundancy, although they do not totally eliminate the problem because the foreign key values can be repeated many times. However, the proper use of foreign keys *minimizes* data redundancies and the chances that destructive data anomalies will develop.

Note

The real test of redundancy is *not* how many copies of a given attribute are stored, *but whether the elimination of an attribute will eliminate information*. Therefore, if you delete an attribute and the original information can still be generated through relational algebra, the inclusion of that attribute would be redundant. Given that view of redundancy, proper foreign keys are clearly not redundant in spite of their multiple occurrences in a table. However, even when you use this less restrictive view of redundancy, keep in mind that *controlled* redundancies are often designed as part of the system to ensure transaction speed and/or information requirements.

You will learn in Chapter 4 that database designers must reconcile three often contradictory requirements: design elegance, processing speed, and information requirements. Also, you will learn in Chapter 13, Business Intelligence and Data Warehouses, that proper data warehousing design requires carefully defined and controlled data redundancies to function properly. Regardless of how you describe data redundancies, the potential for damage is limited by proper implementation and careful control.

As important as it is to control data redundancy, sometimes the level of data redundancy must actually be increased to make the database serve crucial information purposes. You will learn about such redundancies in Chapter 13. Also, data redundancies sometimes *seem* to exist to preserve the historical accuracy of the data. For example, consider a small invoicing system. The system includes the CUSTOMER, who may buy one or more PRODUCTs, thus generating an INVOICE. Because a customer may buy more than one product at a time, an invoice may contain several invoice LINEs, each providing details about the purchased product. The PRODUCT table should contain the product price to provide a consistent pricing input for each product that appears on the invoice. The tables that are part of such a system are shown in Figure 3.29. The system's relational diagram is shown in Figure 3.30.

As you examine the tables and relationships in the two figures, note that you can keep track of typical sales information. For example, by tracing the relationships among the four tables, you discover that customer 10014 (Myron Orlando) bought two items on March 8, 2016, that were written to invoice number 1001: one Houselite chain saw with a 16-inch bar and three rat-tail files. In other words, trace the CUS_CODE number 10014 in the CUSTOMER table to the matching CUS_CODE value in the INVOICE table. Next, trace the INV_NUMBER 1001 to the first two rows in the LINE table. Finally, match the two PROD_CODE values in LINE with the PROD_CODE values in PRODUCT. Application software will be used to write the correct bill by multiplying each invoice line item's LINE_UNITS by its LINE_PRICE, adding the results, and applying appropriate taxes. Later, other application software

FIGURE 3.29 A SMALL INVOICING SYSTEM

Table name: CUSTOMER
Primary key: CUS_CODE
Foreign key: none

Database name: Ch03_SaleCo

CUS_CODE	CUS_LNAME	CUS_FNAME	CUS_INITIAL	CUS_AREACODE	CUS_PHONE
10010	Ramas	Alfred	A	615	844-2573
10011	Dunne	Leona	K	713	894-1238
10012	Smith	Kathy	W	615	894-2285
10013	Olowski	Paul	F	615	894-2180
10014	Orlando	Myron		615	222-1672
10015	O'Brian	Amy	B	713	442-3381
10016	Brown	James	G	615	297-1228
10017	Williams	George		615	290-2556
10018	Farriss	Anne	G	713	382-7185
10019	Smith	Olette	K	615	297-3809

Table name: LINE
Primary key: INV_NUMBER + LINE_NUMBER
Foreign key: INV_NUMBER, PROD_CODE

INV_NUMBER	LINE_NUMBER	PROD_CODE	LINE_UNITS	LINE_PRICE
1001	1	123-21UUY	1	189.99
1001	2	SRE-657UG	3	2.99
1002	1	QER-34256	2	18.63
1003	1	ZZX/3245Q	1	6.79
1003	2	SRE-657UG	1	2.99
1003	3	001278-AB	1	12.95
1004	1	001278-AB	1	12.95
1004	2	SRE-657UG	2	2.99

Table name: INVOICE
Primary key: INV_NUMBER
Foreign key: CUS_CODE

INV_NUMBER	CUS_CODE	INV_DATE
1001	10014	08-Mar-16
1002	10011	08-Mar-16
1003	10012	08-Mar-16
1004	10011	09-Mar-16

Table name: PRODUCT
Primary key: PROD_CODE
Foreign key: none

PROD_CODE	PROD_DESCRIPT	PROD_PRICE	PROD_ON_HAND	VEND_CODE
001278-AB	Claw hammer	12.95	23	232
123-21UUY	Houselite chain saw, 16-in. bar	189.99	4	235
QER-34256	Sledge hammer, 16-lb. head	18.63	6	231
SRE-657UG	Rat-tail file	2.99	15	232
ZZX/3245Q	Steel tape, 12-ft. length	6.79	8	235

FIGURE 3.30 THE RELATIONAL DIAGRAM FOR THE INVOICING SYSTEM

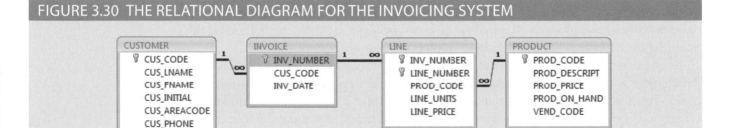

might use the same technique to write sales reports that track and compare sales by week, month, or year.

As you examine the sales transactions in Figure 3.29, you might reasonably suppose that the product price billed to the customer is derived from the PRODUCT table because the product data is stored there. *But why does that same product price occur again in the LINE table? Is that not a data redundancy?* It certainly appears to be, but this time, the apparent redundancy is crucial to the system's success. Copying the product price from the PRODUCT table to the LINE table maintains the *historical accuracy of the transactions*. Suppose, for instance, that you fail to write the LINE_PRICE in the LINE table and that you use the PROD_PRICE from the PRODUCT table to calculate the sales revenue. Now suppose that the PRODUCT table's PROD_PRICE changes, as prices frequently do. This price change will be properly reflected in all subsequent sales revenue calculations. However, the calculations of past sales revenues will also reflect the new product price, which was not in effect when the transaction took place! As a result, the revenue calculations for all past transactions will be incorrect, thus eliminating the possibility of making proper sales comparisons over time. On the other hand, if the price data is copied from the PRODUCT table and stored with the transaction in the LINE table, that price will always accurately reflect the transaction that took place *at that time*. You will discover that such planned "redundancies" are common in good database design.

Finally, you might wonder why the LINE_NUMBER attribute was used in the LINE table in Figure 3.29. Wouldn't the combination of INV_NUMBER and PROD_CODE be a sufficient composite primary key—and, therefore, isn't the LINE_NUMBER redundant? Yes, it is, but this redundancy is common practice on invoicing software that typically generates such line numbers automatically. In this case, the redundancy is not necessary, but given its automatic generation, the redundancy is not a source of anomalies. The inclusion of LINE_NUMBER also adds another benefit: the order of the retrieved invoicing data will always match the order in which the data was entered. If product codes are used as part of the primary key, indexing will arrange those product codes as soon as the invoice is completed and the data is stored. You can imagine the potential confusion when a customer calls and says, "The second item on my invoice has an incorrect price," and you are looking at an invoice whose lines show a different order from those on the customer's copy!

3-8 Indexes

Suppose you want to locate a book in a library. Does it make sense to look through every book until you find the one you want? Of course not; you use the library's catalog, which is indexed by title, topic, and author. The index (in either a manual or computer library catalog) points you to the book's location, making retrieval a quick and simple matter. An **index** is an orderly arrangement used to logically access rows in a table.

Or, suppose you want to find a topic in this book, such as *ER model*. Does it make sense to read through every page until you stumble across the topic? Of course not; it is much simpler to go to the book's index, look up the phrase *ER model*, and read the references that point you to the appropriate page(s). In each case, an index is used to locate a needed item quickly.

Indexes in the relational database environment work like the indexes described in the preceding paragraphs. From a conceptual point of view, an index is composed of an index key and a set of pointers. The **index key** is, in effect, the index's reference point. More formally, an index is an ordered arrangement of keys and pointers. Each key points to the location of the data identified by the key.

For example, suppose you want to look up all of the paintings created by a given painter in the Ch03_Museum database in Figure 3.18. Without an index, you must read each row in the PAINTING table and see if the PAINTER_NUM matches the requested painter. However, if you index the PAINTER table and use the index key PAINTER_NUM, you merely need to look up the appropriate PAINTER_NUM in the index and find the matching pointers. Conceptually speaking, the index would resemble the presentation in Figure 3.31.

As you examine Figure 3.31, note that the first PAINTER_NUM index key value (123) is found in records 1, 2, and 4 of the PAINTING table. The second PAINTER_NUM index key value (126) is found in records 3 and 5 of the PAINTING table.

DBMSs use indexes for many different purposes. You just learned that an index can be used to retrieve data more efficiently, but indexes can also be used by a DBMS

index
An ordered array of index key values and row ID values (pointers). Indexes are generally used to speed up and facilitate data retrieval. Also known as an *index key*.

index key
See *index*.

FIGURE 3.31   COMPONENTS OF AN INDEX

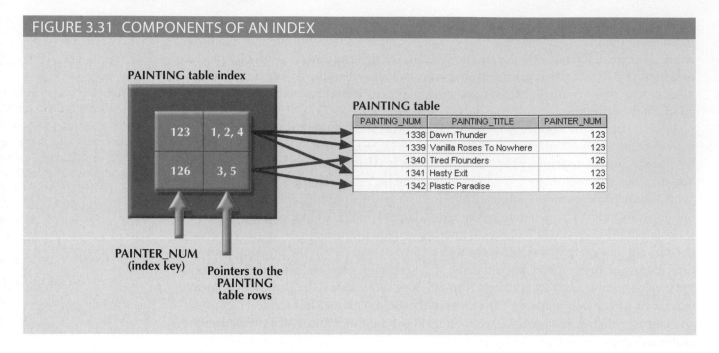

to retrieve data ordered by a specific attribute or attributes. For example, creating an index on a customer's last name will allow you to retrieve the customer data alphabetically by the customer's last name. Also, an index key can be composed of one or more attributes. For example, in Figure 3.29, you can create an index on VEND_CODE and PROD_CODE to retrieve all rows in the PRODUCT table ordered by vendor, and within vendor, ordered by product.

Indexes play an important role in DBMSs for the implementation of primary keys. When you define a table's primary key, the DBMS automatically creates a unique index on the primary key column(s) you declared. For example, in Figure 3.29, when you declare CUS_CODE to be the primary key of the CUSTOMER table, the DBMS automatically creates a unique index on that attribute. In a **unique index**, as its name implies, the index key can have only one pointer value (row) associated with it. (The index in Figure 3.31 is not a unique index because the PAINTER_NUM has multiple pointer values associated with it. For example, painter number 123 points to three rows—1, 2, and 4—in the PAINTING table.)

A table can have many indexes, but each index is associated with only one table. The index key can have multiple attributes (a composite index). Creating an index is easy. You will learn in Chapter 7 that a simple SQL command produces any required index.

3-9 Codd's Relational Database Rules

In 1985, Dr. E. F. Codd published a list of 12 rules to define a relational database system.[1] He published the list out of concern that many vendors were marketing products as "relational" even though those products did not meet minimum relational standards. Dr. Codd's list, shown in Table 3.8, is a frame of reference for what a truly relational database should be. Bear in mind that even the dominant database vendors do not fully support all 12 rules.

unique index
An index in which the index key can have only one associated pointer value (row).

[1]Codd, E., "Is Your DBMS Really Relational?" and "Does Your DBMS Run by the Rules?" *Computerworld*, October 14 and October 21, 1985.

TABLE 3.8

DR. CODD'S 12 RELATIONAL DATABASE RULES

RULE	RULE NAME	DESCRIPTION
1	Information	All information in a relational database must be logically represented as column values in rows within tables.
2	Guaranteed access	Every value in a table is guaranteed to be accessible through a combination of table name, primary key value, and column name.
3	Systematic treatment of nulls	Nulls must be represented and treated in a systematic way, independent of data type.
4	Dynamic online catalog based on the relational model	The metadata must be stored and managed as ordinary data—that is, in tables within the database; such data must be available to authorized users using the standard database relational language.
5	Comprehensive data sublanguage	The relational database may support many languages; however, it must support one well-defined, declarative language as well as data definition, view definition, data manipulation (interactive and by program), integrity constraints, authorization, and transaction management (begin, commit, and rollback).
6	View updating	Any view that is theoretically updatable must be updatable through the system.
7	High-level insert, update, and delete	The database must support set-level inserts, updates, and deletes.
8	Physical data independence	Application programs and ad hoc facilities are logically unaffected when physical access methods or storage structures are changed.
9	Logical data independence	Application programs and ad hoc facilities are logically unaffected when changes are made to the table structures that preserve the original table values (changing order of columns or inserting columns).
10	Integrity independence	All relational integrity constraints must be definable in the relational language and stored in the system catalog, not at the application level.
11	Distribution independence	The end users and application programs are unaware of and unaffected by the data location (distributed vs. local databases).
12	Nonsubversion	If the system supports low-level access to the data, users must not be allowed to bypass the integrity rules of the database.
13	Rule zero	All preceding rules are based on the notion that to be considered relational, a database must use its relational facilities exclusively for management.

Summary

- Tables are the basic building blocks of a relational database. A grouping of related entities, known as an entity set, is stored in a table. Conceptually speaking, the relational table is composed of intersecting rows (tuples) and columns. Each row represents a single entity, and each column represents the characteristics (attributes) of the entities.

- Keys are central to the use of relational tables. Keys define functional dependencies; that is, other attributes are dependent on the key and can therefore be found if the key value is known. A key can be classified as a superkey, a candidate key, a primary key, a secondary key, or a foreign key.

- Each table row must have a primary key. The primary key is an attribute or combination of attributes that uniquely identifies all remaining attributes found in any given row. Because a primary key must be unique, no null values are allowed if entity integrity is to be maintained.

- Although tables are independent, they can be linked by common attributes. Thus, the primary key of one table can appear as the foreign key in another table to which it is linked. Referential integrity dictates that the foreign key must contain values that match the primary key in the related table, or must contain nulls.

- The relational model supports several relational algebra functions, including SELECT, PROJECT, JOIN, INTERSECT, UNION, DIFFERENCE, PRODUCT, and DIVIDE. Understanding the basic mathematical forms of these functions gives a broader understanding of the data manipulation options.

- A relational database performs much of the data manipulation work behind the scenes. For example, when you create a database, the RDBMS automatically produces a structure to house a data dictionary for your database. Each time you create a new table within the database, the RDBMS updates the data dictionary, thereby providing the database documentation.

- Once you know the basics of relational databases, you can concentrate on design. Good design begins by identifying appropriate entities and their attributes and then the relationships among the entities. Those relationships (1:1, 1:M, and M:N) can be represented using ERDs. The use of ERDs allows you to create and evaluate simple logical design. The 1:M relationship is most easily incorporated in a good design; just make sure that the primary key of the "1" is included in the table of the "many."

Key Terms

associative entity	full functional dependence	PRODUCT
attribute domain	functional dependence	PROJECT
bridge entity	homonym	referential integrity
candidate key	index	relational algebra
closure	index key	relvar
composite entity	inner join	RESTRICT
composite key	INTERSECT	right outer join
data dictionary	JOIN	secondary key
dependent	join column(s)	SELECT
determinant	key	set theory
determination	key attribute	superkey
DIFFERENCE	left outer join	synonym
DIVIDE	linking table	system catalog
domain	natural join	theta join
entity integrity	null	tuple
equijoin	outer join	UNION
flags	predicate logic	union-compatible
foreign key (FK)	primary key (PK)	unique index

Online Content

Flashcards and crossword puzzles for key term practice are available at *www.cengagebrain.com*.

Review Questions

1. What is the difference between a database and a table?

2. What does it mean to say that a database displays both entity integrity and referential integrity?

3. Why are entity integrity and referential integrity important in a database?

4. What are the requirements that two relations must satisfy to be considered union-compatible?

5. Which relational algebra operators can be applied to a pair of tables that are not union-compatible?

6. Explain why the data dictionary is sometimes called "the database designer's database."

7. A database user manually notes that "The file contains two hundred records, each record containing nine fields." Use appropriate relational database terminology to "translate" that statement.

Use Figure Q3.8 to answer Questions 8–12.

8. Using the STUDENT and PROFESSOR tables, illustrate the difference between a natural join, an equijoin, and an outer join.

9. Create the table that would result from π_{stu_code} (student).

10. Create the table that would result from $\pi_{stu\_code,\ dept\_code}$ (student $\bowtie$ professor).

Online Content

All of the databases used in the questions and problems are available at *www.cengagebrain.com*. The database names match the database names shown in the figures.

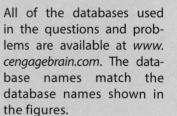

11. Create the basic ERD for the database shown in Figure Q3.8.

12. Create the relational diagram for the database shown in Figure Q3.8.

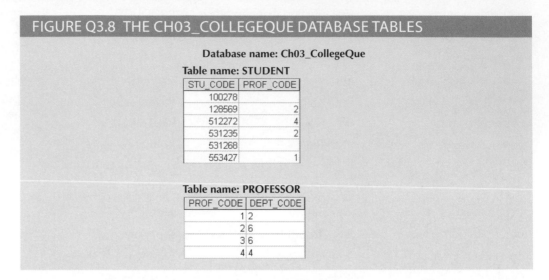

FIGURE Q3.8 THE CH03_COLLEGEQUE DATABASE TABLES

Database name: Ch03_CollegeQue

Table name: STUDENT

STU_CODE	PROF_CODE
100278	
128569	2
512272	4
531235	2
531268	
553427	1

Table name: PROFESSOR

PROF_CODE	DEPT_CODE
1	2
2	6
3	6
4	4

Use Figure Q3.13 to answer Questions 13–17.

FIGURE Q3.13 THE CH03_VENDINGCO DATABASE TABLES

Database name: Ch03_VendingCo

Table name: BOOTH

BOOTH_PRODUCT	BOOTH_PRICE
Chips	1.5
Cola	1.25
Energy Drink	2

Table name: MACHINE

MACHINE_PRODUCT	MACHINE_PRICE
Chips	1.25
Chocolate Bar	1
Energy Drink	2

13. Write the relational algebra formula to apply a UNION relational operator to the tables shown in Figure Q3.13.

14. Create the table that results from applying a UNION relational operator to the tables shown in Figure Q3.13.

15. Write the relational algebra formula to apply an INTERSECT relational operator to the tables shown in Figure Q3.13.

16. Create the table that results from applying an INTERSECT relational operator to the tables shown in Figure Q3.13.

17. Using the tables in Figure Q3.13, create the table that results from MACHINE DIFFERENCE BOOTH.

Use Figure Q3.18 to answer Question 18.

FIGURE Q3.18 THE CROW'S FOOT ERD FOR QUESTION 14

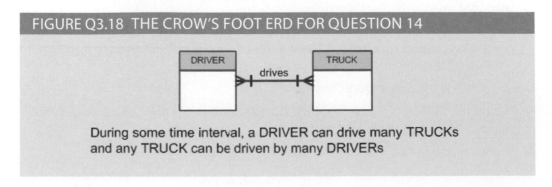

During some time interval, a DRIVER can drive many TRUCKs and any TRUCK can be driven by many DRIVERs

18. Suppose you have the ERD shown in Figure Q3.18. How would you convert this model into an ERM that displays only 1:M relationships? (Make sure you create the revised ERD.)

19. What are homonyms and synonyms, and why should they be avoided in database design?

20. How would you implement a l:M relationship in a database composed of two tables? Give an example.

Use Figure Q3.21 to answer Question 21.

FIGURE Q3.21 THE CH03_NOCOMP DATABASE EMPLOYEE TABLE

Table name: EMPLOYEE   Database name: Ch03_NoComp

EMP_NUM	EMP_LNAME	EMP_INITIAL	EMP_FNAME	DEPT_CODE	JOB_CODE
11234	Friedman	K	Robert	MKTG	12
11238	Olanski	D	Delbert	MKTG	12
11241	Fontein		Juliette	INFS	5
11242	Cruazona	J	Maria	ENG	9
11245	Smithson	B	Bernard	INFS	6
11248	Washington	G	Oleta	ENGR	8
11256	McBride		Randall	ENGR	8
11257	Kachinn	D	Melanie	MKTG	14
11258	Smith	W	William	MKTG	14
11260	Ratula	A	Katrina	INFS	5

21. Identify and describe the components of the table shown in Figure Q3.21, using correct terminology. Use your knowledge of naming conventions to identify the table's probable foreign key(s).

Use the database shown in Figure Q3.22 to answer Questions 22–27.

FIGURE Q3.22 THE CH03_THEATER DATABASE TABLES

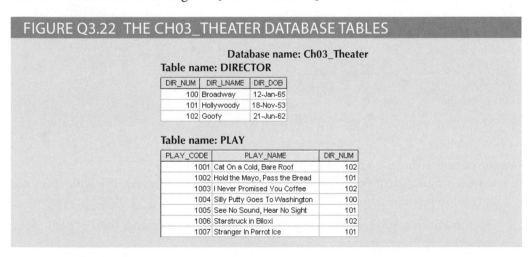

Database name: Ch03_Theater
Table name: DIRECTOR

DIR_NUM	DIR_LNAME	DIR_DOB
100	Broadway	12-Jan-65
101	Hollywoody	18-Nov-53
102	Goofy	21-Jun-62

Table name: PLAY

PLAY_CODE	PLAY_NAME	DIR_NUM
1001	Cat On a Cold, Bare Roof	102
1002	Hold the Mayo, Pass the Bread	101
1003	I Never Promised You Coffee	102
1004	Silly Putty Goes To Washington	100
1005	See No Sound, Hear No Sight	101
1006	Starstruck in Biloxi	102
1007	Stranger In Parrot Ice	101

22. Identify the primary keys.

23. Identify the foreign keys.

24. Create the ERM.

25. Create the relational diagram to show the relationship between DIRECTOR and PLAY.

26. Suppose you wanted quick lookup capability to get a listing of all plays directed by a given director. Which table would be the basis for the INDEX table, and what would be the index key?

27. What would be the conceptual view of the INDEX table described in Question 26? Depict the contents of the conceptual INDEX table.

Problems

FIGURE P3.1 THE CH03_STORECO DATABASE TABLES

Table name: EMPLOYEE   Database name: Ch03_StoreCo

EMP_CODE	EMP_TITLE	EMP_LNAME	EMP_FNAME	EMP_INITIAL	EMP_DOB	STORE_CODE
1	Mr.	Williamson	John	W	21-May-64	3
2	Ms.	Ratula	Nancy		09-Feb-69	2
3	Ms.	Greenboro	Lottie	R	02-Oct-61	4
4	Mrs.	Rumpersfro	Jennie	S	01-Jun-71	5
5	Mr.	Smith	Robert	L	23-Nov-59	3
6	Mr.	Renselaer	Cary	A	25-Dec-65	1
7	Mr.	Ogallo	Roberto	S	31-Jul-62	3
8	Ms.	Johnsson	Elizabeth	I	10-Sep-68	1
9	Mr.	Eindsmar	Jack	W	19-Apr-55	2
10	Mrs.	Jones	Rose	R	06-Mar-66	4
11	Mr.	Broderick	Tom		21-Oct-72	3
12	Mr.	Washington	Alan	Y	08-Sep-74	2
13	Mr.	Smith	Peter	N	25-Aug-64	3
14	Ms.	Smith	Sherry	H	25-May-66	4
15	Mr.	Olenko	Howard	U	24-May-64	5
16	Mr.	Archialo	Barry	V	03-Sep-60	5
17	Ms.	Grimaldo	Jeanine	K	12-Nov-70	4
18	Mr.	Rosenberg	Andrew	D	24-Jan-71	4
19	Mr.	Rosten	Peter	F	03-Oct-68	4
20	Mr.	Mckee	Robert	S	06-Mar-70	1
21	Ms.	Baumann	Jennifer	A	11-Dec-74	3

Table name: STORE

STORE_CODE	STORE_NAME	STORE_YTD_SALES	REGION_CODE	EMP_CODE
1	Access Junction	1003455.76	2	8
2	Database Corner	1421987.39	2	12
3	Tuple Charge	986783.22	1	7
4	Attribute Alley	944568.56	2	3
5	Primary Key Point	2930098.45	1	15

Table name: REGION

REGION_CODE	REGION_DESCRIPT
1	East
2	West

Use the database shown in Figure P3.1 to answer Problems 1–9.

1. For each table, identify the primary key and the foreign key(s). If a table does not have a foreign key, write *None*.

2. Do the tables exhibit entity integrity? Answer yes or no, and then explain your answer.

3. Do the tables exhibit referential integrity? Answer yes or no, and then explain your answer. Write *NA* (Not Applicable) if the table does not have a foreign key.

4. Describe the type(s) of relationship(s) between STORE and REGION.

5. Create the ERD to show the relationship between STORE and REGION.

6. Create the relational diagram to show the relationship between STORE and REGION.

7. Describe the type(s) of relationship(s) between EMPLOYEE and STORE. (*Hint*: Each store employs many employees, one of whom manages the store.)

8. Create the ERD to show the relationships among EMPLOYEE, STORE, and REGION.

9. Create the relational diagram to show the relationships among EMPLOYEE, STORE, and REGION.

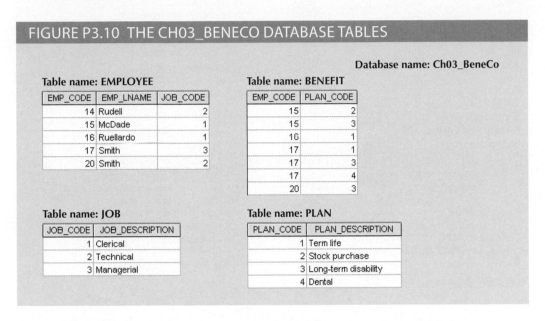

FIGURE P3.10 THE CH03_BENECO DATABASE TABLES

Database name: Ch03_BeneCo

Table name: EMPLOYEE

EMP_CODE	EMP_LNAME	JOB_CODE
14	Rudell	2
15	McDade	1
16	Ruellardo	1
17	Smith	3
20	Smith	2

Table name: BENEFIT

EMP_CODE	PLAN_CODE
15	2
15	3
16	1
17	1
17	3
17	4
20	3

Table name: JOB

JOB_CODE	JOB_DESCRIPTION
1	Clerical
2	Technical
3	Managerial

Table name: PLAN

PLAN_CODE	PLAN_DESCRIPTION
1	Term life
2	Stock purchase
3	Long-term disability
4	Dental

Use the database shown in Figure P3.10 to work Problems 10–16. Note that the database is composed of four tables that reflect these relationships:

- An EMPLOYEE has only one JOB_CODE, but a JOB_CODE can be held by many EMPLOYEEs.

- An EMPLOYEE can participate in many PLANs, and any PLAN can be assigned to many EMPLOYEEs.

Note also that the M:N relationship has been broken down into two 1:M relationships for which the BENEFIT table serves as the composite or bridge entity.

10. For each table in the database, identify the primary key and the foreign key(s). If a table does not have a foreign key, write *None*.

11. Create the ERD to show the relationship between EMPLOYEE and JOB.

12. Create the relational diagram to show the relationship between EMPLOYEE and JOB.

13. Do the tables exhibit entity integrity? Answer yes or no, and then explain your answer.

14. Do the tables exhibit referential integrity? Answer yes or no, and then explain your answer. Write *NA* (Not Applicable) if the table does not have a foreign key.

15. Create the ERD to show the relationships among EMPLOYEE, BENEFIT, JOB, and PLAN.

16. Create the relational diagram to show the relationships among EMPLOYEE, BENEFIT, JOB, and PLAN.

FIGURE P3.17 THE CH03_TRANSCO DATABASE TABLES

Table name: **TRUCK**
Primary key: **TRUCK_NUM**
Foreign key: **BASE_CODE, TYPE_CODE**

Database name: **Ch03_TransCo**

TRUCK_NUM	BASE_CODE	TYPE_CODE	TRUCK_MILES	TRUCK_BUY_DATE	TRUCK_SERIAL_NUM
1001	501	1	32123.5	23-Sep-07	AA-322-12212-W11
1002	502	1	76984.3	05-Feb-06	AC-342-22134-Q23
1003	501	2	12346.6	11-Nov-06	AC-445-78656-Z99
1004		1	2894.3	06-Jan-07	WQ-112-23144-T34
1005	503	2	45673.1	01-Mar-06	FR-998-32245-W12
1006	501	2	193245.7	15-Jul-03	AD-456-00845-R45
1007	502	3	32012.3	17-Oct-04	AA-341-96573-Z84
1008	502	3	44213.6	07-Aug-05	DR-559-22189-D33
1009	503	2	10932.9	12-Feb-08	DE-887-98456-E94

Table name: **BASE**
Primary key: **BASE_CODE**
Foreign key: **none**

BASE_CODE	BASE_CITY	BASE_STATE	BASE_AREA_CODE	BASE_PHONE	BASE_MANAGER
501	Murfreesboro	TN	615	123-4567	Andrea D. Gallager
502	Lexington	KY	568	234-5678	George H. Delarosa
503	Cape Girardeau	MO	456	345-6789	Maria J. Talindo
504	Dalton	GA	901	456-7890	Peter F. McAvee

Table name: **TYPE**
Primary key: **TYPE_CODE**
Foreign key: **none**

TYPE_CODE	TYPE_DESCRIPTION
1	Single box, double-axle
2	Single box, single-axle
3	Tandem trailer, single-axle

Use the database shown in Figure P3.17 to answer Problems 17–23.

17. For each table, identify the primary key and the foreign key(s). If a table does not have a foreign key, write *None*.

18. Do the tables exhibit entity integrity? Answer yes or no, and then explain your answer.

19. Do the tables exhibit referential integrity? Answer yes or no, and then explain your answer. Write *NA* (Not Applicable) if the table does not have a foreign key.

20. Identify the TRUCK table's candidate key(s).

21. For each table, identify a superkey and a secondary key.

22. Create the ERD for this database.

23. Create the relational diagram for this database.

FIGURE P3.24 THE CH03_AVIACO DATABASE TABLES

Table name: **CHARTER**   Database name: **Ch03_AviaCo**

CHAR_TRIP	CHAR_DATE	CHAR_PILOT	CHAR_COPILOT	AC_NUMBER	CHAR_DESTINATION	CHAR_DISTANCE	CHAR_HOURS_FLOWN	CHAR_HOURS_WAIT	CHAR_FUEL_GALLONS	CHAR_OIL_QTS	CUS_CODE
10001	05-Feb-16	104		2289L	ATL	936.0	5.1	2.2	354.1	1	10011
10002	05-Feb-16	101		2778V	BNA	320.0	1.6	0.0	72.6	0	10016
10003	05-Feb-16	105	109	4278Y	GNV	1574.0	7.8	0.0	339.8	2	10014
10004	06-Feb-16	106		1484P	STL	472.0	2.9	4.9	97.2	1	10019
10005	06-Feb-16	101		2289L	ATL	1023.0	5.7	3.5	397.7	2	10011
10006	06-Feb-16	109		4278Y	STL	472.0	2.6	5.2	117.1	0	10017
10007	06-Feb-16	104	105	2778V	GNV	1574.0	7.9	0.0	348.4	2	10012
10008	07-Feb-16	106		1484P	TYS	644.0	4.1	0.0	140.6	1	10014
10009	07-Feb-16	105		2289L	GNV	1574.0	6.6	23.4	459.9	0	10017
10010	07-Feb-16	109		4278Y	ATL	998.0	6.2	3.2	279.7	0	10016
10011	07-Feb-16	101	104	1484P	BNA	352.0	1.9	5.3	66.4	1	10012
10012	08-Feb-16	101		2778V	MOB	884.0	4.8	4.2	215.1	0	10010
10013	08-Feb-16	105		4278Y	TYS	644.0	3.9	4.5	174.3	1	10011
10014	09-Feb-16	106		4278Y	ATL	936.0	6.1	2.1	302.6	0	10017
10015	09-Feb-16	104	101	2289L	GNV	1645.0	6.7	0.0	459.5	2	10016
10016	09-Feb-16	109	105	2778V	MQY	312.0	1.5	0.0	67.2	0	10011
10017	10-Feb-16	101		1484P	STL	508.0	3.1	0.0	105.5	0	10014
10018	10-Feb-16	105	104	4278Y	TYS	644.0	3.8	4.5	167.4	0	10017

The destinations are indicated by standard three-letter airport codes. For example,

STL = St. Louis, MO   ATL = Atlanta, GA   BNA = Nashville, TN

Table name: **AIRCRAFT**

AC_NUMBER	MOD_CODE	AC_TTAF	AC_TTEL	AC_TTER
1484P	PA23-250	1833.1	1833.1	101.8
2289L	C-90A	4243.8	768.9	1123.4
2778V	PA31-350	7992.9	1513.1	789.5
4278Y	PA31-350	2147.3	622.1	243.2

AC-TTAF = Aircraft total time, airframe (hours)
AC-TTEL = Total time, left engine (hours)
AC_TTER = Total time, right engine (hours)

In a fully developed system, such attribute values would be updated by application software when the CHARTER table entries were posted.

Table name: **MODEL**

MOD_CODE	MOD_MANUFACTURER	MOD_NAME	MOD_SEATS	MOD_CHG_MILE
B200	Beechcraft	Super KingAir	10	1.93
C-90A	Beechcraft	KingAir	8	2.67
PA23-250	Piper	Aztec	6	1.93
PA31-350	Piper	Navajo Chieftain	10	2.35

Customers are charged per round-trip mile, using the MOD_CHG_MILE rate. The MOD_SEATS column lists the total number of seats in the airplane, including the pilot and copilot seats. Therefore, a PA31-350 trip that is flown by a pilot and a copilot has eight passenger seats available.

Use the database shown in Figure P3.24 to answer Problems 24–31. AviaCo is an aircraft charter company that supplies on-demand charter flight services using a fleet of four aircraft. Aircraft are identified by a unique registration number. Therefore, the aircraft registration number is an appropriate primary key for the AIRCRAFT table.

FIGURE P3.24   THE CH03_AVIACO DATABASE TABLES (CONTINUED)

Table name: PILOT   Database name: Ch03_AviaCo

EMP_NUM	PIL_LICENSE	PIL_RATINGS	PIL_MED_TYPE	PIL_MED_DATE	PIL_PT135_DATE
101	ATP	ATP/SEL/MEL/Instr/CFII	1	20-Jan-16	11-Jan-16
104	ATP	ATP/SEL/MEL/Instr	1	18-Dec-15	17-Jan-16
105	COM	COMM/SEL/MEL/Instr/CFI	2	05-Jan-16	02-Jan-16
106	COM	COMM/SEL/MEL/Instr	2	10-Dec-15	02-Feb-16
109	COM	ATP/SEL/MEL/SES/Instr/CFII	1	22-Jan-16	15-Jan-16

The pilot licenses shown in the PILOT table include ATP = Airline Transport Pilot and COMM = Commercial Pilot. Businesses that operate on-demand air services are governed by Part 135 of the Federal Air Regulations (FARs), which are enforced by the Federal Aviation Administration (FAA). Such businesses are known as "Part 135 operators." Part 135 operations require that pilots successfully complete flight proficiency checks every six months. The "Part 135" flight proficiency check date is recorded in PIL_PT135_DATE. To fly commercially, pilots must have at least a commercial license and a second-class medical certificate (PIL_MED_TYPE = 2).

The PIL_RATINGS include:
SEL = Single Engine, Land
SES = Single Engine, Sea
CFI = Certified Flight Instructor

MEL = Multiengine, Land
Instr. = Instrument
CFII = Certified Flight Instructor, Instrument

Table name: EMPLOYEE

EMP_NUM	EMP_TITLE	EMP_LNAME	EMP_FNAME	EMP_INITIAL	EMP_DOB	EMP_HIRE_DATE
100	Mr.	Kolmycz	George	D	15-Jun-42	15-Mar-88
101	Ms.	Lewis	Rhonda	G	19-Mar-65	25-Apr-86
102	Mr.	Vandam	Rhett		14-Nov-58	18-May-93
103	Ms.	Jones	Anne	M	11-May-74	26-Jul-99
104	Mr.	Lange	John	P	12-Jul-71	20-Aug-90
105	Mr.	Williams	Robert	D	14-Mar-75	19-Jun-03
106	Mrs.	Duzak	Jeanine	K	12-Feb-68	13-Mar-89
107	Mr.	Diante	Jorge	D	01-May-75	02-Jul-97
108	Mr.	Wiesenbach	Paul	R	14-Feb-66	03-Jun-93
109	Ms.	Travis	Elizabeth	K	18-Jun-61	14-Feb-06
110	Mrs.	Genkazi	Leighla	W	19-May-70	29-Jun-90

Table name: CUSTOMER

CUS_CODE	CUS_LNAME	CUS_FNAME	CUS_INITIAL	CUS_AREACODE	CUS_PHONE	CUS_BALANCE
10010	Ramas	Alfred	A	615	844-2573	0.00
10011	Dunne	Leona	K	713	894-1238	0.00
10012	Smith	Kathy	W	615	894-2285	896.54
10013	Olowski	Paul	F	615	894-2180	1285.19
10014	Orlando	Myron		615	222-1672	673.21
10015	O'Brian	Amy	B	713	442-3381	1014.56
10016	Brown	James	G	615	297-1228	0.00
10017	Williams	George		615	290-2556	0.00
10018	Farriss	Anne	G	713	382-7185	0.00
10019	Smith	Olette	K	615	297-3809	453.98

The nulls in the CHARTER table's CHAR_COPILOT column indicate that a copilot is not required for some charter trips or for some aircraft. Federal Aviation Administration (FAA) rules require a copilot on jet aircraft and on aircraft that have a gross take-off weight over 12,500 pounds. None of the aircraft in the AIRCRAFT table are governed by this requirement; however, some customers may require the presence of a copilot for insurance reasons. All charter trips are recorded in the CHARTER table.

Note

Earlier in the chapter, you were instructed to avoid homonyms and synonyms. In this problem, both the pilot and the copilot are listed in the PILOT table, but EMP_NUM cannot be used for both in the CHARTER table. Therefore, the synonyms CHAR_PILOT and CHAR_COPILOT were used in the CHARTER table.

Although the solution works in this case, it is very restrictive, and it generates nulls when a copilot is not required. Worse, such nulls proliferate as crew requirements change. For example, if the AviaCo charter company grows and starts using larger aircraft, crew requirements may increase to include flight engineers and load masters. The CHARTER table would then have to be modified to include the additional crew assignments; such attributes as CHAR_FLT_ENGINEER and CHAR_LOADMASTER would have to be added to the CHARTER table. Given this change, each time a smaller aircraft flew a charter trip without the number of crew members required in larger aircraft, the missing crew members would yield additional nulls in the CHARTER table.

You will have a chance to correct those design shortcomings in Problem 27. The problem illustrates two important points:

1. Don't use synonyms. If your design requires the use of synonyms, revise the design!

2. To the greatest possible extent, design the database to accommodate growth without requiring structural changes in the database tables. Plan ahead and try to anticipate the effects of change on the database.

24. For each table, identify each of the following when possible:
    a. The primary key
    b. A superkey
    c. A candidate key
    d. The foreign key(s)
    e. A secondary key

25. Create the ERD. (*Hint:* Look at the table contents. You will discover that an AIRCRAFT can fly many CHARTER trips but that each CHARTER trip is flown by one AIRCRAFT, that a MODEL references many AIRCRAFT but that each AIRCRAFT references a single MODEL, and so on.)

26. Create the relational diagram.

27. Modify the ERD you created in Problem 25 to eliminate the problems created by the use of synonyms. (*Hint:* Modify the CHARTER table structure by eliminating the CHAR_PILOT and CHAR_COPILOT attributes; then create a composite table named CREW to link the CHARTER and EMPLOYEE tables. Some crew members, such as flight attendants, may not be pilots. That's why the EMPLOYEE table enters into this relationship.)

28. Create the relational diagram for the design you revised in Problem 27.

You want to see data on charters flown by either Robert Williams (employee number 105) or Elizabeth Travis (employee number 109) as pilot or copilot, but not charters flown by both of them. Complete Problems 29–31 to find this information.

29. Create the table that would result from applying the SELECT and PROJECT relational operators to the CHARTER table to return only the CHAR_TRIP, CHAR_PILOT, and CHAR_COPILOT attributes for charters flown by either employee 105 or employee 109.

30. Create the table that would result from applying the SELECT and PROJECT relational operators to the CHARTER table to return only the CHAR_TRIP, CHAR_PILOT, and CHAR_COPILOT attributes for charters flown by both employee 105 and employee 109.

31. Create the table that would result from applying a DIFFERENCE relational operator of your result from Problem 29 to your result from Problem 30.

Chapter 4

Entity Relationship (ER) Modeling

In this chapter, you will learn:

- The main characteristics of entity relationship components
- How relationships between entities are defined, refined, and incorporated into the database design process
- How ERD components affect database design and implementation
- That real-world database design often requires the reconciliation of conflicting goals

Preview

This chapter expands coverage of the data-modeling aspect of database design. Data modeling is the first step in the database design journey, serving as a bridge between real-world objects and the database model that is implemented in the computer. Therefore, the importance of data-modeling details, expressed graphically through entity relationship diagrams (ERDs), cannot be overstated.

Most of the basic concepts and definitions used in the entity relationship model (ERM) were introduced in Chapter 2, Data Models. For example, the basic components of entities and relationships and their representation should now be familiar to you. This chapter goes much deeper, analyzing the graphic depiction of relationships among the entities and showing how those depictions help you summarize the wealth of data required to implement a successful design.

Finally, the chapter illustrates how conflicting goals can be a challenge in database design and might require design compromises.

Data Files and Available Formats

	MS Access	Oracle	MS SQL	My SQL		MS Access	Oracle	MS SQL	My SQL
CH04_TinyCollege	✓	✓	✓	✓	CH04_Clinic	✓	✓	✓	✓
CH04_TinyCollege_Alt	✓	✓	✓	✓	CH04_PartCo	✓	✓	✓	✓
CH04_ShortCo	✓	✓	✓	✓	CH04_CollegeTry	✓	✓	✓	✓

Data Files Available on cengagebrain.com

Note

Because this book generally focuses on the relational model, you might be tempted to conclude that the ERM is exclusively a relational tool. Actually, conceptual models such as the ERM can be used to understand and design the data requirements of an organization. Therefore, the ERM is independent of the database type. Conceptual models are used in the conceptual design of databases, while relational models are used in the logical design of databases. However, because you are familiar with the relational model from the previous chapter, the relational model is used extensively in this chapter to explain ER constructs and the way they are used to develop database designs.

4-1 The Entity Relationship Model (ERM)

You should remember from Chapter 2, Data Models, and Chapter 3, The Relational Database Model, that the ERM forms the basis of an ERD. The ERD represents the conceptual database as viewed by the end user. ERDs depict the database's main components: entities, attributes, and relationships. Because an entity represents a real-world object, the words *entity* and *object* are often used interchangeably. Thus, the entities (objects) of the Tiny College database design developed in this chapter include students, classes, teachers, and classrooms. The order in which the ERD components are covered in the chapter is dictated by the way the modeling tools are used to develop ERDs that can form the basis for successful database design and implementation.

In Chapter 2, you also learned about the various notations used with ERDs—the original Chen notation and the newer Crow's Foot and UML notations. The first two notations are used at the beginning of this chapter to introduce some basic ER modeling concepts. Some conceptual database modeling concepts can be expressed only using the Chen notation. However, because the emphasis is on *design and implementation* of databases, the Crow's Foot and UML class diagram notations are used for the final Tiny College ER diagram example. Because of its emphasis on implementation, the Crow's Foot notation can represent only what could be implemented. In other words:

- The Chen notation favors conceptual modeling.

- The Crow's Foot notation favors a more implementation-oriented approach.

- The UML notation can be used for both conceptual and implementation modeling.

4-1a Entities

Recall that an entity is an object of interest to the end user. In Chapter 2, you learned that, at the ER modeling level, an entity actually refers to the *entity set* and not to a single entity occurrence. In other words, an *entity* in the ERM corresponds to a table—not to a row—in the relational environment. The ERM refers to a table row as an *entity instance* or *entity occurrence*. In the Chen, Crow's Foot, and UML notations, an entity is represented by a rectangle that contains the entity's name. The entity name, a noun, is usually written in all capital letters.

4-1b Attributes

Attributes are characteristics of entities. For example, the STUDENT entity includes the attributes STU_LNAME, STU_FNAME, and STU_INITIAL, among many others. In the original Chen notation, attributes are represented by ovals and are connected

Online Content

To learn how to create ER diagrams with the help of Microsoft Visio, go to *www.cengagebrain.com*: Appendix A, Designing Databases with Visio Professional: A Tutorial, shows you how to create Crow's Foot ERDs. Appendix H, Unified Modeling Language (UML), shows you how to create UML class diagrams.

to the entity rectangle with a line. Each oval contains the name of the attribute it represents. In the Crow's Foot notation, the attributes are written in the attribute box below the entity rectangle. (See Figure 4.1.) Because the Chen representation consumes more space, software vendors have adopted the Crow's Foot attribute display.

FIGURE 4.1 THE ATTRIBUTES OF THE STUDENT ENTITY: CHEN AND CROW'S FOOT

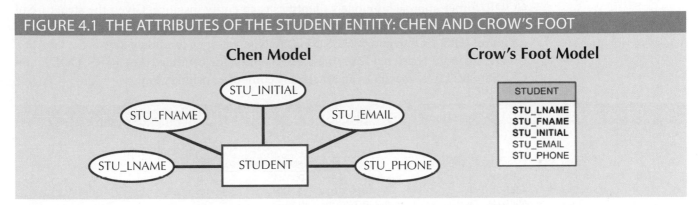

Required and Optional Attributes A **required attribute** is an attribute that must have a value; in other words, it cannot be left empty. As shown in Figure 4.1, the two boldfaced attributes in the Crow's Foot notation indicate that data entry will be required. STU_LNAME and STU_FNAME require data entries because all students are assumed to have a last name and a first name. However, students might not have a middle name, and perhaps they do not yet have a phone number and an email address. Therefore, those attributes are not presented in boldface in the entity box. An **optional attribute** is an attribute that does not require a value; therefore, it can be left empty.

Domains Attributes have a domain. As you learned in Chapter 3, a *domain* is the set of possible values for a given attribute. For example, the domain for a grade point average (GPA) attribute is written (0,4) because the lowest possible GPA value is 0 and the highest possible value is 4. The domain for a gender attribute consists of only two possibilities: M or F (or some other equivalent code). The domain for a company's date of hire attribute consists of all dates that fit in a range (for example, company startup date to current date).

Attributes may share a domain. For instance, a student address and a professor address share the same domain of all possible addresses. In fact, the data dictionary may let a newly declared attribute inherit the characteristics of an existing attribute if the same attribute name is used. For example, the PROFESSOR and STUDENT entities may each have an attribute named ADDRESS and could therefore share a domain.

Identifiers (Primary Keys) The ERM uses **identifiers**—one or more attributes that uniquely identify each entity instance. In the relational model, entities are mapped to tables, and the entity identifier is mapped as the table's primary key (PK). Identifiers are underlined in the ERD. Key attributes are also underlined in a frequently used shorthand notation for the table structure, called a **relational schema**, that uses the following format:

TABLE NAME (**KEY_ATTRIBUTE 1**, ATTRIBUTE 2, ATTRIBUTE 3, … ATTRIBUTE K)

For example, a CAR entity may be represented by:

CAR (**CAR_VIN**, MOD_CODE, CAR_YEAR, CAR_COLOR)

Each car is identified by a unique vehicle identification number, or CAR_VIN.

Composite Identifiers Ideally, an entity identifier is composed of only a single attribute. For example, the table in Figure 4.2 uses a single-attribute primary key named

required attribute
In ER modeling, an attribute that must have a value. In other words, it cannot be left empty.

optional attribute
In ER modeling, an attribute that does not require a value; therefore, it can be left empty.

identifiers
One or more attributes that uniquely identify each entity instance.

relational schema
The organization of a relational database as described by the database administrator.

CLASS_CODE. However, it is possible to use a **composite identifier**, a primary key composed of more than one attribute. For instance, the Tiny College database administrator may decide to identify each CLASS entity instance (occurrence) by using a composite primary key of CRS_CODE and CLASS_SECTION instead of using CLASS_CODE. Either approach uniquely identifies each entity instance. Given the structure of the CLASS table shown in Figure 4.2, CLASS_CODE is the primary key, and the combination of CRS_CODE and CLASS_SECTION is a proper candidate key. If the CLASS_CODE attribute is deleted from the CLASS entity, the candidate key (CRS_CODE and CLASS_SECTION) becomes an acceptable composite primary key.

FIGURE 4.2 THE CLASS TABLE (ENTITY) COMPONENTS AND CONTENTS

Database name: Ch04_TinyCollege

CLASS_CODE	CRS_CODE	CLASS_SECTION	CLASS_TIME	ROOM_CODE	PROF_NUM
10012	ACCT-211	1	MWF 8:00-8:50 a.m.	BUS311	105
10013	ACCT-211	2	MWF 9:00-9:50 a.m.	BUS200	105
10014	ACCT-211	3	TTh 2:30-3:45 p.m.	BUS252	342
10015	ACCT-212	1	MWF 10:00-10:50 a.m.	BUS311	301
10016	ACCT-212	2	Th 6:00-8:40 p.m.	BUS252	301
10017	CIS-220	1	MWF 9:00-9:50 a.m.	KLR209	228
10018	CIS-220	2	MWF 9:00-9:50 a.m.	KLR211	114
10019	CIS-220	3	MWF 10:00-10:50 a.m.	KLR209	228
10020	CIS-420	1	W 6:00-8:40 p.m.	KLR209	162
10021	QM-261	1	MWF 8:00-8:50 a.m.	KLR200	114
10022	QM-261	2	TTh 1:00-2:15 p.m.	KLR200	114
10023	QM-362	1	MWF 11:00-11:50 a.m.	KLR200	162
10024	QM-362	2	TTh 2:30-3:45 p.m.	KLR200	162
10025	MATH-243	1	Th 6:00-8:40 p.m.	DRE155	325

Note

Remember that Chapter 3 made a commonly accepted distinction between COURSE and CLASS. A CLASS constitutes a specific time and place of a COURSE offering. A class is defined by the course description and its time and place, or section. Consider a professor who teaches Database I, Section 2; Database I, Section 5; Database I, Section 8; and Spreadsheet II, Section 6. The professor teaches two courses (Database I and Spreadsheet II), but four classes. Typically, the COURSE offerings are printed in a course catalog, while the CLASS offerings are printed in a class schedule for each term.

composite identifier
In ER modeling, a key composed of more than one attribute.

composite attribute
An attribute that can be further subdivided to yield additional attributes. For example, a phone number such as 615-898-2368 may be divided into an area code (615), an exchange number (898), and a four-digit code (2368). Compare to *simple attribute*.

If the CLASS_CODE in Figure 4.2 is used as the primary key, the CLASS entity may be represented in shorthand form as follows:

CLASS (**CLASS_CODE**, CRS_CODE, CLASS_SECTION, CLASS_TIME, ROOM_CODE, PROF_NUM)

On the other hand, if CLASS_CODE is deleted, and the composite primary key is the combination of CRS_CODE and CLASS_SECTION, the CLASS entity may be represented as follows:

CLASS (**CRS_CODE**, **CLASS_SECTION**, CLASS_TIME, ROOM_CODE, PROF_NUM)

Note that *both* key attributes are underlined in the entity notation.

Composite and Simple Attributes Attributes are classified as simple or composite. A **composite attribute**, not to be confused with a composite key, is an attribute that can be further subdivided to yield additional attributes. For example, the attribute ADDRESS can be subdivided into street, city, state, and zip code. Similarly, the attribute PHONE_NUMBER can be subdivided into area code and exchange number. A **simple attribute** is an attribute that cannot be subdivided. For example, age, sex, and marital status would be classified as simple attributes. To facilitate detailed queries, it is wise to change composite attributes into a series of simple attributes.

The database designer must always be on the lookout for composite attributes. It is common for business rules to use composite attributes to simplify policies, and users often describe entities in their environment using composite attributes. For example, a user at Tiny College might need to know a student's name, address, and phone number. The designer must recognize that these are composite attributes and determine the correct way to decompose the composite into simple attributes.

Single-Valued Attributes A **single-valued attribute** is an attribute that can have only a single value. For example, a person can have only one Social Security number, and a manufactured part can have only one serial number. *Keep in mind that a single-valued attribute is not necessarily a simple attribute.* For instance, a part's serial number (such as SE-08-02-189935) is single-valued, but it is a composite attribute because it can be subdivided into the region in which the part was produced (SE), the plant within that region (08), the shift within the plant (02), and the part number (189935).

Multivalued Attributes **Multivalued attributes** are attributes that can have many values. For instance, a person may have several college degrees, and a household may have several different phones, each with its own number. Similarly, a car's color may be subdivided into many colors for the roof, body, and trim. In the Chen ERM, multivalued attributes are shown by a double line connecting the attribute to the entity. The Crow's Foot notation does not identify multivalued attributes. The ERD in Figure 4.3 contains all of the components introduced thus far; note that CAR_VIN is the primary key, and CAR_COLOR is a multivalued attribute of the CAR entity.

simple attribute
An attribute that cannot be subdivided into meaningful components. Compare to *composite attribute*.

single-valued attribute
An attribute that can have only one value.

multivalued attribute
An attribute that can have many values for a single entity occurrence. For example, an EMP_DEGREE attribute might store the string "BBA, MBA, PHD" to indicate three different degrees held.

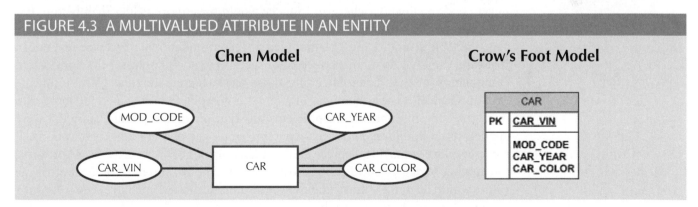

FIGURE 4.3 A MULTIVALUED ATTRIBUTE IN AN ENTITY

Note

In the ERD models in Figure 4.3, the CAR entity's foreign key (FK) has been typed as MOD_CODE. This attribute was manually added to the entity. Actually, proper use of database modeling software will automatically produce the FK when the relationship is defined. In addition, the software will label the FK appropriately and write the FK's implementation details in a data dictionary. Therefore, when you use professional database modeling software, *never type the FK attribute yourself*; let the software handle that task when the relationship between the entities is defined. (You can see how this works in Appendix A, Designing Databases with Visio Professional: A Tutorial, at *www.cengagebrain.com*.)

Implementing Multivalued Attributes Although the conceptual model can handle M:N relationships and multivalued attributes, *you should not implement them in the RDBMS.* Remember from Chapter 3 that in the relational table, each column and row intersection represents a single data value. So, if multivalued attributes exist, the designer must decide on one of two possible courses of action:

1. Within the original entity, create several new attributes, one for each component of the original multivalued attribute. For example, the CAR entity's attribute CAR_COLOR can be split to create the new attributes CAR_TOPCOLOR, CAR_BODY-COLOR, and CAR_TRIMCOLOR, which are then assigned to the CAR entity. (See Figure 4.4.)

FIGURE 4.4 SPLITTING THE MULTIVALUED ATTRIBUTE INTO NEW ATTRIBUTES

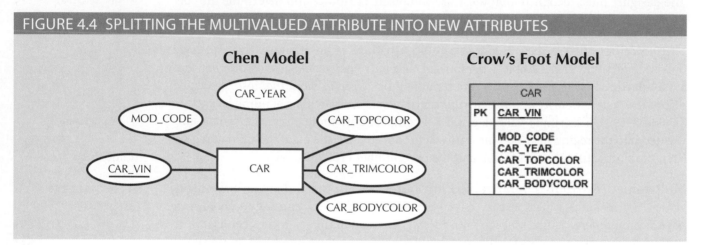

Although this solution seems to work, its adoption can lead to major structural problems in the table. It is only acceptable if every instance will have the same number of values for the multivalued attribute, and no instance will ever have more values. However, even in this case, it is a gamble that new changes in the environment will never create a situation where an instance would have more values than before. For example, if additional color components—such as a logo color—are added for some cars, the table structure must be modified to accommodate the new color section. In that case, cars that do not have such color sections generate nulls for the nonexistent components, or their color entries for those sections are entered as N/A to indicate "not applicable." (The solution in Figure 4.4 is to split a multivalued attribute into new attributes, but imagine the problems this type of solution would cause if it were applied to an employee entity that contains employee degrees and certifications. If some employees have 10 degrees and certifications while most have fewer or none, the number of degree/certification attributes would be 10, and most of those attribute values would be null for most employees.) In short, although you have seen solution 1 applied, it is not always acceptable.

2. Create a new entity composed of the original multivalued attribute's components. This new entity allows the designer to define color for different sections of the car. (See Table 4.1.) Then, this new CAR_COLOR entity is related to the original CAR entity in a 1:M relationship.

Using the approach illustrated in Table 4.1, you even get a fringe benefit: you can now assign as many colors as necessary without having to change the table structure. The ERM shown in Figure 4.5 reflects the components listed in Table 4.1. This is the preferred way to deal with multivalued attributes. Creating a new entity in a 1:M relationship with the original entity yields several benefits: it is a more flexible, expandable solution, and it is compatible with the relational model!

TABLE 4.1

SECTION	COLOR
Top	White
Body	Blue
Trim	Gold
Interior	Blue

COMPONENTS OF THE MULTIVALUED ATTRIBUTE

FIGURE 4.5   A NEW ENTITY SET COMPOSED OF A MULTIVALUED ATTRIBUTE'S COMPONENTS

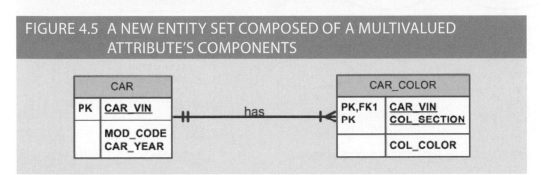

Note

If you are used to looking at relational diagrams such as the ones produced by Microsoft Access, you expect to see the relationship line *in the relational diagram* drawn from the PK to the FK. However, the relational diagram convention is not necessarily reflected in the ERD. In an ERD, the focus is on the entities and the relationships between them, rather than how those relationships are anchored graphically. In a complex ERD that includes both horizontally and vertically placed entities, the placement of the relationship lines is largely dictated by the designer's decision to improve the readability of the design. (Remember that the ERD is used for communication between designers and end users.)

Derived Attributes   Finally, a **derived attribute** is an attribute whose value is calculated (derived) from other attributes. The derived attribute need not be physically stored within the database; instead, it can be derived by using an algorithm. For example, an employee's age, EMP_AGE, may be found by computing the integer value of the difference between the current date and the EMP_DOB. If you use Microsoft Access, you would use the formula INT((DATE() – EMP_DOB)/365). In Microsoft SQL Server, you would use SELECT DATEDIFF("YEAR", EMP_DOB, GETDATE()), where DATEDIFF is a function that computes the difference between dates. The first parameter indicates the measurement (in this case, years). If you use Oracle, you would use SYSDATE instead of DATE(). (You are assuming, of course, that EMP_DOB was stored in the Julian date format.)

Similarly, the total cost of an order can be derived by multiplying the quantity ordered by the unit price. Or, the estimated average speed can be derived by dividing trip distance by the time spent en route. A derived attribute is indicated in the Chen notation by a dashed line that connects the attribute and the entity. (See Figure 4.6.) The Crow's Foot notation does not have a method for distinguishing the derived attribute from other attributes.

Derived attributes are sometimes referred to as *computed attributes*. Computing a derived attribute can be as simple as adding two attribute values located on the same row, or it can be the result of aggregating the sum of values located on many table rows (from the same table or from a different table). The decision to store derived attributes in

derived attribute
An attribute that does not physically exist within the entity and is derived via an algorithm. For example, the Age attribute might be derived by subtracting the birth date from the current date.

database tables depends on the processing requirements and the constraints placed on a particular application. The designer should be able to balance the design in accordance with such constraints. Table 4.2 shows the advantages and disadvantages of storing (or not storing) derived attributes in the database.

FIGURE 4.6 DEPICTION OF A DERIVED ATTRIBUTE

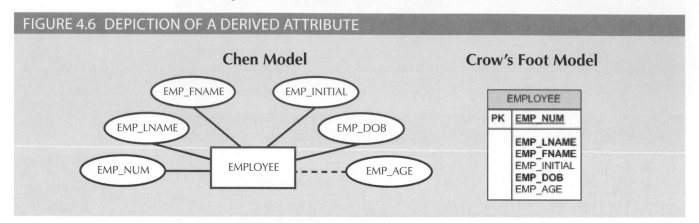

TABLE 4.2

ADVANTAGES AND DISADVANTAGES OF STORING DERIVED ATTRIBUTES

| | DERIVED ATTRIBUTE | |
	STORED	NOT STORED
Advantage	Saves CPU processing cycles Saves data access time Data value is readily available Can be used to keep track of historical data	Saves storage space Computation always yields current value
Disadvantage	Requires constant maintenance to ensure derived value is current, especially if any values used in the calculation change	Uses CPU processing cycles Increases data access time Adds coding complexity to queries

4-1c Relationships

Recall from Chapter 2 that a relationship is an association between entities. The entities that participate in a relationship are also known as **participants**, and each relationship is identified by a name that describes the relationship. The relationship name is an active or passive verb; for example, a STUDENT *takes* a CLASS, a PROFESSOR *teaches* a CLASS, a DEPARTMENT *employs* a PROFESSOR, a DIVISION *is managed by* an EMPLOYEE, and an AIRCRAFT *is flown by* a CREW.

Relationships between entities always operate in both directions. To define the relationship between the entities named CUSTOMER and INVOICE, you would specify that:

- A CUSTOMER may generate many INVOICEs.

- Each INVOICE is generated by one CUSTOMER.

Because you know both directions of the relationship between CUSTOMER and INVOICE, it is easy to see that this relationship can be classified as 1:M.

The relationship classification is difficult to establish if you know only one side of the relationship. For example, if you specify that:

A DIVISION is managed by one EMPLOYEE.

participants
An ER term for entities that participate in a relationship. For example, in the relationship "PROFESSOR teaches CLASS," the *teaches* relationship is based on the participants PROFESSOR and CLASS.

You don't know if the relationship is 1:1 or 1:M. Therefore, you should ask the question "Can an employee manage more than one division?" If the answer is yes, the relationship is 1:M, and the second part of the relationship is then written as:

An EMPLOYEE may manage many DIVISIONs.

If an employee cannot manage more than one division, the relationship is 1:1, and the second part of the relationship is then written as:

An EMPLOYEE may manage only one DIVISION.

4-1d Connectivity and Cardinality

You learned in Chapter 2 that entity relationships may be classified as one-to-one, one-to-many, or many-to-many. You also learned how such relationships were depicted in the Chen and Crow's Foot notations. The term **connectivity** is used to describe the relationship classification.

Cardinality expresses the minimum and maximum number of entity occurrences associated with one occurrence of the related entity. In the ERD, cardinality is indicated by placing the appropriate numbers beside the entities, using the format (x,y). The first value represents the minimum number of associated entities, while the second value represents the maximum number of associated entities. Many database designers who use Crow's Foot modeling notation do not depict the specific cardinalities on the ER diagram itself because the specific limits described by the cardinalities cannot be implemented directly through the database design. Correspondingly, some Crow's Foot ER modeling tools do not print the numeric cardinality range in the diagram; instead, you can add it as text if you want to have it shown. When the specific cardinalities are not included on the diagram in Crow's Foot notation, cardinality is implied by the use of the symbols shown in Figure 4.7, which describe the connectivity and participation (discussed next). The numeric cardinality range has been added using the Microsoft Visio text drawing tool.

FIGURE 4.7 CONNECTIVITY AND CARDINALITY IN AN ERD

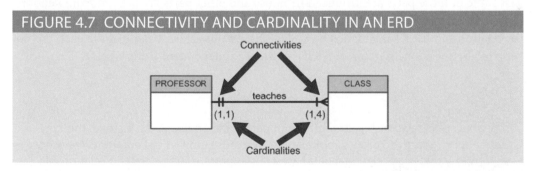

Knowing the minimum and maximum number of entity occurrences is very useful at the application software level. For example, Tiny College might want to ensure that a class is not taught unless it has at least 10 students enrolled. Similarly, if the classroom can hold only 30 students, the application software should use that cardinality to limit enrollment in the class. However, keep in mind that the DBMS cannot handle the implementation of the cardinalities at the table level—that capability is provided by the application software or by triggers. You will learn how to create and execute triggers in Chapter 8, Advanced SQL.

As you examine the Crow's Foot diagram in Figure 4.7, keep in mind that the cardinalities represent the number of occurrences in the *related* entity. For example, the cardinality (1,4) next to the CLASS entity in the "PROFESSOR teaches CLASS" relationship indicates that each professor teaches up to four classes, which means that the PROFESSOR table's primary key value occurs at least once and no more than four times as foreign key values in the CLASS table. If the cardinality had been written as (1,N), there would be no upper limit to the number of classes a professor might teach. Similarly, the cardinality (1,1) next to the PROFESSOR entity indicates that each class is taught by one and only one professor. That is, each CLASS entity occurrence is associated with one and only one entity occurrence in PROFESSOR.

connectivity
The classification of the relationship between entities. Classifications include 1:1, 1:M, and M:N.

cardinality
A property that assigns a specific value to connectivity and expresses the range of allowed entity occurrences associated with a single occurrence of the related entity.

Connectivities and cardinalities are established by concise statements known as business rules, which were introduced in Chapter 2. Such rules, derived from a precise and detailed description of an organization's data environment, also establish the ERM's entities, attributes, relationships, connectivities, cardinalities, and constraints. Because business rules define the ERM's components, making sure that all appropriate business rules are identified is an important part of a database designer's job.

Note

The placement of the cardinalities in the ER diagram is a matter of convention. The Chen notation places the cardinalities on the side of the related entity. The Crow's Foot and UML diagrams place the cardinalities next to the entity to which they apply.

Online Content

Because the careful definition of complete and accurate business rules is crucial to good database design, their derivation is examined in detail in Appendix B, The University Lab: Conceptual Design. The modeling skills you are learning in this chapter are applied in the development of a real database design in Appendix B. The initial design shown in Appendix B is then modified in Appendix C, The University Lab: Conceptual Design Verification, Logical Design, and Implementation. (Both appendixes are available at *www. cengagebrain.com*.)

4-1e Existence Dependence

An entity is said to be **existence-dependent** if it can exist in the database only when it is associated with another related entity occurrence. In implementation terms, an entity is existence-dependent if it has a mandatory foreign key—that is, a foreign key attribute that cannot be null. For example, if an employee wants to claim one or more dependents for tax-withholding purposes, the relationship "EMPLOYEE claims DEPENDENT" would be appropriate. In that case, the DEPENDENT entity is clearly existence-dependent on the EMPLOYEE entity because it is impossible for the dependent to exist apart from the EMPLOYEE in the database.

If an entity can exist apart from all of its related entities, then it is **existence-independent**, and it is referred to as a **strong entity** or **regular entity**. For example, suppose that the XYZ Corporation uses parts to produce its products. Furthermore, suppose that some of those parts are produced in-house and other parts are bought from vendors. In that scenario, it is quite possible for a PART to exist independently from a VENDOR in the relationship "PART is supplied by VENDOR" because at least some of the parts are not supplied by a vendor. Therefore, PART is existence-independent from VENDOR.

Note

The concept of relationship strength is not part of the original ERM. Instead, this concept applies directly to Crow's Foot diagrams. Because Crow's Foot diagrams are used extensively to design relational databases, it is important to understand relationship strength as it affects database implementation. The Chen ERD notation is oriented toward conceptual modeling and therefore does not distinguish between weak and strong relationships.

existence-dependent
A property of an entity whose existence depends on one or more other entities. In such an environment, the existence-independent table must be created and loaded first because the existence-dependent key cannot reference a table that does not yet exist.

4-1f Relationship Strength

The concept of relationship strength is based on how the primary key of a related entity is defined. To implement a relationship, the primary key of one entity (the parent entity, normally on the "one" side of the one-to-many relationship) appears as a foreign key in the related entity (the child entity, mostly the entity on the "many" side of the one-to-many relationship). Sometimes the foreign key also is a primary key component in the related entity. For example, in Figure 4.5, the CAR entity primary key (CAR_VIN) appears as both a primary key component and a foreign key in the CAR_COLOR entity. In this section, you will learn how various relationship strength decisions affect primary key arrangement in database design.

Weak (Non-Identifying) Relationships A **weak relationship**, also known as a **non-identifying relationship**, exists if the primary key of the related entity does not contain a primary key component of the parent entity. By default, relationships are established by having the primary key of the parent entity appear as a foreign key (FK) on the related entity (also known as the child entity). For example, suppose the 1:M relationship between COURSE and CLASS is defined as:

COURSE (**CRS_CODE**, DEPT_CODE, CRS_DESCRIPTION, CRS_CREDIT)

CLASS (**CLASS_CODE**, CRS_CODE, CLASS_SECTION, CLASS_TIME, ROOM_CODE, PROF_NUM)

In this case, a weak relationship exists between COURSE and CLASS because CRS_CODE (the primary key of the parent entity) is only a foreign key in the CLASS entity. In this example, the CLASS primary key did not inherit a primary key component from the COURSE entity.

Figure 4.8 shows how the Crow's Foot notation depicts a weak relationship by placing a dashed relationship line between the entities. The tables shown below the ERD illustrate how such a relationship is implemented.

FIGURE 4.8   A WEAK (NON-IDENTIFYING) RELATIONSHIP BETWEEN COURSE AND CLASS

Table name: **COURSE**   Database name: **Ch04_TinyCollege**

CRS_CODE	DEPT_CODE	CRS_DESCRIPTION	CRS_CREDIT
ACCT-211	ACCT	Accounting I	3
ACCT-212	ACCT	Accounting II	3
CIS-220	CIS	Intro. to Microcomputing	3
CIS-420	CIS	Database Design and Implementation	4
MATH-243	MATH	Mathematics for Managers	3
QM-261	CIS	Intro. to Statistics	3
QM-362	CIS	Statistical Applications	4

Table name: **CLASS**

CLASS_CODE	CRS_CODE	CLASS_SECTION	CLASS_TIME	ROOM_CODE	PROF_NUM
10012	ACCT-211	1	MWF 8:00-8:50 a.m.	BUS311	105
10013	ACCT-211	2	MWF 9:00-9:50 a.m.	BUS200	105
10014	ACCT-211	3	TTh 2:30-3:45 p.m.	BUS252	342
10015	ACCT-212	1	MWF 10:00-10:50 a.m.	BUS311	301
10016	ACCT-212	2	Th 6:00-8:40 p.m.	BUS252	301
10017	CIS-220	1	MWF 9:00-9:50 a.m.	KLR209	228
10018	CIS-220	2	MWF 9:00-9:50 a.m.	KLR211	114
10019	CIS-220	3	MWF 10:00-10:50 a.m.	KLR209	228
10020	CIS-420	1	W 6:00-8:40 p.m.	KLR209	162
10021	QM-261	1	MWF 8:00-8:50 a.m.	KLR200	114
10022	QM-261	2	TTh 1:00-2:15 p.m.	KLR200	114
10023	QM-362	1	MWF 11:00-11:50 a.m.	KLR200	162
10024	QM-362	2	TTh 2:30-3:45 p.m.	KLR200	162
10025	MATH-243	1	Th 6:00-8:40 p.m.	DRE155	325

existence-independent
A property of an entity that can exist apart from one or more related entities. Such a table must be created first when referencing an existence-dependent table.

strong entity
An entity that is existence-independent, that is, it can exist apart from all of its related entities. Also called a *regular entity*.

regular entity
See *strong entity*.

weak (non-identifying) relationship
A relationship in which the primary key of the related entity does not contain a primary key component of the parent entity.

Strong (Identifying) Relationships A **strong (identifying) relationship** exists when the primary key of the related entity contains a primary key component of the parent entity. For example, suppose the 1:M relationship between COURSE and CLASS is defined as:

COURSE (**CRS_CODE**, DEPT_CODE, CRS_DESCRIPTION, CRS_CREDIT)

CLASS (**CRS_CODE, CLASS_SECTION**, CLASS_TIME, ROOM_CODE, PROF_NUM)

In this case, the CLASS entity primary key is composed of CRS_CODE and CLASS_SECTION. Therefore, a strong relationship exists between COURSE and CLASS because CRS_CODE (the primary key of the parent entity) is a primary key component in the CLASS entity. In other words, the CLASS primary key did inherit a primary key component from the COURSE entity. (Note that the CRS_CODE in CLASS is *also* the FK to the COURSE entity.)

The Crow's Foot notation depicts the strong (identifying) relationship with a solid line between the entities, as shown in Figure 4.9.

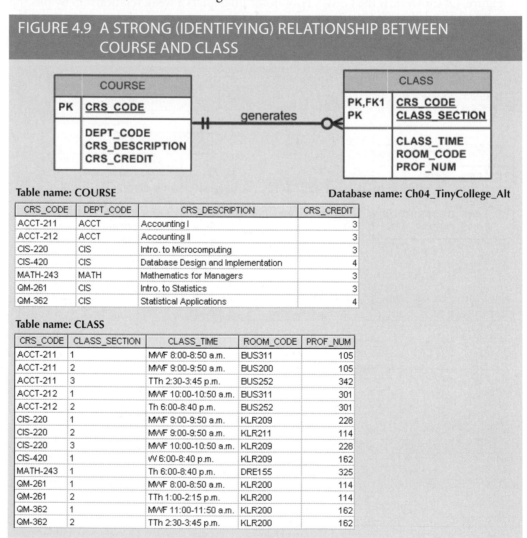

FIGURE 4.9 A STRONG (IDENTIFYING) RELATIONSHIP BETWEEN COURSE AND CLASS

Table name: COURSE   Database name: Ch04_TinyCollege_Alt

CRS_CODE	DEPT_CODE	CRS_DESCRIPTION	CRS_CREDIT
ACCT-211	ACCT	Accounting I	3
ACCT-212	ACCT	Accounting II	3
CIS-220	CIS	Intro. to Microcomputing	3
CIS-420	CIS	Database Design and Implementation	4
MATH-243	MATH	Mathematics for Managers	3
QM-261	CIS	Intro. to Statistics	3
QM-362	CIS	Statistical Applications	4

Table name: CLASS

CRS_CODE	CLASS_SECTION	CLASS_TIME	ROOM_CODE	PROF_NUM
ACCT-211	1	MWF 8:00-8:50 a.m.	BUS311	105
ACCT-211	2	MWF 9:00-9:50 a.m.	BUS200	105
ACCT-211	3	TTh 2:30-3:45 p.m.	BUS252	342
ACCT-212	1	MWF 10:00-10:50 a.m.	BUS311	301
ACCT-212	2	Th 6:00-8:40 p.m.	BUS252	301
CIS-220	1	MWF 9:00-9:50 a.m.	KLR209	228
CIS-220	2	MWF 9:00-9:50 a.m.	KLR211	114
CIS-220	3	MWF 10:00-10:50 a.m.	KLR209	228
CIS-420	1	W 6:00-8:40 p.m.	KLR209	162
MATH-243	1	Th 6:00-8:40 p.m.	DRE155	325
QM-261	1	MWF 8:00-8:50 a.m.	KLR200	114
QM-261	2	TTh 1:00-2:15 p.m.	KLR200	114
QM-362	1	MWF 11:00-11:50 a.m.	KLR200	162
QM-362	2	TTh 2:30-3:45 p.m.	KLR200	162

strong (identifying) relationship
A relationship that occurs when two entities are existence-dependent; from a database design perspective, this relationship exists whenever the primary key of the related entity contains the primary key of the parent entity.

As you examine Figure 4.9, you might wonder what the O symbol next to the CLASS entity signifies. You will discover the meaning of this cardinality in Section 4-1h, Relationship Participation.

In summary, whether the relationship between COURSE and CLASS is strong or weak depends on how the CLASS entity's primary key is defined. Remember that the nature of the relationship is often determined by the database designer, who must use professional

Note

Keep in mind that the *order in which the tables are created and loaded is very important.* For example, in the "COURSE generates CLASS" relationship, the COURSE table must be created before the CLASS table. After all, it would not be acceptable to have the CLASS table's foreign key refer to a COURSE table that did not yet exist. In fact, *you must load the data of the "1" side first in a 1:M relationship to avoid the possibility of referential integrity errors*, regardless of whether the relationships are weak or strong.

judgment to determine which relationship type and strength best suit the database transaction, efficiency, and information requirements. That point will be emphasized in detail!

4-1g Weak Entities

In contrast to the strong or regular entity mentioned in Section 4-1f, a **weak entity** is one that meets two conditions:

1. The entity is existence-dependent; it cannot exist without the entity with which it has a relationship.

2. The entity has a primary key that is partially or totally derived from the parent entity in the relationship.

For example, a company insurance policy insures an employee and any dependents. For the purpose of describing an insurance policy, an EMPLOYEE might or might not have a DEPENDENT, but the DEPENDENT must be associated with an EMPLOYEE. Moreover, the DEPENDENT cannot exist without the EMPLOYEE; that is, a person cannot get insurance coverage as a dependent unless the person is a dependent of an employee. DEPENDENT is the weak entity in the relationship "EMPLOYEE has DEPENDENT." This relationship is shown in Figure 4.10.

FIGURE 4.10 A WEAK ENTITY IN AN ERD

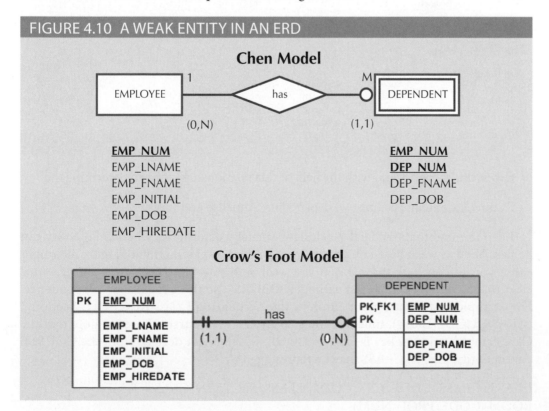

weak entity
An entity that displays existence dependence and inherits the primary key of its parent entity. For example, a DEPENDENT requires the existence of an EMPLOYEE.

Note that the Chen notation in Figure 4.10 identifies the weak entity by using a double-walled entity rectangle. The Crow's Foot notation generated by Visio Professional uses the relationship line and the PK/FK designation to indicate whether the related entity is weak. A strong (identifying) relationship indicates that the related entity is weak. Such a relationship means that both conditions have been met for the weak entity definition—the related entity is existence-dependent, and the PK of the related entity contains a PK component of the parent entity.

Remember that the weak entity inherits part of its primary key from its strong counterpart. For example, at least part of the DEPENDENT entity's key shown in Figure 4.10 was inherited from the EMPLOYEE entity:

EMPLOYEE (**EMP_NUM**, EMP_LNAME, EMP_FNAME, EMP_INITIAL, EMP_DOB, EMP_HIREDATE)

DEPENDENT (**EMP_NUM**, **DEP_NUM**, DEP_FNAME, DEP_DOB)

Figure 4.11 illustrates the implementation of the relationship between the weak entity (DEPENDENT) and its parent or strong counterpart (EMPLOYEE). Note that DEPENDENT's primary key is composed of two attributes, EMP_NUM and DEP_NUM, and that EMP_NUM was inherited from EMPLOYEE.

FIGURE 4.11 A WEAK ENTITY IN A STRONG RELATIONSHIP

Table name: EMPLOYEE   Database name: Ch04_ShortCo

EMP_NUM	EMP_LNAME	EMP_FNAME	EMP_INITIAL	EMP_DOB	EMP_HIREDATE
1001	Callifante	Jeanine	J	12-Mar-64	25-May-97
1002	Smithson	William	K	23-Nov-70	28-May-97
1003	Washington	Herman	H	15-Aug-68	28-May-97
1004	Chen	Lydia	B	23-Mar-74	15-Oct-98
1005	Johnson	Melanie		28-Sep-66	20-Dec-98
1006	Ortega	Jorge	G	12-Jul-79	05-Jan-02
1007	O'Donnell	Peter	D	10-Jun-71	23-Jun-02
1008	Brzenski	Barbara	A	12-Feb-70	01-Nov-03

Table name: DEPENDENT

EMP_NUM	DEP_NUM	DEP_FNAME	DEP_DOB
1001	1	Annelise	05-Dec-97
1001	2	Jorge	30-Sep-02
1003	1	Suzanne	25-Jan-04
1006	1	Carlos	25-May-01
1008	1	Michael	19-Feb-95
1008	2	George	27-Jun-98
1008	3	Katherine	18-Aug-03

Given this scenario, and with the help of this relationship, you can determine that:

Jeanine J. Callifante claims two dependents, Annelise and Jorge.

Keep in mind that the database designer usually determines whether an entity can be described as weak based on the business rules. An examination of Figure 4.8 might cause you to conclude that CLASS is a weak entity to COURSE. After all, it seems clear that a CLASS cannot exist without a COURSE, so there is existence dependence. For example, a student cannot enroll in the Accounting I class ACCT-211, Section 3 (CLASS_CODE 10014), unless there is an ACCT-211 course. However, note that the CLASS table's primary key is CLASS_CODE, which is not derived from the COURSE parent entity. That is, CLASS may be represented by:

CLASS (**CLASS_CODE**, CRS_CODE, CLASS_SECTION, CLASS_TIME, ROOM_CODE, PROF_NUM)

The second weak entity requirement has not been met; therefore, by definition, the CLASS entity in Figure 4.8 may not be classified as weak. On the other hand, if the CLASS entity's primary key had been defined as a composite key composed of the combination CRS_CODE and CLASS_SECTION, CLASS could be represented by:

CLASS (**CRS_CODE**, **CLASS_SECTION**, CLASS_TIME, ROOM_CODE, PROF_NUM)

In that case, as illustrated in Figure 4.9, the CLASS primary key is partially derived from COURSE because CRS_CODE is the COURSE table's primary key. Given this decision, CLASS is a weak entity by definition. (In Visio Professional Crow's Foot terms, the relationship between COURSE and CLASS is classified as strong, or identifying.) In any case, CLASS is always existence-dependent on COURSE, *whether or not it is defined as weak.*

4-1h Relationship Participation

Participation in an entity relationship is either optional or mandatory. Recall that relationships are bidirectional; that is, they operate in both directions. If COURSE is related to CLASS, then by definition, CLASS is related to COURSE. Because of the bidirectional nature of relationships, it is necessary to determine the connectivity of the relationship from COURSE to CLASS and the connectivity of the relationship from CLASS to COURSE. Similarly, the specific maximum and minimum cardinalities must be determined in each direction for the relationship. Once again, you must consider the bidirectional nature of the relationship when determining participation.

Optional participation means that one entity occurrence does not *require* a corresponding entity occurrence in a particular relationship. For example, in the "COURSE generates CLASS" relationship, you noted that at least some courses do not generate a class. In other words, an entity occurrence (row) in the COURSE table does not necessarily require the existence of a corresponding entity occurrence in the CLASS table. (Remember that each entity is implemented as a table.) Therefore, the CLASS entity is considered to be *optional* to the COURSE entity. In Crow's Foot notation, an optional relationship between entities is shown by drawing a small circle (O) on the side of the optional entity, as illustrated in Figure 4.9. The existence of an *optional entity* indicates that its minimum cardinality is 0. (The term *optionality* is used to label any condition in which one or more optional relationships exist.)

Note

Remember that the burden of establishing the relationship is always placed on the entity that contains the foreign key. In most cases, that entity is on the "many" side of the relationship.

Mandatory participation means that one entity occurrence *requires* a corresponding entity occurrence in a particular relationship. If no optionality symbol is depicted with the entity, the entity is assumed to exist in a mandatory relationship with the related entity. If the mandatory participation is depicted graphically, it is typically shown as a small hash mark across the relationship line, similar to the Crow's Foot depiction of a connectivity of 1. The existence of a mandatory relationship indicates that the minimum cardinality is at least 1 for the mandatory entity.

optional participation In ER modeling, a condition in which one entity occurrence does not require a corresponding entity occurrence in a particular relationship.

mandatory participation A relationship in which one entity occurrence must have a corresponding occurrence in another entity. For example, an EMPLOYEE works in a DIVISION. (A person cannot be an employee without being assigned to a company's division.)

Note

You might be tempted to conclude that relationships are weak when they occur between entities in an optional relationship and that relationships are strong when they occur between entities in a mandatory relationship. However, this conclusion is not warranted. Keep in mind that relationship participation and relationship strength do not describe the same thing. You are likely to encounter a strong relationship when one entity is optional to another. For example, the relationship between EMPLOYEE and DEPENDENT is clearly a strong one, but DEPENDENT is clearly optional to EMPLOYEE. After all, you cannot require employees to have dependents. Also, it is just as possible for a weak relationship to be established when one entity is mandatory to another. The relationship strength depends on how the PK of the related entity is formulated, while the relationship participation depends on how the business rule is written. For example, the business rules "Each part must be supplied by a vendor" and "A part may or may not be supplied by a vendor" create different optionalities for the same entities! Failure to understand this distinction may lead to poor design decisions that cause major problems when table rows are inserted or deleted.

When you create a relationship in Microsoft Visio, the default relationship will be mandatory on the "1" side and optional on the "many" side. Table 4.3 shows the various connectivity and participation combinations that are supported by the Crow's Foot notation. Recall that these combinations are often referred to as cardinality in Crow's Foot notation when specific cardinalities are not used.

TABLE 4.3

CROW'S FOOT SYMBOLS

SYMBOL	CARDINALITY	COMMENT
O⋹	(0,N)	Zero or many; the "many" side is optional.
I⋹	(1,N)	One or many; the "many" side is mandatory.
II	(1,1)	One and only one; the "1" side is mandatory.
OⅠ	(0,1)	Zero or one; the "1" side is optional.

Because relationship participation is an important component of database design, you should examine a few more scenarios. Suppose that Tiny College employs some professors who conduct research without teaching classes. If you examine the "PROFESSOR teaches CLASS" relationship, it is quite possible for a PROFESSOR not to teach a CLASS. Therefore, CLASS is *optional* to PROFESSOR. On the other hand, a CLASS must be taught by a PROFESSOR. Therefore, PROFESSOR is *mandatory* to CLASS. Note that the ERD model in Figure 4.12 shows the cardinality next to CLASS to be (0,3), indicating that a professor may teach no classes or as many as three classes. Also, each CLASS table row references one and only one PROFESSOR row—assuming each class is taught by one and only one professor—represented by the (1,1) cardinality next to the PROFESSOR table.

FIGURE 4.12 AN OPTIONAL CLASS ENTITY IN THE RELATIONSHIP "PROFESSOR TEACHES CLASS"

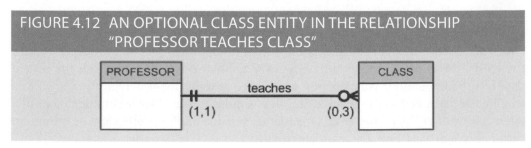

It is important that you clearly understand the distinction between mandatory and optional participation in relationships. Otherwise, you might develop designs in which awkward and unnecessary temporary rows (entity instances) must be created just to accommodate the creation of required entities.

It is also important to understand that the semantics of a problem might determine the type of participation in a relationship. For example, suppose that Tiny College offers several courses; each course has several classes. Note again the distinction between *class* and *course* in this discussion: a CLASS constitutes a specific offering (or section) of a COURSE. Typically, courses are listed in the university's course catalog, while classes are listed in the class schedules that students use to register for their classes.

By analyzing the CLASS entity's contribution to the "COURSE generates CLASS" relationship, it is easy to see that a CLASS cannot exist without a COURSE. Therefore, you can conclude that the COURSE entity is *mandatory* in the relationship. However, two scenarios for the CLASS entity may be written, as shown in Figures 4.13 and 4.14.

FIGURE 4.13 CLASS IS OPTIONAL TO COURSE

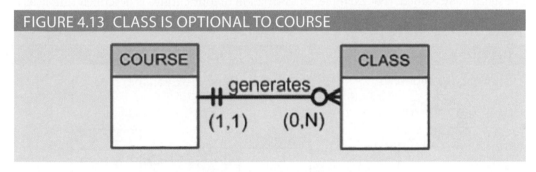

FIGURE 4.14 COURSE AND CLASS IN A MANDATORY RELATIONSHIP

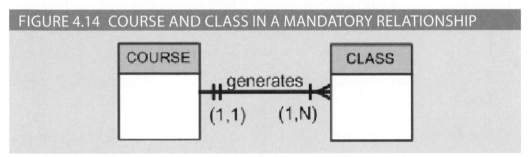

The different scenarios are a function of the problem's semantics; that is, they depend on how the relationship is defined.

1. *CLASS is optional.* It is possible for the department to create the COURSE entity first and then create the CLASS entity after making the teaching assignments. In the real world, such a scenario is very likely; there may be courses for which sections (classes) have not yet been defined. In fact, some courses are taught only once a year and do not generate classes each semester.

2. *CLASS is mandatory.* This condition is created by the constraint imposed by the semantics of the statement "Each COURSE generates one or more CLASSes." In ER terms, each COURSE in the "generates" relationship must have at least one CLASS. Therefore, a CLASS must be created as the COURSE is created to comply with the semantics of the problem.

Keep in mind the practical aspects of the scenario presented in Figure 4.14. Given the semantics of this relationship, the system should not accept a course that is not associated with at least one class section. Is such a rigid environment desirable from an operational point of view? For example, when a new COURSE is created, the database first updates the COURSE table, thereby inserting a COURSE entity that does not yet have a CLASS associated with it.

Naturally, the apparent problem seems to be solved when CLASS entities are inserted into the corresponding CLASS table. However, because of the mandatory relationship, the system will temporarily violate the business rule constraint. For practical purposes, it would be desirable to classify the CLASS as optional to produce a more flexible design.

Finally, as you examine the scenarios in Figures 4.13 and 4.14, keep in mind the role of the DBMS. To maintain data integrity, the DBMS must ensure that the "many" side (CLASS) is associated with a COURSE through the foreign key rules.

4-1i Relationship Degree

A **relationship degree** indicates the number of entities or participants associated with a relationship. A **unary relationship** exists when an association is maintained within a single entity. A **binary relationship** exists when two entities are associated. A **ternary relationship** exists when three entities are associated. Although higher degrees exist, they are rare and are not specifically named. (For example, an association of four entities is described simply as a *four-degree relationship*.) Figure 4.15 shows these types of relationship degrees.

FIGURE 4.15 THREE TYPES OF RELATIONSHIP DEGREE

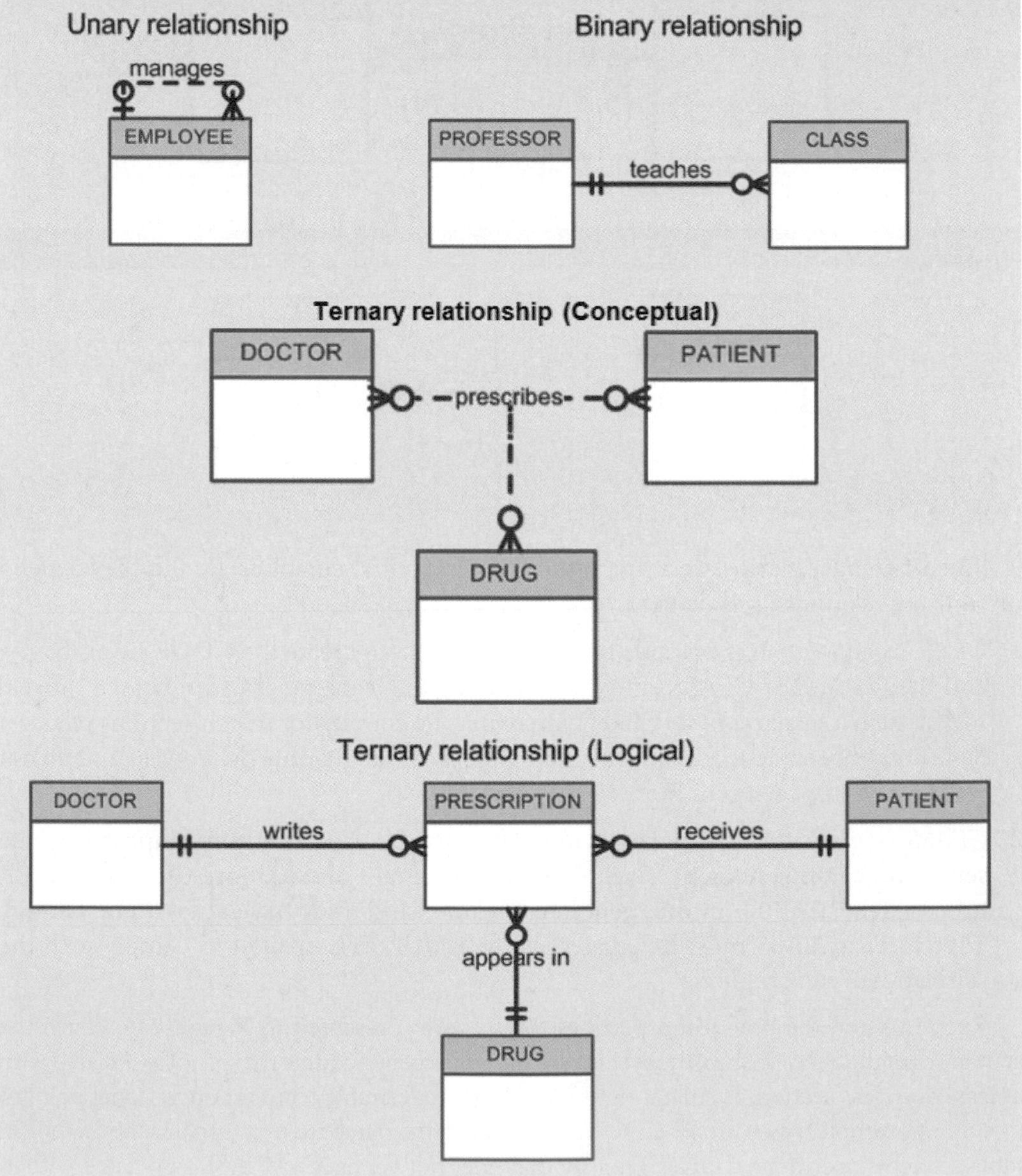

relationship degree
The number of entities or participants associated with a relationship. A relationship degree can be unary, binary, ternary, or higher.

unary relationship
An ER term used to describe an association *within* an entity. For example, an EMPLOYEE might manage another EMPLOYEE.

binary relationship
An ER term for an association (relationship) between two entities. For example, PROFESSOR teaches CLASS.

ternary relationship
An ER term used to describe an association (relationship) between three entities. For example, a DOCTOR prescribes a DRUG for a PATIENT.

Unary Relationships In the case of the unary relationship shown in Figure 4.15, an employee within the EMPLOYEE entity is the manager for one or more employees within that entity. In this case, the existence of the "manages" relationship means that EMPLOYEE requires another EMPLOYEE to be the manager—that is, EMPLOYEE has a relationship with itself. Such a relationship is known as a **recursive relationship**. The various cases of recursive relationships are explained in Section 4-1j.

Binary Relationships A binary relationship exists when two entities are associated in a relationship. Binary relationships are the most common type of relationship. In fact, to simplify the conceptual design, most higher-order (ternary and higher) relationships are decomposed into appropriate equivalent binary relationships whenever possible. In Figure 4.15, "a PROFESSOR teaches one or more CLASSes" represents a binary relationship.

Ternary and Higher-Order Relationships Although most relationships are binary, the use of ternary and higher-order relationships does allow the designer some latitude regarding the semantics of a problem. A ternary relationship implies an association among three different entities. For example, in Figure 4.16, note the relationships and their consequences, which are represented by the following business rules:

- A DOCTOR writes one or more PRESCRIPTIONs.

- A PATIENT may receive one or more PRESCRIPTIONs.

- A DRUG may appear in one or more PRESCRIPTIONs. (To simplify this example, assume that the business rule states that each prescription contains only one drug. In short, if a doctor prescribes more than one drug, a separate prescription must be written for each drug.)

> **recursive relationship**
> A relationship found within a single entity type. For example, an EMPLOYEE is married to an EMPLOYEE or a PART is a component of another PART.

FIGURE 4.16 THE IMPLEMENTATION OF A TERNARY RELATIONSHIP

Database name: Ch04_Clinic

Table name: DRUG

DRUG_CODE	DRUG_NAME	DRUG_PRICE
AF15	Afgapan-15	25.00
AF25	Afgapan-25	35.00
DRO	Droalene Chloride	111.89
DRZ	Druzocholar Cryptolene	18.99
KO15	Koliabar Oxyhexalene	65.75
OLE	Oleander-Drizapan	123.95
TRYP	Tryptolac Heptadimetric	79.45

Table name: PATIENT

PAT_NUM	PAT_TITLE	PAT_LNAME	PAT_FNAME	PAT_INITIAL	PAT_DOB	PAT_AREACODE	PAT_PHONE
100	Mr.	Kolmycz	George	D	15-Jun-1942	615	324-5456
101	Ms.	Lewis	Rhonda	G	19-Mar-2005	615	324-4472
102	Mr.	Vandam	Rhett		14-Nov-1958	901	675-8993
103	Ms.	Jones	Anne	M	16-Oct-1974	615	898-3456
104	Mr.	Lange	John	P	08-Nov-1971	901	504-4430
105	Mr.	Williams	Robert	D	14-Mar-1975	615	890-3220
106	Mrs.	Smith	Jeanine	K	12-Feb-2003	615	324-7883
107	Mr.	Diante	Jorge	D	21-Aug-1974	615	890-4567
108	Mr.	Wiesenbach	Paul	R	14-Feb-1966	615	897-4358
109	Mr.	Smith	George	K	18-Jun-1961	901	504-3339
110	Mrs.	Genkazi	Leighla	W	19-May-1970	901	569-0093
111	Mr.	Washington	Rupert	E	03-Jan-1966	615	890-4925
112	Mr.	Johnson	Edward	E	14-May-1961	615	898-4387
113	Ms.	Smythe	Melanie	P	15-Sep-1970	615	324-9006
114	Ms.	Brandon	Marie	G	02-Nov-1932	901	882-0845
115	Mrs.	Saranda	Hermine	R	25-Jul-1972	615	324-5505
116	Mr.	Smith	George	A	08-Nov-1965	615	890-2984

Table name: DOCTOR

DOC_ID	DOC_LNAME	DOC_FNAME	DOC_INITIAL	DOC_SPECIALTY
29827	Sanchez	Julio	J	Dermatology
32445	Jorgensen	Annelise	G	Neurology
33456	Korenski	Anatoly	A	Urology
33989	LeGrande	George		Pediatrics
34409	Washington	Dennis	F	Orthopaedics
36221	McPherson	Katye	H	Dermatology
36712	Dreifag	Herman	G	Psychiatry
38995	Minh	Tran		Neurology
40004	Chin	Ming	D	Orthopaedics
40028	Feinstein	Denise	L	Gynecology

Table name: PRESCRIPTION

DOC_ID	PAT_NUM	DRUG_CODE	PRES_DOSAGE	PRES_DATE
32445	102	DRZ	2 tablets every four hours -- 50 tablets total	12-Nov-16
32445	113	OLE	1 teaspoon with each meal -- 250 ml total	14-Nov-16
34409	101	KO15	1 tablet every six hours -- 30 tablets total	14-Nov-16
36221	109	DRO	2 tablets with every meal -- 60 tablets total	14-Nov-16
38995	107	KO15	1 tablet every six hours -- 30 tablets total	14-Nov-16

As you examine the table contents in Figure 4.16, note that it is possible to track all transactions. For instance, you can tell that the first prescription was written by doctor 32445 for patient 102, using the drug DRZ.

4-1j Recursive Relationships

As you just learned, a *recursive relationship* is one in which a relationship can exist between occurrences of the same entity set. (Naturally, such a condition is found within a unary relationship.) For example, a 1:M unary relationship can be expressed by "an EMPLOYEE may manage many EMPLOYEEs, and each EMPLOYEE is managed by one EMPLOYEE." Also, as long as polygamy is not legal, a 1:1 unary relationship may be expressed by "an EMPLOYEE may be married to one and only one other EMPLOYEE." Finally, the M:N unary relationship may be expressed by "a COURSE may be a prerequisite to many other COURSEs, and each COURSE may have many other COURSEs as prerequisites." Those relationships are shown in Figure 4.17.

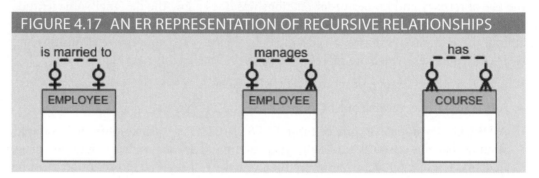

FIGURE 4.17 AN ER REPRESENTATION OF RECURSIVE RELATIONSHIPS

The 1:1 relationship shown in Figure 4.17 can be implemented in the single table shown in Figure 4.18. Note that you can determine that James Ramirez is married to Louise Ramirez, who is married to James Ramirez. Also, Anne Jones is married to Anton Shapiro, who is married to Anne Jones.

FIGURE 4.18 THE 1:1 RECURSIVE RELATIONSHIP "EMPLOYEE IS MARRIED TO EMPLOYEE"

Database name: Ch04_PartCo
Table name: EMPLOYEE_V1

EMP_NUM	EMP_LNAME	EMP_FNAME	EMP_SPOUSE
345	Ramirez	James	347
346	Jones	Anne	349
347	Ramirez	Louise	345
348	Delaney	Robert	
349	Shapiro	Anton	346

Unary relationships are common in manufacturing industries. For example, Figure 4.19 illustrates that a rotor assembly (C-130) is composed of many parts, but each part is used to create only one rotor assembly. Figure 4.19 indicates that a rotor assembly is composed of four 2.5-cm washers, two cotter pins, one 2.5-cm steel shank, four 10.25-cm rotor blades, and two 2.5-cm hex nuts. The relationship implemented in Figure 4.19 thus enables you to track each part within each rotor assembly.

If a part can be used to assemble several different kinds of other parts and is itself composed of many parts, two tables are required to implement the "PART contains PART" relationship. Figure 4.20 illustrates such an environment. Parts tracking is increasingly important as managers become more aware of the legal ramifications of producing more complex output. In many industries, especially those involving aviation, full parts tracking is required by law.

FIGURE 4.19 ANOTHER UNARY RELATIONSHIP: "PART CONTAINS PART"

Table name: PART_V1   Database name: Ch04_PartCo

PART_CODE	PART_DESCRIPTION	PART_IN_STOCK	PART_UNITS_NEEDED	PART_OF_PART
AA21-6	2.5 cm. washer, 1.0 mm. rim	432	4	C-130
AB-121	Cotter pin, copper	1034	2	C-130
C-130	Rotor assembly	36		
E129	2.5 cm. steel shank	128	1	C-130
X10	10.25 cm. rotor blade	345	4	C-130
X34AW	2.5 cm. hex nut	879	2	C-130

FIGURE 4.20 THE IMPLEMENTATION OF THE M:N RECURSIVE RELATIONSHIP "PART CONTAINS PART"

Table name: COMPONENT   Database name: Ch04_PartCo

COMP_CODE	PART_CODE	COMP_PARTS_NEEDED
C-130	AA21-6	4
C-130	AB-121	2
C-130	E129	1
C-131A2	E129	1
C-130	X10	4
C-131A2	X10	1
C-130	X34AW	2
C-131A2	X34AW	2

Table name: PART

PART_CODE	PART_DESCRIPTION	PART_IN_STOCK
AA21-6	2.5 cm. washer, 1.0 mm. rim	432
AB-121	Cotter pin, copper	1034
C-130	Rotor assembly	36
E129	2.5 cm. steel shank	128
X10	10.25 cm. rotor blade	345
X34AW	2.5 cm. hex nut	879

The M:N recursive relationship might be more familiar in a school environment. For instance, note how the M:N "COURSE requires COURSE" relationship illustrated in Figure 4.17 is implemented in Figure 4.21. In this example, MATH-243 is a prerequisite to QM-261 and QM-362, while both MATH-243 and QM-261 are prerequisites to QM-362.

FIGURE 4.21 IMPLEMENTATION OF THE M:N RECURSIVE RELATIONSHIP "COURSE REQUIRES COURSE"

Table name: COURSE   Database name: Ch04_TinyCollege

CRS_CODE	DEPT_CODE	CRS_DESCRIPTION	CRS_CREDIT
ACCT-211	ACCT	Accounting I	3
ACCT-212	ACCT	Accounting II	3
CIS-220	CIS	Intro. to Microcomputing	3
CIS-420	CIS	Database Design and Implementation	4
MATH-243	MATH	Mathematics for Managers	3
QM-261	CIS	Intro. to Statistics	3
QM-362	CIS	Statistical Applications	4

Table name: PREREQ

CRS_CODE	PRE_TAKE
CIS-420	CIS-220
QM-261	MATH-243
QM-362	MATH-243
QM-362	QM-261

Finally, the 1:M recursive relationship "EMPLOYEE manages EMPLOYEE," shown in Figure 4.17, is implemented in Figure 4.22.

One common pitfall when working with unary relationships is to confuse participation with referential integrity. In theory, participation and referential integrity are very different concepts and are normally easy to distinguish in binary relationships. In practical terms, conversely, participation and referential integrity are very similar because they are both implemented through constraints on the same set of attributes. This similarity often leads to confusion when the concepts are applied within the limited structure of a unary relationship. Consider the unary 1:1 spousal relationship between employees, which is described in Figure 4.18. Participation, as described previously, is bidirectional, meaning that it must be addressed in both directions along the relationship. Participation in Figure 4.18 addresses the following questions:

- Must every employee have a spouse who is an employee?

- Must every employee be a spouse to another employee?

FIGURE 4.22 IMPLEMENTATION OF THE 1:M RECURSIVE RELATIONSHIP "EMPLOYEE MANAGES EMPLOYEE"

Database name: Ch04_PartCo

Table name: EMPLOYEE_V2

EMP_CODE	EMP_LNAME	EMP_MANAGER
101	Waddell	102
102	Orincona	
103	Jones	102
104	Reballoh	102
105	Robertson	102
106	Deltona	102

For the data shown in Figure 4.18, the correct answer to both questions is "No." It is possible to be an employee and not have another employee as a spouse. Also, it is possible to be an employee and not be the spouse of another employee.

Referential integrity deals with the correspondence of values in the foreign key with values in the related primary key. Referential integrity is not bidirectional, and therefore answers only one question:

- Must every employee spouse be a valid employee?

For the data shown in Figure 4.18, the correct answer is "Yes." Another way to frame this question is to consider whether every value provided for the EMP_SPOUSE attribute must match some value in the EMP_NUM attribute.

In practical terms, both participation and referential integrity involve the values used as primary keys and foreign keys to implement the relationship. Referential integrity requires that the values in the foreign key correspond to values in the primary key. In one direction, participation considers whether the foreign key can contain a null. In Figure 4.18, for example, employee Robert Delaney is not required to have a value in EMP_SPOUSE. In the other direction, participation considers whether every value in the primary key must appear as a value in the foreign key. In Figure 4.18, for example, employee Robert Delaney's value for EMP_NUM (348) is not required to appear as a value in EMP_SPOUSE for any other employee.

4-1k Associative (Composite) Entities

M:N relationships are a valid construct at the conceptual level, and therefore are found frequently during the ER modeling process. However, implementing the M:N relationship, particularly in the relational model, requires the use of an additional entity, as you learned in Chapter 3. The ER model uses the associative entity to represent an M:N relationship between two or more entities. This associative entity, also called a *composite* or *bridge entity*, is in a 1:M relationship with the parent entities and is composed of the primary key attributes of each parent entity. Furthermore, the associative entity can have additional attributes of its own, as shown by the ENROLL associative entity in Figure 4.23. When using the Crow's Foot notation, the associative entity is identified as a strong (identifying) relationship, as indicated by the solid relationship lines between the parents and the associative entity.

FIGURE 4.23 CONVERTING THE M:N RELATIONSHIP INTO TWO 1:M RELATIONSHIPS

Table name: STUDENT   Database name: Ch04_CollegeTry

STU_NUM	STU_LNAME
321452	Bowser
324257	Smithson

Table name: ENROLL

CLASS_CODE	STU_NUM	ENROLL_GRADE
10014	321452	C
10014	324257	B
10018	321452	A
10018	324257	B
10021	321452	C
10021	324257	C

Table name: CLASS

CLASS_CODE	CRS_CODE	CLASS_SECTION	CLASS_TIME	ROOM_CODE	PROF_NUM
10014	ACCT-211	3	TTh 2:30-3:45 p.m.	BUS252	342
10018	CIS-220	2	MWF 9:00-9:50 a.m.	KLR211	114
10021	QM-261	1	MWF 8:00-8:50 a.m.	KLR200	114

Note that the composite ENROLL entity in Figure 4.23 is existence-dependent on the other two entities; the composition of the ENROLL entity is based on the primary keys of the entities that are connected by the composite entity. The composite entity may also contain additional attributes that play no role in the connective process. For example, although the entity must be composed of at least the STUDENT and CLASS primary keys, it may also include such additional attributes as grades, absences, and other data uniquely identified by the student's performance in a specific class.

Finally, keep in mind that the ENROLL table's key (CLASS_CODE and STU_NUM) is composed entirely of the primary keys of the CLASS and STUDENT tables. Therefore, no null entries are possible in the ENROLL table's key attributes.

Implementing the small database shown in Figure 4.23 requires that you define the relationships clearly. Specifically, you must know the "1" and the "M" sides of each relationship, and you must know whether the relationships are mandatory or optional. For example, note the following point:

- A class may exist (at least at the start of registration) even though it contains no students. Therefore, in Figure 4.24, an optional symbol should appear on the STUDENT side of the M:N relationship between STUDENT and CLASS.

You might argue that to be classified as a STUDENT, a person must be enrolled in at least one CLASS. Therefore, CLASS is mandatory to STUDENT from a purely conceptual point of view. However, when a student is admitted to college, that student has not yet signed up for any classes. Therefore, *at least initially*, CLASS is optional to STUDENT. Note that the practical considerations in the data environment help dictate the use of optionalities.

FIGURE 4.24 THE M:N RELATIONSHIP BETWEEN STUDENT AND CLASS

STUDENT ─enrolls─ CLASS

If CLASS is *not* optional to STUDENT from a database point of view, a class assignment must be made when the student is admitted. However, that's *not* how the process actually works, and the database design must reflect this. In short, the optionality reflects practice.

Because the M:N relationship between STUDENT and CLASS is decomposed into two 1:M relationships through ENROLL, the optionalities must be transferred to ENROLL. (See Figure 4.25.) In other words, it now becomes possible for a class not to occur in ENROLL if no student has signed up for that class. Because a class need not occur in ENROLL, the ENROLL entity becomes optional to CLASS. Also, because the ENROLL entity is created before any students have signed up for a class, the ENROLL entity is also optional to STUDENT, at least initially.

FIGURE 4.25 A COMPOSITE ENTITY IN AN ERD

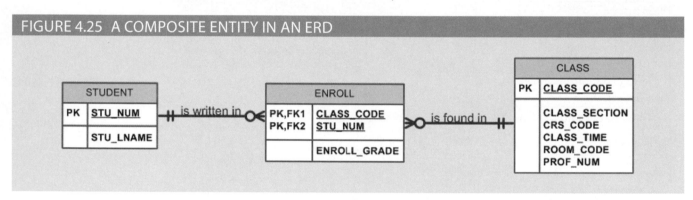

- As students begin to sign up for their classes, they will be entered into the ENROLL entity. Naturally, if a student takes more than one class, that student will occur more than once in ENROLL. For example, note that in the ENROLL table in Figure 4.23, STU_NUM = 321452 occurs three times. On the other hand, each student occurs only once in the STUDENT entity. (Note that the STUDENT table in Figure 4.23 has only one STU_NUM = 321452 entry.) Therefore, in Figure 4.25, the relationship between STUDENT and ENROLL is shown to be 1:M, with the "M" on the ENROLL side.

- As you can see in Figure 4.23, a class can occur more than once in the ENROLL table. For example, CLASS_CODE = 10014 occurs twice. However, CLASS_CODE = 10014 occurs only once in the CLASS table to reflect that the relationship between CLASS and ENROLL is 1:M. Note that in Figure 4.25, the "M" is located on the ENROLL side, while the "1" is located on the CLASS side.

4-2 Developing an ER Diagram

The process of database design is iterative rather than a linear or sequential process. The verb *iterate* means "to do again or repeatedly." Thus, an **iterative process** is based on repetition of processes and procedures. Building an ERD usually involves the following activities:

- Create a detailed narrative of the organization's description of operations.
- Identify the business rules based on the description of operations.

iterative process
A process based on repetition of steps and procedures.

- Identify the main entities and relationships from the business rules.
- Develop the initial ERD.
- Identify the attributes and primary keys that adequately describe the entities.
- Revise and review the ERD.

During the review process, additional objects, attributes, and relationships probably will be uncovered. Therefore, the basic ERM will be modified to incorporate the newly discovered ER components. Subsequently, another round of reviews might yield additional components or clarification of the existing diagram. The process is repeated until the end users and designers agree that the ERD is a fair representation of the organization's activities and functions.

During the design process, the database designer does not depend simply on interviews to help define entities, attributes, and relationships. A surprising amount of information can be gathered by examining the business forms and reports that an organization uses in its daily operations.

To illustrate the use of the iterative process that ultimately yields a workable ERD, start with an initial interview with the Tiny College administrators. The interview process yields the following business rules:

1. Tiny College (TC) is divided into several schools: business, arts and sciences, education, and applied sciences. Each school is administered by a dean who is a professor. Each professor can be the dean of only one school, and a professor is not required to be the dean of any school. Therefore, a 1:1 relationship exists between PROFESSOR and SCHOOL. Note that the cardinality can be expressed by writing (1,1) next to the entity PROFESSOR and (0,1) next to the entity SCHOOL.

2. Each school comprises several departments. For example, the school of business has an accounting department, a management/marketing department, an economics/finance department, and a computer information systems department. Note again the cardinality rules: The smallest number of departments operated by a school is one, and the largest number of departments is indeterminate (N). On the other hand, each department belongs to only a single school; thus, the cardinality is expressed by (1,1). That is, the minimum number of schools to which a department belongs is one, as is the maximum number. Figure 4.26 illustrates these first two business rules.

FIGURE 4.26 THE FIRST TINY COLLEGE ERD SEGMENT

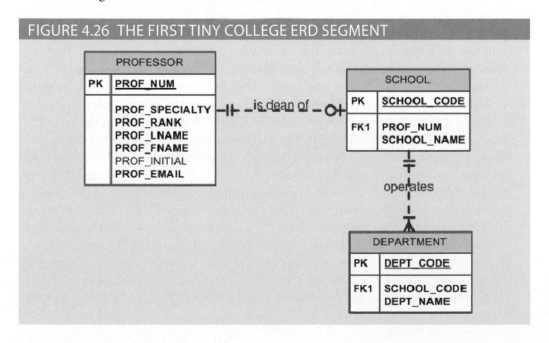

Note

It is again appropriate to evaluate the reason for maintaining the 1:1 relationship between PROFESSOR and SCHOOL in the "PROFESSOR is dean of SCHOOL" relationship. It is worth repeating that the existence of 1:1 relationships often indicates a misidentification of attributes as entities. In this case, the 1:1 relationship could easily be eliminated by storing the dean's attributes in the SCHOOL entity. This solution would also make it easier to answer the queries "Who is the dean?" and "What are the dean's credentials?" The downside of this solution is that it requires the duplication of data that is already stored in the PROFESSOR table, thus setting the stage for anomalies. However, because each school is run by a single dean, the problem of data duplication is rather minor. The selection of one approach over another often depends on information requirements, transaction speed, and the database designer's professional judgment. In short, do not use 1:1 relationships lightly, and make sure that each 1:1 relationship within the database design is defensible.

3. Each department may offer courses. For example, the management/marketing department offers courses such as Introduction to Management, Principles of Marketing, and Production Management. The ERD segment for this condition is shown in Figure 4.27. Note that this relationship is based on the way Tiny College operates. For example, if Tiny College had some departments that were classified as "research only," they would not offer courses; therefore, the COURSE entity would be optional to the DEPARTMENT entity.

FIGURE 4.27 THE SECOND TINY COLLEGE ERD SEGMENT

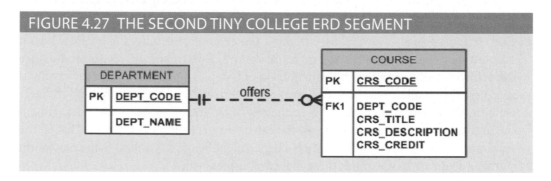

4. The relationship between COURSE and CLASS was illustrated in Figure 4.9. Nevertheless, it is worth repeating that a CLASS is a section of a COURSE. That is, a department may offer several sections (classes) of the same database course. Each of those classes is taught by a professor at a given time in a given place. In short, a 1:M relationship exists between COURSE and CLASS. Additionally, each class is offered during a given semester. SEMESTER defines the year and the term that the class will be offered. Note that this is different from the date when the student actually enrolls in a class. For example, students are able to enroll in summer and fall term classes near the end of the spring term. It is possible that the Tiny College calendar is set with semester beginning and ending dates prior to the creation of the semester class schedule so CLASS is optional to SEMESTER. This design will also help for reporting purposes, for example, you could answer questions such as: what classes were offered X semester? Or, what classes did student Y take on semester X? Because a course may exist in Tiny College's course catalog even when it is not offered as a class in a given semester, CLASS is optional to COURSE. Therefore, the relationships between SEMESTER, COURSE, and CLASS look like Figure 4.28.

FIGURE 4.28 THE THIRD TINY COLLEGE ERD SEGMENT

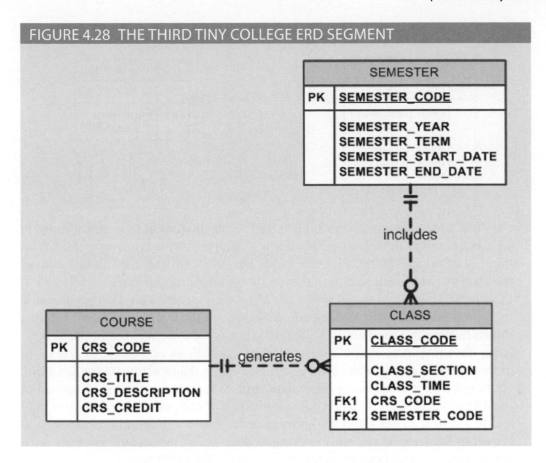

5. Each department should have one or more professors assigned to it. One and only one of those professors chairs the department, and no professor is required to accept the chair position. Therefore, DEPARTMENT is optional to PROFESSOR in the "chairs" relationship. Those relationships are summarized in the ER segment shown in Figure 4.29.

FIGURE 4.29 THE FOURTH TINY COLLEGE ERD SEGMENT

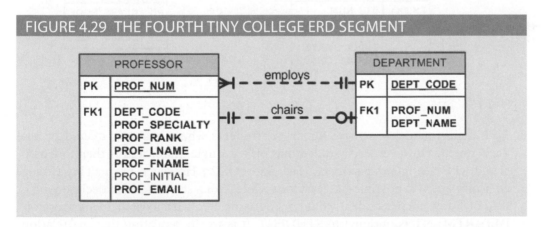

6. Each professor may teach up to four classes; each class is a section of a course. A professor may also be on a research contract and teach no classes at all. The ERD segment in Figure 4.30 depicts those conditions.

7. A student may enroll in several classes but take each class only once during any given enrollment period. For example, during the current enrollment period, a student may decide to take five classes—Statistics, Accounting, English, Database, and History—but that student would not be enrolled in the same Statistics

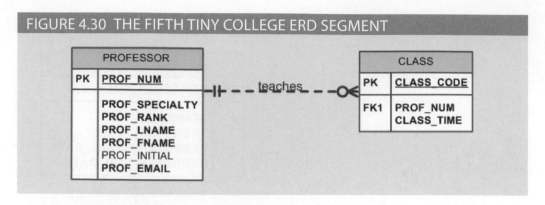

FIGURE 4.30   THE FIFTH TINY COLLEGE ERD SEGMENT

class five times during the enrollment period! Each student may enroll in up to six classes, and each class may have up to 35 students, thus creating an M:N relationship between STUDENT and CLASS. Because a CLASS can initially exist at the start of the enrollment period even though no students have enrolled in it, STUDENT is optional to CLASS in the M:N relationship. This M:N relationship must be divided into two 1:M relationships through the use of the ENROLL entity, shown in the ERD segment in Figure 4.31. However, note that the optional symbol is shown next to ENROLL. If a class exists but has no students enrolled in it, that class does not occur in the ENROLL table. Note also that the ENROLL entity is weak: it is existence-dependent, and its (composite) PK is composed of the PKs of the STUDENT and CLASS entities. You can add the cardinalities (0,6) and (0,35) next to the ENROLL entity to reflect the business rule constraints, as shown in Figure 4.31. (Visio Professional does not automatically generate such cardinalities, but you can use a text box to accomplish that task.)

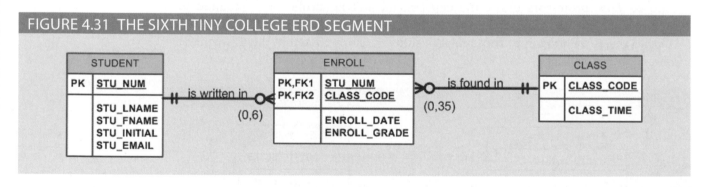

FIGURE 4.31   THE SIXTH TINY COLLEGE ERD SEGMENT

8. Each department has several (or many) students whose major is offered by that department. However, each student has only a single major and is therefore associated with a single department. (See Figure 4.32.) However, in the Tiny College environment, it is possible—at least for a while—for a student not to declare a major field of study. Such a student would not be associated with a department; therefore, DEPARTMENT is optional to STUDENT. It is worth repeating that the relationships between entities and the entities themselves reflect the organization's operating environment. That is, the business rules define the ERD components.

9. Each student has an advisor in his or her department; each advisor counsels several students. An advisor is also a professor, but not all professors advise students. Therefore, STUDENT is optional to PROFESSOR in the "PROFESSOR advises STUDENT" relationship. (See Figure 4.33.)

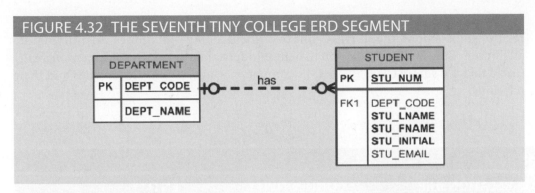

FIGURE 4.32 THE SEVENTH TINY COLLEGE ERD SEGMENT

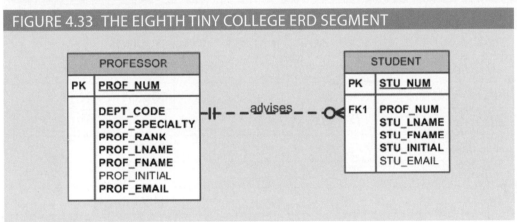

FIGURE 4.33 THE EIGHTH TINY COLLEGE ERD SEGMENT

10. As you can see in Figure 4.34, the CLASS entity contains a ROOM_CODE attribute. Given the naming conventions, it is clear that ROOM_CODE is an FK to another entity. Clearly, because a class is taught in a room, it is reasonable to assume that the ROOM_CODE in CLASS is the FK to an entity named ROOM. In turn, each room is located in a building. So, the last Tiny College ERD is created by observing that a BUILDING can contain many ROOMs, but each ROOM is found in a single BUILDING. In this ERD segment, it is clear that some buildings do not contain (class) rooms. For example, a storage building might not contain any named rooms at all.

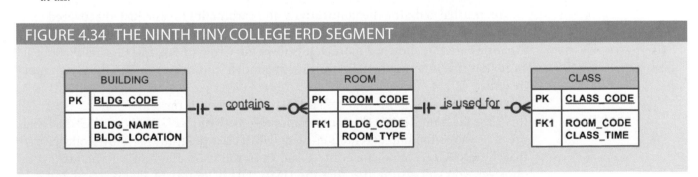

FIGURE 4.34 THE NINTH TINY COLLEGE ERD SEGMENT

Using the preceding summary, you can identify the following entities:

PROFESSOR	SCHOOL	DEPARTMENT
COURSE	CLASS	SEMESTER
STUDENT	BUILDING	ROOM
ENROLL (the associative entity between STUDENT and CLASS)		

Once you have discovered the relevant entities, you can define the initial set of relationships among them. Next, you describe the entity attributes. Identifying the attributes of the entities helps you to better understand the relationships among entities. Table 4.4 summarizes the ERM's components, and names the entities and their relations.

TABLE 4.4

COMPONENTS OF THE ERM

ENTITY	RELATIONSHIP	CONNECTIVITY	ENTITY
SCHOOL	operates	1:M	DEPARTMENT
DEPARTMENT	has	1:M	STUDENT
DEPARTMENT	employs	1:M	PROFESSOR
DEPARTMENT	offers	1:M	COURSE
COURSE	generates	1:M	CLASS
SEMESTER	includes	1:M	CLASS
PROFESSOR	is dean of	1:1	SCHOOL
PROFESSOR	chairs	1:1	DEPARTMENT
PROFESSOR	teaches	1:M	CLASS
PROFESSOR	advises	1:M	STUDENT
STUDENT	enrolls in	M:N	CLASS
BUILDING	contains	1:M	ROOM
ROOM	is used for	1:M	CLASS

Note: ENROLL is the composite entity that implements the M:N relationship "STUDENT enrolls in CLASS."

You must also define the connectivity and cardinality for the just-discovered relations based on the business rules. However, to avoid crowding the diagram, the cardinalities are not shown. Figure 4.35 shows the Crow's Foot ERD for Tiny College. Note that this is an implementation-ready model, so it shows the ENROLL composite entity.

Figure 4.36 shows the conceptual UML class diagram for Tiny College. Note that this class diagram depicts the M:N relationship between STUDENT and CLASS. Figure 4.37 shows the implementation-ready UML class diagram for Tiny College (note that the ENROLL composite entity is shown in this class diagram). If you are a good observer, you will also notice that the UML class diagrams in Figures 4.36 and 4.37 show the entity and attribute names but do not identify the primary key attributes. The reason goes back to UML's roots. UML class diagrams are an object-oriented modeling language, and therefore do not support the notion of "primary or foreign keys" found mainly in the relational world. Rather, in the object-oriented world, objects inherit a unique object identifier at creation time. For more information, see Appendix G, Object-Oriented Databases.

FIGURE 4.35 THE COMPLETED TINY COLLEGE ERD

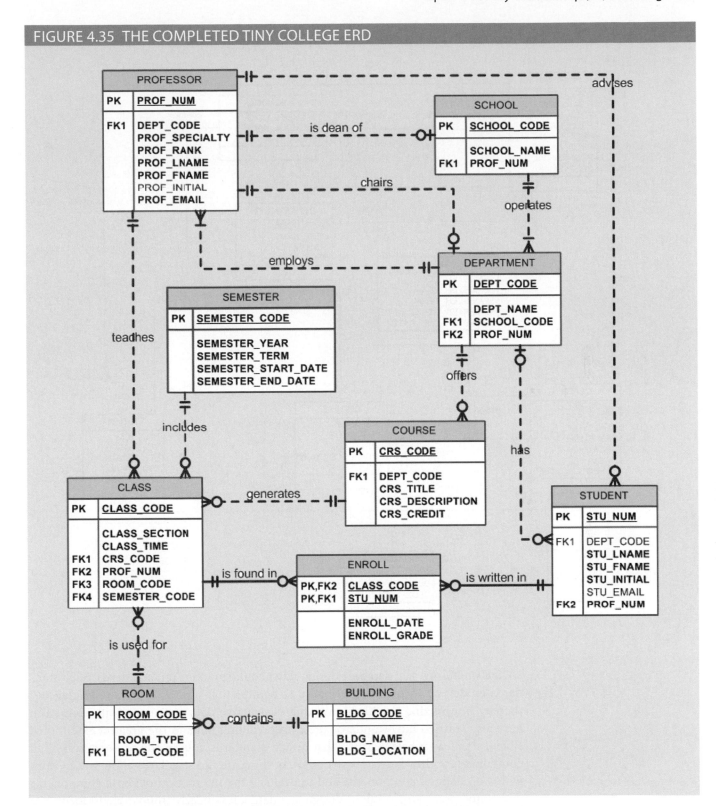

4-3 Database Design Challenges: Conflicting Goals

Database designers must often make design compromises that are triggered by conflicting goals, such as adherence to design standards (design elegance), processing speed, and information requirements.

FIGURE 4.36 THE CONCEPTUAL UML CLASS DIAGRAM FOR TINY COLLEGE

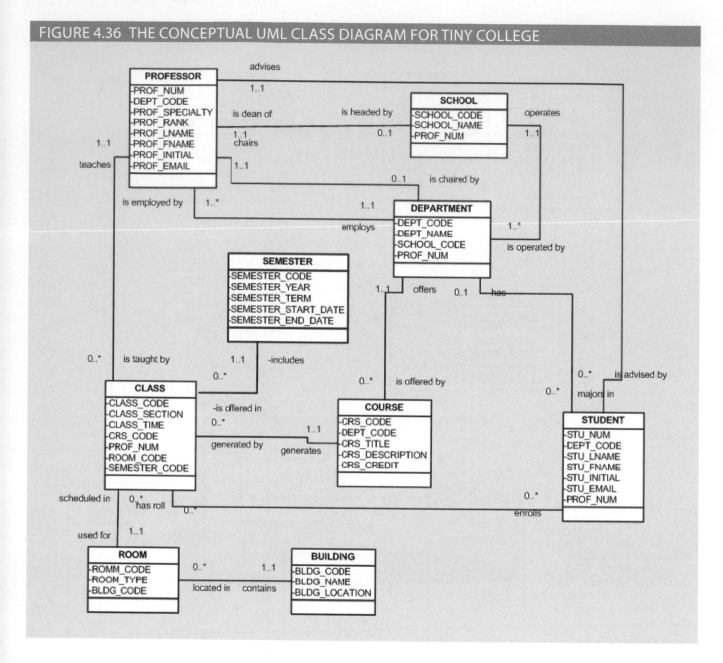

- *Design standards.* The database design must conform to design standards. Such standards guide you in developing logical structures that minimize data redundancies, thereby minimizing the likelihood that destructive data anomalies will occur. You have also learned how standards prescribe avoiding nulls to the greatest extent possible. In fact, you have learned that design standards govern the presentation of all components within the database design. In short, design standards allow you to work with well-defined components and to evaluate the interaction of those components with some precision. Without design standards, it is nearly impossible to formulate a proper design process, to evaluate an existing design, or to trace the likely logical impact of changes in design.

- *Processing speed.* In many organizations, particularly those that generate large numbers of transactions, high processing speeds are often a top priority in database design. High processing speed means minimal access time, which may be achieved by minimizing the number and complexity of logically desirable relationships.

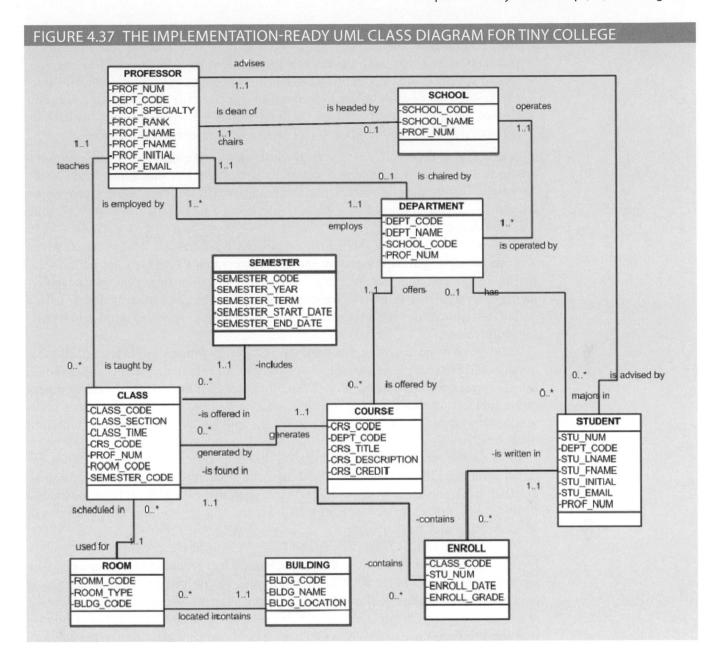

FIGURE 4.37 THE IMPLEMENTATION-READY UML CLASS DIAGRAM FOR TINY COLLEGE

For example, a "perfect" design might use a 1:1 relationship to avoid nulls, while a design that emphasizes higher transaction speed might combine the two tables to avoid the use of an additional relationship, using dummy entries to avoid the nulls. If the focus is on data-retrieval speed, you might also be forced to include derived attributes in the design.

• *Information requirements.* The quest for timely information might be the focus of database design. Complex information requirements may dictate data transformations, and they may expand the number of entities and attributes within the design. Therefore, the database may have to sacrifice some of its "clean" design structures and high transaction speed to ensure maximum information generation. For example, suppose that a detailed sales report must be generated periodically. The sales report includes all invoice subtotals, taxes, and totals; even the invoice lines include subtotals. If the sales report includes hundreds of thousands (or even millions) of invoices, computing the totals, taxes, and subtotals is likely to take some time. If those computations had been made and the results had been stored as derived attributes in the INVOICE and LINE tables at the time of the transaction, the real-time transaction speed might have declined. However, that loss of speed would only be noticeable if there were many simultaneous transactions. The cost of a slight loss of transaction speed at the front end and the addition of multiple derived attributes is likely to pay off when the sales reports are generated (not to mention that it will be simpler to generate the queries).

A design that meets all logical requirements and design conventions is an important goal. However, if this perfect design fails to meet the customer's transaction speed and information requirements, the designer will not have done a proper job from the end user's point of view. Compromises are a fact of life in the real world of database design.

Even while focusing on the entities, attributes, relationships, and constraints, the designer should begin thinking about end-user requirements such as performance, security, shared access, and data integrity. The designer must consider processing requirements and verify that all update, retrieval, and deletion options are available. Finally, a design is of little value unless the end product can deliver all specified query and reporting requirements.

You will probably discover that even the best design process produces an ERD that requires further changes mandated by operational requirements. Such changes should not discourage you from using the process. ER modeling is essential in the development of a sound design that can meet the demands of adjustment and growth. Using ERDs yields perhaps the richest bonus of all: a thorough understanding of how an organization really functions.

Occasionally, design and implementation problems do not yield "clean" implementation solutions. To get a sense of the design and implementation choices a database designer faces, you will revisit the 1:1 recursive relationship "EMPLOYEE is married to EMPLOYEE," first examined in Figure 4.18. Figure 4.38 shows three different ways of implementing such a relationship.

Note that the EMPLOYEE_V1 table in Figure 4.38 is likely to yield data anomalies. For example, if Anne Jones divorces Anton Shapiro, two records must be updated—by setting the respective EMP_SPOUSE values to null—to properly reflect that change. If only one record is updated, inconsistent data occurs. The problem becomes even worse if several of the divorced employees then marry each other. In addition, that implementation also produces undesirable nulls for employees who are *not* married to other employees in the company.

Another approach would be to create a new entity shown as MARRIED_V1 in a 1:M relationship with EMPLOYEE. (See Figure 4.38.) This second implementation does eliminate the nulls for employees who are not married to other employees in the same company. (Such employees would not be entered in the MARRIED_V1 table.) However, this approach still yields possible duplicate values. For example, the marriage between employees 345 and 347 may still appear twice, once as 345,347 and once as 347,345. (Because each of those permutations is unique the first time it appears, the creation of a unique index will not solve the problem.)

As you can see, the first two implementations yield several problems:

- Both solutions use synonyms. The EMPLOYEE_V1 table uses EMP_NUM and EMP_SPOUSE to refer to an employee. The MARRIED_V1 table uses the same synonyms.

- Both solutions are likely to produce redundant data. For example, it is possible to enter employee 345 as married to employee 347 and to enter employee 347 as married to employee 345.

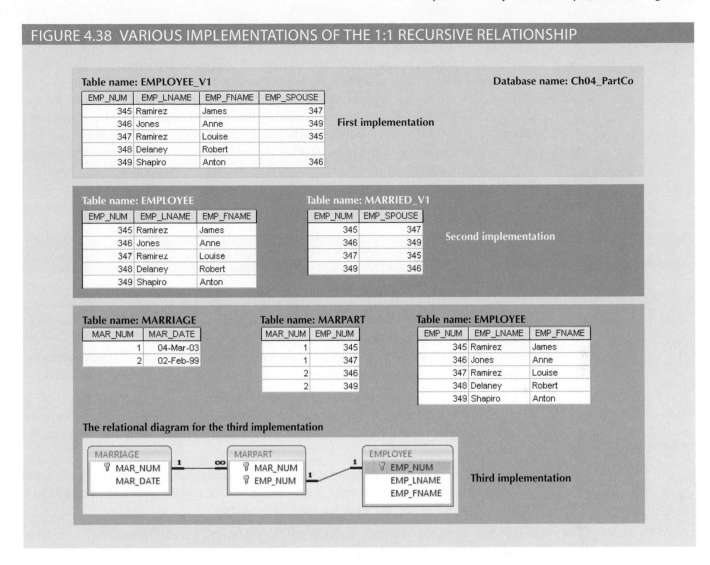

FIGURE 4.38 VARIOUS IMPLEMENTATIONS OF THE 1:1 RECURSIVE RELATIONSHIP

- Both solutions are likely to produce inconsistent data. For example, it is possible to have data pairs such as 345,347 and 348,345 and 347,349, none of which will violate entity integrity requirements because they are all unique. However, this solution would allow any one employee to be married to multiple employees.

A third approach would be to have two new entities, MARRIAGE and MARPART, in a 1:M relationship. MARPART contains the EMP_NUM foreign key to EMPLOYEE. (See the relational diagram in Figure 4.38.) However, even this approach has issues. It requires the collection of additional data regarding the employees' marriage—the marriage date. If the business users do not need this data, then requiring them to collect it would be inappropriate. To ensure that an employee occurs only once in any given marriage, you would have to create a unique index on the EMP_NUM attribute in the MARPART table. Another potential problem with this solution is that the database implementation would theoretically allow more than two employees to "participate" in the same marriage.

As you can see, a recursive 1:1 relationship yields many different solutions with varying degrees of effectiveness and adherence to basic design principles. Any of the preceding solutions would likely involve the creation of program code to help ensure the integrity and consistency of the data. In a later chapter, you will examine the creation of database triggers that can do exactly that. Your job as a database designer is to use your professional judgment to yield a solution that meets the requirements imposed by business rules, processing requirements, and basic design principles.

Finally, document, document, and document! Put all design activities in writing, and then review what you have written. Documentation not only helps you stay on track during the design process, it also enables you and your coworkers to pick up the design thread when the time comes to modify the design. Although the need for documentation should be obvious, one of the most vexing problems in database and systems analysis work is that this need is often ignored in the design and implementation stages. The development of organizational documentation standards is an important aspect of ensuring data compatibility and coherence.

Summary

- The ERM uses ERDs to represent the conceptual database as viewed by the end user. The ERM's main components are entities, relationships, and attributes. The ERD includes connectivity and cardinality notations, and can also show relationship strength, relationship participation (optional or mandatory), and degree of relationship (such as unary, binary, or ternary).

- Connectivity describes the relationship classification (1:1, 1:M, or M:N). Cardinality expresses the specific number of entity occurrences associated with an occurrence of a related entity. Connectivities and cardinalities are usually based on business rules.

- In the ERM, an M:N relationship is valid at the conceptual level. However, when implementing the ERM in a relational database, the M:N relationship must be mapped to a set of 1:M relationships through a composite entity.

- ERDs may be based on many different ERMs. However, regardless of which model is selected, the modeling logic remains the same. Because no ERM can accurately portray all real-world data and action constraints, application software must be used to augment the implementation of at least some of the business rules.

- Unified Modeling Language (UML) class diagrams are used to represent the static data structures in a data model. The symbols used in the UML class and ER diagrams are very similar. The UML class diagrams can be used to depict data models at the conceptual or implementation abstraction levels.

- Database designers, no matter how well they can produce designs that conform to all applicable modeling conventions, are often forced to make design compromises. Those compromises are required when end users have vital transaction-speed and information requirements that prevent the use of "perfect" modeling logic and adherence to all modeling conventions. Therefore, database designers must use their professional judgment to determine how and to what extent the modeling conventions are subject to modification. To ensure that their professional judgments are sound, database designers must have detailed and in-depth knowledge of data-modeling conventions. It is also important to document the design process from beginning to end, which helps keep the design process on track and allows for easy modifications in the future.

Key Terms

binary relationship	mandatory participation	simple attribute
cardinality	multivalued attribute	single-valued attribute
composite attribute	optional attribute	strong entity
composite identifier	optional participation	strong (identifying) relationship
connectivity	participants	ternary relationship
derived attribute	recursive relationship	unary relationship
existence-dependent	regular entity	weak entity
existence-independent	relational schema	weak (non-identifying) relationship
identifier	relationship degree	
iterative process	required attribute	

Online Content

Flashcards and crossword puzzles for key term practice are available at *www.cengagebrain.com*.

Review Questions

1. What two conditions must be met before an entity can be classified as a weak entity? Give an example of a weak entity.

2. What is a strong (or identifying) relationship, and how is it depicted in a Crow's Foot ERD?

3. Given the business rule "an employee may have many degrees," discuss its effect on attributes, entities, and relationships. (*Hint:* Remember what a multivalued attribute is and how it might be implemented.)

4. What is a composite entity, and when is it used?

5. Suppose you are working within the framework of the conceptual model in Figure Q4.5.

FIGURE Q4.5 THE CONCEPTUAL MODEL FOR QUESTION 5

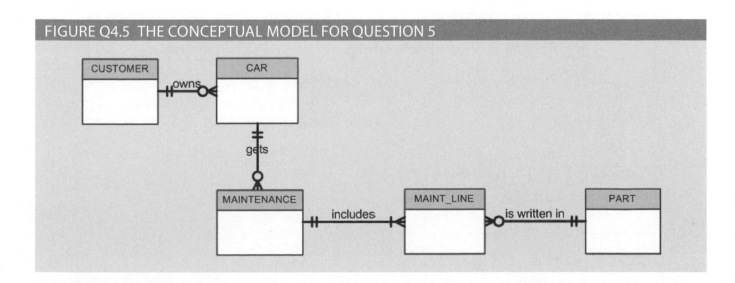

Given the conceptual model in Figure Q4.5:

    a. Write the business rules that are reflected in it.
    b. Identify all of the cardinalities.

6. What is a recursive relationship? Give an example.

7. How would you (graphically) identify each of the following ERM components in a Crow's Foot notation?

    a. an entity
    b. the cardinality (0,N)
    c. a weak relationship
    d. a strong relationship

8. Discuss the difference between a composite key and a composite attribute. How would each be indicated in an ERD?

9. What two courses of action are available to a designer who encounters a multivalued attribute?

10. What is a derived attribute? Give an example.

11. How is a relationship between entities indicated in an ERD? Give an example using the Crow's Foot notation.

12. Discuss two ways in which the 1:M relationship between COURSE and CLASS can be implemented. (*Hint:* Think about relationship strength.)

13. How is a composite entity represented in an ERD, and what is its function? Illustrate the Crow's Foot notation.

14. What three (often conflicting) database requirements must be addressed in database design?

15. Briefly, but precisely, explain the difference between single-valued attributes and simple attributes. Give an example of each.

16. What are multivalued attributes, and how can they be handled within the database design?

Questions 17–20 are based on the ERD in Figure Q4.17.

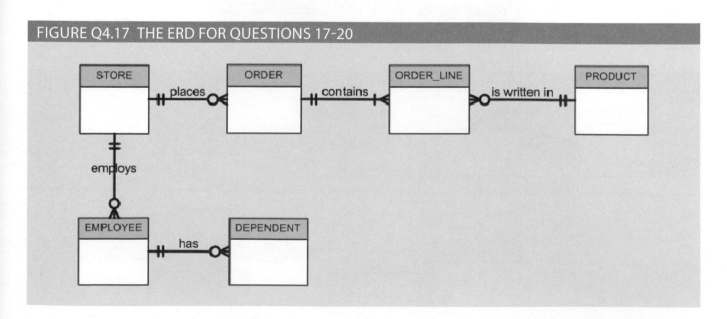

FIGURE Q4.17 THE ERD FOR QUESTIONS 17-20

17. Write the 10 cardinalities that are appropriate for this ERD.

18. Write the business rules reflected in this ERD.

19. What two attributes must be contained in the composite entity between STORE and PRODUCT? Use proper terminology in your answer.

20. Describe precisely the composition of the DEPENDENT weak entity's primary key. Use proper terminology in your answer.

21. The local city youth league needs a database system to help track children who sign up to play soccer. Data needs to be kept on each team, the children who will play on each team, and their parents. Also, data needs to be kept on the coaches for each team.

    Draw a data model with the entities and attributes described here.

    Entities required: Team, Player, Coach, and Parent

    Attributes required:

    Team: Team ID number, Team name, and Team colors

    Player: Player ID number, Player first name, Player last name, and Player age

    Coach: Coach ID number, Coach first name, Coach last name, and Coach home phone number

    Parent: Parent ID number, Parent last name, Parent first name, Home phone number, and Home address (Street, City, State, and Zip code)

    The following relationships must be defined:

    - Team is related to Player.
    - Team is related to Coach.
    - Player is related to Parent.

    Connectivities and participations are defined as follows:

    - A Team may or may not have a Player.
    - A Player must have a Team.
    - A Team may have many Players.
    - A Player has only one Team.
    - A Team may or may not have a Coach.
    - A Coach must have a Team.
    - A Team may have many Coaches.
    - A Coach has only one Team.
    - A Player must have a Parent.
    - A Parent must have a Player.
    - A Player may have many Parents.
    - A Parent may have many Players.

Problems

1. Use the following business rules to create a Crow's Foot ERD. Write all appropriate connectivities and cardinalities in the ERD.

   - A department employs many employees, but each employee is employed by only one department.

   - Some employees, known as "rovers," are not assigned to any department.

   - A division operates many departments, but each department is operated by only one division.

   - An employee may be assigned many projects, and a project may have many employees assigned to it.

   - A project must have at least one employee assigned to it.

   - One of the employees manages each department, and each department is managed by only one employee.

   - One of the employees runs each division, and each division is run by only one employee.

2. Create a complete ERD in Crow's Foot notation that can be implemented in the relational model using the following description of operations. Hot Water (HW) is a small start-up company that sells spas. HW does not carry any stock. A few spas are set up in a simple warehouse so customers can see some of the models available, but any products sold must be ordered at the time of the sale.

   - HW can get spas from several different manufacturers.

   - Each manufacturer produces one or more different brands of spas.

   - Each and every brand is produced by only one manufacturer.

   - Every brand has one or more models.

   - Every model is produced as part of a brand. For example, Iguana Bay Spas is a manufacturer that produces Big Blue Iguana spas, a premium-level brand, and Lazy Lizard spas, an entry-level brand. The Big Blue Iguana brand offers several models, including the BBI-6, an 81-jet spa with two 6-hp motors, and the BBI-10, a 102-jet spa with three 6-hp motors.

   - Every manufacturer is identified by a manufacturer code. The company name, address, area code, phone number, and account number are kept in the system for every manufacturer.

   - For each brand, the brand name and brand level (premium, mid-level, or entry-level) are kept in the system.

   - For each model, the model number, number of jets, number of motors, number of horsepower per motor, suggested retail price, HW retail price, dry weight, water capacity, and seating capacity must be kept in the system.

3. The Jonesburgh County Basketball Conference (JCBC) is an amateur basketball association. Each city in the county has one team as its representative. Each team has a maximum of 12 players and a minimum of 9 players. Each team also has up to 3 coaches (offensive, defensive, and physical training coaches). During the season,

each team plays 2 games (home and visitor) against each of the other teams. Given those conditions, do the following:

- Identify the connectivity of each relationship.
- Identify the type of dependency that exists between CITY and TEAM.
- Identify the cardinality between teams and players and between teams and city.
- Identify the dependency between COACH and TEAM and between TEAM and PLAYER.
- Draw the Chen and Crow's Foot ERDs to represent the JCBC database.
- Draw the UML class diagram to depict the JCBC database.

4. Create an ERD based on the Crow's Foot notation using the following requirements:

- An INVOICE is written by a SALESREP. Each sales representative can write many invoices, but each invoice is written by a single sales representative.
- The INVOICE is written for a single CUSTOMER. However, each customer can have many invoices.
- An INVOICE can include many detail lines (LINE), each of which describes one product bought by the customer.
- The product information is stored in a PRODUCT entity.
- The product's vendor information is found in a VENDOR entity.

5. The Hudson Engineering Group (HEG) has contacted you to create a conceptual model whose application will meet the expected database requirements for the company's training program. The HEG administrator gives you the following description of the training group's operating environment. (*Hint:* Some of the following sentences identify the volume of data rather than cardinalities. Can you tell which ones?)

The HEG has 12 instructors and can handle up to 30 trainees per class. HEG offers 5 Advanced Technology courses, each of which may generate several classes. If a class has fewer than 10 trainees, it will be canceled. Therefore, it is possible for a course not to generate any classes. Each class is taught by one instructor. Each instructor may teach up to 2 classes or may be assigned to do research only. Each trainee may take up to 2 classes per year.

Given that information, do the following:

a. Define all of the entities and relationships. (Use Table 4.4 as your guide.)

b. Describe the relationship between instructor and class in terms of connectivity, cardinality, and existence dependence.

6. Automata, Inc., produces specialty vehicles by contract. The company operates several departments, each of which builds a particular vehicle, such as a limousine, truck, van, or RV.

- Before a new vehicle is built, the department places an order with the purchasing department to request specific components. Automata's purchasing department is interested in creating a database to keep track of orders and to accelerate the process of delivering materials.
- The order received by the purchasing department may contain several different items. An inventory is maintained so the most frequently requested items are delivered almost immediately. When an order comes in, it is checked to determine whether the requested item is in inventory. If an item is not in inventory, it must be ordered from a supplier. Each item may have several suppliers.

Given that functional description of the processes at Automata's purchasing department, do the following:

a. Identify all of the main entities.

b. Identify all of the relations and connectivities among entities.

c. Identify the type of existence dependence in all the relationships.

d. Give at least two examples of the types of reports that can be obtained from the database.

7. United Helpers is a nonprofit organization that provides aid to people after natural disasters. Based on the following brief description of operations, create the appropriate fully labeled Crow's Foot ERD.

- Volunteers carry out the tasks of the organization. The name, address, and telephone number are tracked for each volunteer. Each volunteer may be assigned to several tasks, and some tasks require many volunteers. A volunteer might be in the system without having been assigned a task yet. It is possible to have tasks that no one has been assigned. When a volunteer is assigned to a task, the system should track the start time and end time of that assignment.

- Each task has a task code, task description, task type, and task status. For example, there may be a task with task code "101," a description of "answer the telephone," a type of "recurring," and a status of "ongoing." Another task might have a code of "102," a description of "prepare 5,000 packages of basic medical supplies," a type of "packing," and a status of "open."

- For all tasks of type "packing," there is a packing list that specifies the contents of the packages. There are many packing lists to produce different packages, such as basic medical packages, child-care packages, and food packages. Each packing list has an ID number, a packing list name, and a packing list description, which describes the items that should make up the package. Every packing task is associated with only one packing list. A packing list may not be associated with any tasks, or it may be associated with many tasks. Tasks that are not packing tasks are not associated with any packing list.

- Packing tasks result in the creation of packages. Each individual package of supplies produced by the organization is tracked, and each package is assigned an ID number. The date the package was created and its total weight are recorded. A given package is associated with only one task. Some tasks (such as "answer the phones") will not produce any packages, while other tasks (such as "prepare 5,000 packages of basic medical supplies") will be associated with many packages.

- The packing list describes the *ideal* contents of each package, but it is not always possible to include the ideal number of each item. Therefore, the actual items included in each package should be tracked. A package can contain many different items, and a given item can be used in many different packages.

- Each item that the organization provides has an item ID number, item description, item value, and item quantity on hand stored in the system. Along with tracking the actual items that are placed in each package, the quantity of each item placed in the package must be tracked as well. For example, a packing list may state that basic medical packages should include 100 bandages, 4 bottles of iodine, and 4 bottles of hydrogen peroxide. However, because of the limited supply of items, a given package may include only 10 bandages, 1 bottle of iodine, and no hydrogen peroxide. The fact that the package includes bandages and iodine needs to be recorded along with the quantity of each item included. It is possible for the organization to have items that have not been included in any package yet, but every package will contain at least one item.

8. Using the Crow's Foot notation, create an ERD that can be implemented for a medical clinic using the following business rules:

   - A patient can make many appointments with one or more doctors in the clinic, and a doctor can accept appointments with many patients. However, each appointment is made with only one doctor and one patient.

   - Emergency cases do not require an appointment. However, for appointment management purposes, an emergency is entered in the appointment book as "unscheduled."

   - If kept, an appointment yields a visit with the doctor specified in the appointment. The visit yields a diagnosis and, when appropriate, treatment.

   - With each visit, the patient's records are updated to provide a medical history.

   - Each patient visit creates a bill. Each patient visit is billed by one doctor, and each doctor can bill many patients.

   - Each bill must be paid. However, a bill may be paid in many installments, and a payment may cover more than one bill.

   - A patient may pay the bill directly, or the bill may be the basis for a claim submitted to an insurance company.

   - If the bill is paid by an insurance company, the deductible is submitted to the patient for payment.

9. Create a Crow's Foot notation ERD to support the following business operations:

   - A friend of yours has opened Professional Electronics and Repairs (PEAR) to repair smartphones, laptops, tablets, and MP3 players. She wants you to create a database to help her run her business.

   - When a customer brings a device to PEAR for repair, data must be recorded about the customer, the device, and the repair. The customer's name, address, and a contact phone number must be recorded (if the customer has used the shop before, the information already in the system for the customer is verified as being current). For the device to be repaired, the type of device, model, and serial number are recorded (or verified if the device is already in the system). Only customers who have brought devices into PEAR for repair will be included in this system.

   - Since a customer might sell an older device to someone else who then brings the device to PEAR for repair, it is possible for a device to be brought in for repair by more than one customer. However, each repair is associated with only one customer. When a customer brings in a device to be fixed, it is referred to as a repair request, or just "repair," for short. Each repair request is given a reference number, which is recorded in the system along with the date of the request, and a description of the problem(s) that the customer wants fixed. It is possible for a device to be brought to the shop for repair many different times, and only devices that are brought in for repair are recorded in the system. Each repair request is for the repair of one and only one device. If a customer needs multiple devices fixed, then each device will require its own repair request.

   - There are a limited number of repair services that PEAR can perform. For each repair service, there is a service ID number, description, and charge. "Charge" is how much the customer is charged for the shop to perform the service, including any parts used. The actual repair of a device is the performance of the services necessary to address the problems described by the customer. Completing a repair request may require the performance of many services. Each service can be performed many different times during the repair of different devices, but each service will be performed only once during a given repair request.

- All repairs eventually require the performance of at least one service, but which services will be required may not be known at the time the repair request is made. It is possible for services to be available at PEAR but that have never been required in performing any repair.

- Some services involve only labor activities and no parts are required, but most services require the replacement of one or more parts. The quantity of each part required in the performance of each service should also be recorded. For each part, the part number, part description, quantity in stock, and cost is recorded in the system. The cost indicated is the amount that PEAR pays for the part. Some parts may be used in more than one service, but each part is required for at least one service.

10. Luxury-Oriented Scenic Tours (LOST) provides guided tours to groups of visitors to the Washington D.C. area. In recent years, LOST has grown quickly and is having difficulty keeping up with all of the various information needs of the company. The company's operations are as follows:

- LOST offers many different tours. For each tour, the tour name, approximate length (in hours), and fee charged is needed. Guides are identified by an employee ID, but the system should also record a guide's name, home address, and date of hire. Guides take a test to be qualified to lead specific tours. It is important to know which guides are qualified to lead which tours and the date that they completed the qualification test for each tour. A guide may be qualified to lead many different tours. A tour can have many different qualified guides. New guides may or may not be qualified to lead any tours, just as a new tour may or may not have any qualified guides.

- Every tour must be designed to visit at least three locations. For each location, a name, type, and official description are kept. Some locations (such as the White House) are visited by more than one tour, while others (such as Arlington Cemetery) are visited by a single tour. All locations are visited by at least one tour. The order in which the tour visits each location should be tracked as well.

- When a tour is actually given, that is referred to as an "outing." LOST schedules outings well in advance so they can be advertised and so employees can understand their upcoming work schedules. A tour can have many scheduled outings, although newly designed tours may not have any outings scheduled. Each outing is for a single tour and is scheduled for a particular date and time. All outings must be associated with a tour. All tours at LOST are guided tours, so a guide must be assigned to each outing. Each outing has one and only one guide. Guides are occasionally asked to lead an outing of a tour even if they are not officially qualified to lead that tour. Newly hired guides may not have ever been scheduled to lead any outings. Tourists, called "clients" by LOST, pay to join a scheduled outing. For each client, the name and telephone number are recorded. Clients may sign up to join many different outings, and each outing can have many clients. Information is kept only on clients who have signed up for at least one outing, although newly scheduled outings may not have any clients signed up yet.

a. Create a Crow's Foot notation ERD to support LOST operations.

b. The operations provided state that it is possible for a guide to lead an outing of a tour even if the guide is not officially qualified to lead outings of that tour. Imagine that the business rules instead specified that a guide is never, under any circumstance, allowed to lead an outing unless he or she is qualified to lead outings of that tour. How could the data model in Part a. be modified to enforce this new constraint?

Note

You can use the following cases and additional problems from the Instructor Online Companion as the basis for class projects. These problems illustrate the challenge of translating a description of operations into a set of business rules that will define the components for an ERD you can implement successfully. These problems can also be used as the basis for discussions about the components and contents of a proper description of operations. If you want to create databases that can be successfully implemented, you must learn to separate the generic background material from the details that directly affect database design. You must also keep in mind that many constraints cannot be incorporated into the database design; instead, such constraints are handled by the application software.

Cases

11. The administrators of Tiny College are so pleased with your design and implementation of their student registration and tracking system that they want you to expand the design to include the database for their motor vehicle pool. A brief description of operations follows:

- Faculty members may use the vehicles owned by Tiny College for officially sanctioned travel. For example, the vehicles may be used by faculty members to travel to off-campus learning centers, to travel to locations at which research papers are presented, to transport students to officially sanctioned locations, and to travel for public service purposes. The vehicles used for such purposes are managed by Tiny College's Travel Far But Slowly (TFBS) Center.

- Using reservation forms, each department can reserve vehicles for its faculty, who are responsible for filling out the appropriate trip completion form at the end of a trip. The reservation form includes the expected departure date, vehicle type required, destination, and name of the authorized faculty member. The faculty member who picks up a vehicle must sign a checkout form to log out the vehicle and pick up a trip completion form. (The TFBS employee who releases the vehicle for use also signs the checkout form.) The faculty member's trip completion form includes the faculty member's identification code, the vehicle's identification, the odometer readings at the start and end of the trip, maintenance complaints (if any), gallons of fuel purchased (if any), and the Tiny College credit card number used to pay for the fuel. If fuel is purchased, the credit card receipt must be stapled to the trip completion form. Upon receipt of the trip completion form, the faculty member's department is billed at a mileage rate based on the vehicle type used: sedan, station wagon, panel truck, minivan, or minibus. (*Hint:* Do *not* use more entities than are necessary. Remember the difference between attributes and entities!)

- All vehicle maintenance is performed by TFBS. Each time a vehicle requires maintenance, a maintenance log entry is completed on a prenumbered maintenance log form. The maintenance log form includes the vehicle identification, brief description of the type of maintenance required, initial log entry date, date the maintenance was completed, and name of the mechanic who released the vehicle back into service. (Only mechanics who have an inspection authorization may release a vehicle back into service.)

- As soon as the log form has been initiated, the log form's number is transferred to a maintenance detail form; the log form's number is also forwarded to the parts department manager, who fills out a parts usage form on which the maintenance log number is recorded. The maintenance detail form contains separate lines for each maintenance item performed, for the parts used, and for identification of the mechanic who performed the maintenance. When all maintenance items have been completed, the maintenance detail form is stapled to the maintenance log form, the maintenance log form's completion date is filled out, and the mechanic who releases the vehicle back into service signs the form. The stapled forms are then filed, to be used later as the source for various maintenance reports.

- TFBS maintains a parts inventory, including oil, oil filters, air filters, and belts of various types. The parts inventory is checked daily to monitor parts usage and to reorder parts that reach the "minimum quantity on hand" level. To track parts usage, the parts manager requires each mechanic to sign out the parts that are used to perform each vehicle's maintenance; the parts manager records the maintenance log number under which the part is used.

- Each month TFBS issues a set of reports. The reports include the mileage driven by vehicle, by department, and by faculty members within a department. In addition, various revenue reports are generated by vehicle and department. A detailed parts usage report is also filed each month. Finally, a vehicle maintenance summary is created each month.

Given that brief summary of operations, draw the appropriate (and fully labeled) ERD. Use the Crow's foot methodology to indicate entities, relationships, connectivities, and participations.

12. During peak periods, Temporary Employment Corporation (TEC) places temporary workers in companies. TEC's manager gives you the following description of the business:

- TEC has a file of candidates who are willing to work.

- Any candidate who has worked before has a specific job history. (Naturally, no job history exists if the candidate has never worked.) Each time the candidate works, one additional job history record is created.

- Each candidate has earned several qualifications. Each qualification may be earned by more than one candidate. (For example, more than one candidate may have earned a Bachelor of Business Administration degree or a Microsoft Network Certification, and clearly a candidate may have earned both a BBA and a Microsoft Network Certification.)

- TEC offers courses to help candidates improve their qualifications.

- Every course develops one specific qualification; however, TEC does not offer a course for every qualification. Some qualifications are developed through multiple courses.

- Some courses cover advanced topics that require specific qualifications as prerequisites. Some courses cover basic topics that do not require any prerequisite qualifications. A course can have several prerequisites. A qualification can be a prerequisite for more than one course.

- Courses are taught during training sessions. A training session is the presentation of a single course. Over time, TEC will offer many training sessions for each course; however, new courses may not have any training sessions scheduled right away.

- Candidates can pay a fee to attend a training session. A training session can accommodate several candidates, although new training sessions will not have any candidates registered at first.

- TEC also has a list of companies that request temporaries.

- Each time a company requests a temporary employee, TEC makes an entry in the Openings folder. That folder contains an opening number, a company name, required qualifications, a starting date, an anticipated ending date, and hourly pay.

- Each opening requires only one specific or main qualification.

- When a candidate matches the qualification, the job is assigned, and an entry is made in the Placement Record folder. The folder contains such information as an opening number, candidate number, and total hours worked. In addition, an entry is made in the job history for the candidate.

- An opening can be filled by many candidates, and a candidate can fill many openings.

- TEC uses special codes to describe a candidate's qualifications for an opening. The list of codes is shown in Table P4.12.

TABLE P4.12

CODE	DESCRIPTION
SEC-45	Secretarial work; candidate must type at least 45 words per minute
SEC-60	Secretarial work; candidate must type at least 60 words per minute
CLERK	General clerking work
PRG-VB	Programmer, Visual Basic
PRG-C++	Programmer, C++
DBA-ORA	Database Administrator, Oracle
DBA-DB2	Database Administrator, IBM DB2
DBA-SQLSERV	Database Administrator, MS SQL Server
SYS-1	Systems Analyst, level 1
SYS-2	Systems Analyst, level 2
NW-NOV	Network Administrator, Novell experience
WD-CF	Web Developer, ColdFusion

TEC's management wants to keep track of the following entities:

COMPANY, OPENING, QUALIFICATION, CANDIDATE, JOB_HISTORY, PLACEMENT, COURSE, and SESSION. Given that information, do the following:

a. Draw the Crow's Foot ERDs for this enterprise.

b. Identify all necessary relationships.

c. Identify the connectivity for each relationship.

d. Identify the mandatory and optional dependencies for the relationships.

e. Resolve all M:N relationships.

13. Use the following description of the operations of the RC_Charter2 Company to complete this exercise.

- The RC_Charter2 Company operates a fleet of aircraft under the Federal Air Regulations (FAR) Part 135 (air taxi or charter) certificate, enforced by the FAA. The aircraft are available for air taxi (charter) operations within the United States and Canada.

- Charter companies provide so-called unscheduled operations—that is, charter flights take place only after a customer reserves the use of an aircraft at a designated date and time to fly to one or more designated destinations; the aircraft transports passengers, cargo, or some combination of passengers and cargo. Of course, a customer can reserve many different charter trips during any time frame. However, for billing purposes, each charter trip is reserved by one and only one customer. Some of RC_Charter2's customers do not use the company's charter operations; instead, they purchase fuel, use maintenance services, or use other RC_Charter2 services. However, this database design will focus on the charter operations only.

- Each charter trip yields revenue for the RC_Charter2 Company. This revenue is generated by the charges a customer pays upon the completion of a flight. The charter flight charges are a function of aircraft model used, distance flown, waiting time, special customer requirements, and crew expenses. The distance flown charges are computed by multiplying the round-trip miles by the model's charge per mile. Round-trip miles are based on the actual navigational path flown. The sample route traced in Figure P4.13 illustrates the procedure. Note that the number of round-trip miles is calculated to be 130 + 200 + 180 + 390 = 900.

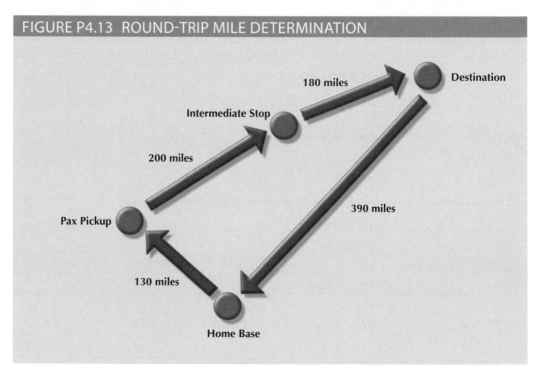

FIGURE P4.13 ROUND-TRIP MILE DETERMINATION

- Depending on whether a customer has RC_Charter2 credit authorization, the customer may do the following:

  a. Pay the entire charter bill upon the completion of the charter flight.

  b. Pay a part of the charter bill and charge the remainder to the account. The charge amount may not exceed the available credit.

c. Charge the entire charter bill to the account. The charge amount may not exceed the available credit.

d. Customers may pay all or part of the existing balance for previous charter trips. Such payments may be made at any time and are not necessarily tied to a specific charter trip. The charter mileage charge includes the expense of the pilot(s) and other crew required by FAR 135. However, if customers request *additional* crew *not* required by FAR 135, those customers are charged for the crew members on an hourly basis. The hourly crew-member charge is based on each crew member's qualifications.

e. The database must be able to handle crew assignments. Each charter trip requires the use of an aircraft, and a crew flies each aircraft. The smaller, piston-engine charter aircraft require a crew consisting of only a single pilot. All jets and other aircraft that have a gross takeoff weight of at least 12,500 pounds require a pilot and a copilot, while some of the larger aircraft used to transport passengers may require flight attendants as part of the crew. Some of the older aircraft require the assignment of a flight engineer, and larger cargo-carrying aircraft require the assignment of a loadmaster. In short, a crew can consist of more than one person, and not all crew members are pilots.

f. The charter flight's aircraft waiting charges are computed by multiplying the hours waited by the model's hourly waiting charge. Crew expenses are limited to meals, lodging, and ground transportation.

The RC_Charter2 database must be designed to generate a monthly summary of all charter trips, expenses, and revenues derived from the charter records. Such records are based on the data that each pilot in command is required to record for each charter trip: trip date(s) and time(s), destination(s), aircraft number, pilot data and other crew data, distance flown, fuel usage, and other data pertinent to the charter flight. Such charter data is then used to generate monthly reports that detail revenue and operating cost information for customers, aircraft, and pilots. All pilots and other crew members are RC_Charter2 Company employees; that is, the company does not use contract pilots and crew.

FAR Part 135 operations are conducted under a strict set of requirements that govern the licensing and training of crew members. For example, pilots must have earned either a commercial license or an Airline Transport Pilot (ATP) license. Both licenses require appropriate ratings, which are specific competency requirements. For example, consider the following:

- To operate a multiengine aircraft designed for takeoffs and landings on land only, the appropriate rating is MEL, or Multiengine Landplane. When a multiengine aircraft can take off and land on water, the appropriate rating is MES, or Multiengine Seaplane.

- The instrument rating is based on a demonstrated ability to conduct all flight operations with sole reference to cockpit instrumentation. The instrument rating is required to operate an aircraft under Instrument Meteorological Conditions (IMC), and all such operations are governed under FAR-specified Instrument Flight Rules (IFR). In contrast, operations conducted under "good weather" or *visual* flight conditions are based on the FAR Visual Flight Rules (VFR).

- The type rating is required for all aircraft with a takeoff weight of more than 12,500 pounds or for aircraft that are purely jet-powered. If an aircraft uses jet engines to drive propellers, that aircraft is said to be turboprop-powered. A turboprop—that is, a turbo-propeller-powered aircraft—does not require a type rating unless it meets the 12,500-pound weight limitation.

- Although pilot licenses and ratings are not time limited, exercising the privilege of the license and ratings under Part 135 requires both *a current medical certificate and a current Part 135 checkride.* The following distinctions are important:

  a. The medical certificate may be Class I or Class II. The Class I medical is more stringent than the Class II, and it must be renewed every six months. The Class II medical must be renewed yearly. If the Class I medical is not renewed during the six-month period, it automatically reverts to a Class II certificate. If the Class II medical is not renewed within the specified period, it automatically reverts to a Class III medical, which is not valid for commercial flight operations.

  b. A Part 135 checkride is a practical flight examination that must be successfully completed every six months. The checkride includes all flight maneuvers and procedures specified in Part 135.

Nonpilot crew members must also have the proper certificates to meet specific job requirements. For example, loadmasters need an appropriate certificate, as do flight attendants. Crew members such as loadmasters and flight attendants may be required in operations that involve large aircraft with a takeoff weight of more than 12,500 pounds and more than 19 passengers; these crew members are also required to pass a written and practical exam periodically. The RC_Charter2 Company is required to keep a complete record of all test types, dates, and results for each crew member, as well as examination dates for pilot medical certificates.

In addition, all flight crew members are required to submit to periodic drug testing; the results must be tracked as well. Note that nonpilot crew members are not required to take pilot-specific tests such as Part 135 checkrides, nor are pilots required to take crew tests such as loadmaster and flight attendant practical exams. However, many crew members have licenses and certifications in several areas. For example, a pilot may have an ATP and a loadmaster certificate. If that pilot is assigned to be a loadmaster on a given charter flight, the loadmaster certificate is required. Similarly, a flight attendant may have earned a commercial pilot's license. Sample data formats are shown in Table P4.13.

Pilots and other crew members must receive recurrency training appropriate to their work assignments. Recurrency training is based on an FAA-approved curriculum that is job specific. For example, pilot recurrency training includes a review of all applicable Part 135 flight rules and regulations, weather data interpretation, company flight operations requirements, and specified flight procedures. The RC_Charter2 Company is required to keep a complete record of all recurrency training for each crew member subject to the training.

The RC_Charter2 Company is required to maintain a detailed record of all crew credentials and all training mandated by Part 135. The company must keep a complete record of each requirement and of all compliance data.

To conduct a charter flight, the company must have a properly maintained aircraft available. A pilot who meets all of the FAA's licensing and currency requirements must fly the aircraft as Pilot in Command (PIC). For aircraft that are powered by piston engines or turboprops and have a gross takeoff weight under 12,500 pounds, single-pilot operations are permitted under Part 135 as long as a properly maintained autopilot is available. However, even if FAR Part 135 permits single-pilot operations, many customers require the presence of a copilot who is capable of conducting the flight operations under Part 135.

TABLE P4.13

PART A TESTS

TEST CODE	TEST DESCRIPTION	TEST FREQUENCY
1	Part 135 Flight Check	6 months
2	Medical, Class I	6 months
3	Medical, Class II	12 months
4	Loadmaster Practical	12 months
5	Flight Attendant Practical	12 months
6	Drug test	Random
7	Operations, written exam	6 months

PART B RESULTS

EMPLOYEE	TEST CODE	TEST DATE	TEST RESULT
101	1	12-Nov-15	Pass-1
103	6	23-Dec-15	Pass-1
112	4	23-Dec-15	Pass-2
103	7	11-Jan-16	Pass-1
112	7	16-Jan-16	Pass-1
101	7	16-Jan-16	Pass-1
101	6	11-Feb-16	Pass-2
125	2	15-Feb-16	Pass-1

PART C LICENSES AND CERTIFICATIONS

LICENSE OR CERTIFICATE	LICENSE OR CERTIFICATE DESCRIPTION
ATP	Airline Transport Pilot
Comm	Commercial license
Med-1	Medical certificate, Class I
Med-2	Medical certificate, Class II
Instr	Instrument rating
MEL	Multiengine Land aircraft rating
LM	Loadmaster
FA	Flight Attendant

EMPLOYEE	LICENSE OR CERTIFICATE	DATE EARNED
101	Comm	12-Nov-93
101	Instr	28-Jun-94
101	MEL	9-Aug-94
103	Comm	21-Dec-95
112	FA	23-Jun-02
103	Instr	18-Jan-96
112	LM	27-Nov-05

The RC_Charter2 operations manager anticipates the lease of turbojet-powered aircraft, which are required to have a crew consisting of a pilot and copilot. Both the pilot and copilot must meet the same Part 135 licensing, ratings, and training requirements.

The company also leases larger aircraft that exceed the 12,500-pound gross takeoff weight. Those aircraft might carry enough passengers to require the presence of one or more flight attendants. If those aircraft carry cargo that weighs more than 12,500 pounds, a loadmaster must be assigned as a crew member to supervise the loading and securing of the cargo. *The database must be designed to meet the anticipated capability for additional charter crew assignments.*

a. Given this incomplete description of operations, write all applicable business rules to establish entities, relationships, optionalities, connectivities, and cardinalities. (*Hint:* Use the following five business rules as examples, and write the remaining business rules in the same format.) A customer may request many charter trips.

- Each charter trip is requested by only one customer.

- Some customers have not yet requested a charter trip.

- An employee may be assigned to serve as a crew member on many charter trips.

- Each charter trip may have many employees assigned to serve as crew members.

b. Draw the fully labeled and implementable Crow's Foot ERD based on the business rules you wrote in Part a. of this problem. Include all entities, relationships, optionalities, connectivities, and cardinalities.

Chapter 5

Advanced Data Modeling

In this chapter, you will learn:
- About the extended entity relationship (EER) model
- How entity clusters are used to represent multiple entities and relationships
- The characteristics of good primary keys and how to select them
- How to use flexible solutions for special data-modeling cases

Preview

In the previous two chapters, you learned how to use entity relationship diagrams (ERDs) to properly create a data model. In this chapter, you will learn about the extended entity relationship (EER) model. The EER model builds on ER concepts and adds support for entity supertypes, subtypes, and entity clustering.

Most current database implementations are based on relational databases. Because the relational model uses keys to create associations among tables, it is essential to learn the characteristics of good primary keys and how to select them. Selecting a good primary key is too important to be left to chance, so this chapter covers the critical aspects of primary key identification and placement.

Focusing on practical database design, this chapter also illustrates some special design cases that highlight the importance of flexible designs, which can be adapted to meet the demands of changing data and information requirements. Data modeling is a vital step in the development of databases that in turn provides a good foundation for successful application development. Remember that good database applications cannot be based on bad database designs, and no amount of outstanding coding can overcome the limitations of poor database design.

Data Files and Available Formats

	MS Access	Oracle	MS SQL	My SQL		MS Access	Oracle	MS SQL	My SQL
CH05_AirCo	✓	✓	✓	✓	CH05_GCSdata	✓	✓	✓	✓
CH05_TinyCollege	✓	✓	✓	✓					

Data Files Available on cengagebrain.com

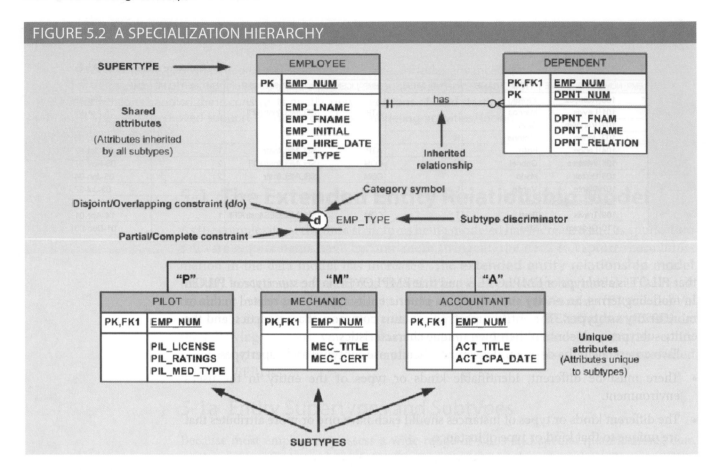

FIGURE 5.2   A SPECIALIZATION HIERARCHY

specialization hierarchy can have many levels of supertype/subtype relationships—that is, you can have a specialization hierarchy in which a supertype has many subtypes. In turn, one of the subtypes is the supertype to other lower-level subtypes.

As you can see in Figure 5.2, the arrangement of entity supertypes and subtypes in a specialization hierarchy is more than a cosmetic convenience. Specialization hierarchies enable the data model to capture additional semantic content (meaning) into the ERD. A specialization hierarchy provides the means to:

- Support attribute inheritance.

- Define a special supertype attribute known as the subtype discriminator.

- Define disjoint/overlapping constraints and complete/partial constraints.

The following sections cover such characteristics and constraints in more detail.

5-1c Inheritance

The property of **inheritance** enables an entity subtype to inherit the attributes and relationships of the supertype. As discussed earlier, a supertype contains attributes that are common to all of its subtypes. In contrast, subtypes contain only the attributes that are unique to the subtype. For example, Figure 5.2 illustrates that pilots, mechanics, and accountants all inherit the employee number, last name, first name, middle initial, and hire date from the EMPLOYEE entity. However, Figure 5.2 also illustrates that pilots have unique attributes; the same is true for mechanics and accountants. *One important inheritance characteristic is that all entity subtypes inherit their primary key attribute from their supertype.* Note in Figure 5.2 that the EMP_NUM attribute is the primary key for each of the subtypes.

At the implementation level, the supertype and its subtype(s) depicted in the specialization hierarchy maintain a 1:1 relationship. For example, the specialization hierarchy

Online Content

This chapter covers only specialization hierarchies. The EER model also supports specialization *lattices*, in which a subtype can have multiple parents (supertypes). However, those concepts are better covered under the object-oriented model in Appendix G, Object-Oriented Databases. The appendix is available at *www.cengagebrain.com*.

inheritance
In the EERD, the property that enables an entity subtype to inherit the attributes and relationships of the entity supertype.

lets you replace the undesirable EMPLOYEE table structure in Figure 5.1 with two tables—one representing the supertype EMPLOYEE and the other representing the subtype PILOT. (See Figure 5.3.)

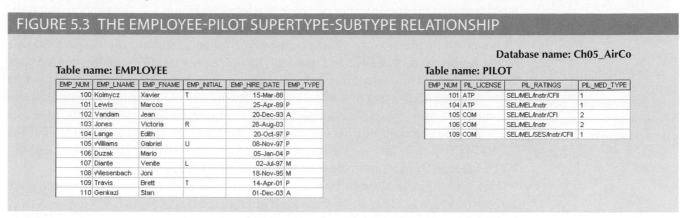

FIGURE 5.3 THE EMPLOYEE-PILOT SUPERTYPE-SUBTYPE RELATIONSHIP

Database name: Ch05_AirCo

Table name: EMPLOYEE

EMP_NUM	EMP_LNAME	EMP_FNAME	EMP_INITIAL	EMP_HIRE_DATE	EMP_TYPE
100	Kolmycz	Xavier	T	15-Mar-88	
101	Lewis	Marcos		25-Apr-89	P
102	Vandam	Jean		20-Dec-93	A
103	Jones	Victoria	R	28-Aug-03	
104	Lange	Edith		20-Oct-97	P
105	Williams	Gabriel	U	08-Nov-97	P
106	Duzak	Mario		05-Jan-04	P
107	Diante	Venite	L	02-Jul-97	M
108	Wiesenbach	Joni		18-Nov-95	M
109	Travis	Brett	T	14-Apr-01	P
110	Genkazi	Stan		01-Dec-03	A

Table name: PILOT

EMP_NUM	PIL_LICENSE	PIL_RATINGS	PIL_MED_TYPE
101	ATP	SEL/MEL/Instr/CFII	1
104	ATP	SEL/MEL/Instr	1
105	COM	SEL/MEL/Instr/CFI	2
106	COM	SEL/MEL/Instr	2
109	COM	SEL/MEL/SES/Instr/CFII	1

Entity subtypes inherit all relationships in which the supertype entity participates. For example, Figure 5.2 shows the EMPLOYEE entity supertype participating in a 1:M relationship with a DEPENDENT entity. Through inheritance, all subtypes also participate in that relationship. In specialization hierarchies with multiple levels of supertype and subtypes, a lower-level subtype inherits all of the attributes and relationships from all of its upper-level supertypes.

Inheriting the relationships of their supertypes does not mean that subtypes cannot have relationships of their own. Figure 5.4 illustrates a 1:M relationship between EMPLOYEE, a subtype of PERSON, and OFFICE. Because only employees and no other type of person will ever have an office within this system, the relationship is modeled with the subtype directly.

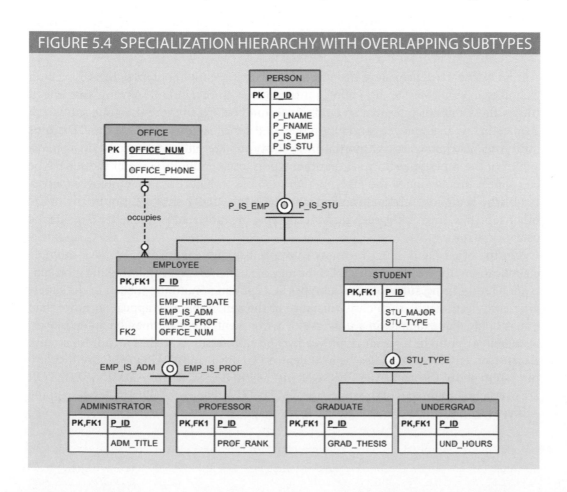

FIGURE 5.4 SPECIALIZATION HIERARCHY WITH OVERLAPPING SUBTYPES

5-1d Subtype Discriminator

A **subtype discriminator** is the attribute in the supertype entity that determines to which subtype the supertype occurrence is related. In Figure 5.2, the subtype discriminator is the employee type (EMP_TYPE).

It is common practice to show the subtype discriminator and its value for each subtype in the ER diagram, as shown in Figure 5.2. However, not all ER modeling tools follow that practice. For example, Microsoft Visio shows the subtype discriminator but not its value. In Figure 5.2, a text tool was used to manually add the discriminator value above the entity subtype, close to the connector line. Using Figure 5.2 as your guide, note that the supertype is related to a PILOT subtype if the EMP_TYPE has a value of "P." If the EMP_TYPE value is "M," the supertype is related to a MECHANIC subtype. If the EMP_TYPE value is "A," the supertype is related to the ACCOUNTANT subtype.

Note that the default comparison condition for the subtype discriminator attribute is the equality comparison. However, in some situations the subtype discriminator is not necessarily based on an equality comparison. For example, based on business requirements, you might create two new pilot subtypes: pilot-in-command (PIC)-qualified and copilot-qualified only. A PIC-qualified pilot must have more than 1,500 PIC flight hours. In this case, the subtype discriminator would be "Flight_Hours," and the criteria would be > 1,500 or <= 1,500, respectively.

Note

In Visio 2010, you select the subtype discriminator when creating a category by using the Category shape from the available shapes. The Category shape is a small circle with a horizontal line underneath that connects the supertype to its subtypes. Visio 2013 does not support specialization hierarchy.

5-1e Disjoint and Overlapping Constraints

subtype discriminator
The attribute in the supertype entity that determines to which entity subtype each supertype occurrence is related.

disjoint subtype
In a specialization hierarchy, a unique and nonoverlapping subtype entity set.

nonoverlapping subtype
See *disjoint subtype*.

overlapping subtype
In a specialization hierarchy, a condition in which each entity instance (row) of the supertype can appear in more than one subtype.

An entity supertype can have disjoint or overlapping entity subtypes. In the aviation example, an employee can be a pilot, a mechanic, or an accountant. Assume that one of the business rules dictates that an employee cannot belong to more than one subtype at a time; that is, an employee cannot be a pilot and a mechanic at the same time. **Disjoint subtypes**, also known as **nonoverlapping subtypes**, are subtypes that contain a *unique* subset of the supertype entity set; in other words, each entity instance of the supertype can appear in only one of the subtypes. For example, in Figure 5.2, an employee (supertype) who is a pilot (subtype) can appear only in the PILOT subtype, not in any of the other subtypes. In an ERD, such disjoint subtypes are indicated by the letter *d* inside the category shape.

On the other hand, if the business rule specifies that employees can have multiple classifications, the EMPLOYEE supertype may contain *overlapping* job classification subtypes. **Overlapping subtypes** are subtypes that contain nonunique subsets of the supertype entity set; that is, each entity instance of the supertype may appear in more than one subtype. For example, in a university environment, a person may be an employee, a student, or both. In turn, an employee may be a professor as well as an administrator. Because an employee may also be a student, STUDENT and EMPLOYEE are overlapping subtypes of the supertype PERSON, just as PROFESSOR and ADMINISTRATOR are overlapping subtypes of the supertype EMPLOYEE. Figure 5.4 illustrates overlapping subtypes with the letter *o* inside the category shape.

It is common practice to show disjoint and overlapping symbols in the ERD. (See Figures 5.2 and 5.4.) However, not all ER modeling tools follow that practice. For example, by default, Visio shows only the subtype discriminator (using the Category shape), but not the disjoint and overlapping symbols. The Visio text tool was used to manually add the *d* and *o* symbols in Figures 5.2 and 5.4.

Note

> Alternative notations exist for representing disjoint and overlapping subtypes. For example, Toby J. Teorey popularized the use of G and Gs to indicate disjoint and overlapping subtypes.

As you learned earlier in this section, the implementation of disjoint subtypes is based on the value of the subtype discriminator attribute in the supertype. However, *implementing* overlapping subtypes requires the use of one discriminator attribute for each subtype. For example, in the case of the Tiny College database design in Chapter 4, Entity Relationship (ER) Modeling, a professor can also be an administrator. Therefore, the EMPLOYEE supertype would have the subtype discriminator attributes and values shown in Table 5.1.

completeness constraint
A constraint that specifies whether each entity supertype occurrence must also be a member of at least one subtype. The completeness constraint can be partial or total.

TABLE 5.1

DISCRIMINATOR ATTRIBUTES WITH OVERLAPPING SUBTYPES		
DISCRIMINATOR ATTRIBUTES		**COMMENT**
PROFESSOR	**ADMINISTRATOR**	
Y	N	The Employee is a member of the Professor subtype.
N	Y	The Employee is a member of the Administrator subtype.
Y	Y	The Employee is both a Professor and an Administrator.

5-1f Completeness Constraint

The **completeness constraint** specifies whether each entity supertype occurrence must also be a member of at least one subtype. The completeness constraint can be partial or total. **Partial completeness** means that not every supertype occurrence is a member of a subtype; some supertype occurrences may not be members of any subtype. **Total completeness** means that every supertype occurrence must be a member of at least one subtype.

The ERDs in Figures 5.2 and 5.4 represent the completeness constraint based on the Visio Category shape. A single horizontal line under the circle represents a partial completeness constraint; a double horizontal line under the circle represents a total completeness constraint.

partial completeness
In a generalization/ specialization hierarchy, a condition in which some supertype occurrences might not be members of any subtype.

total completeness
In a generalization/ specialization hierarchy, a condition in which every supertype occurrence must be a member of at least one subtype.

Note

> Alternative notations exist to represent the completeness constraint. For example, some notations use a single line (partial) or double line (total) to connect the supertype to the Category shape.

Given the disjoint and overlapping subtypes and completeness constraints, it is possible to have the specialization hierarchy constraint scenarios shown in Table 5.2.

TABLE 5.2

SPECIALIZATION HIERARCHY CONSTRAINT SCENARIOS

TYPE	DISJOINT CONSTRAINT	OVERLAPPING CONSTRAINT
Partial	Supertype has optional subtypes. Subtype discriminator can be null. Subtype sets are unique.	Supertype has optional subtypes. Subtype discriminators can be null. Subtype sets are not unique.
Total	Every supertype occurrence is a member of only one subtype. Subtype discriminator cannot be null. Subtype sets are unique.	Every supertype occurrence is a member of at least one subtype. Subtype discriminators cannot be null. Subtype sets are not unique.

5-1g Specialization and Generalization

You can use various approaches to develop entity supertypes and subtypes. For example, you can first identify a regular entity, and then identify all entity subtypes based on their distinguishing characteristics. You can also start by identifying multiple entity types and then later extract the common characteristics of those entities to create a higher-level supertype entity.

Specialization is the top-down process of identifying lower-level, more specific entity subtypes from a higher-level entity supertype. Specialization is based on grouping the unique characteristics and relationships of the subtypes. In the aviation example, you used specialization to identify multiple entity subtypes from the original employee supertype. **Generalization** is the bottom-up process of identifying a higher-level, more generic entity supertype from lower-level entity subtypes. Generalization is based on grouping the common characteristics and relationships of the subtypes. For example, you might identify multiple types of musical instruments: piano, violin, and guitar. Using the generalization approach, you could identify a "string instrument" entity supertype to hold the common characteristics of the multiple subtypes.

5-2 Entity Clustering

Developing an ER diagram entails the discovery of possibly hundreds of entity types and their respective relationships. Generally, the data modeler will develop an initial ERD that contains a few entities. As the design approaches completion, the ERD will contain hundreds of entities and relationships that crowd the diagram to the point of making it unreadable and inefficient as a communication tool. In those cases, you can use entity clusters to minimize the number of entities shown in the ERD.

An **entity cluster** is a "virtual" entity type used to represent multiple entities and relationships in the ERD. An entity cluster is formed by combining multiple interrelated entities into a single, abstract entity object. An entity cluster is considered "virtual" or "abstract" in the sense that it is not actually an entity in the final ERD. Instead, it is a temporary entity used to represent multiple entities and relationships, with the purpose of simplifying the ERD and thus enhancing its readability.

Figure 5.5 illustrates the use of entity clusters based on the Tiny College example in Chapter 4. Note that the ERD contains two entity clusters:

- OFFERING, which groups the SEMESTER, COURSE, and CLASS entities and relationships
- LOCATION, which groups the ROOM and BUILDING entities and relationships

specialization
In a specialization hierarchy, the grouping of unique attributes into a subtype entity.

generalization
In a specialization hierarchy, the grouping of common attributes into a supertype entity.

entity cluster
A "virtual" entity type used to represent multiple entities and relationships in the ERD. An entity cluster is formed by combining multiple interrelated entities into a single abstract entity object. An entity cluster is considered "virtual" or "abstract" because it is not actually an entity in the final ERD.

Note also that the ERD in Figure 5.5 does not show attributes for the entities. When using entity clusters, the key attributes of the combined entities are no longer available. Without the key attributes, primary key inheritance rules change. In turn, the change in the inheritance rules can have undesirable consequences, such as changes in relationships—from identifying to nonidentifying or vice versa—and the loss of foreign key attributes from some entities. To eliminate those problems, the general rule is to *avoid the display of attributes when entity clusters are used.*

FIGURE 5.5 TINY COLLEGE ERD USING ENTITY CLUSTERS

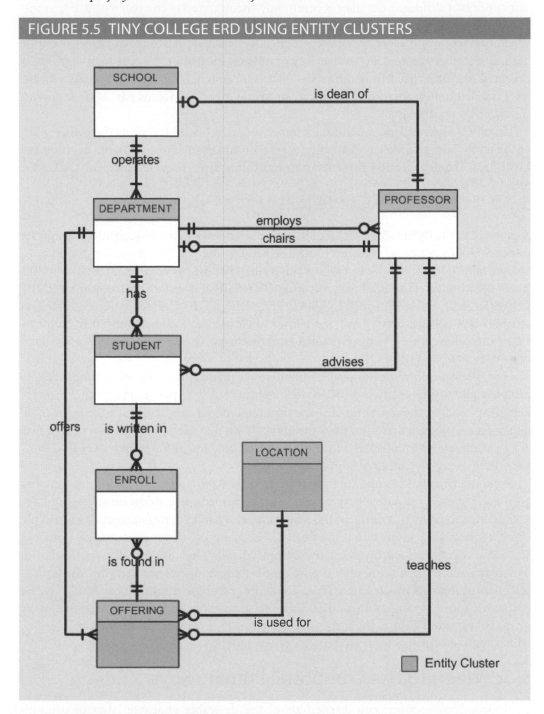

5-3 Entity Integrity: Selecting Primary Keys

Arguably, the most important characteristic of an entity is its primary key (a single attribute or some combination of attributes), which uniquely identifies each entity instance.

The primary key's function is to guarantee entity integrity. Furthermore, primary keys and foreign keys work together to implement relationships in the relational model. Therefore, the importance of properly selecting the primary key has a direct bearing on the efficiency and effectiveness of database implementation.

5-3a Natural Keys and Primary Keys

The concept of a unique identifier is commonly encountered in the real world. For example, you use class or section numbers to register for classes, invoice numbers to identify specific invoices, and account numbers to identify credit cards. Those examples illustrate natural identifiers or keys. A **natural key** or **natural identifier** is a real-world, generally accepted identifier used to distinguish—that is, uniquely identify—real-world objects. As its name implies, a natural key is familiar to end users and forms part of their day-to-day business vocabulary.

Usually, if an entity *has* a natural identifier, a data modeler uses it as the primary key of the entity being modeled. Generally, most natural keys make acceptable primary key identifiers. The next section presents some basic guidelines for selecting primary keys.

5-3b Primary Key Guidelines

A primary key is the attribute or combination of attributes that uniquely identifies entity instances in an entity set. However, can the primary key be based on, for example, 12 attributes? And just how long can a primary key be? In previous examples, why was EMP_NUM selected as a primary key of EMPLOYEE and not a combination of EMP_LNAME, EMP_FNAME, EMP_INITIAL, and EMP_DOB? Can a single, 256-byte text attribute be a good primary key? There is no single answer to those questions, but database experts have built a body of practice over the years. This section examines that body of documented practices.

First, you should understand the function of a primary key. Its main function is to uniquely identify an entity instance or row within a table. In particular, given a primary key value—that is, the determinant—the relational model can determine the value of all dependent attributes that "describe" the entity. Note that identification and description are separate semantic constructs in the model. *The function of the primary key is to guarantee entity integrity, not to "describe" the entity.*

Second, primary keys and foreign keys are used to implement relationships among entities. However, the implementation of such relationships is done mostly behind the scenes, hidden from end users. In the real world, end users identify objects based on the characteristics they know about the objects. For example, when shopping at a grocery store, you select products by taking them from a display shelf and reading the labels, not by looking at the stock number. It is wise for database applications to mimic the human selection process as much as possible. Therefore, database applications should let the end user choose among multiple descriptive narratives of different objects, while using primary key values behind the scenes. Keeping those concepts in mind, look at Table 5.3, which summarizes desirable primary key characteristics.

5-3c When To Use Composite Primary Keys

In the previous section, you learned about the desirable characteristics of primary keys. For example, you learned that the primary key should use the minimum number of attributes possible. However, that does *not* mean that composite primary keys are not permitted in a model. In fact, composite primary keys are particularly useful in two cases:

natural key (natural identifier)
A generally accepted identifier for real-world objects. As its name implies, a natural key is familiar to end users and forms part of their day-to-day business vocabulary.

TABLE 5.3

DESIRABLE PRIMARY KEY CHARACTERISTICS

PK CHARACTERISTIC	RATIONALE
Unique values	The PK must uniquely identify each entity instance. A primary key must be able to guarantee unique values. It cannot contain nulls.
Nonintelligent	The PK should not have embedded semantic meaning other than to uniquely identify each entity instance. An attribute with embedded semantic meaning is probably better used as a descriptive characteristic of the entity than as an identifier. For example, a student ID of 650973 would be preferred over *Smith, Martha L.* as a primary key identifier.
No change over time	If an attribute has semantic meaning, it might be subject to updates, which is why names do not make good primary keys. If *Vickie Smith* is the primary key, what happens if she changes her name when she gets married? If a primary key is subject to change, the foreign key values must be updated, thus adding to the database work load. Furthermore, changing a primary key value means that you are basically changing the identity of an entity. In short, the PK should be permanent and unchangeable.
Preferably single-attribute	A primary key should have the minimum number of attributes possible (irreducible). Single-attribute primary keys are desirable but not required. Single-attribute primary keys simplify the implementation of foreign keys. Having multiple-attribute primary keys can cause primary keys of related entities to grow through the possible addition of many attributes, thus adding to the database workload and making (application) coding more cumbersome.
Preferably numeric	Unique values can be better managed when they are numeric, because the database can use internal routines to implement a counter-style attribute that automatically increments values with the addition of each new row. In fact, most database systems include the ability to use special constructs, such as Autonumber in Microsoft Access, sequence in Oracle, or uniqueidentifier in MS SQL Server to support self-incrementing primary key attributes.
Security-compliant	The selected primary key must not be composed of any attribute(s) that might be considered a security risk or violation. For example, using a Social Security number as a PK in an EMPLOYEE table is not a good idea.

- As identifiers of composite entities, in which each primary key combination is allowed only once in the M:N relationship
- As identifiers of weak entities, in which the weak entity has a strong identifying relationship with the parent entity

To illustrate the first case, assume that you have a STUDENT entity set and a CLASS entity set. In addition, assume that those two sets are related in an M:N relationship via an ENROLL entity set, in which each student/class combination may appear only once in the composite entity. Figure 5.6 shows the ERD to represent such a relationship.

As shown in Figure 5.6, the composite primary key automatically provides the benefit of ensuring that there cannot be duplicate values—that is, it ensures that the same student cannot enroll more than once in the same class.

In the second case, a weak entity in a strong identifying relationship with a parent entity is normally used to represent one of two situations:

1. *A real-world object that is existence-dependent on another real-world object.* Such objects are distinguishable in the real world. A dependent and an employee are two separate people who exist independently of each other. However, such objects can exist in the model only when they relate to each other in a strong identifying relationship. For example, the relationship between EMPLOYEE and DEPENDENT is one of existence dependency, in which the primary key of the dependent entity is a composite key that contains the key of the parent entity.

FIGURE 5.6 THE M:N RELATIONSHIP BETWEEN STUDENT AND CLASS

Database name: Ch05_Tinycollege

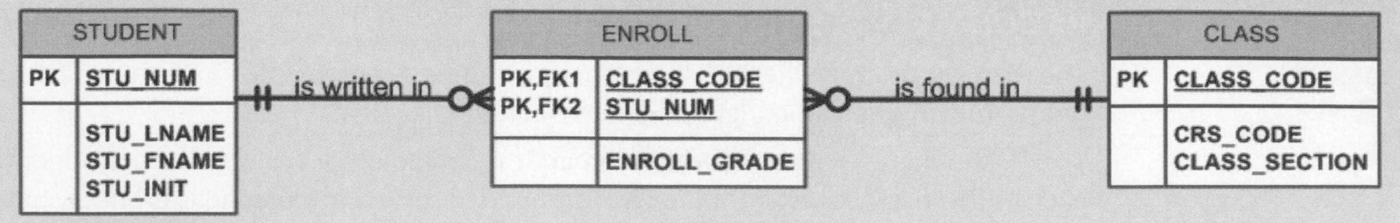

Table name: STUDENT
(first four fields)

STU_NUM	STU_LNAME	STU_FNAME	STU_INIT
321452	Bowser	William	C
324257	Smithson	Anne	K
324258	Brewer	Juliette	
324269	Oblonski	Walter	H
324273	Smith	John	D
324274	Katinga	Raphael	P
324291	Robertson	Gerald	T
324299	Smith	John	B

Table name: ENROLL

CLASS_CODE	STU_NUM	ENROLL_GRADE
10014	321452	C
10014	324257	B
10018	321452	A
10018	324257	B
10021	321452	C
10021	324257	C

Table name: CLASS
(first three fields)

CLASS_CODE	CRS_CODE	CLASS_SECTION
10012	ACCT-211	1
10013	ACCT-211	2
10014	ACCT-211	3
10015	ACCT-212	1
10016	ACCT-212	2
10017	CIS-220	1
10018	CIS-220	2
10019	CIS-220	3
10020	CIS-420	1
10021	QM-261	1
10022	QM-261	2
10023	QM-362	1
10024	QM-362	2
10025	MATH-243	1

2. *A real-world object that is represented in the data model as two separate entities in a strong identifying relationship.* For example, the real-world invoice object is represented by two entities in a data model: INVOICE and LINE. Clearly, the LINE entity does not exist in the real world as an independent object, but as part of an INVOICE.

In both situations, having a strong identifying relationship ensures that the dependent entity can exist only when it is related to the parent entity. In summary, the selection of a composite primary key for composite and weak entity types provides benefits that enhance the integrity and consistency of the model.

5-3d When To Use Surrogate Primary Keys

In some instances a primary key doesn't exist in the real world or the existing natural key might not be a suitable primary key. In these cases, it is standard practice to create a surrogate key. A **surrogate key** is a primary key created by the database designer to simplify the identification of entity instances. The surrogate key has no meaning in the user's environment—it exists only to distinguish one entity instance from another (just like any other primary key). One practical advantage of a surrogate key is that because it has no intrinsic meaning, values for it can be generated by the DBMS to ensure that unique values are always provided.

For example, consider the case of a park recreation facility that rents rooms for small parties. The manager of the facility keeps track of all events, using a folder with the format shown in Table 5.4.

Given the data shown in Table 5.4, you would model the EVENT entity as follows:

EVENT (DATE, TIME_START, TIME_END, ROOM, EVENT_NAME, PARTY_OF)

What primary key would you suggest? In this case, there is no simple natural key that could be used as a primary key in the model. Based on the primary key concepts you learned in previous chapters, you might suggest one of these options:

surrogate key
A system-assigned primary key, generally numeric and auto-incremented.

TABLE 5.4

DATA USED TO KEEP TRACK OF EVENTS					
DATE	**TIME_START**	**TIME_END**	**ROOM**	**EVENT_NAME**	**PARTY_OF**
6/17/2016	11:00a.m.	2:00p.m.	Allure	Burton Wedding	60
6/17/2016	11:00a.m.	2:00p.m.	Bonanza	Adams Office	12
6/17/2016	3:00p.m.	5:30p.m.	Allure	Smith Family	15
6/17/2016	3:30p.m.	5:30p.m.	Bonanza	Adams Office	12
6/18/2016	1:00p.m.	3:00p.m.	Bonanza	Boy Scouts	33
6/18/2016	11:00a.m.	2:00p.m.	Allure	March of Dimes	25
6/18/2016	11:00a.m.	12:30p.m.	Bonanza	Smith Family	12

(**DATE**, **TIME_START**, **ROOM**) or (**DATE**, **TIME_END**, **ROOM**)

Assume that you select the composite primary key (**DATE**, **TIME_START**, **ROOM**) for the EVENT entity. Next, you determine that one EVENT may use many RESOURCEs (such as tables, projectors, PCs, and stands) and that the same RESOURCE may be used for many EVENTs. The RESOURCE entity would be represented by the following attributes:

RESOURCE (**RSC_ID**, RSC_DESCRIPTION, RSC_TYPE, RSC_QTY, RSC_PRICE)

Given the business rules, the M:N relationship between RESOURCE and EVENT would be represented via the EVNTRSC composite entity with a composite primary key as follows:

EVNTRSC (**DATE**, **TIME_START**, **ROOM**, **RSC_ID**, QTY_USED)

You now have a lengthy, four-attribute composite primary key. What would happen if the EVNTRSC entity's primary key were inherited by another existence-dependent entity? At this point, you can see that the composite primary key could make the database implementation and program coding unnecessarily complex.

As a data modeler, you probably noticed that the EVENT entity's selected primary key might not fare well, given the primary key guidelines in Table 5.3. In this case, the EVENT entity's selected primary key contains embedded semantic information and is formed by a combination of date, time, and text data columns. In addition, the selected primary key would cause lengthy primary keys for existence-dependent entities. The preferred alternative is to use a numeric, single-attribute surrogate primary key.

Surrogate primary keys are accepted practice in today's complex data environments. They are especially helpful when there is no natural key, when the selected candidate key has embedded semantic contents, or when the selected candidate key is too long or cumbersome. However, there is a trade-off: if you use a surrogate key, you must ensure that the candidate key of the entity in question performs properly through the use of "unique index" and "not null" constraints.

Note

This example shows a case in which entity integrity is maintained but semantic correctness of business rules is not. For example, you could have two events that overlap and whose primary keys are perfectly compliant. The only way to ensure adherence to this type of business rule (two events cannot overlap—occur on the same room at the same time) would be via application programming code.

5-4 Design Cases: Learning Flexible Database Design

Data modeling and database design require skills that are acquired through experience. In turn, experience is acquired through practice—regular and frequent repetition, applying the concepts learned to specific and different design problems. This section presents four special design cases that highlight the importance of flexible designs, proper identification of primary keys, and placement of foreign keys.

Note

In describing the various modeling concepts throughout this book, the focus is on relational models. Also, given the focus on the practical nature of database design, all design issues are addressed with the implementation goal in mind. Therefore, there is no sharp line of demarcation between design and implementation.

At the pure conceptual stage of the design, foreign keys are not part of an ER diagram. The ERD displays only entities and relationships. Entity instances are distinguished by identifiers that may become primary keys. During design, the modeler attempts to understand and define the entities and relationships. Foreign keys are the mechanism through which the relationship designed in an ERD is implemented in a relational model.

5-4a Design Case 1: Implementing 1:1 Relationships

Foreign keys work with primary keys to properly implement relationships in the relational model. The basic rule is very simple: put the primary key of the "one" side (the parent entity) on the "many" side (the dependent entity) as a foreign key. However, where do you place the foreign key when you are working with a 1:1 relationship? For example, take the case of a 1:1 relationship between EMPLOYEE and DEPARTMENT based on the business rule "one EMPLOYEE is the manager of one DEPARTMENT, and one DEPARTMENT is managed by one EMPLOYEE." In that case, there are two options for selecting and placing the foreign key:

1. *Place a foreign key in both entities*. This option is derived from the basic rule you learned in Chapter 4. Place EMP_NUM as a foreign key in DEPARTMENT, and place DEPT_ID as a foreign key in EMPLOYEE. However, this solution is not recommended because it duplicates work, and it could conflict with other existing relationships. (Remember that DEPARTMENT and EMPLOYEE also participate in a 1:M relationship—one department employs many employees.)

2. *Place a foreign key in one of the entities*. In that case, the primary key of one of the two entities appears as a foreign key in the other entity. That is the preferred solution, but a question remains: *which* primary key should be used as a foreign key? The answer is found in Table 5.5, which shows the rationale for selecting the foreign key in a 1:1 relationship based on the relationship properties in the ERD.

Figure 5.7 illustrates the "EMPLOYEE manages DEPARTMENT" relationship. Note that in this case, EMPLOYEE is mandatory to DEPARTMENT. Therefore, EMP_NUM is placed as the foreign key in DEPARTMENT. Alternatively, you might also argue that the "manager" role is played by the EMPLOYEE in the DEPARTMENT.

As a designer, you must recognize that 1:1 relationships exist in the real world; therefore, they should be supported in the data model. In fact, a 1:1 relationship is used to

TABLE 5.5

	SELECTION OF FOREIGN KEY IN A 1:1 RELATIONSHIP	
CASE	**ER RELATIONSHIP CONSTRAINTS**	**ACTION**
I	One side is mandatory and the other side is optional.	Place the PK of the entity on the mandatory side in the entity on the optional side as a FK, and make the FK mandatory.
II	Both sides are optional.	Select the FK that causes the fewest nulls, or place the FK in the entity in which the (relationship) role is played.
III	Both sides are mandatory.	See Case II, or consider revising your model to ensure that the two entities do not belong together in a single entity.

FIGURE 5.7 THE 1:1 RELATIONSHIP BETWEEN DEPARTMENT AND EMPLOYEE

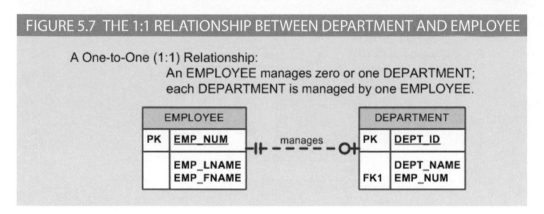

ensure that two entity sets are not placed in the same table. In other words, EMPLOYEE and DEPARTMENT are clearly separate and unique entity types that do not belong together in a single entity. If you grouped them together in one entity, what would you name that entity?

5-4b Design Case 2: Maintaining History of Time-Variant Data

Company managers generally realize that good decision making is based on the information generated through the data stored in databases. Such data reflects both current and past events. Company managers use the data stored in databases to answer questions such as "How do the current company profits compare to those of previous years?" and "What are XYZ product's sales trends?" In other words, the data stored in databases reflects not only current data, but historic data.

Normally, data changes are managed by replacing the existing attribute value with the new value, without regard to the previous value. However, in some situations the history of values for a given attribute must be preserved. From a data-modeling point of view, **time-variant data** refer to data whose values change over time and for which you *must* keep a history of the data changes. You could argue that all data in a database is subject to change over time and is therefore time variant. However, some attribute values, such as your date of birth or your Social Security number, are not time variant. On the other hand, attributes such as your student GPA or your bank account balance are subject to change over time. Sometimes the data changes are externally originated and event driven, such as a product price change. On other occasions, changes are based on well-defined schedules, such as the daily stock quote "open" and "close" values.

time-variant data
Data whose values are a function of time. For example, time-variant data can be seen at work when a company's history of all administrative appointments is tracked.

The storage of time-variant data requires changes in the data model; the type of change depends on the nature of the data. Some time-variant data is equivalent to having a multivalued attribute in your entity. To model this type of time-variant data, you must create a new entity in a 1:M relationship with the original entity. This new entity will contain the new value, the date of the change, and any other attribute that is pertinent to the event being modeled. For example, if you want to track salary histories for each employee, then the EMP_SALARY attribute becomes multivalued, as shown in Figure 5.8. In this case, for each employee, there will be one or more records in the SALARY_HIST entity, which stores the salary amount and the date when the new salary goes into effect.

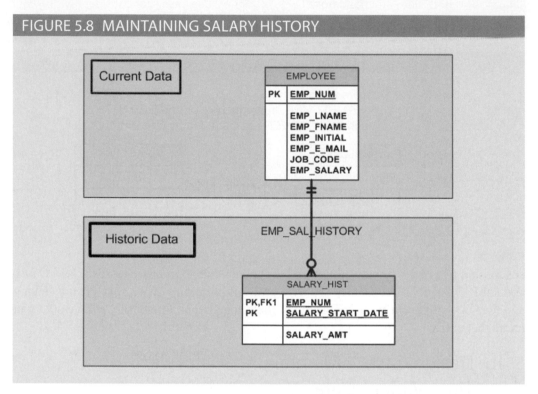

FIGURE 5.8 MAINTAINING SALARY HISTORY

Other time-variant data can turn a 1:M relationship into an M:N relationship. Assume that in addition to employee data, your data model includes data about the different departments in the organization and which employee manages each department. Assuming that each department is managed by only one employee and each employee can manage one department at most, then a 1:1 relationship would exist between EMPLOYEE and DEPARTMENT. This relationship would record the current manager of each department. However, if you want to keep track of the history of all department managers as well as the current manager, you can create the model shown in Figure 5.9.

Note that in Figure 5.9, the MGR_HIST entity has a 1:M relationship with EMPLOYEE and a 1:M relationship with DEPARTMENT to reflect the fact that an employee could be the manager of many different departments over time, and a department could have many different employee managers. Because you are recording time-variant data, you must store the DATE_ASSIGN attribute in the MGR_HIST entity to provide the date that the employee (EMP_NUM) became the department manager. The primary key of MGR_HIST permits the same employee to be the manager of the same department, but on different dates. If that scenario is not the case in your environment—if, for example, an employee is the manager of a department only once—you could make DATE_ASSIGN a nonprime attribute in the MGR_HIST entity.

FIGURE 5.9 MAINTAINING MANAGER HISTORY

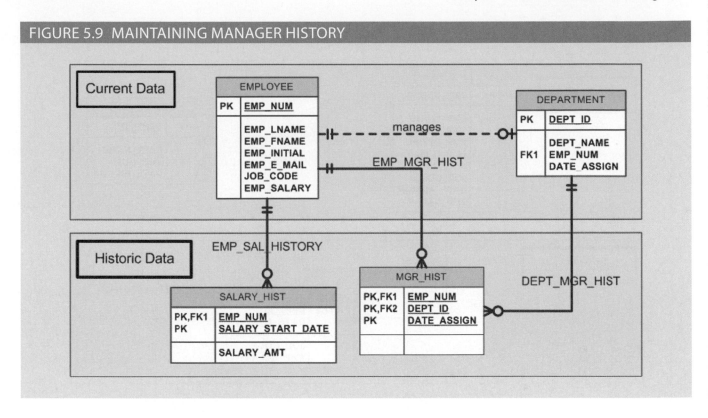

Note in Figure 5.9 that the "manages" relationship is optional in theory and redundant in practice. At any time, you could identify the manager of a department by retrieving the most recent DATE_ASSIGN date from MGR_HIST for a given department. On the other hand, the ERD in Figure 5.9 differentiates between current data and historic data. The *current* manager relationship is implemented by the "manages" relationship between EMPLOYEE and DEPARTMENT. Additionally, the historic data is managed through EMP_MGR_HIST and DEPT_MGR_HIST. The trade-off with that model is that each time a new manager is assigned to a department, there will be two data modifications: one update in the DEPARTMENT entity and one insert in the MGR_HIST entity.

The flexibility of the model proposed in Figure 5.9 becomes more apparent when you add the 1:M "one department employs many employees" relationship. In that case, the PK of the "1" side (DEPT_ID) appears in the "many" side (EMPLOYEE) as a foreign key. Now suppose you would like to keep track of the job history for each of the company's employees—you'd probably want to store the department, the job code, the date assigned, and the salary. To accomplish that task, you could modify the model in Figure 5.9 by adding a JOB_HIST entity. Figure 5.10 shows the use of the new JOB_HIST entity to maintain the employee's history.

Again, it is worth emphasizing that the "manages" and "employs" relationships are theoretically optional and redundant in practice. You can always find out where each employee works by looking at the job history and selecting only the most current data row for each employee. However, as you will discover in Chapter 7, Introduction to Structured Query Language (SQL), and in Chapter 8, Advanced SQL, finding where each employee works is not a trivial task. Therefore, the model represented in Figure 5.10 includes the admittedly redundant but unquestionably useful "manages" and "employs" relationships to separate current data from historic data.

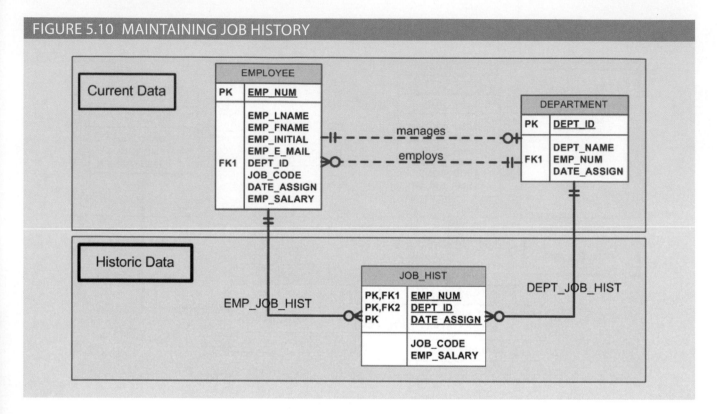

FIGURE 5.10 MAINTAINING JOB HISTORY

5-4c Design Case 3: Fan Traps

Creating a data model requires proper identification of the data relationships among entities. However, due to miscommunication or incomplete understanding of the business rules or processes, it is not uncommon to misidentify relationships among entities. Under those circumstances, the ERD may contain a design trap. A **design trap** occurs when a relationship is improperly or incompletely identified and is therefore represented in a way that is not consistent with the real world. The most common design trap is known as a *fan trap*.

A **fan trap** occurs when you have one entity in two 1:M relationships to other entities, thus producing an association among the other entities that is not expressed in the model. For example, assume that the JCB basketball league has many divisions. Each division has many players, and each division has many teams. Given those "incomplete" business rules, you might create an ERD that looks like the one in Figure 5.11.

As you can see in Figure 5.11, DIVISION is in a 1:M relationship with TEAM and in a 1:M relationship with PLAYER. Although that representation is semantically correct, the relationships are not properly identified. For example, there is no way to identify which players belong to which team. Figure 5.11 also shows a sample instance relationship representation for the ERD. Note that the relationship lines for the DIVISION instances fan out to the TEAM and PLAYER entity instances—thus the "fan trap" label.

Figure 5.12 shows the correct ERD after the fan trap has been eliminated. Note that, in this case, DIVISION is in a 1:M relationship with TEAM. In turn, TEAM is in a 1:M relationship with PLAYER. Figure 5.12 also shows the instance relationship representation after eliminating the fan trap.

Given the design in Figure 5.12, note how easy it is to see which players play for which team. However, to find out which players play in which division, you first need to see what teams belong to each division; then you need to find out which players play on each team. In other words, there is a transitive relationship between DIVISION and PLAYER via the TEAM entity.

design trap
A problem that occurs when a relationship is improperly or incompletely identified and therefore is represented in a way that is not consistent with the real world. The most common design trap is known as a *fan trap*.

fan trap
A design trap that occurs when one entity is in two 1:M relationships with other entities, thus producing an association among the other entities that is not expressed in the model.

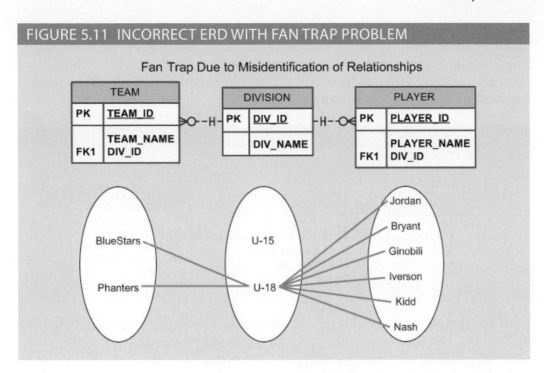

FIGURE 5.11 INCORRECT ERD WITH FAN TRAP PROBLEM

Fan Trap Due to Misidentification of Relationships

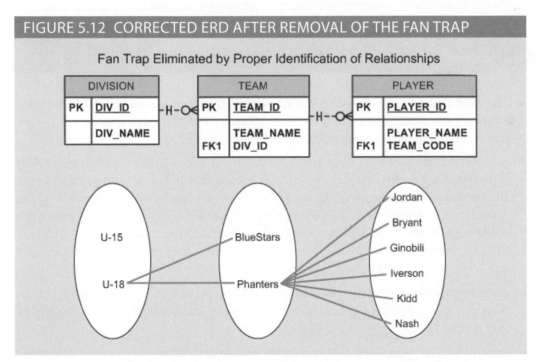

FIGURE 5.12 CORRECTED ERD AFTER REMOVAL OF THE FAN TRAP

Fan Trap Eliminated by Proper Identification of Relationships

5-4d Design Case 4: Redundant Relationships

Although redundancy is often good to have in computer environments (multiple backups in multiple places, for example), redundancy is seldom good in the database environment. (As you learned in Chapter 3, The Relational Database Model, redundancies can cause data anomalies in a database.) Redundant relationships occur when there are multiple relationship paths between related entities. The main concern with redundant relationships is that they remain consistent across the model. However, it is important to note that some designs use redundant relationships as a way to simplify the design.

An example of redundant relationships was first introduced in Figure 5.9 during the discussion of maintaining a history of time-variant data. However, the use of the redundant "manages" and "employs" relationships was justified by the fact that such relationships dealt with current data rather than historic data. Another more specific example of a redundant relationship is represented in Figure 5.13.

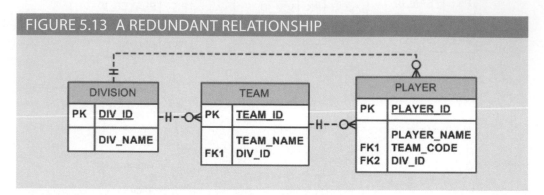

FIGURE 5.13 A REDUNDANT RELATIONSHIP

In Figure 5.13, note the transitive 1:M relationship between DIVISION and PLAYER through the TEAM entity set. Therefore, the relationship that connects DIVISION and PLAYER is redundant, for all practical purposes. In that case, the relationship could be safely deleted without losing any information-generation capabilities in the model.

Summary

- The extended entity relationship (EER) model adds semantics to the ER model via entity supertypes, subtypes, and clusters. An entity supertype is a generic entity type that is related to one or more entity subtypes.

- A specialization hierarchy depicts the arrangement and relationships between entity supertypes and entity subtypes. Inheritance means that an entity subtype inherits the attributes and relationships of the supertype. Subtypes can be disjoint or overlapping. A subtype discriminator is used to determine to which entity subtype the supertype occurrence is related. The subtypes can exhibit partial or total completeness. There are basically two approaches to developing a specialization hierarchy of entity supertypes and subtypes: specialization and generalization.

- An entity cluster is a "virtual" entity type used to represent multiple entities and relationships in the ERD. An entity cluster is formed by combining multiple interrelated entities and relationships into a single, abstract entity object.

- Natural keys are identifiers that exist in the real world. Natural keys sometimes make good primary keys, but not always. Primary keys must have unique values, they should be nonintelligent, they must not change over time, and they are preferably numeric and composed of a single attribute.

- Composite keys are useful to represent M:N relationships and weak (strong identifying) entities.

- Surrogate primary keys are useful when there is no natural key that makes a suitable primary key, when the primary key is a composite primary key with multiple data types, or when the primary key is too long to be usable.

- In a 1:1 relationship, place the PK of the mandatory entity as a foreign key in the optional entity, as an FK in the entity that causes the fewest nulls, or as an FK where the role is played.

- Time-variant data refers to data whose values change over time and require that you keep a history of data changes. To maintain the history of time-variant data, you must create an entity that contains the new value, the date of change, and any other time-relevant data. This entity maintains a 1:M relationship with the entity for which the history is to be maintained.

- A fan trap occurs when you have one entity in two 1:M relationships to other entities, and there is an association among the other entities that is not expressed in the model. Redundant relationships occur when there are multiple relationship paths between related entities. The main concern with redundant relationships is that they remain consistent across the model.

Key Terms

completeness constraint	extended entity relationship model (EERM)	partial completeness
design trap	fan trap	specialization
disjoint subtype	generalization	specialization hierarchy
EER diagram (EERD)	inheritance	subtype discriminator
entity cluster	natural key (natural identifier)	surrogate key
entity subtype	nonoverlapping subtype	time-variant data
entity supertype	overlapping subtype	total completeness

Online Content

Flashcards and crossword puzzles for key term practice are available at *www.cengagebrain.com*.

Review Questions

1. What is an entity supertype, and why is it used?
2. What kinds of data would you store in an entity subtype?
3. What is a specialization hierarchy?
4. What is a subtype discriminator? Give an example of its use.
5. What is an overlapping subtype? Give an example.
6. What is the difference between partial completeness and total completeness?

For Questions 7–9, refer to Figure Q5.7.

7. List all of the attributes of a movie.
8. According to the data model, is it required that every entity instance in the PRODUCT table be associated with an entity instance in the CD table? Why, or why not?
9. Is it possible for a book to appear in the BOOK table without appearing in the PRODUCT table? Why, or why not?
10. What is an entity cluster, and what advantages are derived from its use?
11. What primary key characteristics are considered desirable? Explain *why* each characteristic is considered desirable.
12. Under what circumstances are composite primary keys appropriate?
13. What is a surrogate primary key, and when would you use one?

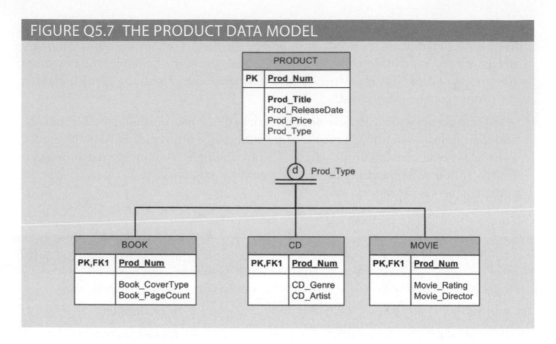

FIGURE Q5.7 THE PRODUCT DATA MODEL

14. When implementing a 1:1 relationship, where should you place the foreign key if one side is mandatory and one side is optional? Should the foreign key be mandatory or optional?

15. What is time-variant data, and how would you deal with such data from a database design point of view?

16. What is the most common design trap, and how does it occur?

Problems

1. Given the following business scenario, create a Crow's Foot ERD using a specialization hierarchy if appropriate. Two-Bit Drilling Company keeps information on employees and their insurance dependents. Each employee has an employee number, name, date of hire, and title. If an employee is an inspector, then the date of certification and certification renewal date should also be recorded in the system. For all employees, the Social Security number and dependent names should be kept. All dependents must be associated with one and only one employee. Some employees will not have dependents, while others will have many dependents.

2. Given the following business scenario, create a Crow's Foot ERD using a specialization hierarchy if appropriate. Tiny Hospital keeps information on patients and hospital rooms. The system assigns each patient a patient ID number. In addition, the patient's name and date of birth are recorded. Some patients are resident patients who spend at least one night in the hospital, and others are outpatients who are treated and released. Resident patients are assigned to a room. Each room is identified by a room number. The system also stores the room type (private or semiprivate) and room fee. Over time, each room will have many patients. Each resident patient will stay in only one room. Every room must have had a patient, and every resident patient must have a room.

3. Given the following business scenario, create a Crow's Foot ERD using a specialization hierarchy if appropriate. Granite Sales Company keeps information on

employees and the departments in which they work. For each department, the department name, internal mail box number, and office phone extension are kept. A department can have many assigned employees, and each employee is assigned to only one department. Employees can be salaried, hourly, or work on contract. All employees are assigned an employee number, which is kept along with the employee's name and address. For hourly employees, hourly wages and target weekly work hours are stored; for example, the company may target 40 hours/ week for some employees, 32 for others, and 20 for others. Some salaried employees are salespeople who can earn a commission in addition to their base salary. For all salaried employees, the yearly salary amount is recorded in the system. For salespeople, their commission percentage on sales and commission percentage on profit are stored in the system. For example, John is a salesperson with a base salary of $50,000 per year plus a 2 percent commission on the sales price for all sales he makes, plus another 5 percent of the profit on each of those sales. For contract employees, the beginning date and end date of their contracts are stored along with the billing rate for their hours.

4. In Chapter 4, you saw the creation of the Tiny College database design, which reflected such business rules as "a professor may advise many students" and "a professor may chair one department." Modify the design shown in Figure 4.36 to include these business rules:

   - An employee could be staff, a professor, or an administrator.

   - A professor may also be an administrator.

   - Staff employees have a work-level classification, such as Level I or Level II.

   - Only professors can chair a department. A department is chaired by only one professor.

   - Only professors can serve as the dean of a college. Each of the university's colleges is served by one dean.

   - A professor can teach many classes.

   - Administrators have a position title.

   Given that information, create the complete ERD that contains all primary keys, foreign keys, and main attributes.

5. Tiny College wants to keep track of the history of all its administrative appointments, including dates of appointment and dates of termination. (*Hint*: Time-variant data is at work.) The Tiny College chancellor may want to know how many deans worked in the College of Business between January 1, 1960, and January 1, 2016, or who the dean of the College of Education was in 1990. Given that information, create the complete ERD that contains all primary keys, foreign keys, and main attributes.

6. Some Tiny College staff employees are information technology (IT) personnel. Some IT personnel provide technology support for academic programs, some provide technology infrastructure support, and some provide support for both. IT personnel are not professors; they are required to take periodic training to retain their technical expertise. Tiny College tracks all IT personnel training by date, type, and results (completed versus not completed). Given that information, create the complete ERD that contains all primary keys, foreign keys, and main attributes.

7. The FlyRight Aircraft Maintenance (FRAM) division of the FlyRight Company (FRC) performs all maintenance for FRC's aircraft. Produce a data model segment that reflects the following business rules:

- All mechanics are FRC employees. Not all employees are mechanics.

- Some mechanics are specialized in engine (EN) maintenance. Others are specialized in airframe (AF) maintenance or avionics (AV) maintenance. (Avionics are the electronic components of an aircraft that are used in communication and navigation.) All mechanics take periodic refresher courses to stay current in their areas of expertise. FRC tracks all courses taken by each mechanic—date, course type, certification (Y/N), and performance.

- FRC keeps an employment history of all mechanics. The history includes the date hired, date promoted, and date terminated.

Given those requirements, create the Crow's Foot ERD segment.

Cases

8. "Martial Arts R Us" (MARU) needs a database. MARU is a martial arts school with hundreds of students. The database must keep track of all the classes that are offered, who is assigned to teach each class, and which students attend each class. Also, it is important to track the progress of each student as they advance. Create a complete Crow's Foot ERD for these requirements:

- Students are given a student number when they join the school. The number is stored along with their name, date of birth, and the date they joined the school.

- All instructors are also students, but clearly not all students are instructors. In addition to the normal student information, for all instructors, the date that they start working as an instructor must be recorded along with their instructor status (compensated or volunteer).

- An instructor may be assigned to teach any number of classes, but each class has one and only one assigned instructor. Some instructors, especially volunteer instructors, may not be assigned to any class.

- A class is offered for a specific level at a specific time, day of the week, and location. For example, one class taught on Mondays at 5:00 p.m. in Room 1 is an intermediate-level class. Another class taught on Mondays at 6:00 p.m. in Room 1 is a beginner-level class. A third class taught on Tuesdays at 5:00 p.m. in Room 2 is an advanced-level class.

- Students may attend any class of the appropriate level during each week, so there is no expectation that any particular student will attend any particular class session. Therefore, the attendance of students at each individual class meeting must be tracked.

- A student will attend many different class meetings, and each class meeting is normally attended by many students. Some class meetings may not be attended by any students. New students may not have attended any class meetings yet.

- At any given meeting of a class, instructors other than the assigned instructor may show up to help. Therefore, a given class meeting may have a head instructor and many assistant instructors, but it will always have at least the one instructor who is assigned to that class. For each class meeting, the date of the class and the instructors' roles (head instructor or assistant instructor) need to be recorded. For example, Mr. Jones is assigned to teach the Monday, 5:00 p.m., intermediate class in Room 1. During a particular meeting of that class, Mr. Jones was the head instructor and Ms. Chen served as an assistant instructor.

- Each student holds a rank in the martial arts. The rank name, belt color, and rank requirements are stored. Most ranks have numerous rank requirements, but each requirement is associated with only one particular rank. All ranks except white belt have at least one requirement.

- A given rank may be held by many students. While it is customary to think of a student as having a single rank, it is necessary to track each student's progress through the ranks. Therefore, every rank that a student attains is kept in the system. New students joining the school are automatically given the rank of white belt. The date that a student is awarded each rank should be kept in the system. All ranks have at least one student who has achieved that rank at some time.

9. The *Journal of E-commerce Research Knowledge* is a prestigious information systems research journal. It uses a peer-review process to select manuscripts for publication. Only about 10 percent of the manuscripts submitted to the journal are accepted for publication. A new issue of the journal is published each quarter. Create a complete ERD to support the business needs described below.

   - Unsolicited manuscripts are submitted by authors. When a manuscript is received, the editor assigns it a number and records some basic information about it in the system, including the title of the manuscript, the date it was received, and a manuscript status of "received." Information about the author(s) is also recorded, including each author's name, mailing address, email address, and affiliation (the author's school or company). Every manuscript must have an author. Only authors who have submitted manuscripts are kept in the system. It is typical for a manuscript to have several authors. A single author may have submitted many different manuscripts to the journal. Additionally, when a manuscript has multiple authors, it is important to record the order in which the authors are listed in the manuscript credits.

   - At his or her earliest convenience, the editor will briefly review the topic of the manuscript to ensure that its contents fall within the scope of the journal. If the content is not appropriate for the journal, the manuscript's status is changed to "rejected," and the author is notified via email. If the content is within the scope of the journal, then the editor selects three or more reviewers to review the manuscript. Reviewers work for other companies or universities and read manuscripts to ensure their scientific validity. For each reviewer, the system records a reviewer number, name, email address, affiliation, and areas of interest. Areas of interest are predefined areas of expertise that the reviewer has specified. An area of interest is identified by an IS code and includes a description (for example, IS2003 is the code for "database modeling"). A reviewer can have many areas of interest, and an area of interest can be associated with many reviewers. All reviewers must specify at least one area of interest. It is unusual, but possible, to have an area of interest for which the journal has no reviewers. The editor will change the status of the manuscript to "under review" and record which reviewers received the manuscript and the date it was sent to each reviewer. A reviewer will typically receive several manuscripts to review each year, although new reviewers may not have received any manuscripts yet.

   - The reviewers will read the manuscript at their earliest convenience and provide feedback to the editor. The feedback from each reviewer includes rating the manuscript on a 10-point scale for appropriateness, clarity, methodology, and contribution to the field, as well as a recommendation for publication (accept or reject). The editor will record all of this information in the system for each review received, along with the date the feedback was received. Once all of the reviewers have provided their evaluations, the editor will decide whether to publish the manuscript and change its status to "accepted" or "rejected." If the manuscript will be published, the date of acceptance is recorded.

- Once a manuscript has been accepted for publication, it must be scheduled. For each issue of the journal, the publication period (fall, winter, spring, or summer), publication year, volume, and number are recorded. An issue will contain many manuscripts, although the issue may be created in the system before it is known which manuscripts will be published in that issue. An accepted manuscript appears in only one issue of the journal. Each manuscript goes through a typesetting process that formats the content, including fonts, font size, line spacing, justification, and so on. Once the manuscript has been typeset, its number of pages is recorded in the system. The editor will then decide which issue each accepted manuscript will appear in and the order of manuscripts within each issue. The order and the beginning page number for each manuscript must be stored in the system. Once the manuscript has been scheduled for an issue, the status of the manuscript is changed to "scheduled." Once an issue is published, the print date for the issue is recorded, and the status of each manuscript in that issue is changed to "published."

10. Global Unified Technology Sales (GUTS) is moving toward a "bring your own device" (BYOD) model for employee computing. Employees can use traditional desktop computers in their offices. They can also use a variety of personal mobile computing devices such as tablets, smartphones, and laptops. The new computing model introduces some security risks that GUTS is attempting to address. The company wants to ensure that any devices connecting to their servers are properly registered and approved by the Information Technology department. Create a complete ERD to support the business needs described below:

   - Every employee works for a department that has a department code, name, mail box number, and phone number. The smallest department currently has 5 employees, and the largest department has 40 employees. This system will only track in which department an employee is currently employed. Very rarely, a new department can be created within the company. At such times, the department may exist temporarily without any employees. For every employee, an employee number and name (first, last, and middle initial) are recorded in the system. It is also necessary to keep each employee's title.

   - An employee can have many devices registered in the system. Each device is assigned an identification number when it is registered. Most employees have at least one device, but newly hired employees might not have any devices registered initially. For each device, the brand and model need to be recorded. Only devices that are registered to an employee will be in the system. While unlikely, it is possible that a device could transfer from one employee to another. However, if that happens, only the employee who currently owns the device is tracked in the system. When a device is registered in the system, the date of that registration needs to be recorded.

   - Devices can be either desktop systems that reside in a company office or mobile devices. Desktop devices are typically provided by the company and are intended to be a permanent part of the company network. As such, each desktop device is assigned a static IP address, and the MAC address for the computer hardware is kept in the system. A desktop device is kept in a static location (building name and office number). This location should also be kept in the system so that, if the device becomes compromised, the IT department can dispatch someone to remediate the problem.

   - For mobile devices, it is important to also capture the device's serial number, which operating system (OS) it is using, and the version of the OS. The IT department is also verifying that each mobile device has a screen lock enabled and has encryption enabled for data. The system should support storing information on whether or not each mobile device has these capabilities enabled.

- Once a device is registered in the system, and the appropriate capabilities are enabled if it is a mobile device, the device may be approved for connections to one or more servers. Not all devices meet the requirements to be approved at first so the device might be in the system for a period of time before it is approved to connect to any server. GUTS has a number of servers, and a device must be approved for each server individually. Therefore, it is possible for a single device to be approved for several servers but not for all servers.

- Each server has a name, brand, and IP address. Within the IT department's facilities are a number of climate-controlled server rooms where the physical servers can be located. Which room each server is in should also be recorded. Further, it is necessary to track which operating system is being used on each server. Some servers are virtual servers and some are physical servers. If a server is a virtual server, then the system should track which physical server it is running on. A single physical server can host many virtual servers, but each virtual server is hosted on only one physical server. Only physical servers can host a virtual server. In other words, one virtual server cannot host another virtual server. Not all physical servers host a virtual server.

- A server will normally have many devices that are approved to access the server, but it is possible for new servers to be created that do not yet have any approved devices. When a device is approved for connection to a server, the date of that approval should be recorded. It is also possible for a device that was approved for a server to lose its approval. If that happens, the date that the approval was removed should be recorded. If a device loses its approval, it may regain that approval at a later date if whatever circumstance that lead to the removal is resolved.

- A server can provide many user services, such as email, chat, homework managers, and others. Each service on a server has a unique identification number and name. The date that GUTS began offering that service should be recorded. Each service runs on only one server although new servers might not offer any services initially. Client-side services are not tracked in this system so every service must be associated with a server.

- Employees must get permission to access a service before they can use it. Most employees have permissions to use a wide array of services, but new employees might not have permission on any service. Each service can support multiple approved employees as users, but new services might not have any approved users at first. The date on which the employee is approved to use a service is tracked by the system. The first time an employee is approved to access a service, the employee must create a username and password. This will be the same username and password that the employee will use for every service for which the employee is eventually approved.

11. Global Computer Solutions (GCS) is an information technology consulting company with many offices throughout the United States. The company's success is based on its ability to maximize its resources—that is, its ability to match highly skilled employees with projects according to region. To better manage its projects, GCS has contacted you to design a database so GCS managers can keep track of their customers, employees, projects, project schedules, assignments, and invoices.

    The GCS database must support all of GCS's operations and information requirements. A basic description of the main entities follows:

    - The *employees* of GCS must have an employee ID, a last name, a middle initial, a first name, a region, and a date of hire recorded in the system.

- Valid *regions* are as follows: Northwest (NW), Southwest (SW), Midwest North (MN), Midwest South (MS), Northeast (NE), and Southeast (SE).

- Each employee has many skills, and many employees have the same skill.

- Each *skill* has a skill ID, description, and rate of pay. Valid skills are as follows: Data Entry I, Data Entry II, Systems Analyst I, Systems Analyst II, Database Designer I, Database Designer II, Cobol I, Cobol II, C++ I, C++ II, VB I, VB II, ColdFusion I, ColdFusion II, ASP I, ASP II, Oracle DBA, MS SQL Server DBA, Network Engineer I, Network Engineer II, Web Administrator, Technical Writer, and Project Manager. Table P5.11a shows an example of the Skills Inventory.

TABLE P5.11A

SKILL	EMPLOYEE
Data Entry I	Seaton Amy; Williams Josh; Underwood Trish
Data Entry II	Williams Josh; Seaton Amy
Systems Analyst I	Craig Brett; Sewell Beth; Robbins Erin; Bush Emily; Zebras Steve
Systems Analyst II	Chandler Joseph; Burklow Shane; Robbins Erin
DB Designer I	Yarbrough Peter; Smith Mary
DB Designer II	Yarbrough Peter; Pascoe Jonathan
Cobol I	Kattan Chris; Ephanor Victor; Summers Anna; Ellis Maria
Cobol II	Kattan Chris; Ephanor Victor; Batts Melissa
C++ I	Smith Jose; Rogers Adam; Cope Leslie
C++ II	Rogers Adam; Bible Hanah
VB I	Zebras Steve; Ellis Maria
VB II	Zebras Steve; Newton Christopher
ColdFusion I	Duarte Miriam; Bush Emily
ColdFusion II	Bush Emily; Newton Christopher
ASP I	Duarte Miriam; Bush Emily
ASP II	Duarte Miriam; Newton Christopher
Oracle DBA	Smith Jose; Pascoe Jonathan
SQL Server DBA	Yarbrough Peter; Smith Jose
Network Engineer I	Bush Emily; Smith Mary
Network Engineer II	Bush Emily; Smith Mary
Web Administrator	Bush Emily; Smith Mary; Newton Christopher
Technical Writer	Kilby Surgena; Bender Larry
Project Manager	Paine Brad; Mudd Roger; Kenyon Tiffany; Connor Sean

- GCS has many *customers*. Each customer has a customer ID, name, phone number, and region.

- GCS works by *projects*. A project is based on a contract between the customer and GCS to design, develop, and implement a computerized solution. Each project has specific characteristics such as the project ID, the customer to which the project belongs, a brief description, a project date (the date the contract was signed), an estimated project start date and end date, an estimated project budget, an actual start date, an actual end date, an actual cost, and one employee assigned as the manager of the project.

- The actual cost of the project is updated each Friday by adding that week's cost to the actual cost. The week's cost is computed by multiplying the hours each employee worked by the rate of pay for that skill.

- The employee who is the manager of the project must complete a *project schedule*, which effectively is a design and development plan. In the project schedule (or plan), the manager must determine the tasks that will be performed to take the project from beginning to end. Each task has a task ID, a brief task description, starting and ending dates, the types of skills needed, and the number of employees (with the required skills) needed to complete the task. General tasks are the initial interview, database and system design, implementation, coding, testing, and final evaluation and sign-off. For example, GCS might have the project schedule shown in Table P5.11b.

TABLE P5.11B

PROJECT ID: 1   **DESCRIPTION: SALES MANAGEMENT SYSTEM**
COMPANY: SEE ROCKS   **CONTRACT DATE: 2/12/2016**   **REGION: NW**
START DATE: 3/1/2016   **END DATE: 7/1/2016**   **BUDGET: $15,500**

START DATE	END DATE	TASK DESCRIPTION	SKILL(S) REQUIRED	QUANTITY REQUIRED
3/1/16	3/6/16	Initial interview	Project Manager Systems Analyst II DB Designer I	1 1 1
3/11/16	3/15/16	Database design	DB Designer I	1
3/11/16	4/12/16	System design	Systems Analyst II Systems Analyst I	1 2
3/18/16	3/22/16	Database implementation	Oracle DBA	1
3/25/16	5/20/16	System coding and testing	Cobol I Cobol II Oracle DBA	2 1 1
3/25/16	6/7/16	System documentation	Technical Writer	1
6/10/16	6/14/16	Final evaluation	Project Manager Systems Analyst II DB Designer I Cobol II	1 1 1 1
6/17/16	6/21/16	On-site system online and data loading	Project Manager Systems Analyst II DB Designer I Cobol II	1 1 1 1
7/1/16	7/1/16	Sign-off	Project Manager	1

- GCS pools all of its employees by region; from this pool, employees are assigned to a specific task scheduled by the project manager. For example, in the first project's schedule, you know that a Systems Analyst II, Database Designer I, and Project Manager are needed for the period from 3/1/16 to 3/6/16. The project manager is assigned when the project is created and remains for the duration of the project. Using that information, GCS searches the employees who are located in the same region as the customer, matches the skills required, and assigns the employees to the project task.

- Each project schedule task can have many employees assigned to it, and a given employee can work on multiple project tasks. However, an employee can work on only one project task at a time. For example, if an employee is already assigned to work on a project task from 2/20/16 to 3/3/16, the employee cannot work on another task until the current assignment is closed (ends). The date that an assignment is closed does not necessarily match the ending date of the project schedule task because a task can be completed ahead of or behind schedule.

- Given all of the preceding information, you can see that the assignment associates an employee with a project task, using the project schedule. Therefore, to keep track of the *assignment*, you require at least the following information: assignment ID, employee, project schedule task, assignment start date, and assignment end date. The end date could be any date, as some projects run ahead of or behind schedule. Table P5.11c shows a sample assignment form.

TABLE P5.11C

PROJECT ID: 1			DESCRIPTION: SALES MANAGEMENT SYSTEM			
COMPANY: SEE ROCKS			CONTRACT DATE: 2/12/2016		AS OF: 03/29/16	
SCHEDULED				ACTUAL ASSIGNMENTS		
PROJECT TASK	START DATE	END DATE	SKILL	EMPLOYEE	START DATE	END DATE
Initial interview	3/1/16	3/6/16	Project Mgr. Sys. Analyst II DB Designer I	101-Connor S. 102-Burklow S. 103-Smith M.	3/1/16 3/1/16 3/1/16	3/6/16 3/6/16 3/6/16
Database design	3/11/16	3/15/16	DB Designer I	104-Smith M.	3/11/16	3/14/16
System design	3/11/16	4/12/16	Sys. Analyst II Sys. Analyst I Sys. Analyst I	105-Burklow S. 106-Bush E. 107-Zebras S.	3/11/16 3/11/16 3/11/16	
Database implementation	3/18/16	3/22/16	Oracle DBA	108-Smith J.	3/15/16	3/19/16
System coding and testing	3/25/16	5/20/16	Cobol I Cobol I Cobol II Oracle DBA	109-Summers A. 110-Ellis M. 111-Ephanor V. 112-Smith J.	3/21/16 3/21/16 3/21/16 3/21/16	
System documentation	3/25/16	6/7/16	Tech. Writer	113-Kilby S.	3/25/16	
Final evaluation	6/10/16	6/14/16	Project Mgr. Sys. Analyst II DB Designer I Cobol II			
On-site system online and data loading	6/17/16	6/21/16	Project Mgr. Sys. Analyst II DB Designer I Cobol II			
Sign-off	7/1/16	7/1/16	Project Mgr.			

(*Note:* The assignment number is shown as a prefix of the employee name—for example, 101 or 102.) Assume that the assignments shown previously are the only ones as of the date of this design. The assignment number can be any number that matches your database design.

- Employee work hours are kept in a *work log*, which contains a record of the actual hours worked by employees on a given assignment. The work log is a form that the employee fills out at the end of each week (Friday) or at the end of each month. The form contains the date, which is either the current Friday of the month or the last workday of the month if it does not fall on a Friday. The form also contains the assignment ID, the total hours worked either that week or up to the end of the month, and the bill number to which the work-log entry is charged. Obviously, each work-log entry can be related to only one bill. A sample list of the current work-log entries for the first sample project is shown in Table P5.11d.

TABLE P5.11D

EMPLOYEE NAME	WEEK ENDING	ASSIGNMENT NUMBER	HOURS WORKED	BILL NUMBER
Burklow S.	3/1/16	1-102	4	xxx
Connor S.	3/1/16	1-101	4	xxx
Smith M.	3/1/16	1-103	4	xxx
Burklow S.	3/8/16	1-102	24	xxx
Connor S.	3/8/16	1-101	24	xxx
Smith M.	3/8/16	1-103	24	xxx
Burklow S.	3/15/16	1-105	40	xxx
Bush E.	3/15/16	1-106	40	xxx
Smith J.	3/15/16	1-108	6	xxx
Smith M.	3/15/16	1-104	32	xxx
Zebras S.	3/15/16	1-107	35	xxx
Burklow S.	3/22/16	1-105	40	
Bush E.	3/22/16	1-106	40	
Ellis M.	3/22/16	1-110	12	
Ephanor V.	3/22/16	1-111	12	
Smith J.	3/22/16	1-108	12	
Smith J.	3/22/16	1-112	12	
Summers A.	3/22/16	1-109	12	
Zebras S.	3/22/16	1-107	35	
Burklow S.	3/29/16	1-105	40	
Bush E.	3/29/16	1-106	40	
Ellis M.	3/29/16	1-110	35	
Ephanor V.	3/29/16	1-111	35	
Kilby S.	3/29/16	1-113	40	
Smith J.	3/29/16	1-112	35	
Summers A.	3/29/16	1-109	35	
Zebras S.	3/29/16	1-107	35	

Note: xxx represents the bill ID. Use the one that matches the bill number in your database.

- Finally, every 15 days, a *bill* is written and sent to the customer for the total hours worked on the project during that period. When GCS generates a bill, it uses the bill number to update the work-log entries that are part of the bill. In summary, a bill can refer to many work-log entries, and each work-log entry can be related to only one bill. GCS sent one bill on 3/15/16 for the first project (SEE ROCKS), totaling the hours worked between 3/1/16 and 3/15/16. Therefore, you can safely assume that there is only one bill in this table and that the bill covers the work-log entries shown in the preceding form.

Your assignment is to create a database that fulfills the operations described in this problem. The minimum required entities are employee, skill, customer, region, project, project schedule, assignment, work log, and bill. (There are additional required entities that are not listed.)

- Create all of the required tables and required relationships.
- Create the required indexes to maintain entity integrity when using surrogate primary keys.
- Populate the tables as needed, as indicated in the sample data and forms.

Chapter 6

Normalization of Database Tables

In this chapter, you will learn:

• What normalization is and what role it plays in the database design process

• About the normal forms 1NF, 2NF, 3NF, BCNF, and 4NF

• How normal forms can be transformed from lower normal forms to higher normal forms

• That normalization and ER modeling are used concurrently to produce a good database design

• That some situations require denormalization to generate information efficiently

Preview

Good database design must be matched to good table structures. In this chapter, you will learn to evaluate and design good table structures to control data redundancies, thereby avoiding data anomalies. The process that yields such desirable results is known as normalization.

To recognize and appreciate the characteristics of a good table structure, it is useful to examine a poor one. Therefore, the chapter begins by examining the characteristics of a poor table structure and the problems it creates. You then learn how to correct the table structure. This methodology will yield important dividends: you will know how to design a good table structure and how to repair a poor one.

You will discover not only that data anomalies can be eliminated through normalization, but that a properly normalized set of table structures is actually less complicated to use than an unnormalized set. In addition, you will learn that the normalized set of table structures more faithfully reflects an organization's real operations.

Data Files and Available Formats

	MS Access	Oracle	MS SQL	My SQL		MS Access	Oracle	MS SQL	My SQL
CH06_ConstructCo	✓	✓	✓	✓	CH06_Service	✓	✓	✓	✓
CH06_Eval	✓	✓	✓	✓					

Data Files Available on cengagebrain.com

6-1 Database Tables and Normalization

Having good relational database software is not enough to avoid the data redundancy discussed in Chapter 1, Database Systems. If the database tables are treated as though they are files in a file system, the relational database management system (RDBMS) never has a chance to demonstrate its superior data-handling capabilities.

The table is the basic building block of database design. Consequently, the table's structure is of great interest. Ideally, the database design process explored in Chapter 4, Entity Relationship (ER) Modeling, yields good table structures. Yet, it is possible to create poor table structures even in a good database design. How do you recognize a poor table structure, and how do you produce a good table? The answer to both questions involves normalization. **Normalization** is a process for evaluating and correcting table structures to minimize data redundancies, thereby reducing the likelihood of data anomalies. The normalization process involves assigning attributes to tables based on the concept of determination you learned in Chapter 3, The Relational Database Model.

Normalization works through a series of stages called normal forms. The first three stages are described as first normal form (1NF), second normal form (2NF), and third normal form (3NF). From a structural point of view, 2NF is better than 1NF, and 3NF is better than 2NF. For most purposes in business database design, 3NF is as high as you need to go in the normalization process. However, you will discover that properly designed 3NF structures also meet the requirements of fourth normal form (4NF).

Although normalization is a very important ingredient in database design, you should not assume that the highest level of normalization is always the most desirable. Generally, the higher the normal form, the more relational join operations you need to produce a specified output. Also, more resources are required by the database system to respond to end-user queries. A successful design must also consider end-user demand for fast performance. Therefore, you will occasionally need to *denormalize* some portions of a database design to meet performance requirements. **Denormalization** produces a lower normal form; that is, a 3NF will be converted to a 2NF through denormalization. However, the price you pay for increased performance through denormalization is greater data redundancy.

normalization
A process that assigns attributes to entities so that data redundancies are reduced or eliminated.

denormalization
A process by which a table is changed from a higher-level normal form to a lower-level normal form, usually to increase processing speed. Denormalization potentially yields data anomalies.

prime attribute
A key attribute; that is, an attribute that is part of a key or is the whole key. See also *key attributes*.

key attributes
The attributes that form a primary key. See also *prime attribute*.

nonprime attribute
An attribute that is not part of a key.

nonkey attribute
See *nonprime attribute*.

Note

Although the word *table* is used throughout this chapter, formally, normalization is concerned with relations. In Chapter 3 you learned that the terms *table* and *relation* are frequently used interchangeably. In fact, you can say that a table is the implementation view of a logical relation that meets some specific conditions. (See Table 3.1.) However, being more rigorous, the mathematical relation does not allow duplicate tuples; whereas they could exist in tables (see Section 6-5). Also, in normalization terminology, any attribute that is at least part of a key is known as a **prime attribute** instead of the more common term **key attribute**, which was introduced earlier. Conversely, a **nonprime attribute**, or a **nonkey attribute**, is not part of any candidate key.

6-2 The Need For Normalization

Normalization is typically used in conjunction with the entity relationship modeling that you learned in the previous chapters. Database designers commonly use normalization in two situations. When designing a new database structure based on the business requirements of the end users, the database designer will construct a data model using a technique such as Crow's Foot notation ERDs. After the initial design is complete, the designer can use normalization to analyze the relationships among the attributes within each entity and determine if the structure can be improved through normalization. Alternatively, database designers are often asked to modify existing data structures that can be in the form of flat files, spreadsheets, or older database structures. Again, by analyzing relationships among the attributes or fields in the data structure, the database designer can use the normalization process to improve the existing data structure and create an appropriate database design. Whether you are designing a new database structure or modifying an existing one, the normalization process is the same.

To get a better idea of the normalization process, consider the simplified database activities of a construction company that manages several building projects. Each project has its own project number, name, assigned employees, and so on. Each employee has an employee number, name, and job classification, such as engineer or computer technician.

The company charges its clients by billing the hours spent on each contract. The hourly billing rate is dependent on the employee's position. For example, one hour of computer technician time is billed at a different rate than one hour of engineer time. Periodically, a report is generated that contains the information displayed in Table 6.1.

The total charge in Table 6.1 is a derived attribute and is not stored in the table at this point.

The easiest short-term way to generate the required report might seem to be a table whose contents correspond to the reporting requirements. (See Figure 6.1.)

FIGURE 6.1 TABULAR REPRESENTATION OF THE REPORT FORMAT

Table name: **RPT_FORMAT**   Database name: **Ch06_ConstructCo**

PROJ_NUM	PROJ_NAME	EMP_NUM	EMP_NAME	JOB_CLASS	CHG_HOUR	HOURS
15	Evergreen	103	June E. Arbough	Elect. Engineer	84.50	23.8
		101	John G. News	Database Designer	105.00	19.4
		105	Alice K. Johnson *	Database Designer	105.00	35.7
		106	William Smithfield	Programmer	35.75	12.6
		102	David H. Senior	Systems Analyst	96.75	23.8
18	Amber Wave	114	Annelise Jones	Applications Designer	48.10	24.6
		118	James J. Frommer	General Support	18.36	45.3
		104	Anne K. Ramoras *	Systems Analyst	96.75	32.4
		112	Darlene M. Smithson	DSS Analyst	45.95	44.0
22	Rolling Tide	105	Alice K. Johnson	Database Designer	105.00	64.7
		104	Anne K. Ramoras	Systems Analyst	96.75	48.4
		113	Delbert K. Joenbrood *	Applications Designer	48.10	23.6
		111	Geoff B. Wabash	Clerical Support	26.87	22.0
		106	William Smithfield	Programmer	35.75	12.8
25	Starflight	107	Maria D. Alonzo	Programmer	35.75	24.6
		115	Travis B. Bawangi	Systems Analyst	96.75	45.8
		101	John G. News *	Database Designer	105.00	56.3
		114	Annelise Jones	Applications Designer	48.10	33.1
		108	Ralph B. Washington	Systems Analyst	96.75	23.6
		118	James J. Frommer	General Support	18.36	30.5
		112	Darlene M. Smithson	DSS Analyst	45.95	41.4

Note that the data in Figure 6.1 reflects the assignment of employees to projects. Apparently, an employee can be assigned to more than one project. For example, Darlene Smithson (EMP_NUM = 112) has been assigned to two projects: Amber Wave and Starflight. Given the structure of the dataset, each project includes only a single occurrence of any one employee. Therefore, knowing the PROJ_NUM and EMP_NUM values will let you find the job classification and its hourly charge. In addition, you will know the total number of hours each employee worked on each project. (The total charge—a derived attribute whose value can be computed by multiplying the hours billed and the charge per hour—has not been included in Figure 6.1. No structural harm is done if this derived attribute is included.)

TABLE 6.1

A SAMPLE REPORT LAYOUT

PROJECT NUMBER	PROJECT NAME	EMPLOYEE NUMBER	EMPLOYEE NAME	JOB CLASS	CHARGE/ HOUR	HOURS BILLED	TOTAL CHARGE
15	Evergreen	103	June E. Arbough	Elec. Engineer	$ 84.50	23.8	$ 2,011.10
		101	John G. News	Database Designer	$105.00	19.4	$ 2,037.00
		105	Alice K. Johnson *	Database Designer	$105.00	35.7	$ 3,748.50
		106	William Smithfield	Programmer	$ 35.75	12.6	$ 450.45
		102	David H. Senior	Systems Analyst	$ 96.75	23.8	$ 2,302.65
				Subtotal			$10,549.70
18	Amber Wave	114	Annelise Jones	Applications Designer	$ 48.10	24.6	$ 1,183.26
		118	James J. Frommer	General Support	$ 18.36	45.3	$ 831.71
		104	Anne K. Ramoras *	Systems Analyst	$ 96.75	32.4	$ 3,134.70
		112	Darlene M. Smithson	DSS Analyst	$ 45.95	44.0	$ 2,021.80
				Subtotal			$ 7,171.47
22	Rolling Tide	105	Alice K. Johnson	Database Designer	$105.00	64.7	$ 6,793.50
		104	Anne K. Ramoras	Systems Analyst	$ 96.75	48.4	$ 4,682.70
		113	Delbert K. Joenbrood *	Applications Designer	$ 48.10	23.6	$ 1,135.16
		111	Geoff B. Wabash	Clerical Support	$ 26.87	22.0	$ 591.14
		106	William Smithfield	Programmer	$ 35.75	12.8	$ 457.60
				Subtotal			$13,660.10
25	Starflight	107	Maria D. Alonzo	Programmer	$ 35.75	24.6	$ 879.45
		115	Travis B. Bawangi	Systems Analyst	$ 96.75	45.8	$ 4,431.15
		101	John G. News *	Database Designer	$105.00	56.3	$ 5,911.50
		114	Annelise Jones	Applications Designer	$ 48.10	33.1	$ 1,592.11
		108	Ralph B. Washington	Systems Analyst	$ 96.75	23.6	$ 2,283.30
		118	James J. Frommer	General Support	$ 18.36	30.5	$ 559.98
		112	Darlene M. Smithson	DSS Analyst	$ 45.95	41.4	$ 1,902.33
				Subtotal			$17,559.82
				Total			$48,941.09

Note: A * indicates the project leader.

Unfortunately, the structure of the dataset in Figure 6.1 does not conform to the requirements discussed in Chapter 3, nor does it handle data very well. Consider the following deficiencies:

1. The project number (PROJ_NUM) is apparently intended to be a primary key (PK) or at least a part of a PK, but it contains nulls. Given the preceding discussion, you know that PROJ_NUM + EMP_NUM will define each row.

2. The table entries invite data inconsistencies. For example, the JOB_CLASS value "Elect. Engineer" might be entered as "Elect.Eng." in some cases, "El. Eng." in others, and "EE" in still others.

3. The table displays data redundancies that yield the following anomalies:

   a. *Update anomalies.* Modifying the JOB_CLASS for employee number 105 requires many potential alterations, one for each EMP_NUM = 105.

   b. *Insertion anomalies.* Just to complete a row definition, a new employee must be assigned to a project. If the employee is not yet assigned, a phantom project must be created to complete the employee data entry.

   c. *Deletion anomalies.* Suppose that only one employee is associated with a given project. If that employee leaves the company and the employee data is deleted, the project information will also be deleted. To prevent the loss of the project information, a fictitious employee must be created.

In spite of those structural deficiencies, the table structure *appears* to work; the report is generated with ease. Unfortunately, the report might yield varying results depending on what data anomaly has occurred. For example, if you want to print a report to show the total "hours worked" value by the job classification "Database Designer," that report will not include data for "DB Design" and "Database Design" data entries. Such reporting anomalies cause a multitude of problems for managers—and cannot be fixed through application programming.

Even if careful data-entry auditing can eliminate most of the reporting problems (at a high cost), it is easy to demonstrate that even a simple data entry becomes inefficient. Given the existence of update anomalies, suppose Darlene M. Smithson is assigned to work on the Evergreen project. The data-entry clerk must update the PROJECT file with the following entry:

| 15 | Evergreen | 112 | Darlene M. Smithson | DSS Analyst | $45.95 | 0.0 |

to match the attributes PROJ_NUM, PROJ_NAME, EMP_NUM, EMP_NAME, JOB_CLASS, CHG_HOUR, and HOURS. (If Smithson has just been assigned to the project, the total number of hours worked is 0.0.)

Note

Remember that the naming convention makes it easy to see what each attribute stands for and its likely origin. For example, PROJ_NAME uses the prefix PROJ to indicate that the attribute is associated with the PROJECT table, while the NAME component is self-documenting as well. However, keep in mind that name length is also an issue, especially in the prefix designation. For that reason, the prefix CHG was used rather than CHARGE. (Given the database's context, it is not likely that the prefix will be misunderstood.)

Each time another employee is assigned to a project, some data entries (such as PROJ_NAME, EMP_NAME, and CHG_HOUR) are unnecessarily repeated. Imagine the data-entry chore when 200 or 300 table entries must be made! The entry of the employee number should be sufficient to identify Darlene M. Smithson, her job description, and her hourly charge. Because only one person is identified by the number 112, that person's characteristics (name, job classification, and so on) should not have to be entered each time the main file is updated. Unfortunately, the structure displayed in Figure 6.1 does not make allowances for that possibility.

The data redundancy evident in Figure 6.1 leads to wasted data storage space. Even worse, data redundancy produces data anomalies. For example, suppose the data-entry clerk had entered the data as:

| 15 | Evergeen | 112 | Darla Smithson | DCS Analyst | $45.95 | 0.0 |

At first glance, the data entry appears to be correct. But is *Evergeen* the same project as *Evergreen*? And is *DCS Analyst* supposed to be *DSS Analyst*? Is Darla Smithson the same person as Darlene M. Smithson? Such confusion is a data integrity problem because the data entry failed to conform to the rule that all copies of redundant data must be identical.

The possibility of introducing data integrity problems caused by data redundancy must be considered during database design. The relational database environment is especially well suited to help the designer overcome those problems.

6-3 The Normalization Process

In this section, you will learn how to use normalization to produce a set of normalized tables to store the data that will be used to generate the required information. The objective of normalization is to ensure that each table conforms to the concept of well-formed relations—in other words, tables that have the following characteristics:

- Each table represents a single subject. For example, a COURSE table will contain only data that directly pertain to courses. Similarly, a STUDENT table will contain only student data.

- No data item will be *unnecessarily* stored in more than one table (in short, tables have minimum controlled redundancy). The reason for this requirement is to ensure that the data is updated in only one place.

- All nonprime attributes in a table are dependent on the primary key—the entire primary key and nothing but the primary key. The reason for this requirement is to ensure that the data is uniquely identifiable by a primary key value.

- Each table is void of insertion, update, or deletion anomalies, which ensures the integrity and consistency of the data.

To accomplish the objective, the normalization process takes you through the steps that lead to successively higher normal forms. The most common normal forms and their basic characteristic are listed in Table 6.2. You will learn the details of these normal forms in the indicated sections.

The concept of keys is central to the discussion of normalization. Recall from Chapter 3 that a candidate key is a minimal (irreducible) superkey. The primary key is the candidate key selected to be the primary means used to identify the rows in the table. Although normalization is typically presented from the perspective of candidate keys, this initial discussion assumes for the sake of simplicity that each table has only one candidate key; therefore, that candidate key is the primary key.

From the data modeler's point of view, the objective of normalization is to ensure that all tables are at least in third normal form (3NF). Even higher-level normal forms exist.

TABLE 6.2

NORMAL FORMS

NORMAL FORM	CHARACTERISTIC	SECTION
First normal form (1NF)	Table format, no repeating groups, and PK identified	6-3a
Second normal form (2NF)	1NF and no partial dependencies	6-3b
Third normal form (3NF)	2NF and no transitive dependencies	6-3c
Boyce-Codd normal form (BCNF)	Every determinant is a candidate key (special case of 3NF)	6-6a
Fourth normal form (4NF)	3NF and no independent multivalued dependencies	6-6b

However, normal forms such as the fifth normal form (5NF) and domain-key normal form (DKNF) are not likely to be encountered in a business environment and are mainly of theoretical interest. Such higher normal forms usually increase joins, which slows performance without adding any value in the elimination of data redundancy. Some very specialized applications, such as statistical research, might require normalization beyond the 4NF, but those applications fall outside the scope of most business operations. Because this book focuses on practical applications of database techniques, the higher-level normal forms are not covered.

Functional Dependence Before outlining the normalization process, it is a good idea to review the concepts of determination and functional dependence that were covered in detail in Chapter 3. Table 6.3 summarizes the main concepts.

TABLE 6.3

FUNCTIONAL DEPENDENCE CONCEPTS

CONCEPT	DEFINITION
Functional dependence	The attribute B is fully functionally dependent on the attribute A if each value of A determines one and only one value of B. Example: PROJ_NUM $\rightarrow$ PROJ_NAME (read as *PROJ_NUM functionally determines PROJ_NAME*) In this case, the attribute PROJ_NUM is known as the determinant attribute, and the attribute PROJ_NAME is known as the dependent attribute.
Functional dependence (generalized definition)	Attribute A determines attribute B (that is, B is functionally dependent on A) if all (generalized definition) of the rows in the table that agree in value for attribute A also agree in value for attribute B.
Fully functional dependence (composite key)	If attribute B is functionally dependent on a composite key A but not on any subset of that composite key, the attribute B is fully functionally dependent on A.

It is crucial to understand these concepts because they are used to derive the set of functional dependencies for a given relation. The normalization process works one relation at a time, identifying the dependencies on that relation and normalizing the relation. As you will see in the following sections, normalization starts by identifying the dependencies of a given relation and progressively breaking up the relation (table) into a set of new relations (tables) based on the identified dependencies.

Two types of functional dependencies that are of special interest in normalization are partial dependencies and transitive dependencies. A **partial dependency** exists when there is a functional dependence in which the determinant is only part of the primary key (remember the assumption that there is only one candidate key). For example, if (A, B) $\rightarrow$ (C, D), B $\rightarrow$ C, and (A, B) is the primary key, then the functional

partial dependency
A condition in which an attribute is dependent on only a portion (subset) of the primary key.

dependence B → C is a partial dependency because only part of the primary key (B) is needed to determine the value of C. Partial dependencies tend to be straightforward and easy to identify.

A **transitive dependency** exists when there are functional dependencies such that X → Y, Y → Z, and X is the primary key. In that case, the dependency X → Z is a transitive dependency because X determines the value of Z via Y. Unlike partial dependencies, transitive dependencies are more difficult to identify among a set of data. Fortunately, there is an effective way to identify transitive dependencies: they occur only when a functional dependence exists among nonprime attributes. In the previous example, the actual transitive dependency is X → Z. However, the dependency Y → Z signals that a transitive dependency exists. Hence, throughout the discussion of the normalization process, the existence of a functional dependence among nonprime attributes will be considered a sign of a transitive dependency. To address the problems related to transitive dependencies, changes to the table structure are made based on the functional dependence that signals the transitive dependency's existence. Therefore, to simplify the description of normalization, from this point forward the signaling dependency will be called the *transitive dependency*.

6-3a Conversion To First Normal Form

Because the relational model views data as part of a table or a collection of tables in which all key values must be identified, the data depicted in Figure 6.1 might not be stored as shown. Note that Figure 6.1 contains what is known as repeating groups. A **repeating group** derives its name from the fact that a group of multiple entries of the same type can exist for any *single* key attribute occurrence. In Figure 6.1, note that each single project number (PROJ_NUM) occurrence can reference a group of related data entries. For example, the Evergreen project (PROJ_NUM = 15) shows five entries at this point—and those entries are related because they each share the PROJ_NUM = 15 characteristic. Each time a new record is entered for the Evergreen project, the number of entries in the group grows by one.

A relational table must not contain repeating groups. The existence of repeating groups provides evidence that the RPT_FORMAT table in Figure 6.1 fails to meet even the lowest normal form requirements, thus reflecting data redundancies.

Normalizing the table structure will reduce the data redundancies. If repeating groups do exist, they must be eliminated by making sure that each row defines a single entity. In addition, the dependencies must be identified to diagnose the normal form. Identification of the normal form lets you know where you are in the normalization process. Normalization starts with a simple three-step procedure.

Step 1: Eliminate the Repeating Groups   Start by presenting the data in a tabular format, where each cell has a single value and there are no repeating groups. To eliminate the repeating groups, eliminate the nulls by making sure that each repeating group attribute contains an appropriate data value. That change converts the table in Figure 6.1 to 1NF in Figure 6.2.

Step 2: Identify the Primary Key   The layout in Figure 6.2 represents more than a mere cosmetic change. Even a casual observer will note that PROJ_NUM is not an adequate primary key because the project number does not uniquely identify all of the remaining entity (row) attributes. For example, the PROJ_NUM value 15 can identify any one of five employees. To maintain a proper primary key that will *uniquely* identify any attribute value, the new key must be composed of a *combination* of PROJ_NUM and EMP_NUM. For example, using the data shown in Figure 6.2, if you know that

transitive dependency
A condition in which an attribute is dependent on another attribute that is not part of the primary key.

repeating group
In a relation, a characteristic describing a group of multiple entries of the same type for a single key attribute occurrence. For example, a car can have multiple colors for its top, interior, bottom, trim, and so on.

FIGURE 6.2 A TABLE IN FIRST NORMAL FORM

Table name: **DATA_ORG_1NF**   Database name: **Ch06_ConstructCo**

PROJ_NUM	PROJ_NAME	EMP_NUM	EMP_NAME	JOB_CLASS	CHG_HOUR	HOURS
15	Evergreen	103	June E. Arbough	Elect. Engineer	84.50	23.8
15	Evergreen	101	John G. News	Database Designer	105.00	19.4
15	Evergreen	105	Alice K. Johnson *	Database Designer	105.00	35.7
15	Evergreen	106	William Smithfield	Programmer	35.75	12.6
15	Evergreen	102	David H. Senior	Systems Analyst	96.75	23.8
18	Amber Wave	114	Annelise Jones	Applications Designer	48.10	24.6
18	Amber Wave	118	James J. Frommer	General Support	18.36	45.3
18	Amber Wave	104	Anne K. Ramoras *	Systems Analyst	96.75	32.4
18	Amber Wave	112	Darlene M. Smithson	DSS Analyst	45.95	44.0
22	Rolling Tide	105	Alice K. Johnson	Database Designer	105.00	64.7
22	Rolling Tide	104	Anne K. Ramoras	Systems Analyst	96.75	48.4
22	Rolling Tide	113	Delbert K. Joenbrood *	Applications Designer	48.10	23.6
22	Rolling Tide	111	Geoff B. Wabash	Clerical Support	26.87	22.0
22	Rolling Tide	106	William Smithfield	Programmer	35.75	12.8
25	Starflight	107	Maria D. Alonzo	Programmer	35.75	24.6
25	Starflight	115	Travis B. Bawangi	Systems Analyst	96.75	45.8
25	Starflight	101	John G. News *	Database Designer	105.00	56.3
25	Starflight	114	Annelise Jones	Applications Designer	48.10	33.1
25	Starflight	108	Ralph B. Washington	Systems Analyst	96.75	23.6
25	Starflight	118	James J. Frommer	General Support	18.36	30.5
25	Starflight	112	Darlene M. Smithson	DSS Analyst	45.95	41.4

PROJ_NUM = 15 and EMP_NUM = 103, the entries for the attributes PROJ_NAME, EMP_NAME, JOB_CLASS, CHG_HOUR, and HOURS must be Evergreen, June E. Arbough, Elect. Engineer, $84.50, and 23.8, respectively.

Step 3: Identify All Dependencies The identification of the PK in Step 2 means that you have already identified the following dependency:

PROJ_NUM, EMP_NUM → PROJ_NAME, EMP_NAME, JOB_CLASS, CHG_HOUR, HOURS

That is, the PROJ_NAME, EMP_NAME, JOB_CLASS, CHG_HOUR, and HOURS values are all dependent on—they are determined by—the combination of PROJ_NUM and EMP_NUM. There are additional dependencies. For example, the project number identifies (determines) the project name. In other words, the project name is dependent on the project number. You can write that dependency as:

PROJ_NUM → PROJ_NAME

Also, if you know an employee number, you also know that employee's name, job classification, and charge per hour. Therefore, you can identify the dependency shown next:

EMP_NUM → EMP_NAME, JOB_CLASS, CHG_HOUR

In simpler terms, an employee has the following attributes: a number, a name, a job classification, and a charge per hour. However, by further studying the data in Figure 6.2, you can see that knowing the job classification means knowing the charge per hour for that job classification. (Notice that all "System Analyst" or "Programmer" positions have the same charge per hour regardless of the project or employee.) In other words, the charge per hour depends on the job classification, not the employee. Therefore, you can identify one last dependency:

JOB_CLASS → CHG_HOUR

This dependency exists between two nonprime attributes; therefore, it is a signal that a transitive dependency exists, and we will refer to it as a transitive dependency. The dependencies you have just examined can also be depicted with the help of the diagram shown in Figure 6.3. Because such a diagram depicts all dependencies found within a given table structure, it is known as a **dependency diagram**. Dependency diagrams are very helpful in getting a bird's-eye view of all the relationships among a table's attributes, and their use makes it less likely that you will overlook an important dependency.

FIGURE 6.3 FIRST NORMAL FORM (1NF) DEPENDENCY DIAGRAM

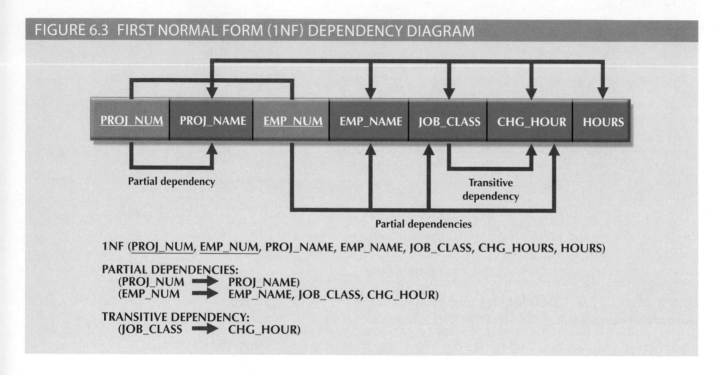

1NF (<u>PROJ_NUM</u>, <u>EMP_NUM</u>, PROJ_NAME, EMP_NAME, JOB_CLASS, CHG_HOURS, HOURS)

PARTIAL DEPENDENCIES:
   (PROJ_NUM ➡ PROJ_NAME)
   (EMP_NUM ➡ EMP_NAME, JOB_CLASS, CHG_HOUR)

TRANSITIVE DEPENDENCY:
   (JOB_CLASS ➡ CHG_HOUR)

As you examine Figure 6.3, note the following features of a dependency diagram:

1. The primary key attributes are bold, underlined, and in a different color.

2. The arrows above the attributes indicate all desirable dependencies—that is, dependencies based on the primary key. In this case, note that the entity's attributes are dependent on the *combination* of **PROJ_NUM** and **EMP_NUM**.

3. The arrows below the dependency diagram indicate less desirable dependencies. Two types of such dependencies exist:

   a. *Partial dependencies.* You need to know only the PROJ_NUM to determine the PROJ_NAME; that is, the PROJ_NAME is dependent on only part of the primary key. Also, you need to know only the EMP_NUM to find the EMP_NAME, the JOB_CLASS, and the CHG_HOUR. A dependency based on only a part of a composite primary key is a partial dependency.

   b. *Transitive dependencies.* Note that CHG_HOUR is dependent on JOB_CLASS. Because neither CHG_HOUR nor JOB_CLASS is a prime attribute—that is, neither attribute is at least part of a key—the condition is a transitive dependency. In other words, a transitive dependency is a dependency of one nonprime attribute on another nonprime attribute. The problem with transitive dependencies is that they still yield data anomalies.

Figure 6.3 includes the relational schema for the table in 1NF and a textual notation for each identified dependency.

dependency diagram
A representation of all data dependencies (primary key, partial, or transitive) within a table.

Note

The term **first normal form (1NF)** describes the tabular format in which:

- All of the key attributes are defined.

- There are no repeating groups in the table. In other words, each row/column intersection contains one and only one value, not a set of values.

- All attributes are dependent on the primary key.

All relational tables satisfy the 1NF requirements. The problem with the 1NF table structure shown in Figure 6.3 is that it contains partial dependencies—dependencies based on only a part of the primary key.

While partial dependencies are sometimes used for performance reasons, they should be used with caution. Such caution is warranted because a table that contains partial dependencies is still subject to data redundancies, and therefore to various anomalies. The data redundancies occur because every row entry requires duplication of data. For example, if Alice K. Johnson submits her work log, then the user would have to make multiple entries during the course of a day. For each entry, the EMP_NAME, JOB_CLASS, and CHG_HOUR must be entered each time, even though the attribute values are identical for each row entered. Such duplication of effort is very inefficient, and it helps create data anomalies; nothing prevents the user from typing slightly different versions of the employee name, the position, or the hourly pay. For instance, the employee name for EMP_NUM = 102 might be entered as *Dave Senior* or *D. Senior*. The project name might also be entered correctly as *Evergreen* or misspelled as *Evergeen*. Such data anomalies violate the relational database's integrity and consistency rules.

6-3b Conversion To Second Normal Form

Conversion to 2NF occurs only when the 1NF has a composite primary key. If the 1NF has a single-attribute primary key, then the table is automatically in 2NF. The 1NF-to-2NF conversion is simple. Starting with the 1NF format displayed in Figure 6.3, you take the following steps:

Step 1: Make New Tables to Eliminate Partial Dependencies For each component of the primary key that acts as a determinant in a partial dependency, create a new table with a copy of that component as the primary key. While these components are placed in the new tables, it is important that they also remain in the original table as well. The determinants must remain in the original table because they will be the foreign keys for the relationships needed to relate these new tables to the original table. To construct the revised dependency diagram, write each key component on a separate line and then write the original (composite) key on the last line. For example:

PROJ_NUM

EMP_NUM

PROJ_NUM EMP_NUM

Each component will become the key in a new table. In other words, the original table is now divided into three tables (PROJECT, EMPLOYEE, and ASSIGNMENT).

first normal form (1NF)
The first stage in the normalization process. It describes a relation depicted in tabular format, with no repeating groups and a primary key identified. All nonkey attributes in the relation are dependent on the primary key.

Step 2: Reassign Corresponding Dependent Attributes Use Figure 6.3 to determine attributes that are dependent in the partial dependencies. The dependencies for the original key components are found by examining the arrows below the dependency diagram shown in Figure 6.3. The attributes that are dependent in a partial dependency are removed from the original table and placed in the new table with the dependency's determinant. Any attributes that are not dependent in a partial dependency will remain in the original table. In other words, the three tables that result from the conversion to 2NF are given appropriate names (PROJECT, EMPLOYEE, and ASSIGNMENT) and are described by the following relational schemas:

PROJECT (**PROJ_NUM**, PROJ_NAME)

EMPLOYEE (**EMP_NUM**, EMP_NAME, JOB_CLASS, CHG_HOUR)

ASSIGNMENT (**PROJ_NUM, EMP_NUM**, ASSIGN_HOURS)

Because the number of hours spent on each project by each employee is dependent on both PROJ_NUM and EMP_NUM in the ASSIGNMENT table, you leave those hours in the ASSIGNMENT table as ASSIGN_HOURS. Notice that the ASSIGNMENT table contains a composite primary key composed of the attributes PROJ_NUM and EMP_NUM. Notice that by leaving the determinants in the original table as well as making them the primary keys of the new tables, primary key/foreign key relationships have been created. For example, in the EMPLOYEE table, EMP_NUM is the primary key. In the ASSIGNMENT table, EMP_NUM is part of the composite primary key (PROJ_NUM, EMP_NUM) and is a foreign key relating the EMPLOYEE table to the ASSIGNMENT table.

The results of Steps 1 and 2 are displayed in Figure 6.4. At this point, most of the anomalies discussed earlier have been eliminated. For example, if you now want to add, change, or delete a PROJECT record, you need to go only to the PROJECT table and make the change to only one row.

Because a partial dependency can exist only when a table's primary key is composed of several attributes, a table whose primary key consists of only a single attribute is automatically in 2NF once it is in 1NF.

Figure 6.4 still shows a transitive dependency, which can generate anomalies. For example, if the charge per hour changes for a job classification held by many employees, that change must be made for *each* of those employees. If you forget to update some of the employee records that are affected by the charge per hour change, different employees with the same job description will generate different hourly charges.

> **second normal form (2NF)**
> The second stage in the normalization process, in which a relation is in 1NF and there are no partial dependencies (dependencies in only part of the primary key).

Note

A table is in **second normal form (2NF)** when:

- It is in 1NF.

and

- It includes no partial dependencies; that is, no attribute is dependent on only a portion of the primary key.

It is still possible for a table in 2NF to exhibit transitive dependency. That is, the primary key may rely on one or more nonprime attributes to functionally determine other nonprime attributes, as indicated by a functional dependence among the nonprime attributes.

FIGURE 6.4 SECOND NORMAL FORM (2NF) CONVERSION RESULTS

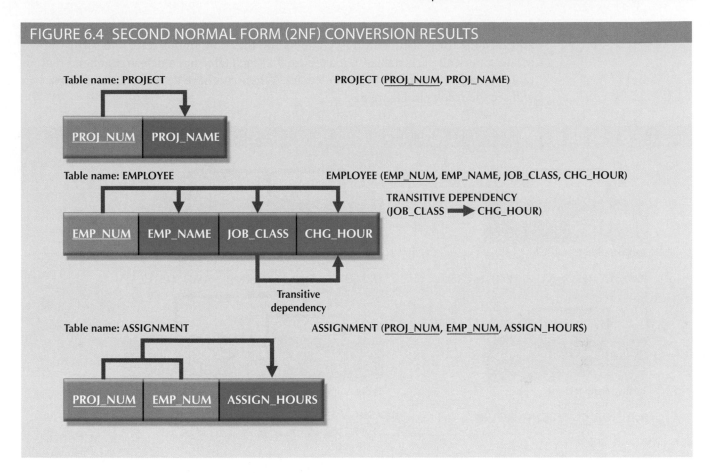

Table name: PROJECT

PROJECT (<u>PROJ_NUM</u>, PROJ_NAME)

Table name: EMPLOYEE

EMPLOYEE (<u>EMP_NUM</u>, EMP_NAME, JOB_CLASS, CHG_HOUR)

TRANSITIVE DEPENDENCY
(JOB_CLASS ➡ CHG_HOUR)

Transitive dependency

Table name: ASSIGNMENT

ASSIGNMENT (<u>PROJ_NUM</u>, <u>EMP_NUM</u>, ASSIGN_HOURS)

6-3c Conversion To Third Normal Form

The data anomalies created by the database organization shown in Figure 6.4 are easily eliminated by completing the following two steps:

Step 1: Make New Tables to Eliminate Transitive Dependencies For every transitive dependency, write a copy of its determinant as a primary key for a new table. A **determinant** is any attribute whose value determines other values within a row. If you have three different transitive dependencies, you will have three different determinants. As with the conversion to 2NF, it is important that the determinant remain in the original table to serve as a foreign key. Figure 6.4 shows only one table that contains a transitive dependency. Therefore, write the determinant for this transitive dependency as:

JOB_CLASS

Step 2: Reassign Corresponding Dependent Attributes Using Figure 6.4, identify the attributes that are dependent on each determinant identified in Step 1. Place the dependent attributes in the new tables with their determinants and remove them from their original tables. In this example, eliminate CHG_HOUR from the EMPLOYEE table shown in Figure 6.4 to leave the EMPLOYEE table dependency definition as:

EMP_NUM → EMP_NAME, JOB_CLASS

determinant
Any attribute in a specific row whose value directly determines other values in that row.

Draw a new dependency diagram to show all of the tables you have defined in Steps 1 and 2. Name the table to reflect its contents and function. In this case, JOB seems appropriate. Check all of the tables to make sure that each table has a determinant and that no table contains inappropriate dependencies. When you have completed these steps, you will see the results in Figure 6.5.

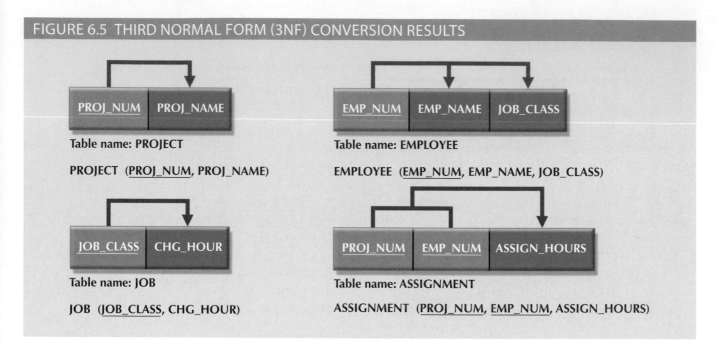

FIGURE 6.5 THIRD NORMAL FORM (3NF) CONVERSION RESULTS

Table name: PROJECT

PROJECT (**PROJ_NUM**, PROJ_NAME)

Table name: EMPLOYEE

EMPLOYEE (**EMP_NUM**, EMP_NAME, JOB_CLASS)

Table name: JOB

JOB (**JOB_CLASS**, CHG_HOUR)

Table name: ASSIGNMENT

ASSIGNMENT (**PROJ_NUM, EMP_NUM**, ASSIGN_HOURS)

In other words, after the 3NF conversion has been completed, your database will contain four tables:

PROJECT (**PROJ_NUM**, PROJ_NAME)

EMPLOYEE (**EMP_NUM**, EMP_NAME, JOB_CLASS)

JOB (**JOB_CLASS**, CHG_HOUR)

ASSIGNMENT (**PROJ_NUM, EMP_NUM**, ASSIGN_HOURS)

> **third normal form (3NF)**
> A table is in 3NF when it is in 2NF and no nonkey attribute is functionally dependent on another nonkey attribute; that is, it cannot include transitive dependencies.

Note that this conversion has eliminated the original EMPLOYEE table's transitive dependency. The tables are now said to be in third normal form (3NF).

Note

A table is in **third normal form (3NF)** when:

- It is in 2NF.

and

- It contains no transitive dependencies.

It is interesting to note the similarities between resolving 2NF and 3NF problems. To convert a table from 1NF to 2NF, it is necessary to remove the partial dependencies. To convert a table from 2NF to 3NF, it is necessary to remove the transitive dependencies. No matter whether the "problem" dependency is a partial dependency or a transitive dependency, the solution is the same: create a new table for each problem dependency. The determinant of the problem dependency remains in the original table and is placed as the primary key of the new table. The dependents of the problem dependency are removed from the original table and placed as nonprime attributes in the new table.

Be aware, however, that while the technique is the same, it is imperative that 2NF be achieved before moving on to 3NF; be certain to resolve the partial dependencies before resolving the transitive dependencies. Also, recall the assumption that was made at the beginning of the normalization discussion—that each table has only one candidate key, which is the primary key. If a table has multiple candidate keys, then the overall process remains the same, but there are additional considerations.

For example, if a table has multiple candidate keys and one of them is a composite key, the table can have partial dependencies based on this composite candidate key, even when the primary key chosen is a single attribute. In those cases, following the process described above, those dependencies would be perceived as transitive dependencies and would not be resolved until 3NF. The simplified process described above will allow the designer to achieve the correct result, but through practice, you should recognize all candidate keys and their dependencies as such, and resolve them appropriately. The existence of multiple candidate keys can also influence the identification of transitive dependencies. Previously, a transitive dependency was defined to exist when one nonprime attribute determined another nonprime attribute. In the presence of multiple candidate keys, the definition of a nonprime attribute as an attribute that is not a part of any candidate key is critical. If the determinant of a functional dependence is not the primary key but is a part of another candidate key, then it is not a nonprime attribute and does not signal the presence of a transitive dependency.

6-4 Improving the Design

Now that the table structures have been cleaned up to eliminate the troublesome partial and transitive dependencies, you can focus on improving the database's ability to provide information and on enhancing its operational characteristics. In the next few paragraphs, you will learn about the various types of issues you need to address to produce a good normalized set of tables. Note that for space issues, each section presents just one example—the designer must apply the principle to all remaining tables in the design. Remember that normalization cannot, by itself, be relied on to make good designs. Instead, normalization is valuable because its use helps eliminate data redundancies.

Evaluate PK Assignments Each time a new employee is entered into the EMPLOYEE table, a JOB_CLASS value must be entered. Unfortunately, it is too easy to make data-entry errors that lead to referential integrity violations. For example, entering *DB Designer* instead of *Database Designer* for the JOB_CLASS attribute in the EMPLOYEE table will trigger such a violation. Therefore, it would be better to add a JOB_CODE attribute to create a unique identifier. The addition of a JOB_CODE attribute produces the following dependency:

JOB_CODE → JOB_CLASS, CHG_HOUR

If you assume that the JOB_CODE is a proper primary key, this new attribute does produce the following dependency:

JOB_CLASS → CHG_HOUR

However, this dependency is not a transitive dependency because the determinant is a candidate key. Further, the presence of JOB_CODE greatly decreases the likelihood of referential integrity violations. Note that the new JOB table now has two candidate keys—JOB_CODE and JOB_CLASS. In this case, JOB_CODE is the chosen primary key as well as a surrogate key. A surrogate key, as you should recall, is an artificial PK introduced by the designer with the purpose of simplifying the assignment of primary keys to tables. Surrogate keys are usually numeric, they are often generated automatically by the DBMS, they are free of semantic content (they have no special meaning), and they are usually hidden from the end users.

Evaluate Naming Conventions It is best to adhere to the naming conventions outlined in Chapter 2, Data Models. Therefore, CHG_HOUR will be changed to JOB_CHG_HOUR to indicate its association with the JOB table. In addition, the attribute name JOB_CLASS does not quite describe entries such as *Systems Analyst*, *Database Designer*, and so on; the label JOB_DESCRIPTION fits the entries better. Also, you might have noticed that HOURS was changed to ASSIGN_HOURS in the conversion from 1NF to 2NF. That change lets you associate the hours worked with the ASSIGNMENT table.

Refine Attribute Atomicity It is generally good practice to pay attention to the *atomicity* requirement. An **atomic attribute** is one that cannot be further subdivided. Such an attribute is said to display **atomicity**. Clearly, the use of the EMP_NAME in the EMPLOYEE table is not atomic because EMP_NAME can be decomposed into a last name, a first name, and an initial. By improving the degree of atomicity, you also gain querying flexibility. For example, if you use EMP_LNAME, EMP_FNAME, and EMP_INITIAL, you can easily generate phone lists by sorting last names, first names, and initials. Such a task would be very difficult if the name components were within a single attribute. In general, designers prefer to use simple, single-valued attributes, as indicated by the business rules and processing requirements.

Identify New Attributes If the EMPLOYEE table were used in a real-world environment, several other attributes would have to be added. For example, year-to-date gross salary payments, Social Security payments, and Medicare payments would be desirable. An employee hire date attribute (EMP_HIREDATE) could be used to track an employee's job longevity, and it could serve as a basis for awarding bonuses to long-term employees and for other morale-enhancing measures. The same principle must be applied to all other tables in your design.

Identify New Relationships According to the original report, the users need to track which employee is acting as the manager of each project. This can be implemented as a relationship between EMPLOYEE and PROJECT. From the original report, it is clear that each project has only one manager. Therefore, the system's ability to supply detailed information about each project's manager is ensured by using the EMP_NUM as a foreign key in PROJECT. That action ensures that you can access the details of each PROJECT's manager data without producing unnecessary and undesirable data duplication. The designer must take care to place the right attributes in the right tables by using normalization principles.

Refine Primary Keys as Required for Data Granularity Granularity refers to the level of detail represented by the values stored in a table's row. Data stored at its lowest level of granularity is said to be *atomic data*, as explained earlier. In Figure 6.5, the ASSIGNMENT table in 3NF uses the ASSIGN_HOURS attribute to represent the hours worked by a given employee on a given project. However, are those values recorded at

atomic attribute
An attribute that cannot be further subdivided to produce meaningful components. For example, a person's last name attribute cannot be meaningfully subdivided.

atomicity
Not being able to be divided into smaller units.

granularity
The level of detail represented by the values stored in a table's row. Data stored at its lowest level of granularity is said to be *atomic data*.

their lowest level of granularity? In other words, does ASSIGN_HOURS represent the *hourly* total, *daily* total, *weekly* total, *monthly* total, or *yearly* total? Clearly, ASSIGN_ HOURS requires more careful definition. In this case, the relevant question would be as follows: for what time frame—hour, day, week, month, and so on—do you want to record the ASSIGN_HOURS data?

For example, assume that the combination of EMP_NUM and PROJ_NUM is an acceptable (composite) primary key in the ASSIGNMENT table. That primary key is useful in representing only the total number of hours an employee worked on a project since its start. Using a surrogate primary key such as ASSIGN_NUM provides lower granularity and yields greater flexibility. For example, assume that the EMP_NUM and PROJ_NUM combination is used as the primary key, and then an employee makes two "hours worked" entries in the ASSIGNMENT table. That action violates the entity integrity requirement. Even if you add the ASSIGN_DATE as part of a composite PK, an entity integrity violation is still generated if any employee makes two or more entries for the same project on the same day. (The employee might have worked on the project for a few hours in the morning and then worked on it again later in the day.) The same data entry yields no problems when ASSIGN_ NUM is used as the primary key.

Note

In an ideal database design, the level of desired granularity would be determined during the conceptual design or while the requirements were being gathered. However, as you have already seen in this chapter, many database designs involve the refinement of existing data requirements, thus triggering design modifications. In a real-world environment, changing granularity requirements might dictate changes in primary key selection, and those changes might ultimately require the use of surrogate keys.

Maintain Historical Accuracy Writing the job charge per hour into the ASSIGN-MENT table is crucial to maintaining the historical accuracy of the table's data. It would be appropriate to name this attribute ASSIGN_CHG_HOUR. Although this attribute would appear to have the same value as JOB_CHG_HOUR, this is true *only* if the JOB_ CHG_HOUR value remains the same forever. It is reasonable to assume that the job charge per hour will change over time. However, suppose that the charges to each project were calculated and billed by multiplying the hours worked from the ASSIGNMENT table by the charge per hour from the JOB table. Those charges would always show the current charge per hour stored in the JOB table rather than the charge per hour that was in effect at the time of the assignment.

Evaluate Using Derived Attributes Finally, you can use a derived attribute in the ASSIGNMENT table to store the actual charge made to a project. That derived attribute, named ASSIGN_CHARGE, is the result of multiplying ASSIGN_HOURS by ASSIGN_ CHG_HOUR. This creates a transitive dependency such that:

(ASSIGN_CHARGE + ASSIGN_HOURS) → ASSIGN_CHG_HOUR

From a system functionality point of view, such derived attribute values can be calculated when they are needed to write reports or invoices. However, storing the derived attribute in the table makes it easy to write the application software to produce the desired results. Also, if many transactions must be reported and/or summarized, the availability of the derived attribute will save reporting time. (If the calculation is done at the time of data entry, it will be completed when the end user presses the Enter key, thus speeding up the process.) Review Chapter 4 for a discussion of the implications of storing derived attributes in a database table.

The enhancements described in the preceding sections are illustrated in the tables and dependency diagrams shown in Figure 6.6.

FIGURE 6.6 THE COMPLETED DATABASE

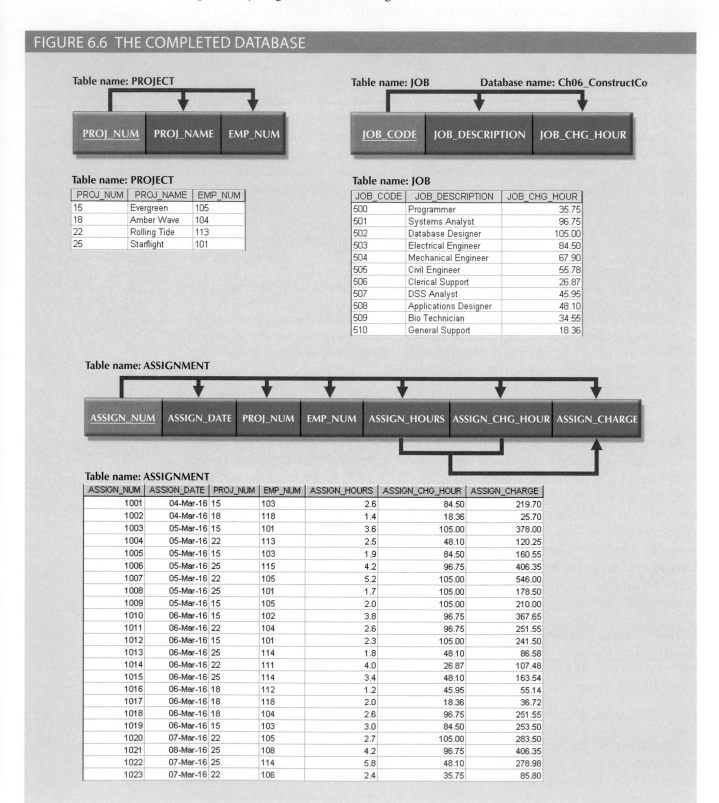

Table name: PROJECT

PROJ_NUM	PROJ_NAME	EMP_NUM
15	Evergreen	105
18	Amber Wave	104
22	Rolling Tide	113
25	Starflight	101

Database name: Ch06_ConstructCo

Table name: JOB

JOB_CODE	JOB_DESCRIPTION	JOB_CHG_HOUR
500	Programmer	35.75
501	Systems Analyst	96.75
502	Database Designer	105.00
503	Electrical Engineer	84.50
504	Mechanical Engineer	67.90
505	Civil Engineer	55.78
506	Clerical Support	26.87
507	DSS Analyst	45.95
508	Applications Designer	48.10
509	Bio Technician	34.55
510	General Support	18.36

Table name: ASSIGNMENT

ASSIGN_NUM	ASSIGN_DATE	PROJ_NUM	EMP_NUM	ASSIGN_HOURS	ASSIGN_CHG_HOUR	ASSIGN_CHARGE
1001	04-Mar-16	15	103	2.6	84.50	219.70
1002	04-Mar-16	18	118	1.4	18.36	25.70
1003	05-Mar-16	15	101	3.6	105.00	378.00
1004	05-Mar-16	22	113	2.5	48.10	120.25
1005	05-Mar-16	15	103	1.9	84.50	160.55
1006	05-Mar-16	25	115	4.2	96.75	406.35
1007	05-Mar-16	22	105	5.2	105.00	546.00
1008	05-Mar-16	25	101	1.7	105.00	178.50
1009	05-Mar-16	15	105	2.0	105.00	210.00
1010	06-Mar-16	15	102	3.8	96.75	367.65
1011	06-Mar-16	22	104	2.6	96.75	251.55
1012	06-Mar-16	15	101	2.3	105.00	241.50
1013	06-Mar-16	25	114	1.8	48.10	86.58
1014	06-Mar-16	22	111	4.0	26.87	107.48
1015	06-Mar-16	25	114	3.4	48.10	163.54
1016	06-Mar-16	18	112	1.2	45.95	55.14
1017	06-Mar-16	18	118	2.0	18.36	36.72
1018	06-Mar-16	18	104	2.6	96.75	251.55
1019	06-Mar-16	15	103	3.0	84.50	253.50
1020	07-Mar-16	22	105	2.7	105.00	283.50
1021	08-Mar-16	25	108	4.2	96.75	406.35
1022	07-Mar-16	25	114	5.8	48.10	278.98
1023	07-Mar-16	22	106	2.4	35.75	85.80

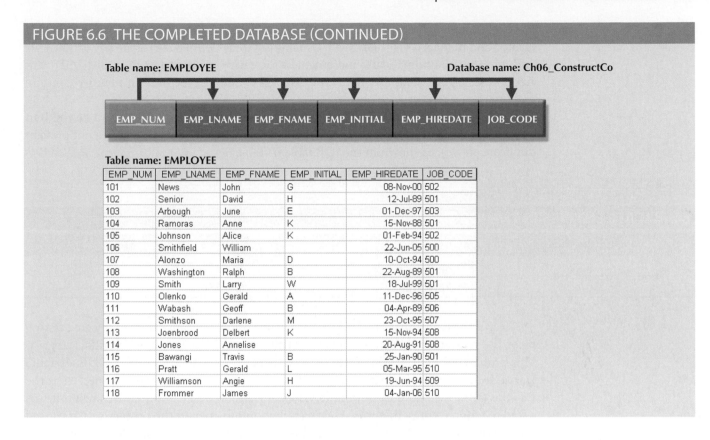

FIGURE 6.6 THE COMPLETED DATABASE (CONTINUED)

Table name: EMPLOYEE   Database name: Ch06_ConstructCo

Table name: EMPLOYEE

EMP_NUM	EMP_LNAME	EMP_FNAME	EMP_INITIAL	EMP_HIREDATE	JOB_CODE
101	News	John	G	08-Nov-00	502
102	Senior	David	H	12-Jul-89	501
103	Arbough	June	E	01-Dec-97	503
104	Ramoras	Anne	K	15-Nov-88	501
105	Johnson	Alice	K	01-Feb-94	502
106	Smithfield	William		22-Jun-05	500
107	Alonzo	Maria	D	10-Oct-94	500
108	Washington	Ralph	B	22-Aug-89	501
109	Smith	Larry	W	18-Jul-99	501
110	Olenko	Gerald	A	11-Dec-96	505
111	Wabash	Geoff	B	04-Apr-89	506
112	Smithson	Darlene	M	23-Oct-95	507
113	Joenbrood	Delbert	K	15-Nov-94	508
114	Jones	Annelise		20-Aug-91	508
115	Bawangi	Travis	B	25-Jan-90	501
116	Pratt	Gerald	L	05-Mar-95	510
117	Williamson	Angie	H	19-Jun-94	509
118	Frommer	James	J	04-Jan-06	510

Figure 6.6 is a vast improvement over the original database design. If the application software is designed properly, the most active table (ASSIGNMENT) requires the entry of only the PROJ_NUM, EMP_NUM, and ASSIGN_HOURS values. The values for the attributes ASSIGN_NUM and ASSIGN_DATE can be generated by the application. For example, the ASSIGN_NUM can be created by using a counter, and the ASSIGN_DATE can be the system date read by the application and automatically entered into the ASSIGNMENT table. In addition, the application software can automatically insert the correct ASSIGN_CHG_HOUR value by writing the appropriate JOB table's JOB_CHG_HOUR value into the ASSIGNMENT table. (The JOB and ASSIGNMENT tables are related through the JOB_CODE attribute.) If the JOB table's JOB_CHG_HOUR value changes, the next insertion of that value into the ASSIGNMENT table will reflect the change automatically. The table structure thus minimizes the need for human intervention. In fact, if the system requires the employees to enter their own work hours, they can scan their EMP_NUM into the ASSIGNMENT table by using a magnetic card reader that enters their identity. Thus, the ASSIGNMENT table's structure can set the stage for maintaining some desired level of security.

6-5 Surrogate Key Considerations

Although this design meets the vital entity and referential integrity requirements, the designer must still address some concerns. For example, a composite primary key might become too cumbersome to use as the number of attributes grows. (It becomes difficult to create a suitable foreign key when the related table uses a composite primary key. In addition, a composite primary key makes it more difficult to write search routines.) Or, a primary key attribute might simply have too much descriptive content to be usable—which is why the JOB_CODE attribute was added to the JOB table to serve as its primary key. When the primary key is considered to be unsuitable for some reason, designers use surrogate keys, as discussed in the previous chapter.

At the implementation level, a surrogate key is a system-defined attribute generally created and managed via the DBMS. Usually, a system-defined surrogate key is numeric, and its value is automatically incremented for each new row. For example, Microsoft Access uses an AutoNumber data type, Microsoft SQL Server uses an identity column, and Oracle uses a sequence object.

Recall from Section 6-4 that the JOB_CODE attribute was designated to be the JOB table's primary key. However, remember that the JOB_CODE attribute does not prevent duplicate entries, as shown in the JOB table in Table 6.4.

TABLE 6.4

DUPLICATE ENTRIES IN THE JOB TABLE

JOB_CODE	JOB_DESCRIPTION	JOB_CHG_HOUR
511	Programmer	$35.75
512	Programmer	$35.75

Clearly, the data entries in Table 6.4 are inappropriate because they duplicate existing records—yet there has been no violation of either entity integrity or referential integrity. This problem of multiple duplicate records was created when the JOB_CODE attribute was added as the PK. (When the JOB_DESCRIPTION was initially designated to be the PK, the DBMS would ensure unique values for all job description entries when it was asked to enforce entity integrity. However, that option created the problems that caused the use of the JOB_CODE attribute in the first place!) In any case, if JOB_CODE is to be the surrogate PK, you still must ensure the existence of unique values in the JOB_DESCRIPTION *through the use of a unique index*.

Note that all of the remaining tables (PROJECT, ASSIGNMENT, and EMPLOYEE) are subject to the same limitations. For example, if you use the EMP_NUM attribute in the EMPLOYEE table as the PK, you can make multiple entries for the same employee. To avoid that problem, you might create a unique index for EMP_LNAME, EMP_FNAME, and EMP_INITIAL, but how would you then deal with two employees named Joe B. Smith? In that case, you might use another (preferably externally defined) attribute to serve as the basis for a unique index.

It is worth repeating that database design often involves trade-offs and the exercise of professional judgment. In a real-world environment, you must strike a balance between design integrity and flexibility. For example, you might design the ASSIGNMENT table to use a unique index on PROJ_NUM, EMP_NUM, and ASSIGN_DATE if you want to limit an employee to only one ASSIGN_HOURS entry per date. That limitation would ensure that employees could not enter the same hours multiple times for any given date. Unfortunately, that limitation is likely to be undesirable from a managerial point of view. After all, if an employee works several different times on a project during any given day, it must be possible to make multiple entries for that same employee and the same project during that day. In that case, the best solution might be to add a new externally defined attribute—such as a stub, voucher, or ticket number—to ensure uniqueness. In any case, frequent data audits would be appropriate.

6-6 Higher-Level Normal Forms

Tables in 3NF will perform suitably in business transactional databases. However, higher normal forms are sometimes useful. In this section, you will learn about a special case of 3NF, known as Boyce-Codd normal form, and about fourth normal form (4NF).

6-6a   The Boyce-Codd Normal Form

A table is in Boyce-Codd normal form (BCNF) when every determinant in the table is a candidate key. (Recall from Chapter 3 that a candidate key has the same characteristics as a primary key, but for some reason, it was not chosen to be the primary key.) Clearly, when a table contains only one candidate key, the 3NF and the BCNF are equivalent. In other words, BCNF can be violated only when the table contains more than one candidate key. In the previous normal form examples, tables with only one candidate key were used to simplify the explanations. Remember, however, that multiple candidate keys are always possible, and normalization rules focus on candidate keys, not just the primary key. Consider the table structure shown in Figure 6.7.

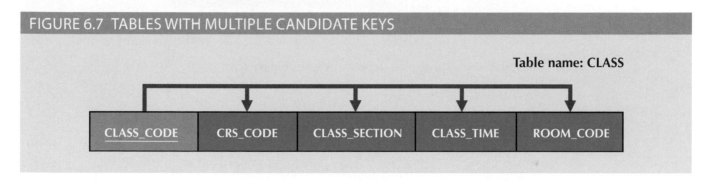

FIGURE 6.7   TABLES WITH MULTIPLE CANDIDATE KEYS

Table name: **CLASS**

| CLASS_CODE | CRS_CODE | CLASS_SECTION | CLASS_TIME | ROOM_CODE |

The CLASS table has two candidate keys:

- CLASS_CODE
- CRS_CODE + CLASS_SECTION

The table is in 1NF because the key attributes are defined and all nonkey attributes are determined by the key. This is true for both candidate keys. Both candidate keys have been identified, and all of the other attributes can be determined by either candidate key. The table is in 2NF because it is in 1NF and there are no partial dependencies on either candidate key. Since CLASS_CODE is a single attribute candidate key, the issue of partial dependencies doesn't apply. However, the composite candidate key of CRS_CODE + CLASS_SECTION could potentially have a partial dependency so 2NF must be evaluated for that candidate key. In this case, there are no partial dependencies involving the composite key. Finally, the table is in 3NF because there are no transitive dependencies. Remember, because CRS_CODE + CLASS_SECTION is a candidate key, the fact that this composite can determine the CLASS_TIME and ROOM_CODE is not a transitive dependency. A transitive dependency exists when a *nonkey* attribute can determine another nonkey attribute, and CRS_CODE + CLASS_SECTION is a key.

Note

A table is in **Boyce-Codd normal form (BCNF)** when every determinant in the table is a candidate key.

Boyce-Codd normal form (BCNF)
A special type of third normal form (3NF) in which every determinant is a candidate key. A table in BCNF must be in 3NF. See also *determinant*.

Most designers consider the BCNF to be a special case of the 3NF. In fact, if the techniques shown in this chapter are used, most tables conform to the BCNF requirements once the 3NF is reached. So, how can a table be in 3NF and not be in BCNF? To answer that question, you must keep in mind that a transitive dependency exists when one nonprime attribute is dependent on another nonprime attribute.

In other words, a table is in 3NF when it is in 2NF and there are no transitive dependencies, but what about a case in which one key attribute is the determinant of another key attribute? That condition does not violate 3NF, yet it fails to meet the BCNF requirements (see Figure 6.8) because BCNF requires that every determinant in the table be a candidate key.

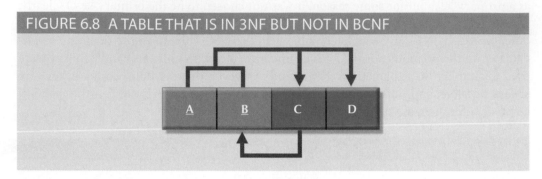

FIGURE 6.8 A TABLE THAT IS IN 3NF BUT NOT IN BCNF

Note these functional dependencies in Figure 6.8:

$A + B \rightarrow C, D$

$A + C \rightarrow B, D$

$C \rightarrow B$

Notice that this structure has two candidate keys: (A + B) and (A + C). The table structure shown in Figure 6.8 has no partial dependencies, nor does it contain transitive dependencies. (The condition $C \rightarrow B$ indicates that *one key attribute determines part of the primary key*—and *that* dependency *is not* transitive or partial because the dependent is a prime attribute!) Thus, the table structure in Figure 6.8 meets the 3NF requirements, although the condition $C \rightarrow B$ causes the table to fail to meet the BCNF requirements.

To convert the table structure in Figure 6.8 into table structures that are in 3NF and in BCNF, first change the primary key to A + C. This change is appropriate because the dependency $C \rightarrow B$ means that C is effectively a superset of B. At this point, the table is in 1NF because it contains a partial dependency, $C \rightarrow B$. Next, follow the standard decomposition procedures to produce the results shown in Figure 6.9.

To see how this procedure can be applied to an actual problem, examine the sample data in Table 6.5.

Table 6.5 reflects the following conditions:

- Each CLASS_CODE identifies a class uniquely. This condition illustrates the case in which a course might generate many classes. For example, a course labeled INFS 420 might be taught in two classes (sections), each identified by a unique code to facilitate registration. Thus, the CLASS_CODE 32456 might identify INFS 420, class section 1, while the CLASS_CODE 32457 might identify INFS 420, class section 2. Or, the CLASS_CODE 28458 might identify QM 362, class section 5.

- A student can take many classes. Note, for example, that student 125 has taken both 21334 and 32456, earning the grades A and C, respectively.

- A staff member can teach many classes, but each class is taught by only one staff member. Note that staff member 20 teaches the classes identified as 32456 and 28458.

The structure shown in Table 6.5 is reflected in Panel A of Figure 6.10:

$STU\_ID + STAFF\_ID \rightarrow CLASS\_CODE, ENROLL\_GRADE$

$CLASS\_CODE \rightarrow STAFF\_ID$

FIGURE 6.9   DECOMPOSITION TO BCNF

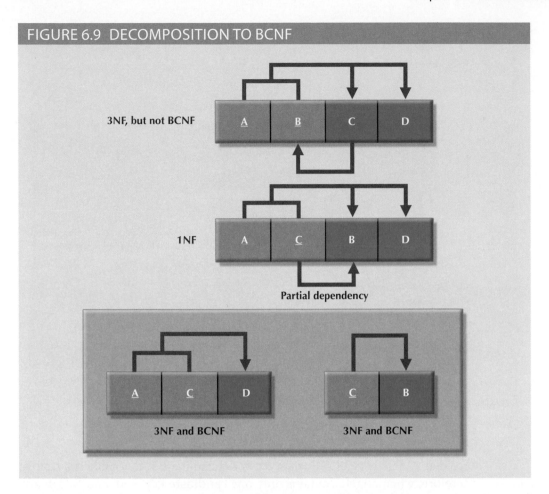

3NF, but not BCNF

1NF

Partial dependency

3NF and BCNF

3NF and BCNF

TABLE 6.5

SAMPLE DATA FOR A BCNF CONVERSION

STU_ID	STAFF_ID	CLASS_CODE	ENROLL_GRADE
125	25	21334	A
125	20	32456	C
135	20	28458	B
144	25	27563	C
144	20	32456	B

Panel A of Figure 6.10 shows a structure that is clearly in 3NF, but the table represented by this structure has a major problem because it is trying to describe two things: staff assignments to classes and student enrollment information. Such a dual-purpose table structure will cause anomalies. For example, if a different staff member is assigned to teach class 32456, two rows will require updates, thus producing an update anomaly. Also, if student 135 drops class 28458, information about who taught that class is lost, thus producing a deletion anomaly. The solution to the problem is to decompose the table structure, following the procedure outlined earlier. The decomposition of Panel B shown in Figure 6.10 yields two table structures that conform to both 3NF and BCNF requirements.

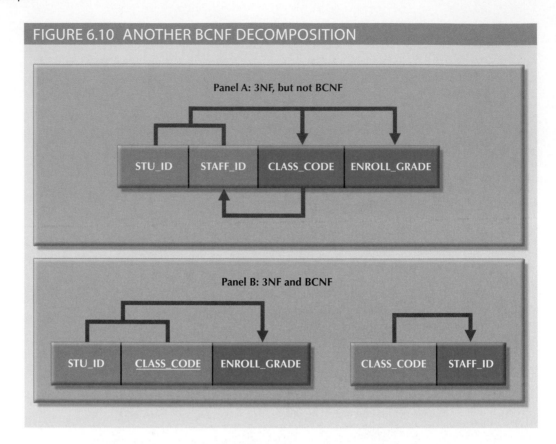

FIGURE 6.10 ANOTHER BCNF DECOMPOSITION

Panel A: 3NF, but not BCNF

| STU_ID | STAFF_ID | CLASS_CODE | ENROLL_GRADE |

Panel B: 3NF and BCNF

| STU_ID | CLASS_CODE | ENROLL_GRADE |

| CLASS_CODE | STAFF_ID |

Remember that a table is in BCNF when every determinant in that table is a candidate key. Therefore, when a table contains only one candidate key, 3NF and BCNF are equivalent.

6-6b Fourth Normal Form (4NF)

You might encounter poorly designed databases, or you might be asked to convert spreadsheets into a database format in which multiple multivalued attributes exist. For example, consider the possibility that an employee can have multiple assignments and can also be involved in multiple service organizations. Suppose employee 10123 volunteers for the Red Cross and United Way. In addition, the same employee might be assigned to work on three projects: 1, 3, and 4. Figure 6.11 illustrates how that set of facts can be recorded in very different ways.

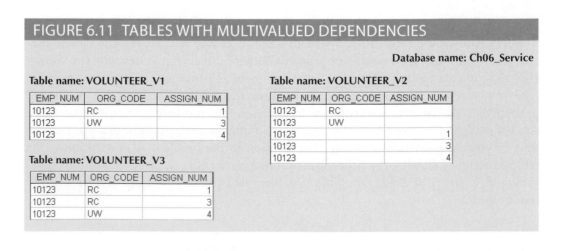

FIGURE 6.11 TABLES WITH MULTIVALUED DEPENDENCIES

Database name: Ch06_Service

Table name: VOLUNTEER_V1

EMP_NUM	ORG_CODE	ASSIGN_NUM
10123	RC	1
10123	UW	3
10123		4

Table name: VOLUNTEER_V2

EMP_NUM	ORG_CODE	ASSIGN_NUM
10123	RC	
10123	UW	
10123		1
10123		3
10123		4

Table name: VOLUNTEER_V3

EMP_NUM	ORG_CODE	ASSIGN_NUM
10123	RC	1
10123	RC	3
10123	UW	4

There is a problem with the tables in Figure 6.11. The attributes ORG_CODE and ASSIGN_NUM each may have many different values. In normalization terminology, this situation is referred to as a multivalued dependency, which occurs when one key determines multiple values of two other attributes and those attributes are independent of each other. (One employee can have many service entries and many assignment entries. Therefore, one EMP_NUM can determine multiple values of ORG_CODE and multiple values of ASSIGN_NUM; however, ORG_CODE and ASSIGN_NUM are independent of each other.) The presence of a multivalued dependency means that if table versions 1 and 2 are implemented, the tables are likely to contain quite a few null values; in fact, the tables do not even have a viable candidate key. (The EMP_NUM values are not unique, so they cannot be PKs. No combination of the attributes in table versions 1 and 2 can be used to create a PK because some of them contain nulls.) Such a condition is not desirable, especially when there are thousands of employees, many of whom may have multiple job assignments and many service activities. Version 3 at least has a PK, but it is composed of all the attributes in the table. In fact, version 3 meets 3NF requirements, yet it contains many redundancies that are clearly undesirable.

The solution is to eliminate the problems caused by the multivalued dependency. You do this by creating new tables for the components of the multivalued dependency. In this example, the multivalued dependency is resolved and eliminated by creating the ASSIGNMENT and SERVICE_V1 tables depicted in Figure 6.12. Those tables are said to be in 4NF.

FIGURE 6.12 A SET OF TABLES IN 4NF

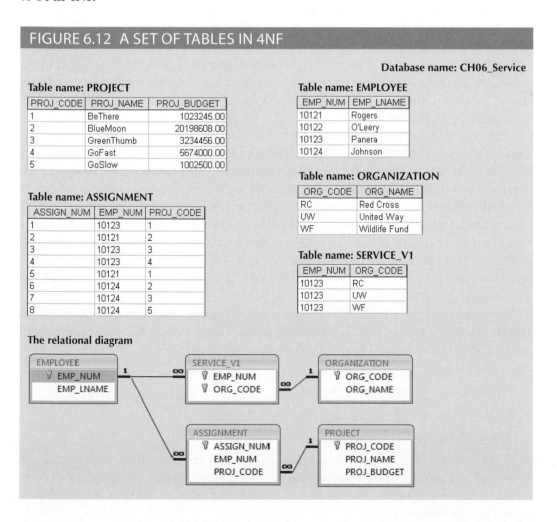

Database name: CH06_Service

Table name: PROJECT

PROJ_CODE	PROJ_NAME	PROJ_BUDGET
1	BeThere	1023245.00
2	BlueMoon	20198608.00
3	GreenThumb	3234456.00
4	GoFast	5674000.00
5	GoSlow	1002500.00

Table name: EMPLOYEE

EMP_NUM	EMP_LNAME
10121	Rogers
10122	O'Leery
10123	Panera
10124	Johnson

Table name: ORGANIZATION

ORG_CODE	ORG_NAME
RC	Red Cross
UW	United Way
WF	Wildlife Fund

Table name: ASSIGNMENT

ASSIGN_NUM	EMP_NUM	PROJ_CODE
1	10123	1
2	10121	2
3	10123	3
4	10123	4
5	10121	1
6	10124	2
7	10124	3
8	10124	5

Table name: SERVICE_V1

EMP_NUM	ORG_CODE
10123	RC
10123	UW
10123	WF

The relational diagram

If you follow the proper design procedures illustrated in this book, you should not encounter the problem shown in Figure 6.11. Specifically, the discussion of 4NF is largely academic if you make sure that your tables conform to the following two rules:

1. All attributes must be dependent on the primary key, but they must be independent of each other.

2. No row may contain two or more multivalued facts about an entity.

Note

A table is in **fourth normal form** (4NF) when it is in 3NF and has no multivalued dependencies.

6-7 Normalization and Database Design

The tables shown in Figure 6.6 illustrate how normalization procedures can be used to produce good tables from poor ones. You will likely have ample opportunity to put this skill into practice when you begin to work with real-world databases. *Normalization should be part of the design process.* Therefore, make sure that proposed entities meet the required normal form *before* the table structures are created. Keep in mind that if you follow the design procedures discussed in Chapters 3 and 4, the likelihood of data anomalies will be small. However, even the best database designers are known to make occasional mistakes that come to light during normalization checks. Also, many of the real-world databases you encounter will have been improperly designed or burdened with anomalies if they were improperly modified over the course of time. That means you might be asked to redesign and modify existing databases that are, in effect, anomaly traps. Therefore, you should be aware of good design principles and procedures as well as normalization procedures.

First, an ERD is created through an iterative process. You begin by identifying relevant entities, their attributes, and their relationships. Then you use the results to identify additional entities and attributes. The ERD provides the big picture, or macro view, of an organization's data requirements and operations.

Second, normalization focuses on the characteristics of specific entities; that is, normalization represents a micro view of the entities within the ERD. Also, as you learned in the previous sections of this chapter, the normalization process might yield additional entities and attributes to be incorporated into the ERD. Therefore, it is difficult to separate normalization from ER modeling; the two techniques are used in an iterative and incremental process.

To understand the proper role of normalization in the design process, you should reexamine the operations of the contracting company whose tables were normalized in the preceding sections. Those operations can be summarized by using the following business rules:

- The company manages many projects.

- Each project requires the services of many employees.

- An employee may be assigned to several different projects.

- Some employees are not assigned to a project and perform duties not specifically related to a project. Some employees are part of a labor pool, to be shared by all project teams. For example, the company's executive secretary would not be assigned to any one particular project.

fourth normal form (4NF)
A table is in 4NF if it is in 3NF and contains no multiple independent sets of multivalued dependencies.

- Each employee has a single primary job classification, which determines the hourly billing rate.

- Many employees can have the same job classification. For example, the company employs more than one electrical engineer.

Given that simple description of the company's operations, two entities and their attributes are initially defined:

- PROJECT (**PROJ_NUM**, PROJ_NAME)

- EMPLOYEE (**EMP_NUM**, EMP_LNAME, EMP_FNAME, EMP_INITIAL, JOB_DESCRIPTION, JOB_CHG_HOUR)

Those two entities constitute the initial ERD shown in Figure 6.13.

FIGURE 6.13 INITIAL CONTRACTING COMPANY ERD

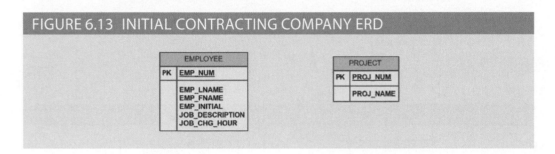

After creating the initial ERD shown in Figure 6.13, the normal forms are defined:

- PROJECT is in 3NF and needs no modification at this point.

- EMPLOYEE requires additional scrutiny. The JOB_DESCRIPTION attribute defines job classifications such as Systems Analyst, Database Designer, and Programmer. In turn, those classifications determine the billing rate, JOB_CHG_HOUR. Therefore, EMPLOYEE contains a transitive dependency.

The removal of EMPLOYEE's transitive dependency yields three entities:

- PROJECT (**PROJ_NUM**, PROJ_NAME)

- EMPLOYEE (**EMP_NUM**, EMP_LNAME, EMP_FNAME, EMP_INITIAL, JOB_CODE)

- JOB (**JOB_CODE**, JOB_DESCRIPTION, JOB_CHG_HOUR)

Because the normalization process yields an additional entity (JOB), the initial ERD is modified as shown in Figure 6.14.

To represent the M:N relationship between EMPLOYEE and PROJECT, you might think that two 1:M relationships could be used—an employee can be assigned to many projects, and each project can have many employees assigned to it. (See Figure 6.15.) Unfortunately, that representation yields a design that cannot be correctly implemented.

Because the M:N relationship between EMPLOYEE and PROJECT cannot be implemented, the ERD in Figure 6.15 must be modified to include the ASSIGNMENT entity to track the assignment of employees to projects, thus yielding the ERD shown in Figure 6.16. The ASSIGNMENT entity in Figure 6.16 uses the primary keys from the entities PROJECT and EMPLOYEE to serve as its foreign keys. However, note that in this implementation, the ASSIGNMENT entity's surrogate primary key is ASSIGN_NUM, to avoid the use of a composite primary key. Therefore, the "enters" relationship between EMPLOYEE and ASSIGNMENT and the "requires" relationship between PROJECT and ASSIGNMENT are shown as weak or nonidentifying.

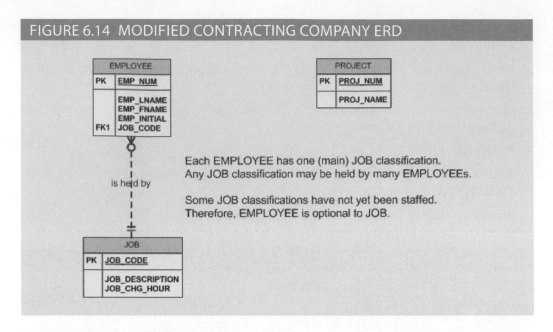

FIGURE 6.14 MODIFIED CONTRACTING COMPANY ERD

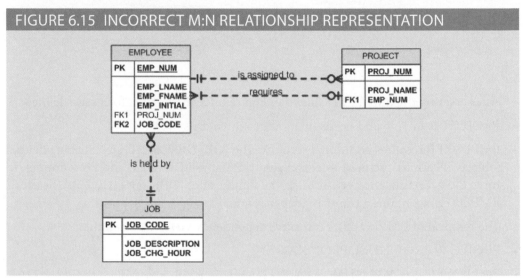

FIGURE 6.15 INCORRECT M:N RELATIONSHIP REPRESENTATION

In Figure 6.16, the ASSIGN_HOURS attribute is assigned to the composite entity named ASSIGNMENT. Because you will likely need detailed information about each project's manager, the creation of a "manages" relationship is useful. The "manages" relationship is implemented through the foreign key in PROJECT. Finally, some additional attributes may be created to improve the system's ability to generate additional information. For example, you may want to include the date the employee was hired (EMP_HIREDATE) to keep track of worker longevity. Based on this last modification, the model should include four entities and their attributes:

PROJECT (**PROJ_NUM**, PROJ_NAME, EMP_NUM)

EMPLOYEE (**EMP_NUM**, EMP_LNAME, EMP_FNAME, EMP_INITIAL, EMP_HIREDATE, JOB_CODE)

JOB (**JOB_CODE**, JOB_DESCRIPTION, JOB_CHG_HOUR)

ASSIGNMENT (**ASSIGN_NUM**, ASSIGN_DATE, PROJ_NUM, EMP_NUM, ASSIGN_HOURS, ASSIGN_CHG_HOUR, ASSIGN_CHARGE)

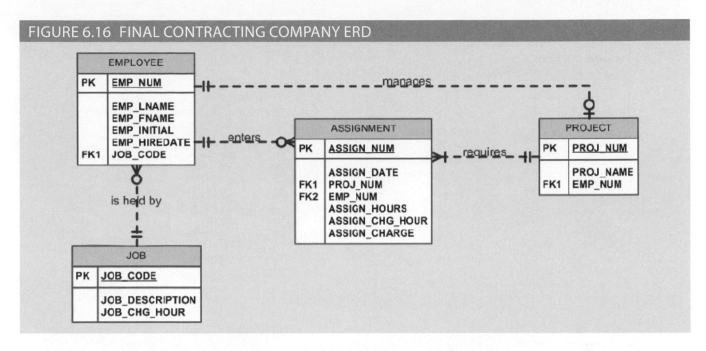

FIGURE 6.16 FINAL CONTRACTING COMPANY ERD

The design process is now on the right track. The ERD represents the operations accurately, and the entities now reflect their conformance to 3NF. The combination of normalization and ER modeling yields a useful ERD, whose entities may now be translated into appropriate table structures. In Figure 6.15, note that PROJECT is optional to EMPLOYEE in the "manages" relationship. This optionality exists because not all employees manage projects. The final database contents are shown in Figure 6.17.

6-8 Denormalization

It is important to remember that the optimal relational database implementation requires that all tables be at least in third normal form (3NF). A good relational DBMS excels at managing normalized relations—that is, relations void of any unnecessary redundancies that might cause data anomalies. Although the creation of normalized relations is an important database design goal, it is only one of many such goals. Good database design also considers processing (or reporting) requirements and processing speed. The problem with normalization is that as tables are decomposed to conform to normalization requirements, the number of database tables expands. Therefore, in order to generate information, data must be put together from various tables. Joining a large number of tables takes additional input/output (I/O) operations and processing logic, thereby reducing system speed. Most relational database systems are able to handle joins very efficiently. However, rare and occasional circumstances may allow some degree of denormalization so processing speed can be increased.

Keep in mind that the advantage of higher processing speed must be carefully weighed against the disadvantage of data anomalies. On the other hand, some anomalies are of only theoretical interest. For example, should people in a real-world database environment worry that a ZIP_CODE determines CITY in a CUSTOMER table whose primary key is the customer number? Is it really practical to produce a separate table for

ZIP (**ZIP_CODE**, CITY)

to eliminate a transitive dependency from the CUSTOMER table? (Perhaps your answer to that question changes if you are in the business of producing mailing lists.) As explained earlier, the problem with denormalized relations and redundant data is that data integrity could be compromised due to the possibility of insert, update, and

FIGURE 6.17 THE IMPLEMENTED DATABASE

Table name: EMPLOYEE

Database name: Ch06_ConstructCo

EMP_NUM	EMP_LNAME	EMP_FNAME	EMP_INITIAL	EMP_HIREDATE	JOB_CODE
101	News	John	G	08-Nov-00	502
102	Senior	David	H	12-Jul-89	501
103	Arbough	June	E	01-Dec-97	503
104	Ramoras	Anne	K	15-Nov-88	501
105	Johnson	Alice	K	01-Feb-94	502
106	Smithfield	William		22-Jun-05	500
107	Alonzo	Maria	D	10-Oct-94	500
108	Washington	Ralph	B	22-Aug-89	501
109	Smith	Larry	W	18-Jul-99	501
110	Olenko	Gerald	A	11-Dec-96	505
111	Wabash	Geoff	B	04-Apr-89	506
112	Smithson	Darlene	M	23-Oct-95	507
113	Joenbrood	Delbert	K	15-Nov-94	508
114	Jones	Annelise		20-Aug-91	508
115	Bawangi	Travis	B	25-Jan-90	501
116	Pratt	Gerald	L	05-Mar-95	510
117	Williamson	Angie	H	19-Jun-94	509
118	Frommer	James	J	04-Jan-06	510

Table name: JOB

JOB_CODE	JOB_DESCRIPTION	JOB_CHG_HOUR
500	Programmer	35.75
501	Systems Analyst	96.75
502	Database Designer	105.00
503	Electrical Engineer	84.50
504	Mechanical Engineer	67.90
505	Civil Engineer	55.78
506	Clerical Support	26.87
507	DSS Analyst	45.95
508	Applications Designer	48.10
509	Bio Technician	34.55
510	General Support	18.36

Table name: PROJECT

PROJ_NUM	PROJ_NAME	EMP_NUM
15	Evergreen	105
18	Amber Wave	104
22	Rolling Tide	113
25	Starflight	101

Table name: ASSIGNMENT

ASSIGN_NUM	ASSIGN_DATE	PROJ_NUM	EMP_NUM	ASSIGN_HOURS	ASSIGN_CHG_HOUR	ASSIGN_CHARGE
1001	04-Mar-16	15	103	2.6	84.50	219.70
1002	04-Mar-16	18	118	1.4	18.36	25.70
1003	05-Mar-16	15	101	3.6	105.00	378.00
1004	05-Mar-16	22	113	2.5	48.10	120.25
1005	05-Mar-16	15	103	1.9	84.50	160.55
1006	05-Mar-16	25	115	4.2	96.75	406.35
1007	05-Mar-16	22	105	5.2	105.00	546.00
1008	05-Mar-16	25	101	1.7	105.00	178.50
1009	05-Mar-16	15	105	2.0	105.00	210.00
1010	06-Mar-16	15	102	3.8	96.75	367.65
1011	06-Mar-16	22	104	2.6	96.75	251.55
1012	06-Mar-16	15	101	2.3	105.00	241.50
1013	06-Mar-16	25	114	1.8	48.10	86.58
1014	06-Mar-16	22	111	4.0	26.87	107.48
1015	06-Mar-16	25	114	3.4	48.10	163.54
1016	06-Mar-16	18	112	1.2	45.95	55.14
1017	06-Mar-16	18	118	2.0	18.36	36.72
1018	06-Mar-16	18	104	2.6	96.75	251.55
1019	06-Mar-16	15	103	3.0	84.50	253.50
1020	07-Mar-16	22	105	2.7	105.00	283.50
1021	08-Mar-16	25	108	4.2	96.75	406.35
1022	07-Mar-16	25	114	5.8	48.10	278.98
1023	07-Mar-16	22	106	2.4	35.75	85.80

deletion anomalies. The advice is simple: use common sense during the normalization process.

Furthermore, the database design process could, in some cases, introduce some small degree of redundant data in the model, as seen in the previous example. This, in effect, creates "denormalized" relations. Table 6.6 shows some common examples of data redundancy that are generally found in database implementations.

A more comprehensive example of the need for denormalization due to reporting requirements is the case of a faculty evaluation report in which each row lists the scores obtained during the last four semesters taught. (See Figure 6.18.)

Although this report seems simple enough, the problem is that the data is stored in a normalized table in which each row represents a different score for a given faculty member in a given semester. (See Figure 6.19.)

The difficulty of transposing multirow data to multicolumn data is compounded by the fact that the last four semesters taught are not necessarily the same for all faculty members. Some might have taken sabbaticals, some might have had research appointments, some might be new faculty with only two semesters on the job, and so on. To generate this report, the two tables in Figure 6.18 were used. The EVALDATA table is

TABLE 6.6

COMMON DENORMALIZATION EXAMPLES

CASE	EXAMPLE	RATIONALE AND CONTROLS
Redundant data	Storing ZIP and CITY attributes in the AGENT table when ZIP determines CITY (see Figure 2.2)	Avoid extra join operations Program can validate city (drop-down box) based on the zip code
Derived data	Storing STU_HRS and STU_CLASS (student classification) when STU_HRS determines STU_CLASS (see Figure 3.28)	Avoid extra join operations Program can validate classification (lookup) based on the student hours
Preaggregated data (also derived data)	Storing the student grade point average (STU_GPA) aggregate value in the STUDENT table when this can be calculated from the ENROLL and COURSE tables (see Figure 3.28)	Avoid extra join operations Program computes the GPA every time a grade is entered or updated STU_GPA can be updated only via administrative routine
Information requirements	Using a temporary denormalized table to hold report data; this is required when creating a tabular report in which the columns represent data that are stored in the table as rows (see Figures 6.17 and 6.18)	Impossible to generate the data required by the report using plain SQL No need to maintain table Temporary table is deleted once report is done Processing speed is not an issue

FIGURE 6.18 THE FACULTY EVALUATION REPORT

Faculty Evaluation Report

Instructor	Department	I Semester	I Mean	II Semester	II Mean	III Semester	III Mean	IV Semester	IV Mean	Last Two Sem. Avg.
Alton	INFS	2015S	2.91	2014F	2.84	2014S	2.55	2013F	2.51	2.875
Ames	INFS	2015S	3.24	2014F	3.26	2014S	3.31	2013F	3.19	3.250
Crandon	INFS	2015S	3.93	2014F	3.95	2014S	3.91	2013F	3.88	3.940
Dumas	MGMT	2014F	3.66	2014S	3.69	2013F	3.56	2013S	3.72	3.675
Landon	BMOM	2015S	3.57	2014F	3.64	2014S	3.39	2013F	3.57	3.605
Lohar	ECON	2009F	3.53	2008F	3.53					3.530
Rolman	INFS	2006S	3.50							3.500

the master data table containing the evaluation scores for each faculty member for each semester taught; this table is normalized. The FACHIST table contains the last four data points—that is, evaluation score and semester—for each faculty member. The FACHIST table is a temporary denormalized table created from the EVALDATA table via a series of queries. (The FACHIST table is the basis for the report shown in Figure 6.18.)

As shown in the faculty evaluation report, the conflicts between design efficiency, information requirements, and performance are often resolved through compromises that may include denormalization. In this case, and assuming there is enough storage space, the designer's choices could be narrowed down to:

- Store the data in a permanent denormalized table. This is not the recommended solution because the denormalized table is subject to data anomalies (insert, update, and delete). This solution is viable only if performance is an issue.

- Create a temporary denormalized table from the permanent normalized table(s). The denormalized table exists only as long as it takes to generate the report; it disappears after the report is produced. Therefore, there are no data anomaly problems. This solution is practical only if performance is not an issue and there are no other viable processing options.

As shown, *normalization purity is often difficult to sustain in the modern database environment.* You will learn in Chapter 13, Business Intelligence and Data Warehouses, that lower

FIGURE 6.19 THE EVALDATA AND FACHIST TABLES

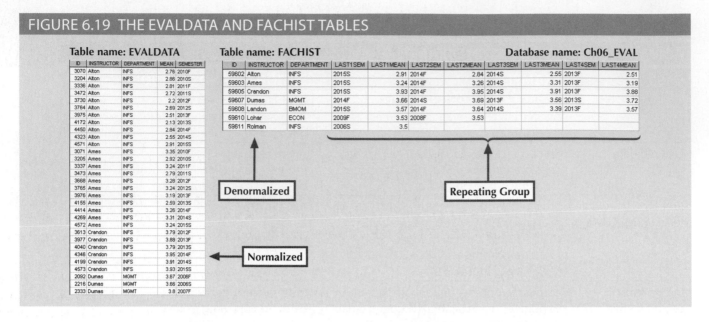

Table name: EVALDATA

ID	INSTRUCTOR	DEPARTMENT	MEAN	SEMESTER
3070	Alton	INFS	2.76	2010F
3204	Alton	INFS	2.86	2010S
3336	Alton	INFS	2.81	2011F
3472	Alton	INFS	2.72	2011S
3730	Alton	INFS	2.2	2012F
3764	Alton	INFS	2.69	2012S
3975	Alton	INFS	2.51	2013F
4172	Alton	INFS	2.13	2013S
4450	Alton	INFS	2.84	2014F
4323	Alton	INFS	2.55	2014S
4571	Alton	INFS	2.91	2015S
3071	Ames	INFS	3.35	2010F
3205	Ames	INFS	2.92	2010S
3337	Ames	INFS	3.24	2011F
3473	Ames	INFS	2.79	2011S
3668	Ames	INFS	3.28	2012F
3765	Ames	INFS	3.24	2012S
3976	Ames	INFS	3.19	2013F
4155	Ames	INFS	2.59	2013S
4414	Ames	INFS	3.26	2014F
4269	Ames	INFS	3.31	2014S
4572	Ames	INFS	3.24	2015S
3613	Crandon	INFS	3.79	2012F
3977	Crandon	INFS	3.88	2013F
4040	Crandon	INFS	3.79	2013S
4348	Crandon	INFS	3.95	2014F
4199	Crandon	INFS	3.91	2014S
4573	Crandon	INFS	3.93	2015S
2092	Dumas	MGMT	3.67	2006F
2216	Dumas	MGMT	3.66	2006S
2333	Dumas	MGMT	3.8	2007F

Table name: FACHIST   Database name: Ch06_EVAL

ID	INSTRUCTOR	DEPARTMENT	LAST1SEM	LAST1MEAN	LAST2SEM	LAST2MEAN	LAST3SEM	LAST3MEAN	LAST4SEM	LAST4MEAN
59602	Alton	INFS	2015S	2.91	2014F	2.84	2014S	2.55	2013F	2.51
59603	Ames	INFS	2015S	3.24	2014F	3.26	2014S	3.31	2013F	3.19
59605	Crandon	INFS	2015S	3.93	2014F	3.95	2014S	3.91	2013F	3.88
59607	Dumas	MGMT	2014F	3.66	2014S	3.69	2013F	3.58	2013S	3.72
59608	Landon	BMOM	2015S	3.57	2014F	3.64	2014S	3.39	2013F	3.57
59610	Lohar	ECON	2009F	3.53	2008F	3.53				
59611	Rolman	INFS	2006S	3.5						

Denormalized

Repeating Group

Normalized

normalization forms occur (and are even required) in specialized databases known as data warehouses. Such specialized databases reflect the ever-growing demand for greater scope and depth in the data on which decision support systems increasingly rely. You will discover that the data warehouse routinely uses 2NF structures in its complex, multilevel, multisource data environment. In short, although normalization is very important, especially in the so-called production database environment, 2NF is no longer disregarded as it once was.

Although 2NF tables cannot always be avoided, the problem of working with tables that contain partial and/or transitive dependencies in a production database environment should not be minimized. Aside from the possibility of troublesome data anomalies being created, unnormalized tables in a production database tend to suffer from these defects:

- Data updates are less efficient because programs that read and update tables must deal with larger tables.

- Indexing is more cumbersome. It is simply not practical to build all of the indexes required for the many attributes that might be located in a single unnormalized table.

- Unnormalized tables yield no simple strategies for creating virtual tables known as *views*. You will learn how to create and use views in Chapter 8, Advanced SQL.

Remember that good design cannot be created in the application programs that use a database. Also keep in mind that unnormalized database tables often lead to various data redundancy disasters in production databases, such as the problems examined thus far. In other words, use denormalization cautiously and make sure that you can explain why the unnormalized tables are a better choice in certain situations than their normalized counterparts.

6-9 Data-Modeling Checklist

In the chapters of Part 2, you have learned how data modeling translates a specific real-world environment into a data model that represents the real-world data, users, processes, and interactions. The modeling techniques you have learned thus far give you the tools needed to produce successful database designs. However, just as any good pilot uses a checklist to ensure that all is in order for a successful flight, the data-modeling checklist shown in Table 6.7 will help ensure that you perform data-modeling tasks successfully based on the concepts and tools you have learned in this text.

Note

You can also find this data-modeling checklist on the inside front cover of this book for easy reference.

TABLE 6.7

DATA-MODELING CHECKLIST

BUSINESS RULES

- Properly document and verify all business rules with the end users.
- Ensure that all business rules are written precisely, clearly, and simply. The business rules must help identify entities, attributes, relationships, and constraints.
- Identify the source of all business rules, and ensure that each business rule is justified, dated, and signed off by an approving authority.

DATA MODELING

Naming conventions: All names should be limited in length (database-dependent size).
- Entity names:
  - Should be nouns that are familiar to business and should be short and meaningful
  - Should document abbreviations, synonyms, and aliases for each entity
  - Should be unique within the model
  - For composite entities, may include a combination of abbreviated names of the entities linked through the composite entity
- Attribute names:
  - Should be unique within the entity
  - Should use the entity abbreviation as a prefix
  - Should be descriptive of the characteristic
  - Should use suffixes such as _ID, _NUM, or _CODE for the PK attribute
  - Should not be a reserved word
  - Should not contain spaces or special characters such as @, !, or &
- Relationship names:
  - Should be active or passive verbs that clearly indicate the nature of the relationship

Entities:
- Each entity should represent a single subject.
- Each entity should represent a set of distinguishable entity instances.
- All entities should be in 3NF or higher. Any entities below 3NF should be justified.
- The granularity of the entity instance should be clearly defined.
- The PK should be clearly defined and support the selected data granularity.

Attributes:
- Should be simple and single-valued (atomic data)
- Should document default values, constraints, synonyms, and aliases
- Derived attributes should be clearly identified and include source(s)
- Should not be redundant unless this is required for transaction accuracy, performance, or maintaining a history
- Nonkey attributes must be fully dependent on the PK attribute

Relationships:
- Should clearly identify relationship participants
- Should clearly define participation, connectivity, and document cardinality

ER model:
- Should be validated against expected processes: inserts, updates, and deletions
- Should evaluate where, when, and how to maintain a history
- Should not contain redundant relationships except as required (see attributes)
- Should minimize data redundancy to ensure single-place updates
- Should conform to the minimal data rule: All that is needed is there, and all that is there is needed.

Summary

- Normalization is a technique used to design tables in which data redundancies are minimized. The first three normal forms (1NF, 2NF, and 3NF) are the most common. From a structural point of view, higher normal forms are better than lower normal forms because higher normal forms yield relatively fewer data redundancies in the database. Almost all business designs use 3NF as the ideal normal form. A special, more restricted 3NF known as Boyce-Codd normal form, or BCNF, is also used.

- A table is in 1NF when all key attributes are defined and all remaining attributes are dependent on the primary key. However, a table in 1NF can still contain both partial and transitive dependencies. A partial dependency is one in which an attribute is functionally dependent on only a part of a multiattribute primary key. A transitive dependency is one in which an attribute is functionally dependent on another nonkey attribute. A table with a single-attribute primary key cannot exhibit partial dependencies.

- A table is in 2NF when it is in 1NF and contains no partial dependencies. Therefore, a 1NF table is automatically in 2NF when its primary key is based on only a single attribute. A table in 2NF may still contain transitive dependencies.

- A table is in 3NF when it is in 2NF and contains no transitive dependencies. Given that definition, the Boyce-Codd normal form (BCNF) is merely a special 3NF case in which all determinant keys are candidate keys. When a table has only a single candidate key, a 3NF table is automatically in BCNF.

- A table that is not in 3NF may be split into new tables until all of the tables meet the 3NF requirements.

- Normalization is an important part—but only a part—of the design process. As entities and attributes are defined during the ER modeling process, subject each entity (set) to normalization checks and form new entities (sets) as required. Incorporate the normalized entities into the ERD and continue the iterative ER process until all entities and their attributes are defined and all equivalent tables are in 3NF.

- A table in 3NF might contain multivalued dependencies that produce either numerous null values or redundant data. Therefore, it might be necessary to convert a 3NF table to the fourth normal form (4NF) by splitting the table to remove the multivalued dependencies. Thus, a table is in 4NF when it is in 3NF and contains no multivalued dependencies.

- The larger the number of tables, the more additional I/O operations and processing logic you need to join them. Therefore, tables are sometimes denormalized to yield less I/O in order to increase processing speed. Unfortunately, with larger tables, you pay for the increased processing speed by making the data updates less efficient, by making indexing more cumbersome, and by introducing data redundancies that are likely to yield data anomalies. In the design of production databases, use denormalization sparingly and cautiously.

- The data-modeling checklist provides a way for the designer to check that the ERD meets a set of minimum requirements.

Key Term

atomic attribute	first normal form (1NF)	partial dependency
atomicity	fourth normal form (4NF)	prime attribute
Boyce-Codd normal form (BCNF)	granularity	repeating group
	key attribute	second normal form (2NF)
denormalization	nonkey attribute	third normal form (3NF)
dependency diagram	nonprime attribute	transitive dependency
determinant	normalization	

Online Content

Flashcards and crossword puzzles for key term practice are available at *www.cengagebrain.com*.

Review Questions

1. What is normalization?

2. When is a table in 1NF?

3. When is a table in 2NF?

4. When is a table in 3NF?

5. When is a table in BCNF?

6. Given the dependency diagram shown in Figure Q6.6, answer Items 6a–6c.

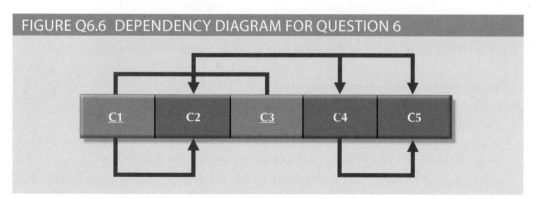

FIGURE Q6.6   DEPENDENCY DIAGRAM FOR QUESTION 6

a. Identify and discuss each of the indicated dependencies.

b. Create a database whose tables are at least in 2NF, showing the dependency diagrams for each table.

c. Create a database whose tables are at least in 3NF, showing the dependency diagrams for each table.

7. The dependency diagram in Figure Q6.7 indicates that authors are paid royalties for each book they write for a publisher. The amount of the royalty can vary by author, by book, and by edition of the book.

FIGURE Q6.7 BOOK ROYALTY DEPENDENCY DIAGRAM

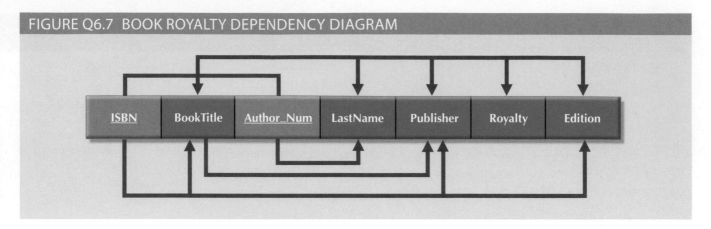

a. Based on the dependency diagram, create a database whose tables are at least in 2NF, showing the dependency diagram for each table.

b. Create a database whose tables are at least in 3NF, showing the dependency diagram for each table.

8. The dependency diagram in Figure Q6.8 indicates that a patient can receive many prescriptions for one or more medicines over time. Based on the dependency diagram, create a database whose tables are in at least 2NF, showing the dependency diagram for each table.

FIGURE Q6.8 PRESCRIPTION DEPENDENCY DIAGRAM

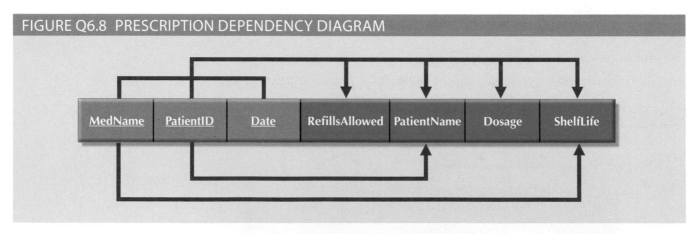

9. What is a partial dependency? With what normal form is it associated?

10. What three data anomalies are likely to be the result of data redundancy? How can such anomalies be eliminated?

11. Define and discuss the concept of transitive dependency.

12. What is a surrogate key, and when should you use one?

13. Why is a table whose primary key consists of a single attribute automatically in 2NF when it is in 1NF?

14. How would you describe a condition in which one attribute is dependent on another attribute when neither attribute is part of the primary key?

15. Suppose someone tells you that an attribute that is part of a composite primary key is also a candidate key. How would you respond to that statement?

16. A table is in _____ normal form when it is in _____ and there are no transitive dependencies.

Problems

1. Using the descriptions of the attributes given in the figure, convert the ERD shown in Figure P6.1 into a dependency diagram that is in at least 3NF.

FIGURE P6.1   APPOINTMENT ERD FOR PROBLEM 1

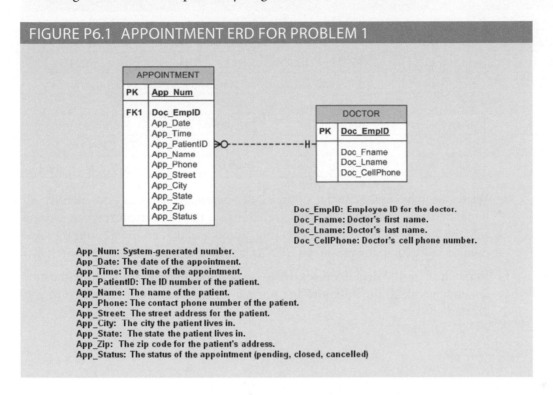

App_Num: System-generated number.
App_Date: The date of the appointment.
App_Time: The time of the appointment.
App_PatientID: The ID number of the patient.
App_Name: The name of the patient.
App_Phone: The contact phone number of the patient.
App_Street: The street address for the patient.
App_City: The city the patient lives in.
App_State: The state the patient lives in.
App_Zip: The zip code for the patient's address.
App_Status: The status of the appointment (pending, closed, cancelled)

Doc_EmpID: Employee ID for the doctor.
Doc_Fname: Doctor's first name.
Doc_Lname: Doctor's last name.
Doc_CellPhone: Doctor's cell phone number.

2. Using the descriptions of the attributes given in the figure, convert the ERD shown in Figure P6.2 into a dependency diagram that is in at least 3NF.

FIGURE P6.2   PRESENTATION ERD FOR PROBLEM 2

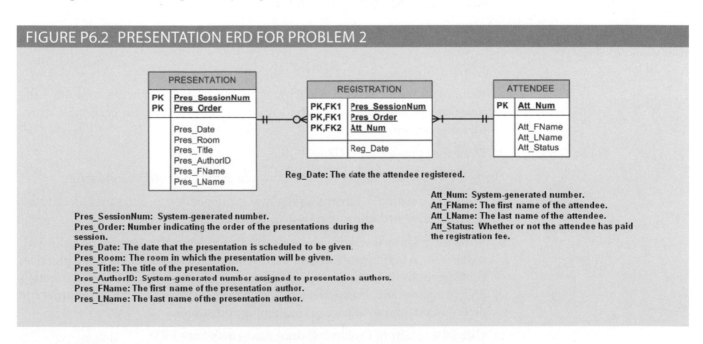

Reg_Date: The date the attendee registered.

Pres_SessionNum: System-generated number.
Pres_Order: Number indicating the order of the presentations during the session.
Pres_Date: The date that the presentation is scheduled to be given.
Pres_Room: The room in which the presentation will be given.
Pres_Title: The title of the presentation.
Pres_AuthorID: System-generated number assigned to presentation authors.
Pres_FName: The first name of the presentation author.
Pres_LName: The last name of the presentation author.

Att_Num: System-generated number.
Att_FName: The first name of the attendee.
Att_LName: The last name of the attendee.
Att_Status: Whether or not the attendee has paid the registration fee.

3. Using the INVOICE table structure shown in Table P6.3, do the following:

TABLE P6.3

ATTRIBUTE NAME	SAMPLE VALUE	SAMPLE VALUE	SAMPLE VALUE	SAMPLE VALUE	SAMPLE VALUE
INV_NUM	211347	211347	211347	211348	211349
PROD_NUM	AA-E3422QW	QD-300932X	RU-995748G	AA-E3422QW	GH-778345P
SALE_DATE	15-Jan-2016	15-Jan-2016	15-Jan-2016	15-Jan-2016	16-Jan-2016
PROD_LABEL	Rotary sander	0.25-in. drill bit	Band saw	Rotary sander	Power drill
VEND_CODE	211	211	309	211	157
VEND_NAME	NeverFail, Inc.	NeverFail, Inc.	BeGood, Inc.	NeverFail, Inc.	ToughGo, Inc.
QUANT_SOLD	1	8	1	2	1
PROD_PRICE	$49.95	$3.45	$39.99	$49.95	$87.75

a. Write the relational schema, draw its dependency diagram, and identify all dependencies, including all partial and transitive dependencies. You can assume that the table does not contain repeating groups and that an invoice number references more than one product. (*Hint:* This table uses a composite primary key.)

b. Remove all partial dependencies, write the relational schema, and draw the new dependency diagrams. Identify the normal forms for each table structure you created.

Note

You can assume that any given product is supplied by a single vendor, but a vendor can supply many products. Therefore, it is proper to conclude that the following dependency exists:

PROD_NUM → PROD_LABEL, PROD_PRICE, VEND_CODE, VEND_NAME

(*Hint:* Your actions should produce three dependency diagrams.)

c. Remove all transitive dependencies, write the relational schema, and draw the new dependency diagrams. Also identify the normal forms for each table structure you created.

d. Draw the Crow's Foot ERD.

4. Using the STUDENT table structure shown in Table P6.4, do the following:

a. Write the relational schema and draw its dependency diagram. Identify all dependencies, including all transitive dependencies.

b. Write the relational schema and draw the dependency diagram to meet the 3NF requirements to the greatest practical extent possible. If you believe that practical considerations dictate using a 2NF structure, explain why your decision to retain 2NF is appropriate. If necessary, add or modify attributes to create appropriate determinants and to adhere to the naming conventions.

c. Using the results of Problem 4, draw the Crow's Foot ERD.

TABLE P6.4

ATTRIBUTE NAME	SAMPLE VALUE	SAMPLE VALUE	SAMPLE VALUE	SAMPLE VALUE	SAMPLE VALUE
STU_NUM	211343	200128	199876	198648	223456
STU_LNAME	Stephanos	Smith	Jones	Ortiz	McKulski
STU_MAJOR	Accounting	Accounting	Marketing	Marketing	Statistics
DEPT_CODE	ACCT	ACCT	MKTG	MKTG	MATH
DEPT_NAME	Accounting	Accounting	Marketing	Marketing	Mathematics
DEPT_PHONE	4356	4356	4378	4378	3420
COLLEGE_NAME	Business Admin	Business Admin	Business Admin	Business Admin	Arts & Sciences
ADVISOR_LNAME	Grastrand	Grastrand	Gentry	Tillery	Chen
ADVISOR_OFFICE	T201	T201	T228	T356	J331
ADVISOR_BLDG	Torre Building	Torre Building	Torre Building	Torre Building	Jones Building
ADVISOR_PHONE	2115	2115	2123	2159	3209
STU_GPA	3.87	2.78	2.31	3.45	3.58
STU_HOURS	75	45	117	113	87
STU_CLASS	Junior	Sophomore	Senior	Senior	Junior

Note

Although the completed student hours (STU_HOURS) do determine the student classification (STU_CLASS), this dependency is not as obvious as you might initially assume it to be. For example, a student is considered a junior if the student has completed between 61 and 90 credit hours.

5. To keep track of office furniture, computers, printers, and other office equipment, the FOUNDIT Company uses the table structure shown in Table P6.5.

TABLE P6.5

ATTRIBUTE NAME	SAMPLE VALUE	SAMPLE VALUE	SAMPLE VALUE
ITEM_ID	231134-678	342245-225	254668-449
ITEM_LABEL	HP DeskJet 895Cse	HP Toner	DT Scanner
ROOM_NUMBER	325	325	123
BLDG_CODE	NTC	NTC	CSF
BLDG_NAME	Nottooclear	Nottooclear	Canseefar
BLDG_MANAGER	I. B. Rightonit	I. B. Rightonit	May B. Next

a. Given that information, write the relational schema and draw the dependency diagram. Make sure that you label the transitive and/or partial dependencies.

b. Write the relational schema and create a set of dependency diagrams that meet 3NF requirements. Rename attributes to meet the naming conventions, and create new entities and attributes as necessary.

c. Draw the Crow's Foot ERD.

6. The table structure shown in Table P6.6 contains many unsatisfactory components and characteristics. For example, there are several multivalued attributes, naming conventions are violated, and some attributes are not atomic.

TABLE P6.6

EMP_NUM	1003	1018	1019	1023
EMP_LNAME	Willaker	Smith	McGuire	McGuire
EMP_EDUCATION	BBA, MBA	BBA		BS, MS, Ph.D.
JOB_CLASS	SLS	SLS	JNT	DBA
EMP_DEPENDENTS	Gerald (spouse), Mary (daughter), John (son)		JoAnne (spouse)	George (spouse) Jill (daughter)
DEPT_CODE	MKTG	MKTG	SVC	INFS
DEPT_NAME	Marketing	Marketing	General Service	Info. Systems
DEPT_MANAGER	Jill H. Martin	Jill H. Martin	Hank B. Jones	Carlos G. Ortez
EMP_TITLE	Sales Agent	Sales Agent	Janitor	DB Admin
EMP_DOB	23-Dec-1968	28-Mar-1979	18-May-1982	20-Jul-1959
EMP_HIRE_DATE	14-Oct-1997	15-Jan-2006	21-Apr-2003	15-Jul-1999
EMP_TRAINING	L1, L2	L1	L1	L1, L3, L8, L15
EMP_BASE_SALARY	$38,255.00	$30,500.00	$19,750.00	$127,900.00
EMP_COMMISSION_RATE	0.015	0.010		

a. Given the structure shown in Table P6.6, write the relational schema and draw its dependency diagram. Label all transitive and/or partial dependencies.

b. Draw the dependency diagrams that are in 3NF. (*Hint:* You might have to create a few new attributes. Also make sure that the new dependency diagrams contain attributes that meet proper design criteria; i.e., make sure there are no multivalued attributes, that the naming conventions are met, and so on.)

c. Draw the relational diagram.

d. Draw the Crow's Foot ERD.

7. Suppose you are given the following business rules to form the basis for a database design. The database must enable the manager of a company dinner club to mail invitations to the club's members, to plan the meals, to keep track of who attends the dinners, and so on.

- Each dinner serves many members, and each member may attend many dinners.

- A member receives many invitations, and each invitation is mailed to many members.

- A dinner is based on a single entree, but an entree may be used as the basis for many dinners. For example, a dinner may be composed of a fish entree, rice, and corn, or the dinner may be composed of a fish entree, a baked potato, and string beans.

Because the manager is not a database expert, the first attempt at creating the database uses the structure shown in Table P6.7.

a. Given the table structure illustrated in Table P6.7, write the relational schema and draw its dependency diagram. Label all transitive and/or partial dependencies. (*Hint:* This structure uses a composite primary key.)

TABLE P6.7

ATTRIBUTE NAME	SAMPLE VALUE	SAMPLE VALUE	SAMPLE VALUE
MEMBER_NUM	214	235	214
MEMBER_NAME	Alice B. VanderVoort	Gerald M. Gallega	Alice B. VanderVoort
MEMBER_ADDRESS	325 Meadow Park	123 Rose Court	325 Meadow Park
MEMBER_CITY	Murkywater	Highlight	Murkywater
MEMBER_ZIPCODE	12345	12349	12345
INVITE_NUM	8	9	10
INVITE_DATE	23-Feb-2016	12-Mar-2016	23-Feb-2016
ACCEPT_DATE	27-Feb-2016	15-Mar-2016	27-Feb-2016
DINNER_DATE	15-Mar-2016	17-Mar-2016	15-Mar-2016
DINNER_ATTENDED	Yes	Yes	No
DINNER_CODE	DI5	DI5	DI2
DINNER_DESCRIPTION	Glowing Sea Delight	Glowing Sea Delight	Ranch Superb
ENTREE_CODE	EN3	EN3	EN5
ENTREE_DESCRIPTION	Stuffed crab	Stuffed crab	Marinated steak
DESSERT_CODE	DE8	DE5	DE2
DESSERT_DESCRIPTION	Chocolate mousse with raspberry sauce	Cherries jubilee	Apple pie with honey crust

b. Break up the dependency diagram you drew in Problem 7a to produce dependency diagrams that are in 3NF, and write the relational schema. (*Hint:* You might have to create a few new attributes. Also, make sure that the new dependency diagrams contain attributes that meet proper design criteria; i.e., make sure there are no multivalued attributes, that the naming conventions are met, and so on.)

c. Using the results of Problem 7b, draw the Crow's Foot ERD.

8. Use the dependency diagram shown in Figure P6.8 to work the following problems.

a. Break up the dependency diagram shown in Figure P6.8 to create two new dependency diagrams: one in 3NF and one in 2NF.

b. Modify the dependency diagrams you created in Problem 8a to produce a set of dependency diagrams that are in 3NF. (*Hint:* One of your dependency diagrams should be in 3NF but not in BCNF.)

c. Modify the dependency diagrams you created in Problem 8b to produce a collection of dependency diagrams that are in 3NF and BCNF.

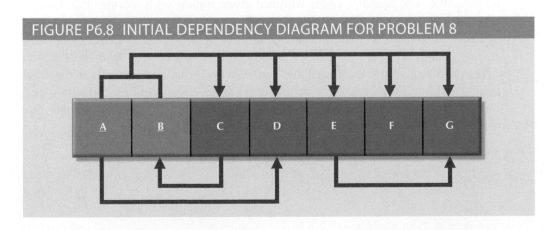

FIGURE P6.8   INITIAL DEPENDENCY DIAGRAM FOR PROBLEM 8

9. Suppose you have been given the table structure and data shown in Table P6.9, which was imported from an Excel spreadsheet. The data reflects that a professor can have multiple advisees, can serve on multiple committees, and can edit more than one journal.

TABLE P6.9

ATTRIBUTE NAME	SAMPLE VALUE	SAMPLE VALUE	SAMPLE VALUE	SAMPLE VALUE
EMP_NUM	123	104	118	
PROF_RANK	Professor	Asst. Professor	Assoc. Professor	Assoc. Professor
EMP_NAME	Ghee	Rankin	Ortega	Smith
DEPT_CODE	CIS	CHEM	CIS	ENG
DEPT_NAME	Computer Info. Systems	Chemistry	Computer Info. Systems	English
PROF_OFFICE	KDD-567	BLF-119	KDD-562	PRT-345
ADVISEE	1215, 2312, 3233, 2218, 2098	3102, 2782, 3311, 2008, 2876, 2222, 3745, 1783, 2378	2134, 2789, 3456, 2002, 2046, 2018, 2764	2873, 2765, 2238, 2901, 2308
COMMITTEE_CODE	PROMO, TRAF, APPL, DEV	DEV	SPR, TRAF	PROMO, SPR, DEV
JOURNAL_CODE	JMIS, QED, JMGT		JCIS, JMGT	

Given the information in Table P6.9:

a. Draw the dependency diagram.

b. Identify the multivalued dependencies.

c. Create the dependency diagrams to yield a set of table structures in 3NF.

d. Eliminate the multivalued dependencies by converting the affected table structures to 4NF.

e. Draw the Crow's Foot ERD to reflect the dependency diagrams you drew in Problem 9c. (*Note:* You might have to create additional attributes to define the proper PKs and FKs. Make sure that all of your attributes conform to the naming conventions.)

10. The manager of a consulting firm has asked you to evaluate a database that contains the table structure shown in Table P6.10.

Table P6.10 was created to enable the manager to match clients with consultants. The objective is to match a client within a given region with a consultant in that region and to make sure that the client's need for specific consulting services is properly matched to the consultant's expertise. For example, if the client needs help with database design and is located in the Southeast, the objective is to make a match with a consultant who is located in the Southeast and whose expertise is in database design. (Although the consulting company manager tries to match consultant and client locations to minimize travel expense, it is not always possible to do so.) The following basic business rules are maintained:

- Each client is located in one region.

- A region can contain many clients.

- Each consultant can work on many contracts.
- Each contract might require the services of many consultants.
- A client can sign more than one contract, but each contract is signed by only one client.
- Each contract might cover multiple consulting classifications. (For example, a contract may list consulting services in database design and networking.)

TABLE P6.10

ATTRIBUTE NAME	SAMPLE VALUE	SAMPLE VALUE	SAMPLE VALUE
CLIENT_NUM	298	289	289
CLIENT_NAME	Marianne R. Brown	James D. Smith	James D. Smith
CLIENT_REGION	Midwest	Southeast	Southeast
CONTRACT_DATE	10-Feb-2016	15-Feb-2016	12-Mar-2016
CONTRACT_NUMBER	5841	5842	5843
CONTRACT_AMOUNT	$2,985,000.00	$670,300.00	$1,250,000.00
CONSULT_CLASS_1	Database Administration	Internet Services	Database Design
CONSULT_CLASS_2	Web Applications		Database Administration
CONSULT_CLASS_3			Network Installation
CONSULT_CLASS_4			
CONSULTANT_NUM_1	29	34	25
CONSULTANT_NAME_1	Rachel G. Carson	Gerald K. Ricardo	Angela M. Jamison
CONSULTANT_REGION_1	Midwest	Southeast	Southeast
CONSULTANT_NUM_2	56	38	34
CONSULTANT_NAME_2	Karl M. Spenser	Anne T. Dimarco	Gerald K. Ricardo
CONSULTANT_REGION_2	Midwest	Southeast	Southeast
CONSULTANT_NUM_3	22	45	
CONSULTANT_NAME_3	Julian H. Donatello	Geraldo J. Rivera	
CONSULTANT_REGION_3	Midwest	Southeast	
CONSULTANT_NUM_4		18	
CONSULTANT_NAME_4		Donald Chen	
CONSULTANT_REGION_4		West	

- Each consultant is located in one region.
- A region can contain many consultants.
- Each consultant has one or more areas of expertise (class). For example, a consultant might be classified as an expert in both database design and networking.
- Each area of expertise (class) can have many consultants. For example, the consulting company might employ many consultants who are networking experts.

a. Given this brief description of the requirements and the business rules, write the relational schema and draw the dependency diagram for the preceding (and very poor) table structure. Label all transitive and/or partial dependencies.

b. Break up the dependency diagram you drew in Problem 10a to produce dependency diagrams that are in 3NF and write the relational schema. (*Hint:* You might have to create a few new attributes. Also make sure that the new dependency diagrams contain attributes that meet proper design criteria; that is, make sure there are no multivalued attributes, that the naming conventions are met, and so on.)

c. Using the results of Problem 10b, draw the Crow's Foot ERD.

11. Given the sample records in the CHARTER table shown in Table P6.11, do the following:

a. Write the relational schema and draw the dependency diagram for the table structure. Make sure that you label all dependencies. CHAR_PAX indicates the number of passengers carried. The CHAR_MILES entry is based on round-trip miles, including pickup points. (*Hint:* Look at the data values to determine the nature of the relationships. For example, note that employee Melton has flown two charter trips as pilot and one trip as copilot.)

b. Decompose the dependency diagram you drew to solve Problem 11a to create table structures that are in 3NF and write the relational schema.

c. Draw the Crow's Foot ERD to reflect the properly decomposed dependency diagrams you created in Problem 11b. Make sure the ERD yields a database that can track all of the data shown in Problem 11. Show all entities, relationships, connectivities, optionalities, and cardinalities.

TABLE P6.11

ATTRIBUTE NAME	SAMPLE VALUE	SAMPLE VALUE	SAMPLE VALUE	SAMPLE VALUE
CHAR_TRIP	10232	10233	10234	10235
CHAR_DATE	15-Jan-2016	15-Jan-2016	16-Jan-2016	17-Jan-2016
CHAR_CITY	STL	MIA	TYS	ATL
CHAR_MILES	580	1,290	524	768
CUST_NUM	784	231	544	784
CUST_LNAME	Brown	Hanson	Bryana	Brown
CHAR_PAX	5	12	2	5
CHAR_CARGO	235 lbs.	18,940 lbs.	348 lbs.	155 lbs.
PILOT	Melton	Chen	Henderson	Melton
COPILOT		Henderson	Melton	
FLT_ENGINEER		O'Shaski		
LOAD_MASTER		Benkasi		
AC_NUMBER	1234Q	3456Y	1234Q	2256W
MODEL_CODE	PA31-350	CV-580	PA31-350	PA31-350
MODEL_SEATS	10	38	10	10
MODEL_CHG_MILE	$2.79	$23.36	$2.79	$2.79

PART 3
Advanced Design and Implementation

Introduction to Structured Query Language (SQL)

In this chapter, you will learn:

• The basic commands and functions of SQL
• How to use SQL for data administration (to create tables and indexes)
• How to use SQL for data manipulation (to add, modify, delete, and retrieve data)
• How to use SQL to query a database for useful information

Preview

In this chapter, you will learn the basics of Structured Query Language (SQL). SQL, which is pronounced *S-Q-L* or *sequel*, is composed of commands that enable users to create database and table structures, perform various types of data manipulation and data administration, and query the database to extract useful information. All relational DBMS software supports SQL, and many software vendors have developed extensions to the basic SQL command set.

Although it is quite useful and powerful, SQL is not meant to stand alone in the applications arena. Data entry with SQL is possible but awkward, as are data corrections and additions. SQL itself does not create menus, special report forms, overlays, pop-ups, or other features that end users usually expect. Instead, those features are available as vendor-supplied enhancements. SQL focuses on data definition (creating tables and indexes) and data manipulation (adding, modifying, deleting, and retrieving data). This chapter covers these basic functions. In spite of its limitations, SQL is a powerful tool for extracting information and managing data.

Data Files and Available Formats

	MS Access	Oracle	MS SQL	My SQL		MS Access	Oracle	MS SQL	My SQL
CH07_SaleCo	✓	✓	✓	✓	CH07_ConstructCo	✓	✓	✓	✓
					CH07_LargeCo	✓	✓	✓	✓
					Ch07_Fact	✓	✓	✓	✓

Data Files Available on cengagebrain.com

Note

Although you can use the MS Access databases and SQL script files for creating the tables and loading the data supplied online, it is strongly suggested that you create your own database structures so you can practice the SQL commands illustrated in this chapter.

How you connect to your database depends on how the software was installed on your computer. Follow the instructions provided by your instructor or school.

7-1 **Introduction to SQL**

Ideally, a database language allows you to create database and table structures, perform basic data management chores (add, delete, and modify), and perform complex queries designed to transform the raw data into useful information. Moreover, a database language must perform such basic functions with minimal user effort, and its command structure and syntax must be easy to learn. Finally, it must be portable; that is, it must conform to some basic standard so a person does not have to relearn the basics when moving from one RDBMS to another. SQL meets those ideal database language requirements well.

SQL functions fit into two broad categories:

- It is a *data definition language (DDL)*. SQL includes commands to create database objects such as tables, indexes, and views, as well as commands to define access rights to those database objects. Some common data definition commands you will learn are listed in Table 7.1.

- It is a *data manipulation language (DML)*. SQL includes commands to insert, update, delete, and retrieve data within the database tables. The data manipulation commands you will learn in this chapter are listed in Table 7.2.

TABLE 7.1

SQL DATA DEFINITION COMMANDS

COMMAND OR OPTION	DESCRIPTION
CREATE SCHEMA AUTHORIZATION	Creates a database schema
CREATE TABLE	Creates a new table in the user's database schema
NOT NULL	Ensures that a column will not have null values
UNIQUE	Ensures that a column will not have duplicate values
PRIMARY KEY	Defines a primary key for a table
FOREIGN KEY	Defines a foreign key for a table
DEFAULT	Defines a default value for a column (when no value is given)
CHECK	Validates data in an attribute
CREATE INDEX	Creates an index for a table
CREATE VIEW	Creates a dynamic subset of rows and columns from one or more tables (see Chapter 8, Advanced SQL)
ALTER TABLE	Modifies a table's definition (adds, modifies, or deletes attributes or constraints)
CREATE TABLE AS	Creates a new table based on a query in the user's database schema
DROP TABLE	Permanently deletes a table (and its data)
DROP INDEX	Permanently deletes an index
DROP VIEW	Permanently deletes a view

SQL is relatively easy to learn. Its basic command set has a vocabulary of fewer than 100 words. Better yet, SQL is a nonprocedural language: you merely command *what* is to be done; you do not have to worry about *how*. For example, a single command creates the complex table structures required to store and manipulate data successfully; end users and programmers do not need to know the physical data storage format or the complex activities that take place when a SQL command is executed.

TABLE 7.2

SQL DATA MANIPULATION COMMANDS

COMMAND OR OPTION	DESCRIPTION
INSERT	Inserts row(s) into a table
SELECT	Selects attributes from rows in one or more tables or views
WHERE	Restricts the selection of rows based on a conditional expression
GROUP BY	Groups the selected rows based on one or more attributes
HAVING	Restricts the selection of grouped rows based on a condition
ORDER BY	Orders the selected rows based on one or more attributes
UPDATE	Modifies an attribute's values in one or more table's rows
DELETE	Deletes one or more rows from a table
COMMIT	Permanently saves data changes
ROLLBACK	Restores data to its original values
Comparison operators	
=, <, >, <=, >=, <>, !=	Used in conditional expressions
Logical operators	
AND/OR/NOT	Used in conditional expressions
Special operators	Used in conditional expressions
BETWEEN	Checks whether an attribute value is within a range
IS NULL	Checks whether an attribute value is null
LIKE	Checks whether an attribute value matches a given string pattern
IN	Checks whether an attribute value matches any value within a value list
EXISTS	Checks whether a subquery returns any rows
DISTINCT	Limits values to unique values
Aggregate functions	Used with SELECT to return mathematical summaries on columns
COUNT	Returns the number of rows with non-null values for a given column
MIN	Returns the minimum attribute value found in a given column
MAX	Returns the maximum attribute value found in a given column
SUM	Returns the sum of all values for a given column
AVG	Returns the average of all values for a given column

The American National Standards Institute (ANSI) prescribes a standard SQL. The ANSI SQL standards are also accepted by the International Organization for Standardization (ISO), a consortium composed of national standards bodies of more than 150 countries. Although adherence to the ANSI/ISO SQL standard is usually required in commercial and government contract database specifications, many RDBMS vendors add their own special enhancements. Consequently, it is seldom possible to move a SQL-based application from one RDBMS to another without making some changes.

However, even though there are several different SQL "dialects," their differences are minor. Whether you use Oracle, Microsoft SQL Server, MySQL, IBM's DB2, Microsoft Access, or any other well-established RDBMS, a software manual should be sufficient to get you up to speed if you know the material presented in this chapter.

At the heart of SQL is the query. In Chapter 1, Database Systems, you learned that a query is a spur-of-the-moment question. Actually, in the SQL environment, the word *query* covers both questions and actions. Most SQL queries are used to answer questions such as these: "What products currently held in inventory are priced over $100, and what is the quantity on hand for each of those products?" or "How many employees have been hired since January 1, 2016, by each of the company's departments?" However, many SQL queries are used to perform actions such as adding or deleting table rows or changing attribute values within tables. Still other SQL queries create new tables or indexes. In short, for a DBMS, a query is simply a SQL statement that must be executed. However, before you can use SQL to query a database, you must define the database environment for SQL with its data definition commands.

7-2 Data Definition Commands

Before you examine the SQL syntax for creating and defining tables and other elements, first examine a simple database model and the database tables that form the basis for the many SQL examples you will explore in this chapter.

7-2a The Database Model

A simple database composed of the following tables is used to illustrate the SQL commands in this chapter: CUSTOMER, INVOICE, LINE, PRODUCT, and VENDOR. This database model is shown in Figure 7.1.

The database model in Figure 7.1 reflects the following business rules:

- A customer may generate many invoices. Each invoice is generated by one customer.

- An invoice contains one or more invoice lines. Each invoice line is associated with one invoice.

- Each invoice line references one product. A product may be found in many invoice lines. (You can sell more than one hammer to more than one customer.)

FIGURE 7.1 THE DATABASE MODEL

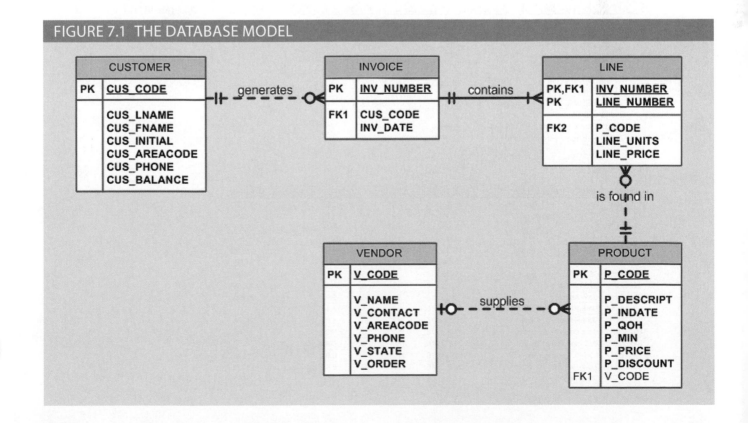

- A vendor *may* supply many products. Some vendors do not yet supply products. For example, a vendor list may include *potential* vendors.

- If a product is vendor-supplied, it is supplied by only a single vendor.

- Some products are not supplied by a vendor. For example, some products may be produced in-house or bought on the open market.

As you can see in Figure 7.1, the database model contains many tables. However, to illustrate the initial set of data definition commands, the focus of attention will be the PRODUCT and VENDOR tables. You will have the opportunity to use the remaining tables later in this chapter and in the Problems section.

To give you a point of reference for understanding the effect of the SQL queries, the contents of the PRODUCT and VENDOR tables are listed in Figure 7.2. In the tables, note the following features, which correspond to the business rules reflected in the ERD shown in Figure 7.1:

- The VENDOR table contains vendors who are not referenced in the PRODUCT table. Database designers note that possibility by saying that PRODUCT is optional to VENDOR; a vendor may exist without a reference to a product. You examined such optional relationships in detail in Chapter 4, Entity Relationship (ER) Modeling.

- Existing V_CODE values in the PRODUCT table must (and do) have a match in the VENDOR table to ensure referential integrity.

- A few products are supplied factory-direct, a few are made in-house, and a few may have been bought in a warehouse sale. In other words, a product is not necessarily supplied by a vendor. Therefore, VENDOR is optional to PRODUCT.

FIGURE 7.2 THE VENDOR AND PRODUCT TABLES

Table name: VENDOR   Database name: Ch07_SaleCo

V_CODE	V_NAME	V_CONTACT	V_AREACODE	V_PHONE	V_STATE	V_ORDER
21225	Bryson, Inc.	Smithson	615	223-3234	TN	Y
21226	SuperLoo, Inc.	Flushing	904	215-8995	FL	N
21231	D&E Supply	Singh	615	228-3245	TN	Y
21344	Gomez Bros.	Ortega	615	889-2546	KY	N
22567	Dome Supply	Smith	901	678-1419	GA	N
23119	Randsets Ltd.	Anderson	901	678-3998	GA	Y
24004	Brackman Bros.	Browning	615	228-1410	TN	N
24288	ORDVA, Inc.	Hakford	615	898-1234	TN	Y
25443	B&K, Inc.	Smith	904	227-0093	FL	N
25501	Damal Supplies	Smythe	615	890-3529	TN	N
25595	Rubicon Systems	Orton	904	456-0092	FL	Y

Table name: PRODUCT

P_CODE	P_DESCRIPT	P_INDATE	P_QOH	P_MIN	P_PRICE	P_DISCOUNT	V_CODE
11QER/31	Power painter, 15 psi., 3-nozzle	03-Nov-15	8	5	109.99	0.00	25595
13-Q2/P2	7.25-in. pwr. saw blade	13-Dec-15	32	15	14.99	0.05	21344
14-Q1/L3	9.00-in. pwr. saw blade	13-Nov-15	18	12	17.49	0.00	21344
1546-QQ2	Hrd. cloth, 1/4-in., 2x50	15-Jan-16	15	8	39.95	0.00	23119
1558-QW1	Hrd. cloth, 1/2-in., 3x50	15-Jan-16	23	5	43.99	0.00	23119
2232/QTY	B&D jigsaw, 12-in. blade	30-Dec-15	8	5	109.92	0.05	24288
2232/QWE	B&D jigsaw, 8-in. blade	24-Dec-15	6	5	99.87	0.05	24288
2238/QPD	B&D cordless drill, 1/2-in.	20-Jan-16	12	5	38.95	0.05	25595
23109-HB	Claw hammer	20-Jan-16	23	10	9.95	0.10	21225
23114-AA	Sledge hammer, 12 lb.	02-Jan-16	8	5	14.40	0.05	
54778-2T	Rat-tail file, 1/8-in. fine	15-Dec-15	43	20	4.99	0.00	21344
89-WRE-Q	Hicut chain saw, 16 in.	07-Feb-16	11	5	256.99	0.05	24288
PVC23DRT	PVC pipe, 3.5-in., 8-ft	20-Feb-16	188	75	5.87	0.00	
SM-18277	1.25-in. metal screw, 25	01-Mar-16	172	75	6.99	0.00	21225
SW-23116	2.5-in. wd. screw, 50	24-Feb-16	237	100	8.45	0.00	21231
WR3/TT3	Steel matting, 4'x8'x1/6", .5" mesh	17-Jan-16	18	5	119.95	0.10	25595

A few of the conditions just described were made for the sake of illustrating specific SQL features. For example, null V_CODE values were used in the PRODUCT table to illustrate how you can track such nulls using SQL.

7-2b Creating The Database

Before you can use a new RDBMS, you must complete two tasks: create the database structure and create the tables that will hold the end-user data. To complete the first task, the RDBMS creates the physical files that will hold the database. When you create a new database, the RDBMS automatically creates the data dictionary tables in which to store the metadata and creates a default database administrator. Creating the physical files that will hold the database means interacting with the operating system and the file systems supported by the operating system. Therefore, creating the database structure is the one feature that tends to differ substantially from one RDBMS to another. However, it is relatively easy to create a database structure, regardless of which RDBMS you use.

If you use Microsoft Access, creating the database is simple: start Access, click the FILE tab, click New in the left pane, and then click Blank desktop database in the right pane. Specify the folder in which you want to store the database, and then name the database. However, if you work in a database environment typically used by larger organizations, you will probably use an enterprise RDBMS such as Oracle, MS SQL Server, MySQL, or DB2. Given their security requirements and greater complexity, creating a database with these products is a more elaborate process. (See Appendix N, Creating a New Database Using Oracle 11g, for specific instructions to create a database structure in Oracle.)

With the exception of creating the database, most RDBMS vendors use SQL that deviates little from the ANSI standard SQL. For example, most RDBMSs require each SQL command to end with a semicolon. However, some SQL implementations do not use a semicolon. Important syntax differences among implementations will be highlighted in the Note boxes in this chapter.

If you are using an enterprise RDBMS, you must be authenticated by the RDBMS before you can start creating tables. **Authentication** is the process the DBMS uses to verify that only registered users access the database. To be authenticated, you must log on to the RDBMS using a user ID and a password created by the database administrator. In an enterprise RDBMS, every user ID is associated with a database schema.

7-2c The Database Schema

In the SQL environment, a **schema** is a logical group of database objects—such as tables and indexes—that are related to each other. Usually, the schema belongs to a single user or application. A single database can hold multiple schemas that belong to different users or applications. Schemas are useful in that they group tables by owner (or function) and enforce a first level of security by allowing each user to see only the tables that belong to that user.

ANSI SQL standards define a command to create a database schema:

CREATE SCHEMA AUTHORIZATION {creator};

Therefore, if the creator is JONES, the following command is used:

CREATE SCHEMA AUTHORIZATION JONES;

Most enterprise RDBMSs support that command. However, the command is seldom used directly—that is, from the command line. (When a user is created,

authentication
The process through which a DBMS verifies that only registered users can access the database.

schema
A logical grouping of database objects, such as tables, indexes, views, and queries, that are related to each other. Usually, a schema belongs to a single user or application.

the DBMS automatically assigns a schema to that user.) When the DBMS *is* used, the CREATE SCHEMA AUTHORIZATION command must be issued by the user who owns the schema. That is, if you log on as JONES, you can only use CREATE SCHEMA AUTHORIZATION JONES.

For most RDBMSs, the CREATE SCHEMA AUTHORIZATION command is optional, which is why this chapter focuses on the ANSI SQL commands required to create and manipulate tables.

7-2d Data Types

In the data dictionary in Table 7.3, note the data types selected. Keep in mind that data-type selection is usually dictated by the nature and intended use of the data. For example:

- P_PRICE clearly requires some kind of numeric data type; defining it as a character field is not acceptable.

- Just as clearly, a vendor name is an obvious candidate for a character data type. For example, VARCHAR(35) fits well because vendor names are variable-length character strings, and in this case, such strings may be up to 35 characters long.

- At first glance, it might seem logical to select a numeric data type for V_AREACODE because it contains only digits. However, adding and subtracting area codes does not yield meaningful results. Therefore, selecting a character data type is more appropriate. This is true for many common attributes found in business data models. For example, even though zip codes contain all digits, they must be defined as character data because some zip codes begin with the digit zero (0), and a numeric data type would cause the leading zero to be dropped.

- U.S. state abbreviations are always two characters, so CHAR(2) is a logical choice.

- Selecting P_INDATE to be a (Julian) DATE field rather than a character field is desirable because Julian dates allow you to make simple date comparisons and perform date arithmetic. For instance, if you have used DATE fields, you can determine the number of days between dates.

If you use DATE fields, you can also determine a future date using a simple command. For example, you can determine the date that is 60 days from a given P_INDATE by using P_INDATE + 60 in most DBMSs. MySQL requires a function for adding dates. For example, the AddDate() function used in "AddDate(P_INDATE, 60)" determines the date that is 60 days from the P_INDATE. Or, you can use the RDBMS's system date—SYSDATE in Oracle, SYSDATE() or NOW() in MySQL, GETDATE() in MS SQL Server, and Date() in Access—to answer questions such as "What will be the date 60 days from today?" For example, you might use SYSDATE + 60 in Oracle, AddDate(SYSDATE(), 60) in MySQL, GETDATE() + 60 in MS SQL Server, or Date() + 60 in Access.

Note

Oracle uses DATE data types to store complete dates, that is, a date and time. Access uses Date/Time as the data type to store these types of values. MySQL and MS SQL Server use the DATE data type to store only dates without a time component. Storing a complete date with time component in MySQL or MS SQL Server requires the DATETIME data type.

Date arithmetic capability is particularly useful in billing. Perhaps you want your system to start charging interest on a customer balance 60 days after the invoice is generated. Such simple date arithmetic would be impossible if you used a character data type.

TABLE 7.3

DATA DICTIONARY FOR THE CH07_SALECO DATABASE

TABLE NAME	ATTRIBUTE NAME	CONTENTS	TYPE	FORMAT	RANGE	REQUIRED	PK OR FK	FK REFERENCED TABLE
PRODUCT	P_CODE	Product code	VARCHAR(10)	XXXXXXXXXX	NA	Y	PK	
	P_DESCRIPT	Product description	VARCHAR(35)	Xxxxxxxxxxx	NA	Y		
	P_INDATE	Stocking date	DATE	DD-MON-YYYY	NA	Y		
	P_QOH	Units available	SMALLINT	####	0–9999	Y		
	P_MIN	Minimum units	SMALLINT	####	0–9999	Y		
	P_PRICE	Product price	NUMBER(8,2)	####.##	0.00–9999.00	Y		
	P_DISCOUNT	Discount rate	NUMBER(5,2)	0.##	0.00–0.20	Y		
	V_CODE	Vendor code	INTEGER	###	100–999		FK	VENDOR
VENDOR	V_CODE	Vendor code	INTEGER	#####	1000–9999	Y	PK	
	V_NAME	Vendor name	VARCHAR(35)	Xxxxxxxxxxx	NA	Y		
	V_CONTACT	Contact person	VARCHAR(25)	Xxxxxxxxxxx	NA	Y		
	V_AREACODE	Area code	CHAR(3)	999	NA	Y		
	V_PHONE	Phone number	CHAR(8)	999-9999	NA	Y		
	V_STATE	State	CHAR(2)	XX	NA	Y		
	V_ORDER	Previous order	CHAR(1)	X	Y or N	Y		

FK   = Foreign key
PK   = Primary key
CHAR   = Fixed-length character data, 1 to 255 characters
VARCHAR = Variable-length character data, 1 to 2,000 characters. VARCHAR is automatically converted to VARCHAR2 in Oracle.
NUMBER   = Numeric data. NUMBER(9,2) is used to specify numbers that have two decimal places and are up to nine digits long, including the decimal places.
           Some RDBMSs permit the use of a MONEY or a CURRENCY data type.
NUMERIC = Numeric data. DBMSs that do not support the NUMBER data type typically use NUMERIC instead.
INT   = Integer values only. INT is automatically converted to NUMBER in Oracle.
SMALLINT = Small integer values only. SMALLINT is automatically converted to NUMBER in Oracle.
DATE formats vary. Commonly accepted formats are DD-MON-YYYY, DD-MON-YY, MM/DD/YYYY, and MM/DD/YY.

*Not all the ranges shown here will be illustrated in this chapter. However, you can use these constraints to practice writing your own.

Data-type selection sometimes requires professional judgment. For example, you must make a decision about the V_CODE's data type as follows:

- If you want the computer to generate new vendor codes by adding 1 to the largest recorded vendor code, you must classify V_CODE as a numeric attribute. (You cannot perform mathematical procedures on character data.) The designation INTEGER will ensure that only the counting numbers (integers) can be used. Most SQL implementations also permit the use of SMALLINT for integer values up to six digits.

- If you do not want to perform mathematical procedures based on V_CODE, you should classify it as a character attribute, even though it is composed entirely of numbers. When there is no need to perform mathematical procedures on the attribute, store it as a character attribute.

The first option is used to demonstrate the SQL procedures in this chapter.

When you define the attribute's data type, you must pay close attention to the expected use of the attributes for sorting and data-retrieval purposes. For example, in a real estate application, an attribute that represents the numbers of bathrooms in a home (H_BATH_NUM) could be assigned the CHAR(3) data type because the application will probably not do any addition, multiplication, or division with the number of bathrooms. Based on the CHAR(3) data-type definition, valid H_BATH_NUM values would be '2','1','2.5','10'. However, this data-type decision creates potential problems. For example, if an application sorts the homes by number of bathrooms, a query would "see" the value '10' as less than '2', which is clearly incorrect. So, you must consider the expected use of the data to properly define the attribute data type.

The data dictionary in Table 7.3 contains only a few of the data types supported by SQL. For teaching purposes, the selection of data types is limited to ensure that almost any RDBMS can be used to implement the examples. If your RDBMS is fully compliant with ANSI SQL, it will support many more data types than those shown in Table 7.4. Also, many RDBMSs support data types beyond the ones specified in ANSI SQL.

TABLE 7.4

SOME COMMON SQL DATA TYPES

DATA TYPE	FORMAT	COMMENTS
Numeric	NUMBER(L,D) or NUMERIC(L,D)	The declaration NUMBER(7,2) or NUMERIC(7,2) indicates that numbers will be stored with two decimal places and may be up to seven digits long, including the sign and the decimal place (for example, 12.32 or −134.99).
	INTEGER	May be abbreviated as INT. Integers are (whole) counting numbers, so they cannot be used if you want to store numbers that require decimal places.
	SMALLINT	Like INTEGER but limited to integer values up to six digits. If your integer values are relatively small, use SMALLINT instead of INT.
	DECIMAL(L,D)	Like the NUMBER specification, but the storage length is a *minimum* specification. That is, greater lengths are acceptable, but smaller ones are not. DECIMAL(9,2), DECIMAL(9), and DECIMAL are all acceptable.
Character	CHAR(L)	Fixed-length character data for up to 255 characters. If you store strings that are not as long as the CHAR parameter value, the remaining spaces are left unused. Therefore, if you specify CHAR(25), strings such as *Smith* and *Katzenjammer* are each stored as 25 characters. However, a U.S. area code is always three digits long, so CHAR(3) would be appropriate if you wanted to store such codes.
	VARCHAR(L) or VARCHAR2(L)	Variable-length character data. The designation VARCHAR2(25) or VARCHAR(25) will let you store characters up to 25 characters long. However, unlike CHAR, VARCHAR will not leave unused spaces. Oracle automatically converts VARCHAR to VARCHAR2.
Date	DATE	Stores dates in the Julian date format.

In addition to the data types shown in Table 7.4, SQL supports several other data types, including TIME, TIMESTAMP, REAL, DOUBLE, and FLOAT, and intervals, such as INTERVAL DAY TO HOUR. Many RDBMSs have also expanded the list to include other types of data, such as LOGICAL, CURRENCY, and AutoNumber (Access). However, because this chapter is designed to introduce the basics of SQL, the discussion is limited to the data types summarized in Table 7.4.

7-2e Creating Table Structures

Now you are ready to implement the PRODUCT and VENDOR table structures with the help of SQL, using the **CREATE TABLE** syntax shown next.

CREATE TABLE *tablename* (

column1	*data type*	[*constraint*] [,
column2	*data type*	[*constraint*]] [,
PRIMARY KEY	(*column1*	[, *column2*])] [,
FOREIGN KEY	(*column1*	[, *column2*]) REFERENCES *tablename*] [,
CONSTRAINT	*constraint*]);	

To make the SQL code more readable, most SQL programmers use one line per column (attribute) definition. In addition, spaces are used to line up the attribute characteristics and constraints. Finally, both table and attribute names are fully capitalized. Those conventions are used in the following examples that create VENDOR and PRODUCT tables and subsequent tables throughout the book.

Online Content

All the SQL commands used in this chapter are located in script files at *www.cengagebrain.com*. You can copy and paste the SQL commands into your SQL program. Script files are provided for Oracle, MS SQL Server, and MySQL users.

Note

SQL Syntax

Syntax notation for SQL commands used in this book:

CAPITALS	Required SQL command keywords
italics	A parameter provided by the end user (generally required)
{a	b	..}	A mandatory parameter; use one option from the list separated by	
[......]	An optional parameter—anything inside square brackets is optional
Tablename	The name of a table
Column	The name of an attribute in a table
data type	A valid data-type definition
constraint	A valid constraint definition
condition	A valid conditional expression (evaluates to true or false)
columnlist	One or more column names or expressions separated by commas
tablelist	One or more table names separated by commas
conditionlist	One or more conditional expressions separated by logical operators
expression	A simple value (such as 76 or Married) or a formula (such as P_PRICE – 10)

CREATE TABLE
A SQL command that creates a table's structures using the characteristics and attributes given.

```
CREATE TABLE VENDOR (
V_CODE          INTEGER       NOT NULL      UNIQUE,
V_NAME          VARCHAR(35)   NOT NULL,
V_CONTACT       VARCHAR(25)   NOT NULL,
V_AREACODE      CHAR(3)       NOT NULL,
V_PHONE         CHAR(8)       NOT NULL,
V_STATE         CHAR(2)       NOT NULL,
V_ORDER         CHAR(1)       NOT NULL,
PRIMARY KEY (V_CODE));
```

Note

- Because the PRODUCT table contains a foreign key that references the VENDOR table, create the VENDOR table first. (In fact, the "M" side of a relationship always references the "1" side. Therefore, in a 1:M relationship, you must *always* create the table for the "1" side first.)

- If your RDBMS does not support the VARCHAR2 and FCHAR format, use CHAR.

- Oracle accepts the VARCHAR data type and automatically converts it to VARCHAR2.

- If your RDBMS does not support SINT or SMALLINT, use INTEGER or INT. If INTEGER is not supported, use NUMBER (Oracle or Access) or NUMERIC (MS SQL Server or MySQL).

- If you use Access, you can use the NUMBER data type, but you cannot use the number delimiters at the SQL level. For example, using NUMBER(8,2) to indicate numbers with up to eight digits with two digits to the right of the decimal place is fine in Oracle, but you cannot use it in Access—you must use NUMBER without the delimiters.

- If your RDBMS does not support primary and foreign key designations or the UNIQUE specification, delete them from the SQL code shown here.

- If you use the PRIMARY KEY designation in Oracle, you do not need the NOT NULL and UNIQUE specifications.

- The ON UPDATE CASCADE clause is part of the ANSI standard, but it may not be supported by your RDBMS. In that case, delete the ON UPDATE CASCADE clause.

```
CREATE TABLE PRODUCT (
P_CODE          VARCHAR(10)   NOT NULL      UNIQUE,
P_DESCRIPT      VARCHAR(35)   NOT NULL,
P_INDATE        DATE          NOT NULL,
P_QOH           SMALLINT      NOT NULL,
P_MIN           SMALLINT      NOT NULL,
P_PRICE         NUMBER(8,2)   NOT NULL,
P_DISCOUNT      NUMBER(5,2)   NOT NULL,
V_CODE          INTEGER,
PRIMARY KEY (P_CODE),
FOREIGN KEY (V_CODE) REFERENCES VENDOR ON UPDATE CASCADE);
```

Note

Note to MySQL Users

MySQL was originally designed to handle very rapid retrieval of data. To improve retrieval speed, the developers sacrificed many features that ensure data integrity. As MySQL has become more robust, many of those features, such as referential integrity, have been added. To provide developers with options for database behavior, MySQL still supports "nontransaction-safe" tables that do not enable some of the features for data integrity, as well as "transaction-safe" tables that do. MySQL storage engines allow the developer to specify which type of tables to use. MySQL defaults to the MyISAM storage engine, which produces nontransaction-safe tables. The InnoDB storage engine produces transaction-safe tables. The storage engine is specified at the end of the CREATE TABLE command as shown below:

```
CREATE TABLE PRODUCT (
P_CODE          VARCHAR(10)    NOT NULL        UNIQUE,
P_DESCRIPT      VARCHAR(35)    NOT NULL,
P_INDATE        DATE           NOT NULL,
P_QOH           SMALLINT       NOT NULL,
P_MIN           SMALLINT       NOT NULL,
P_PRICE         NUMBER(8,2)    NOT NULL,
P_DISCOUNT      NUMBER(5,2)    NOT NULL,
V_CODE          INTEGER,
PRIMARY KEY (P_CODE),
FOREIGN KEY (V_CODE) REFERENCES VENDOR (V_CODE) ON UPDATE CASCADE);
;
```

Transaction-safe tables provide improved support for data integrity, implementation of database transactions and transaction logs (as discussed in Chapter 10, Transaction Management and Concurrency Control), and improved backup and recovery options.

As you examine the preceding SQL table-creating command sequences, note the following features:

- The NOT NULL specifications for the attributes ensure that a data entry will be made. When it is crucial to have the data available, the NOT NULL specification will not allow the end user to leave the attribute empty (with no data entry at all). Because this specification is made at the table level and stored in the data dictionary, application programs can use this information to create the data dictionary validation automatically.

- The UNIQUE specification creates a unique index in the respective attribute. Use it to avoid having duplicated values in a column.

- The primary key attributes contain both a NOT NULL and UNIQUE specification, which enforce the entity integrity requirements. If the NOT NULL and UNIQUE specifications are not supported, use PRIMARY KEY without the specifications. (For example, if you designate the PK in MS Access, the NOT NULL and UNIQUE specifications are automatically assumed and are not spelled out.)

- The entire table definition is enclosed in parentheses. A comma is used to separate each table element definition (attributes, primary key, and foreign key).

Note

If you are working with a composite primary key, all of the primary key's attributes are contained within the parentheses and are separated with commas. For example, the LINE table in Figure 7.1 has a primary key that consists of the two attributes INV_NUMBER and LINE_NUMBER. Therefore, you would define the primary key by typing the following:

PRIMARY KEY (INV_NUMBER, LINE_NUMBER),

The order of the primary key components is important because the indexing starts with the first mentioned attribute, then proceeds with the next attribute, and so on. In this example, the line numbers would be ordered within each of the invoice numbers:

INV_NUMBER	LINE_NUMBER
1001	1
1001	2
1002	1
1003	1
1003	2

reserved words
Words used by a system that cannot be used for any other purpose. For example, in Oracle SQL, the word INITIAL cannot be used to name tables or columns.

- The ON UPDATE CASCADE specification ensures that if you make a change in any VENDOR's V_CODE that change is automatically applied to all foreign key references throughout the system to ensure that referential integrity is maintained. (Although the ON UPDATE CASCADE clause is part of the ANSI standard, some RDBMSs, such as Oracle, do not support it. If your RDBMS does not support the clause, delete it from the code shown here.)

- An RDBMS automatically enforces referential integrity for foreign keys. That is, you cannot have an invalid entry in the foreign key column; at the same time, you cannot delete a vendor row as long as a product row references that vendor.

- The command sequence ends with a semicolon. (Remember that your RDBMS may require you to omit the semicolon.)

Note

Note About Column Names

Do *not* use mathematical symbols such as +, −, and / in your column names; instead, use an underscore to separate words, if necessary. For example, PER-NUM might generate an error message, but PER_NUM is acceptable. Also, do *not* use reserved words. **Reserved words** are words used by SQL to perform specific functions. For example, in some RDBMSs, the column name INITIAL will generate the message "invalid column name."

Note

Note to Oracle Users

When you press Enter after typing each line, a line number is automatically generated as long as you do not type a semicolon before pressing Enter. For example, Oracle's execution of the CREATE TABLE command will look like the following:

```
CREATE TABLE PRODUCT (
2          P_CODE              VARCHAR2(10)
3          CONSTRAINT          PRODUCT_P_CODE_PK          PRIMARY KEY,
4          P_DESCRIPT          VARCHAR2(35)               NOT NULL,
5          P_INDATE            DATE                       NOT NULL,
6          P_QOH               NUMBER                     NOT NULL,
7          P_MIN               NUMBER                     NOT NULL,
8          P_PRICE             NUMBER(8,2)                NOT NULL,
9          P_DISCOUNT          NUMBER(5,2)                NOT NULL,
10         V_CODE              NUMBER,
11         CONSTRAINT          PRODUCT_V_CODE_FK
12         FOREIGN KEY         V_CODE REFERENCES VENDOR
13         ;
```

In the preceding SQL command sequence, note the following:

- The attribute definition for P_CODE starts in line 2 and ends with a comma at the end of line 3.

- The CONSTRAINT clause (line 3) allows you to define and name a constraint in Oracle. You can name the constraint to meet your own naming conventions. In this case, the constraint was named PRODUCT_P_CODE_PK.

- Examples of constraints are NOT NULL, UNIQUE, PRIMARY KEY, FOREIGN KEY, and CHECK. Additional details about constraints are explained as follows.

- To define a PRIMARY KEY constraint, you could also use the following syntax: P_CODE VARCHAR2(10) PRIMARY KEY.

- In this case, Oracle would automatically name the constraint.

- Lines 11 and 12 define a FOREIGN KEY constraint named PRODUCT_V_CODE_FK for the attribute V_CODE. The CONSTRAINT clause is generally used at the end of the CREATE TABLE command sequence.

- If you do not name the constraints yourself, Oracle will automatically assign a name. Unfortunately, the Oracle-assigned name makes sense only to Oracle, so you will have a difficult time deciphering it later. You should assign a name that makes sense to human beings!

7-2f SQL Constraints

In Chapter 3, The Relational Database Model, you learned that adherence to rules for entity integrity and referential integrity is crucial in a relational database environment. Fortunately, most SQL implementations support both integrity rules. Entity integrity is enforced automatically when the primary key is specified in the CREATE TABLE command sequence. For example, you can create the VENDOR table structure and set the stage for the enforcement of entity integrity rules by using the following:

PRIMARY KEY (V_CODE)

In the PRODUCT table's CREATE TABLE sequence, note that referential integrity has been enforced by specifying the following in the PRODUCT table:

FOREIGN KEY (V_CODE) REFERENCES VENDOR ON UPDATE CASCADE

Online Content

For a more detailed discussion of the options for using the ON DELETE and ON UPDATE clauses, see Appendix D, Converting the ER Model into a Database Structure, Section D.2, General Rules Governing Relationships Among Tables. Appendix D is available at *www.cengagebrain.com*.

The foreign key constraint definition ensures that:

- You cannot delete a vendor from the VENDOR table if at least one product row references that vendor. This is the default behavior for the treatment of foreign keys.

- On the other hand, if a change is made in an existing VENDOR table's V_CODE, that change must be reflected automatically in any PRODUCT table V_CODE reference (ON UPDATE CASCADE). That restriction makes it impossible for a V_CODE value to exist in the PRODUCT table if it points to a nonexistent VENDOR table V_CODE value. In other words, the ON UPDATE CASCADE specification ensures the preservation of referential integrity. (Oracle does not support ON UPDATE CASCADE.)

In general, ANSI SQL permits the use of ON DELETE and ON UPDATE clauses to cover CASCADE, SET NULL, or SET DEFAULT.

Note

Note about Referential Constraint Actions

The support for the referential constraint's actions varies from product to product. For example:

- MySQL requires the InnoDB storage engine to enforce referential integrity.

- MS Access, SQL Server, MySQL, and Oracle support ON DELETE CASCADE.

- MS Access, MySQL, and SQL Server support ON UPDATE CASCADE.

- Oracle does not support ON UPDATE CASCADE.

- Oracle and MySQL support SET NULL.

- MS Access and SQL Server do not support SET NULL.

- Refer to your product manuals for additional information on referential constraints.

While MS Access does not support ON DELETE CASCADE or ON UPDATE CASCADE at the SQL command-line level, it does support them through the relationship window interface. In fact, whenever you try to establish a relationship between two tables in Access, the relationship window interface will automatically pop up.

Besides the PRIMARY KEY and FOREIGN KEY constraints, the ANSI SQL standard also defines the following constraints:

- The NOT NULL constraint ensures that a column does not accept nulls.

- The UNIQUE constraint ensures that all values in a column are unique.

- The DEFAULT constraint assigns a value to an attribute when a new row is added to a table. The end user may, of course, enter a value other than the default value.

- The CHECK constraint is used to validate data when an attribute value is entered. The CHECK constraint does precisely what its name suggests: it checks to see that a specified condition exists. Examples of such constraints include the following:

  – The minimum order value must be at least 10.

  – The date must be after April 15, 2016.

  If the CHECK constraint is met for the specified attribute (that is, the condition is true), the data is accepted for that attribute. If the condition is found to be false, an error message is generated and the data is not accepted.

Note that the CREATE TABLE command lets you define constraints in two different places:

- When you create the column definition (known as a *column constraint*)
- When you use the CONSTRAINT keyword (known as a *table constraint*)

A column constraint applies to just one column; a table constraint may apply to many columns. Those constraints are supported at varying levels of compliance by enterprise RDBMSs.

In this chapter, Oracle is used to illustrate SQL constraints. For example, note that the following SQL command sequence uses the DEFAULT and CHECK constraints to define the table named CUSTOMER.

```
CREATE TABLE CUSTOMER (
CUS_CODE        NUMBER          PRIMARY KEY,
CUS_LNAME       VARCHAR(15)     NOT NULL,
CUS_FNAME       VARCHAR(15)     NOT NULL,
CUS_INITIAL     CHAR(1),
CUS_AREACODE    CHAR(3)         DEFAULT '615'    NOT NULL
                                CHECK(CUS_AREACODE IN
                                ('615','713','931')),
CUS_PHONE       CHAR(8)         NOT NULL,
CUS_BALANCE     NUMBER(9,2)     DEFAULT 0.00,
CONSTRAINT CUS_UI1 UNIQUE (CUS_LNAME, CUS_FNAME));
```

In this case, the CUS_AREACODE attribute is assigned a default value of '615'. Therefore, if a new CUSTOMER table row is added and the end user makes no entry for the area code, the '615' value will be recorded. Also, the CHECK condition restricts the values for the customer's area code to 615, 713, and 931; any other values will be rejected.

It is important to note that the DEFAULT value applies only when new rows are added to a table, and then only when no value is entered for the customer's area code. (The default value is not used when the table is modified.) In contrast, the CHECK condition is validated whether a customer row is added or *modified*. However, while the CHECK condition may include any valid expression, it applies only to the attributes in the table being checked. If you want to check for conditions that include attributes in other tables, you must use triggers. (See Chapter 8, Advanced SQL.) Finally, the last line of the CREATE TABLE command sequence creates a unique index constraint (named CUS_UI1) on the customer's last name and first name. The index will prevent the entry of two customers with the same last name and first name. (This index merely illustrates the process. Clearly, it should be possible to have more than one person named John Smith in the CUSTOMER table.)

Note

Note to MS Access and MySQL Users

MS Access does not accept the DEFAULT or CHECK constraints. However, MS Access will accept the CONSTRAINT CUS_UI1 UNIQUE (CUS_LNAME, CUS_FNAME) line and create the unique index.

MySQL will allow CHECK constraints in the table definition for compatibility, but it does not enforce them. MySQL does allow DEFAULT constraints, but the DEFAULT value cannot be a function. Therefore, it is not possible to set the default value for a date field to be the current date using SYSDATE() or NOW() because they are both functions.

In the following SQL command to create the INVOICE table, the DEFAULT constraint assigns a default date to a new invoice, and the CHECK constraint validates that the invoice date is greater than January 1, 2016.

```
CREATE TABLE INVOICE (
INV_NUMBER    NUMBER    PRIMARY KEY,
CUS_CODE      NUMBER    NOT NULL REFERENCES CUSTOMER(CUS_CODE),
INV_DATE      DATE      DEFAULT SYSDATE NOT NULL,
CONSTRAINT INV_CK1 CHECK (INV_DATE > TO_DATE('01-JAN-2016',
'DD-MON-YYYY')));
```

In this case, notice the following:

- The CUS_CODE attribute definition contains REFERENCES CUSTOMER (CUS_CODE) to indicate that the CUS_CODE is a foreign key. This is another way to define a foreign key.

- The DEFAULT constraint uses the SYSDATE special function. This function always returns today's date.

- The invoice date (INV_DATE) attribute is automatically given today's date (returned by SYSDATE) when a new row is added and no value is given for the attribute.

- A CHECK constraint is used to validate that the invoice date is greater than 'January 1, 2016'. When comparing a date to a manually entered date in a CHECK clause, Oracle requires the use of the TO_DATE function. The TO_DATE function takes two parameters: the literal date and the date format used.

The final SQL command sequence creates the LINE table. The LINE table has a composite primary key (INV_NUMBER, LINE_NUMBER) and uses a UNIQUE constraint in INV_NUMBER and P_CODE to ensure that the same product is not ordered twice in the same invoice.

```
CREATE TABLE LINE (
INV_NUMBER      NUMBER          NOT NULL,
LINE_NUMBER     NUMBER(2,0)     NOT NULL,
P_CODE          VARCHAR(10)     NOT NULL,
LINE_UNITS      NUMBER(9,2)     DEFAULT 0.00    NOT NULL,
LINE_PRICE      NUMBER(9,2)     DEFAULT 0.00    NOT NULL,
PRIMARY KEY (INV_NUMBER, LINE_NUMBER),
FOREIGN KEY (INV_NUMBER) REFERENCES INVOICE ON DELETE CASCADE,
FOREIGN KEY (P_CODE) REFERENCES PRODUCT(P_CODE),
CONSTRAINT LINE_UI1 UNIQUE(INV_NUMBER, P_CODE));
```

In the creation of the LINE table, note that a UNIQUE constraint is added to prevent the duplication of an invoice line. A UNIQUE constraint is enforced through the creation of a unique index. Also note that the ON DELETE CASCADE foreign key enforces referential integrity. The use of ON DELETE CASCADE is recommended for weak entities to ensure that the deletion of a row in the strong entity automatically triggers the deletion of the corresponding rows in the dependent weak entity. In that case, the deletion of an INVOICE row will automatically delete all of the LINE rows related to the invoice. In the following section, you will learn more about indexes and how to use SQL commands to create them.

7-2g SQL Indexes

You learned in Chapter 3 that indexes can be used to improve the efficiency of searches and to avoid duplicate column values. In the previous section, you saw how to declare unique indexes on selected attributes when the table is created. In fact, when you declare a primary key, the DBMS automatically creates a unique index. Even with this feature, you often need additional indexes. The ability to create indexes quickly and efficiently is important. Using the **CREATE INDEX** command, SQL indexes can be created on the basis of any selected attribute. The syntax is:

CREATE [UNIQUE]INDEX *indexname* ON *tablename(column1 [, column2])*

For example, based on the attribute P_INDATE stored in the PRODUCT table, the following command creates an index named P_INDATEX:

CREATE INDEX P_INDATEX ON PRODUCT(P_INDATE);

SQL does not let you write over an existing index without warning you first, thus preserving the index structure within the data dictionary. Using the UNIQUE index qualifier, you can even create an index that prevents you from using a value that has been used before. Such a feature is especially useful when the index attribute is a candidate key whose values must not be duplicated:

CREATE UNIQUE INDEX P_CODEX ON PRODUCT(P_CODE);

If you now try to enter a duplicate P_CODE value, SQL produces the error message "duplicate value in index." Many RDBMSs, including Access, automatically create a unique index on the PK attribute(s) when you declare the PK.

A common practice is to create an index on any field that is used as a search key, in comparison operations in a conditional expression, or when you want to list rows in a specific order. For example, if you want to create a report of all products by vendor, it would be useful to create an index on the V_CODE attribute in the PRODUCT table. Remember that a vendor can supply many products. Therefore, you should not create a UNIQUE index in this case. Better yet, to make the search as efficient as possible, using a composite index is recommended.

Unique composite indexes are often used to prevent data duplication. For example, consider the case illustrated in Table 7.5, in which required employee test scores are stored. (An employee can take a test only once on a given date.) Given the structure of Table 7.5, the PK is EMP_NUM + TEST_NUM. The third test entry for employee 111 meets entity integrity requirements—the combination 111,3 is unique—yet the WEA test entry is clearly duplicated.

CREATE INDEX
A SQL command that creates indexes on the basis of a selected attribute or attributes.

TABLE 7.5

A DUPLICATED TEST RECORD

EMP_NUM	TEST_NUM	TEST_CODE	TEST_DATE	TEST_SCORE
110	1	WEA	15-Jan-2016	93
110	2	WEA	12-Jan-2016	87
111	1	HAZ	14-Dec-2015	91
111	2	WEA	18-Feb-2016	95
111	3	WEA	18-Feb-2016	95
112	1	CHEM	17-Aug-2015	91

Such duplication could have been avoided through the use of a unique composite index, using the attributes EMP_NUM, TEST_CODE, and TEST_DATE:

CREATE UNIQUE INDEX EMP_TESTDEX ON TEST(EMP_NUM, TEST_CODE, TEST_DATE);

By default, all indexes produce results that are listed in ascending order, but you can create an index that yields output in descending order. For example, if you routinely print a report that lists all products ordered by price from highest to lowest, you could create an index named PROD_PRICEX by typing:

CREATE INDEX PROD_PRICEX ON PRODUCT(P_PRICE DESC);

To delete an index, use the **DROP INDEX** command:

DROP INDEX *indexname*

For example, if you want to eliminate the PROD_PRICEX index, type:

DROP INDEX PROD_PRICEX;

After creating the tables and some indexes, you are ready to start entering data. The following sections use two tables (VENDOR and PRODUCT) to demonstrate most of the data manipulation commands.

7-3 Data Manipulation Commands

In this section, you will learn how to use the basic SQL data manipulation commands INSERT, SELECT, COMMIT, UPDATE, ROLLBACK, and DELETE.

7-3a Adding Table Rows

SQL requires the use of the **INSERT** command to enter data into a table. The INSERT command's basic syntax looks like this:

INSERT INTO *tablename* VALUES (*value1*, *value2*, …, *valuen*)

Because the PRODUCT table uses its V_CODE to reference the VENDOR table's V_CODE, an integrity violation will occur if the VENDOR table V_CODE values do not yet exist. Therefore, you need to enter the VENDOR rows before the PRODUCT rows. Given the VENDOR table structure defined earlier and the sample VENDOR data shown in Figure 7.2, you would enter the first two data rows as follows:

INSERT INTO VENDOR
    VALUES (21225,'Bryson, Inc.','Smithson','615','223-3234','TN','Y');
INSERT INTO VENDOR
    VALUES (21226,'Superloo, Inc.','Flushing','904','215-8995','FL','N');

and so on, until all of the VENDOR table records have been entered.

(To see the contents of the VENDOR table, use the SELECT * FROM VENDOR; command.)

The PRODUCT table rows would be entered in the same fashion, using the PROD-UCT data shown in Figure 7.2. For example, the first two data rows would be entered as follows, pressing Enter at the end of each line:

DROP INDEX
A SQL command used to delete database objects such as tables, views, indexes, and users.

INSERT
A SQL command that allows the insertion of one or more data rows into a table using a subquery.

INSERT INTO PRODUCT
      VALUES ('11QER/31','Power painter, 15 psi., 3-nozzle','03-Nov-15',8,5,109.99,0.00,25595);
INSERT INTO PRODUCT
      VALUES ('13-Q2/P2','7.25-in. pwr. saw blade','13-Dec-15',32,15,14.99, 0.05,
      21344);

(To see the contents of the PRODUCT table, use the SELECT * FROM PRODUCT;
command.)

Note

Date entry is a function of the date format expected by the DBMS. For example, March 25, 2016, might be shown as 25-Mar-2016 in Access and Oracle, 2016-03-25 in MySQL, or it might be displayed in other presentation formats in another RDBMS. MS Access requires the use of # delimiters when performing any computations or comparisons based on date attributes, as in P_INDATE >= #25-Mar-16#. Date data and the functions for manipulating it in various DBMS products is discussed in more detail in Chapter 8.

In the preceding data-entry lines, observe that:

- The row contents are entered between parentheses. Note that the first character after VALUES is a parenthesis and that the last character in the command sequence is also a parenthesis.

- Character (string) and date values must be entered between apostrophes (').

- Numerical entries are *not* enclosed in apostrophes.

- Attribute entries are separated by commas.

- A value is required for each column in the table.

This version of the INSERT command adds one table row at a time.

Inserting Rows with Null Attributes Thus far, you have entered rows in which all of the attribute values are specified. But what do you do if a product does not have a vendor or if you do not yet know the vendor code? In those cases, you would want to leave the vendor code null. To enter a null, use the following syntax:

INSERT INTO PRODUCT
      VALUES ('BRT-345','Titanium drill bit','18-Oct-15', 75, 10, 4.50, 0.06, NULL);

Incidentally, note that the NULL entry is accepted only because the V_CODE attribute is optional—the NOT NULL declaration was not used in the CREATE TABLE statement for this attribute.

Inserting Rows with Optional Attributes Sometimes, more than one attribute is optional. Rather than declaring each attribute as NULL in the INSERT command, you can indicate just the attributes that have required values. You do that by listing the attribute names inside parentheses after the table name. For the purpose of this example, assume that the only required attributes for the PRODUCT table are P_CODE and P_DESCRIPT:

INSERT INTO PRODUCT(P_CODE, P_DESCRIPT) VALUES ('BRT-345','Titanium drill bit');

Note

When inserting rows interactively, omitting the attribute list in the INSERT command is acceptable if the programmer intends to provide a value for each attribute. However, if an INSERT command is embedded inside a program for later use, the attribute list should *always* be used, even if the programmer provides a value for every attribute. The reason is that the structure of the database table may change over time. The programs that are created today become the legacy systems of tomorrow. These applications may be expected to have a very long, useful life. If the structure of the table changes over time as new business requirements develop, an INSERT without an attribute list may inadvertently insert data into the wrong columns if the order of the columns in the table changes, or the INSERT command may generate an error because the command does not provide enough values if new columns are subsequently added to the table.

7-3b Saving Table Changes

Any changes made to the table contents are not saved on disk until you close the database, close the program you are using, or use the **COMMIT** command. If the database is open and a power outage or some other interruption occurs before you issue the COMMIT command, your changes will be lost and only the original table contents will be retained. The syntax for the COMMIT command is:

COMMIT [WORK]

COMMIT
The SQL command that permanently saves data changes to a database.

The COMMIT command permanently saves all changes—such as rows added, attributes modified, and rows deleted—made to any table in the database. Therefore, if you intend to make your changes to the PRODUCT table permanent, it is a good idea to save those changes by using the following command:

COMMIT;

Note

Note to MS Access and MySQL Users

MS Access does not support the COMMIT command because it automatically saves changes after the execution of each SQL command. By default, MySQL also automatically commits changes with each command. However, if START TRANSACTION or BEGIN is placed at the beginning of a series of commands, MySQL will delay committing the commands until the COMMIT or ROLLBACK command is issued.

SELECT
A SQL command that yields the values of all rows or a subset of rows in a table. The SELECT statement is used to retrieve data from tables.

wildcard character
A symbol that can be used as a general substitute for: (1) all columns in a table (*) when used in an attribute list of a SELECT statement or, (2) zero or more characters in a SQL LIKE clause condition (% and _).

However, the COMMIT command's purpose is not just to save changes. In fact, the ultimate purpose of the COMMIT and ROLLBACK commands (see Section 7-3e) is to ensure database update integrity in transaction management. (You will see how such issues are addressed in Chapter 10, Transaction Management and Concurrency Control.)

7-3c Listing Table Rows

The **SELECT** command is used to list the contents of a table. The syntax of the SELECT command is as follows:

SELECT   *columnlist*   FROM   *tablename*

The SELECT clause of the query specifies the columns to be retrieved as a column list. The *columnlist* represents one or more attributes, separated by commas. You could use the asterisk (*) as a wildcard character to list all attributes. A **wildcard character** is a symbol that can be used as a general substitute for other characters or commands. For example, to list all attributes and all rows of the PRODUCT table, use the following:

SELECT * FROM PRODUCT;

The **FROM** clause of the query specifies the table or tables from which the data is to be retrieved. Figure 7.3 shows the output generated by that command. (Figure 7.3 shows all of the rows in the PRODUCT table that serve as the basis for subsequent discussions. If you entered only the PRODUCT table's first two records, as shown in the preceding section, the output of the preceding SELECT command would show only the rows you entered. Don't worry about the difference between your SELECT output and the output shown in Figure 7.3. When you complete the work in this section, you will have created and populated your VENDOR and PRODUCT tables with the correct rows for use in future sections.)

FIGURE 7.3 THE CONTENTS OF THE PRODUCT TABLE

P_CODE	P_DESCRIPT	P_INDATE	P_QOH	P_MIN	P_PRICE	P_DISCOUNT	V_CODE
11QER/31	Power painter, 15 psi., 3-nozzle	03-Nov-15	8	5	109.99	0.00	25595
13-Q2/P2	7.25-in. pwr. saw blade	13-Dec-15	32	15	14.99	0.05	21344
14-Q1/L3	9.00-in. pwr. saw blade	13-Nov-15	18	12	17.49	0.00	21344
1546-QQ2	Hrd. cloth, 1/4-in., 2x50	15-Jan-16	15	8	39.95	0.00	23119
1558-QW1	Hrd. cloth, 1/2-in., 3x50	15-Jan-16	23	5	43.99	0.00	23119
2232/QTY	B&D jigsaw, 12-in. blade	30-Dec-15	8	5	109.92	0.05	24288
2232/QWE	B&D jigsaw, 8-in. blade	24-Dec-15	6	5	99.87	0.05	24288
2238/QPD	B&D cordless drill, 1/2-in.	20-Jan-16	12	5	38.95	0.05	25595
23109-HB	Claw hammer	20-Jan-16	23	10	9.95	0.10	21225
23114-AA	Sledge hammer, 12 lb.	02-Jan-16	8	5	14.40	0.05	
54778-2T	Rat-tail file, 1/8-in. fine	15-Dec-15	43	20	4.99	0.00	21344
89-WRE-Q	Hicut chain saw, 16 in.	07-Feb-16	11	5	256.99	0.05	24288
PVC23DRT	PVC pipe, 3.5-in., 8-ft	20-Feb-16	188	75	5.87	0.00	
SM-18277	1.25-in. metal screw, 25	01-Mar-16	172	75	6.99	0.00	21225
SW-23116	2.5-in. wd. screw, 50	24-Feb-16	237	100	8.45	0.00	21231
WR3/TT3	Steel matting, 4'x8'x1/6", .5" mesh	17-Jan-16	18	5	119.95	0.10	25595

Note

Your listing might not be in the order shown in Figure 7.3. The listings shown in the figure are the result of system-controlled primary-key-based index operations. You will learn later how to control the output so that it conforms to the order you have specified.

Note

Note to Oracle Users

Some SQL implementations (such as Oracle's) cut the attribute labels to fit the width of the column. However, Oracle lets you set the width of the display column to show the complete attribute name. You can also change the display format, regardless of how the data is stored in the table. For example, if you want to display dollar symbols and commas in the P_PRICE output, you can declare:

COLUMN P_PRICE FORMAT $99,999.99

to change the output 12347.67 to $12,347.67.

In the same manner, to display only the first 12 characters of the P_DESCRIPT attribute, use the following:

COLUMN P_DESCRIPT FORMAT A12 TRUNCATE

FROM
A SQL clause that specifies the table or tables from which data is to be retrieved.

Although SQL commands can be grouped together on a single line, complex command sequences are best shown on separate lines, with space between the SQL command and the command's components. Using that formatting convention makes it much easier to see the components of the SQL statements, which in turn makes it easy to trace the SQL logic and make corrections if necessary. The number of spaces used in the indention is up to you. For example, note the following format for a more complex statement:

```
SELECT  P_CODE, P_DESCRIPT, P_INDATE, P_QOH, P_MIN, P_PRICE,
        P_DISCOUNT, V_CODE
FROM    PRODUCT;
```

When you run a SELECT command on a table, the RDBMS returns a set of one or more rows that have the same characteristics as a relational table. In addition, the SELECT command lists all rows from the table you specified in the FROM clause. This is a very important characteristic of SQL commands. By default, most SQL data manipulation commands operate over an entire table (or relation), which is why SQL commands are said to be *set-oriented* commands. A SQL set-oriented command works over a set of rows. The set may include one or more columns and zero or more rows from one or more tables.

Note

Just as with INSERT commands, omitting the column list by specifying " * " for all columns is acceptable when querying the database interactively. However, if the SELECT query is embedded in a program for later use, the column list should always be included even if every column in the table is being included in the result because the structure of the table might change over time. In real-world business applications, SELECT * commands embedded in programs are often considered bugs waiting to happen.

7-3d Updating Table Rows

Use the **UPDATE** command to modify data in a table. The syntax for this command is as follows:

```
UPDATE       tablename
SET          columnname = expression [, columnname = expression]
[WHERE       conditionlist ];
```

For example, if you want to change P_INDATE from December 13, 2015, to January 18, 2016, in the second row of the PRODUCT table (see Figure 7.3), use the primary key (13-Q2/P2) to locate the correct row. Therefore, type:

```
UPDATE       PRODUCT
SET          P_INDATE = '18-JAN-2016'
WHERE        P_CODE = '13-Q2/P2';
```

If more than one attribute is to be updated in the row, separate the corrections with commas:

UPDATE
A SQL command that allows attribute values to be changed in one or more rows of a table.

```
UPDATE       PRODUCT
SET          P_INDATE = '18-JAN-2016', P_PRICE = 17.99, P_MIN = 10
WHERE        P_CODE = '13-Q2/P2';
```

What would have happened if the previous UPDATE command had not included the WHERE condition? The P_INDATE, P_PRICE, and P_MIN values would have been changed in *all* rows of the PRODUCT table. Remember, the UPDATE command is a set-oriented operator. Therefore, if you do not specify a WHERE condition, the UPDATE command will apply the changes to *all* rows in the specified table.

Confirm the correction(s) by using the following SELECT command to check the PRODUCT table's listing:

SELECT * FROM PRODUCT;

7-3e Restoring Table Contents

If you have not yet used the COMMIT command to store the changes permanently in the database, you can restore the database to its previous condition with the **ROLLBACK** command. ROLLBACK undoes any changes since the last COMMIT command and brings all of the data back to the values that existed before the changes were made. To restore the data to its "prechange" condition, type:

ROLLBACK;

and then press Enter. Use the SELECT statement again to verify that the ROLLBACK restored the data to its original values.

COMMIT and ROLLBACK work only with data manipulation commands that add, modify, or delete table rows. For example, assume that you perform these actions:

1. CREATE a table called SALES.

2. INSERT 10 rows in the SALES table.

3. UPDATE two rows in the SALES table.

4. Execute the ROLLBACK command.

Will the SALES table be removed by the ROLLBACK command? No, the ROLLBACK command will undo *only* the results of the INSERT and UPDATE commands. All data definition commands (CREATE TABLE) are automatically committed to the data dictionary and cannot be rolled back. The COMMIT and ROLLBACK commands are examined in greater detail in Chapter 10.

Note

Note to MS Access Users

MS Access does not support the ROLLBACK command.

Some RDBMSs, such as Oracle, automatically COMMIT data changes when issuing data definition commands. For example, if you had used the CREATE INDEX command after updating the two rows in the previous example, all previous changes would have been committed automatically; doing a ROLLBACK afterward would not have undone anything. *Check your RDBMS manual to understand these subtle differences.*

7-3f Deleting Table Rows

It is easy to delete a table row using the **DELETE** statement. The syntax is:

DELETE FROM   *tablename*
[WHERE   *conditionlist*];

ROLLBACK
A SQL command that restores the database table contents to the condition that existed after the last COMMIT statement.

DELETE
A SQL command that allows data rows to be deleted from a table.

For example, if you want to delete the product you added earlier whose code (P_CODE) is 'BRT-345', use the following command:

DELETE FROM   PRODUCT
WHERE   P_CODE = 'BRT-345';

In this example, the primary key value lets SQL find the exact record to be deleted from the PRODUCT table. However, deletions are not limited to a primary key match; any attribute may be used. For example, in your PRODUCT table, you will see several products for which the P_MIN attribute is equal to 5. Use the following command to delete all rows from the PRODUCT table for which the P_MIN is equal to 5:

DELETE FROM   PRODUCT
WHERE   P_MIN = 5;

Check the PRODUCT table's contents again to verify that all products with P_MIN equal to 5 have been deleted.

Finally, remember that DELETE is a set-oriented command, and that the WHERE condition is optional. Therefore, if you do not specify a WHERE condition, *all* rows from the specified table will be deleted!

Note

Note to MySQL Users

By default MySQL is set for "safe mode" for updates and deletes. This means that users cannot update or delete rows from a table unless the UPDATE or DELETE command includes a WHERE clause that provides a value for the primary key. To disable safe mode temporarily, set the sql_safe_updates variable to 0. Safe mode can be re-enabled by setting the variable back to 1. For example, to complete the DELETE command shown above, the following sequence could be used:

SET SQL_SAFE_UPDATES = 0;

DELETE FROM PRODUCT WHERE P_MIN = 5;

SET SQL_SAFE_UPDATES = 1;

To permanently disable safe mode, uncheck the safe mode option in MySQL Workbench under the Edit → Preferences window.

subquery
A query that is embedded (or nested) inside another query. Also known as a *nested query* or an *inner query*.

nested query
In SQL, a query that is embedded in another query. See *subquery*.

inner query
A query that is embedded or nested inside another query. Also known as a *nested query* or a *subquery*.

7-3g Inserting Table Rows with a Select Subquery

You learned in Section 7-3a how to use the INSERT statement to add rows to a table. In that section, you added rows one at a time. In this section, you will learn how to add multiple rows to a table, using another table as the source of the data. The syntax for the INSERT statement is:

INSERT INTO *tablename*   SELECT *columnlist*   FROM *tablename*;

In this case, the INSERT statement uses a SELECT subquery. A **subquery**, also known as a **nested query** or an **inner query**, is a query that is embedded (or nested) inside another query. The inner query is always executed first by the RDBMS. Given the previous SQL statement, the INSERT portion represents the outer query, and the SELECT portion represents the subquery. You can nest queries (place queries inside queries) many levels deep. In every case, the output of the inner query is used as the input for the outer (higher-level) query. In Chapter 8 you will learn more about the various types of subqueries.

The values returned by the SELECT subquery should match the attributes and data types of the table in the INSERT statement. If the table into which you are inserting rows has one date attribute, one number attribute, and one character attribute, the SELECT subquery should return one or more rows in which the first column has date values, the second column has number values, and the third column has character values.

7-4 SELECT Queries

In this section, you will learn how to fine-tune the SELECT command by adding restrictions to the search criteria. When coupled with appropriate search conditions, SELECT is an incredibly powerful tool that enables you to transform data into information. For example, in the following sections, you will learn how to create queries that can answer questions such as these: "What products were supplied by a particular vendor?", "Which products are priced below $10?", and "How many products supplied by a given vendor were sold between January 5, 2016, and March 20, 2016?"

7-4a Selecting Rows with Conditional Restrictions

You can select partial table contents by placing restrictions on the rows to be included in the output. Use the **WHERE** clause to add conditional restrictions to the SELECT statement that limit the rows returned by the query. The following syntax enables you to specify which rows to select:

SELECT	*columnlist*
FROM	*tablelist*
[WHERE	*conditionlist*];

The SELECT statement retrieves all rows that match the specified condition(s)—also known as the conditional criteria—you specified in the WHERE clause. The *conditionlist* in the WHERE clause of the SELECT statement is represented by one or more conditional expressions, separated by logical operators. The WHERE clause is optional. If no rows match the specified criteria in the WHERE clause, you see a blank screen or a message that tells you no rows were retrieved. For example, consider the following query:

SELECT	P_DESCRIPT, P_INDATE, P_PRICE, V_CODE
FROM	PRODUCT
WHERE	V_CODE = 21344;

This query returns the description, date, and price of products with a vendor code of 21344, as shown in Figure 7.4.

MS Access users can use the Access QBE (query by example) query generator. Although the Access QBE generates its own "native" version of SQL, you can also elect to type standard SQL in the Access SQL window, as shown at the bottom of Figure 7.5. The figure shows the Access QBE screen, the SQL window's QBE-generated SQL, and the listing of the modified SQL.

Online Content

Before you execute the commands in the following sections, you *must* do the following:

- If you are using Oracle, MySQL, or MS SQL Server, run the respective sqlintrodbinit.sql script file at *www.cengagebrain.com* to create all tables and load the data in the database.

- If you are using Access, copy the original Ch07_SaleCo.mdb file from *www.cengagebrain.com.*

WHERE
A SQL clause that adds conditional restrictions to a SELECT statement that limit the rows returned by the query.

FIGURE 7.4 SELECTED PRODUCT TABLE ATTRIBUTES FOR VENDOR CODE 21344

P_DESCRIPT	P_INDATE	P_PRICE	V_CODE
7.25-in. pwr. saw blade	13-Dec-15	14.99	21344
9.00-in. pwr. saw blade	13-Nov-15	17.49	21344
Rat-tail file, 1/8-in. fine	15-Dec-15	4.99	21344

FIGURE 7.5 THE MICROSOFT ACCESS QBE AND ITS SQL

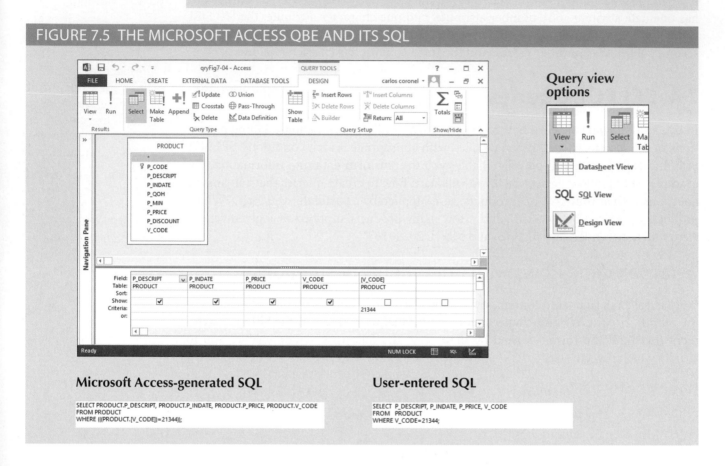

Microsoft Access-generated SQL

```
SELECT PRODUCT.P_DESCRIPT, PRODUCT.P_INDATE, PRODUCT.P_PRICE, PRODUCT.V_CODE
FROM PRODUCT
WHERE (((PRODUCT.[V_CODE])=21344));
```

User-entered SQL

```
SELECT  P_DESCRIPT, P_INDATE, P_PRICE, V_CODE
FROM  PRODUCT
WHERE V_CODE=21344;
```

Note

Note to MS Access Users

The MS Access QBE interface automatically designates the data source by using the table name as a prefix. You will discover later that the table name prefix is used to avoid ambiguity when the same column name appears in multiple tables. For example, both the VENDOR and PRODUCT tables contain the V_CODE attribute. Therefore, if both tables are used (as they would be in a join), the source of the V_CODE attribute must be specified.

Numerous conditional restrictions can be placed on the selected table contents. For example, the comparison operators shown in Table 7.6 can be used to restrict output.

The following example uses the "not equal to" operator:

```
SELECT      P_DESCRIPT, P_QOH, P_PRICE, V_CODE
FROM        PRODUCT
WHERE       V_CODE <> 21344;
```

The output, shown in Figure 7.6, lists all of the rows for which the vendor code is *not* 21344.

TABLE 7.6

COMPARISON OPERATORS

SYMBOL	MEANING
=	Equal to
<	Less than
<=	Less than or equal to
>	Greater than
>=	Greater than or equal to
<> or !=	Not equal to

Note that, in Figure 7.6, rows with nulls in the V_CODE column (see Figure 7.3) are not included in the SELECT command's output.

FIGURE 7.6   PRODUCT TABLE ATTRIBUTES FOR VENDOR CODES OTHER THAN 21344

P_DESCRIPT	P_QOH	P_PRICE	V_CODE
Power painter, 15 psi., 3-nozzle	8	109.99	25595
Hrd. cloth, 1/4-in., 2x50	15	39.95	23119
Hrd. cloth, 1/2-in., 3x50	23	43.99	23119
B&D jigsaw, 12-in. blade	8	109.92	24288
B&D jigsaw, 8-in. blade	6	99.87	24288
B&D cordless drill, 1/2-in.	12	38.95	25595
Claw hammer	23	9.95	21225
Hicut chain saw, 16 in.	11	256.99	24288
1.25-in. metal screw, 25	172	6.99	21225
2.5-in. wd. screw, 50	237	8.45	21231
Steel matting, 4'x8'x1/6", .5" mesh	18	119.95	25595

The following command sequence:

SELECT   P_DESCRIPT, P_QOH, P_MIN, P_PRICE
FROM   PRODUCT
WHERE   P_PRICE <= 10;

yields the output shown in Figure 7.7.

FIGURE 7.7   SELECTED PRODUCT TABLE ATTRIBUTES WITH A P_PRICE RESTRICTION

P_DESCRIPT	P_QOH	P_MIN	P_PRICE
Claw hammer	23	10	9.95
Rat-tail file, 1/8-in. fine	43	20	4.99
PVC pipe, 3.5-in., 8-ft	188	75	5.87
1.25-in. metal screw, 25	172	75	6.99
2.5-in. wd. screw, 50	237	100	8.45

Using Comparison Operators on Character Attributes Because computers identify all characters by their numeric American Standard Code for Information Interchange (ASCII) codes, comparison operators may even be used to place restrictions on character-based attributes. Therefore, the command:

```
SELECT        P_CODE, P_DESCRIPT, P_QOH, P_MIN, P_PRICE
FROM          PRODUCT
WHERE         P_CODE < '1558-QW1';
```

would be correct and would yield a list of all rows in which the P_CODE is alphabetically less than 1558-QW1. (Because the ASCII code value for the letter B is greater than the value of the letter A, it follows that A is less than B.) Therefore, the output will be generated as shown in Figure 7.8.

FIGURE 7.8 SELECTED PRODUCT TABLE ATTRIBUTES: THE ASCII CODE EFFECT

P_CODE	P_DESCRIPT	P_QOH	P_MIN	P_PRICE
11QER/31	Power painter, 15 psi., 3-nozzle	8	5	109.99
13-Q2/P2	7.25-in. pwr. saw blade	32	15	14.99
14-Q1/L3	9.00-in. pwr. saw blade	18	12	17.49
1546-QQ2	Hrd. cloth, 1/4-in., 2x50	15	8	39.95

String (character) comparisons are made from left to right. This left-to-right comparison is especially useful when attributes such as names are to be compared. For example, the string "Ardmore" would be judged *greater than* the string "Aarenson" but *less than* the string "Brown"; such results may be used to generate alphabetical listings like those in a phone directory. If the characters 0–9 are stored as strings, the same left-to-right string comparisons can lead to apparent anomalies. For example, the ASCII code for the character "5" is *greater than* the ASCII code for the character "4," as expected. Yet, the same "5" will also be judged *greater than* the string "44" because the *first* character in the string "44" is less than the string "5." For that reason, you may get some unexpected results from comparisons when dates or other numbers are stored in character format. For example, the left-to-right ASCII character comparison would force the conclusion that the date "01/01/2016" occurred *before* "12/31/2015." Because the leftmost character "0" in "01/01/2016" is *less than* the leftmost character "1" in "12/31/2015," "01/01/2016" is *less than* "12/31/2015." Naturally, if date strings are stored in a yyyy/mm/dd format, the comparisons will yield appropriate results, but this is a nonstandard date presentation. Therefore, all current RDBMSs support date data types; you should use them. In addition, using date data types gives you the benefit of date arithmetic.

Using Comparison Operators on Dates Date procedures are often more software-specific than other SQL procedures. For example, the query to list all of the rows in which the inventory stock dates occur on or after January 20, 2016, looks like this:

```
SELECT        P_DESCRIPT, P_QOH, P_MIN, P_PRICE, P_INDATE
FROM          PRODUCT
WHERE         P_INDATE >= '20-Jan-2016';
```

Remember that MS Access users must use the # delimiters for dates. For example, you would use #20-Jan-16# in the preceding WHERE clause. The date-restricted output is shown in Figure 7.9. In MySQL, the expected date format is yyyy-mm-dd, so the WHERE clause would be written as:

```
WHERE         P_INDATE >= '2016-01-20'
```

Using Computed Columns and Column Aliases Suppose that you want to determine the total value of each of the products currently held in inventory. Logically, that determination requires the multiplication of each product's quantity on hand by its current price. You can accomplish this task with the following command:

```
SELECT          P_DESCRIPT, P_QOH, P_PRICE, P_QOH * P_PRICE
FROM            PRODUCT;
```

FIGURE 7.9 SELECTED PRODUCT TABLE ATTRIBUTES: DATE RESTRICTION

P_DESCRIPT	P_QOH	P_MIN	P_PRICE	P_INDATE
B&D cordless drill, 1/2-in.	12	5	38.95	20-Jan-16
Claw hammer	23	10	9.95	20-Jan-16
Hicut chain saw, 16 in.	11	5	256.99	07-Feb-16
PVC pipe, 3.5-in., 8-ft	188	75	5.87	20-Feb-16
1.25-in. metal screw, 25	172	75	6.99	01-Mar-16
2.5-in. wd. screw, 50	237	100	8.45	24-Feb-16

Entering the SQL command in Access generates the output shown in Figure 7.10.

FIGURE 7.10 SELECT STATEMENT WITH A COMPUTED COLUMN

P_DESCRIPT	P_QOH	P_PRICE	Expr1
Power painter, 15 psi., 3-nozzle	8	109.99	879.92
7.25-in. pwr. saw blade	32	14.99	479.68
9.00-in. pwr. saw blade	18	17.49	314.82
Hrd. cloth, 1/4-in., 2x50	15	39.95	599.25
Hrd. cloth, 1/2-in., 3x50	23	43.99	1011.77
B&D jigsaw, 12-in. blade	8	109.92	879.36
B&D jigsaw, 8-in. blade	6	99.87	599.22
B&D cordless drill, 1/2-in.	12	38.95	467.40
Claw hammer	23	9.95	228.85
Sledge hammer, 12 lb.	8	14.40	115.20
Rat-tail file, 1/8-in. fine	43	4.99	214.57
Hicut chain saw, 16 in.	11	256.99	2826.89
PVC pipe, 3.5-in., 8-ft	188	5.87	1103.56
1.25-in. metal screw, 25	172	6.99	1202.28
2.5-in. wd. screw, 50	237	8.45	2002.65
Steel matting, 4'x8'x1/6", .5" mesh	18	119.95	2159.10

SQL accepts any valid expressions (or formulas) in the computed columns. Such formulas can contain any valid mathematical operators and functions that are applied to attributes in any of the tables specified in the FROM clause of the SELECT statement. Note also that Access automatically adds an *Expr* label to all computed columns. (The first computed column would be labeled Expr1; the second, Expr2; and so on.) Oracle uses the actual formula text as the label for the computed column.

To make the output more readable, the SQL standard permits the use of aliases for any column in a SELECT statement. An **alias** is an alternate name given to a column or table in any SQL statement.

For example, you can rewrite the previous SQL statement as follows:

```
SELECT          P_DESCRIPT, P_QOH, P_PRICE, P_QOH * P_PRICE AS
                TOTVALUE
FROM            PRODUCT;
```

The output of the command is shown in Figure 7.11.

You could also use a computed column, an alias, and date arithmetic in a single query. For example, assume that you want to get a list of out-of-warranty products that have been stored more than 90 days. In that case, the P_INDATE is at least 90 days less than the current (system) date. The MS Access version of this query is:

```
SELECT          P_CODE, P_INDATE, DATE() - 90 AS CUTDATE
FROM            PRODUCT
WHERE           P_INDATE <= DATE() - 90;
```

alias
An alternative name for a column or table in a SQL statement.

FIGURE 7.11 SELECT STATEMENT WITH A COMPUTED COLUMN AND AN ALIAS

P_DESCRIPT	P_QOH	P_PRICE	TOTVALUE
Power painter, 15 psi., 3-nozzle	8	109.99	879.92
7.25-in. pwr. saw blade	32	14.99	479.68
9.00-in. pwr. saw blade	18	17.49	314.82
Hrd. cloth, 1/4-in., 2x50	15	39.95	599.25
Hrd. cloth, 1/2-in., 3x50	23	43.99	1011.77
B&D jigsaw, 12-in. blade	8	109.92	879.36
B&D jigsaw, 8-in. blade	6	99.87	599.22
B&D cordless drill, 1/2-in.	12	38.95	467.40
Claw hammer	23	9.95	228.85
Sledge hammer, 12 lb.	8	14.40	115.20
Rat-tail file, 1/8-in. fine	43	4.99	214.57
Hicut chain saw, 16 in.	11	256.99	2826.89
PVC pipe, 3.5-in., 8-ft	188	5.87	1103.56
1.25-in. metal screw, 25	172	6.99	1202.28
2.5-in. wd. screw, 50	237	8.45	2002.65
Steel matting, 4'x8'x1/6", .5" mesh	18	119.95	2159.10

The Oracle version of the same query is shown here:

```
SELECT          P_CODE, P_INDATE, SYSDATE - 90 AS CUTDATE
FROM            PRODUCT
WHERE           P_INDATE <= SYSDATE - 90;
```

Note that DATE() and SYSDATE are special functions that return the current date in MS Access and Oracle, respectively. You can use the DATE() and SYSDATE functions anywhere a date literal is expected, such as in the value list of an INSERT statement, in an UPDATE statement when changing the value of a date attribute, or in a SELECT statement, as shown here. Of course, the previous query output would change based on the current date.

Suppose that a manager wants a list of all products, the dates they were received, and the warranty expiration date (90 days from receiving the product). To generate that list, type:

```
SELECT          P_CODE, P_INDATE, P_INDATE + 90 AS EXPDATE
FROM            PRODUCT;
```

Note that you can use all arithmetic operators with date attributes as well as with numeric attributes.

7-4b Arithmetic Operators: The Rule of Precedence

As you saw in the previous example, you can use arithmetic operators with table attributes in a column list or in a conditional expression. In fact, SQL commands are often used in conjunction with the arithmetic operators shown in Table 7.7.

TABLE 7.7

THE ARITHMETIC OPERATORS

OPERATOR	DESCRIPTION
+	Add
-	Subtract
*	Multiply
/	Divide
^	Raise to the power of (some applications use ** instead of ^)

Do not confuse the multiplication symbol (*) with the wildcard symbol used by some SQL implementations, such as MS Access. The wildcard symbol is used only in string comparisons, while the multiplication symbol is used in conjunction with mathematical procedures.

As you perform mathematical operations on attributes, remember the mathematical rules of precedence. As the name suggests, the **rules of precedence** are the rules that establish the order in which computations are completed. For example, note the order of the following computational sequence:

1. Perform operations within parentheses.
2. Perform power operations.
3. Perform multiplications and divisions.
4. Perform additions and subtractions.

The application of the rules of precedence will tell you that $8 + 2 * 5 = 8 + 10 = 18$, but $(8 + 2)$ $* 5 = 10 * 5 = 50$. Similarly, $4 + 5^2 * 3 = 4 + 25 * 3 = 79$, but $(4 + 5)^2 * 3 = 81 * 3 = 243$, while the operation expressed by $(4 + 5^2) * 3$ yields the answer $(4 + 25) * 3 = 29 * 3 = 87$.

7-4c Logical Operators: AND, OR, and NOT

In the real world, a search of data normally involves multiple conditions. For example, when you are buying a new house, you look for a certain area, a certain number of bedrooms, bathrooms, stories, and so on. In the same way, SQL allows you to include multiple conditions in a query through the use of logical operators. The logical operators are AND, OR, and NOT. For example, if you want a list of the table contents for either the V_CODE = 21344 *or* the V_CODE = 24288, you can use the **OR** logical operator, as in the following command sequence:

```
SELECT      P_DESCRIPT, P_INDATE, P_PRICE, V_CODE
FROM        PRODUCT
WHERE       V_CODE = 21344 OR V_CODE = 24288;
```

This command generates the six rows shown in Figure 7.12 that match the logical restriction.

FIGURE 7.12 SELECTED PRODUCT TABLE ATTRIBUTES: THE LOGICAL OR

P_DESCRIPT	P_INDATE	P_PRICE	V_CODE
7.25-in. pwr. saw blade	13-Dec-15	14.99	21344
9.00-in. pwr. saw blade	13-Nov-15	17.49	21344
B&D jigsaw, 12-in. blade	30-Dec-15	109.92	24288
B&D jigsaw, 8-in. blade	24-Dec-15	99.87	24288
Rat-tail file, 1/8-in. fine	15-Dec-15	4.99	21344
Hicut chain saw, 16 in.	07-Feb-16	256.99	24288

The logical operator **AND** has the same SQL syntax requirement as OR. The following command generates a list of all rows for which P_PRICE is less than $50 *and* for which P_INDATE is a date occurring after January 15, 2016:

```
SELECT      P_DESCRIPT, P_INDATE, P_PRICE, V_CODE
FROM        PRODUCT
WHERE       P_PRICE < 50
AND         P_INDATE > '15-Jan-2016';
```

rules of precedence
Basic algebraic rules that specify the order in which operations are performed. For example, operations within parentheses are executed first, so in the equation $2 + (3 \times 5)$, the multiplication portion is calculated first, making the correct answer 17.

OR
The SQL logical operator used to link multiple conditional expressions in a WHERE or HAVING clause. It requires only one of the conditional expressions to be true.

AND
The SQL logical operator used to link multiple conditional expressions in a WHERE or HAVING clause. It requires that all conditional expressions evaluate to true.

This command produces the output shown in Figure 7.13.

FIGURE 7.13 SELECTED PRODUCT TABLE ATTRIBUTES: THE LOGICAL AND

P_DESCRIPT	P_INDATE	P_PRICE	V_CODE
B&D cordless drill, 1/2-in.	20-Jan-16	38.95	25595
Claw hammer	20-Jan-16	9.95	21225
PVC pipe, 3.5-in., 8-ft	20-Feb-16	5.87	
1.25-in. metal screw, 25	01-Mar-16	6.99	21225
2.5-in. wd. screw, 50	24-Feb-16	8.45	21231

You can combine the logical OR with the logical AND to place further restrictions on the output. For example, suppose that you want a table listing for the following conditions:

- The P_INDATE is after January 15, 2016, *and* the P_PRICE is less than $50.
- *Or* the V_CODE is 24288.

The required listing can be produced by using the following:

```
SELECT      P_DESCRIPT, P_INDATE, P_PRICE, V_CODE
FROM        PRODUCT
WHERE       (P_PRICE < 50 AND P_INDATE > '15-Jan-2016')
OR          V_CODE = 24288;
```

Note the use of parentheses to combine logical restrictions. Where you place the parentheses depends on how you want the logical restrictions to be executed. Conditions listed within parentheses are always executed first. The preceding query yields the output shown in Figure 7.14.

FIGURE 7.14 SELECTED PRODUCT TABLE ATTRIBUTES: THE LOGICAL AND AND OR

P_DESCRIPT	P_INDATE	P_PRICE	V_CODE
B&D jigsaw, 12-in. blade	30-Dec-15	109.92	24288
B&D jigsaw, 8-in. blade	24-Dec-15	99.87	24288
B&D cordless drill, 1/2-in.	20-Jan-16	38.95	25595
Claw hammer	20-Jan-16	9.95	21225
Hicut chain saw, 16 in.	07-Feb-16	256.99	24288
PVC pipe, 3.5-in., 8-ft	20-Feb-16	5.87	
1.25-in. metal screw, 25	01-Mar-16	6.99	21225
2.5-in. wd. screw, 50	24-Feb-16	8.45	21231

Note that the three rows with the V_CODE = 24288 are included regardless of the P_INDATE and P_PRICE entries for those rows.

The use of the logical operators OR and AND can become quite complex when numerous restrictions are placed on the query. In fact, a specialty field in mathematics known as **Boolean algebra** is dedicated to the use of logical operators.

The logical operator **NOT** is used to negate the result of a conditional expression. That is, in SQL, all conditional expressions evaluate to true or false. If an expression is true, the row is selected; if an expression is false, the row is not selected. The NOT logical operator is typically used to find the rows that *do not* match a certain condition. For example, if you want to see a listing of all rows for which the vendor code is *not* 21344, use the following command sequence:

Boolean algebra
A branch of mathematics that uses the logical operators OR, AND, and NOT.

NOT
A SQL logical operator that negates a given predicate.

```
SELECT      *
FROM        PRODUCT
WHERE       NOT (V_CODE = 21344);
```

Note that the condition is enclosed in parentheses; that practice is optional, but it is highly recommended for clarity. The logical operator NOT can be combined with AND and OR.

Note

If your SQL version does not support the logical NOT, you can generate the required output by using the following condition:

WHERE V_CODE <> 21344

   If your version of SQL does not support <>, use:

WHERE V_CODE != 21344

7-4d Special Operators

ANSI-standard SQL allows the use of special operators in conjunction with the WHERE clause. These special operators include:

BETWEEN: Used to check whether an attribute value is within a range

IS NULL: Used to check whether an attribute value is null

LIKE: Used to check whether an attribute value matches a given string pattern

IN: Used to check whether an attribute value matches any value within a value list

EXISTS: Used to check whether a subquery returns any rows

The BETWEEN Special Operator If you use software that implements a standard SQL, the operator BETWEEN may be used to check whether an attribute value is within a range of values. For example, if you want to see a listing for all products whose prices are *between* $50 and $100, use the following command sequence:

```
SELECT          *
FROM            PRODUCT
WHERE           P_PRICE BETWEEN 50.00 AND 100.00;
```

Note

Note to Oracle Users

When using the BETWEEN special operator, always specify the lower-range value first. The WHERE clause of the command above is interpreted as:

WHERE P_PRICE >= 50 AND P_PRICE <= 100

   If you list the higher-range value first, the DBMS will return an empty result set because the WHERE clause will be interpreted as:

WHERE P_PRICE >= 100 and P_PRICE <= 50

   Clearly, no product can have a price that is both greater than 100 and simultaneously less than 50, Therefore, no rows can possibly match the criteria.

BETWEEN
In SQL, a special comparison operator used to check whether a value is within a range of specified values.

IS NULL
In SQL, a comparison operator used to check whether an attribute has a value.

LIKE
In SQL, a comparison operator used to check whether an attribute's text value matches a specified string pattern.

IN
In SQL, a comparison operator used to check whether a value is among a list of specified values.

EXISTS
In SQL, a comparison operator that checks whether a subquery returns any rows.

If your DBMS does not support BETWEEN, you can use:

```
SELECT          *
FROM            PRODUCT
WHERE           P_PRICE => 50.00 AND P_PRICE <= 100.00;
```

The IS NULL Special Operator Standard SQL allows the use of IS NULL to check for a null attribute value. For example, suppose that you want to list all products that do not have a vendor assigned (V_CODE is null). Such a null entry could be found by using the following command sequence:

```
SELECT          P_CODE, P_DESCRIPT, V_CODE
FROM            PRODUCT
WHERE           V_CODE IS NULL;
```

Similarly, if you want to check a null date entry, the command sequence is:

```
SELECT          P_CODE, P_DESCRIPT, P_INDATE
FROM            PRODUCT
WHERE           P_INDATE IS NULL;
```

Note that SQL uses a special operator to test for nulls. Why? Couldn't you just enter a condition such as "V_CODE = NULL"? No. Technically, NULL is not a "value" the way the number 0 or the blank space is; instead, a NULL is a special property of an attribute that represents the absence of any value.

The LIKE Special Operator The LIKE special operator is used in conjunction with wildcards to find patterns within string attributes. Standard SQL allows you to use the percent sign (%) and underscore (_) wildcard characters to make matches when the entire string is not known:

- % means any and all *following* or *preceding* characters are eligible. For example:

  'J%' includes Johnson, Jones, Jernigan, July, and J-231Q.

  'Jo%' includes Johnson and Jones.

  '%n' includes Johnson and Jernigan.

- _ means any *one* character may be substituted for the underscore. For example:

  '_23-456-6789' includes 123-456-6789, 223-456-6789, and 323-456-6789.

  '_23-_56-678_' includes 123-156-6781, 123-256-6782, and 823-956-6788.

  '_o_es' includes Jones, Cones, Cokes, totes, and roles.

Note

Some RDBMSs, such as Microsoft Access, use the wildcard characters * and ? instead of % and _.

For example, the following query would find all VENDOR rows for contacts whose last names begin with *Smith*.

```
SELECT          V_NAME, V_CONTACT, V_AREACODE, V_PHONE
FROM            VENDOR
WHERE           V_CONTACT LIKE 'Smith%';
```

If you check the original VENDOR data in Figure 7.2 again, you'll see that this SQL query yields three records: two Smiths and one Smithson.

Keep in mind that most SQL implementations yield case-sensitive searches. For example, Oracle will not yield a result that includes *Jones* if you use the wildcard search delimiter 'jo%' in a search for last names; *Jones* begins with a capital *J*, and your wildcard search starts with a lowercase *j*. On the other hand, MS Access searches are not case sensitive.

For example, suppose that you typed the following query in Oracle:

```
SELECT          V_NAME, V_CONTACT, V_AREACODE, V_PHONE
FROM            VENDOR
WHERE           V_CONTACT LIKE 'SMITH%';
```

No rows will be returned because character-based queries may be case sensitive. That is, an uppercase character has a different ASCII code than a lowercase character, causing *SMITH*, *Smith*, and *smith* to be evaluated as different (unequal) entries. Because the table contains no vendor whose last name begins with *SMITH* (all uppercase), the 'SMITH%' used in the query cannot be matched. Matches can be made only when the query entry is written exactly like the table entry.

Some RDBMSs, such as Microsoft Access, automatically make the necessary conversions to eliminate case sensitivity. Others, such as Oracle, provide a special UPPER function to convert both table and query character entries to uppercase. (The conversion is done in the computer's memory only; the conversion has no effect on how the value is actually stored in the table.) So, if you want to avoid a no-match result based on case sensitivity, and if your RDBMS allows the use of the UPPER function, you can generate the same results by using the following query:

```
SELECT          V_NAME, V_CONTACT, V_AREACODE, V_PHONE
FROM            VENDOR
WHERE           UPPER(V_CONTACT) LIKE 'SMITH%';
```

The preceding query produces a list that includes all rows containing a last name that begins with *Smith*, regardless of uppercase or lowercase letter combinations such as *Smith*, *smith*, and *SMITH*.

The logical operators may be used in conjunction with the special operators. For instance, the following query:

```
SELECT          V_NAME, V_CONTACT, V_AREACODE, V_PHONE
FROM            VENDOR
WHERE           V_CONTACT NOT LIKE 'Smith%';
```

will yield an output of all vendors whose names do not start with *Smith*.

Suppose that you do not know whether a person's name is spelled *Johnson* or *Johnsen*. The wildcard character _ lets you find a match for either spelling. The proper search would be instituted by the following query:

```
SELECT          *
FROM            VENDOR
WHERE           V_CONTACT LIKE 'Johns_n';
```

Thus, the wildcards allow you to make matches when only approximate spellings are known. Wildcard characters may be used in combinations. For example, the wildcard search based on the string '_l%' can yield the strings "Al", "Alton", "Elgin", "Blakeston", "blank", "bloated", and "eligible".

The IN Special Operator Many queries that would require the use of the logical OR can be more easily handled with the help of the special operator IN. For example, the following query:

```
SELECT          *
FROM            PRODUCT
WHERE           V_CODE = 21344
OR              V_CODE = 24288;
```

can be handled more efficiently with:

```
SELECT          *
FROM            PRODUCT
WHERE           V_CODE IN (21344, 24288);
```

Note that the IN operator uses a value list. All of the values in the list must be of the same data type. Each of the values in the value list is compared to the attribute—in this case, V_CODE. If the V_CODE value matches any of the values in the list, the row is selected. In this example, the rows selected will be only those in which the V_CODE is either 21344 or 24288.

If the attribute used is of a character data type, the list values must be enclosed in single quotation marks. For instance, if the V_CODE had been defined as CHAR(5) when the table was created, the preceding query would have read:

```
SELECT          *
FROM            PRODUCT
WHERE           V_CODE IN ('21344', '24288');
```

The IN operator is especially valuable when it is used in conjunction with subqueries. For example, suppose that you want to list the V_CODE and V_NAME of only those vendors who provide products. In that case, you could use a subquery within the IN operator to automatically generate the value list. The query would be:

```
SELECT          V_CODE, V_NAME
FROM            VENDOR
WHERE           V_CODE IN (SELECT V_CODE FROM PRODUCT);
```

The preceding query will be executed in two steps:

1. The inner query or subquery will generate a list of V_CODE values from the PRODUCT tables. Those V_CODE values represent the vendors who supply products.

2. The IN operator will compare the values generated by the subquery to the V_CODE values in the VENDOR table, and will select only the rows with matching values—that is, the vendors who provide products.

The IN special operator will receive additional attention in Chapter 8, where you will learn more about subqueries.

The EXISTS Special Operator The EXISTS special operator can be used whenever there is a requirement to execute a command based on the result of another query. That is, if a subquery returns any rows, run the main query; otherwise, do not. For example, the following query will list all vendors, but only if there are products to order:

SELECT   *
FROM   VENDOR
WHERE   EXISTS (SELECT * FROM PRODUCT WHERE P_QOH <= P_MIN);

The EXISTS special operator is used in the following example to list all vendors, but only if there are products with the quantity on hand, and less than double the minimum quantity:

SELECT   *
FROM   VENDOR
WHERE   EXISTS (SELECT * FROM PRODUCT WHERE P_QOH < P_MIN * 2);

The EXISTS special operator will receive additional attention in Chapter 8, where you will learn more about subqueries.

7-5 Additional Data Definition Commands

In this section, you will learn how to change table structures by changing attribute characteristics and by adding columns. Then you will learn how to make advanced data updates to the new columns. Finally, you will learn how to copy tables or parts of tables and how to delete tables.

All changes in the table structure are made by using the **ALTER TABLE** command followed by a keyword that produces the specific change you want to make. Three options are available: ADD, MODIFY, and DROP. You use ADD to add a column, MODIFY to change column characteristics, and DROP to delete a column from a table. Most RDBMSs do not allow you to delete a column unless the column does not contain any values; otherwise, such an action might delete crucial data used by other tables. The basic syntax to add or modify columns is:

ALTER TABLE *tablename*
    {ADD | MODIFY} (*columnname datatype* [{ADD | MODIFY}
    columnname datatype]);

The ALTER TABLE command can also be used to add table constraints. In those cases, the syntax would be:

ALTER TABLE *tablename*
    ADD *constraint* [ADD *constraint*];

where *constraint* refers to a constraint definition similar to those you learned in Section 7-2f.

You could also use the ALTER TABLE command to remove a column or table constraint. The syntax would be as follows:

ALTER TABLE *tablename*
    DROP {PRIMARY KEY | COLUMN *columnname* | CONSTRAINT
    constraintname };

ALTER TABLE
The SQL command used to make changes to table structure. When the command is followed by a keyword (ADD or MODIFY), it adds a column or changes column characteristics.

Notice that when removing a constraint, you need to specify it by name, which is one reason you should always name constraints in your CREATE TABLE or ALTER TABLE statement.

7-5a Changing a Column's Data Type

Using the ALTER syntax, the integer V_CODE in the PRODUCT table can be changed to a character V_CODE by using the following command:

```
ALTER TABLE PRODUCT
    MODIFY (V_CODE CHAR(5));
```

Some RDBMSs, such as Oracle, do not let you change data types unless the column to be changed is empty. For example, if you want to change the V_CODE field from the current number definition to a character definition, the preceding command will yield an error message because the V_CODE column already contains data. The error message is easily explained. Remember that the V_CODE in PRODUCT references the V_CODE in VENDOR. If you change the V_CODE data type, the data types do not match, and there is a referential integrity violation, which triggers the error message. If the V_CODE column does not contain data, the preceding command sequence will alter the table structure as expected (if the foreign key reference was not specified during the creation of the PRODUCT table).

7-5b Changing a Column's Data Characteristics

If the column to be changed already contains data, you can make changes in the column's characteristics if those changes do not alter the data type. For example, if you want to increase the width of the P_PRICE column to nine digits, use the following command:

```
ALTER TABLE PRODUCT
    MODIFY (P_PRICE DECIMAL(9,2));
```

If you now list the table contents, you can see that the column width of P_PRICE has increased by one digit.

Note

Some DBMSs impose limitations on when it is possible to change attribute characteristics. For example, Oracle lets you increase (but not decrease) the size of a column because an attribute modification will affect the integrity of the data in the database. In fact, some attribute changes can be made only when there is no data in any rows for the affected attribute.

7-5c Adding a Column

You can alter an existing table by adding one or more columns. In the following example, you add the column named P_SALECODE to the PRODUCT table. (This column will be used later to determine whether goods that have been in inventory for a certain length of time should be placed on special sale.)

Suppose that you expect the P_SALECODE entries to be 1, 2, or 3. Because no arithmetic will be performed with the P_SALECODE, the P_SALECODE will be classified as a single-character attribute. Note the inclusion of all required information in the following ALTER command:

ALTER TABLE PRODUCT
     ADD (P_SALECODE CHAR(1));

When adding a column, be careful not to include the NOT NULL clause for the new column. Doing so will cause an error message; if you add a new column to a table that already has rows, the existing rows will default to a value of null for the new column. Therefore, it is not possible to add the NOT NULL clause for this new column. (Of course, you can add the NOT NULL clause to the table structure after all the data for the new column has been entered and the column no longer contains nulls.)

7-5d Dropping a Column

Occasionally, you might want to modify a table by deleting a column. Suppose that you want to delete the V_ORDER attribute from the VENDOR table. You would use the following command:

ALTER TABLE VENDOR
     DROP COLUMN V_ORDER;

Again, some RDBMSs impose restrictions on attribute deletion. For example, you may not drop attributes that are involved in foreign key relationships, nor may you delete an attribute if it is the only one in a table.

7-5e Advanced Data Updates

To make changes to data in the columns of existing rows, use the UPDATE command. Do not confuse the INSERT and UPDATE commands: INSERT creates new rows in the table, while UPDATE changes rows that already exist. For example, to enter the P_SALE-CODE value '2' in the fourth row, use the UPDATE command together with the primary key P_CODE '1546-QQ2'. Enter the value by using the following command sequence:

UPDATE   PRODUCT
SET   P_SALECODE = '2'
WHERE   P_CODE = '1546-QQ2';

Subsequent data can be entered the same way, defining each entry location by its primary key (P_CODE) and its column location (P_SALECODE). For example, if you want to enter the P_SALECODE value '1' for the P_CODE values '2232/QWE' and '2232/QTY', you use:

UPDATE   PRODUCT
SET   P_SALECODE = '1'
WHERE   P_CODE IN ('2232/QWE', '2232/QTY');

If your RDBMS does not support IN, use the following command:

UPDATE   PRODUCT
SET   P_SALECODE = '1'
WHERE   P_CODE = '2232/QWE' OR P_CODE = '2232/QTY';

Online Content

If you are using the MS Access databases provided at *www.cengagebrain. com*, you can track each of the updates in the following sections. For example, look at the copies of the PRODUCT table in the Ch07_SaleCo database, one named PROD-UCT_2 and one named PRODUCT_3. Each of the two copies includes the new P_SALECODE column. If you want to see the *cumulative* effect of all UPDATE commands, you can continue using the PRODUCT table with the P_SALECODE modification and all of the changes you will make in the following sections. (You might even want to use both options, first to examine the individual effects of the update queries and then to examine the cumulative effects.)

You can check the results of your efforts by using the following commands:

```
SELECT          P_CODE, P_DESCRIPT, P_INDATE, P_PRICE, P_SALECODE
FROM            PRODUCT;
```

Although the UPDATE sequences just shown allow you to enter values into specified table cells, the process is very cumbersome. Fortunately, if a relationship can be established between the entries and the existing columns, the relationship can be used to assign values to their appropriate slots. For example, suppose that you want to place sales codes into the table based on the P_INDATE using the following schedule:

P_INDATE	P_SALECODE
before December 25, 2015	2
between January 16, 2016 and February 10, 2016	1

Using the PRODUCT table, the following two command sequences make the appropriate assignments:

```
UPDATE          PRODUCT
SET             P_SALECODE = '2'
WHERE           P_INDATE < '25-Dec-2015';
UPDATE          PRODUCT
SET             P_SALECODE = '1'
WHERE           P_INDATE >= '16-Jan-2016' AND P_INDATE <='10-Feb-2016';
```

To check the results of those two command sequences, use:

```
SELECT          P_CODE, P_DESCRIPT, P_INDATE, P_PRICE, P_SALECODE
FROM            PRODUCT;
```

If you have made *all* of the updates shown in this section using Oracle, your PRODUCT table should look like Figure 7.15. *Make sure that you issue a COMMIT statement to save these changes.*

The arithmetic operators are particularly useful in data updates. For example, if the quantity on hand in your PRODUCT table has dropped below the minimum desirable value, you will order more of the product. Suppose, for example, that you have ordered 20 units of product 2232/QWE. When the 20 units arrive, you will want to add them to inventory using the following commands:

```
UPDATE          PRODUCT
SET             P_QOH = P_QOH + 20
WHERE           P_CODE = '2232/QWE';
```

If you want to add 10 percent to the price for all products that have current prices below $50, you can use:

```
UPDATE          PRODUCT
SET             P_PRICE = P_PRICE * 1.10
WHERE           P_PRICE < 50.00;
```

If you are using Oracle, issue a ROLLBACK command to undo the changes made by the last two UPDATE statements.

FIGURE 7.15 THE CUMULATIVE EFFECT OF THE MULTIPLE UPDATES IN THE PRODUCT TABLE

```
 SQL Plus                                                            _  □  X

SQL> SELECT P_CODE, P_DESCRIPT, P_INDATE, P_PRICE, P_SALECODE FROM PRODUCT;

P_CODE     P_DESCRIPT                          P_INDATE     P_PRICE P
---------  ----------------------------------  ----------   ------- -
11QER/31   Power painter, 15 psi., 3-nozzle    03-NOV-15     109.99 2
13-Q2/P2   7.25-in. pwr. saw blade             13-DEC-15      14.99 2
14-Q1/L3   9.00-in. pwr. saw blade             13-NOV-15      17.49 2
1546-QQ2   Hrd. cloth, 1/4-in., 2x50           15-JAN-16      39.95 2
1558-QW1   Hrd. cloth, 1/2-in., 3x50           15-JAN-16      43.99
2232/QTY   B\&D jigsaw, 12-in. blade           30-DEC-15     109.92 1
2232/QWE   B\&D jigsaw, 8-in. blade            24-DEC-15      99.87 2
2238/QPD   B\&D cordless drill, 1/2-in.        20-JAN-16      38.95 1
23109-HB   Claw hammer                         20-JAN-16       9.95 1
23114-AA   Sledge hammer, 12 lb.               02-JAN-16      14.4
54778-2T   Rat-tail file, 1/8-in. fine         15-DEC-15       4.99 2
89-WRE-Q   Hicut chain saw, 16 in.             07-FEB-16     256.99 1
PVC23DRT   PVC pipe, 3.5-in., 8-ft             20-FEB-16       5.87
SM-18277   1.25-in. metal screw, 25            01-MAR-16       6.99
SW-23116   2.5-in. wd. screw, 50               24-FEB-16       8.45
WR3/TT3    Steel matting, 4'x8'x1/6", .5" mesh 17-JAN-16     119.95 1

16 rows selected.

SQL>
```

Note

If you fail to roll back the changes of the preceding UPDATE queries, the output of the subsequent queries will not match the results shown in the figures. Therefore:

- If you are using Oracle, use the ROLLBACK command to restore the database to its previous state.
- If you are using Access, copy the original Ch07_SaleCo.mdb file from *www.cengage brain.com*.

7-5f Copying Parts of Tables

As you will discover in later chapters on database design, sometimes it is necessary to break up a table structure into several component parts (or smaller tables). Fortunately, SQL allows you to copy the contents of selected table columns so that the data need not be re-entered manually into the newly created table(s). For example, if you want to copy P_CODE, P_DESCRIPT, P_PRICE, and V_CODE from the PRODUCT table to a new table named PART, you create the PART table structure first, as follows:

```
CREATE TABLE PART(
PART_CODE               CHAR(8),
PART_DESCRIPT           CHAR(35),
PART_PRICE              DECIMAL(8,2),
V_CODE                  INTEGER,
PRIMARY KEY (PART_CODE));
```

Note that the PART column names need not be identical to those of the original table and that the new table need not have the same number of columns as the original table. In this case, the first column in the PART table is PART_CODE, rather than the original P_CODE in the PRODUCT table. Also, the PART table contains only four columns rather than the eight columns in the PRODUCT table. However, column characteristics must match; you cannot copy a character-based attribute into a numeric structure, and vice versa.

Next, you need to add the rows to the new PART table, using the PRODUCT table rows and the INSERT command you learned in Section 7–3g. The syntax is:

INSERT INTO   *target_tablename*[(*target_columnlist*)]
SELECT   *source_columnlist*
FROM   *source_tablename*;

Note that the target column list is required if the source column list does not match all of the attribute names and characteristics of the target table (including the order of the columns). Otherwise, you do not need to specify the target column list. In this example, you must specify the target column list in the following INSERT command because the column names of the target table are different:

INSERT INTO PART   (PART_CODE, PART_DESCRIPT, PART_PRICE, V_CODE)
SELECT   P_CODE, P_DESCRIPT, P_PRICE, V_CODE FROM
                   PRODUCT;

The contents of the PART table can now be examined by using the following query to generate the new PART table's contents, shown in Figure 7.16:

SELECT   * FROM PART;

FIGURE 7.16 PART TABLE ATTRIBUTES COPIED FROM THE PRODUCT TABLE

PART_CODE	PART_DESCRIPT	PART_PRICE	V_CODE
11QER/31	Power painter, 15 psi., 3-nozzle	109.99	25595
13-Q2/P2	7.25-in. pwr. saw blade	14.99	21344
14-Q1/L3	9.00-in. pwr. saw blade	17.49	21344
1546-QQ2	Hrd. cloth, 1/4-in., 2x50	39.95	23119
1558-QW1	Hrd. cloth, 1/2-in., 3x50	43.99	23119
2232/QTY	B&D jigsaw, 12-in. blade	109.92	24288
2232/QWE	B&D jigsaw, 8-in. blade	99.87	24288
2238/QPD	B&D cordless drill, 1/2-in.	38.95	25595
23109-HB	Claw hammer	9.95	21225
23114-AA	Sledge hammer, 12 lb.	14.4	
54778-2T	Rat-tail file, 1/8-in. fine	4.99	21344
89-WRE-Q	Hicut chain saw, 16 in.	256.99	24288
PVC23DRT	PVC pipe, 3.5-in., 8-ft	5.87	
SM-18277	1.25-in. metal screw, 25	6.99	21225
SW-23116	2.5-in. wd. screw, 50	8.45	21231
WR3/TT3	Steel matting, 4'x8'x1/6", .5" mesh	119.95	25595

SQL provides another way to rapidly create a new table based on selected columns and rows of an existing table. In this case, the new table will copy the attribute names, data characteristics, and rows of the original table. The Oracle version of the command is:

CREATE TABLE PART AS
SELECT   P_CODE AS PART_CODE, P_DESCRIPT AS PART_DESCRIPT,
             P_PRICE AS PART_PRICE, V_CODE
FROM   PRODUCT;

If the PART table already exists, Oracle will not let you overwrite the existing table. To run this command, you must first delete the existing PART table. (See Section 7-5h.)

The MS Access version of this command is:

```
SELECT          P_CODE AS PART_CODE, P_DESCRIPT AS PART_DESCRIPT,
                P_PRICE AS PART_PRICE, V_CODE INTO PART
FROM            PRODUCT;
```

If the PART table already exists, MS Access will ask if you want to delete the existing table and continue with the creation of the new PART table.

The SQL command just shown creates a new PART table with PART_CODE, PART_DESCRIPT, PART_PRICE, and V_CODE columns. In addition, all of the data rows for the selected columns will be copied automatically. *However, note that no entity integrity (primary key) or referential integrity (foreign key) rules are automatically applied to the new table.* In the next section, you will learn how to define the PK to enforce entity integrity and the FK to enforce referential integrity.

7-5g Adding Primary and Foreign Key Designations

When you create a new table based on another table, the new table does not include integrity rules from the old table. In particular, there is no primary key. To define the primary key for the new PART table, use the following command:

```
ALTER TABLE     PART
    ADD         PRIMARY KEY (PART_CODE);
```

Several other scenarios could leave you without entity and referential integrity. For example, you might have forgotten to define the primary and foreign keys when you created the original tables. Or, if you imported tables from a different database, you might have discovered that the importing procedure did not transfer the integrity rules. In any case, you can re-establish the integrity rules by using the ALTER command. For example, if the PART table's foreign key has not yet been designated, it can be designated by:

```
ALTER TABLE     PART
    ADD         FOREIGN KEY (V_CODE) REFERENCES VENDOR;
```

Alternatively, if neither the PART table's primary key nor its foreign key has been designated, you can incorporate both changes at once:

```
ALTER TABLE     PART
    ADD         PRIMARY KEY (PART_CODE)
    ADD         FOREIGN KEY (V_CODE) REFERENCES VENDOR;
```

Even composite primary keys and multiple foreign keys can be designated in a single SQL command. For example, if you want to enforce the integrity rules for the LINE table shown in Figure 7.1, you can use:

```
ALTER TABLE     LINE
    ADD         PRIMARY KEY (INV_NUMBER, LINE_NUMBER)
    ADD         FOREIGN KEY (INV_NUMBER) REFERENCES INVOICE
    ADD         FOREIGN KEY (P_CODE) REFERENCES PRODUCT;
```

7-5h Deleting a Table from the Database

A table can be deleted from the database using the **DROP TABLE** command. For example, you can delete the PART table you just created with the following command:

DROP TABLE PART;

You can drop a table only if it is not the "one" side of any relationship. If you try to drop a table otherwise, the RDBMS will generate an error message indicating that a foreign key integrity violation has occurred.

7-6 **Additional SELECT Query Keywords**

One of the most important advantages of SQL is its ability to produce complex free-form queries. The logical operators that were introduced earlier to update table contents work just as well in the query environment. In addition, SQL provides useful functions that count, find minimum and maximum values, calculate averages, and so on. Better yet, SQL allows the user to limit queries to only those entries that have no duplicates or entries whose duplicates can be grouped.

7-6a Ordering a Listing

The **ORDER BY** clause is especially useful when the listing order is important to you. The syntax is:

SELECT	*columnlist*
FROM	*tablelist*
[WHERE	*conditionlist*]
[ORDER BY	*columnlist* [ASC | DESC]];

Although you have the option of declaring the order type—ascending or descending—the default order is ascending. For example, if you want the contents of the PRODUCT table to be listed by P_PRICE in ascending order, use the following commands:

SELECT	P_CODE, P_DESCRIPT, P_QOH, P_PRICE
FROM	PRODUCT
ORDER BY	P_PRICE;

The output is shown in Figure 7.17. Note that ORDER BY yields an ascending price listing.

Comparing the listing in Figure 7.17 to the actual table contents shown earlier in Figure 7.2, you will see that the lowest-priced product is listed first in Figure 7.17, followed by the next lowest-priced product, and so on. However, although ORDER BY produces a sorted output, the actual table contents are unaffected by the ORDER BY command.

To produce the list in descending order, you would enter:

SELECT	P_CODE, P_DESCRIPT, P_INDATE, P_PRICE
FROM	PRODUCT
ORDER BY	P_PRICE DESC;

DROP TABLE
A SQL command used to delete database objects such as tables, views, indexes, and users.

ORDER BY
A SQL clause that is useful for ordering the output of a SELECT query (for example, in ascending or descending order).

FIGURE 7.17 SELECTED PRODUCT TABLE ATTRIBUTES: ORDERED BY (ASCENDING) P_PRICE

P_CODE	P_DESCRIPT	P_QOH	P_PRICE
54778-2T	Rat-tail file, 1/8-in. fine	43	4.99
PVC23DRT	PVC pipe, 3.5-in., 8-ft	188	5.87
SM-18277	1.25-in. metal screw, 25	172	6.99
SW-23116	2.5-in. wd. screw, 50	237	8.45
23109-HB	Claw hammer	23	9.95
23114-AA	Sledge hammer, 12 lb.	8	14.40
13-Q2/P2	7.25-in. pwr. saw blade	32	14.99
14-Q1/L3	9.00-in. pwr. saw blade	18	17.49
2238/QPD	B&D cordless drill, 1/2-in.	12	38.95
1546-QQ2	Hrd. cloth, 1/4-in., 2x50	15	39.95
1558-QW1	Hrd. cloth, 1/2-in., 3x50	23	43.99
2232/QWE	B&D jigsaw, 8-in. blade	6	99.87
2232/QTY	B&D jigsaw, 12-in. blade	8	109.92
11QER/31	Power painter, 15 psi., 3-nozzle	8	109.99
WR3/TT3	Steel matting, 4'x8'x1/6", .5" mesh	18	119.95
89-WRE-Q	Hicut chain saw, 16 in.	11	256.99

Ordered listings are used frequently. For example, suppose that you want to create a phone directory. It would be helpful if you could produce an ordered sequence (last name, first name, initial) in three stages:

1. ORDER BY last name.

2. Within the last names, ORDER BY first name.

3. Within the first and last names, ORDER BY middle initial.

Such a multilevel ordered sequence is known as a **cascading order sequence**, and it can be created easily by listing several attributes, separated by commas, after the ORDER BY clause.

The cascading order sequence is the basis for any telephone directory. To illustrate a cascading order sequence, use the following SQL command on the EMPLOYEE table:

```
SELECT      EMP_LNAME, EMP_FNAME, EMP_INITIAL, EMP_AREACODE,
            EMP_PHONE
FROM        EMPLOYEE
ORDER BY    EMP_LNAME, EMP_FNAME, EMP_INITIAL;
```

This command yields the results shown in Figure 7.18.

The ORDER BY clause is useful in many applications, especially because the DESC qualifier can be invoked. For example, listing the most recent items first is a standard procedure. Typically, invoice due dates are listed in descending order. Or, if you want to examine budgets, it is probably useful to list the largest budget line items first.

You can use the ORDER BY clause in conjunction with other SQL commands, too. For example, note the use of restrictions on date and price in the following command sequence:

```
SELECT      P_DESCRIPT, V_CODE, P_INDATE, P_PRICE
FROM        PRODUCT
WHERE       P_INDATE < '21-Jan-2016'
AND         P_PRICE <= 50.00
ORDER BY    V_CODE, P_PRICE DESC;
```

cascading order sequence
A nested ordering sequence for a set of rows, such as a list in which all last names are alphabetically ordered and, within the last names, all first names are ordered.

FIGURE 7.18 TELEPHONE LIST QUERY RESULTS

EMP_LNAME	EMP_FNAME	EMP_INITIAL	EMP_AREACODE	EMP_PHONE
Brandon	Marie	G	901	882-0845
Diante	Jorge	D	615	890-4567
Genkazi	Leighla	W	901	569-0093
Johnson	Edward	E	615	898-4387
Jones	Anne	M	615	898-3456
Kolmycz	George	D	615	324-5456
Lange	John	P	901	504-4430
Lewis	Rhonda	G	615	324-4472
Saranda	Hermine	R	615	324-5505
Smith	George	A	615	890-2984
Smith	George	K	901	504-3339
Smith	Jeanine	K	615	324-7883
Smythe	Melanie	P	615	324-9006
Vandam	Rhett		901	675-8993
Washington	Rupert	E	615	890-4925
Wiesenbach	Paul	R	615	897-4358
Williams	Robert	D	615	890-3220

The output is shown in Figure 7.19. Note that within each V_CODE, the P_PRICE values are in descending order.

FIGURE 7.19 A QUERY BASED ON MULTIPLE RESTRICTIONS

P_DESCRIPT	V_CODE	P_INDATE	P_PRICE
Sledge hammer, 12 lb.		02-Jan-16	14.40
Claw hammer	21225	20-Jan-16	9.95
9.00-in. pwr. saw blade	21344	13-Nov-15	17.49
7.25-in. pwr. saw blade	21344	13-Dec-15	14.99
Rat-tail file, 1/8-in. fine	21344	15-Dec-15	4.99
Hrd. cloth, 1/2-in., 3x50	23119	15-Jan-16	43.99
Hrd. cloth, 1/4-in., 2x50	23119	15-Jan-16	39.95
B&D cordless drill, 1/2-in.	25595	20-Jan-16	38.95

7-6b Listing Unique Values

How many *different* vendors are currently represented in the PRODUCT table? A simple listing (SELECT) is not very useful if the table contains several thousand rows and you have to sift through the vendor codes manually. Fortunately, SQL's **DISTINCT** clause produces a list of only those values that are different from one another. For example, the command

SELECT   DISTINCT V_CODE
FROM   PRODUCT;

yields only the different vendor codes (V_CODE) in the PRODUCT table, as shown in Figure 7.20. Note that the first output row shows the null. The placement of nulls does not affect the list contents. In Oracle, you could use ORDER BY V_CODE NULLS FIRST to place nulls at the top of the list.

DISTINCT
A SQL clause that produces only a list of values that are different from one another.

7-6c Aggregate Functions

SQL can perform various mathematical summaries for you, such as counting the number of rows that contain a specified condition, finding the minimum or maximum values

FIGURE 7.20   A LISTING OF DISTINCT V_CODE VALUES IN THE PRODUCT TABLE

V_CODE
21225
21231
21344
23119
24288
25595

for a specified attribute, summing the values in a specified column, and averaging the values in a specified column. Those aggregate functions are shown in Table 7.8.

Note

> If the ordering column has nulls, they are listed either first or last, depending on the RDBMS. The ORDER BY clause must always be listed last in the SELECT command sequence.

TABLE 7.8

SOME BASIC SQL AGGREGATE FUNCTIONS

FUNCTION	OUTPUT
COUNT	The number of rows containing non-null values
MIN	The minimum attribute value encountered in a given column
MAX	The maximum attribute value encountered in a given column
SUM	The sum of all values for a given column
AVG	The arithmetic mean (average) for a specified column

To illustrate another standard SQL command format, most of the remaining input and output sequences are presented using the Oracle RDBMS.

COUNT The **COUNT** function is used to tally the number of non-null values of an attribute. COUNT can be used in conjunction with the DISTINCT clause. For example, suppose that you want to find out how many different vendors are in the PRODUCT table. The answer, generated by the first SQL code set shown in Figure 7.21, is 6. Note that the nulls are not counted as V_CODE values.

The aggregate functions can be combined with the SQL commands explored earlier. For example, the second SQL command set in Figure 7.21 supplies the answer to the question, "How many vendors referenced in the PRODUCT table have supplied products with prices that are less than or equal to $10?" The answer is that three vendors' products meet the price specification.

The COUNT aggregate function uses one parameter within parentheses, generally a column name such as COUNT(V_CODE) or COUNT(P_CODE). The parameter may also be an expression such as COUNT(DISTINCT V_CODE) or COUNT(P_PRICE+10). Using that syntax, COUNT always returns the number of non-null values in the given column. (Whether the column values are computed or show stored table row values is immaterial.) In contrast, the syntax COUNT(*) returns the number of total rows from the query, including

COUNT
A SQL aggregate function that outputs the number of rows containing not null values for a given column or expression, sometimes used in conjunction with the DISTINCT clause.

FIGURE 7.21 COUNT FUNCTION OUTPUT EXAMPLES

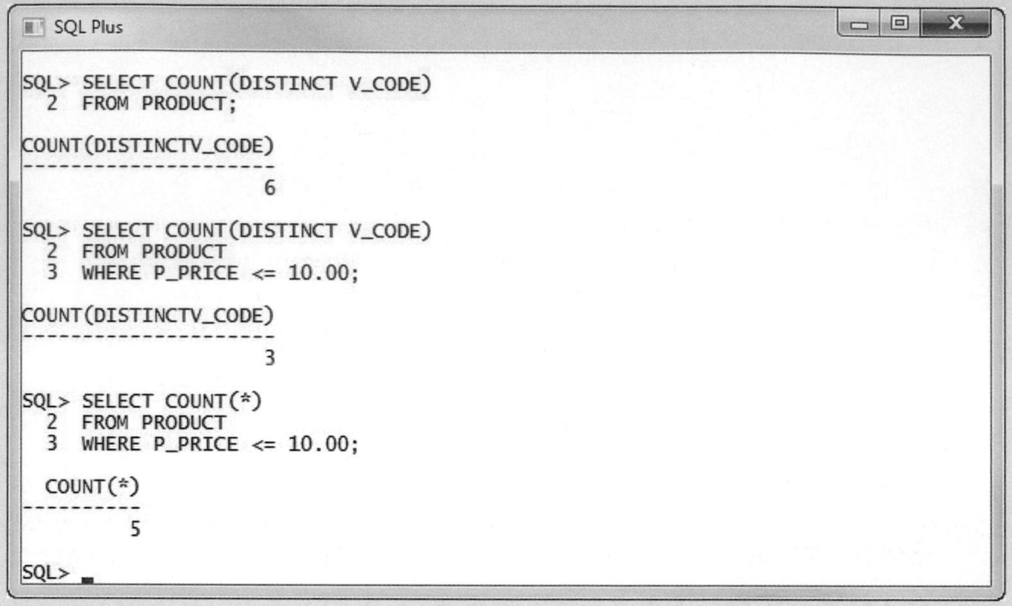

```
SQL Plus                                              □ ▣  X

SQL> SELECT COUNT(DISTINCT V_CODE)
  2  FROM PRODUCT;

COUNT(DISTINCTV_CODE)
--------------------
                   6

SQL> SELECT COUNT(DISTINCT V_CODE)
  2  FROM PRODUCT
  3  WHERE P_PRICE <= 10.00;

COUNT(DISTINCTV_CODE)
--------------------
                   3

SQL> SELECT COUNT(*)
  2  FROM PRODUCT
  3  WHERE P_PRICE <= 10.00;

  COUNT(*)
----------
         5

SQL> ▄
```

the rows that contain nulls. In the example in Figure 7.21, SELECT COUNT(P_CODE) FROM PRODUCT and SELECT COUNT(*) FROM PRODUCT will yield the same answer because there are no null values in the P_CODE primary key column.

Note that the third SQL command set in Figure 7.21 uses the COUNT(*) command to answer the question, "How many rows in the PRODUCT table have a P_PRICE value less than or equal to $10?" The answer indicates that five products have a listed price that meets the specification. The COUNT(*) aggregate function is used to count rows in a query result set. In contrast, the COUNT(*column*) aggregate function counts the number of non-null values in a given column. For example, in Figure 7.20, the COUNT(*) function would return a value of 7 to indicate seven rows returned by the query. The COUNT(V_CODE) function would return a value of 6 to indicate the six non-null vendor code values.

Note

Note to MS Access Users

MS Access does not support the use of COUNT with the DISTINCT clause. If you want to use such queries in MS Access, you must create subqueries with DISTINCT and NOT NULL clauses. For example, the equivalent MS Access queries for the first two queries shown in Figure 7.21 are:

```
SELECT     COUNT(*)
FROM       (SELECT DISTINCT V_CODE FROM PRODUCT WHERE V_CODE IS NOT NULL)
```

and

```
SELECT     COUNT(*)
FROM       (SELECT DISTINCT(V_CODE)
              FROM (SELECT V_CODE, P_PRICE FROM PRODUCT
                WHERE V_CODE IS NOT NULL AND P_PRICE<10))
```

The two queries are available at *www.cengagebrain.com* in the Ch07_SaleCo (Access) database. MS Access does add a trailer at the end of the query after you have executed it, but you can delete that trailer the next time you use the query. Subqueries are covered in detail in Chapter 8, Advanced SQL.

MAX and MIN The **MAX** and **MIN** functions help you find answers to problems such as the highest and lowest (maximum and minimum) prices in the PRODUCT table. The highest price, $256.99, is supplied by the first SQL command set in Figure 7.22. The second SQL command set shown in Figure 7.22 yields the minimum price of $4.99.

FIGURE 7.22 MAX AND MIN OUTPUT EXAMPLES

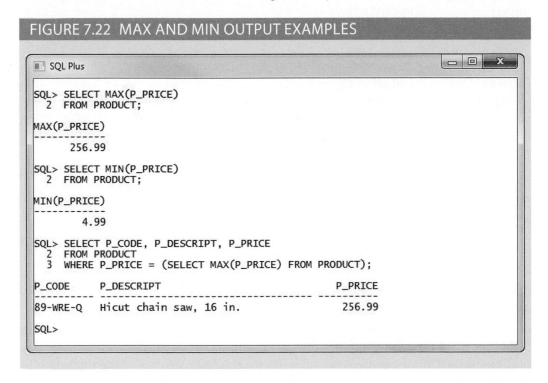

```
SQL> SELECT MAX(P_PRICE)
  2  FROM PRODUCT;

MAX(P_PRICE)
------------
      256.99

SQL> SELECT MIN(P_PRICE)
  2  FROM PRODUCT;

MIN(P_PRICE)
------------
        4.99

SQL> SELECT P_CODE, P_DESCRIPT, P_PRICE
  2  FROM PRODUCT
  3  WHERE P_PRICE = (SELECT MAX(P_PRICE) FROM PRODUCT);

P_CODE     P_DESCRIPT                               P_PRICE
---------- ---------------------------------------- ----------
89-WRE-Q   Hicut chain saw, 16 in.                     256.99

SQL>
```

The third SQL command set in Figure 7.22 demonstrates that the numeric functions can be used in conjunction with more complex queries. However, you must remember that *the numeric functions yield only one value* based on all the values found in the table: a single maximum value, a single minimum value, a single count, or a single average value. *It is easy to overlook this warning.* For example, examine the question, "Which product has the highest price?"

Although that query seems simple enough, the SQL command sequence:

```
SELECT        P_CODE, P_DESCRIPT, P_PRICE
FROM          PRODUCT
WHERE         P_PRICE = MAX(P_PRICE);
```

does not yield the expected results because the use of MAX(P_PRICE) on the right side of a comparison operator is incorrect, thus producing an error message. The aggregate function MAX(*columnname*) can be used only in the column list of a SELECT statement. Also, in a comparison that uses an equality symbol, you can use only a single value to the right of the equals sign.

MAX
A SQL aggregate function that yields the maximum attribute value in a given column.

MIN
A SQL aggregate function that yields the minimum attribute value in a given column.

To answer the question, therefore, you must compute the maximum price first, then compare it to each price returned by the query. To do that, you need a nested query. In this case, the nested query is composed of two parts:

- The *inner query*, which is executed first.
- The *outer query*, which is executed last. (Remember that the outer query is always the first SQL command you encounter—in this case, SELECT.)

Using the following command sequence as an example, note that the inner query first finds the maximum price value, which is stored in memory. Because the outer query now has a value to which to compare each P_PRICE value, the query executes properly.

```
SELECT      P_CODE, P_DESCRIPT, P_PRICE
FROM        PRODUCT
WHERE       P_PRICE = (SELECT MAX(P_PRICE) FROM PRODUCT);
```

The execution of the nested query yields the correct answer, shown below the third (nested) SQL command set in Figure 7.22.

The MAX and MIN aggregate functions can also be used with date columns. For example, to find out which product has the oldest date, you would use MIN(P_INDATE). In the same manner, to find out the most recent product, you would use MAX(P_INDATE).

Note

You can use expressions anywhere a column name is expected. Suppose that you want to know what product has the highest inventory value. To find the answer, you can write the following query:

```
SELECT *
FROM PRODUCT
WHERE P_QOH*P_PRICE = (SELECT MAX(P_QOH*P_PRICE) FROM PRODUCT);
```

SUM The **SUM** function computes the total sum for any specified attribute, using any condition(s) you have imposed. For example, if you want to compute the total amount owed by your customers, you could use the following command:

```
SELECT      SUM(CUS_BALANCE) AS TOTBALANCE
FROM        CUSTOMER;
```

You could also compute the sum total of an expression. For example, if you want to find the total value of all items carried in inventory, you could use the following:

```
SELECT      SUM(P_QOH * P_PRICE) AS TOTVALUE
FROM        PRODUCT;
```

SUM
A SQL aggregate function that yields the sum of all values for a given column or expression.

AVG
A SQL aggregate function that outputs the mean average for a specified column or expression.

The total value is the sum of the product of the quantity on hand and the price for all items. (See Figure 7.23.)

AVG The **AVG** function format is similar to those of MIN and MAX and is subject to the same operating restrictions. The first SQL command set in Figure 7.24 shows how a simple average P_PRICE value can be generated to yield the computed average price of 56.42125. The second SQL command set in Figure 7.24 produces five output lines that describe products whose prices exceed the average product price. Note that the second query uses nested SQL commands and the ORDER BY clause examined earlier.

FIGURE 7.23 THE TOTAL VALUE OF ALL ITEMS IN THE PRODUCT TABLE

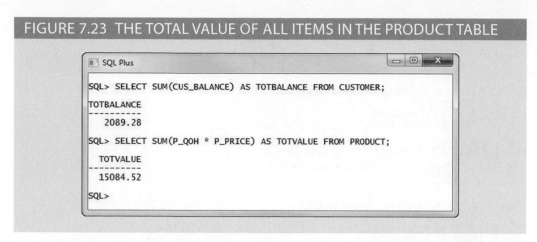

FIGURE 7.24 AVG FUNCTION OUTPUT EXAMPLES

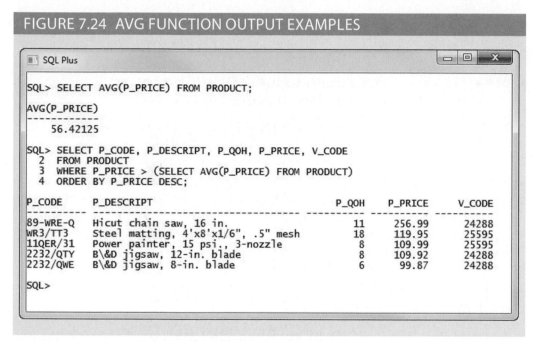

7-6d Grouping Data

In the previous examples, the aggregate functions summarized data across all rows in the given tables. Sometimes, however, you do not want to treat the entire table as a single collection of data for summarizing. Rows can be grouped into smaller collections quickly and easily using the **GROUP BY** clause within the SELECT statement. The aggregate functions will then summarize the data within each smaller collection. The syntax is

SELECT	*columnlist*
FROM	*tablelist*
[WHERE	*conditionlist*]
[GROUP BY	*columnlist*]
[HAVING	*conditionlist*]
[ORDER BY	*columnlist* [ASC | DESC]];

The GROUP BY clause is generally used when you have attribute columns combined with aggregate functions in the SELECT statement. For example, to determine the minimum price for each sales code, use the first SQL command set shown in Figure 7.25.

GROUP BY
A SQL clause used to create frequency distributions when combined with any of the aggregate functions in a SELECT statement.

FIGURE 7.25 GROUP BY CLAUSE OUTPUT EXAMPLES

```
SQL> SELECT P_SALECODE, MIN(P_PRICE)
  2  FROM PRODUCT
  3  GROUP BY P_SALECODE
  4  ORDER BY P_SALECODE;

P MIN(P_PRICE)
- ------------
1         9.95
2         4.99
          5.87

SQL> SELECT P_SALECODE, AVG(P_PRICE)
  2  FROM PRODUCT
  3  GROUP BY P_SALECODE
  4  ORDER BY P_SALECODE;

P AVG(P_PRICE)
- ------------
1      107.152
2        47.88
         15.94

SQL> _
```

The second SQL command set in Figure 7.25 generates the average price within each sales code. Note that the P_SALECODE nulls are included within the grouping.

The GROUP BY clause is valid only when used in conjunction with one of the SQL aggregate functions, such as COUNT, MIN, MAX, AVG, and SUM. For example, as shown in the first command set in Figure 7.26, if you try to group the output by using

```
SELECT      V_CODE, P_CODE, P_DESCRIPT, P_PRICE
FROM        PRODUCT
GROUP BY    V_CODE;
```

you generate a "not a GROUP BY expression" error. However, if you write the preceding SQL command sequence in conjunction with an aggregate function, the GROUP BY clause works properly. The second SQL command sequence in Figure 7.26 properly answers the question, "How many products are supplied by each vendor?" because it uses a COUNT aggregate function.

Note that the last output line in Figure 7.26 shows a null for the V_CODE, indicating that two products were not supplied by a vendor. Perhaps those products were produced in-house, or they might have been bought without the use of a vendor, or the person who entered the data might have merely forgotten to enter a vendor code. (Remember that nulls can be the result of many things.)

Note

When using the GROUP BY clause with a SELECT statement:

- The SELECT's *columnlist* must include a combination of column names and aggregate functions.
- The GROUP BY clause's *columnlist* must include all nonaggregate function columns specified in the SELECT's *columnlist*. If required, you could also group by any aggregate function columns that appear in the SELECT's *columnlist*.
- The GROUP BY clause *columnlist* can include any columns from the tables in the FROM clause of the SELECT statement, even if they do not appear in the SELECT's *columnlist*.

FIGURE 7.26 INCORRECT AND CORRECT USE OF THE GROUP BY CLAUSE

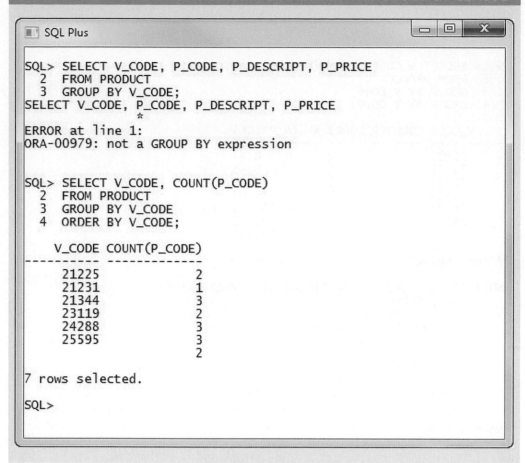

The GROUP BY Feature's HAVING Clause A particularly useful extension of the GROUP BY feature is the **HAVING** clause. The HAVING clause operates very much like the WHERE clause in the SELECT statement. However the WHERE clause applies to columns and expressions for individual rows, while the HAVING clause is applied to the output of a GROUP BY operation. For example, suppose that you want to generate a listing of the number of products in the inventory supplied by each vendor. However, this time you want to limit the listing to products whose prices average less than $10. The first part of that requirement is satisfied with the help of the GROUP BY clause, as illustrated in the first SQL command set in Figure 7.27. Note that the HAVING clause is used in conjunction with the GROUP BY clause in the second SQL command set in Figure 7.27 to generate the desired result.

If you use the WHERE clause instead of the HAVING clause, the second SQL command set in Figure 7.27 will produce an error message.

You can also combine multiple clauses and aggregate functions. For example, consider the following SQL statement:

```
SELECT        V_CODE, SUM(P_QOH * P_PRICE) AS TOTCOST
FROM          PRODUCT
GROUP BY      V_CODE
HAVING        (SUM(P_QOH * P_PRICE) > 500)
ORDER BY      SUM(P_QOH * P_PRICE) DESC;
```

HAVING
A clause applied to the output of a GROUP BY operation to restrict selected rows.

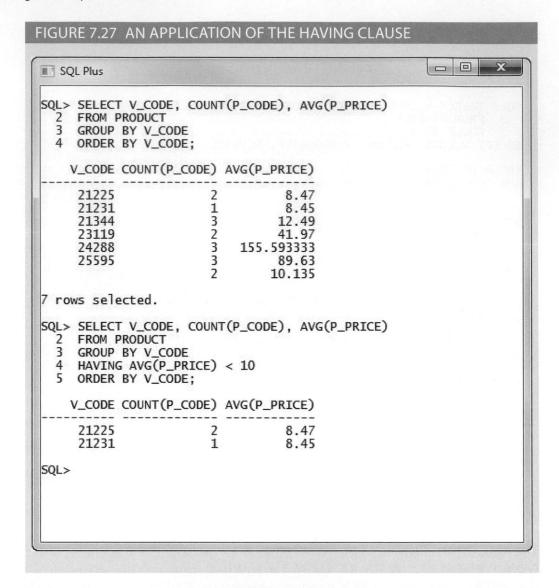

FIGURE 7.27 AN APPLICATION OF THE HAVING CLAUSE

```
SQL> SELECT V_CODE, COUNT(P_CODE), AVG(P_PRICE)
  2  FROM PRODUCT
  3  GROUP BY V_CODE
  4  ORDER BY V_CODE;

    V_CODE COUNT(P_CODE) AVG(P_PRICE)
---------- ------------- ------------
     21225             2         8.47
     21231             1         8.45
     21344             3        12.49
     23119             2        41.97
     24288             3   155.593333
     25595             3        89.63
                       2       10.135

7 rows selected.

SQL> SELECT V_CODE, COUNT(P_CODE), AVG(P_PRICE)
  2  FROM PRODUCT
  3  GROUP BY V_CODE
  4  HAVING AVG(P_PRICE) < 10
  5  ORDER BY V_CODE;

    V_CODE COUNT(P_CODE) AVG(P_PRICE)
---------- ------------- ------------
     21225             2         8.47
     21231             1         8.45

SQL>
```

This statement will do the following:

- Aggregate the total cost of products grouped by V_CODE.
- Select only the rows with totals that exceed $500.
- List the results in descending order by the total cost.

Note the syntax used in the HAVING and ORDER BY clauses; in both cases, you must specify the column expression (formula) used in the SELECT statement's column list, rather than the column alias (TOTCOST). Some RDBMSs allow you to replace the column expression with the column alias, while others do not.

7-7 Joining Database Tables

The ability to combine, or join, tables on common attributes is perhaps the most important distinction between a relational database and other databases. A join is performed when data is retrieved from more than one table at a time. If necessary, review the join definitions and examples in Chapter 3, The Relational Database Model.

To join tables, you simply list the tables in the FROM clause of the SELECT statement. The DBMS will create the Cartesian product of every table in the FROM clause. However, to get the correct result—that is, a natural join—you must select only the rows in which the common attribute values match. To do this, use the WHERE clause to indicate the common attributes used to link the tables; this WHERE clause is sometimes referred to as the join condition.

The join condition is generally composed of an equality comparison between the foreign key and the primary key of related tables. For example, suppose that you want to join the two tables VENDOR and PRODUCT. Because V_CODE is the foreign key in the PRODUCT table and the primary key in the VENDOR table, the link is established on V_CODE. (See Table 7.9.)

TABLE 7.9

CREATING LINKS THROUGH FOREIGN KEYS

TABLE	ATTRIBUTES TO BE SHOWN	LINKING ATTRIBUTE
PRODUCT	P_DESCRIPT, P_PRICE	V_CODE
VENDOR	V_NAME, V_CONTACT, V_AREACODE, V_PHONE	V_CODE

When the same attribute name appears in more than one of the joined tables, the source table of the attributes listed in the SELECT command sequence must be defined. To join the PRODUCT and VENDOR tables, you would use the following, which produces the output shown in Figure 7.28:

FIGURE 7.28 THE RESULTS OF A JOIN

P_DESCRIPT	P_PRICE	V_NAME	V_CONTACT	V_AREACODE	V_PHONE
Claw hammer	9.95	Bryson, Inc.	Smithson	615	223-3234
1.25-in. metal screw, 25	6.99	Bryson, Inc.	Smithson	615	223-3234
2.5-in. wd. screw, 50	8.45	D&E Supply	Singh	615	228-3245
7.25-in. pwr. saw blade	14.99	Gomez Bros.	Ortega	615	889-2546
9.00-in. pwr. saw blade	17.49	Gomez Bros.	Ortega	615	889-2546
Rat-tail file, 1/8-in. fine	4.99	Gomez Bros.	Ortega	615	889-2546
Hrd. cloth, 1/4-in., 2x50	39.95	Randsets Ltd.	Anderson	901	678-3998
Hrd. cloth, 1/2-in., 3x50	43.99	Randsets Ltd.	Anderson	901	678-3998
B&D jigsaw, 12-in. blade	109.92	ORDVA, Inc.	Hakford	615	898-1234
B&D jigsaw, 8-in. blade	99.87	ORDVA, Inc.	Hakford	615	898-1234
Hicut chain saw, 16 in.	256.99	ORDVA, Inc.	Hakford	615	898-1234
Power painter, 15 psi., 3-nozzle	109.99	Rubicon Systems	Orton	904	456-0092
B&D cordless drill, 1/2-in.	38.95	Rubicon Systems	Orton	904	456-0092
Steel matting, 4'x8'x1/6", .5" mesh	119.95	Rubicon Systems	Orton	904	456-0092

SELECT   P_DESCRIPT, P_PRICE, V_NAME, V_CONTACT,
                 V_AREACODE, V_PHONE
FROM   PRODUCT, VENDOR
WHERE   PRODUCT.V_CODE = VENDOR.V_CODE;

Your output might be presented in a different order because the SQL command produces a listing in which the order of the rows is not relevant. In fact, you are likely to get a different order of the same listing the next time you execute the command. However, you can generate a more predictable list by using an ORDER BY clause:

```
SELECT          PRODUCT.P_DESCRIPT, PRODUCT.P_PRICE,
                VENDOR.V_NAME, VENDOR.V_CONTACT,
                VENDOR.V_AREACODE, VENDOR.V_PHONE
FROM            PRODUCT, VENDOR
WHERE           PRODUCT.V_CODE = VENDOR.V_CODE
ORDER BY        PRODUCT.P_PRICE;
```

In this case, your listing will always be arranged from the lowest price to the highest price.

Note

Table names were used as prefixes in the preceding SQL command sequence. For example, PRODUCT.P_PRICE was used rather than P_PRICE. Most current-generation RDBMSs do not require table names to be used as prefixes unless the same attribute name occurs in several of the tables being joined. In the previous case, V_CODE is used as a foreign key in PROD-UCT and as a primary key in VENDOR; therefore, you must use the table names as prefixes in the WHERE clause. In other words, you can write the previous query as:

```
SELECT P_DESCRIPT, P_PRICE, V_NAME, V_CONTACT, V_AREACODE,
V_PHONE
FROM PRODUCT, VENDOR WHERE PRODUCT.V_CODE = VENDOR.V_CODE ORDER BY
P_PRICE;
```

Naturally, if an attribute name occurs in several places, its origin (table) must be specified. If you fail to provide such a specification, SQL will generate an error message to indicate that you have been ambiguous about the attribute's origin.

The preceding SQL command sequence joins a row in the PRODUCT table with a row in the VENDOR table, in which the V_CODE values of these rows are the same, as indicated in the WHERE clause's condition. Because any vendor can deliver any number of ordered products, the PRODUCT table might contain multiple V_CODE entries for each V_CODE entry in the VENDOR table. In other words, each V_CODE in VENDOR can be matched with many V_CODE rows in PRODUCT.

If you do not specify the WHERE clause, the result will be the Cartesian product of PRODUCT and VENDOR. Because the PRODUCT table contains 16 rows and the VENDOR table contains 11 rows, the Cartesian product will yield a listing of (16 × 11) = 176 rows. (Each row in PRODUCT will be joined to each row in the VENDOR table.)

All of the SQL commands can be used on the joined tables. For example, the following command sequence is quite acceptable in SQL and produces the output shown in Figure 7.29.

```
SELECT          P_DESCRIPT, P_PRICE, V_NAME, V_CONTACT,
                V_AREACODE, V_PHONE
FROM            PRODUCT, VENDOR
WHERE           PRODUCT.V_CODE = VENDOR.V_CODE
AND             P_INDATE > '15-Jan-2016';
```

FIGURE 7.29 AN ORDERED AND LIMITED LISTING AFTER A JOIN

P_DESCRIPT	P_PRICE	V_NAME	V_CONTACT	V_AREACODE	V_PHONE
1.25-in. metal screw, 25	6.99	Bryson, Inc.	Smithson	615	223-3234
2.5-in. wd. screw, 50	8.45	D&E Supply	Singh	615	228-3245
Claw hammer	9.95	Bryson, Inc.	Smithson	615	223-3234
B&D cordless drill, 1/2-in.	38.95	Rubicon Systems	Orton	904	456-0092
Steel matting, 4'x8'x1/6", .5" mesh	119.95	Rubicon Systems	Orton	904	456-0092
Hicut chain saw, 16 in.	256.99	ORDVA, Inc.	Hakford	615	898-1234

When joining three or more tables, you need to specify a join condition for each pair of tables. The number of join conditions will always be $n-1$, where n represents the number of tables listed in the FROM clause. For example, if you have three tables, you must have two join conditions; if you have five tables, you must have four join conditions; and so on.

Remember, the join condition will match the foreign key of a table to the primary key of the related table. For example, using Figure 7.1, if you want to list the customer last name, invoice number, invoice date, and product descriptions for all invoices for customer 10014, you must type the following:

```
SELECT      CUS_LNAME, INVOICE.INV_NUMBER, INV_DATE, P_DESCRIPT
FROM        CUSTOMER, INVOICE, LINE, PRODUCT
WHERE       CUSTOMER.CUS_CODE = INVOICE.CUS_CODE
AND         INVOICE.INV_NUMBER = LINE.INV_NUMBER
AND         LINE.P_CODE = PRODUCT.P_CODE
AND         CUSTOMER.CUS_CODE = 10014
ORDER BY    INV_NUMBER;
```

Finally, be careful not to create circular join conditions. For example, if Table A is related to Table B, Table B is related to Table C, and Table C is also related to Table A, create only two join conditions: join A with B and B with C. Do not join C with A!

7-7a Joining Tables with an Alias

An alias may be used to identify the source table from which the data is taken. The aliases P and V are used to label the PRODUCT and VENDOR tables in the next command sequence. Any legal table name may be used as an alias. (Also notice that there are no table name prefixes because the attribute listing contains no duplicate names in the SELECT statement.)

```
SELECT      P_DESCRIPT, P_PRICE, V_NAME, V_CONTACT, V_AREACODE,
            V_PHONE
FROM        PRODUCT P, VENDOR V
WHERE       P.V_CODE = V.V_CODE
ORDER BY    P_PRICE;
```

7-7b Recursive Joins

An alias is especially useful when a table must be joined to itself in a **recursive query**. For example, suppose that you are working with the EMP table shown in Figure 7.30. Using the data in the EMP table, you can generate a list of all employees with their managers'

recursive query
A nested query that joins a table to itself.

FIGURE 7.30 THE CONTENTS OF THE EMP TABLE

EMP_NUM	EMP_TITLE	EMP_LNAME	EMP_FNAME	EMP_INITIAL	EMP_DOB	EMP_HIRE_DATE	EMP_AREACODE	EMP_PHONE	EMP_MGR
100	Mr.	Kolmycz	George	D	15-Jun-42	15-Mar-85	615	324-5456	
101	Ms.	Lewis	Rhonda	G	19-Mar-65	25-Apr-86	615	324-4472	100
102	Mr.	Vandam	Rhett		14-Nov-58	20-Dec-90	901	675-8993	100
103	Ms.	Jones	Anne	M	16-Oct-74	28-Aug-94	615	898-3456	100
104	Mr.	Lange	John	P	08-Nov-71	20-Oct-94	901	504-4430	105
105	Mr.	Williams	Robert	D	14-Mar-75	08-Nov-98	615	890-3220	
106	Mrs.	Smith	Jeanine	K	12-Feb-68	05-Jan-89	615	324-7883	105
107	Mr.	Diante	Jorge	D	21-Aug-74	02-Jul-94	615	890-4567	105
108	Mr.	Wiesenbach	Paul	R	14-Feb-66	18-Nov-92	615	897-4358	
109	Mr.	Smith	George	K	18-Jun-61	14-Apr-89	901	504-3339	108
110	Mrs.	Genkazi	Leighla	W	19-May-70	01-Dec-90	901	569-0093	108
111	Mr.	Washington	Rupert	E	03-Jan-66	21-Jun-93	615	890-4925	105
112	Mr	Johnson	Edward	E	14-May-61	01-Dec-83	615	898-4387	100
113	Ms.	Smythe	Melanie	P	15-Sep-70	11-May-99	615	324-9006	105
114	Ms.	Brandon	Marie	G	02-Nov-56	15-Nov-79	901	882-0845	108
115	Mrs.	Saranda	Hermine	R	25-Jul-72	23-Apr-93	615	324-5505	105
116	Mr.	Smith	George	A	08-Nov-65	10-Dec-88	615	890-2984	108

names by joining the EMP table to itself. In that case, you would also use aliases to differentiate the table from itself. The SQL command sequence would look like this:

```
SELECT      E.EMP_NUM, E.EMP_LNAME, E.EMP_MGR,
            M.EMP_LNAME
FROM        EMP E, EMP M
WHERE       E.EMP_MGR=M.EMP_NUM
ORDER BY    E.EMP_MGR;
```

Online Content

For a complete walk-through example of converting an ER model into a database structure and using SQL commands to create tables, see Appendix D, Converting the ER Model into a Database Structure at *www.cengagebrain.com*.

The output of the preceding command sequence is shown in Figure 7.31.

FIGURE 7.31 USING AN ALIAS TO JOIN A TABLE TO ITSELF

EMP_NUM	E.EMP_LNAME	EMP_MGR	M.EMP_LNAME
112	Johnson	100	Kolmycz
103	Jones	100	Kolmycz
102	Vandam	100	Kolmycz
101	Lewis	100	Kolmycz
115	Saranda	105	Williams
113	Smythe	105	Williams
111	Washington	105	Williams
107	Diante	105	Williams
106	Smith	105	Williams
104	Lange	105	Williams
116	Smith	108	Wiesenbach
114	Brandon	108	Wiesenbach
110	Genkazi	108	Wiesenbach
109	Smith	108	Wiesenbach

Note

In MS Access, you would add AS to the previous SQL command sequence. For example:

```
SELECT      E.EMP_NUM,E.EMP_LNAME,E.EMP_MGR,M.EMP_LNAME
FROM        EMP AS E, EMP AS M
WHERE       E.EMP_MGR = M.EMP_NUM
ORDER BY    E.EMP_MGR;
```

Summary

- SQL commands can be divided into two overall categories: data definition language (DDL) commands and data manipulation language (DML) commands.

- The ANSI standard data types are supported by all RDBMS vendors in different ways. The basic data types are NUMBER, NUMERIC, INTEGER, CHAR, VARCHAR, and DATE.

- The basic data definition commands allow you to create tables and indexes. Many SQL constraints can be used with columns. The commands are CREATE TABLE, CREATE INDEX, ALTER TABLE, DROP TABLE, and DROP INDEX.

- DML commands allow you to add, modify, and delete rows from tables. The basic DML commands are SELECT, INSERT, UPDATE, DELETE, COMMIT, and ROLLBACK.

- The INSERT command is used to add new rows to tables. The UPDATE command is used to modify data values in existing rows of a table. The DELETE command is used to delete rows from tables. The COMMIT and ROLLBACK commands are used to permanently save or roll back changes made to the rows. Once you COMMIT the changes, you cannot undo them with a ROLLBACK command.

- The SELECT statement is the main data retrieval command in SQL. A SELECT statement has the following syntax:

SELECT	*columnlist*
FROM	*tablelist*
[WHERE	*conditionlist*]
[GROUP BY	*columnlist*]
[HAVING	*conditionlist*]
[ORDER BY	*columnlist* [ASC | DESC]];

- The column list represents one or more column names separated by commas. The column list may also include computed columns, aliases, and aggregate functions. A computed column is represented by an expression or formula (for example, P_PRICE * P_QOH). The FROM clause contains a list of table names.

- The WHERE clause can be used with the SELECT, UPDATE, and DELETE statements to restrict the rows affected by the DDL command. The condition list represents one or more conditional expressions separated by logical operators (AND, OR, and NOT). The conditional expression can contain any comparison operators (=, >, <, >=, <=, and <>) as well as special operators (BETWEEN, IS NULL, LIKE, IN, and EXISTS).

- Aggregate functions (COUNT, MIN, MAX, and AVG) are special functions that perform arithmetic computations over a set of rows. The aggregate functions are usually used in conjunction with the GROUP BY clause to group the output of aggregate computations by one or more attributes. The HAVING clause is used to restrict the output of the GROUP BY clause by selecting only the aggregate rows that match a given condition.

- The ORDER BY clause is used to sort the output of a SELECT statement. The ORDER BY clause can sort by one or more columns and can use either ascending or descending order.

- You can join the output of multiple tables with the SELECT statement. The join operation is performed every time you specify two or more tables in the FROM clause and use a join condition in the WHERE clause to match the foreign key of one table to the primary key of the related table. If you do not specify a join condition, the DBMS will automatically perform a Cartesian product of the tables you specify in the FROM clause.

- The natural join uses the join condition to match only rows with equal values in the specified columns.

Key Terms

 Online Content

Flashcards and crossword puzzles for key term practice are available at *www.cengagebrain.com*.

alias	DROP INDEX	NOT
ALTER TABLE	DROP TABLE	OR
AND	EXISTS	ORDER BY
authentication	FROM	recursive query
AVG	GROUP BY	reserved words
BETWEEN	HAVING	ROLLBACK
Boolean algebra	IN	rules of precedence
cascading order sequence	inner query	schema
COMMIT	INSERT	SELECT
COUNT	IS NULL	subquery
CREATE INDEX	LIKE	SUM
CREATE TABLE	MAX	UPDATE
DELETE	MIN	WHERE
DISTINCT	nested query	wildcard character

Review Questions

1. In a SELECT query, what is the difference between a WHERE clause and a HAVING clause?

2. Explain why the following command would create an error and what changes could be made to fix the error:

   SELECT V_CODE, SUM(P_QOH) FROM PRODUCT;

3. What type of integrity is enforced when a primary key is declared?

4. Explain why it might be more appropriate to declare an attribute that contains only digits as a character data type instead of a numeric data type.

5. What is the difference between a column constraint and a table constraint?

6. What are "referential constraint actions"?

7. Rewrite the following WHERE clause without the use of the IN special operator:

   WHERE V_STATE IN ('TN', 'FL', 'GA')

8. Explain the difference between an ORDER BY clause and a GROUP BY clause.

9. Explain why the following two commands produce different results:

   SELECT DISTINCT COUNT (V_CODE) FROM PRODUCT;
   SELECT COUNT (DISTINCT V_CODE) FROM PRODUCT;

10. What is the difference between the COUNT aggregate function and the SUM aggregate function?

11. Explain why it would be preferable to use a DATE data type to store date data instead of a character data type.

12. What is a recursive join?

Online Content

Problems 1–25 are based on the Ch07_ConstructCo database at *www.cengagebrain.com*. This database is stored in Microsoft Access format. Oracle, MySQL, and MS SQL Server script files are available at *www.cengagebrain.com*.

Problems

The Ch07_ConstructCo database stores data for a consulting company that tracks all charges to projects. The charges are based on the hours each employee works on each project. The structure and contents of the Ch07_ConstructCo database are shown in Figure P7.1.

FIGURE P7.1 THE CH07_CONSTRUCTCO DATABASE

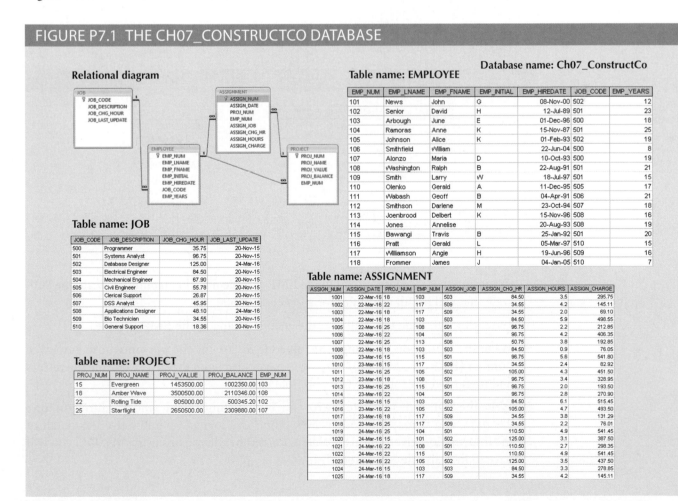

Database name: Ch07_ConstructCo

Relational diagram

Table name: JOB

JOB_CODE	JOB_DESCRIPTION	JOB_CHG_HOUR	JOB_LAST_UPDATE
500	Programmer	35.75	20-Nov-15
501	Systems Analyst	96.75	20-Nov-15
502	Database Designer	125.00	24-Mar-16
503	Electrical Engineer	84.50	20-Nov-15
504	Mechanical Engineer	67.90	20-Nov-15
505	Civil Engineer	55.78	20-Nov-15
506	Clerical Support	26.87	20-Nov-15
507	DSS Analyst	45.95	20-Nov-15
508	Applications Designer	48.10	24-Mar-16
509	Bio Technician	34.55	20-Nov-15
510	General Support	18.36	20-Nov-15

Table name: PROJECT

PROJ_NUM	PROJ_NAME	PROJ_VALUE	PROJ_BALANCE	EMP_NUM
15	Evergreen	1453500.00	1002350.00	103
18	Amber Wave	3500500.00	2110346.00	108
22	Rolling Tide	805000.00	500345.20	102
25	Starflight	2650500.00	2309880.00	107

Table name: EMPLOYEE

EMP_NUM	EMP_LNAME	EMP_FNAME	EMP_INITIAL	EMP_HIREDATE	JOB_CODE	EMP_YEARS
101	News	John	G	08-Nov-00	502	12
102	Senior	David	H	12-Jul-89	501	23
103	Arbough	June	E	01-Dec-96	500	18
104	Ramoras	Anne	K	15-Nov-87	501	25
105	Johnson	Alice	K	01-Feb-93	502	19
106	Smithfield	William		22-Jun-04	500	8
107	Alonzo	Maria	D	10-Oct-93	500	19
108	Washington	Ralph	B	22-Aug-91	501	21
109	Smith	Larry	W	18-Jul-97	501	15
110	Olenko	Gerald	A	11-Dec-95	505	17
111	Wabash	Geoff	B	04-Apr-91	506	21
112	Smithson	Darlene	M	23-Oct-94	507	18
113	Joenbrood	Delbert	K	15-Nov-96	508	16
114	Jones	Annelise		20-Aug-93	508	19
115	Bawangi	Travis	B	25-Jan-92	501	20
116	Pratt	Gerald	L	05-Mar-97	510	15
117	Williamson	Angie	H	19-Jun-96	509	16
118	Frommer	James	J	04-Jan-05	510	7

Table name: ASSIGNMENT

ASSIGN_NUM	ASSIGN_DATE	PROJ_NUM	EMP_NUM	ASSIGN_JOB	ASSIGN_CHG_HR	ASSIGN_HOURS	ASSIGN_CHARGE
1001	22-Mar-16	18	103	503	84.50	3.5	295.75
1002	22-Mar-16	22	117	509	34.55	4.2	145.11
1003	22-Mar-16	18	117	509	34.55	2.0	69.10
1004	22-Mar-16	18	103	503	84.50	5.9	498.55
1005	22-Mar-16	25	108	501	96.75	2.2	212.85
1006	22-Mar-16	22	104	501	96.75	4.2	406.35
1007	22-Mar-16	25	113	508	50.75	3.8	192.85
1008	22-Mar-16	18	103	503	84.50	0.9	76.05
1009	23-Mar-16	15	115	501	96.75	5.6	541.80
1010	23-Mar-16	15	117	509	34.55	2.4	82.92
1011	23-Mar-16	25	105	502	105.00	4.3	451.50
1012	23-Mar-16	18	108	501	96.75	3.4	328.95
1013	23-Mar-16	25	115	501	96.75	2.0	193.50
1014	23-Mar-16	22	104	501	96.75	2.8	270.90
1015	23-Mar-16	15	103	503	84.50	6.1	515.45
1016	23-Mar-16	22	105	502	105.00	4.7	493.50
1017	23-Mar-16	18	117	509	34.55	3.8	131.29
1018	23-Mar-16	25	117	509	34.55	2.2	76.01
1019	24-Mar-16	25	104	501	110.50	4.9	541.45
1020	24-Mar-16	15	101	502	125.00	3.1	387.50
1021	24-Mar-16	22	108	501	110.50	2.7	298.35
1022	24-Mar-16	22	115	501	110.50	4.9	541.45
1023	24-Mar-16	22	105	502	125.00	3.5	437.50
1024	24-Mar-16	15	103	503	84.50	3.3	278.85
1025	24-Mar-16	18	117	509	34.55	4.2	145.11

Note that the ASSIGNMENT table in Figure P7.1 stores the JOB_CHG_HOUR values as an attribute (ASSIGN_CHG_HR) to maintain historical accuracy of the data. The JOB_CHG_HOUR values are likely to change over time. In fact, a JOB_CHG_HOUR change will be reflected in the ASSIGNMENT table. Naturally, the employee primary job assignment might also change, so the ASSIGN_JOB is also stored. Because those attributes are required to maintain the historical accuracy of the data, they are *not* redundant.

Given the structure and contents of the Ch07_ConstructCo database shown in Figure P7.1, use SQL commands to answer Problems 1–25.

1. Write the SQL code that will create the table structure for a table named EMP_1. This table is a subset of the EMPLOYEE table. The basic EMP_1 table structure is summarized in the following table. (Note that the JOB_CODE is the FK to JOB.)

ATTRIBUTE (FIELD) NAME	DATA DECLARATION
EMP_NUM	CHAR(3)
EMP_LNAME	VARCHAR(15)
EMP_FNAME	VARCHAR(15)
EMP_INITIAL	CHAR(1)
EMP_HIREDATE	DATE
JOB_CODE	CHAR(3)

2. Having created the table structure in Problem 1, write the SQL code to enter the first two rows for the table shown in Figure P7.2.

FIGURE P7.2   THE CONTENTS OF THE EMP_1 TABLE

EMP_NUM	EMP_LNAME	EMP_FNAME	EMP_INITIAL	EMP_HIREDATE	JOB_CODE
101	News	John	G	08-Nov-00	502
102	Senior	David	H	12-Jul-89	501
103	Arbough	June	E	01-Dec-96	500
104	Ramoras	Anne	K	15-Nov-87	501
105	Johnson	Alice	K	01-Feb-93	502
106	Smithfield	William		22-Jun-04	500
107	Alonzo	Maria	D	10-Oct-93	500
108	Washington	Ralph	B	22-Aug-91	501
109	Smith	Larry	W	18-Jul-97	501

3. Assuming that the data shown in the EMP_1 table have been entered, write the SQL code that will list all attributes for a job code of 502.

4. Write the SQL code that will save the changes made to the EMP_1 table.

5. Write the SQL code to change the job code to 501 for the person whose employee number (EMP_NUM) is 107. After you have completed the task, examine the results and then reset the job code to its original value.

6. Write the SQL code to delete the row for William Smithfield, who was hired on June 22, 2004, and whose job code is 500. (*Hint:* Use logical operators to include all of the information given in this problem. Remember, if you are using MySQL, you will have to first disable "safe mode.")

7. Write the SQL code that will restore the data to its original status; that is, the table should contain the data that existed before you made the changes in Problems 5 and 6.

8. Write the SQL code to create a copy of EMP_1, naming the copy EMP_2. Then write the SQL code that will add the attributes EMP_PCT and PROJ_NUM to the structure. The EMP_PCT is the bonus percentage to be paid to each employee. The new attribute characteristics are:

EMP_PCT   NUMBER(4,2)

PROJ_NUM   CHAR(3)

[*Note:* If your SQL implementation allows it, you may use DECIMAL(4,2) or NUMERIC(4,2) rather than NUMBER(4,2).]

9. Write the SQL code to change the EMP_PCT value to 3.85 for the person whose employee number (EMP_NUM) is 103. Next, write the SQL command sequences to change the EMP_PCT values, as shown in Figure P7.9.

FIGURE P7.9 THE EMP_2 TABLE AFTER THE MODIFICATIONS

EMP_NUM	EMP_LNAME	EMP_FNAME	EMP_INITIAL	EMP_HIREDATE	JOB_CODE	EMP_PCT	PROJ_NUM
101	News	John	G	08-Nov-00	502	5.00	
102	Senior	David	H	12-Jul-89	501	8.00	
103	Arbough	June	E	01-Dec-96	500	3.85	
104	Ramoras	Anne	K	15-Nov-87	501	10.00	
105	Johnson	Alice	K	01-Feb-93	502	5.00	
106	Smithfield	William		22-Jun-04	500	6.20	
107	Alonzo	Maria	D	10-Oct-93	500	5.15	
108	Washington	Ralph	B	22-Aug-91	501	10.00	
109	Smith	Larry	W	18-Jul-97	501	2.00	

10. Using a single command sequence, write the SQL code that will change the project number (PROJ_NUM) to 18 for all employees whose job classification (JOB_CODE) is 500.

11. Using a single command sequence, write the SQL code that will change the project number (PROJ_NUM) to 25 for all employees whose job classification (JOB_CODE) is 502 or higher. When you finish Problems 10 and 11, the EMP_2 table will contain the data shown in Figure P7.11. (You may assume that the table has been saved again at this point.)

FIGURE P7.11 THE EMP_2 TABLE CONTENTS AFTER THE MODIFICATIONS

EMP_NUM	EMP_LNAME	EMP_FNAME	EMP_INITIAL	EMP_HIREDATE	JOB_CODE	EMP_PCT	PROJ_NUM
101	News	John	G	08-Nov-00	502	5.00	25
102	Senior	David	H	12-Jul-89	501	8.00	
103	Arbough	June	E	01-Dec-96	500	3.85	18
104	Ramoras	Anne	K	15-Nov-87	501	10.00	
105	Johnson	Alice	K	01-Feb-93	502	5.00	25
106	Smithfield	William		22-Jun-04	500	6.20	18
107	Alonzo	Maria	D	10-Oct-93	500	5.15	18
108	Washington	Ralph	B	22-Aug-91	501	10.00	
109	Smith	Larry	W	18-Jul-97	501	2.00	

12. Write the SQL code that will change the PROJ_NUM to 14 for employees who were hired before January 1, 1994, and whose job code is at least 501. (You may assume that the table will be restored to its condition preceding this question.)

13. Write the two SQL command sequences required to:

    a. Create a temporary table named TEMP_1 whose structure is composed of the EMP_2 attributes EMP_NUM and EMP_PCT.

    b. Copy the matching EMP_2 values into the TEMP_1 table.

14. Write the SQL command that will delete the newly created TEMP_1 table from the database.

15. Write the SQL code required to list all employees whose last names start with *Smith*. In other words, the rows for both Smith and Smithfield should be included in the listing. Assume case sensitivity.

16. Using the EMPLOYEE, JOB, and PROJECT tables in the Ch07_ConstructCo database (see Figure P7.1), write the SQL code that will produce the results shown in Figure P7.16.

FIGURE P7.16 THE QUERY RESULTS FOR PROBLEM 16

PROJ_NAME	PROJ_VALUE	PROJ_BALANCE	EMP_LNAME	EMP_FNAME	EMP_INITIAL	JOB_CODE	JOB_DESCRIPTION	JOB_CHG_HOUR
Rolling Tide	805000.00	500345.20	Senior	David	H	501	Systems Analyst	96.75
Evergreen	1453500.00	1002350.00	Arbough	June	E	500	Programmer	35.75
Starflight	2650500.00	2309880.00	Alonzo	Maria	D	500	Programmer	35.75
Amber Wave	3500500.00	2110346.00	Washington	Ralph	B	501	Systems Analyst	96.75

17. Write the SQL code that will produce the same information that was shown in Problem 16, but sorted by the employee's last name.

18. Write the SQL code to find the average bonus percentage in the EMP_2 table you created in Problem 8.

19. Write the SQL code that will produce a listing for the data in the EMP_2 table in ascending order by the bonus percentage.

20. Write the SQL code that will list only the distinct project numbers in the EMP_2 table.

21. Write the SQL code to calculate the ASSIGN_CHARGE values in the ASSIGNMENT table in the Ch07_ConstructCo database. (See Figure P7.1.) Note that ASSIGN_CHARGE is a derived attribute that is calculated by multiplying ASSIGN_CHG_HR by ASSIGN_HOURS.

22. Using the data in the ASSIGNMENT table, write the SQL code that will yield the total number of hours worked for each employee and the total charges stemming from those hours worked. The results of running that query are shown in Figure P7.22.

FIGURE P7.22 TOTAL HOURS AND CHARGES BY EMPLOYEE

EMP_NUM	EMP_LNAME	SumOfASSIGN_HOURS	SumOfASSIGN_CHARGE
101	News	3.1	387.50
103	Arbough	19.7	1664.65
104	Ramoras	11.9	1218.70
105	Johnson	12.5	1382.50
108	Washington	8.3	840.15
113	Joenbrood	3.8	192.85
115	Bawangi	12.5	1276.75
117	Williamson	18.8	649.54

23. Write a query to produce the total number of hours and charges for each of the projects represented in the ASSIGNMENT table. The output is shown in Figure P7.23.

FIGURE P7.23 TOTAL HOUR AND CHARGES BY PROJECT

PROJ_NUM	SumOfASSIGN_HOURS	SumOfASSIGN_CHARGE
15	20.5	1806.52
18	23.7	1544.80
22	27.0	2593.16
25	19.4	1668.16

24. Write the SQL code to generate the total hours worked and the total charges made by all employees. The results are shown in Figure P7.24. (*Hint:* This is a nested query. If you use Microsoft Access, you can use the output shown in Figure P7.22 as the basis for the query that will produce the output shown in Figure P7.24.)

FIGURE P7.24 TOTAL HOURS AND CHARGES, ALL EMPLOYEES

SumOfSumOfASSIGN_HOURS	SumOfSumOfASSIGN_CHARGE
90.6	7612.64

25. Write the SQL code to generate the total hours worked and the total charges made to all projects. The results should be the same as those shown in Figure P7.24. (*Hint:* This is a nested query. If you use Microsoft Access, you can use the output shown in Figure P7.23 as the basis for this query.)

    The structure and contents of the Ch07_SaleCo database are shown in Figure P7.26. Use this database to answer the following problems. Save each query as Q*XX*, where *XX* is the problem number.

26. Write a query to count the number of invoices.

27. Write a query to count the number of customers with a balance of more than $500.

28. Generate a listing of all purchases made by the customers, using the output shown in Figure P7.28 as your guide. (*Hint:* Use the ORDER BY clause to order the resulting rows shown in Figure P7.28.)

Online Content

Problems 26–43 are based on the Ch07_SaleCo database, which is available at *www. cengagebrain.com*. This database is stored in Microsoft Access format. Oracle, MySQL, and MS SQL Server script files are available at *www.cengage brain.com*.

FIGURE P7.26 THE CH07_SALECO DATABASE

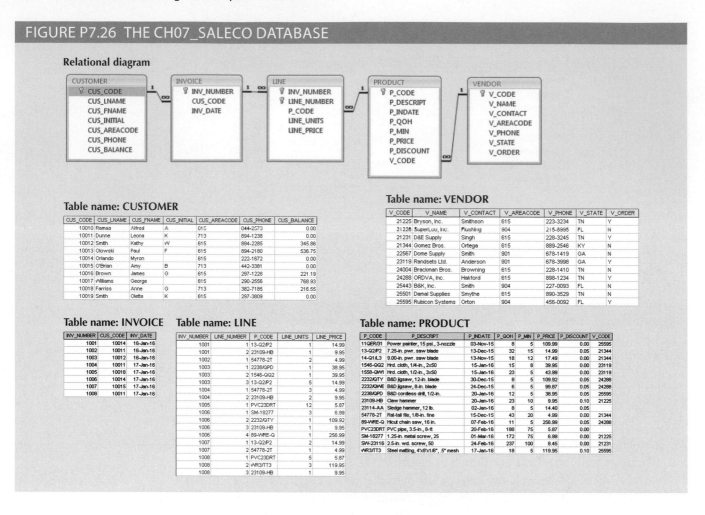

Relational diagram

Table name: CUSTOMER

CUS_CODE	CUS_LNAME	CUS_FNAME	CUS_INITIAL	CUS_AREACODE	CUS_PHONE	CUS_BALANCE
10010	Ramas	Alfred	A	015	044-2573	0.00
10011	Dunne	Leona	K	713	894-1238	0.00
10012	Smith	Kathy	W	615	894-2285	345.86
10013	Olowski	Paul	F	615	894-2180	536.75
10014	Orlando	Myron		615	222-1672	0.00
10015	O'Brian	Amy	B	713	442-3381	0.00
10016	Brown	James	G	615	297-1228	221.19
10017	Williams	George		615	290-2556	768.93
10018	Farriss	Anne	G	713	382-7185	216.55
10019	Smith	Olette	K	615	297-3809	0.00

Table name: VENDOR

V_CODE	V_NAME	V_CONTACT	V_AREACODE	V_PHONE	V_STATE	V_ORDER
21225	Bryson, Inc.	Smithson	615	223-3234	TN	Y
21226	SuperLoo, Inc.	Flushing	904	215-8995	FL	N
21231	D&E Supply	Singh	615	228-3245	TN	Y
21344	Gomez Bros.	Ortega	615	889-2546	KY	N
22567	Dome Supply	Smith	901	678-1419	GA	N
23119	Randsets Ltd.	Anderson	901	678-3998	GA	Y
24004	Brackman Bros.	Browning	615	228-1410	TN	N
24288	ORDVA, Inc.	Hakford	615	898-1234	TN	Y
25443	B&K, Inc.	Smith	904	227-0093	FL	N
25501	Damal Supplies	Smythe	615	890-3529	TN	N
25595	Rubicon Systems	Orton	904	456-0092	FL	Y

Table name: INVOICE

INV_NUMBER	CUS_CODE	INV_DATE
1001	10014	16-Jan-16
1002	10011	16-Jan-16
1003	10012	16-Jan-16
1004	10011	17-Jan-16
1005	10018	17-Jan-16
1006	10014	17-Jan-16
1007	10015	17-Jan-16
1008	10011	17-Jan-16

Table name: LINE

INV_NUMBER	LINE_NUMBER	P_CODE	LINE_UNITS	LINE_PRICE
1001	1	13-Q2/P2	1	14.99
1001	2	23109-HB	1	9.95
1002	1	54778-2T	2	4.99
1003	1	2238/QPD	1	38.95
1003	2	1546-QQ2	1	39.95
1003	3	13-Q2/P2	5	14.99
1004	1	54778-2T	3	4.99
1004	2	23109-HB	2	9.95
1005	1	PVC23DRT	12	5.87
1006	1	SM-18277	3	6.99
1006	2	2232/QTY	1	109.92
1006	3	23109-HB	1	9.95
1006	4	89-WRE-Q	1	256.99
1007	1	13-Q2/P2	2	14.99
1007	2	54778-2T	1	4.99
1008	1	PVC23DRT	5	5.87
1008	2	WR3/TT3	3	119.95
1008	3	23109-HB	1	9.95

Table name: PRODUCT

P_CODE	P_DESCRIPT	P_INDATE	P_QOH	P_MIN	P_PRICE	P_DISCOUNT	V_CODE
11QER/31	Power painter, 15 psi., 3-nozzle	03-Nov-15	8	5	109.99	0.00	25595
13-Q2/P2	7.25-in. pwr. saw blade	13-Dec-15	32	15	14.99	0.05	21344
14-Q1/L3	9.00-in. pwr. saw blade	13-Nov-15	18	12	17.49	0.00	21344
1546-QQ2	Hrd. cloth, 1/4-in., 2x50	15-Jan-16	15	8	39.95	0.00	23119
1558-QW1	Hrd. cloth, 1/2-in., 3x50	15-Jan-16	23	5	43.99	0.00	23119
2232/QTY	B&D jigsaw, 12-in. blade	30-Dec-15	8	5	109.92	0.05	24288
2232/QWE	B&D jigsaw, 8-in. blade	24-Dec-15	6	5	99.87	0.05	24288
2238/QPD	B&D cordless drill, 1/2-in.	20-Jan-16	12	5	38.95	0.05	25595
23109-HB	Claw hammer	20-Jan-16	23	10	9.95	0.10	21225
23114-AA	Sledge hammer, 12 lb.	02-Jan-16	8	5	14.40	0.05	
54778-2T	Rat-tail file, 1/8-in. fine	15-Dec-15	43	20	4.99	0.00	21344
89-WRE-Q	Hicut chain saw, 16 in.	07-Feb-16	11	5	256.99	0.05	24288
PVC23DRT	PVC pipe, 3.5-in., 8-ft	20-Feb-16	188	75	5.87	0.00	
SM-18277	1.25-in. metal screw, 25	01-Mar-16	172	75	6.99	0.00	21225
SW-23116	2.5-in. wd. screw, 50	24-Feb-16	237	100	8.45	0.00	21231
WR3/TT3	Steel matting, 4'x8'x1/6", .5" mesh	17-Jan-16	18	5	119.95	0.10	25595

FIGURE P7.28 LIST OF CUSTOMER PURCHASES

CUS_CODE	INV_NUMBER	INV_DATE	P_DESCRIPT	LINE_UNITS	LINE_PRICE
10011	1002	16-Jan-16	Rat-tail file, 1/8-in. fine	2	4.99
10011	1004	17-Jan-16	Claw hammer	2	9.95
10011	1004	17-Jan-16	Rat-tail file, 1/8-in. fine	3	4.99
10011	1008	17-Jan-16	Claw hammer	1	9.95
10011	1008	17-Jan-16	PVC pipe, 3.5-in., 8-ft	5	5.87
10011	1008	17-Jan-16	Steel matting, 4'x8'x1/6", .5" mesh	3	119.95
10012	1003	16-Jan-16	7.25-in. pwr. saw blade	5	14.99
10012	1003	16-Jan-16	B&D cordless drill, 1/2-in.	1	38.95
10012	1003	16-Jan-16	Hrd. cloth, 1/4-in., 2x50	1	39.95
10014	1001	16-Jan-16	7.25-in. pwr. saw blade	1	14.99
10014	1001	16-Jan-16	Claw hammer	1	9.95
10014	1006	17-Jan-16	1.25-in. metal screw, 25	3	6.99
10014	1006	17-Jan-16	B&D jigsaw, 12-in. blade	1	109.92
10014	1006	17-Jan-16	Claw hammer	1	9.95
10014	1006	17-Jan-16	Hicut chain saw, 16 in.	1	256.99
10015	1007	17-Jan-16	7.25-in. pwr. saw blade	2	14.99
10015	1007	17-Jan-16	Rat-tail file, 1/8-in. fine	1	4.99
10018	1005	17-Jan-16	PVC pipe, 3.5-in., 8-ft	12	5.87

29. Using the output shown in Figure P7.29 as your guide, generate a list of customer purchases, including the subtotals for each of the invoice line numbers. (*Hint:* Modify the query format used to produce the list of customer purchases in Problem 28, delete the INV_DATE column, and add the derived attribute LINE_UNITS * LINE_PRICE to calculate the subtotals.)

FIGURE P7.29 SUMMARY OF CUSTOMER PURCHASES WITH SUBTOTALS

CUS_CODE	INV_NUMBER	P_DESCRIPT	Units Bought	Unit Price	Subtotal
10011	1002	Rat-tail file, 1/8-in. fine	2	4.99	9.98
10011	1004	Claw hammer	2	9.95	19.90
10011	1004	Rat-tail file, 1/8-in. fine	3	4.99	14.97
10011	1008	Claw hammer	1	9.95	9.95
10011	1008	PVC pipe, 3.5-in., 8-ft	5	5.87	29.35
10011	1008	Steel matting, 4'x8'x1/6", .5" mesh	3	119.95	359.85
10012	1003	7.25-in. pwr. saw blade	5	14.99	74.95
10012	1003	B&D cordless drill, 1/2-in.	1	38.95	38.95
10012	1003	Hrd. cloth, 1/4-in., 2x50	1	39.95	39.95
10014	1001	7.25-in. pwr. saw blade	1	14.99	14.99
10014	1001	Claw hammer	1	9.95	9.95
10014	1006	1.25-in. metal screw, 25	3	6.99	20.97
10014	1006	B&D jigsaw, 12-in. blade	1	109.92	109.92
10014	1006	Claw hammer	1	9.95	9.95
10014	1006	Hicut chain saw, 16 in.	1	256.99	256.99
10015	1007	7.25-in. pwr. saw blade	2	14.99	29.98
10015	1007	Rat-tail file, 1/8-in. fine	1	4.99	4.99
10018	1005	PVC pipe, 3.5-in., 8-ft	12	5.87	70.44

30. Modify the query used in Problem 29 to produce the summary shown in Figure P7.30.

FIGURE P7.30 CUSTOMER PURCHASE SUMMARY

CUS_CODE	CUS_BALANCE	Total Purchases
10011	0.00	444.00
10012	345.86	153.85
10014	0.00	422.77
10015	0.00	34.97
10018	216.55	70.44

31. Modify the query in Problem 30 to include the number of individual product purchases made by each customer. (In other words, if the customer's invoice is based on three products, one per LINE_NUMBER, you count three product purchases. Note that in the original invoice data, customer 10011 generated three invoices, which contained a total of six lines, each representing a product purchase.) Your output values must match those shown in Figure P7.31.

FIGURE P7.31 CUSTOMER TOTAL PURCHASE AMOUNTS AND NUMBER OF PURCHASES

CUS_CODE	CUS_BALANCE	Total Purchases	Number of Purchases
10011	0.00	444.00	6
10012	345.86	153.85	3
10014	0.00	422.77	6
10015	0.00	34.97	2
10018	216.55	70.44	1

32. Use a query to compute the average purchase amount per product made by each customer. (*Hint:* Use the results of Problem 31 as the basis for this query.) Your output values must match those shown in Figure P7.32. Note that the average purchase amount is equal to the total purchases divided by the number of purchases per customer.

FIGURE P7.32 AVERAGE PURCHASE AMOUNT BY CUSTOMER

CUS_CODE	CUS_BALANCE	Total Purchases	Number of Purchases	Average Purchase Amount
10011	0.00	444.00	6	74.00
10012	345.86	153.85	3	51.28
10014	0.00	422.77	6	70.46
10015	0.00	34.97	2	17.48
10018	216.55	70.44	1	70.44

33. Create a query to produce the total purchase per invoice, generating the results shown in Figure P7.33. The invoice total is the sum of the product purchases in the LINE that corresponds to the INVOICE.

FIGURE P7.33 INVOICE TOTALS

INV_NUMBER	Invoice Total
1001	24.94
1002	9.98
1003	153.85
1004	34.87
1005	70.44
1006	397.83
1007	34.97
1008	399.15

34. Use a query to show the invoices and invoice totals in Figure P7.34. (*Hint:* Group by the CUS_CODE.)

FIGURE P7.34 INVOICE TOTALS BY CUSTOMER

CUS_CODE	INV_NUMBER	Invoice Total
10011	1002	9.98
10011	1004	34.87
10011	1008	399.15
10012	1003	153.85
10014	1001	24.94
10014	1006	397.83
10015	1007	34.97
10018	1005	70.44

35. Write a query to produce the number of invoices and the total purchase amounts by customer, using the output shown in Figure P7.35 as your guide. (Compare this summary to the results shown in Problem 34.)

FIGURE P7.35 NUMBER OF INVOICES AND TOTAL PURCHASE AMOUNTS BY CUSTOMER

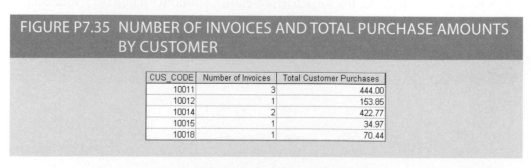

CUS_CODE	Number of Invoices	Total Customer Purchases
10011	3	444.00
10012	1	153.85
10014	2	422.77
10015	1	34.97
10018	1	70.44

36. Using the query results in Problem 35 as your basis, write a query to generate the total number of invoices, the invoice total for all of the invoices, the smallest of the customer purchase amounts, the largest of the customer purchase amounts, and the average of all the customer purchase amounts. (*Hint:* Check the figure output in Problem 35.) Your output must match Figure P7.36.

FIGURE P7.36 NUMBER OF INVOICES, INVOICE TOTALS, MINIMUM, MAXIMUM, AND AVERAGE SALES

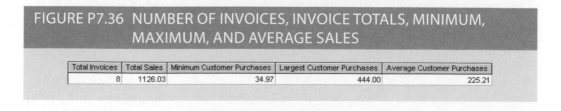

Total Invoices	Total Sales	Minimum Customer Purchases	Largest Customer Purchases	Average Customer Purchases
8	1126.03	34.97	444.00	225.21

37. List the balances of customers who have made purchases during the current invoice cycle—that is, for the customers who appear in the INVOICE table. The results of this query are shown in Figure P7.37.

FIGURE P7.37 BALANCES FOR CUSTOMERS WHO MADE PURCHASES

CUS_CODE	CUS_BALANCE
10011	0.00
10012	345.86
10014	0.00
10015	0.00
10018	216.55

38. Using the results of the query created in Problem 37, provide a summary of customer balance characteristics, as shown in Figure P7.38.

FIGURE P7.38 BALANCE SUMMARY OF CUSTOMERS WHO MADE PURCHASES

Minimum Balance	Maximum Balance	Average Balance
0	345.86	112.48

39. Create a query to find the balance characteristics for all customers, including the total of the outstanding balances. The results of this query are shown in Figure P7.39.

FIGURE P7.39 BALANCE SUMMARY FOR ALL CUSTOMERS

Total Balances	Minimum Balance	Maximum Balance	Average Balance
2089.28	0.00	768.93	208.93

40. Find the listing of customers who did not make purchases during the invoicing period. Your output must match the output shown in Figure P7.40.

FIGURE P7.40 BALANCES OF CUSTOMERS WHO DID NOT MAKE PURCHASES

CUS_CODE	CUS_BALANCE
10010	0.00
10013	536.75
10016	221.19
10017	768.93
10019	0.00

41. Find the customer balance summary for all customers who have not made purchases during the current invoicing period. The results are shown in Figure P7.41.

FIGURE P7.41 SUMMARY OF CUSTOMER BALANCES FOR CUSTOMERS WHO DID NOT MAKE PURCHASES

Total Balance	Minimum Balance	Maximum Balance	Average Balance
1526.87	0.00	768.93	305.37

42. Create a query that summarizes the value of products currently in inventory. Note that the value of each product is a result of multiplying the units currently in inventory by the unit price. Use the ORDER BY clause to match the order shown in Figure P7.42.

FIGURE P7.42 VALUE OF PRODUCTS CURRENTLY IN INVENTORY

P_DESCRIPT	P_QOH	P_PRICE	Subtotal
Hicut chain saw, 16 in.	11	256.99	2826.89
Steel matting, 4'x8'x1/6", .5" mesh	18	119.95	2159.10
2.5-in. wd. screw, 50	237	8.45	2002.65
1.25-in. metal screw, 25	172	6.99	1202.28
PVC pipe, 3.5-in., 8-ft	188	5.87	1103.56
Hrd. cloth, 1/2-in., 3x50	23	43.99	1011.77
Power painter, 15 psi., 3-nozzle	8	109.99	879.92
B&D jigsaw, 12-in. blade	8	109.92	879.36
Hrd. cloth, 1/4-in., 2x50	15	39.95	599.25
B&D jigsaw, 8-in. blade	6	99.87	599.22
7.25-in. pwr. saw blade	32	14.99	479.68
B&D cordless drill, 1/2-in.	12	38.95	467.40
9.00-in. pwr. saw blade	18	17.49	314.82
Claw hammer	23	9.95	228.85
Rat-tail file, 1/8-in. fine	43	4.99	214.57
Sledge hammer, 12 lb.	8	14.40	115.20

43. Using the results of the query created in Problem 42, find the total value of the product inventory. The results are shown in Figure P7.43.

FIGURE P7.43 TOTAL VALUE OF ALL PRODUCTS IN INVENTORY

Total Value of Inventory
15084.52

Online Content

Problems 44–64 are based on the Ch07_LargeCo database, which is available at *www.cengagebrain.com*. This database is stored in Microsoft Access format. Oracle, MySQL, and MS SQL Server script files are available at *www.cengagebrain. com*.

The Ch07_LargeCo database (see Figure P7.44) stores data for a company that sells paint products. The company tracks the sale of products to customers. The database keeps data on customers (LGCUSTOMER), sales (LGINVOICE), products (LGPRODUCT), which products are on which invoices (LGLINE), employees (LGEMPLOYEE), the salary history of each employee (LGSALARY_ HISTORY), departments (LGDEPARTMENT), product brands (LGBRAND), vendors (LGVENDOR), and which vendors supply each product (LGSUPPLIES). Some of the tables contain only a few rows of data, while other tables are quite large; for example, there are only eight departments, but more than 3,300 invoices containing over 11,000 invoice lines. For Problems 45–64, a figure of the correct output for each problem is provided. If the output of the query is very large, only the first several rows of the output are shown.

FIGURE P7.44   THE CH07_LARGECO ERD

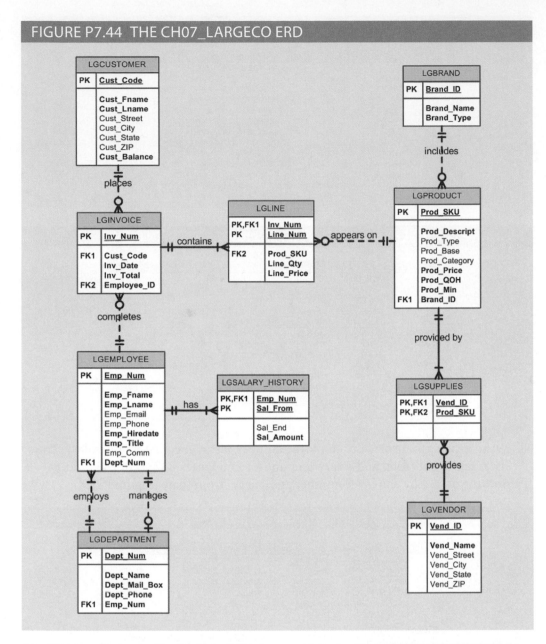

44. Write a query to display the eight departments in the LGDEPARTMENT table.

45. Write a query to display the SKU (stock keeping unit), description, type, base, category, and price for all products that have a PROD_BASE of water and a PROD_CATEGORY of sealer (Figure P7.45).

FIGURE P7.45   WATER-BASED SEALERS

PROD_SKU	PROD_DESCRIPT	PROD_TYPE	PROD_BASE	PROD_CATEGORY	PROD_PRICE
1403-TUY	Sealer, Water Based, for Concrete Floors	Interior	Water	Sealer	42.99

46. Write a query to display the first name, last name, and email address of employees hired from January 1, 2003, to December 31, 2012. Sort the output by last name and then by first name (Figure P7.46).

47. Write a query to display the first name, last name, phone number, title, and department number of employees who work in department 300 or have the title "CLERK I." Sort the output by last name and then by first name (Figure P7.47).

FIGURE P7.46 EMPLOYEES HIRED FROM 2003 – 2012

EMP_FNAME	EMP_LNAME	EMP_EMAIL
SUE	ASH	S.ASH98@LGCOMPANY.COM
ALIDA	BLACKWELL	A.BLACKW99@LGCOMPANY.COM
PHOEBE	BLEDSOE	P.BLEDSO99@LGCOMPANY.COM
VALARIE	BLEDSOE	V.BLEDSO99@LGCOMPANY.COM
WILFORD	BURGOS	W.BURGOS6@LGCOMPANY.COM
KASEY	CASH	K.CASH0@LGCOMPANY.COM
DANICA	CASTLE	C.DANICA99@LGCOMPANY.COM
DOUG	CAUDILL	C.DOUG0@LGCOMPANY.COM
LUCIO	CAUDILL	L.CAUDIL4@LGCOMPANY.COM
HANNAH	COLEMAN	H.COLEMA7@LGCOMPANY.COM
PHILLIS	CONKLIN	P.CONKLI4@LGCOMPANY.COM
LEE	CONNOR	L.CONNOR99@LGCOMPANY.COM

FIGURE P7.47 CLERKS AND EMPLOYEES IN DEPARTMENT 300

EMP_FNAME	EMP_LNAME	EMP_PHONE	EMP_TITLE	DEPT_NUM
LAVINA	ACEVEDO	862-6787	ASSOCIATE	300
LAUREN	AVERY	550-2270	SENIOR ASSOCIATE	300
ROSALBA	BAKER	632-8197	ASSOCIATE	300
FERN	CARPENTER	735-4820	PURCHASING SPECIALIST	300
LEEANN	CLINTON	616-9615	CLERK I	600
TANIKA	CRANE	449-6336	PURCHASING SPECIALIST	300
SAMMY	DIGGS	525-2101	SENIOR ASSOCIATE	300
LANA	DOWDY	471-8795	SENIOR ASSOCIATE	300
STEPHAINE	DUNLAP	618-8203	BUYER - RAW MATERIALS	300
HAL	FISHER	676-3662	SENIOR ASSOCIATE	300
LINDSAY	GOOD	337-9570	CLERK I	600
LEEANN	HORN	828-4361	SENIOR ASSOCIATE	300

48. Write a query to display the employee number, last name, first name, salary "from" date, salary end date, and salary amount for employees 83731, 83745, and 84039. Sort the output by employee number and salary "from" date (Figure P7.48).

FIGURE P7.48 SALARY HISTORY FOR SELECTED EMPLOYEES

EMP_NUM	EMP_LNAME	EMP_FNAME	SAL_FROM	SAL_END	SAL_AMOUNT
83731	VARGAS	SHERON	7/15/2012	7/14/2013	43740
83731	VARGAS	SHERON	7/14/2013	7/13/2014	48110
83731	VARGAS	SHERON	7/14/2014	7/14/2015	49550
83731	VARGAS	SHERON	7/15/2015		51040
83745	SPICER	DWAIN	8/2/2009	8/1/2010	56020
83745	SPICER	DWAIN	8/2/2010	8/2/2011	57700
83745	SPICER	DWAIN	8/3/2011	8/1/2012	63470
83745	SPICER	DWAIN	8/2/2012	8/1/2013	68550
83745	SPICER	DWAIN	8/1/2013	7/31/2014	71980
83745	SPICER	DWAIN	8/1/2014	8/1/2015	74140
83745	SPICER	DWAIN	8/2/2015		76360
84039	COLEMAN	HANNAH	6/28/2012	6/27/2013	47380
84039	COLEMAN	HANNAH	6/27/2013	6/26/2014	51170
84039	COLEMAN	HANNAH	6/27/2014	6/27/2015	52700
84039	COLEMAN	HANNAH	6/28/2015		54280

49. Write a query to display the first name, last name, street, city, state, and zip code of any customer who purchased a Foresters Best brand top coat between July 15, 2015, and July 31, 2015. If a customer purchased more than one such product, display the customer's information only once in the output. Sort the output by state, last name, and then first name (Figure P7.49).

50. Write a query to display the employee number, last name, email address, title, and department name of each employee whose job title ends in the word "ASSOCIATE." Sort the output by department name and employee title (Figure P7.50).

51. Write a query to display a brand name and the number of products of that brand that are in the database. Sort the output by the brand name (Figure P7.51).

FIGURE P7.49 CUSTOMERS WHO PURCHASED FORESTERS BEST TOP COAT

CUST_FNAME	CUST_LNAME	CUST_STREET	CUST_CITY	CUST_STATE	CUST_ZIP
LUPE	SANTANA	1292 WEST 70TH PLACE	Phenix City	AL	36867
HOLLIS	STILES	1493 DOLLY MADISON CIRCLE	Snow Hill	AL	36778
LISETTE	WHITTAKER	339 NORTHPARK DRIVE	Montgomery	AL	36197
DEANDRE	JAMISON	1571 HANES STREET	Miami	FL	33169
CATHLEEN	WHITMAN	1712 NORTHFIELD DRIVE	Marshallville	GA	31057
SHERIE	STOVER	640 MOUNTAIN VIEW DRIVE	Parksville	KY	40464
BRYCE	HOGAN	1860 IMLACH DRIVE	Newbury	MA	01951
SHELBY	SALAS	486 SUSITNA VIEW COURT	North Tisbury	MA	02568
JERMAINE	HANCOCK	1627 SAUNDERS ROAD	Ellicott City	MD	21041
WHITNEY	WHITFIELD	1259 RHONE STREET	Phippsburg	ME	04567
MONROE	ALLISON	272 SCHODDE STREET	Kalamazoo	MI	49002
DARLEEN	PARRA	561 COLLIE HILL WAY	Madison	MS	39130
CLINTON	AGUIRRE	1651 VANGUARD DRIVE	Franklinville	NC	27248
TOMMIE	PALMER	933 ELCADORE CIRCLE	Arapahoe	NC	28510
JEFFEREY	MCBRIDE	1043 ROCKRIDGE DRIVE	Glenwood	NJ	07418
SIDNEY	GARZA	772 SHEPPARD DRIVE	Fair Harbor	NY	11706
TAMELA	GUIDRY	1873 BAXTER ROAD	Brooklyn	NY	11252
KAREN	LEVINE	1534 PALMER COURT	Cincinnati	OH	45218
STEPHENIE	MCKENZIE	1039 DELAWARE PLACE	Wilkes Barre	PA	18763
LAN	NICHOLS	367 LAKEVIEW DRIVE	Pittsburgh	PA	15262
KASEY	SOSA	975 WEST 96TH AVENUE	Kinzers	PA	17535
SHELBY	THAYER	1634 RUANE ROAD	Bordeaux	SC	29835
WILSON	BELL	1127 CUNNINGHAM STREET	Louisville	TN	37777
RENATE	LADD	652 LEWIS STREET	Crystal City	VA	22202
MELONIE	JIMENEZ	848 DOWNEY FINCH LANE	East Monkton	VT	05443

FIGURE P7.50 EMPLOYEES WITH THE TITLE OF ASSOCIATE

EMP_NUM	EMP_LNAME	EMP_EMAIL	EMP_TITLE	DEPT_NAME
84526	LASSITER	F.LASSIT8@LGCOMPANY.COM	ASSOCIATE	ACCOUNTING
83517	ALBRIGHT	SO.ALBRI96@LGCOMPANY.COM	ASSOCIATE	ACCOUNTING
84386	RIVERA	D.RIVERA76@LGCOMPANY.COM	ASSOCIATE	ACCOUNTING
83378	DUNHAM	F.DUNHAM5@LGCOMPANY.COM	ASSOCIATE	ACCOUNTING
83583	ROLLINS	M.ROLLIN99@LGCOMPANY.COM	ASSOCIATE	ACCOUNTING
83661	FINN	D.FINN87@LGCOMPANY.COM	ASSOCIATE	ACCOUNTING
84383	WASHINGTON	L.WASHIN98@LGCOMPANY.COM	ASSOCIATE	CUSTOMER SERVICE
84206	HEALY	N.HEALY82@LGCOMPANY.COM	ASSOCIATE	CUSTOMER SERVICE
83451	ELLIS	R.ELLIS81@LGCOMPANY.COM	ASSOCIATE	CUSTOMER SERVICE
84442	GREGORY	A.GREGOR95@LGCOMPANY.COM	ASSOCIATE	CUSTOMER SERVICE
84459	GILLIAM	E.GILLIA10@LGCOMPANY.COM	ASSOCIATE	CUSTOMER SERVICE
84300	SEAY	A.SEAY75@LGCOMPANY.COM	ASSOCIATE	CUSTOMER SERVICE

FIGURE P7.51 NUMBER OF PRODUCTS OF EACH BRAND

BRAND_NAME	NUMPRODUCTS
BINDER PRIME	27
BUSTERS	25
FORESTERS BEST	15
HOME COMFORT	36
LE MODE	36
LONG HAUL	41
OLDE TYME QUALITY	27
STUTTENFURST	27
VALU-MATTE	18

52. Write a query to display the number of products in each category that have a water base (Figure P7.52).

53. Write a query to display the number of products within each base and type combination (Figure P7.53).

54. Write a query to display the total inventory—that is, the sum of all products on hand for each brand ID. Sort the output by brand ID in descending order (Figure P7.54).

FIGURE P7.52 NUMBER OF WATER-BASED PRODUCTS IN EACH CATEGORY

PROD_CATEGORY	NUMPRODUCTS
Cleaner	2
Filler	2
Primer	16
Sealer	1
Top Coat	81

FIGURE P7.53 NUMBER OF PRODUCTS OF EACH BASE AND TYPE

PROD_BASE	PROD_TYPE	NUMPRODUCTS
Solvent	Exterior	67
Solvent	Interior	83
Water	Exterior	39
Water	Interior	63

FIGURE P7.54 TOTAL INVENTORY OF EACH BRAND OF PRODUCTS

BRAND_ID	TOTALINVENTORY
35	2431
33	2158
31	1117
30	3012
29	1735
28	2200
27	2596
25	1829
23	1293

55. Write a query to display the brand ID, brand name, and average price of products of each brand. Sort the output by brand name. (Results are shown with the average price rounded to two decimal places.) (Figure P7.55.)

FIGURE P7.55 AVERAGE PRICE OF PRODUCTS OF EACH BRAND

BRAND_ID	BRAND_NAME	AVGPRICE
33	BINDER PRIME	16.12
29	BUSTERS	22.59
23	FORESTERS BEST	20.94
27	HOME COMFORT	21.8
35	LE MODE	19.22
30	LONG HAUL	20.12
28	OLDE TYME QUALITY	18.33
25	STUTTENFURST	16.47
31	VALU-MATTE	16.84

56. Write a query to display the department number and most recent employee hire date for each department. Sort the output by department number (Figure P7.56).

FIGURE P7.56 MOST RECENT HIRE IN EACH DEPARTMENT

DEPT_NUM	MOSTRECENT
200	6/8/2005
250	12/15/2015
280	4/16/2014
300	12/12/2014
400	1/26/2015
500	4/26/2015
550	10/22/2015
600	10/2/2015

57. Write a query to display the employee number, first name, last name, and largest salary amount for each employee in department 200. Sort the output by largest salary in descending order (Figure P7.57).

FIGURE P7.57   LARGEST SALARY AMOUNT FOR EACH EMPLOYEE IN DEPARTMENT 200

EMP_NUM	EMP_FNAME	EMP_LNAME	LARGESTSALARY
83509	FRANKLYN	STOVER	210000
83705	JOSE	BARR	147000
83537	CLEO	ENGLISH	136000
83565	LOURDES	ABERNATHY	133000
83593	ROSANNE	NASH	129000
83621	FONDA	GONZALEZ	126000
83649	DELMA	JACOB	123000
83677	HERB	MANNING	120000
83936	BRADFORD	BRAY	117000
83734	INEZ	ROCHA	112000
84049	LANE	BRANDON	110000
83763	JAIME	FELTON	107000

58. Write a query to display the customer code, first name, last name, and sum of all invoice totals for customers with cumulative invoice totals greater than $1,500. Sort the output by the sum of invoice totals in descending order (Figure P7.58).

FIGURE P7.58   LIST OF CUSTOMERS WITH CUMULATIVE PURCHASES OF MORE THAN $1,500

CUST_CODE	CUST_FNAME	CUST_LNAME	TOTALINVOICES
215	CHARMAINE	BRYAN	3134.15
98	VALENTIN	MARINO	3052.46
152	LISETTE	WHITTAKER	3042.78
117	KARON	MATA	3009.63
97	ERWIN	ANDERSON	2895.49
112	LAN	NICHOLS	2867.14
118	JESSE	HICKS	2786.55
220	ABRAHAM	PLATT	2187.26
103	CORRINA	GIFFORD	2122.07
302	SHIRLENE	FITCH	2046.31
173	INGRID	HARDY	2040.31
132	JANIS	DUBOIS	2015.62

59. Write a query to display the department number, department name, department phone number, employee number, and last name of each department manager. Sort the output by department name (Figure P7.59).

FIGURE P7.59   DEPARTMENT MANAGERS

DEPT_NUM	DEPT_NAME	DEPT_PHONE	EMP_NUM	EMP_LNAME
600	ACCOUNTING	555-2333	84583	YAZZIE
250	CUSTOMER SERVICE	555-5555	84001	FARMER
500	DISTRIBUTION	555-3624	84052	FORD
280	MARKETING	555-8500	84042	PETTIT
300	PURCHASING	555-4873	83746	RANKIN
200	SALES	555-2824	83509	STOVER
550	TRUCKING	555-0057	83683	STONE
400	WAREHOUSE	555-1003	83759	CHARLES

60. Write a query to display the vendor ID, vendor name, brand name, and number of products of each brand supplied by each vendor. Sort the output by vendor name and then by brand name (Figure P7.60).

FIGURE P7.60 NUMBER OF PRODUCTS OF EACH BRAND SUPPLIED BY EACH VENDOR

VEND_ID	VEND_NAME	BRAND_NAME	NUMPRODUCTS
8	Baltimore Paints Consolidated	BINDER PRIME	27
8	Baltimore Paints Consolidated	FORESTERS BEST	1
8	Baltimore Paints Consolidated	HOME COMFORT	36
8	Baltimore Paints Consolidated	LE MODE	3
8	Baltimore Paints Consolidated	LONG HAUL	3
8	Baltimore Paints Consolidated	VALU-MATTE	1
13	Boykin Chemical Workshop	BUSTERS	1
13	Boykin Chemical Workshop	LE MODE	2
13	Boykin Chemical Workshop	LONG HAUL	2
13	Boykin Chemical Workshop	OLDE TYME QUALITY	2
13	Boykin Chemical Workshop	STUTTENFURST	1
13	Boykin Chemical Workshop	VALU-MATTE	1

61. Write a query to display the employee number, last name, first name, and sum of invoice totals for all employees who completed an invoice. Sort the output by employee last name and then by first name (Figure P7.61).

FIGURE P7.61 TOTAL VALUE OF INVOICES COMPLETED BY EACH EMPLOYEE

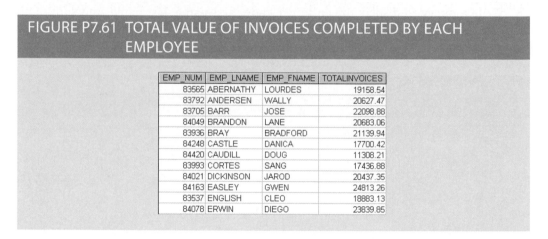

EMP_NUM	EMP_LNAME	EMP_FNAME	TOTALINVOICES
83565	ABERNATHY	LOURDES	19158.54
83792	ANDERSEN	WALLY	20627.47
83705	BARR	JOSE	22098.88
84049	BRANDON	LANE	20683.06
83936	BRAY	BRADFORD	21139.94
84248	CASTLE	DANICA	17700.42
84420	CAUDILL	DOUG	11308.21
83993	CORTES	SANG	17436.88
84021	DICKINSON	JAROD	20437.35
84163	EASLEY	GWEN	24813.26
83537	ENGLISH	CLEO	18883.13
84078	ERWIN	DIEGO	23839.85

62. Write a query to display the largest average product price of any brand (Figure P7.62).

FIGURE P7.62 LARGEST AVERAGE BRAND PRICE

LARGEST AVERAGE
22.59

63. Write a query to display the brand ID, brand name, brand type, and average price of products for the brand that has the largest average product price (Figure P7.63).

FIGURE P7.63 BRAND WITH HIGHEST AVERAGE PRICE

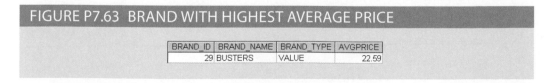

BRAND_ID	BRAND_NAME	BRAND_TYPE	AVGPRICE
29	BUSTERS	VALUE	22.59

64. Write a query to display the manager name, department name, department phone number, employee name, customer name, invoice date, and invoice total for the department manager of the employee who made a sale to a customer whose last name is Hagan on May 18, 2015 (Figure P7.64).

FIGURE P7.64 MANAGER OF EMPLOYEE MAKING A SALE TO CUSTOMER HAGAN

Manager FName	Manager LName	DEPT_NAME	DEPT_PHONE	Employee Fname	Employee Lname	Customer FName	Customer LName	INV_DATE	INV_TOTAL
FRANKLYN	STOVER	SALES	555-2824	THURMAN	WILKINSON	DARELL	HAGAN	5/18/2015	315.04

The CIS Department at Tiny College maintains the Free Access to Current Technology (FACT) library of ebooks. FACT is a collection of current technology ebooks for use by faculty and students. Agreements with the publishers allow patrons to electronically check out a book, which gives them exclusive access to the book online through the FACT website, but only one patron at a time can have access to a book. A book must have at least one author but can have many. An author must have written at least one book to be included in the system, but may have written many. A book may have never been checked out, but can be checked out many times by the same patron or different patrons over time. Because all faculty and staff in the department are given accounts at the online library, a patron may have never checked out a book or they may have checked out many books over time. To simplify determining which patron currently has a given book checked out, a redundant relationship between BOOK and PATRON is maintained. The ERD for this system is shown in Figure P7.65 and should be used to answer Problems 65–95. For Problems 66–95, a figure of the correct output is provided for each problem. If the output of the query is very large, only the first several rows of the output are shown.

65. Write a query that displays the book title, cost and year of publication for every book in the system.

FIGURE P7.65 THE CH07_FACT ERD

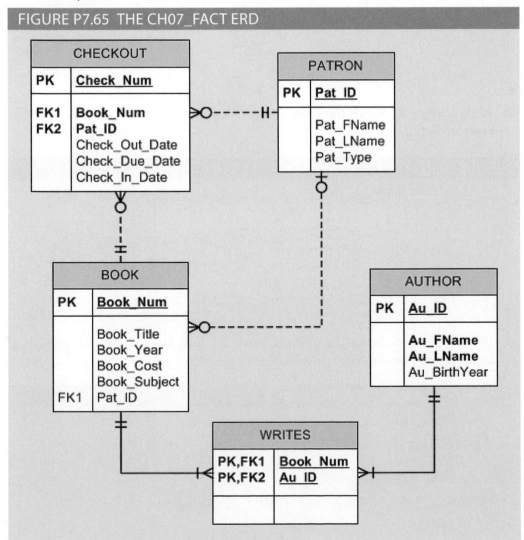

66. Write a query that displays the first and last name of every patron (Figure P7.66). (50 rows)

FIGURE P7.66 ALL PATRON NAMES

PAT_FNAME	PAT_LNAME
robert	carter
Kelsey	Koch
Cedric	Baldwin
Vera	Alvarado
Alan	Martin
Cory	Barry
Peggy	Marsh
Tony	Miles
Betsy	Malone

67. Write a query to display the checkout number, checkout date, and due date for every book that has been checked out (Figure P7.67). (68 rows)

FIGURE P7.67 ALL CHECKOUTS

CHECK_NUM	CHECK_OUT_DATE	CHECK_DUE_DATE
91001	3/31/2015	4/14/2015
91002	3/31/2015	4/7/2015
91003	3/31/2015	4/14/2015
91004	3/31/2015	4/14/2015
91005	3/31/2015	4/7/2015
91006	4/5/2015	4/12/2015
91007	4/5/2015	4/12/2015
91008	4/5/2015	4/12/2015
91009	4/5/2015	4/19/2015
91010	4/5/2015	4/19/2015
91011	4/5/2015	4/12/2015

68. Write a query to display the book number, book title, and year of publication for every book (Figure P7.68).

FIGURE P7.68 TITLE AND YEAR FOR ALL BOOKS

BOOK_NUM	TITLE	Year Published
5235	Beginner's Guide to JAVA	2012
5236	Database in the Cloud	2012
5237	Mastering the database environment	2013
5238	Conceptual Programming	2013
5239	J++ in Mobile Apps	2013
5240	iOS Programming	2013
5241	JAVA First Steps	2013
5242	C# in Middleware Deployment	2013
5243	DATABASES in Theory	2013
5244	Cloud-based Mobile Applications	2013

69. Write a query to display the different years in which books have been published. Include each year only once (Figure P7.69).

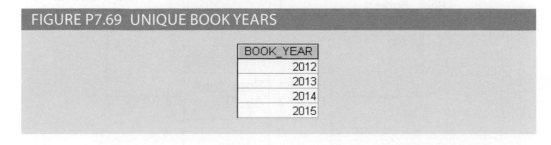

FIGURE P7.69 UNIQUE BOOK YEARS

BOOK_YEAR
2012
2013
2014
2015

70. Write a query to display the different subjects on which FACT has books. Include each subject only once (Figure P7.70).

FIGURE P7.70 UNIQUE BOOK SUBJECTS

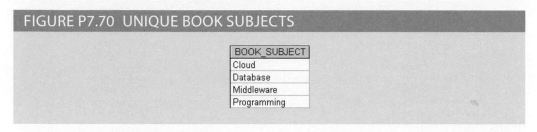

BOOK_SUBJECT
Cloud
Database
Middleware
Programming

71. Write a query to display the book number, title, and cost of each book (Figure P7.71).

FIGURE P7.71 TITLE AND REPLACEMENT COST FOR BOOKS

BOOK_NUM	BOOK_TITLE	Replacement Cost
5235	Beginner's Guide to JAVA	59.95
5236	Database in the Cloud	79.95
5237	Mastering the database environment	89.95
5238	Conceptual Programming	59.95
5239	J++ in Mobile Apps	49.95
5240	iOS Programming	79.95
5241	JAVA First Steps	49.95
5242	C# in Middleware Deployment	59.95
5243	DATABASES in Theory	129.95
5244	Cloud-based Mobile Applications	69.95
5245	The Golden Road to Platform independence	119.95
5246	Capture the Cloud	69.95
5247	Shining Through the Cloud: Sun Programming	109.95
5248	What You Always Wanted to Know About Database, But Were Afraid to Ask	49.95

72. Write a query to display the checkout number, book number, patron ID, checkout date, and due date for every checkout that has ever occurred in the system. Sort the results by checkout date in descending order (Figure P7.72). (68 rows)

FIGURE P7.72 CHECKOUTS BY DATE

CHECK_NUM	BOOK_NUM	PAT_ID	CHECK_OUT_DATE	CHECK_DUE_DATE
91067	5252	1229	5/24/2015	5/31/2015
91068	5238	1229	5/24/2015	5/31/2015
91066	5242	1228	5/19/2015	5/26/2015
91065	5244	1210	5/17/2015	5/24/2015
91064	5236	1183	5/17/2015	5/31/2015
91060	5235	1209	5/15/2015	5/22/2015
91063	5243	1223	5/15/2015	5/22/2015
91062	5254	1223	5/15/2015	5/22/2015
91061	5246	1172	5/15/2015	5/22/2015
91056	5254	1224	5/10/2015	5/17/2015

73. Write a query to display the book title, year, and subject for every book. Sort the results by book subject in ascending order, year in descending order, and then title in ascending order (Figure P7.73).

74. Write a query to display the book number, title, and year of publication for all books published in 2012 (Figure P7.74).

75. Write a query to display the book number, title, and year of publication for all books in the "Database" subject (Figure P7.75).

FIGURE P7.73 BOOKS BY CASCADING SORT

BOOK_TITLE	BOOK_YEAR	BOOK_SUBJECT
Capture the Cloud	2014	Cloud
Starlight Applications	2014	Cloud
Cloud-based Mobile Applications	2013	Cloud
Database in the Cloud	2012	Cloud
Beyond the Database Veil	2014	Database
What You Always Wanted to Know About Database, But Were Afraid to Ask	2014	Database
DATABASES in Theory	2013	Database
Mastering the database environment	2013	Database
Reengineering the Middle Tier	2014	Middleware

FIGURE P7.74 BOOKS PUBLISHED IN 2012

BOOK_NUM	BOOK_TITLE	BOOK_YEAR
5235	Beginner's Guide to JAVA	2012
5236	Database in the Cloud	2012

FIGURE P7.75 DATABASE BOOKS

BOOK_NUM	BOOK_TITLE	BOOK_YEAR
5237	Mastering the database environment	2013
5243	DATABASES in Theory	2013
5248	What You Always Wanted to Know About Database, But Were Afraid to Ask	2014
5252	Beyond the Database Veil	2014

76. Write a query to display the checkout number, book number, and checkout date of all books checked out before April 5, 2015 (Figure P7.76).

FIGURE P7.76 CHECKOUTS BEFORE APRIL 5TH

CHECK_NUM	BOOK_NUM	CHECK_OUT_DATE
91001	5235	3/31/2015
91002	5238	3/31/2015
91003	5240	3/31/2015
91004	5237	3/31/2015
91005	5236	3/31/2015

77. Write a query to display the book number, title, and year of publication of all books published after 2013 and on the "Programming" subject (Figure P7.77).

FIGURE P7.77 NEWER BOOKS ON PROGRAMMING

BOOK_NUM	BOOK_TITLE	BOOK_YEAR
5247	Shining Through the Cloud: Sun Programming	2014
5251	Thoughts on Revitalizing Ruby	2014
5253	Virtual Programming for Virtual Environments	2014
5254	Coding Style for Maintenance	2015

78. Write a query to display the book number, title, year of publication, subject, and cost for all books that are on the subjects of "Middleware" or "Cloud," and that cost more than $70 (Figure P7.78).

FIGURE P7.78 EXPENSIVE MIDDLEWARE OR CLOUD BOOKS

BOOK_NUM	BOOK_TITLE	BOOK_YEAR	BOOK_SUBJECT	BOOK_COST
5236	Database in the Cloud	2012	Cloud	79.95
5245	The Golden Road to Platform independence	2014	Middleware	119.95
5250	Reengineering the Middle Tier	2014	Middleware	89.95

79. Write a query to display the author ID, first name, last name, and year of birth for all authors born in the decade of the 1980s (Figure P7.79).

FIGURE P7.79 AUTHORS BORN IN THE 1980S

AU_ID	AU_FNAME	AU_LNAME	AU_BIRTHYEAR
218	Rachel	Beatney	1983
383	Neal	Walsh	1980
394	Robert	Lake	1982
438	Perry	Pearson	1986
460	Connie	Paulsen	1983
581	Manish	Aggerwal	1984
603	Julia	Palca	1988

80. Write a query to display the book number, title, and year of publication for all books that contain the word "Database" in the title, regardless of how it is capitalized (Figure P7.80).

FIGURE P7.80 BOOK TITLES CONTAINING DATABASE

BOOK_NUM	BOOK_TITLE	BOOK_YEAR
5236	Database in the Cloud	2012
5237	Mastering the database environment	2013
5243	DATABASES in Theory	2013
5248	What You Always Wanted to Know About Database, But Were Afraid to Ask	2014
5252	Beyond the Database Veil	2014

81. Write a query to display the patron ID, first and last name of all patrons who are students (Figure P7.81). (44 rows)

FIGURE P7.81 STUDENT PATRONS

PAT_ID	PAT_FNAME	PAT_LNAME
1166	Vera	Alvarado
1171	Peggy	Marsh
1172	Tony	Miles
1174	Betsy	Malone
1180	Nadine	Blair
1181	Allen	Horne
1182	Jamal	Melendez
1184	Jimmie	Love
1185	Sandra	Yang
1200	Lorenzo	Torres

82. Write a query to display the patron ID, first and last name, and patron type for all patrons whose last name begins with the letter "C" (Figure P7.82).

FIGURE P7.82 PATRONS WHOSE LAST NAME STARTS WITH 'C'

PAT_ID	PAT_FNAME	PAT_LNAME	PAT_TYPE
1160	robert	carter	Faculty
1208	Ollie	Cantrell	Student
1210	Keith	Cooley	STUdent

83. Write a query to display the author ID, first and last name of all authors whose year of birth is unknown (Figure P7.83).

FIGURE P7.83 AUTHORS WITH UNKNOWN BIRTH YEAR

AU_ID	AU_FNAME	AU_LNAME
229	Carmine	Salvadore
262	Xia	Chiang
559	Rachel	McGill

84. Write a query to display the author ID, first and last name of all authors whose year of birth is known (Figure P7.84).

FIGURE P7.84 AUTHORS WITH KNOWN BIRTH YEAR

AU_ID	AU_FNAME	AU_LNAME
185	Benson	Reeves
218	Rachel	Beatney
251	Hugo	Bruer
273	Reba	Durante
284	Trina	Tankersly
383	Neal	Walsh
394	Robert	Lake
438	Perry	Pearson
460	Connie	Paulsen
581	Manish	Aggerwal
592	Lawrence	Sheel
603	Julia	Palca

85. Write a query to display the checkout number, book number, patron ID, checkout date, and due date for all checkouts that have not yet been returned. Sort the results by book number (Figure P7.85).

FIGURE P7.85 UNRETURNED CHECKOUTS

CHECK_NUM	BOOK_NUM	PAT_ID	CHECK_OUT_DATE	CHECK_DUE_DATE
91068	5238	1229	5/24/2015	5/31/2015
91053	5240	1212	5/9/2015	5/16/2015
91066	5242	1228	5/19/2015	5/26/2015
91061	5246	1172	5/15/2015	5/22/2015
91059	5249	1207	5/10/2015	5/17/2015
91067	5252	1229	5/24/2015	5/31/2015

86. Write a query to display the author ID, first name, last name, and year of birth for all authors. Sort the results in descending order by year of birth, and then in ascending order by last name (Figure P7.86). (*Note:* Some DBMS sort NULLs as being large and some DBMS sort NULLs as being small.)

FIGURE P7.86 AUTHORS BY BIRTH YEAR

AU_ID	AU_FNAME	AU_LNAME	AU_BIRTHYEAR
185	Benson	Reeves	1990
603	Julia	Palca	1988
438	Perry	Pearson	1986
581	Manish	Aggerwal	1984
218	Rachel	Beatney	1983
460	Connie	Paulsen	1983
394	Robert	Lake	1982
383	Neal	Walsh	1980
592	Lawrence	Sheel	1976
251	Hugo	Bruer	1972
273	Reba	Durante	1969
284	Trina	Tankersly	1961
262	Xia	Chiang	
559	Rachel	McGill	
229	Carmine	Salvadore	

87. Write a query to display the number of books in the FACT system (Figure P7.87).

FIGURE P7.87 NUMBER OF BOOKS

Number of Books
20

88. Write a query to display the number of different book subjects in the FACT system (Figure P7.88).

FIGURE P7.88 NUMBER OF DIFFERENT SUBJECTS

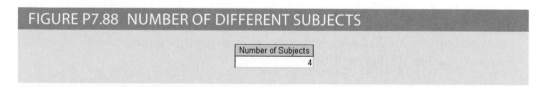

Number of Subjects
4

89. Write a query to display the number of books that are available (not currently checked out) (Figure P7.89).

FIGURE P7.89 NUMBER OF BOOKS NOT CURRENTLY CHECKED OUT

Available Books
14

90. Write a query to display the highest book cost in the system (Figure P7.90).

FIGURE P7.90 MOST EXPENSIVE BOOK PRICE

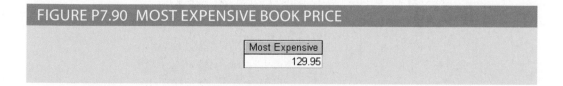

Most Expensive
129.95

91. Write a query to display the lowest book cost in the system (Figure P7.91).

FIGURE P7.91 LEAST EXPENSIVE BOOK PRICE

Least Expensive
49.95

92. Write a query to display the number of different patrons who have ever checked out a book (Figure P7.92).

FIGURE P7.92 DIFFERENT PATRONS TO CHECKOUT A BOOK

DIFFERENT PATRONS
33

93. Write a query to display the subject and the number of books in each subject. Sort the results by the number of books in descending order, then by subject name in ascending order (Figure P7.93).

FIGURE P7.93 NUMBER OF BOOKS PER SUBJECT

BOOK_SUBJECT	Books In Subject
Programming	9
Cloud	4
Database	4
Middleware	3

94. Write a query to display the author ID and the number of books written by that author. Sort the results in descending order by number of books, then in ascending order by author ID (Figure P7.94).

FIGURE P7.94 NUMBER OF BOOKS PER AUTHOR

AU_ID	Books Written
262	3
460	3
185	2
229	2
251	2
383	2
394	2
559	2
218	1
273	1
284	1
438	1
581	1
592	1
603	1

95. Write a query to display the total value of all books in the library (Figure P7.95).

FIGURE P7.95   TOTAL OF ALL BOOKS

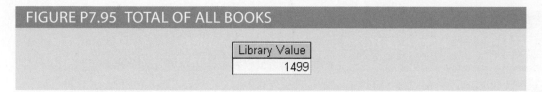

Library Value
1499

Cases

EliteVideo is a startup company providing concierge DVD kiosk service in upscale neighborhoods. EliteVideo can own several copies (VIDEO) of each movie (MOVIE). For example, a kiosk may have 10 copies of the movie *Twist in the Wind*. In the database, *Twist in the Wind* would be one MOVIE, and each copy would be a VIDEO. A rental transaction (RENTAL) involves one or more videos being rented to a member (MEMBERSHIP). A video can be rented many times over its lifetime; therefore, there is an M:N relationship between RENTAL and VIDEO. DETAILRENTAL is the bridge table to resolve this relationship. The complete ERD is provided in Figure P7.96.

FIGURE P7.96   THE CH07_MOVIECO ERD

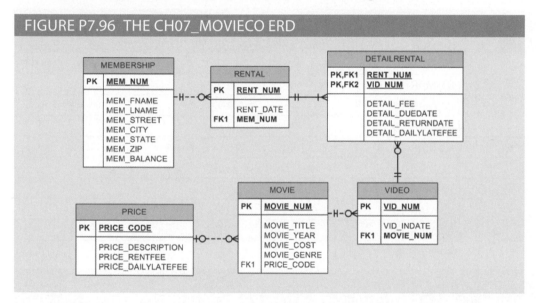

96. Write the SQL code to create the table structures for the entities shown in Figure P7.96. The structures should contain the attributes specified in the ERD. Use data types that are appropriate for the data that will need to be stored in each attribute. Enforce primary key and foreign key constraints as indicated by the ERD.

97. The following tables provide a very small portion of the data that will be kept in the database. The data needs to be inserted into the database for testing purposes. Write the INSERT commands necessary to place the following data in the tables that were created in Problem 96.

MEMBERSHIP

MEM NUM	MEM_ FNAME	MEM_ LNAME	MEM_STREET	MEM_CITY	MEM_ STATE	MEM_ ZIP	MEM_ BALANCE
102	Tami	Dawson	2632 Takli Circle	Norene	TN	37136	11
103	Curt	Knight	4025 Cornell Court	Flatgap	KY	41219	6
104	Jamal	Melendez	788 East 145th Avenue	Quebeck	TN	38579	0
105	Iva	Mcclain	6045 Musket Ball Circle	Summit	KY	42783	15
106	Miranda	Parks	4469 Maxwell Place	Germantown	TN	38183	0
107	Rosario	Elliott	7578 Danner Avenue	Columbia	TN	38402	5
108	Mattie	Guy	4390 Evergreen Street	Lily	KY	40740	0
109	Clint	Ochoa	1711 Elm Street	Greeneville	TN	37745	10
110	Lewis	Rosales	4524 Southwind Circle	Counce	TN	38326	0
111	Stacy	Mann	2789 East Cook Avenue	Murfreesboro	TN	37132	8
112	Luis	Trujillo	7267 Melvin Avenue	Heiskell	TN	37754	3
113	Minnie	Gonzales	6430 Vasili Drive	Williston	TN	38076	0

RENTAL

RENT_NUM	RENT_DATE	MEM_NUM
1001	01-MAR-16	103
1002	01-MAR-16	105
1003	02-MAR-16	102
1004	02-MAR-16	110
1005	02-MAR-16	111
1006	02-MAR-16	107
1007	02-MAR-16	104
1008	03-MAR-16	105
1009	03-MAR-16	111

DETAILRENTAL

RENT_ NUM	VID_NUM	DETAIL_FEE	DETAIL_DUEDATE	DETAIL_RETURNDATE	DETAIL_ DAILYLATEFEE
1001	34342	2	04-MAR-16	02-MAR-16	
1001	61353	2	04-MAR-16	03-MAR-16	1
1002	59237	3.5	04-MAR-16	04-MAR-16	3
1003	54325	3.5	04-MAR-16	09-MAR-16	3
1003	61369	2	06-MAR-16	09-MAR-16	1
1003	61388	0	06-MAR-16	09-MAR-16	1
1004	44392	3.5	05-MAR-16	07-MAR-16	3
1004	34367	3.5	05-MAR-16	07-MAR-16	3
1004	34341	2	07-MAR-16	07-MAR-16	1
1005	34342	2	07-MAR-16	05-MAR-16	1
1005	44397	3.5	05-MAR-16	05-MAR-16	3

DETAILRENTAL

RENT_NUM	VID_NUM	DETAIL_FEE	DETAIL_DUEDATE	DETAIL_RETURNDATE	DETAIL_DAILYLATEFEE
1006	34366	3.5	05-MAR-16	04-MAR-16	3
1006	61367	2	07-MAR-16		1
1007	34368	3.5	05-MAR-16		3
1008	34369	3.5	05-MAR-16	05-MAR-16	3
1009	54324	3.5	05-MAR-16		3
1001	34366	3.5	04-MAR-16	02-MAR-16	3

VIDEO

VID_NUM	VID_INDATE	MOVIE_NUM
54321	18-JUN-15	1234
54324	18-JUN-15	1234
54325	18-JUN-15	1234
34341	22-JAN-14	1235
34342	22-JAN-14	1235
34366	02-MAR-16	1236
34367	02-MAR-16	1236
34368	02-MAR-16	1236
34369	02-MAR-16	1236
44392	21-OCT-15	1237
44397	21-OCT-15	1237
59237	14-FEB-16	1237
61388	25-JAN-14	1239
61353	28-JAN-13	1245
61354	28-JAN-13	1245
61367	30-JUL-15	1246
61369	30-JUL-15	1246

MOVIE

MOVIE_NUM	MOVIE_TITLE	MOVIE_YEAR	MOVIE_COST	MOVIE_GENRE	PRICE_CODE
1234	The Cesar Family Christmas	2014	39.95	FAMILY	2
1235	Smokey Mountain Wildlife	2011	59.95	ACTION	1
1236	Richard Goodhope	2015	59.95	DRAMA	2
1237	Beatnik Fever	2014	29.95	COMEDY	2
1238	Constant Companion	2015	89.95	DRAMA	
1239	Where Hope Dies	2005	25.49	DRAMA	3
1245	Time to Burn	2012	45.49	ACTION	1
1246	What He Doesn't Know	2013	58.29	COMEDY	1

PRICE			
PRICE_CODE	PRICE_DESCRIPTION	PRICE_RENTFEE	PRICE_DAILYLATEFEE
1	Standard	2	1
2	New Release	3.5	3
3	Discount	1.5	1
4	Weekly Special	1	.5

For Questions 98–127, use the tables that were created in Problem 96 and the data that was loaded into those tables in Problem 97.

98. Write the SQL command to save the rows inserted in Problem 97.

99. Write the SQL command to change the movie year for movie number 1245 to 2013.

100. Write the SQL command to change the price code for all action movies to price code 3.

101. Write a single SQL command to increase all price rental fee values in the PRICE table by $0.50.

102. Write the SQL command to save the changes made to the PRICE and MOVIE tables in Problems 98–101.

103. Write a query to display the movie title, movie year, and movie genre for all movies. (The results are shown in Figure P7.103.)

FIGURE P7.103 ALL MOVIES

Movie_Title	Movie_Year	Movie_Genre
The Cesar Family Christmas	2014	FAMILY
Smokey Mountain Wildlife	2011	ACTION
Richard Goodhope	2015	DRAMA
Beatnik Fever	2014	COMEDY
Constant Companion	2015	DRAMA
Where Hope Dies	2005	DRAMA
Time to Burn	2013	ACTION
What He Doesn't Know	2013	COMEDY

104. Write a query to display the movie year, movie title, and movie cost sorted by movie year in descending order. (The results are shown in Figure P7.104.)

FIGURE P7.104 MOVIES BY YEAR

Movie_Year	Movie_Title	Movie_Cost
2015	Constant Companion	89.95
2015	Richard Goodhope	59.95
2014	Beatnik Fever	29.95
2014	The Cesar Family Christmas	39.95
2013	What He Doesn't Know	58.29
2013	Time to Burn	45.49
2011	Smokey Mountain Wildlife	59.95
2005	Where Hope Dies	25.49

105. Write a query to display the movie title, movie year, and movie genre for all movies sorted by movie genre in ascending order, then sorted by movie year in descending order within genre. (The results are shown in Figure P7.105.)

FIGURE P7.105 MOVIES WITH MULTICOLUMN SORT

Movie_Title	Movie_Year	Movie_Genre
Time to Burn	2013	ACTION
Smokey Mountain Wildlife	2011	ACTION
Beatnik Fever	2014	COMEDY
What He Doesn't Know	2013	COMEDY
Constant Companion	2015	DRAMA
Richard Goodhope	2015	DRAMA
Where Hope Dies	2005	DRAMA
The Cesar Family Christmas	2014	FAMILY

106. Write a query to display the movie number, movie title, and price code for all movies with a title that starts with the letter *R*. (The results are shown in Figure P7.106.)

FIGURE P7.106 MOVIES STARTING WITH *R*

Movie_Num	Movie_Title	Price_Code
1236	Richard Goodhope	2

107. Write a query to display the movie title, movie year, and movie cost for all movies that contain the word *hope* in the title. Sort the results in ascending order by title. (The results are shown in Figure P7.107.)

FIGURE P7.107 MOVIES WITH "HOPE"

Movie_Title	Movie_Year	Movie_Cost
Richard Goodhope	2015	59.95
Where Hope Dies	2005	25.49

108. Write a query to display the movie title, movie year, and movie genre for all action movies. (The results are shown in Figure P7.108.)

FIGURE P7.108 ACTION MOVIES

Movie_Title	Movie_Year	Movie_Genre
Smokey Mountain Wildlife	2011	ACTION
Time to Burn	2013	ACTION

109. Write a query to display the movie number, movie title, and movie cost for all movies that cost more than $40. (The results are shown in Figure P7.109.)

FIGURE P7.109 MOVIES COSTING MORE THAN $40

Movie_Num	Movie_Title	Movie_Cost
1235	Smokey Mountain Wildlife	59.95
1236	Richard Goodhope	59.95
1238	Constant Companion	89.95
1245	Time to Burn	45.49
1246	What He Doesn't Know	58.29

110. Write a query to display the movie number, movie title, movie cost, and movie genre for all action or comedy movies that cost less than $50. Sort the results in ascending order by genre. (The results are shown in Figure P7.110.)

FIGURE P7.110 ACTION OR COMEDY MOVIES LESS THAN $50

Movie_Num	Movie_Title	Movie_Cost	Movie_Genre
1245	Time to Burn	45.49	ACTION
1237	Beatnik Fever	29.95	COMEDY

111. Write a query to display the membership number, name, street, state, and balance for all members in Tennessee (TN), with a balance less than $5, and whose street name ends in "Avenue". (The results are shown in Figure P7.111.)

FIGURE P7.111 MEMBERS WITH MULTIPLE RESTRICTIONS

Mem_Num	Mem_FName	Mem_LName	Mem_Street	Mem_State	Mem_Balance
104	Jamal	Melendez	788 East 145th Avenue	TN	0
112	Luis	Trujillo	7267 Melvin Avenue	TN	3

112. Write a query to display the movie genre and the number of movies in each genre. (The results are shown in Figure P7.112.)

FIGURE P7.112 NUMBER OF MOVIES IN GENRE

Movie_Genre	Number of Movies
ACTION	2
COMEDY	2
DRAMA	3
FAMILY	1

113. Write a query to display the average cost of all the movies. (The results are shown in Figure P7.113.)

FIGURE P7.113 AVERAGE MOVIE COST

Average Movie Cost
51.1275

114. Write a query to display the movie genre and average cost of movies in each genre. (The results are shown in Figure P7.114.)

FIGURE P7.114 AVERAGE COST BY GENRE

Movie_Genre	Average Cost
ACTION	52.72
COMEDY	44.12
DRAMA	58.46
FAMILY	39.95

115. Write a query to display the movie title, movie genre, price description, and price rental fee for all movies with a price code. (The results are shown in Figure P7.115.)

FIGURE P7.115 RENTAL FEES FOR MOVIES

Movie_Title	Movie_Genre	Price_Description	Price_RentFee
What He Doesn't Know	COMEDY	Standard	2.5
The Cesar Family Christmas	FAMILY	New Release	4
Richard Goodhope	DRAMA	New Release	4
Beatnik Fever	COMEDY	New Release	4
Smokey Mountain Wildlife	ACTION	Discount	2
Where Hope Dies	DRAMA	Discount	2
Time to Burn	ACTION	Discount	2

116. Write a query to display the movie genre and average rental fee for movies in each genre that have a price. (The results are shown in Figure P7.116.)

FIGURE P7.116 AVERAGE RENTAL FEE BY GENRE

Movie_Genre	Average Rental Fee
ACTION	2
COMEDY	3.25
DRAMA	3
FAMILY	4

117. Write a query to display the movie title and breakeven amount for each movie that has a price. The breakeven amount is the movie cost divided by the price rental fee for each movie that has a price; it determines the number of rentals needed to break even on the purchase of the movie. (The results are shown in Figure P7.117.)

FIGURE P7.117 BREAKEVEN RENTALS

Movie_Title	Breakeven Rentals
What He Doesn't Know	23.32
The Cesar Family Christmas	9.99
Richard Goodhope	14.99
Beatnik Fever	7.49
Smokey Mountain Wildlife	29.98
Where Hope Dies	12.75
Time to Burn	22.75

118. Write a query to display the movie title and movie year for all movies that have a price code. (The results are shown in Figure P7.118.)

FIGURE P7.118 MOVIES WITH A PRICE

Movie_Title	Movie_Year
The Cesar Family Christmas	2014
Smokey Mountain Wildlife	2011
Richard Goodhope	2015
Beatnik Fever	2014
Where Hope Dies	2005
Time to Burn	2013
What He Doesn't Know	2013

119. Write a query to display the movie title, movie genre, and movie cost for all movies that cost between $44.99 and $49.99. (The results are shown in Figure P7.119.)

FIGURE P7.119 MOVIES COSTS WITHIN A RANGE

Movie_Title	Movie_Genre	Movie_Cost
Time to Burn	ACTION	45.49

120. Write a query to display the movie title, price description, price rental fee, and genre for all movies that are in the genres of family, comedy, or drama. (The results are shown in Figure P7.120.)

FIGURE P7.120 MOVIES WITHIN SPECIFIC GENRES

Movie_Title	Price_Description	Price_RentFee	Movie_Genre
The Cesar Family Christmas	New Release	4	FAMILY
Richard Goodhope	New Release	4	DRAMA
Beatnik Fever	New Release	4	COMEDY
Where Hope Dies	Discount	2	DRAMA
What He Doesn't Know	Standard	2.5	COMEDY

121. Write a query to display the membership number, first name, last name, and balance of the memberships that have a rental. (The results are shown in Figure P7.121.)

FIGURE P7.121 BALANCES OF MEMBERSHIPS WITH RENTALS

Mem_Num	Mem_FName	Mem_LName	Mem_Balance
102	Tami	Dawson	11
103	Curt	Knight	6
104	Jamal	Melendez	0
105	Iva	McClain	15
107	Rosario	Elliott	5
110	Lewis	Rosales	0
111	Stacy	Mann	8

122. Write a query to display the minimum balance, maximum balance, and average balance for memberships that have a rental. (The results are shown in Figure P7.122.)

FIGURE P7.122 MINIMUM, MAXIMUM, AND AVERAGE BALANCES

Minimum Balance	Maximum Balance	Average Balance
0	15	6.43

123. Write a query to display the rental number, rental date, video number, movie title, due date, and return date for all videos that were returned after the due date. Sort the results by rental number and movie title. (The results are shown in Figure P7.123.)

FIGURE P7.123 LATE VIDEO RETURNS

Rent_Num	Rent_Date	Vid_Num	Movie_Title	Detail_DueDate	Detail_ReturnDate
1003	02-Mar-16	54325	The Cesar Family Christmas	04-Mar-16	09-Mar-16
1003	02-Mar-16	61369	What He Doesn't Know	06-Mar-16	09-Mar-16
1003	02-Mar-16	61388	Where Hope Dies	06-Mar-16	09-Mar-16
1004	02-Mar-16	44392	Beatnik Fever	05-Mar-16	07-Mar-16
1004	02-Mar-16	34367	Richard Goodhope	05-Mar-16	07-Mar-16

124. Write a query to display the rental number, rental date, movie title, and detail fee for each movie that was returned on or before the due date. (The results are shown in Figure P7.124.)

FIGURE P7.124 ACTUAL RENTAL FEES CHARGED

Rent_Num	Rent_Date	Movie_Title	Detail_Fee
1001	01-Mar-16	Smokey Mountain Wildlife	2
1001	01-Mar-16	Time to Burn	2
1002	01-Mar-16	Beatnik Fever	3.5
1004	02-Mar-16	Smokey Mountain Wildlife	2
1005	02-Mar-16	Smokey Mountain Wildlife	2
1005	02-Mar-16	Beatnik Fever	3.5
1006	02-Mar-16	Richard Goodhope	3.5
1008	03-Mar-16	Richard Goodhope	3.5
1001	01-Mar-16	Richard Goodhope	3.5

125. Write a query to display the movie number, movie genre, average cost of movies in that genre, cost of the individual movie, and the percentage difference between the average movie cost and the individual movie cost. The results are shown in Figure P7.125. The percentage difference is the cost of the individual movie minus the average cost of movies in that genre, divided by the average cost of movies in that genre multiplied by 100. For example, if the average cost of movies in the family genre is $25 and a given family movie costs $26, then the calculation would be [(26 − 25) / 25 * 100], or 4.00 percent. In this case, the individual movie costs 4 percent more than the average family movie.

FIGURE P7.125 MOVIE DIFFERENCES FROM GENRE AVERAGE

Movie_Num	Movie_Genre	Average Cost	Movie_Cost	Percent Difference
1234	FAMILY	39.95	39.95	0.00
1235	ACTION	52.72	59.95	13.71
1236	DRAMA	58.46	59.95	2.54
1237	COMEDY	44.12	29.95	-32.12
1238	DRAMA	58.46	89.95	53.86
1239	DRAMA	58.46	25.49	-56.40
1245	ACTION	52.72	45.49	-13.71
1246	COMEDY	44.12	58.29	32.12

Chapter 8

Advanced SQL

In this chapter, you will learn:

• How to use the advanced SQL JOIN operator syntax

• About the different types of subqueries and correlated queries

• How to use SQL functions to manipulate dates, strings, and other data

• About the relational set operators UNION, UNION ALL, INTERSECT, and MINUS

• How to create and use views and updatable views

• How to create and use triggers and stored procedures

• How to create embedded SQL

Preview

In Chapter 7, Introduction to Structured Query Language (SQL), you learned the basic SQL data definition and data manipulation commands. In this chapter, you build on that knowledge and learn how to use more advanced SQL features.

You will learn about the SQL relational set operators (UNION, INTERSECT, and MINUS) and learn how they are used to merge the results of multiple queries. Joins are at the heart of SQL, so you must learn how to use the SQL JOIN statement to extract information from multiple tables. You will also learn about the different styles of subqueries that you can implement in a SELECT statement and about more of SQL's many functions to extract information from data, including manipulation of dates and strings and computations based on stored or even derived data.

Finally, you will learn how to use triggers and stored procedures to perform actions when a specific event occurs. You will also see how SQL facilitates the application of business procedures when it is embedded in a programming language such as Visual Basic .NET, C#, or COBOL.

Data Files and Available Formats

	MS Access	Oracle	MS SQL	My SQL		MS Access	Oracle	MS SQL	My SQL
CH08_SaleCo	✓	✓	✓	✓	CH08_SimpleCo	✓	✓	✓	✓
CH08_UV	✓	✓	✓	✓	CH08_LargeCo	✓	✓	✓	✓
					CH08_SaleCo2	✓	✓	✓	✓
					CH08_AviaCo	✓	✓	✓	✓
					CH08_Fact	✓	✓	✓	✓

Data Files Available on cengagebrain.com

8-1 SQL Join Operators

The relational join operation merges rows from two tables and returns the rows with one of the following conditions:

- Have common values in common columns (natural join).

- Meet a given join condition (equality or inequality).

- Have common values in common columns or have no matching values (outer join).

In Chapter 7, you learned how to use the SELECT statement in conjunction with the WHERE clause to join two or more tables. For example, you can join the PRODUCT and VENDOR tables through their common V_CODE by writing the following:

```
SELECT       P_CODE, P_DESCRIPT, P_PRICE, V_NAME
FROM         PRODUCT, VENDOR
WHERE        PRODUCT.V_CODE = VENDOR.V_CODE;
```

The preceding SQL join syntax is sometimes referred to as an "old-style" join. Note that the FROM clause contains the tables being joined and that the WHERE clause contains the condition(s) used to join the tables.

Note the following points about the preceding query:

- The FROM clause indicates which tables are to be joined. If three or more tables are included, the join operation takes place two tables at a time, from left to right. For example, if you are joining tables T1, T2, and T3, the first join is table T1 with T2; the results of that join are then joined to table T3.

- The join condition in the WHERE clause tells the SELECT statement which rows will be returned. In this case, the SELECT statement returns all rows for which the V_CODE values in the PRODUCT and VENDOR tables are equal.

- The number of join conditions is always equal to the number of tables being joined minus one. For example, if you join three tables (T1, T2, and T3), you *will* have two join conditions (j1 and j2). All join conditions are connected through an AND logical operator. The first join condition (j1) defines the join criteria for T1 and T2. The second join condition (j2) defines the join criteria for the output of the first join and T3.

- Generally, the join condition will be an equality comparison of the primary key in one table and the related foreign key in the second table.

Join operations can be classified as inner joins and outer joins. The **inner join** is the traditional join in which only rows that meet a given criterion are selected. The join criterion can be an equality condition (also called a natural join or an *equijoin*) or an inequality condition (also called a *theta join*). An **outer join** returns not only the matching rows but the rows with unmatched attribute values for one table or both tables to be joined. The SQL standard also introduces a special type of join, called a *cross join*, that returns the same result as the Cartesian product of two sets or tables.

In this section, you will learn various ways to express join operations that meet the ANSI SQL standard, as outlined in Table 8.1. Remember that not all DBMS vendors provide the same level of SQL support and that some do not support the join styles shown in this section. Oracle 12c is used to demonstrate the following queries; refer to your DBMS manual if you are using a different DBMS.

inner join
A join operation in which only rows that meet a given criterion are selected. The join criterion can be an equality condition (natural join or equijoin) or an inequality condition (theta join). The inner join is the most commonly used type of join. Contrast with *outer join*.

outer join
A join operation that produces a table in which all unmatched pairs are retained; unmatched values in the related table are left null. Contrast with *inner join*.

TABLE 8.1

SQL JOIN EXPRESSION STYLES

JOIN CLASSIFICATION	JOIN TYPE	SQL SYNTAX EXAMPLE	DESCRIPTION
CROSS	CROSS JOIN	SELECT * FROM T1, T2	Returns the Cartesian product of T1 and T2 (old style)
		SELECT * FROM T1 CROSS JOIN T2	Returns the Cartesian product of T1 and T2
INNER	Old-style JOIN	SELECT * FROM T1, T2 WHERE T1.C1=T2.C1	Returns only the rows that meet the join condition in the WHERE clause (old style); only rows with matching values are selected
	NATURAL JOIN	SELECT * FROM T1 NATURAL JOIN T2	Returns only the rows with matching values in the matching columns; the matching columns must have the same names and similar data types
	JOIN USING	SELECT * FROM T1 JOIN T2 USING (C1)	Returns only the rows with matching values in the columns indicated in the USING clause
	JOIN ON	SELECT * FROM T1 JOIN T2 ON T1.C1=T2.C1	Returns only the rows that meet the join condition indicated in the ON clause
OUTER	LEFT JOIN	SELECT * FROM T1 LEFT OUTER JOIN T2 ON T1.C1=T2.C1	Returns rows with matching values and includes all rows from the left table (T1) with unmatched values
	RIGHT JOIN	SELECT * FROM T1 RIGHT OUTER JOIN T2 ON T1.C1=T2.C1	Returns rows with matching values and includes all rows from the right table (T2) with unmatched values
	FULL JOIN	SELECT * FROM T1 FULL OUTER JOIN T2 ON T1.C1=T2.C1	Returns rows with matching values and includes all rows from both tables (T1 and T2) with unmatched values

8-1a Cross Join

A **cross join** performs a relational product (also known as the *Cartesian product*) of two tables. The cross join syntax is:

SELECT *column-list* FROM *table1* CROSS JOIN *table2*

For example, the following command:

SELECT * FROM INVOICE CROSS JOIN LINE;

performs a cross join of the INVOICE and LINE tables that generates 144 rows. (There are 8 invoice rows and 18 line rows, yielding 8 × 18 = 144 rows.)

You can also perform a cross join that yields only specified attributes. For example, you can specify:

SELECT   INVOICE.INV_NUMBER, CUS_CODE, INV_DATE, P_CODE
FROM   INVOICE CROSS JOIN LINE;

The results generated through that SQL statement can also be generated by using the following syntax:

SELECT   INVOICE.INV_NUMBER, CUS_CODE, INV_DATE, P_CODE
FROM   INVOICE, LINE;

cross join
A join that performs a relational product (or Cartesian product) of two tables.

Note

Unlike Oracle, MS SQL Server, and MySQL, Access does not support the CROSS JOIN command. However, all DBMSs support producing a cross join by placing a comma between the tables in the FROM clause.

8-1b Natural Join

Recall from Chapter 3 that a natural join returns all rows with matching values in the matching columns and eliminates duplicate columns. This style of query is used when the tables share one or more common attributes with common names. The natural join syntax is:

SELECT *column-list* FROM *table1* NATURAL JOIN *table2*

The natural join will perform the following tasks:

- Determine the common attribute(s) by looking for attributes with identical names and compatible data types.
- Select only the rows with common values in the common attribute(s).
- If there are no common attributes, return the relational product of the two tables.

The following example performs a natural join of the CUSTOMER and INVOICE tables and returns only selected attributes:

SELECT   CUS_CODE, CUS_LNAME, INV_NUMBER, INV_DATE
FROM   CUSTOMER NATURAL JOIN INVOICE;

The SQL code and its results are shown at the top of Figure 8.1.

You are not limited to two tables when performing a natural join. For example, you can perform a natural join of the INVOICE, LINE, and PRODUCT tables and project only selected attributes by writing the following:

SELECT   INV_NUMBER, P_CODE, P_DESCRIPT, LINE_UNITS, LINE_PRICE
FROM   INVOICE NATURAL JOIN LINE NATURAL JOIN PRODUCT;

The SQL code and its results are shown at the bottom of Figure 8.1.

One important difference between the natural join and the old-style join syntax is that the natural join does not require the use of a table qualifier for the common attributes. In the first natural join example, you projected CUS_CODE. However, the projection did not require any table qualifier, even though the CUS_CODE attribute appears in both the CUSTOMER and INVOICE tables. The same can be said of the INV_NUMBER attribute in the second natural join example.

Note

Although natural joins are common in theoretical discussions of databases and DBMS functionality, they are typically discouraged in most development environments. Natural joins do not document the join condition in the code, so they are harder to maintain, and many developers do not like the DBMS "guessing" about how the tables should be joined. Oracle and MySQL support NATURAL JOIN, but MS SQL Server and Access do not.

FIGURE 8.1 NATURAL JOIN RESULTS

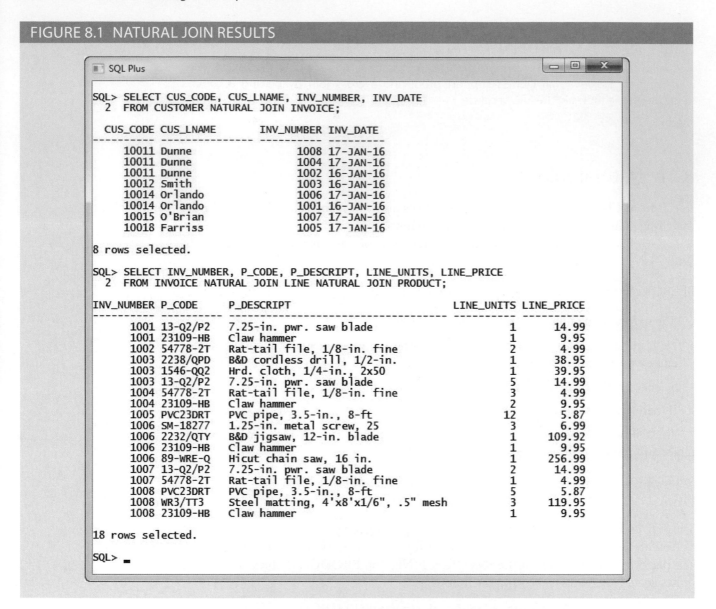

```
SQL> SELECT CUS_CODE, CUS_LNAME, INV_NUMBER, INV_DATE
  2  FROM CUSTOMER NATURAL JOIN INVOICE;

  CUS_CODE CUS_LNAME         INV_NUMBER INV_DATE
---------- ---------------- ----------- ---------
     10011 Dunne                   1008 17-JAN-16
     10011 Dunne                   1004 17-JAN-16
     10011 Dunne                   1002 16-JAN-16
     10012 Smith                   1003 16-JAN-16
     10014 Orlando                 1006 17-JAN-16
     10014 Orlando                 1001 16-JAN-16
     10015 O'Brian                 1007 17-JAN-16
     10018 Farriss                 1005 17-JAN-16

8 rows selected.

SQL> SELECT INV_NUMBER, P_CODE, P_DESCRIPT, LINE_UNITS, LINE_PRICE
  2  FROM INVOICE NATURAL JOIN LINE NATURAL JOIN PRODUCT;

INV_NUMBER P_CODE    P_DESCRIPT                              LINE_UNITS LINE_PRICE
---------- --------- --------------------------------------- ---------- ----------
      1001 13-Q2/P2  7.25-in. pwr. saw blade                          1      14.99
      1001 23109-HB  Claw hammer                                      1       9.95
      1002 54778-2T  Rat-tail file, 1/8-in. fine                      2       4.99
      1003 2238/QPD  B&D cordless drill, 1/2-in.                      1      38.95
      1003 1546-QQ2  Hrd. cloth, 1/4-in., 2x50                        1      39.95
      1003 13-Q2/P2  7.25-in. pwr. saw blade                          5      14.99
      1004 54778-2T  Rat-tail file, 1/8-in. fine                      3       4.99
      1004 23109-HB  Claw hammer                                      2       9.95
      1005 PVC23DRT  PVC pipe, 3.5-in., 8-ft                         12       5.87
      1006 SM-18277  1.25-in. metal screw, 25                         3       6.99
      1006 2232/QTY  B&D jigsaw, 12-in. blade                         1     109.92
      1006 23109-HB  Claw hammer                                      1       9.95
      1006 89-WRE-Q  Hicut chain saw, 16 in.                          1     256.99
      1007 13-Q2/P2  7.25-in. pwr. saw blade                          2      14.99
      1007 54778-2T  Rat-tail file, 1/8-in. fine                      1       4.99
      1008 PVC23DRT  PVC pipe, 3.5-in., 8-ft                          5       5.87
      1008 WR3/TT3   Steel matting, 4'x8'x1/6", .5" mesh              3     119.95
      1008 23109-HB  Claw hammer                                      1       9.95

18 rows selected.

SQL>
```

8-1c JOIN USING Clause

A second way to express a join is through the USING keyword. The query returns only the rows with matching values in the column indicated in the USING clause—and that column must exist in both tables. The syntax is:

SELECT *column-list* FROM *table1* JOIN *table2* USING (*common-column*)

To see the JOIN USING query in action, perform a join of the INVOICE and LINE tables by writing the following:

SELECT   INV_NUMBER, P_CODE, P_DESCRIPT, LINE_UNITS, LINE_PRICE
FROM   INVOICE JOIN LINE USING (INV_NUMBER) JOIN PRODUCT
                 USING (P_CODE);

The SQL statement produces the results shown in Figure 8.2.

FIGURE 8.2 JOIN USING RESULTS

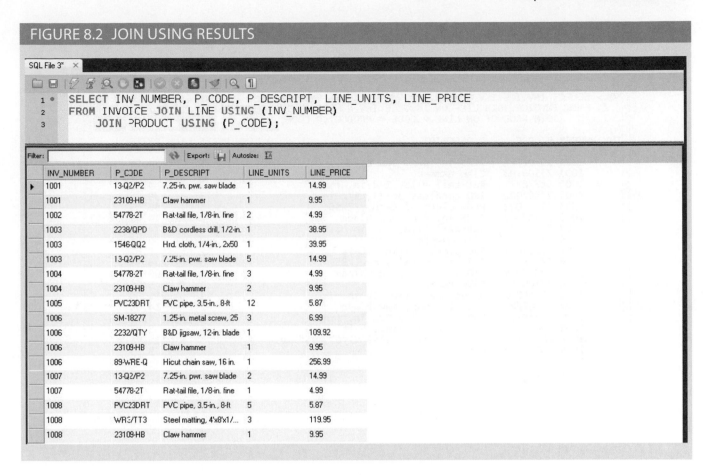

As with the NATURAL JOIN command, the JOIN USING operand does not require table qualifiers and will only return one copy of the common attribute.

Note

Oracle and MySQL support the JOIN USING syntax. MS SQL Server and Access do not. If JOIN USING is used in Oracle, then table qualifiers cannot be used with the common attribute anywhere within the query. MySQL will allow table qualifiers on the common attribute anywhere except in the USING clause itself.

8-1d JOIN ON Clause

The previous two join styles use common attribute names in the joining tables. Another way to express a join when the tables have no common attribute names is to use the JOIN ON operand. The query will return only the rows that meet the indicated join condition. The join condition will typically include an equality comparison expression of two columns. (The columns may or may not share the same name, but obviously they must have comparable data types.) The syntax is:

SELECT *column-list* FROM *table1* JOIN *table2* ON *join-condition*

The following example performs a join of the INVOICE and LINE tables using the ON clause. The result is shown in Figure 8.3.

FIGURE 8.3 JOIN ON RESULTS

```
SQL Plus                                                    □ ▣ X

SQL> SELECT INVOICE.INV_NUMBER, PRODUCT.P_CODE, P_DESCRIPT, LINE_UNITS, LINE_PRICE
  2  FROM INVOICE JOIN LINE ON INVOICE.INV_NUMBER = LINE.INV_NUMBER
  3       JOIN PRODUCT ON LINE.P_CODE = PRODUCT.P_CODE;

INV_NUMBER P_CODE     P_DESCRIPT                         LINE_UNITS LINE_PRICE
---------- ---------- -------------------------------    ---------- ----------
      1001 13-Q2/P2   7.25-in. pwr. saw blade                     1      14.99
      1001 23109-HB   Claw hammer                                 1       9.95
      1002 54778-2T   Rat-tail file, 1/8-in. fine                 2       4.99
      1003 2238/QPD   B&D cordless drill, 1/2-in.                 1      38.95
      1003 1546-QQ2   Hrd. cloth, 1/4-in., 2x50                   1      39.95
      1003 13-Q2/P2   7.25-in. pwr. saw blade                     5      14.99
      1004 54778-2T   Rat-tail file, 1/8-in. fine                 3       4.99
      1004 23109-HB   Claw hammer                                 2       9.95
      1005 PVC23DRT   PVC pipe, 3.5-in., 8-ft                    12       5.87
      1006 SM-18277   1.25-in. metal screw, 25                    3       6.99
      1006 2232/QTY   B&D jigsaw, 12-in. blade                    1     109.92
      1006 23109-HB   Claw hammer                                 1       9.95
      1006 89-WRE-Q   Hicut chain saw, 16 in.                     1     256.99
      1007 13-Q2/P2   7.25-in. pwr. saw blade                     2      14.99
      1007 54778-2T   Rat-tail file, 1/8-in. fine                 1       4.99
      1008 PVC23DRT   PVC pipe, 3.5-in., 8-ft                     5       5.87
      1008 WR3/TT3    Steel matting, 4'x8'x1/6", .5" mesh         3     119.95
      1008 23109-HB   Claw hammer                                 1       9.95

18 rows selected.

SQL> _
```

SELECT	INVOICE.INV_NUMBER, PRODUCT.P_CODE, P_DESCRIPT, LINE_UNITS, LINE_PRICE
FROM	INVOICE JOIN LINE ON INVOICE.INV_NUMBER = LINE.INV_NUMBER JOIN PRODUCT ON LINE.P_CODE = PRODUCT.P_CODE;

Unlike the NATURAL JOIN and JOIN USING operands, the JOIN ON clause requires a table qualifier for the common attributes. If you do not specify the table qualifier, you will get a "column ambiguously defined" error message.

Keep in mind that the JOIN ON syntax lets you perform a join even when the tables do not share a common attribute name. For example, to generate a list of all employees with the managers' names, you can use the following (recursive) query:

SELECT	E.EMP_MGR, M.EMP_LNAME, E.EMP_NUM, E.EMP_LNAME
FROM	EMP E JOIN EMP M ON E.EMP_MGR = M.EMP_NUM
ORDER BY	E.EMP_MGR;

Note

Oracle, MS SQL Server, MySQL, and Access all support the JOIN ON syntax. In many environments, including the SQL code generated by Access when queries are created using the QBE window, it is common to include the optional word INNER to the join syntax. For example,

SELECT	P.P_CODE, P.P_DESCRIPT, V.V_CODE, V.V_NAME
FROM	PRODUCT P INNER JOIN VENDOR V ON P.V_CODE = V.V_CODE;

8-1e Outer Joins

An outer join returns not only the rows matching the join condition (that is, rows with matching values in the common columns), it returns the rows with unmatched values. The ANSI standard defines three types of outer joins: left, right, and full. The left and right designations reflect the order in which the tables are processed by the DBMS. Remember that join operations take place two tables at a time. The first table named in the FROM clause will be the left side, and the second table named will be the right side. If three or more tables are being joined, the result of joining the first two tables becomes the left side, and the third table becomes the right side.

The left outer join returns not only the rows matching the join condition (that is, rows with matching values in the common column), it returns the rows in the left table with unmatched values in the right table. The syntax is:

SELECT   column-list
FROM   table1 LEFT [OUTER] JOIN table2 ON join-condition

For example, the following query lists the product code, vendor code, and vendor name for all products and includes those vendors with no matching products:

SELECT   P_CODE, VENDOR.V_CODE, V_NAME
FROM   VENDOR LEFT JOIN PRODUCT ON VENDOR.
              V_CODE = PRODUCT.V_CODE;

The preceding SQL code and its results are shown in Figure 8.4.

FIGURE 8.4 LEFT JOIN RESULTS

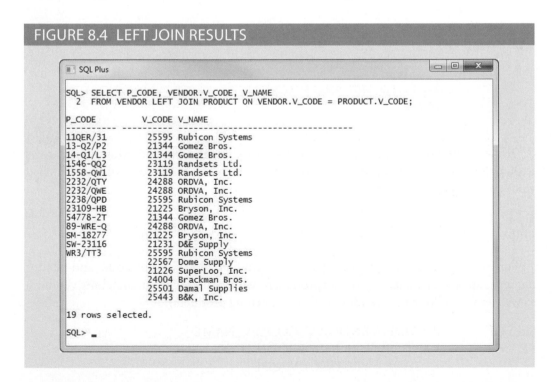

The right outer join returns not only the rows matching the join condition (that is, rows with matching values in the common column), it returns the rows in the right table with unmatched values in the left table. The syntax is:

SELECT   column-list
FROM   table1 RIGHT [OUTER] JOIN table2 ON join-condition

For example, the following query lists the product code, vendor code, and vendor name for all products and includes products that do not have a matching vendor code:

SELECT   P_CODE, VENDOR.V_CODE, V_NAME
FROM   VENDOR RIGHT JOIN PRODUCT ON VENDOR.
                    V_CODE = PRODUCT.V_CODE;

The SQL code and its output are shown in Figure 8.5.

FIGURE 8.5 RIGHT JOIN RESULTS

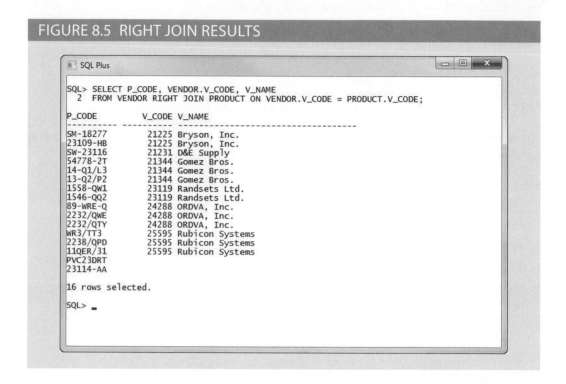

The full outer join returns not only the rows matching the join condition (that is, rows with matching values in the common column), it returns all of the rows with unmatched values in the table on either side. The syntax is:

SELECT   column-list
FROM   *table1* FULL [OUTER] JOIN *table2* ON *join-condition*

For example, the following query lists the product code, vendor code, and vendor name for all products and includes all product rows (products without matching vendors) as well as all vendor rows (vendors without matching products):

SELECT   P_CODE, VENDOR.V_CODE, V_NAME
FROM   VENDOR FULL JOIN PRODUCT ON VENDOR.
                    V_CODE = PRODUCT.V_CODE;

The SQL code and its results are shown in Figure 8.6.

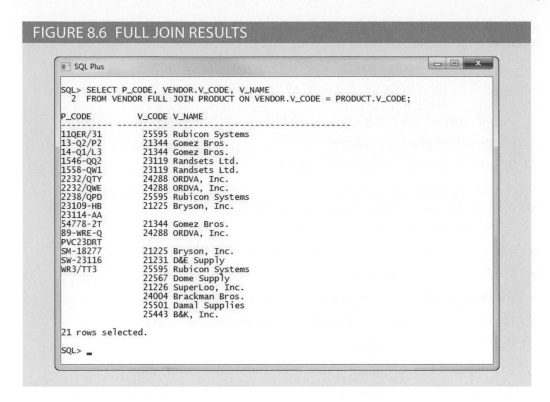

FIGURE 8.6 FULL JOIN RESULTS

Note

Oracle and MS SQL Server support the FULL JOIN syntax. MySQL and Access do not.

8-2 Subqueries and Correlated Queries

The use of joins in a relational database allows you to get information from two or more tables. For example, the following query allows you to get customer data with its respective invoices by joining the CUSTOMER and INVOICE tables.

```
SELECT      INV_NUMBER, INVOICE.CUS_CODE, CUS_LNAME,
            CUS_FNAME
FROM        CUSTOMER, INVOICE
WHERE       CUSTOMER.CUS_CODE = INVOICE.CUS_CODE;
```

In the previous query, the data from both tables (CUSTOMER and INVOICE) is processed at once, matching rows with shared CUS_CODE values.

However, it is often necessary to process data based on *other* processed data. For example, suppose that you want to generate a list of vendors who do not provide products. (Recall that not all vendors in the VENDOR table have provided products—some are only *potential* vendors.) In Chapter 7, you learned that you could generate such a list by writing the following query:

```
SELECT      V_CODE, V_NAME FROM VENDOR
WHERE       V_CODE NOT IN (SELECT V_CODE FROM PRODUCT);
```

Similarly, to generate a list of all products with a price greater than or equal to the average product price, you can write the following query:

```
SELECT      P_CODE, P_PRICE FROM PRODUCT
WHERE       P_PRICE >= (SELECT AVG(P_PRICE) FROM PRODUCT);
```

In both queries, you needed to get information that was not previously known:

- What vendors provide products?
- What is the average price of all products?

In both cases, you used a subquery to generate the required information, which could then be used as input for the originating query. You learned how to use subqueries in Chapter 7; review their basic characteristics:

- A subquery is a query (SELECT statement) inside another query.
- A subquery is normally expressed inside parentheses.
- The first query in the SQL statement is known as the outer query.
- The query inside the SQL statement is known as the inner query.
- The inner query is executed first.
- The output of an inner query is used as the input for the outer query.
- The entire SQL statement is sometimes referred to as a nested query.

In this section, you learn more about the practical use of subqueries. You already know that a subquery is based on the use of the SELECT statement to return one or more values to another query, but subqueries have a wide range of uses. For example, you can use a subquery within a SQL data manipulation language (DML) statement such as INSERT, UPDATE, or DELETE, in which a value or list of values (such as multiple vendor codes or a table) is expected. Table 8.2 uses simple examples to summarize the use of SELECT subqueries in DML statements.

TABLE 8.2

SELECT SUBQUERY EXAMPLES

SELECT SUBQUERY EXAMPLES	EXPLANATION
INSERT INTO PRODUCT SELECT * FROM P;	Inserts all rows from Table P into the PRODUCT table. Both tables must have the same attributes. The subquery returns all rows from Table P.

UPDATE PRODUCT
SET P_PRICE = (SELECT AVG(P_PRICE)
        FROM PRODUCT)
WHERE V_CODE IN (SELECT V_CODE
        FROM VENDOR
        WHERE V_AREACODE = '615')
	Updates the product price to the average product price, but only for products provided by vendors who have an area code equal to 615. The first subquery returns the average price; the second subquery returns the list of vendors with an area code equal to 615.

DELETE FROM PRODUCT
WHERE V_CODE IN (SELECT V_CODE
        FROM VENDOR
        WHERE V_AREACODE = '615')
	Deletes the PRODUCT table rows provided by vendors with an area code equal to 615. The subquery returns the list of vendor codes with an area code equal to 615.

Using the examples in Table 8.2, note that the subquery is always on the right side of a comparison or assigning expression. Also, a subquery can return one or more values. To be precise, the subquery can return the following:

- *One single value (one column and one row).* This subquery is used anywhere a single value is expected, as in the right side of a comparison expression. An example is the preceding UPDATE subquery, in which you assigned the average price to the product's price. Obviously, when you assign a value to an attribute, you are assigning a single value, not a list of them. Therefore, the subquery must return only one value (one column, one row). If the query returns multiple values, the DBMS will generate an error.

- *A list of values (one column and multiple rows).* This type of subquery is used anywhere a list of values is expected, such as when using the IN clause—for example, when comparing the vendor code to a list of vendors. Again, in this case, there is only one column of data with multiple value instances. This type of subquery is used frequently in combination with the IN operator in a WHERE conditional expression.

- *A virtual table (multicolumn, multirow set of values).* This type of subquery can be used anywhere a table is expected, such as when using the FROM clause. You will see an example later in this chapter.

It is important to note that a subquery can return no values at all; it is a NULL. In such cases, the output of the outer query might result in an error or a null empty set, depending on where the subquery is used (in a comparison, an expression, or a table set).

In the following sections, you will learn how to write subqueries within the SELECT statement to retrieve data from the database.

8-2a WHERE Subqueries

The most common type of subquery uses an inner SELECT subquery on the right side of a WHERE comparison expression. For example, to find all products with a price greater than or equal to the average product price, you write the following query:

```
SELECT       P_CODE, P_PRICE FROM PRODUCT
WHERE        P_PRICE >= (SELECT AVG(P_PRICE) FROM PRODUCT);
```

The output of the preceding query is shown in Figure 8.7. Note that this type of query, when used in a >, <, =, >=, or <= conditional expression, requires a subquery that returns only one value (one column, one row). The value generated by the subquery must be of a comparable data type; if the attribute to the left of the comparison symbol is a character type, the subquery must return a character string. Also, if the query returns more than a single value, the DBMS will generate an error.

Subqueries can also be used in combination with joins. For example, the following query lists all customers who ordered a claw hammer:

```
SELECT       DISTINCT CUS_CODE, CUS_LNAME, CUS_FNAME
FROM         CUSTOMER     JOIN INVOICE USING (CUS_CODE)
                          JOIN LINE USING (INV_NUMBER)
                          JOIN PRODUCT USING (P_CODE)
WHERE        P_CODE = (SELECT P_CODE FROM PRODUCT WHERE
             P_DESCRIPT = 'Claw hammer');
```

The result of the query is shown in Figure 8.7.

FIGURE 8.7 WHERE SUBQUERY EXAMPLES

```
SQL Plus

SQL> SELECT P_CODE, P_PRICE FROM PRODUCT
  2  WHERE P_PRICE >= (SELECT AVG(P_PRICE) FROM PRODUCT);

P_CODE        P_PRICE
----------    ----------
11QER/31       109.99
2232/QTY       109.92
2232/QWE        99.87
89-WRE-Q       256.99
WR3/TT3        119.95

SQL> SELECT DISTINCT CUS_CODE, CUS_LNAME, CUS_FNAME
  2  FROM CUSTOMER JOIN INVOICE USING (CUS_CODE)
  3               JOIN LINE USING (INV_NUMBER)
  4               JOIN PRODUCT USING (P_CODE)
  5  WHERE P_CODE = (SELECT P_CODE FROM PRODUCT WHERE P_DESCRIPT = 'Claw hammer');

  CUS_CODE CUS_LNAME        CUS_FNAME
---------- ---------------- ----------------
     10011 Dunne            Leona
     10014 Orlando          Myron

SQL>
```

In the preceding example, the inner query finds the P_CODE for the claw hammer. The P_CODE is then used to restrict the selected rows to those in which the P_CODE in the LINE table matches the P_CODE for "Claw hammer." Note that the previous query could have been written this way:

```
SELECT      DISTINCT CUS_CODE, CUS_LNAME, CUS_FNAME
FROM        CUSTOMER     JOIN INVOICE USING (CUS_CODE)
                         JOIN LINE USING (INV_NUMBER)
                         JOIN PRODUCT USING (P_CODE)
WHERE       P_DESCRIPT = 'Claw hammer';
```

If the original query encounters the "Claw hammer" string in more than one product description, you get an error message. To compare one value to a list of values, you must use an IN operand, as shown in the next section.

8-2b IN Subqueries

What if you wanted to find all customers who purchased a hammer or any kind of saw or saw blade? Note that the product table has two different types of hammers: a claw hammer and a sledge hammer. Also, there are multiple occurrences of products that contain "saw" in their product descriptions, including saw blades and jigsaws. In such cases, you need to compare the P_CODE not to one product code (a single value), but to a list of product code values. When you want to compare a single attribute to a list of values, you use the IN operator. When the P_CODE values are not known beforehand, but they can be derived using a query, you must use an IN subquery. The following example lists all customers who have purchased hammers, saws, or saw blades.

```
SELECT      DISTINCT CUS_CODE, CUS_LNAME, CUS_FNAME
FROM        CUSTOMER    JOIN INVOICE USING (CUS_CODE)
                        JOIN LINE USING (INV_NUMBER)
                        JOIN PRODUCT USING (P_CODE)
WHERE       P_CODE IN   (SELECT P_CODE FROM PRODUCT
                        WHERE P_DESCRIPT LIKE '%hammer%'
                        OR P_DESCRIPT LIKE '%saw%');
```

The result of the query is shown in Figure 8.8.

FIGURE 8.8 IN SUBQUERY EXAMPLE

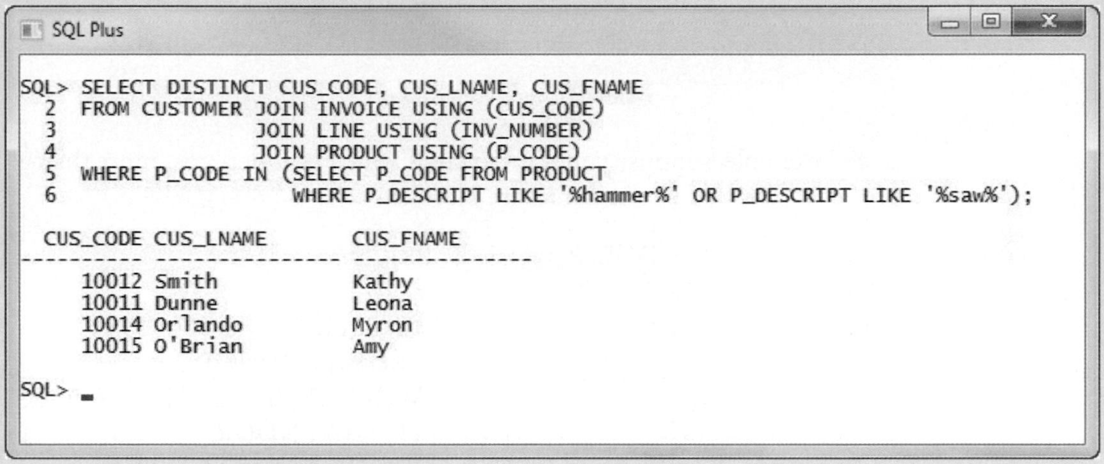

```
SQL> SELECT DISTINCT CUS_CODE, CUS_LNAME, CUS_FNAME
  2  FROM CUSTOMER JOIN INVOICE USING (CUS_CODE)
  3                JOIN LINE USING (INV_NUMBER)
  4                JOIN PRODUCT USING (P_CODE)
  5  WHERE P_CODE IN (SELECT P_CODE FROM PRODUCT
  6                   WHERE P_DESCRIPT LIKE '%hammer%' OR P_DESCRIPT LIKE '%saw%');

CUS_CODE CUS_LNAME        CUS_FNAME
-------- ---------------- ----------------
   10012 Smith            Kathy
   10011 Dunne            Leona
   10014 Orlando          Myron
   10015 O'Brian          Amy

SQL>
```

8-2c HAVING Subqueries

Just as you can use subqueries with the WHERE clause, you can use a subquery with a HAVING clause. The HAVING clause is used to restrict the output of a GROUP BY query by applying conditional criteria to the grouped rows. For example, to list all products with a total quantity sold greater than the average quantity sold, you would write the following query:

```
SELECT      P_CODE, SUM(LINE_UNITS)
FROM        LINE
GROUP BY    P_CODE
HAVING      SUM(LINE_UNITS) > (SELECT AVG(LINE_UNITS) FROM LINE);
```

The result of the query is shown in Figure 8.9.

8-2d Multirow Subquery Operators: ANY and ALL

So far, you have learned that you must use an IN subquery to compare a value to a list of values. However, the IN subquery uses an equality operator; that is, it selects only those rows that are equal to at least one of the values in the list. What happens if you need to make an inequality comparison (> or <) of one value to a list of values?

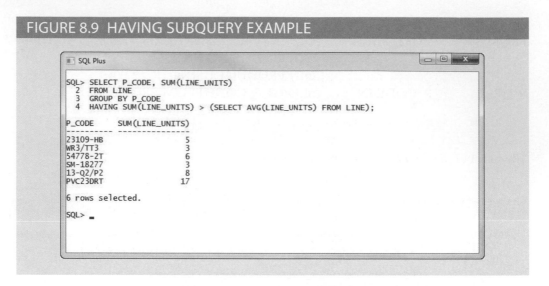

FIGURE 8.9 HAVING SUBQUERY EXAMPLE

For example, suppose you want to know which products cost more than all individual products provided by vendors from Florida:

```
SELECT      P_CODE, P_QOH * P_PRICE
FROM        PRODUCT
WHERE       P_QOH * P_PRICE > ALL (SELECT P_QOH * P_PRICE
                                   FROM PRODUCT
                                   WHERE V_CODE IN      (SELECT V_CODE
                                   FROM VENDOR
                                   WHERE V_STATE = 'FL'));
```

The result of the query is shown in Figure 8.10.

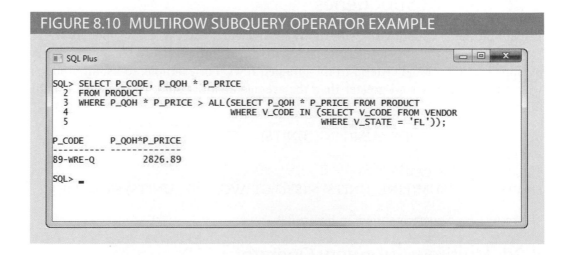

FIGURE 8.10 MULTIROW SUBQUERY OPERATOR EXAMPLE

It is important to note the following points about the query and its output in Figure 8.10:

- The query is a typical example of a nested query.
- The query has one outer SELECT statement with a SELECT subquery (call it sqA) that contains a second SELECT subquery (call it sqB).

- The last SELECT subquery (sqB) is executed first and returns a list of all vendors from Florida.

- The first SELECT subquery (sqA) uses the output of the second SELECT subquery (sqB). The sqA subquery returns the list of costs for all products provided by vendors from Florida.

- The use of the ALL operator allows you to compare a single value (P_QOH * P_PRICE) with a list of values returned by the first subquery (sqA) using a comparison operator other than equals.

- For a row to appear in the result set, it has to meet the criterion P_QOH * P_PRICE > ALL of the individual values returned by the subquery sqA. The values returned by sqA are a list of product costs. In fact, "greater than ALL" is equivalent to "greater than the highest product cost of the list." In the same way, a condition of "less than ALL" is equivalent to "less than the lowest product cost of the list."

Another powerful operator is the ANY multirow operator, which you can consider the cousin of the ALL multirow operator. The ANY operator allows you to compare a single value to a list of values and select only the rows for which the inventory cost is greater than or less than any value in the list. You could use the equal to ANY operator, which would be the equivalent of the IN operator.

8-2e FROM Subqueries

So far you have seen how the SELECT statement uses subqueries within WHERE, HAVING, and IN statements, and how the ANY and ALL operators are used for multirow subqueries. In all of those cases, the subquery was part of a conditional expression, and it always appeared at the right side of the expression. In this section, you will learn how to use subqueries in the FROM clause.

As you already know, the FROM clause specifies the table(s) from which the data will be drawn. Because the output of a SELECT statement is another table (or more precisely, a "virtual" table), you could use a SELECT subquery in the FROM clause. For example, assume that you want to know all customers who have purchased products 13-Q2/P2 *and* 23109-HB. All product purchases are stored in the LINE table, so you can easily find out who purchased any given product by searching the P_CODE attribute in the LINE table. In this case, however, you want to know all customers who purchased both products, not just one. You could write the following query:

```
SELECT     DISTINCT CUSTOMER.CUS_CODE, CUSTOMER.CUS_LNAME
FROM       CUSTOMER,
           (SELECT INVOICE.CUS_CODE FROM INVOICE NATURAL JOIN LINE
           WHERE P_CODE = '13-Q2/P2') CP1,
           (SELECT INVOICE.CUS_CODE FROM INVOICE NATURAL JOIN LINE
           WHERE P_CODE = '23109-HB') CP2
WHERE      CUSTOMER.CUS_CODE = CP1.CUS_CODE AND
           CP1.CUS_CODE = CP2.CUS_CODE;
```

The result of the query is shown in Figure 8.11.

Note in Figure 8.11 that the first subquery returns all customers who purchased product 13-Q2/P2, while the second subquery returns all customers who purchased

FIGURE 8.11 FROM SUBQUERY EXAMPLE

```
SQL Plus

SQL> SELECT DISTINCT CUSTOMER.CUS_CODE, CUSTOMER.CUS_LNAME
  2  FROM CUSTOMER,
  3        (SELECT INVOICE.CUS_CODE FROM INVOICE NATURAL JOIN LINE WHERE P_CODE = '13-Q2/P2') CP1,
  4        (SELECT INVOICE.CUS_CODE FROM INVOICE NATURAL JOIN LINE WHERE P_CODE = '23109-HB') CP2
  5  WHERE CUSTOMER.CUS_CODE = CP1.CUS_CODE AND
  6        CP1.CUS_CODE = CP2.CUS_CODE;

 CUS_CODE CUS_LNAME
---------- ---------------
    10014 Orlando

SQL>
```

product 23109-HB. So, in this FROM subquery, you are joining the CUSTOMER table with two virtual tables. The join condition selects only the rows with matching CUS_CODE values in each table (base or virtual).

8-2f Attribute List Subqueries

The SELECT statement uses the attribute list to indicate what columns to project in the resulting set. Those columns can be attributes of base tables, computed attributes, or the result of an aggregate function. The attribute list can also include a subquery expression, also known as an *inline subquery*. A subquery in the attribute list must return one value; otherwise, an error code is raised. For example, a simple inline query can be used to list the difference between each product's price and the average product price:

```
SELECT      P_CODE, P_PRICE, (SELECT AVG(P_PRICE) FROM PRODUCT)
            AS AVGPRICE,
            P_PRICE – (SELECT AVG(P_PRICE) FROM PRODUCT) AS DIFF
FROM        PRODUCT;
```

Figure 8.12 shows the result of the query.

In Figure 8.12, note that the inline query output returns one value (the average product's price) and that the value is the same in every row. Note also that the query uses the full expression instead of the column aliases when computing the difference. In fact, if you try to use the alias in the difference expression, you will get an error message. The column alias cannot be used in computations in the attribute list when the alias is defined in the same attribute list. That DBMS requirement is the result of the way the DBMS parses and executes queries.

Another example will help you understand the use of attribute list subqueries and column aliases. For example, suppose that you want to know the product code, the total sales by product, and the contribution by employee of each product's sales. To get the sales by product, you need to use only the LINE table. To compute the contribution by employee, you need to know the number of employees (from the EMPLOYEE table). As you study the tables' structures, you can see that the LINE and EMPLOYEE tables do not share a common attribute. In fact, you do not need a common attribute. You only need to know the total number of employees, not the

FIGURE 8.12 INLINE SUBQUERY EXAMPLE

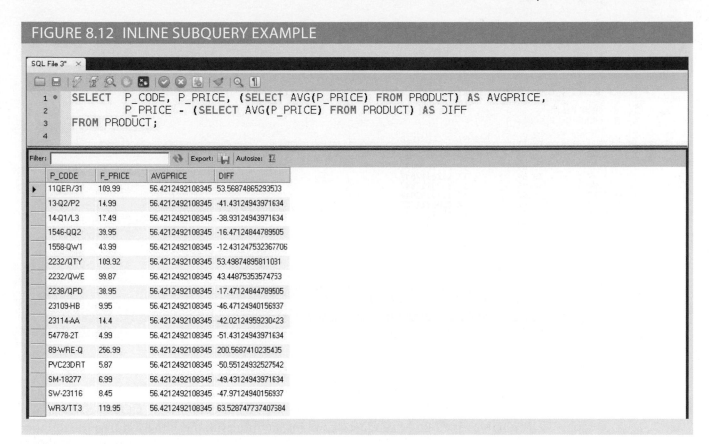

```
SELECT  P_CODE, P_PRICE, (SELECT AVG(P_PRICE) FROM PRODUCT) AS AVGPRICE,
        P_PRICE - (SELECT AVG(P_PRICE) FROM PRODUCT) AS DIFF
FROM PRODUCT;
```

P_CODE	P_PRICE	AVGPRICE	DIFF
11QER/31	109.99	56.4212492108345	53.5687486529353
13-Q2/P2	14.99	56.4212492108345	-41.43124943971634
14-Q1/L3	17.49	56.4212492108345	-38.93124943971634
1546-QQ2	39.95	56.4212492108345	-16.47124844789505
1558-QW1	43.99	56.4212492108345	-12.431247532367706
2232/QTY	109.92	56.4212492108345	53.49874895811031
2232/QWE	99.87	56.4212492108345	43.44875353574753
2238/QPD	38.95	56.4212492108345	-17.47124844789505
23109-HB	9.95	56.4212492108345	-46.47124940156937
23114-AA	14.4	56.4212492108345	-42.02124959230423
54778-2T	4.99	56.4212492108345	-51.43124943971634
89-WRE-Q	256.99	56.4212492108345	200.5687410235405
PVC23DRT	5.87	56.4212492108345	-50.55124932527542
SM-18277	6.99	56.4212492108345	-49.43124943971634
SW-23116	8.45	56.4212492108345	-47.97124940156937
WR3/TT3	119.95	56.4212492108345	63.528747737407384

total employees related to each product. So, to answer the query, you would write the following code:

```
SELECT      P_CODE, SUM(LINE_UNITS * LINE_PRICE) AS SALES,
            (SELECT COUNT(*) FROM EMPLOYEE) AS ECOUNT,
            SUM(LINE_UNITS * LINE_PRICE)/(SELECT COUNT(*) FROM
            EMPLOYEE) AS CONTRIB
FROM        LINE
GROUP BY    P_CODE;
```

The result of the query is shown in Figure 8.13.

As you can see in Figure 8.13, the number of employees remains the same for each row in the result set. The use of this type of subquery is limited to certain instances when you need to include data from other tables that is not directly related to a main table or tables in the query. The value will remain the same for each row, like a constant in a programming language. (You will learn another use of inline subqueries in Section 8-2g, Correlated Subqueries.) Note that you cannot use an alias in the attribute list to write the expression that computes the contribution per employee.

Another way to write the same query by using column aliases requires the use of a subquery in the FROM clause, as follows:

```
SELECT      P_CODE, SALES, ECOUNT, SALES/ECOUNT AS CONTRIB
FROM        (SELECT P_CODE, SUM(LINE_UNITS * LINE_PRICE) AS SALES,
                    (SELECT COUNT(*) FROM EMPLOYEE) AS ECOUNT
            FROM        LINE
            GROUP BY    P_CODE);
```

FIGURE 8.13 ANOTHER EXAMPLE OF AN INLINE SUBQUERY

```
SQL Plus                                                                    _ □ x

SQL> SELECT P_CODE, SUM(LINE_UNITS * LINE_PRICE) AS SALES, (SELECT COUNT(*) FROM EMPLOYEE) AS ECOUNT,
  2        SUM(LINE_UNITS * LINE_PRICE) / (SELECT COUNT(*) FROM EMPLOYEE) AS CONTRIB
  3  FROM LINE
  4  GROUP BY P_CODE;

P_CODE          SALES      ECOUNT     CONTRIB
----------  ----------  ----------  ----------
23109-HB        49.75          17  2.92647059
WR3/TT3        359.85          17  21.1676471
54778-2T        29.94          17  1.76117647
2238/QPD        38.95          17  2.29117647
SM-18277        20.97          17  1.23352941
13-Q2/P2       119.92          17  7.05411765
89-WRE-Q       256.99          17  15.1170588
2232/QTY       109.92          17  6.46588235
PVC23DRT        99.79          17        5.87
1546-QQ2        39.95          17        2.35

10 rows selected.

SQL> _
```

In this case, you are actually using two subqueries. The subquery in the FROM clause executes first and returns a virtual table with three columns: P_CODE, SALES, and ECOUNT. The FROM subquery contains an inline subquery that returns the number of employees as ECOUNT. Because the outer query receives the output of the inner query, you can now refer to the columns in the outer subquery by using the column aliases.

8-2g Correlated Subqueries

Until now, all subqueries you have learned execute independently. That is, each subquery in a command sequence executes in a serial fashion, one after another. The inner subquery executes first; its output is used by the outer query, which then executes until the last outer query finishes (the first SQL statement in the code).

In contrast, a **correlated subquery** is a subquery that executes once for each row in the outer query. The process is similar to the typical nested loop in a programming language. For example:

FOR X = 1 TO 2
    FOR Y = 1 TO 3
        PRINT "X = "X, "Y = "Y
    END
END

will yield the following output:

X = 1   Y = 1
X = 1   Y = 2
X = 1   Y = 3
X = 2   Y = 1
X = 2   Y = 2
X = 2   Y = 3

correlated subquery
A subquery that executes once for each row in the outer query.

Note that the outer loop X = 1 TO 2 begins the process by setting X = 1, and then the inner loop Y = 1 TO 3 is completed for each X outer loop value. The relational DBMS uses the same sequence to produce correlated subquery results:

1. It initiates the outer query.
2. For each row of the outer query result set, it executes the inner query by passing the outer row to the inner query.

This process is the opposite of that of the subqueries, as you have already seen. The query is called a *correlated* subquery because the inner query is *related* to the outer query; the inner query references a column of the outer subquery.

To see the correlated subquery in action, suppose that you want to know all product sales in which the units sold value is greater than the average units sold value *for that product* (as opposed to the average for *all* products). In that case, the following procedure must be completed:

1. Compute the average units sold for a product.
2. Compare the average computed in Step 1 to the units sold in each sale row, and then select only the rows in which the number of units sold is greater.

The following correlated query completes the preceding two-step process:

```
SELECT      INV_NUMBER, P_CODE, LINE_UNITS
FROM        LINE LS
WHERE       LS.LINE_UNITS > (SELECT AVG(LINE_UNITS)
                            FROM LINE LA
                            WHERE LA.P_CODE = LS.P_CODE);
```

The first example in Figure 8.14 shows the result of the query.

FIGURE 8.14 CORRELATED SUBQUERY EXAMPLES

```
SQL> SELECT INV_NUMBER, P_CODE, LINE_UNITS
  2  FROM LINE LS
  3  WHERE LS.LINE_UNITS > (SELECT AVG(LINE_UNITS)
  4                         FROM LINE LA
  5                         WHERE LA.P_CODE = LS.P_CODE);

INV_NUMBER P_CODE     LINE_UNITS
---------- ---------- ----------
      1003 13-Q2/P2            5
      1004 54778-2T           3
      1004 23109-HB           2
      1005 PVC23DRT          12

SQL> SELECT INV_NUMBER, P_CODE, LINE_UNITS,
  2         (SELECT AVG(LINE_UNITS) FROM LINE LX WHERE LX.P_CODE = LS.P_CODE) AS AVG
  3  FROM LINE LS
  4  WHERE LS.LINE_UNITS > (SELECT AVG(LINE_UNITS)
  5                         FROM LINE LA
  6                         WHERE LA.P_CODE = LS.P_CODE);

INV_NUMBER P_CODE     LINE_UNITS        AVG
---------- ---------- ---------- ----------
      1004 23109-HB           2       1.25
      1004 54778-2T           3          2
      1003 13-Q2/P2           5 2.66666667
      1005 PVC23DRT          12        8.5

SQL> _
```

In the top query and its result in Figure 8.14, note that the LINE table is used more than once, so you must use table aliases. In this case, the inner query computes the average units sold of the product that matches the P_CODE of the outer query P_CODE. That is, the inner query runs once, using the first product code found in the outer LINE table, and it returns the average sale for that product. When the number of units sold in the outer LINE row is greater than the average computed, the row is added to the output. Then the inner query runs again, this time using the second product code found in the outer LINE table. The process repeats until the inner query has run for all rows in the outer LINE table. In this case, the inner query will be repeated as many times as there are rows in the outer query.

To verify the results and to provide an example of how you can combine subqueries, you can add a correlated inline subquery to the previous query. (See the second query and its results in Figure 8.14.) As you can see, the new query contains a correlated inline subquery that computes the average units sold for each product. You not only get an answer, you can also verify that the answer is correct.

Correlated subqueries can also be used with the EXISTS special operator. For example, suppose that you want to know the names of all customers who have placed an order lately. In that case, you could use a correlated subquery like the first one shown in Figure 8.15.

FIGURE 8.15 EXISTS CORRELATED SUBQUERY EXAMPLES

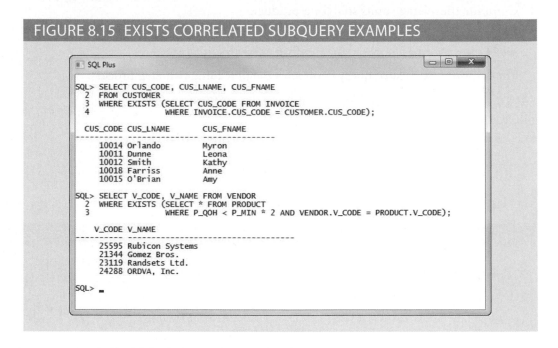

```
SELECT      CUS_CODE, CUS_LNAME, CUS_FNAME
FROM        CUSTOMER
WHERE       EXISTS      (SELECT      CUS_CODE FROM INVOICE
                         WHERE       INVOICE.CUS_CODE =
                                     CUSTOMER.CUS_CODE);
```

The second example in Figure 8.15 will help you understand how to use correlated queries. For example, suppose that you want to know what vendors you must contact to order products that are approaching the minimum quantity-on-hand value. In particular, you want to know the vendor code and vendor name for products with a quantity on hand that is less than double the minimum quantity. The query that answers the question is as follows:

```
SELECT      V_CODE, V_NAME
FROM        VENDOR
WHERE       EXISTS      (SELECT      *
                         FROM        PRODUCT
                         WHERE       P_QOH < P_MIN * 2
                         AND         VENDOR.V_CODE = PRODUCT.V_CODE);
```

In the second query in Figure 8.15, note that:

1. The inner correlated subquery runs using the first vendor.

2. If any products match the condition (the quantity on hand is less than double the minimum quantity), the vendor code and name are listed in the output.

3. The correlated subquery runs using the second vendor, and the process repeats itself until all vendors are used.

8-3 SQL Functions

The data in databases is the basis of critical business information. Generating information from data often requires many data manipulations. Sometimes such data manipulation involves the decomposition of data elements. For example, an employee's date of birth can be subdivided into a day, a month, and a year. A product manufacturing code (for example, SE-05-2-09-1234-1-3/12/16-19:26:48) can be designed to record the manufacturing region, plant, shift, production line, employee number, date, and time. For years, conventional programming languages have had special functions that enabled programmers to perform data transformations like the preceding data decompositions. If you know a modern programming language, it is very likely that the SQL functions in this section will look familiar.

SQL functions are very useful tools. You'll need to use functions when you want to list all employees ordered by year of birth, or when your marketing department wants you to generate a list of all customers ordered by zip code and the first three digits of their telephone numbers. In both of these cases, you'll need to use data elements that are not present as such in the database. Instead, you will need a SQL function that can be derived from an existing attribute. Functions always use a numerical, date, or string value. The value may be part of the command itself (a constant or literal) or it may be an attribute located in a table. Therefore, a function may appear anywhere in a SQL statement where a value or an attribute can be used.

There are many types of SQL functions, such as arithmetic, trigonometric, string, date, and time functions. This section will not explain all of these functions in detail, but it will give you a brief overview of the most useful ones.

Note

Although the main DBMS vendors support the SQL functions covered here, the syntax or degree of support will probably differ. In fact, DBMS vendors invariably add their own functions to products to lure new customers. The functions covered in this section represent just a small portion of functions supported by your DBMS. Read your DBMS SQL reference manual for a complete list of available functions.

8-3a Date and Time Functions

All SQL-standard DBMSs support date and time functions. All date functions take one parameter of a date or character data type and return a value (character, numeric, or date type). Unfortunately, date/time data types are implemented differently by different DBMS vendors. The problem occurs because the ANSI SQL standard defines date data types, but it does not specify how those data types are to be stored. Instead, it lets the vendor deal with that issue.

Because date/time functions differ from vendor to vendor, this section will cover basic date/time functions for MS Access, SQL Server, and Oracle. Table 8.3 shows a list of selected MS Access and SQL Server date/time functions.

TABLE 8.3

SELECTED MS ACCESS AND SQL SERVER DATA/TIME FUNCTIONS

FUNCTION	EXAMPLE(S)
CONVERT (MS SQL Server) Convert can be used to perform a wide array of data type conversions as discussed next. It can also be used to format date data. Syntax: CONVERT(varchar(length), date_value, fmt_code) fmt_code = format used; can be: 1: MM/DD/YY 101: MM/DD/YYYY 2: YY.MM.DD 102: YYYY.MM.DD 3: DD/MM/YY 103: DD/MM/YYYY	Displays the product code and date the product was last received into stock for all products: SELECT P_CODE, CONVERT(VARCHAR(8), P_INDATE, 1) FROM PRODUCT; SELECT P_CODE, CONVERT(VARCHAR(10), P_INDATE, 102) FROM PRODUCT;
YEAR Returns a four-digit year Syntax: YEAR(date_value)	Lists all employees born in 1982: SELECT EMP_LNAME, EMP_FNAME, EMP_DOB, YEAR(EMP_DOB) AS YEAR FROM EMPLOYEE WHERE YEAR(EMP_DOB) = 1982;
MONTH Returns a two-digit month code Syntax: MONTH(date_value)	Lists all employees born in November: SELECT EMP_LNAME, EMP_FNAME, EMP_DOB, MONTH(EMP_DOB) AS MONTH FROM EMPLOYEE WHERE MONTH(EMP_DOB) = 11;
DAY Returns the number of the day Syntax: DAY(date_value)	Lists all employees born on the 14th day of the month: SELECT EMP_LNAME, EMP_FNAME, EMP_DOB, DAY(EMP_DOB) AS DAY FROM EMPLOYEE WHERE DAY(EMP_DOB) = 14;
DATE() MS Access **GETDATE() SQL Server** Returns today's date	Lists how many days are left until Christmas: SELECT #25-Dec-2016# – DATE(); Note two features: • There is no FROM clause, which is acceptable in Access and MS SQL Server. • The Christmas date is enclosed in number signs (#) because you are doing date arithmetic. In MS SQL Server: Use GETDATE() to get the current system date. To compute the difference between dates, use the DATEDIFF function (see below).

TABLE 8.3   (CONTINUED)

SELECTED MS ACCESS AND SQL SERVER DATA/TIME FUNCTIONS

DATEADD SQL Server Adds a number of selected time periods to a date Syntax: DATEADD(datepart, number, date)	Adds a number of dateparts to a given date. Dateparts can be minutes, hours, days, weeks, months, quarters, or years. For example: SELECT DATEADD(day,90, P_INDATE) AS DueDate FROM PRODUCT; The preceding example adds 90 days to P_INDATE. In MS Access, use the following: SELECT P_INDATE+90 AS DueDate FROM PRODUCT;
DATEDIFF SQL Server Subtracts two dates Syntax: DATEDIFF(datepart, startdate, enddate)	Returns the difference between two dates expressed in a selected datepart. For example: SELECT DATEDIFF(day, P_INDATE, GETDATE()) AS DaysAgo FROM PRODUCT; In MS Access, use the following: SELECT DATE() - P_INDATE AS DaysAgo FROM PRODUCT;

Table 8.4 shows the equivalent date/time functions used in Oracle. Note that Oracle uses the same function (TO_CHAR) to extract the various parts of a date. Also, another function (TO_DATE) is used to convert character strings to a valid Oracle date format that can be used in date arithmetic.

TABLE 8.4

SELECTED ORACLE DATE/TIME FUNCTIONS

FUNCTION	EXAMPLE(S)

TO_CHAR
Returns a character string or a formatted string from a date value
Syntax:
TO_CHAR(date_value, fmt)
fmt = format used; can be:
MONTH: name of month
MON: three-letter month name
MM: two-digit month name
D: number for day of week
DD: number for day of month
DAY: name of day of week
YYYY: four-digit year value
YY: two-digit year value
	Lists all employees born in 1982:
	SELECT EMP_LNAME, EMP_FNAME, EMP_DOB,
	                TO_CHAR(EMP_DOB, 'YYYY') AS YEAR
	FROM EMPLOYEE
	WHERE TO_CHAR(EMP_DOB, 'YYYY') = '1982';
	Lists all employees born in November:
	SELECT EMP_LNAME, EMP_FNAME, EMP_DOB,
	                TO_CHAR(EMP_DOB, 'MM') AS MONTH
	FROM EMPLOYEE
	WHERE TO_CHAR(EMP_DOB, 'MM') = '11';
	Lists all employees born on the 14th day of the month:
	SELECT EMP_LNAME, EMP_FNAME, EMP_DOB,
	                TO_CHAR(EMP_DOB, 'DD') AS DAY
	FROM EMPLOYEE
	WHERE TO_CHAR(EMP_DOB, 'DD') = '14';

TABLE 8.4 (CONTINUED)

SELECTED ORACLE DATE/TIME FUNCTIONS

TO_DATE
Returns a date value using a character string and a date format mask; also used to translate a date between formats
Syntax:
TO_DATE(char_value, fmt)
fmt = format used; can be:
MONTH: name of month
MON: three-letter month name
MM: two-digit month name
D: number for day of week
DD: number for day of month
DAY: name of day of week
YYYY: four-digit year value
YY: two-digit year value
	Lists the approximate age of employees on the company's tenth anniversary date (11/25/2016):
	SELECT EMP_LNAME, EMP_FNAME,
	            EMP_DOB, '11/25/2016' AS ANIV_DATE,
	            (TO_DATE('11/25/2004','MM/DD/YYYY') - EMP_DOB)/365 AS YEARS
	FROM EMPLOYEE
	ORDER BY YEARS;
	Note the following:
	• '11/25/2016' is a text string, not a date.
	• The TO_DATE function translates the text string to a valid Oracle date used in date arithmetic.
	How many days are there between Thanksgiving and Christmas 2016?
	SELECT TO_DATE('2016/12/25','YYYY/MM/DD') –
	          TO_DATE('NOVEMBER 27, 2016','MONTH DD, YYYY')
	FROM DUAL;
	Note the following:
	• The TO_DATE function translates the text string to a valid Oracle date used in date arithmetic.
	• DUAL is Oracle's pseudo-table, used only for cases in which a table is not really needed.

SYSDATE Returns today's date	Lists how many days are left until Christmas: SELECT TO_DATE('25-Dec-2016','DD-MON-YYYY') - SYSDATE FROM DUAL; Notice two things: • DUAL is Oracle's pseudo-table, used only for cases in which a table is not really needed. • The Christmas date is enclosed in a TO_DATE function to translate the date to a valid date format.
ADD_MONTHS Adds a number of months or years to a date Syntax: ADD_MONTHS(date_value, n) n = number of months	Lists all products with their expiration date (two years from the purchase date): SELECT P_CODE, P_INDATE, ADD_MONTHS(P_INDATE,24) FROM PRODUCT ORDER BY ADD_MONTHS(P_INDATE,24);
LAST_DAY Returns the date of the last day of the month given in a date Syntax: LAST_DAY(date_value)	Lists all employees who were hired within the last seven days of a month: SELECT EMP_LNAME, EMP_FNAME, EMP_HIRE_DATE FROM EMPLOYEE WHERE EMP_HIRE_DATE >=LAST_DAY(EMP_HIRE_DATE)-7;

Table 8.5 shows the equivalent functions for MySQL.

TABLE 8.5

SELECTED MYSQL DATE/TIME FUNCTIONS

FUNCTION	EXAMPLE(S)
Date_Format Returns a character string or a formatted string from a date value Syntax: DATE_FORMAT(date_value, fmt) fmt = format used; can be: %M: name of month %m: two-digit month number %b: abbreviated month name %d: number of day of month %W: weekday name %a: abbreviated weekday name %Y: four-digit year %y: two-digit year	Displays the product code and date the product was last received into stock for all products: SELECT P_CODE, DATE_FORMAT(P_INDATE, '%m/%d/%y') FROM PRODUCT; SELECT P_CODE, DATE_FORMAT(P_INDATE, '%M %d, %Y') FROM PRODUCT;
YEAR Returns a four-digit year Syntax: YEAR(date_value)	Lists all employees born in 1982: SELECT EMP_LNAME, EMP_FNAME, EMP_DOB, YEAR(EMP_DOB) AS YEAR FROM EMPLOYEE WHERE YEAR(EMP_DOB) = 1982;
MONTH Returns a two-digit month code Syntax: MONTH(date_value)	Lists all employees born in November: SELECT EMP_LNAME, EMP_FNAME, EMP_DOB, MONTH(EMP_DOB) AS MONTH FROM EMPLOYEE WHERE MONTH(EMP_DOB) = 11;
DAY Returns the number of the day Syntax: DAY(date_value)	Lists all employees born on the 14th day of the month: SELECT EMP_LNAME, EMP_FNAME, EMP_DOB, DAY(EMP_DOB) AS DAY FROM EMPLOYEE WHERE DAY(EMP_DOB) = 14;
ADDDATE Adds a number of days to a date Syntax: ADDDATE(date_value, n) n = number of days **DATE_ADD** Adds a number of days, months, or years to a date. This is similar to ADDDATE except it is more robust. It allows the user to specify the date unit to add. Syntax: DATE_ADD(date, INTERVAL n unit) n = number to add unit = date unit, can be: DAY: add n days WEEK: add n weeks MONTH: add n months YEAR: add n years	List all products with the date they will have been on the shelf for 30 days. SELECT P_CODE, P_INDATE, ADDDATE(P_INDATE, 30) FROM PRODUCT ORDER BY ADDDATE(P_INDATE, 30); Lists all products with their expiration date (two years from the purchase date): SELECT P_CODE, P_INDATE, DATE_ADD(P_INDATE, INTERVAL 2 YEAR) FROM PRODUCT ORDER BY DATE_ADD(P_INDATE, INTERVAL 2 YEAR);
LAST_DAY Returns the date of the last day of the month given in a date Syntax: LAST_DAY(date_value)	Lists all employees who were hired within the last seven days of a month: SELECT EMP_LNAME, EMP_FNAME, EMP_HIRE_DATE FROM EMPLOYEE WHERE EMP_HIRE_DATE >= DATE_ADD(LAST_DAY (EMP_HIRE_DATE), INTERVAL -7 DAY);

8-3b Numeric Functions

Numeric functions can be grouped in many different ways, such as algebraic, trigonometric, and logarithmic. In this section, you will learn two very useful functions. Do not confuse the SQL aggregate functions you saw in the previous chapter with the numeric functions in this section. The first group operates over a set of values (multiple rows—hence, the name *aggregate functions*), while the numeric functions covered here operate over a single row. Numeric functions take one numeric parameter and return one value. Table 8.6 shows a selected group of available numeric functions.

TABLE 8.6

SELECTED NUMERIC FUNCTIONS

FUNCTION	EXAMPLE(S)
ABS Returns the absolute value of a number Syntax: ABS(numeric_value)	In Oracle, use the following: SELECT 1.95, −1.93, ABS(1.95), ABS(−1.93) FROM DUAL; In MS Access, MySQL, and MS SQL Server, use the following: SELECT 1.95, −1.93, ABS(1.95), ABS(−1.93);

ROUND
Rounds a value to a specified precision (number of digits)
Syntax:
ROUND(numeric_value, p)
p = precision
	Lists the product prices rounded to one and zero decimal places:
	SELECT P_CODE, P_PRICE,
	               ROUND(P_PRICE,1) AS PRICE1,
	               ROUND(P_PRICE,0) AS PRICE0
	FROM PRODUCT;

CEIL/CEILING/FLOOR Returns the smallest integer greater than or equal to a number or returns the largest integer equal to or less than a number, respectively Syntax: CEIL(numeric_value) Oracle or MySQL CEILING(numeric_value) MS SQL Server or MySQL FLOOR(numeric_value)	Lists the product price, the smallest integer greater than or equal to the product price, and the largest integer equal to or less than the product price. In Oracle or MySQL, use the following: SELECT P_PRICE, CEIL(P_PRICE), FLOOR(P_PRICE) FROM PRODUCT; In MS SQL Server or MySQL, use the following: SELECT P_PRICE, CEILING(P_PRICE), FLOOR(P_PRICE) FROM PRODUCT; MS Access does not support these functions. Note that MySQL supports both CEIL and CEILING.

8-3c String Functions

String manipulations are among the most-used functions in programming. If you have ever created a report using any programming language, you know the importance of properly concatenating strings of characters, printing names in uppercase, or knowing the length of a given attribute. Table 8.7 shows a subset of useful string manipulation functions.

TABLE 8.7

SELECTED STRING FUNCTIONS

FUNCTION	EXAMPLE(S)
Concatenation **|| Oracle** **+ Access and MS SQL Server** **& Access** **CONCAT() MySQL** Concatenates data from two different character columns and returns a single column. Syntax: strg_value || strg_value strg_value + strg_value strg_value & strg_value CONCAT(strg_value, strg_value) The CONCAT function can only accept two string values so nested CONCAT functions are required when more than two values are to be concatenated.	Lists all employee names (concatenated). In Oracle, use the following: SELECT EMP_LNAME || ', ' || EMP_FNAME AS NAME FROM EMPLOYEE; In Access and MS SQL Server, use the following: SELECT EMP_LNAME + ', ' + EMP_FNAME AS NAME FROM EMPLOYEE; In MySQL, use the following: SELECT CONCAT(CONCAT(EMP_LNAME, ', '), EMP_FNAME AS NAME FROM EMPLOYEE;

UPPER Oracle, MS SQL Server, and MySQL
UCASE MySQL and Access
LOWER Oracle, MS SQL Server, and MySQL
LCASE MySQL and Access
Returns a string in all capital or all lowercase letters
Syntax:
UPPER(strg_value)
UCASE(strg_value)
LOWER(strg_value)
LCASE(strg_value)
	Lists all employee names in all capital letters (concatenated).
	In Oracle, use the following:
	SELECT UPPER(EMP_LNAME || ', ' || EMP_FNAME) AS NAME
	FROM EMPLOYEE;
	In MS SQL Server, use the following:
	SELECT UPPER(EMP_LNAME + ', ' + EMP_FNAME) AS NAME
	FROM EMPLOYEE;
	In Access, use the following:
	SELECT UCASE(EMP_LNAME & ', ' & EMP_FNAME) AS NAME
	FROM EMPLOYEE;
	In MySQL, use the following:
	SELECT UPPER(CONCAT(CONCAT(EMP_LNAME, ', '),
	                      EMP_FNAME AS NAME
	FROM EMPLOYEE;
	Lists all employee names in all lowercase letters (concatenated).
	In Oracle, use the following:
	SELECT LOWER(EMP_LNAME || ', ' || EMP_FNAME) AS NAME
	FROM EMPLOYEE;
	In MS SQL Server, use the following:
	SELECT LOWER(EMP_LNAME + ', ' + EMP_FNAME) AS NAME
	FROM EMPLOYEE;
	In Access, use the following:
	SELECT LCASE(EMP_LNAME & ', ' & EMP_FNAME) AS NAME
	FROM EMPLOYEE;
	In MySQL, use the following:
	SELECT LOWER(CONCAT(CONCAT(EMP_LNAME, ', '),
	                      EMP_FNAME AS NAME
	FROM EMPLOYEE;

TABLE 8.7 (CONTINUED)

SELECTED STRING FUNCTIONS

FUNCTION	EXAMPLE(S)
SUBSTRING Returns a substring or part of a given string parameter Syntax: SUBSTR(strg_value, p, l) Oracle and MySQL SUBSTRING(strg_value,p,l) MS SQL Server and MySQL MID(strg_value,p,l) Access p = start position l = length of characters If the length of characters is omitted, the functions will return the remainder of the string value.	Lists the first three characters of all employee phone numbers. In Oracle or MySQL, use the following: SELECT EMP_PHONE, SUBSTR(EMP_PHONE,1,3) AS PREFIX FROM EMPLOYEE; In MS SQL Server or MySQL, use the following: SELECT EMP_PHONE, SUBSTRING(EMP_PHONE,1,3) AS PREFIX FROM EMPLOYEE; In Access, use the following: SELECT EMP_PHONE, MID(EMP_PHONE, 1,3) AS PREFIX FROM EMPLOYEE;
LENGTH Returns the number of characters in a string value Syntax: LENGTH(strg_value) Oracle and MySQL LEN(strg_value) MS SQL Server and Access	Lists all employee last names and the length of their names in descending order by last name length. In Oracle and MySQL, use the following: SELECT EMP_LNAME, LENGTH(EMP_LNAME) AS NAMESIZE FROM EMPLOYEE; In MS Access and SQL Server, use the following: SELECT EMP_LNAME, LEN(EMP_LNAME) AS NAMESIZE FROM EMPLOYEE;

8-3d Conversion Functions

Conversion functions allow you to take a value of a given data type and convert it to the equivalent value in another data type. In Section 8-3a, you learned about two basic Oracle SQL conversion functions: TO_CHAR and TO_DATE. Note that the TO_CHAR function takes a date value and returns a character string representing a day, a month, or a year. In the same way, the TO_DATE function takes a character string representing a date and returns an actual date in Oracle format. SQL Server uses the CAST and CONVERT functions to convert one data type to another. A summary of the selected functions is shown in Table 8.8.

TABLE 8.8

SELECTED CONVERSION FUNCTIONS

FUNCTION	EXAMPLE(S)

Numeric or Date to Character:
TO_CHAR Oracle
CAST Oracle, MS SQL Server, MySQL
CONVERT MS SQL Server, MySQL
CSTR Access
Returns a character string from a numeric or date value.
Syntax:
TO_CHAR(value-to-convert, fmt)
fmt = format used; can be:
9 = displays a digit
0 = displays a leading zero
, = displays the comma
. = displays the decimal point
$= displays the dollar sign
B = leading blank
S = leading sign
MI = trailing minus sign
CAST (value-to-convert AS char(length))
Note that Oracle and MS SQL Server can use CAST to convert the numeric data into fixed length or variable length character data type.
MySQL cannot CAST into variable length character data, only fixed length.
MS SQL Server:
CONVERT(varchar(length), value-to-convert)
MySQL:
CONVERT(value-to-convert, char(length))
The primary difference between CAST and CONVERT is that CONVERT can also be used to change the character set of the data.
CSTR(value-to-convert)
	Lists all product prices, product received date, and percent discount using formatted values.
	TO_CHAR:
	SELECT P_CODE,
	    TO_CHAR(P_PRICE,'999.99') AS PRICE,
	    TO_CHAR(P_INDATE, 'MM/DD/YYYY') AS INDATE,
	    TO_CHAR(P_DISCOUNT,'0.99') AS DISC
	FROM PRODUCT;
	CAST in Oracle and MS SQL Server:
	SELECT P_CODE, CAST(P_PRICE AS VARCHAR(8)) AS PRICE,
	    CAST(P_INDATE AS VARCHAR(20)) AS INDATE,
	    CAST(P_DISCOUNT AS VARCHAR(4)) AS DISC
	FROM PRODUCT;
	CAST in MySQL:
	SELECT P_CODE, CAST(P_PRICE AS CHAR(8)) AS PRICE,
	    CAST(P_INDATE AS CHAR(20)) AS INDATE,
	    CAST(P_DISCOUNT AS CHAR(4)) AS DISC
	FROM PRODUCT;
	CONVERT in MS SQL Server:
	SELECT P_CODE, CONVERT(VARCHAR(8), P_PRICE) AS PRICE,
	    CONVERT(VARCHAR(20), P_INDATE) AS INDATE,
	    CONVERT(VARCHAR(4), P_DISC) AS DISC
	FROM PRODUCT;
	CONVERT in MySQL:
	SELECT P_CODE, CONVERT(P_PRICE, CHAR(8)) AS PRICE,
	    CONVERT(P_INDATE, CHAR(20)) AS INDATE,
	    CONVERT(P_DISC, CHAR(4)) AS DISC
	FROM PRODUCT;
	CSTR in Access:
	SELECT P_CODE, CSTR(P_PRICE) AS PRICE,
	    CSTR(P_INDATE) AS INDATE,
	    CSTR(P_DISC) AS DISCOUNT
	FROM PRODUCT;

TABLE 8.8 (CONTINUED)

SELECTED CONVERSION FUNCTIONS

FUNCTION	EXAMPLE(S)

String to Number:
TO_NUMBER Oracle
CAST Oracle, MS SQL Server, MySQL
CONVERT MS SQL Server, MySQL
CINT Access
CDEC Access
Returns a number from a character string
Syntax:
Oracle:
TO_NUMBER(char_value, fmt)
fmt = format used; can be:
9 = indicates a digit
B = leading blank
S = leading sign
MI = trailing minus sign
CAST (value-to-convert as numeric-data type) Note that in addition to the INTEGER and DECIMAL(l,d) data types, Oracle supports NUMBER and MS SQL Server supports NUMERIC.
MS SQL Server:
CONVERT(value-to-convert, decimal(l,d))
MySQL:
CONVERT(value-to-convert, decimal(l,d))
Other than the data type to be converted into, these functions operate the same as described above.
CINT in Access returns the number in the integer data type, while CDEC returns decimal data type.
	Converts text strings to numeric values when importing data to a table from another source in text format; for example, the query shown here uses the TO_NUMBER function to convert text formatted to Oracle default numeric values using the format masks given.
	TO_NUMBER:
	SELECT TO_NUMBER('−123.99', 'S999.99'),
	           TO_NUMBER('99.78−','B999.99MI')
	FROM DUAL;
	CAST:
	SELECT CAST('−123.99' AS DECIMAL(8,2)),
	           CAST('−99.78' AS DECIMAL(8,2));
	The CAST function does not support the trailing sign on the character string.
	CINT and CDEC:
	SELECT CINT('−123'), CDEC('−123.99');

CASE Oracle, MS SQL Server, MySQL
DECODE Oracle
SWITCH Access
Compares an attribute or expression with a series of values and returns an associated value or a default value if no match is found
Syntax:
DECODE:
DECODE(e, x, y, d)
e = attribute or expression
x = value with which to compare e
y = value to return in e = x
d = default value to return if e is not equal to x
CASE:
CASE When condition
THEN value1 ELSE value2 END
SWITCH:
SWITCH(e1, x, e2, y, TRUE, d)
e1 = comparison expression
x = value to return if e1 is true
e2 = comparison expression
y = value to return if e2 is true
TRUE = keyword indicating the next value is the default
d = default value to return if none of the expressions were true
	The following example returns the sales tax rate for specified states:
	Compares V_STATE to 'CA'; if the values match, it returns .08.
	Compares V_STATE to 'FL'; if the values match, it returns .05.
	Compares V_STATE to 'TN'; if the values match, it returns .085.
	If there is no match, it returns 0.00 (the default value).
	SELECT V_CODE, V_STATE,
	           DECODE(V_STATE,'CA',.08,'FL',.05, 'TN',.085, 0.00)
	           AS TAX
	FROM VENDOR;
	CASE:
	SELECT V_CODE, V_STATE,
	CASE WHEN V_STATE = 'CA' THEN .08
	WHEN V_STATE = 'FL' THEN .05
	WHEN V_STATE = 'TN' THEN .085
	           ELSE 0.00 END AS TAX
	FROM VENDOR
	SWITCH:
	SELECT V_CODE, V_STATE,
	           SWITCH(V_STATE ='CA',.08,
	           V_STATE = 'FL',.05,
	           V_STATE = 'TN',.085,
	           TRUE, 0.00) AS TAX
	FROM VENDOR;

8-4 **Relational Set Operators**

In Chapter 3, The Relational Database Model, you learned about the eight general relational operators. In this section, you will learn how to use three SQL commands—UNION, INTERSECT, and EXCEPT (MINUS)—to implement the union, intersection, and difference relational operators.

In previous chapters, you learned that SQL data manipulation commands are **set-oriented**; that is, they operate over entire sets of rows and columns (tables) at once. You can combine two or more sets to create new sets (or relations). That is precisely what the UNION, INTERSECT, and EXCEPT (MINUS) statements do. In relational database terms, you can use the words *sets*, *relations*, and *tables* interchangeably because they all provide a conceptual view of the data set as it is presented to the relational database user.

Note

The SQL standard defines the operations that all DBMSs must perform on data, but it leaves the implementation details to the DBMS vendors. Therefore, some advanced SQL features might not work on all DBMS implementations. Also, some DBMS vendors might implement additional features not found in the SQL standard. The SQL standard defines UNION, INTERSECT, and EXCEPT as the keywords for the UNION, INTERSECT, and DIFFERENCE relational operators, and these are the names used in MS SQL Server. However, Oracle uses MINUS as the name of the DIFFERENCE operator instead of EXCEPT. Other RDBMS vendors might use a different command name or might not implement a given command at all. For example, Access and MySQL do not have direct support for INTERSECT or DIFFERENCE operations because that functionality can be achieved using combinations of joins and subqueries. To learn more about the ANSI/ISO SQL standards and find out how to obtain the latest standard documents in electronic form, check the ANSI website (*www.ansi.org*).

UNION, INTERSECT, and EXCEPT (MINUS) work properly only if relations are **union-compatible**, which means that the number of attributes must be the same and their corresponding data types must be alike. In practice, some RDBMS vendors require the data types to be compatible but not exactly the same. For example, compatible data types are VARCHAR (35) and CHAR (15). Both attributes store character (string) values; the only difference is the string size. Another example of compatible data types is NUMBER and SMALLINT. Both data types are used to store numeric values.

Note

Some DBMS products might require union-compatible tables to have *identical* data types.

8-4a UNION

Suppose that SaleCo has bought another company. SaleCo's management wants to make sure that the acquired company's customer list is properly merged with its own customer list. Because some customers might have purchased goods from both companies, the two lists might contain common customers. SaleCo's management wants to make sure that customer records are not duplicated when the two customer lists are merged. The UNION query is a perfect tool for generating a combined listing of customers—one that excludes duplicate records.

set-oriented
Dealing with or related to sets, or groups of things. In the relational model, SQL operators are set-oriented because they operate over entire sets of rows and columns at once.

union-compatible
Two or more tables that share the same number of columns and have columns with compatible data types or domains.

The UNION statement combines rows from two or more queries *without including duplicate rows*. The syntax of the UNION statement is:

query UNION *query*

In other words, the UNION statement combines the output of two SELECT queries. (Remember that the SELECT statements must be union-compatible. That is, they must return the same number of attributes and similar data types.)

To demonstrate the use of the UNION statement in SQL, use the CUSTOMER and CUSTOMER_2 tables in the Ch08_SaleCo database. To show the combined CUSTOMER and CUSTOMER_2 records without duplicates, the UNION query is written as follows:

```
SELECT      CUS_LNAME, CUS_FNAME, CUS_INITIAL, CUS_AREACODE,
            CUS_PHONE
FROM        CUSTOMER
UNION
SELECT      CUS_LNAME, CUS_FNAME, CUS_INITIAL, CUS_AREACODE,
            CUS_PHONE
FROM        CUSTOMER_2;
```

Figure 8.16 shows the contents of the CUSTOMER and CUSTOMER_2 tables and the result of the UNION query. Although MS Access is used to show the results here, similar results can be obtained with Oracle, MS SQL Server, and MySQL.

FIGURE 8.16 UNION QUERY RESULTS

Database name: Ch08_SaleCo

Table name: CUSTOMER

CUS_CODE	CUS_LNAME	CUS_FNAME	CUS_INITIAL	CUS_AREACODE	CUS_PHONE	CUS_BALANCE
10010	Ramas	Alfred	A	615	844-2573	0.00
10011	Dunne	Leona	K	713	894-1238	0.00
10012	Smith	Kathy	W	615	894-2285	345.86
10013	Olowski	Paul	F	615	894-2180	536.75
10014	Orlando	Myron		615	222-1672	0.00
10015	O'Brian	Amy	B	713	442-3381	0.00
10016	Brown	James	G	615	297-1228	221.19
10017	Williams	George		615	290-2556	768.93
10018	Farriss	Anne	G	713	382-7185	216.55
10019	Smith	Olette	K	615	297-3809	0.00

Query name: qryUNION-of-CUSTOMER-and-CUSTOMER_2

CUS_LNAME	CUS_FNAME	CUS_INITIAL	CUS_AREACODE	CUS_PHONE
Brown	James	G	615	297-1228
Dunne	Leona	K	713	894-1238
Farriss	Anne	G	713	382-7185
Hernandez	Carlos	J	723	123-7654
Lewis	Marie	J	734	332-1789
McDowell	George		723	123-7768
O'Brian	Amy	B	713	442-3381
Olowski	Paul	F	615	894-2180
Orlando	Myron		615	222-1672
Ramas	Alfred	A	615	844-2573
Smith	Kathy	W	615	894-2285
Smith	Olette	K	615	297-3809
Terrell	Justine	H	615	322-9870
Tirpin	Khaleed	G	723	123-9876
Williams	George		615	290-2556

Table name: CUSTOMER_2

CUS_CODE	CUS_LNAME	CUS_FNAME	CUS_INITIAL	CUS_AREACODE	CUS_PHONE
345	Terrell	Justine	H	615	322-9870
347	Olowski	Paul	F	615	894-2180
351	Hernandez	Carlos	J	723	123-7654
352	McDowell	George		723	123-7768
365	Tirpin	Khaleed	G	723	123-9876
368	Lewis	Marie	J	734	332-1789
369	Dunne	Leona	K	713	894-1238

Note the following in Figure 8.16:

- The CUSTOMER table contains 10 rows, while the CUSTOMER_2 table contains seven rows.

- Customers Dunne and Olowski are included in the CUSTOMER table as well as the CUSTOMER_2 table.

- The UNION query yields 15 records because the duplicate records of customers Dunne and Olowski are not included. In short, the UNION query yields a unique set of records.

Note

The UNION statement can be used to unite more than just two queries. For example, assume that you have four union-compatible queries named T1, T2, T3, and T4. With the UNION statement, you can combine the output of all four queries into a single result set. The SQL statement will be similar to this:

```
SELECT column-list FROM T1
UNION
SELECT column-list FROM T2
UNION
SELECT column-list FROM T3
UNION
SELECT column-list FROM T4;
```

8-4b UNION ALL

If SaleCo's management wants to know how many customers are on *both* the CUSTOMER and CUSTOMER_2 lists, a UNION ALL query can be used to produce a relation that retains the duplicate rows. Therefore, the following query will keep all rows from both queries (including the duplicate rows) and return 17 rows.

```
SELECT      CUS_LNAME, CUS_FNAME, CUS_INITIAL, CUS_AREACODE,
            CUS_PHONE
FROM        CUSTOMER
UNION ALL
SELECT      CUS_LNAME, CUS_FNAME, CUS_INITIAL, CUS_AREACODE,
            CUS_PHONE
FROM        CUSTOMER_2;
```

Running the preceding UNION ALL query produces the result shown in Figure 8.17. Like the UNION statement, the UNION ALL statement can be used to unite more than just two queries.

8-4c INTERSECT

If SaleCo's management wants to know which customer records are duplicated in the CUSTOMER and CUSTOMER_2 tables, the INTERSECT statement can be used to combine rows from two queries, returning only the rows that appear in both sets. The syntax for the INTERSECT statement is:

query INTERSECT *query*

To generate the list of duplicate customer records, you can use the following commands:

```
SELECT      CUS_LNAME, CUS_FNAME, CUS_INITIAL, CUS_AREACODE,
            CUS_PHONE
FROM        CUSTOMER
```

```
            INTERSECT
SELECT          CUS_LNAME, CUS_FNAME, CUS_INITIAL, CUS_AREACODE,
                CUS_PHONE
FROM            CUSTOMER_2;
```

FIGURE 8.17 UNION ALL QUERY RESULTS

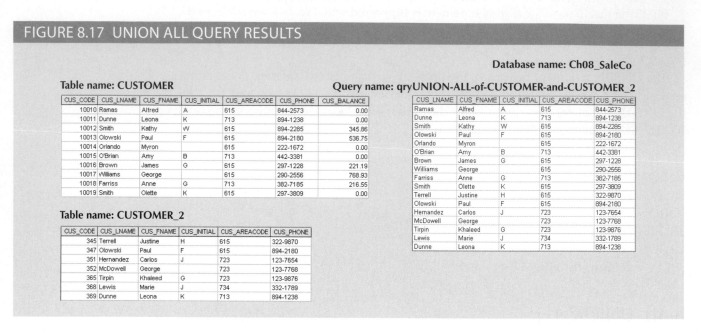

Database name: Ch08_SaleCo

Table name: CUSTOMER

Query name: qryUNION-ALL-of-CUSTOMER-and-CUSTOMER_2

CUS_CODE	CUS_LNAME	CUS_FNAME	CUS_INITIAL	CUS_AREACODE	CUS_PHONE	CUS_BALANCE
10010	Ramas	Alfred	A	615	844-2573	0.00
10011	Dunne	Leona	K	713	894-1238	0.00
10012	Smith	Kathy	W	615	894-2285	345.86
10013	Olowski	Paul	F	615	894-2180	536.75
10014	Orlando	Myron		615	222-1672	0.00
10015	O'Brian	Amy	B	713	442-3381	0.00
10016	Brown	James	G	615	297-1228	221.19
10017	Williams	George		615	290-2556	768.93
10018	Farriss	Anne	G	713	382-7185	216.55
10019	Smith	Olette	K	615	297-3809	0.00

CUS_LNAME	CUS_FNAME	CUS_INITIAL	CUS_AREACODE	CUS_PHONE
Ramas	Alfred	A	615	844-2573
Dunne	Leona	K	713	894-1238
Smith	Kathy	W	615	894-2285
Olowski	Paul	F	615	894-2180
Orlando	Myron		615	222-1672
O'Brian	Amy	B	713	442-3381
Brown	James	G	615	297-1228
Williams	George		615	290-2556
Farriss	Anne	G	713	382-7185
Smith	Olette	K	615	297-3809
Terrell	Justine	H	615	322-9870
Olowski	Paul	F	615	894-2180
Hernandez	Carlos	J	723	123-7654
McDowell	George		723	123-7768
Tirpin	Khaleed	G	723	123-9876
Lewis	Marie	J	734	332-1789
Dunne	Leona	K	713	894-1238

Table name: CUSTOMER_2

CUS_CODE	CUS_LNAME	CUS_FNAME	CUS_INITIAL	CUS_AREACODE	CUS_PHONE
345	Terrell	Justine	H	615	322-9870
347	Olowski	Paul	F	615	894-2180
351	Hernandez	Carlos	J	723	123-7654
352	McDowell	George		723	123-7768
365	Tirpin	Khaleed	G	723	123-9876
368	Lewis	Marie	J	734	332-1789
369	Dunne	Leona	K	713	894-1238

The INTERSECT statement can be used to generate additional useful customer information. For example, the following query returns the customer codes for all customers who are in area code 615 and who have made purchases. (If a customer has made a purchase, there must be an invoice record for that customer.)

```
SELECT          CUS_CODE FROM CUSTOMER WHERE CUS_AREACODE = '615'
INTERSECT
SELECT          DISTINCT CUS_CODE FROM INVOICE;
```

Figure 8.18 shows both sets of SQL statements and their output.

FIGURE 8.18 INTERSECT QUERY RESULTS

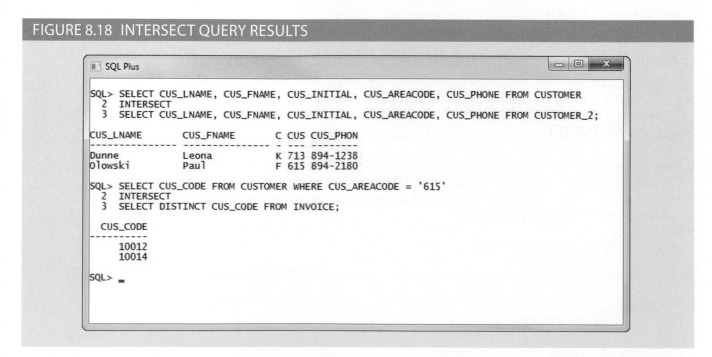

Note

Access and MySQL do not support the INTERSECT query. These DBMSs are able to give the desired results using alternative query formats. For example, INTERSECT results can also be produced in Access and MySQL through an inner join that includes all of the attributes to be returned in the join condition. The query:

SELECT CUS_AREACODE FROM CUSTOMER

INTERSECT

SELECT V_AREACODE FROM VENDOR;

can also be produced without the INTERSECT command with the query:

SELECT DISTINCT CUS_AREACODE

FROM CUSTOMER JOIN VENDOR ON CUS_AREACODE = V_AREACODE;

8-4d EXCEPT (MINUS)

The EXCEPT statement in SQL combines rows from two queries and returns only the rows that appear in the first set but not in the second. The syntax for the EXCEPT statement in MS SQL Server and the MINUS statement in Oracle is:

query EXCEPT *query*

and

query MINUS *query*

For example, if the SaleCo managers want to know which customers in the CUSTOMER table are not found in the CUSTOMER_2 table, they can use the following commands in Oracle:

```
SELECT      CUS_LNAME, CUS_FNAME, CUS_INITIAL, CUS_AREACODE,
            CUS_PHONE
FROM        CUSTOMER
MINUS
SELECT      CUS_LNAME, CUS_FNAME, CUS_INITIAL, CUS_AREACODE,
            CUS_PHONE
FROM        CUSTOMER_2;
```

If the managers want to know which customers in the CUSTOMER_2 table are not found in the CUSTOMER table, they merely switch the table designations:

```
SELECT      CUS_LNAME, CUS_FNAME, CUS_INITIAL, CUS_AREACODE,
            CUS_PHONE
FROM        CUSTOMER_2
MINUS
SELECT      CUS_LNAME, CUS_FNAME, CUS_INITIAL, CUS_AREACODE,
            CUS_PHONE
FROM        CUSTOMER;
```

Users of MS SQL Server would substitute the keyword EXCEPT in place of MINUS, but otherwise the syntax is exactly the same. You can extract useful information by combining MINUS with various clauses such as WHERE. For example, the following query returns the customer codes for all customers in area code 615 minus the ones who have made purchases, leaving the customers in area code 615 who have not made purchases.

SELECT   CUS_CODE FROM CUSTOMER WHERE CUS_AREACODE = '615'
MINUS
SELECT   DISTINCT CUS_CODE FROM INVOICE;

Figure 8.19 shows the preceding three SQL statements and their output.

FIGURE 8.19 MINUS QUERY RESULTS

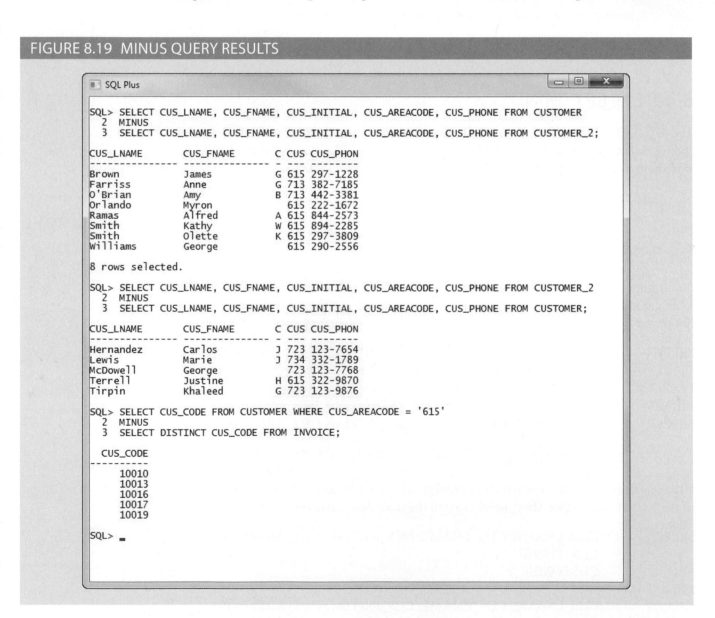

8-4e Syntax Alternatives

If your DBMS does not support the INTERSECT or EXCEPT (MINUS) statements, you can use IN and NOT IN subqueries to obtain similar results. For example, the following query will produce the same results as the INTERSECT query shown in Section 8-4c:

```
SELECT      CUS_CODE FROM CUSTOMER
WHERE       CUS_AREACODE = '615' AND
            CUS_CODE IN (SELECT DISTINCT CUS_CODE FROM INVOICE);
```

Figure 8.20 shows the use of the INTERSECT alternative.

FIGURE 8.20 INTERSECT ALTERNATIVE

Database name: Ch08_SaleCo

Table name: CUSTOMER

CUS_CODE	CUS_LNAME	CUS_FNAME	CUS_INITIAL	CUS_AREACODE	CUS_PHONE	CUS_BALANCE
10010	Ramas	Alfred	A	615	844-2573	0.00
10011	Dunne	Leona	K	713	894-1238	0.00
10012	Smith	Kathy	W	615	894-2285	345.86
10013	Olowski	Paul	F	615	894-2180	536.75
10014	Orlando	Myron		615	222-1672	0.00
10015	O'Brian	Amy	B	713	442-3381	0.00
10016	Brown	James	G	615	297-1228	221.19
10017	Williams	George		615	290-2556	768.93
10018	Farriss	Anne	G	713	382-7185	216.55
10019	Smith	Olette	K	615	297-3809	0.00

Table name: INVOICE

INV_NUMBER	CUS_CODE	INV_DATE
1001	10014	16-Jan-14
1002	10011	16-Jan-14
1003	10012	16-Jan-14
1004	10011	17-Jan-14
1005	10018	17-Jan-14
1006	10014	17-Jan-14
1007	10015	17-Jan-14
1008	10011	17-Jan-14

Query name: qry-INTERSECT-Alternative

CUS_CODE
10012
10014

Using the same alternative to the MINUS statement, you can generate the output for the third MINUS query shown in Section 8-4d by entering the following:

```
SELECT      CUS_CODE FROM CUSTOMER
WHERE       CUS_AREACODE = '615' AND
            CUS_CODE NOT IN (SELECT DISTINCT CUS_CODE FROM
            INVOICE);
```

The results of the query are shown in Figure 8.21. Note that the query output includes only the customers in area code 615 who have not made any purchases and therefore have not generated invoices.

8-5 Virtual Tables: Creating a View

As you learned earlier, the output of a relational operator such as SELECT is another relation (or table). Suppose that at the end of each day, you would like to have a list of all products to reorder—that is, products with a quantity on hand that is less than or equal to the minimum quantity. Instead of typing the same query at the end of each day, wouldn't it be better to permanently save that query in the database? That is the function of a relational view. A **view** is a virtual table based on a SELECT query. The query can contain columns, computed columns, aliases, and aggregate functions from one or more tables. The tables on which the view is based are called **base tables**.

view
A virtual table based on a SELECT query that is saved as an object in the database.

base table
The table on which a view is based.

FIGURE 8.21 MINUS ALTERNATIVE

Database name: Ch08_SaleCo

Table name: CUSTOMER

CUS_CODE	CUS_LNAME	CUS_FNAME	CUS_INITIAL	CUS_AREACODE	CUS_PHONE	CUS_BALANCE
10010	Ramas	Alfred	A	615	844-2573	0.00
10011	Dunne	Leona	K	713	894-1238	0.00
10012	Smith	Kathy	W	615	894-2285	345.86
10013	Olowski	Paul	F	615	894-2180	536.75
10014	Orlando	Myron		615	222-1672	0.00
10015	O'Brian	Amy	B	713	442-3381	0.00
10016	Brown	James	G	615	297-1228	221.19
10017	Williams	George		615	290-2556	768.93
10018	Farriss	Anne	G	713	382-7185	216.55
10019	Smith	Olette	K	615	297-3809	0.00

Table name: INVOICE

INV_NUMBER	CUS_CODE	INV_DATE
1001	10014	16-Jan-14
1002	10011	16-Jan-14
1003	10012	16-Jan-14
1004	10011	17-Jan-14
1005	10018	17-Jan-14
1006	10014	17-Jan-14
1007	10015	17-Jan-14
1008	10011	17-Jan-14

Query name: qry-MINUS-Alternative

CUS_CODE
10010
10013
10016
10017
10019

You can create a view by using the **CREATE VIEW** command:

CREATE VIEW *viewname* AS SELECT *query*

The CREATE VIEW statement is a data definition command that stores the subquery specification—the SELECT statement used to generate the virtual table—in the data dictionary.

The first SQL command set in Figure 8.22 shows the syntax used to create a view named PRICEGT50. This view contains only the designated three attributes (P_DESCRIPT, P_QOH, and P_PRICE) and only rows in which the price is over $50. The second SQL command sequence in Figure 8.22 shows the rows that make up the view.

FIGURE 8.22 CREATING A VIRTUAL TABLE WITH THE CREATE VIEW COMMAND

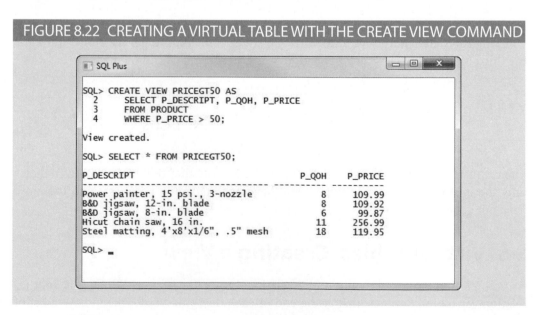

CREATE VIEW
A SQL command that creates a logical, "virtual" table. The view can be treated as a real table.

Note

Note to MS Access Users

The CREATE VIEW command is not directly supported in MS Access. To create a view in MS Access, you simply create a SQL query and then save it.

A relational view has several special characteristics:

- You can use the name of a view anywhere a table name is expected in a SQL statement.

- Views are dynamically updated. That is, the view is re-created on demand each time it is invoked. Therefore, if new products are added or deleted to meet the criterion P_PRICE > 50.00, those new products will automatically appear or disappear in the PRICEGT50 view the next time the view is invoked.

- Views provide a level of security in the database because they can restrict users to seeing only specified columns and rows in a table. For example, if you have a company with hundreds of employees in several departments, you could give each department secretary a view of certain attributes only for the employees who belong to that secretary's department.

- Views may also be used as the basis for reports. For example, if you need a report that shows a summary of total product cost and quantity-on-hand statistics grouped by vendor, you could create a PROD_STATS view as:

```
CREATE VIEW PROD_STATS AS
SELECT      V_CODE, SUM(P_QOH*P_PRICE) AS TOTCOST, MAX(P_QOH)
            AS MAXQTY, MIN(P_QOH) AS MINQTY, AVG(P_QOH) AS
            AVGQTY
FROM        PRODUCT
GROUP BY    V_CODE;
```

8-5a Updatable Views

One of the most common operations in production database environments is to use batch update routines to update a master table attribute (field) with transaction data. As the name implies, a **batch update routine** pools multiple transactions into a single batch to update a master table field in a *single operation*. For example, a batch update routine is commonly used to update a product's quantity on hand based on summary sales transactions. Such routines are typically run as overnight batch jobs to update the quantity on hand of products in inventory. For example, the sales transactions performed by traveling salespeople can be entered during periods when the system is offline.

To perform a batch update routine, begin by defining the master product table (PRODMASTER) and the product monthly sales totals table (PRODSALES) shown in Figure 8.23. Note the 1:1 relationship between the two tables.

Online Content

The PRODMASTER and PRODSALES tables are in the Ch08_UV databases for the different DBMSs, which are available at *www.cengagebrain.com*.

FIGURE 8.23 THE PRODMASTER AND PRODSALES TABLES

Database name: Ch08_UV

Table name: PRODMASTER

PROD_ID	PROD_DESC	PROD_QOH
A123	SCREWS	67
BX34	NUTS	37
C583	BOLTS	50

Table name: PRODSALES

PROD_ID	PS_QTY
A123	7
BX34	3

batch update routine
A routine that pools transactions into a single group to update a master table in a single operation.

Using the tables in Figure 8.23, update the PRODMASTER table by subtracting the PRODSALES table's product monthly sales quantity (PS_QTY) from the PRODMASTER table's PROD_QOH. To produce the required update, the update query would be written like this:

```
UPDATE       PRODMASTER, PRODSALES
SET          PRODMASTER.PROD_QOH = PROD_QOH – PS_QTY
WHERE        PRODMASTER.PROD_ID = PRODSALES.PROD_ID;
```

Note that the update statement reflects the following sequence of events:

- Join the PRODMASTER and PRODSALES tables.

- Update the PROD_QOH attribute (using the PS_QTY value in the PRODSALES table) for each row of the PRODMASTER table with matching PROD_ID values in the PRODSALES table.

Note

Updating using multiple tables in MS SQL Server requires the UPDATE FROM syntax. The above code would be written in MS SQL Server as the following:

```
UPDATE PRODMASTER
SET PROD_QOH = PROD_QOH – PS_QTY
FROM PRODMASTER JOIN PRODSALES ON PRODMASTER.PROD_ID = PRODSALES.
PROD_ID;
```

To be used in a batch update, the PRODSALES data must be stored in a base table rather than in a view. The query will work in MySQL and Access, but Oracle will return the error message shown in Figure 8.24.

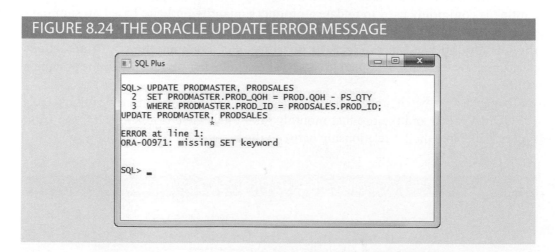

FIGURE 8.24 THE ORACLE UPDATE ERROR MESSAGE

Oracle produced the error message because it expected to find a single table name in the UPDATE statement. In fact, you cannot join tables in the UPDATE statement in Oracle. To solve that problem, you have to create an *updatable* view. As its name suggests, an **updatable view** can be used to update attributes in any base table(s) used in the view. You must realize that *not all views are updatable*. Actually, several restrictions govern updatable views, and some of them are vendor-specific.

updatable view
A view that can update attributes in base tables that are used in the view.

Note

While the examples in this section are generated in Oracle, the same code and techniques also work in MS SQL Server, MySQL, and Access. To see what additional restrictions are placed on updatable views by the DBMS you are using, check the appropriate DBMS documentation.

The most common updatable view restrictions are as follows:

- GROUP BY expressions or aggregate functions cannot be used.
- You cannot use set operators such as UNION, INTERSECT, and MINUS.
- Most restrictions are based on the use of JOINs or group operators in views. More specifically, the base table to be updated must be key-preserved, meaning that the values of the primary key of the base table must still be unique by definition in the view.

An updatable view named PSVUPD has been created, as shown in Figure 8.25.

FIGURE 8.25 CREATING AN UPDATABLE VIEW

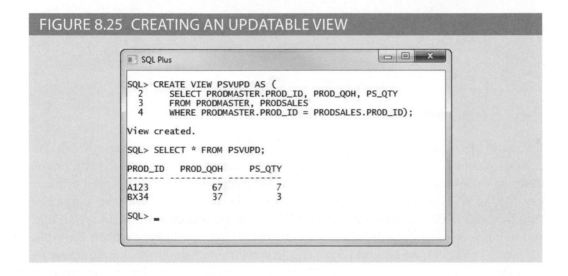

One easy way to determine whether a view can be used to update a base table is to examine the view's output. If the primary key columns of the base table you want to update still have unique values in the view, the base table is updatable. For example, if the PROD_ID column of the view returns the A123 or BX34 values more than once, the PRODMASTER table cannot be updated through the view.

After creating the updatable view shown in Figure 8.25, you can use the UPDATE command to update the view, thereby updating the PRODMASTER table. Figure 8.26 shows how the UPDATE command is used and shows the final contents of the PRODMASTER table after the UPDATE has been executed.

Although the batch update procedure just illustrated meets the goal of updating a master table with data from a transaction table, the preferred real-world solution to the update problem is to use procedural SQL, which you will learn about later in this chapter.

FIGURE 8.26 PRODMASTER TABLE UPDATE, USING AN UPDATABLE VIEW

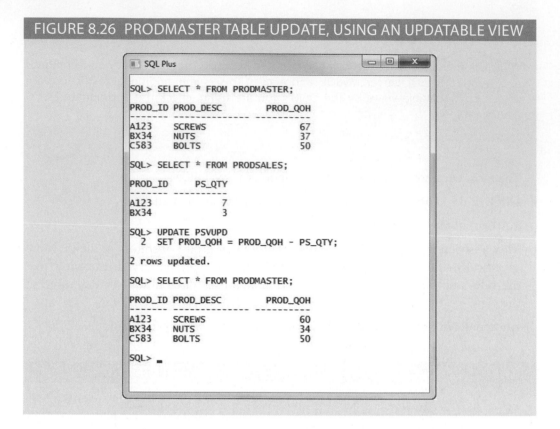

8-6 Sequences

If you use MS Access, you might be familiar with the AutoNumber data type, which you can use to define a column in your table that will be automatically populated with unique numeric values. In fact, if you create a table in MS Access and forget to define a primary key, MS Access will offer to create a primary key column; if you accept, you will notice that MS Access creates a column named "ID" with an AutoNumber data type. After you define a column as an AutoNumber type, every time you insert a row in the table, MS Access will automatically add a value to that column, starting with 1 and increasing the value by 1 in every new row you add. Also, you cannot include that column in your INSERT statements—Access will not let you edit that value at all. MS SQL Server traditionally has used the Identity column property to serve a similar purpose. In MS SQL Server, a table can have at most one column defined as an Identity column. This column behaves similarly to an MS Access column with the AutoNumber data type. MySQL uses the AUTO_INCREMENT property during table creation to indicate that values for an attribute should be generated in the same fashion. AUTO_INCREMENT can be adjusted to start with a value other than 1. Similar to IDENTITY columns in MS SQL Server, only one column in a table can have AUTO_INCREMENT specified, and that column must also be defined as the primary key of the table.

Oracle does not support the AutoNumber data type, or Auto_Increment column properties. Traditionally, Oracle uses a **sequence** to assign values to a column on a table. However, beginning in Oracle 12c, Oracle has added support for Identity columns, and beginning in MS SQL Server 2012, SQL Server supports sequences. There are many similarities in the use of sequences across these DBMS so a database programmer who is comfortable with one should be able to easily transition to the other. However, a sequence is very different from the Access AutoNumber data type and deserves closer scrutiny:

- Sequences are an independent object in the database. (Sequences are not a data type.)
- Sequences have a name.
- Sequences can be used anywhere a value is expected.
- Sequences are not tied to a table or a column.
- Sequences generate a numeric value that can be assigned to any column in any table.
- The table attribute to which you assigned a value based on a sequence can be edited and modified.

The basic syntax to create a sequence is as follows:

CREATE SEQUENCE *name* [START WITH *n*] [INCREMENT BY *n*]
[CACHE | NOCACHE]

where

- *name* is the name of the sequence.
- *n* is an integer value that can be positive or negative.
- *START WITH* specifies the initial sequence value. (The default value is 1.)
- *INCREMENT BY* determines the value by which the sequence is incremented. (The default increment value is 1. The sequence increment can be positive or negative to enable you to create ascending or descending sequences.)
- The *CACHE* or *NOCACHE/NO CACHE* clause indicates whether the DBMS will preallocate sequence numbers in memory. Oracle uses *NOCACHE* as one word and preallocates 20 values by default. SQL Server uses *NO CACHE* as two words. If a cache size is not specified in SQL Server, then the DBMS will determine a default cache size that is not guaranteed to be consistent across different databases.

For example, you could create a sequence to automatically assign values to the customer code each time a new customer is added, and create another sequence to automatically assign values to the invoice number each time a new invoice is added. The SQL code to accomplish those tasks is:

CREATE SEQUENCE CUS_CODE_SEQ START WITH 20010 NOCACHE;

CREATE SEQUENCE INV_NUMBER_SEQ START WITH 4010 NOCACHE;

Note

Remember, SQL Server uses NO CACHE as two words so the corresponding commands in SQL Server would be:

```
CREATE SEQUENCE CUS_CODE_SEQ START WITH 20010 NO CACHE;
CREATE SEQUENCE INV_NUMBER_SEQ START WITH 4010 NO CACHE;
```

You can check all of the sequences you have created by using the following SQL command, as illustrated in Figure 8.27.

SELECT * FROM USER_SEQUENCES;

FIGURE 8.27 ORACLE SEQUENCE

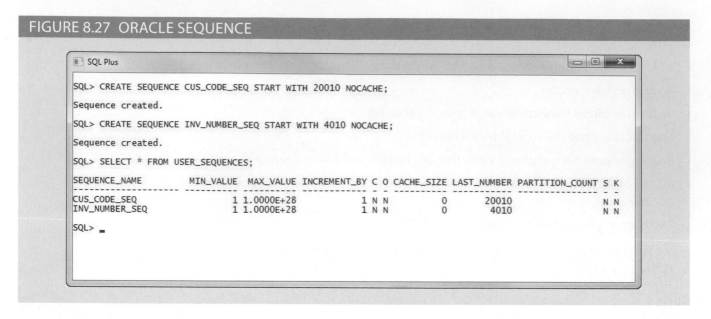

```
SQL> CREATE SEQUENCE CUS_CODE_SEQ START WITH 20010 NOCACHE;

Sequence created.

SQL> CREATE SEQUENCE INV_NUMBER_SEQ START WITH 4010 NOCACHE;

Sequence created.

SQL> SELECT * FROM USER_SEQUENCES;

SEQUENCE_NAME        MIN_VALUE MAX_VALUE INCREMENT_BY C O CACHE_SIZE LAST_NUMBER PARTITION_COUNT S K
-------------------- --------- --------- ------------ - - ---------- ----------- --------------- - -
CUS_CODE_SEQ                 1 1.0000E+28            1 N N          0       20010                 N N
INV_NUMBER_SEQ              1 1.0000E+28            1 N N          0        4010                 N N

SQL> _
```

Note

SQL Server also stores sequences as schema level objects so they can be viewed with the command

SELECT * FROM SYS.SEQUENCES;

Oracle and SQL Server differ slightly in the syntax for retrieving a value from the sequence. In SQL Server, the phrase NEXT VALUE FOR <sequence_name> causes the sequence to generate and return the next value. In Oracle, you must use two special pseudo-columns: NEXTVAL and CURRVAL. NEXTVAL retrieves the next available value from a sequence, and CURRVAL retrieves the current value of a sequence. For example, you can use the following code to enter a new customer in SQL Server:

INSERT INTO CUSTOMER

VALUES (NEXT VALUE FOR CUS_CODE_SEQ, 'Connery', 'Sean', NULL, '615', '898-2007', 0.00);

In Oracle, you would use:

INSERT INTO CUSTOMER

VALUES (CUS_CODE_SEQ.NEXTVAL, 'Connery', 'Sean', NULL, '615', '898-2007', 0.00);

The preceding SQL statement adds a new customer to the CUSTOMER table and assigns the value 20010 to the CUS_CODE attribute. Examine some important sequence characteristics:

- CUS_CODE_SEQ.NEXTVAL retrieves the next available value from the sequence.
- Each time you use NEXTVAL, the sequence is incremented.
- Once a sequence value is used (through NEXTVAL), it cannot be used again. If your SQL statement rolls back for some reason, *the sequence value does not roll back*. If you issue another SQL statement (with another NEXTVAL), the next available sequence value will be returned to the user—it will look like the sequence skips a number.
- You can issue an INSERT statement without using the sequence.

In Oracle, CURRVAL retrieves the current value of a sequence—that is, the last sequence number used, which was generated with a NEXTVAL. You cannot use CURRVAL unless a NEXTVAL was issued previously in the same session. The main use for CURRVAL is to enter rows in dependent tables. For example, the INVOICE and LINE tables are related in a one-to-many relationship through the INV_NUMBER attribute. You can use the INV_NUMBER_SEQ sequence to automatically generate invoice numbers. Then, using CURRVAL, you can get the latest INV_NUMBER used and assign it to the related INV_NUMBER foreign key attribute in the LINE table. For example:

```
INSERT INTO INVOICE VALUES (INV_NUMBER_SEQ.NEXTVAL,
                            20010, SYSDATE);
INSERT INTO LINE    VALUES (INV_NUMBER_SEQ.CURRVAL,
                            1,'13-Q2/P2', 1, 14.99);
INSERT INTO LINE    VALUES (INV_NUMBER_SEQ.CURRVAL,
                            2,'23109-HB', 1, 9.95);
COMMIT;
```

The results are shown in Figure 8.28.

FIGURE 8.28 ORACLE SEQUENCE EXAMPLES

```
SQL Plus

SQL> INSERT INTO CUSTOMER
  2  VALUES (CUS_CODE_SEQ.NEXTVAL, 'Connery', 'Sean', NULL, '615', '898-2007', 0.00);

1 row created.

SQL> SELECT * FROM CUSTOMER WHERE CUS_CODE = 20010;

  CUS_CODE CUS_LNAME       CUS_FNAME       C CUS CUS_PHON CUS_BALANCE
---------- --------------- --------------- - --- -------- -----------
     20010 Connery         Sean              615 898-2007           0

SQL> INSERT INTO INVOICE
  2  VALUES (INV_NUMBER_SEQ.NEXTVAL, 20010, SYSDATE);

1 row created.

SQL> SELECT * FROM INVOICE WHERE INV_NUMBER = 4010;

INV_NUMBER   CUS_CODE INV_DATE
---------- ---------- ---------
      4010      20010 08-JUL-15

SQL> INSERT INTO LINE
  2  VALUES (INV_NUMBER_SEQ.CURRVAL, 1, '13-Q2/P2', 1, 14.99);

1 row created.

SQL> INSERT INTO LINE
  2  VALUES (INV_NUMBER_SEQ.CURRVAL, 2, '23109-HB', 1, 9.95);

1 row created.

SQL> SELECT * FROM LINE WHERE INV_NUMBER = 4010;

INV_NUMBER LINE_NUMBER P_CODE     LINE_UNITS LINE_PRICE
---------- ----------- ---------- ---------- ----------
      4010           1 13-Q2/P2            1      14.99
      4010           2 23109-HB            1       9.95

SQL> COMMIT;

Commit complete.

SQL> _
```

Note

The reason that CURRVAL can only be used in the same session as a NEXTVAL is because the value returned by CURRVAL is not actually provided by the sequence. Remember, the sequence will not generate a value twice. Whenever NEXTVAL is called, Oracle makes a note of the call, which sequence was called, and what number was provided in the user's session information. When CURRVAL is invoked, the DBMS will look in the session information to see what the last value provided to that user by that sequence was. This is very powerful in a multiuser environment. For example, imagine that Maria and Zameer are working in the same database application and with the same data. When Maria calls on a sequence with NEXTVAL, she is provided a new value. If Zameer calls the same sequence with NEXTVAL, he is provided a number different from Maria's number. If, in her same session, Maria now calls on the sequence with CURRVAL, she is not provided with the last number that the sequence generated (which was given to Zameer), but she is given the last number that the sequence provided to her session! Similarly, CURRVAL would provide Zameer with the last value provided to his session. Since this information about the last value provided by the sequence to each user is kept in the user's session information, when Maria disconnects from the database, ending her session, that information is lost. If she reconnects to the database, she will be starting a new session. If she immediately calls on CURRVAL, she will get an error because the DBMS does not have a record of that session being provided any values from the sequence.

In the example shown in Figure 8.28, INV_NUMBER_SEQ.NEXTVAL retrieves the next available sequence number (4010) and assigns it to the INV_NUMBER column in the INVOICE table. Also note the use of the SYSDATE attribute to automatically insert the current date in the INV_DATE attribute. Next, the following two INSERT statements add the products being sold to the LINE table. In this case, INV_NUMBER_SEQ.CURRVAL refers to the last-used INV_NUMBER_SEQ sequence number (4010). In this way, the relationship between INVOICE and LINE is established automatically. The COMMIT statement at the end of the command sequence makes the changes permanent. Of course, you can also issue a ROLLBACK statement, in which case the rows you inserted in the INVOICE and LINE tables would be rolled back (but remember that the sequence number would not). Once you use a sequence number with NEXTVAL, there is no way to reuse it! This "no-reuse" characteristic is designed to guarantee that the sequence will always generate unique values.

Note

At this writing, SQL Server does not provide a direct equivalent to Oracle's CURRVAL. If you wish to find the last number generated by a sequence in SQL Server, you can retrieve it by querying the metadata, but this will only give the last number generated by the sequence for any user.

Remember these points when you think about sequences:

- The use of sequences is optional. You can enter the values manually.
- A sequence is not associated with a table. As in the examples in Figure 8.28, two distinct sequences were created (one for customer code values and one for invoice number values), but you could have created just one sequence and used it to generate unique values for both tables.

Note

The SQL standard defines the use of Identity columns and sequence objects. However, some DBMS vendors might not adhere to the standard. Check your DBMS documentation.

Finally, you can drop a sequence from a database with a DROP SEQUENCE command. For example, to drop the sequences created earlier, you would type:

DROP SEQUENCE CUS_CODE_SEQ;
DROP SEQUENCE INV_NUMBER_SEQ;

Dropping a sequence does not delete the values you assigned to table attributes (CUS_CODE and INV_NUMBER); it deletes only the sequence object from the database. The *values* you assigned to the table columns (CUS_CODE and INV_NUMBER) remain in the database.

Because the CUSTOMER and INVOICE tables are used in the following examples, you will want to keep the original data set. Therefore, you can delete the customer, invoice, and line rows you just added by using the following commands:

DELETE FROM INVOICE WHERE INV_NUMBER = 4010;
DELETE FROM CUSTOMER WHERE CUS_CODE = 20010;
COMMIT;

Those commands delete the recently added invoice, all of the invoice line rows associated with the invoice (the LINE table's INV_NUMBER foreign key was defined with the ON DELETE CASCADE option), and the recently added customer. The COMMIT statement saves all changes to permanent storage.

Note

At this point, you need to re-create the CUS_CODE_SEQ and INV_NUMBER_SEQ sequences, as they will be used again later in the chapter. Enter:

CREATE SEQUENCE CUS_CODE_SEQ START WITH 20010 NOCACHE;
CREATE SEQUENCE INV_NUMBER_SEQ START WITH 4010 NOCACHE;

8-7 Procedural SQL

Thus far, you have learned to use SQL to read, write, and delete data in the database. For example, you learned to update values in a record, to add records, and to delete records. Unfortunately, SQL does not support the *conditional* execution of procedures that are typically supported by a programming language using the general format:

IF <condition>
            THEN <perform procedure>
                        ELSE <perform alternate procedure>
END IF

SQL also fails to support looping operations in programming languages that permit the execution of repetitive actions typically encountered in a programming environment. The typical format is:

DO WHILE

> <perform procedure>

END DO

Traditionally, if you wanted to perform a conditional or looping type of operation (that is, a procedural type of programming using an IF-THEN-ELSE or DO-WHILE statement), you would use a programming language such as Visual Basic .NET, C#, or Java. Although that approach is still common, it usually involves the duplication of application code in many programs. Therefore, when procedural changes are required, modifications must be made in many different programs. An environment characterized by such redundancies often creates data-management problems.

A better approach is to isolate critical code and then have all application programs call the shared code. The advantage of this modular approach is that the application code is isolated in a single program, thus yielding better maintenance and logic control. In any case, the rise of distributed databases and object-oriented databases required that more application code be stored and executed within the database. (For more information on these databases, see Chapter 12, Distributed Database Management Systems, and Appendix G, Object-Oriented Databases, at *www.cengagebrain.com*, respectively.) To meet that requirement, most RDBMS vendors created numerous programming language extensions. Those extensions include:

- Flow-control procedural programming structures (IF-THEN-ELSE, DO-WHILE) for logic representation

- Variable declaration and designation within the procedures

- Error management

To remedy the lack of procedural functionality in SQL and to provide some standardization within the many vendor offerings, the SQL-99 standard defined the use of persistent stored modules. A **persistent stored module (PSM)** is a block of code containing standard SQL statements and procedural extensions that is stored and executed at the DBMS server. The PSM represents business logic that can be encapsulated, stored, and shared among multiple database users. A PSM lets an administrator assign specific access rights to a stored module to ensure that only authorized users can use it. Support for PSMs is left to each vendor to implement. In fact, for many years, some RDBMSs (such as Oracle, SQL Server, and DB2) supported stored procedure modules within the database before the official standard was promulgated.

MS SQL Server implements PSMs via Transact-SQL and other language extensions, the most notable of which are the .NET family of programming languages. Oracle implements PSMs through its procedural SQL language. MySQL uses a procedural version of SQL that is similar in many respects to the Oracle procedural language. **Procedural Language SQL (PL/SQL)** is a language that makes it possible to use and store procedural code and SQL statements within the database and to merge SQL and traditional programming constructs, such as variables, conditional processing (IF-THEN-ELSE), basic loops (FOR and WHILE loops), and error trapping. The procedural code is executed as a unit by the DBMS when it is invoked (directly or indirectly) by the end user. End users can use PL/SQL to create:

- Anonymous PL/SQL blocks

- Triggers (covered in Section 8-7a)

persistent stored module (PSM)
A block of code with standard SQL statements and procedural extensions that is stored and executed at the DBMS server.

Procedural Language SQL (PL/SQL)
An Oracle-specific programming language based on SQL with procedural extensions designed to run inside the Oracle database.

- Stored procedures (covered in Section 8-7b and Section 8-7c)
- PL/SQL functions (covered in Section 8-7d)

Do not confuse PL/SQL functions with SQL's built-in aggregate functions such as MIN and MAX. SQL built-in functions can be used only within SQL statements, while PL/SQL functions are mainly invoked within PL/SQL programs such as triggers and stored procedures. Functions can also be called within SQL statements, provided that they conform to very specific rules that are dependent on your DBMS environment.

Note

PL/SQL, triggers, and stored procedures are illustrated within the context of an Oracle DBMS. All examples in the following sections assume the use of Oracle RDBMS.

Using Oracle SQL*Plus, you can write a PL/SQL code block by enclosing the commands inside BEGIN and END clauses. For example, the following PL/SQL block inserts a new row in the VENDOR table, as shown in Figure 8.29.

FIGURE 8.29 ANONYMOUS PL/SQL BLOCK EXAMPLES

```
SQL Plus

SQL> BEGIN
  2      INSERT INTO VENDOR
  3      VALUES (25678, 'Microsoft Corp.', 'Bill Gates', '765', '546-8484', 'WA', 'N');
  4  END;
  5  /

PL/SQL procedure successfully completed.

SQL> SET SERVEROUTPUT ON
SQL>
SQL> BEGIN
  2      INSERT INTO VENDOR
  3      VALUES (25772, 'Clue Store', 'Issac Hayes', '456', '323-2009', 'VA', 'N');
  4      DBMS_OUTPUT.PUT_LINE('New Vendor Added! ');
  5  END;
  6  /
New Vendor Added!

PL/SQL procedure successfully completed.

SQL> SELECT * FROM VENDOR;

    V_CODE V_NAME                              V_CONTACT         V_A V_PHONE  V_ V
---------- ----------------------------------- ----------------- --- -------- -- -
     21225 Bryson, Inc.                        Smithson          615 223-3234 TN Y
     21226 SuperLoo, Inc.                      Flushing          904 215-8995 FL N
     21231 D&E Supply                          Singh             615 228-3245 TN Y
     21344 Gomez Bros.                         Ortega            615 889-2546 KY N
     22567 Dome Supply                         Smith             901 678-1419 GA N
     23119 Randsets Ltd.                       Anderson          901 678-3998 GA Y
     24004 Brackman Bros.                      Browning          615 228-1410 TN N
     24288 ORDVA, Inc.                         Hakford           615 898-1234 TN Y
     25443 B&K, Inc.                           Smith             904 227-0093 FL N
     25501 Damal Supplies                      Smythe            615 890-3529 TN N
     25595 Rubicon Systems                     Orton             904 456-0092 FL Y
     25678 Microsoft Corp.                     Bill Gates        765 546-8484 WA N
     25772 Clue Store                          Issac Hayes       456 323-2009 VA N

13 rows selected.

SQL>
```

```
          BEGIN
                    INSERT INTO VENDOR
                    VALUES (25678,'Microsoft Corp.', 'Bill Gates','765','546-8484','WA','N');
          END;
          /
```

The PL/SQL block shown in Figure 8.29 is known as an **anonymous PL/SQL block** because it has not been given a specific name. The block's last line uses a forward slash (/) to indicate the end of the command-line entry. This type of PL/SQL block executes as soon as you press Enter after typing the forward slash. Following the PL/SQL block's execution, you will see the message "PL/SQL procedure successfully completed."

Suppose that you want a more specific message displayed on the SQL*Plus screen after a procedure is completed, such as "New Vendor Added." To produce a more specific message, you must do two things:

1. At the SQL > prompt, type SET SERVEROUTPUT ON. This SQL*Plus command enables the client console (SQL*Plus) to receive messages from the server side (Oracle DBMS). Remember, just like standard SQL, the PL/SQL code (anonymous blocks, triggers, and procedures) are executed at the server side, not at the client side. To stop receiving messages from the server, you would enter SET SERVEROUTPUT OFF.

2. To send messages from the PL/SQL block to the SQL*Plus console, use the DBMS_OUTPUT.PUT_LINE function.

The following anonymous PL/SQL block inserts a row in the VENDOR table and displays the message "New Vendor Added!" (see Figure 8.29).

```
          BEGIN
                    INSERT INTO VENDOR
                    VALUES (25772, 'Clue Store', 'Issac Hayes', '456','323-2009', 'VA', 'N');
                    DBMS_OUTPUT.PUT_LINE('New Vendor Added!');
          END;
          /
```

In Oracle, you can use the SQL*Plus command SHOW ERRORS to help you diagnose errors found in PL/SQL blocks. The SHOW ERRORS command yields additional debugging information whenever you generate an error after creating or executing a PL/SQL block.

The following example of an anonymous PL/SQL block demonstrates several of the constructs supported by the procedural language. Remember that the exact syntax of the language is vendor-dependent; in fact, many vendors enhance their products with proprietary features.

```
DECLARE
W_P1 NUMBER(3) := 0;
W_P2 NUMBER(3) := 10;
W_NUM NUMBER(2) := 0;
BEGIN
WHILE W_P2 < 300 LOOP
          SELECT COUNT(P_CODE) INTO W_NUM FROM PRODUCT
          WHERE P_PRICE BETWEEN W_P1 AND W_P2;
          DBMS_OUTPUT.PUT_LINE('There are ' || W_NUM || ' Products with
                    price between ' || W_P1 || ' and ' || W_P2);
          W_P1 := W_P2 + 1;
          W_P2 := W_P2 + 50;
END LOOP;
END;
/
```

anonymous PL/SQL block
A PL/SQL block that has not been given a specific name.

The block's code and execution are shown in Figure 8.30.

FIGURE 8.30 ANONYMOUS PL/SQL BLOCK WITH VARIABLES AND LOOPS

```
SQL Plus

SQL> DECLARE
  2   W_P1   NUMBER(3) := 0;
  3   W_P2   NUMBER(3) := 10;
  4   W_NUM NUMBER(2) := 0;
  5   BEGIN
  6   WHILE W_P2 < 300 LOOP
  7      SELECT COUNT(P_CODE) INTO W_NUM FROM PRODUCT
  8      WHERE P_PRICE BETWEEN W_P1 AND W_P2;
  9      DBMS_OUTPUT.PUT_LINE('There are ' || W_NUM || ' Products with price between ' || W_P1 || ' and ' || W_P2);
 10      W_P1 := W_P1 + 1;
 11      W_P2 := W_P2 + 50;
 12   END LOOP;
 13   END;
 14   /
There are 5 Products with price between 0 and 10
There are 6 Products with price between 11 and 60
There are 3 Products with price between 61 and 110
There are 1 Products with price between 111 and 160
There are 0 Products with price between 161 and 210
There are 1 Products with price between 211 and 260

PL/SQL procedure successfully completed.

SQL>
```

The PL/SQL block shown in Figure 8.30 has the following characteristics:

- The PL/SQL block starts with the DECLARE section, in which you declare the variable names, the data types, and, if desired, an initial value. Supported data types are shown in Table 8.9.

- A WHILE loop is used. Note the following syntax:

  WHILE *condition* LOOP

        PL/SQL statements;

  END LOOP

- The SELECT statement uses the INTO keyword to assign the output of the query to a PL/SQL variable. You can use the INTO keyword only inside a PL/SQL block of code. If the SELECT statement returns more than one value, you will get an error.

- Note the use of the string concatenation symbol (||) to display the output.

- Each statement inside the PL/SQL code must end with a semicolon (;).

The most useful feature of PL/SQL blocks is that they let you create code that can be named, stored, and executed—either implicitly or explicitly—by the DBMS. That capability is especially desirable when you need to use triggers and stored procedures, which you will explore next.

TABLE 8.9

PL/SQL BASIC DATA TYPES

DATA TYPE	DESCRIPTION
CHAR	Character values of a fixed length; for example: W_ZIP CHAR(5)
VARCHAR2	Variable-length character values; for example: W_FNAME VARCHAR2(15)
NUMBER	Numeric values; for example: W_PRICE NUMBER(6,2)
DATE	Date values; for example: W_EMP_DOB DATE
%TYPE	Inherits the data type from a variable that you declared previously or from an attribute of a database table; for example: W_PRICE PRODUCT.P_PRICE%TYPE Assigns W_PRICE the same data type as the P_PRICE column in the PRODUCT table

Note

PL/SQL blocks can contain only standard SQL data manipulation language (DML) commands such as SELECT, INSERT, UPDATE, and DELETE. The use of data definition language (DDL) commands is not directly supported in a PL/SQL block.

8-7a Triggers

Automating business procedures and automatically maintaining data integrity and consistency are critical in a modern business environment. One of the most critical business procedures is proper inventory management. For example, you want to make sure that current product sales can be supported with sufficient product availability. Therefore, you must ensure that a product order is written to a vendor when that product's inventory drops below its minimum allowable quantity on hand. Better yet, how about ensuring that the task is completed automatically?

To automate product ordering, you first must make sure the product's quantity on hand reflects an up-to-date and consistent value. After the appropriate product availability requirements have been set, two key issues must be addressed:

1. Business logic requires an update of the product quantity on hand each time there is a sale of that product.

2. If the product's quantity on hand falls below its minimum allowable inventory level, the product must be reordered.

To accomplish these two tasks, you could write multiple SQL statements: one to update the product quantity on hand and another to update the product reorder flag. Next, you would have to run each statement in the correct order each time there was a new sale. Such a multistage process would be inefficient because a series of SQL statements must be written and executed each time a product is sold. Even worse, this SQL environment requires that someone must remember to perform the SQL tasks.

A **trigger** is procedural SQL code that is *automatically* invoked by the RDBMS upon the occurrence of a given data manipulation event. It is useful to remember that:

- A trigger is invoked before or after a data row is inserted, updated, or deleted.
- A trigger is associated with a database table.
- Each database table may have one or more triggers.
- A trigger is executed as part of the transaction that triggered it.

Triggers are critical to proper database operation and management. For example:

- Triggers can be used to enforce constraints that cannot be enforced at the DBMS design and implementation levels.
- Triggers add functionality by automating critical actions and providing appropriate warnings and suggestions for remedial action. In fact, one of the most common uses for triggers is to facilitate the enforcement of referential integrity.
- Triggers can be used to update table values, insert records in tables, and call other stored procedures.

Triggers play a critical role in making the database truly useful; they also add processing power to the RDBMS and to the database system as a whole. Oracle recommends triggers for:

- Auditing purposes (creating audit logs)
- Automatic generation of derived column values
- Enforcement of business or security constraints
- Creation of replica tables for backup purposes

To see how a trigger is created and used, examine a simple inventory management problem. For example, if a product's quantity on hand is updated when the product is sold, the system should automatically check whether the quantity on hand falls below its minimum allowable quantity. To demonstrate that process, use the PRODUCT table in Figure 8.31. Note the use of the minimum order quantity (P_MIN_ORDER) and product reorder flag (P_REORDER) columns. The P_MIN_ORDER indicates the minimum quantity for restocking an order. The P_REORDER column is a numeric field that indicates whether the product needs to be reordered (1 = Yes, 0 = No). The initial P_REORDER values are set to 0 (No) to serve as the basis for the initial trigger development.

trigger
A procedural SQL code that is automatically invoked by the relational database management system when a data manipulation event occurs.

FIGURE 8.31 THE PRODUCT TABLE

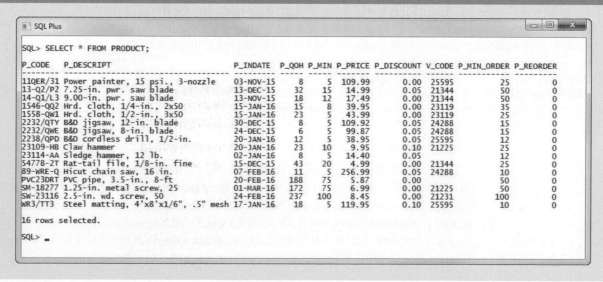

```
SQL Plus

SQL> SELECT * FROM PRODUCT;

P_CODE    P_DESCRIPT                            P_INDATE  P_QOH P_MIN P_PRICE P_DISCOUNT V_CODE P_MIN_ORDER P_REORDER
--------  ------------------------------------  --------  ----- ----- ------- ---------- ------ ----------- ---------
11QER/31  Power painter, 15 psi., 3-nozzle      03-NOV-15     8     5  109.99       0.00  25595          25         0
13-Q2/P2  7.25-in. pwr. saw blade               13-DEC-15    32    15   14.99       0.05  21344          50         0
14-Q1/L3  9.00-in. pwr. saw blade               13-NOV-15    18    12   17.49       0.00  21344          50         0
1546-QQ2  Hrd. cloth, 1/4-in., 2x50             15-JAN-16    15     8   39.95       0.00  23119          35         0
1558-QW1  Hrd. cloth, 1/2-in., 3x50             15-JAN-16    23     5   43.99       0.00  23119          25         0
2232/QTY  B&D jigsaw, 12-in. blade              30-DEC-15     8     5  109.92       0.05  24288          15         0
2232/QWE  B&D jigsaw, 8-in. blade               24-DEC-15     6     5   99.87       0.05  24288          15         0
2238/QPD  B&D cordless drill, 1/2-in.           20-JAN-16    12     5   38.95       0.05  25595          12         0
23109-HB  Claw hammer                           20-JAN-16    23    10    9.95       0.10  21225          25         0
23114-AA  Sledge hammer, 12 lb.                 02-JAN-16     8     5   14.40       0.05                 12         0
54778-2T  Rat-tail file, 1/8-in. fine           15-DEC-15    43    20    4.99       0.00  21344          25         0
89-WRE-Q  Hicut chain saw, 16 in.               07-FEB-16    11     5  256.99       0.05  24288          10         0
PVC23DRT  PVC pipe, 3.5-in., 8-ft               20-FEB-16   188    75    5.87       0.00                 50         0
SM-18277  1.25-in. metal screw, 25              01-MAR-16   172    75    6.99       0.00  21225          50         0
SW-23116  2.5-in. wd. screw, 50                 24-FEB-16   237   100    8.45       0.00  21231         100         0
WR3/TT3   Steel matting, 4'x8'x1/6", .5" mesh   17-JAN-16    18     5  119.95       0.10  25595          10         0

16 rows selected.

SQL> _
```

statement-level trigger
A SQL trigger that is assumed if the FOR EACH ROW keywords are omitted. This type of trigger is executed once, before or after the triggering statement completes, and is the default case.

row-level trigger
A trigger that is executed once for each row affected by the triggering SQL statement. A row-level trigger requires the use of the FOR EACH ROW keywords in the trigger declaration.

Given the PRODUCT table listing shown in Figure 8.31, create a trigger to evaluate the product's quantity on hand, P_QOH. If the quantity on hand is below the minimum quantity shown in P_MIN, the trigger will set the P_REORDER column to 1, which represents "Yes." The syntax to create a trigger in Oracle is as follows:

```
CREATE OR REPLACE TRIGGER trigger_name
[BEFORE / AFTER] [DELETE / INSERT / UPDATE OF column_name] ON table_name
[FOR EACH ROW]
[DECLARE]
[variable_namedata type[:=initial_value] ]
BEGIN
PL/SQL instructions;
…
END;
```

As you can see, a trigger definition contains the following parts:

- *The triggering timing*: BEFORE or AFTER. This timing indicates when the trigger's PL/SQL code executes—in this case, before or after the triggering statement is completed.

- *The triggering event*: The statement that causes the trigger to execute (INSERT, UPDATE, or DELETE).

  - *The triggering level*: The two types of triggers are statement-level triggers and row-level triggers. A **statement-level trigger** is assumed if you omit the FOR EACH ROW keywords. This type of trigger is executed once, before or after the triggering statement is completed. This is the default case.

  - A **row-level trigger** requires use of the FOR EACH ROW keywords. This type of trigger is executed once for each row affected by the triggering statement. (In other words, if you update 10 rows, the trigger executes 10 times.)

- *The triggering action:* The PL/SQL code enclosed between the BEGIN and END keywords. Each statement inside the PL/SQL code must end with a semicolon (;).

Note

Oracle and MS SQL Server allow a trigger to include multiple triggering conditions; that is, any combination of INSERT, UPDATE, and/or DELETE. MySQL allows only one triggering condition per trigger. Therefore, if a certain set of actions should be taken in the case of multiple events, for example, during an UPDATE or an INSERT, then two separate triggers are required in MySQL. To reduce having duplicate code in both triggers, it is a common practice to create a stored procedure that performs the common actions, then have both triggers call the same stored procedure.

Previously, Access did not support triggers for tables. However, starting with Access 2013, "Table Events" have been added that provide trigger functionality. A table can have events before and/or after rows are inserted, updated, or deleted.

In the PRODUCT table's case, you will create a statement-level trigger that is implicitly executed AFTER an UPDATE of the P_QOH attribute for an existing row or AFTER an INSERT of a new row in the PRODUCT table. The trigger action executes an UPDATE statement that compares the P_QOH with the P_MIN column. If the value of P_QOH is equal to or less than P_MIN, the trigger updates the P_REORDER to 1. To create the trigger, Oracle's SQL*Plus will be used. The trigger code is shown in Figure 8.32.

FIGURE 8.32   CREATING THE TRG_PRODUCT_REORDER TRIGGER

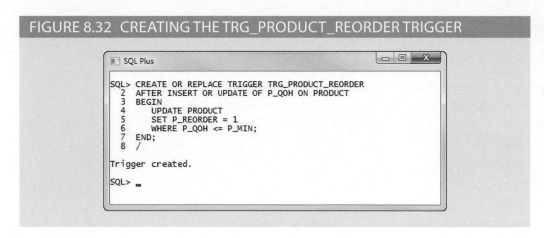

```
SQL> CREATE OR REPLACE TRIGGER TRG_PRODUCT_REORDER
  2    AFTER INSERT OR UPDATE OF P_QOH ON PRODUCT
  3    BEGIN
  4      UPDATE PRODUCT
  5        SET P_REORDER = 1
  6        WHERE P_QOH <= P_MIN;
  7    END;
  8  /

Trigger created.

SQL>
```

To test the TRG_PRODUCT_REORDER trigger, update the quantity on hand of product '11QER/31' to 4. After the UPDATE completes, the trigger is automatically fired and the UPDATE statement inside the trigger code sets the P_REORDER to 1 for all products that are below the minimum. (See Figure 8.33.)

FIGURE 8.33   VERIFYING THE TRG_PRODUCT_REORDER TRIGGER EXECUTION

```
SQL> SELECT P_CODE, P_DESCRIPT, P_QOH, P_MIN, P_MIN_ORDER, P_REORDER
  2    FROM PRODUCT
  3    WHERE P_CODE = '11QER/31';

P_CODE   P_DESCRIPT                             P_QOH P_MIN P_MIN_ORDER P_REORDER
-------- ------------------------------------- ----- ----- ----------- ---------
11QER/31 Power painter, 15 psi., 3-nozzle          8     5          25         0

SQL> UPDATE PRODUCT
  2    SET P_QOH = 4
  3    WHERE P_CODE = '11QER/31';

1 row updated.

SQL> SELECT P_CODE, P_DESCRIPT, P_QOH, P_MIN, P_MIN_ORDER, P_REORDER
  2    FROM PRODUCT
  3    WHERE P_CODE = '11QER/31';

P_CODE   P_DESCRIPT                             P_QOH P_MIN P_MIN_ORDER P_REORDER
-------- ------------------------------------- ----- ----- ----------- ---------
11QER/31 Power painter, 15 psi., 3-nozzle          4     5          25         1

SQL>
```

The trigger shown in Figure 8.33 seems to work, but what happens if you reduce the minimum quantity of product '2232/QWE'? Figure 8.34 shows that when you update the minimum quantity, the quantity on hand of the product '2232/QWE' falls below the new minimum, but the reorder flag is still 0. Why?

The answer is simple: you updated the P_MIN column, but the trigger is never executed. TRG_PRODUCT_ REORDER executes only after an update of the P_QOH column! To avoid that inconsistency, you must modify the trigger event to execute after an update of the P_MIN field, too. The updated trigger code is shown in Figure 8.35.

FIGURE 8.34 THE P_REORDER VALUE MISMATCH AFTER UPDATE OF THE P_MIN ATTRIBUTE

```
SQL Plus

SQL> SELECT P_CODE, P_DESCRIPT, P_QOH, P_MIN, P_MIN_ORDER, P_REORDER
  2  FROM PRODUCT
  3  WHERE P_CODE = '2232/QWE';

P_CODE    P_DESCRIPT                              P_QOH P_MIN P_MIN_ORDER P_REORDER
--------  ------------------------------------    ----- ----- ----------- ---------
2232/QWE  B&D jigsaw, 8-in. blade                   6     5        15         0

SQL> UPDATE PRODUCT
  2  SET P_MIN = 7
  3  WHERE P_CODE = '2232/QWE';

1 row updated.

SQL> SELECT P_CODE, P_DESCRIPT, P_QOH, P_MIN, P_MIN_ORDER, P_REORDER
  2  FROM PRODUCT
  3  WHERE P_CODE = '2232/QWE';

P_CODE    P_DESCRIPT                              P_QOH P_MIN P_MIN_ORDER P_REORDER
--------  ------------------------------------    ----- ----- ----------- ---------
2232/QWE  B&D jigsaw, 8-in. blade                   6     7        15         0

SQL> _
```

FIGURE 8.35 SECOND VERSION OF THE TRG_PRODUCT_REORDER TRIGGER

```
SQL Plus

SQL> CREATE OR REPLACE TRIGGER TRG_PRODUCT_REORDER
  2    AFTER INSERT OR UPDATE OF P_QOH, P_MIN ON PRODUCT
  3  BEGIN
  4      UPDATE PRODUCT
  5      SET P_REORDER = 1
  6      WHERE P_QOH <= P_MIN;
  7  END;
  8  /

Trigger created.

SQL> _
```

To test this new trigger version, change the minimum quantity for product '23114-AA' to 10. After that update, the trigger makes sure that the reorder flag is properly set for all of the products in the PRODUCT table. (See Figure 8.36.)

This second version of the trigger seems to work well, but nothing happens if you change the P_QOH value for product '11QER/31', as shown in Figure 8.37! (Note that the reorder flag is *still* set to 1.) Why didn't the trigger change the reorder flag to 0?

The answer is that the trigger does not consider all possible cases. Examine the second version of the TRG_PRODUCT_REORDER trigger code (Figure 8.35) in more detail:

- The trigger fires after the triggering statement is completed. Therefore, the DBMS always executes two statements (INSERT plus UPDATE or UPDATE plus UPDATE). That is, after you update P_MIN or P_QOH or you insert a new row in the PRODUCT table, the trigger executes another UPDATE statement automatically.

- The triggering action performs an UPDATE of *all* the rows in the PRODUCT table, *even if the triggering statement updates just one row!* This can affect the performance of

FIGURE 8.36 SUCCESSFUL TRIGGER EXECUTION AFTER THE P_MIN VALUE IS UPDATED

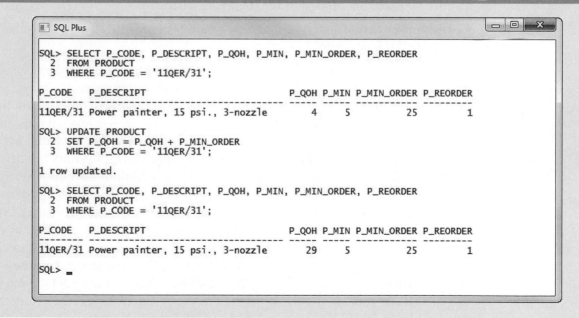

FIGURE 8.37 THE P_REORDER VALUE MISMATCH AFTER INCREASING THE P_QOH VALUE

the database. Imagine what will happen if you have a PRODUCT table with 519,128 rows and you insert just one product. The trigger will update all 519,129 rows, including the rows that do not need an update!

- The trigger sets the P_REORDER value only to 1; it does not reset the value to 0, even if such an action is clearly required when the inventory level is back to a value greater than the minimum value.

In short, the second version of the TRG_PRODUCT_REORDER trigger still does not complete all of the necessary steps. Now modify the trigger to handle all update scenarios, as shown in Figure 8.38.

FIGURE 8.38 THE THIRD VERSION OF THE TRG_PRODUCT_REORDER TRIGGER

```
SQL Plus                                                              [ - ] [ □ ] [ X ]

SQL> CREATE OR REPLACE TRIGGER TRG_PRODUCT_REORDER
  2  BEFORE INSERT OR UPDATE OF P_QOH, P_MIN ON PRODUCT
  3  FOR EACH ROW
  4  BEGIN
  5      IF :NEW.P_QOH <= :NEW.P_MIN THEN
  6          :NEW.P_REORDER := 1;
  7      ELSE
  8          :NEW.P_REORDER := 0;
  9      END IF;
 10  END;
 11  /

Trigger created.

SQL> _
```

The trigger in Figure 8.38 sports several new features:

- The trigger is executed *before* the actual triggering statement is completed. In Figure 8.38, the triggering timing is defined in line 2, BEFORE INSERT OR UPDATE. This clearly indicates that the triggering statement is executed before the INSERT or UPDATE completes, unlike the previous trigger examples.

- The trigger is a row-level trigger instead of a statement-level trigger. The FOR EACH ROW keywords make the trigger a row-level trigger. Therefore, this trigger executes once for each row affected by the triggering statement.

- The trigger action uses the :NEW attribute reference to change the value of the P_REORDER attribute.

The use of the :NEW attribute references deserves a more detailed explanation. To understand its use, you must first consider a basic computing tenet: *all changes are done first in primary memory, then transferred to permanent memory*. In other words, the computer cannot change anything directly in permanent storage (on disk). It must first read the data from permanent storage to primary memory, then make the change in primary memory, and finally write the changed data back to permanent memory (on disk).

The DBMS operates in the same way, with one addition. Because ensuring data integrity is critical, the DBMS makes two copies of every row being changed by a DML (INSERT, UPDATE, or DELETE) statement. You will learn more about this in Chapter 10, Transaction Management and Concurrency Control. The first copy contains the original ("old") values of the attributes before the changes. The second copy contains the changed ("new") values of the attributes that will be permanently saved to the database after any changes made by an INSERT, UPDATE, or DELETE. You can use :OLD to refer to the original values; you can use :NEW to refer to the changed values (the values that will be stored in the table). You can use :NEW and :OLD attribute references only within the PL/SQL code of a database trigger action. For example:

- IF :NEW.P_QOH < = :NEW.P_MIN compares the quantity on hand with the minimum quantity of a product. Remember that this is a row-level trigger. Therefore, this comparison is made for each row that is updated by the triggering statement.

- Although the trigger is a BEFORE trigger, this does not mean that the triggering statement has not executed yet. To the contrary, the triggering statement has already taken place; otherwise, the trigger would not have fired and the :NEW values would not exist. Remember, BEFORE means *before* the changes are permanently saved to disk, but *after* the changes are made in memory.

- The trigger uses the :NEW reference to assign a value to the P_REORDER column before the UPDATE or INSERT results are permanently stored in the table.

The assignment is always made to the :NEW value (never to the :OLD value), and the assignment always uses the := assignment operator. The :OLD values are *read-only* values; you cannot change them. Note that :NEW.P_REORDER := 1; assigns the value 1 to the P_REORDER column and :NEW.P_REORDER := 0; assigns the value 0 to the P_REORDER column.

- This new trigger version does not use any DML statements!

Before testing the new trigger, note that product '11QER/31' currently has a quantity on hand that is above the minimum quantity, yet the reorder flag is set to 1. Given that condition, the reorder flag must be 0. After creating the new trigger, you can execute an UPDATE statement to fire it, as shown in Figure 8.39.

FIGURE 8.39 EXECUTION OF THE THIRD TRIGGER VERSION

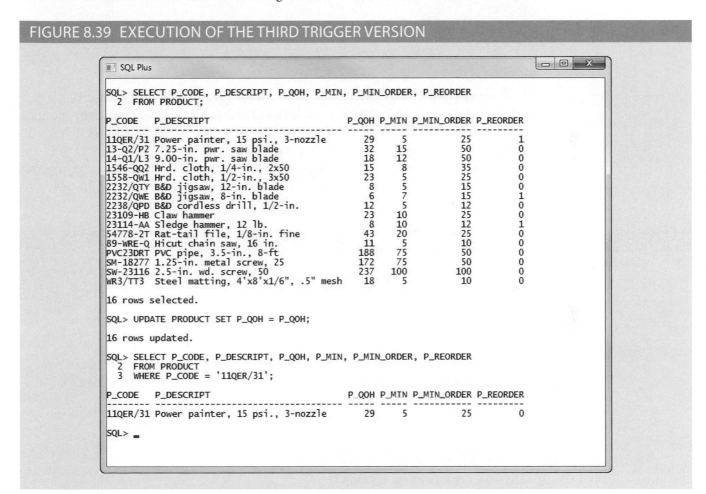

Note the following important features of the code in Figure 8.39:

- The trigger is automatically invoked for each affected row—in this case, all rows of the PRODUCT table. If your triggering statement would have affected only three rows, not all PRODUCT rows would have the correct P_REORDER value set, which is why the triggering statement was set up as shown in Figure 8.38.

- The trigger will run only if you insert a new product row or update P_QOH or P_MIN. If you update any other attribute, the trigger will not run.

You can also use a trigger to update an attribute in a table other than the one being modified. For example, suppose that you would like to create a trigger that automatically reduces the quantity on hand of a product with every sale. To accomplish that task, you must create a trigger for the LINE table that updates a row in the PRODUCT table. The sample code for that trigger is shown in Figure 8.40.

FIGURE 8.40 TRG_LINE_PROD TRIGGER TO UPDATE THE PRODUCT QUANTITY ON HAND

```
SQL Plus                                                    _ □ X

SQL> CREATE OR REPLACE TRIGGER TRG_LINE_PROD
  2    AFTER INSERT ON LINE
  3    FOR EACH ROW
  4    BEGIN
  5       UPDATE PRODUCT
  6       SET P_QOH = P_QOH - :NEW.LINE_UNITS
  7       WHERE PRODUCT.P_CODE = :NEW.P_CODE;
  8    END;
  9    /

Trigger created.

SQL> _
```

Note that the TRG_LINE_PROD row-level trigger executes after inserting a new invoice's LINE and reduces the quantity on hand of the recently sold product by the number of units sold. This row-level trigger updates a row in a different table (PRODUCT), using the :NEW values of the recently added LINE row.

A third trigger example shows the use of variables within a trigger. In this case, you want to update the customer balance (CUS_BALANCE) in the CUSTOMER table after inserting every new LINE row. This trigger code is shown in Figure 8.41.

FIGURE 8.41 TRG_LINE_CUS TRIGGER TO UPDATE THE CUSTOMER BALANCE

```
SQL Plus                                                    _ □ X

SQL> CREATE OR REPLACE TRIGGER TRG_LINE_CUS
  2    AFTER INSERT ON LINE
  3    FOR EACH ROW
  4    DECLARE
  5       W_CUS    CHAR(5);
  6       W_TOT    NUMBER := 0;    --to compute total cost
  7    BEGIN
  8       -- this trigger fires up after an INSERT of a LINE
  9       -- it will update the CUS_BALANCE in CUSTOMER
 10
 11       --1) get the CUS_CODE
 12       SELECT CUS_CODE INTO W_CUS
 13       FROM INVOICE
 14       WHERE INVOICE.INV_NUMBER = :NEW.INV_NUMBER;
 15
 16       --2) compute the total of the current line
 17       W_TOT := :NEW.LINE_PRICE * :NEW.LINE_UNITS;
 18
 19       --3) update the CUS_BALANCE in CUSTOMER
 20       UPDATE CUSTOMER
 21       SET CUS_BALANCE = CUS_BALANCE + W_TOT
 22       WHERE CUS_CODE = W_CUS;
 23
 24       DBMS_OUTPUT.PUT_LINE('*** Balance updated for customer: ' || W_CUS);
 25
 26    END;
 27    /

Trigger created.

SQL>
```

Carefully examine the trigger in Figure 8.41.

- The trigger is a row-level trigger that executes after each new LINE row is inserted.
- The DECLARE section in the trigger is used to declare any variables used inside the trigger code.

- You can declare a variable by assigning a name, a data type, and (optionally) an initial value, as in the case of the W_TOT variable.

- The first step in the trigger code is to get the customer code (CUS_CODE) from the related INVOICE table. Note that the SELECT statement returns only one attribute (CUS_CODE) from the INVOICE table. Also note that the attribute returns only one value as specified by the use of the WHERE clause, *to restrict the query output to a single value.*

- Note the use of the INTO clause within the SELECT statement. You use the INTO clause to assign a value from a SELECT statement to a variable (W_CUS) used within a trigger.

- The second step in the trigger code computes the total of the line by multiplying :NEW. LINE_UNITS by :NEW.LINE_PRICE and assigning the result to the W_TOT variable.

- The final step updates the customer balance by using an UPDATE statement and the W_TOT and W_CUS trigger variables.

- Double dashes (--) are used to indicate comments within the PL/SQL block.

To summarize the triggers created in this section:

- TRG_PRODUCT_REORDER is a row-level trigger that updates P_REORDER in PRODUCT when a new product is added or when the P_QOH or P_MIN columns are updated.

- TRG_LINE_PROD is a row-level trigger that automatically reduces the P_QOH in PRODUCT when a new row is added to the LINE table.

- TRG_LINE_CUS is a row-level trigger that automatically increases the CUS_ BALANCE in CUSTOMER when a new row is added in the LINE table.

The use of triggers facilitates the automation of multiple data management tasks. Although triggers are independent objects, they are associated with database tables. When you delete a table, all its trigger objects are deleted with it. However, if you needed to delete a trigger without deleting the table, you could use the following command:

DROP TRIGGER *trigger_name*

Trigger Action Based on Conditional DML Predicates You could also create triggers whose actions depend on the type of DML statement (INSERT, UPDATE, or DELETE) that fires the trigger. For example, you could create a trigger that executes after an INSERT, an UPDATE, or a DELETE on the PRODUCT table. But how do you know which one of the three statements caused the trigger to execute? In those cases, you could use the following syntax:

IF INSERTING THEN … END IF;
IF UPDATING THEN … END IF;
IF DELETING THEN … END IF;

8-7b Stored Procedures

A **stored procedure** is a named collection of procedural and SQL statements. Just like database triggers, stored procedures are stored in the database. One of the major advantages of stored procedures is that they can be used to encapsulate and represent business transactions. For example, you can create a stored procedure to represent a product sale, a credit update, or the addition of a new customer. By doing that, you can encapsulate SQL statements within a single stored procedure and execute them as a single transaction. There are two clear advantages to the use of stored procedures:

- Stored procedures substantially reduce network traffic and increase performance. Because the procedure is stored at the server, there is no transmission of individual

stored procedure
(1) A named collection of procedural and SQL statements. (2) Business logic stored on a server in the form of SQL code or another DBMS-specific procedural language.

SQL statements over the network. The use of stored procedures improves system performance because all transactions are executed locally on the RDBMS, so each SQL statement does not have to travel over the network.

- Stored procedures help reduce code duplication by means of code isolation and code sharing (creating unique PL/SQL modules that are called by application programs), thereby minimizing the chance of errors and the cost of application development and maintenance.

To create a stored procedure, you use the following syntax:

CREATE OR REPLACE PROCEDURE *procedure_name* [(*argument* [IN/OUT] *data-type*, ...)]

[IS/AS]
[*variable_namedata type*[:=*initial_value*]]

BEGIN

PL/SQL or SQL statements;
...

END;

Note the following important points about stored procedures and their syntax:

- *argument* specifies the parameters that are passed to the stored procedure. A stored procedure could have zero or more arguments or parameters.

- *IN/OUT* indicates whether the parameter is for input, output, or both.

- *data-type* is one of the procedural SQL data types used in the RDBMS. The data types normally match those used in the RDBMS table creation statement.

- Variables can be declared between the keywords IS and BEGIN. You must specify the variable name, its data type, and (optionally) an initial value.

To illustrate stored procedures, assume that you want to create a procedure (PRC_PROD_DISCOUNT) to assign an additional 5 percent discount for all products when the quantity on hand is more than or equal to twice the minimum quantity. Figure 8.42 shows how the stored procedure is created.

FIGURE 8.42 CREATING THE PRC_PROD_DISCOUNT STORED PROCEDURE

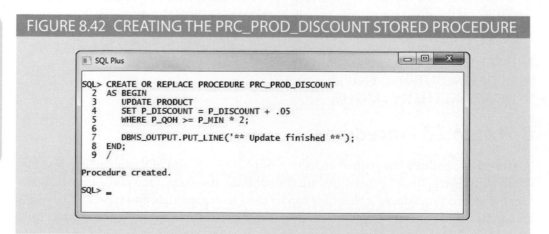

```
SQL> CREATE OR REPLACE PROCEDURE PRC_PROD_DISCOUNT
  2   AS BEGIN
  3      UPDATE PRODUCT
  4      SET P_DISCOUNT = P_DISCOUNT + .05
  5      WHERE P_QOH >= P_MIN * 2;
  6
  7      DBMS_OUTPUT.PUT_LINE('** Update finished **');
  8  END;
  9  /

Procedure created.

SQL>
```

Note in Figure 8.42 that the PRC_PROD_DISCOUNT stored procedure uses the DBMS_OUTPUT.PUT_LINE function to display a message when the procedure executes. (This action assumes that you previously ran SET SERVEROUTPUT ON.)

To execute the stored procedure, you must use the following syntax:

EXEC *procedure_name*[(*parameter_list*)];

For example, to see the results of running the PRC_PROD_DISCOUNT stored procedure, you can use the EXEC PRC_PROD_DISCOUNT command shown in Figure 8.43.

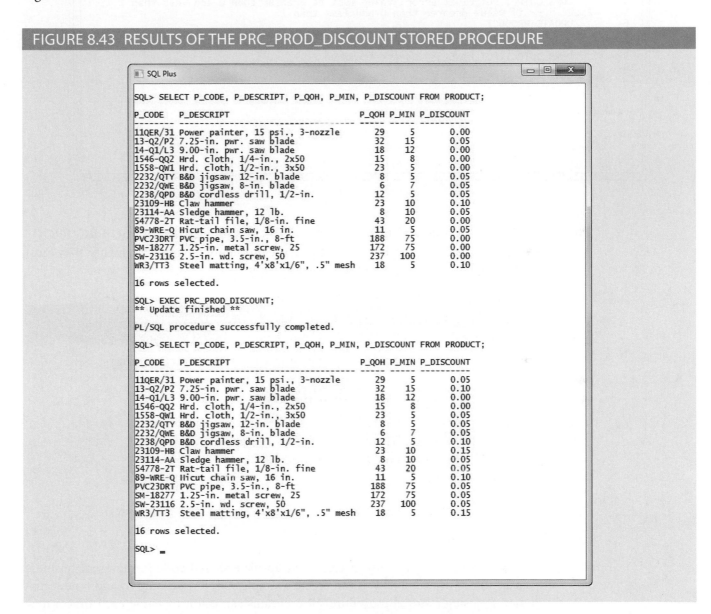

Using Figure 8.43 as your guide, you can see how the product discount attribute was increased by 5 percent for all products with a quantity on hand that was more than or equal to twice the minimum quantity. (Compare the first PRODUCT table listing to the second PRODUCT table listing.)

One of the main advantages of procedures is that you can pass values to them. For example, the previous PRC_PROD_DISCOUNT procedure worked well, but what if you want to make the percentage increase an input variable? In that case, you can pass an argument to represent the rate of increase to the procedure. Figure 8.44 shows the code for that procedure.

FIGURE 8.44 SECOND VERSION OF THE PRC_PROD_DISCOUNT STORED PROCEDURE

```
SQL Plus

SQL> CREATE OR REPLACE PROCEDURE PRC_PROD_DISCOUNT (WPI IN NUMBER) AS
  2    BEGIN
  3      IF ((WPI <= 0) OR (WPI >= 1)) THEN    --validate WPI parameter
  4        DBMS_OUTPUT.PUT_LINE('Error: Value must be greater than 0 and less than 1');
  5      ELSE     -- if value greater than 0 and less than 1
  6        UPDATE PRODUCT
  7        SET P_DISCOUNT = P_DISCOUNT + WPI
  8        WHERE P_QOH >= P_MIN * 2;
  9        DBMS_OUTPUT.PUT_LINE('** Update finished **');
 10      END IF;
 11    END;
 12    /

Procedure created.

SQL> _
```

Figure 8.45 shows the execution of the second version of the PRC_PROD_ DISCOUNT stored procedure. Note that if the procedure requires arguments, they must be enclosed in parentheses and separated by commas.

FIGURE 8.45 RESULTS OF THE SECOND VERSION OF THE PRC_PROD_ DISCOUNT STORED PROCEDURE

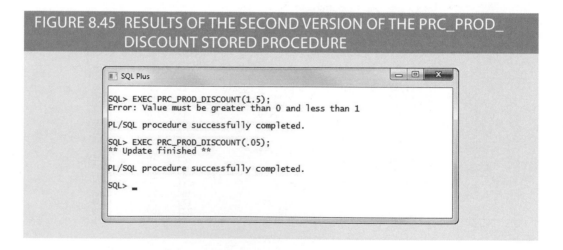

```
SQL Plus

SQL> EXEC PRC_PROD_DISCOUNT(1.5);
Error: Value must be greater than 0 and less than 1

PL/SQL procedure successfully completed.

SQL> EXEC PRC_PROD_DISCOUNT(.05);
** Update finished **

PL/SQL procedure successfully completed.

SQL> _
```

Stored procedures are also useful to encapsulate shared code to represent business transactions. For example, you can create a simple stored procedure to add a new customer. By using a stored procedure, all programs can call it by name each time a new customer is added. Naturally, if new customer attributes are added later, you will need to modify the stored procedure. However, the programs that use the stored procedure will not need to know the name of the newly added attribute; they will need to add only a new parameter to the procedure call. (Notice the PRC_CUS_ADD stored procedure shown in Figure 8.46.)

As you examine Figure 8.46, note these features:

- The PRC_CUS_ADD procedure uses several parameters, one for each required attribute in the CUSTOMER table.

- The stored procedure uses the CUS_CODE_SEQ sequence to generate a new customer code.

FIGURE 8.46 THE PRC_CUS_ADD STORED PROCEDURE

```
SQL Plus                                                              _  □  X

SQL> CREATE OR REPLACE PROCEDURE PRC_CUS_ADD
  2  (W_LN IN VARCHAR, W_FN IN VARCHAR, W_INIT IN VARCHAR, W_AC IN VARCHAR, W_PH IN VARCHAR)
  3  AS
  4  BEGIN
  5      --note this procedure uses the CUS_CODE_SEQ sequence created earlier
  6      --attribute names are required when not giving values for all table attributes
  7      INSERT INTO CUSTOMER(CUS_CODE, CUS_LNAME, CUS_FNAME, CUS_INITIAL, CUS_AREACODE, CUS_PHONE)
  8      VALUES (CUS_CODE_SEQ.NEXTVAL, W_LN, W_FN, W_INIT, W_AC, W_PH);
  9      DBMS_OUTPUT.PUT_LINE('Customer ' || W_LN || ', ' || W_FN || ' added.');
 10  END;
 11  /

Procedure created.

SQL> EXEC PRC_CUS_ADD('Walker', 'James', NULL, '615', '84-HORSE');
Customer Walker, James added.

PL/SQL procedure successfully completed.

SQL> SELECT * FROM CUSTOMER WHERE CUS_LNAME = 'Walker';

  CUS_CODE CUS_LNAME        CUS_FNAME        C CUS CUS_PHON CUS_BALANCE
---------- ---------------  ---------------  - --- -------- -----------
     20012 Walker           James              615 84-HORSE           0

SQL> EXEC PRC_CUS_ADD ('Lowery', 'Denise', NULL, NULL, NULL);
BEGIN PRC_CUS_ADD ('Lowery', 'Denise', NULL, NULL, NULL); END;

*
ERROR at line 1:
ORA-01400: cannot insert NULL into ("TEST"."CUSTOMER"."CUS_AREACODE")
ORA-06512: at "TEST.PRC_CUS_ADD", line 7
ORA-06512: at line 1


SQL> _
```

- The required parameters—those specified in the table definition—must be included and can be null *only* when the table specifications permit nulls for that parameter. For example, note that the second customer addition was unsuccessful because the CUS_AREACODE is a required attribute and cannot be null.

- The procedure displays a message in the SQL*Plus console to let the user know that the customer was added.

The next two examples further illustrate the use of sequences within stored procedures. In this case, create two stored procedures:

1. The PRC_INV_ADD procedure adds a new invoice.

2. The PRC_LINE_ADD procedure adds a new product line row for a given invoice.

Both procedures are shown in Figure 8.47. Note the use of a variable in the PRC_LINE_ADD procedure to get the product price from the PRODUCT table.

To test the procedures shown in Figure 8.47:

1. Call the PRC_INV_ADD procedure with the new invoice data as arguments.

2. Call the PRC_LINE_ADD procedure and pass the product line arguments.

That process is illustrated in Figure 8.48.

FIGURE 8.47 THE PRC_INV_ADD AND PRC_LINE_ADD STORED PROCEDURES

```
SQL Plus

SQL> CREATE OR REPLACE PROCEDURE PRC_INV_ADD (W_CUS_CODE IN VARCHAR2, W_DATE IN DATE) AS
  2   BEGIN
  3      INSERT INTO INVOICE (INV_NUMBER, CUS_CODE, INV_DATE)
  4      VALUES (INV_NUMBER_SEQ.NEXTVAL, W_CUS_CODE, W_DATE);
  5      DBMS_OUTPUT.PUT_LINE('Invoice added.');
  6   END;
  7   /

Procedure created.

SQL> CREATE OR REPLACE PROCEDURE PRC_LINE_ADD (W_LN IN NUMBER, W_P_CODE IN VARCHAR2, W_LU IN NUMBER) AS
  2      W_LP  NUMBER  := 0;
  3   BEGIN
  4      --get the product price
  5      SELECT P_PRICE INTO W_LP
  6      FROM PRODUCT
  7      WHERE PRODUCT.P_CODE = W_P_CODE;
  8
  9      --adds the new line row
 10      INSERT INTO LINE (INV_NUMBER, LINE_NUMBER, P_CODE, LINE_UNITS, LINE_PRICE)
 11      VALUES (INV_NUMBER_SEQ.CURRVAL, W_LN, W_P_CODE, W_LU, W_LP);
 12
 13      DBMS_OUTPUT.PUT_LINE('Invoice line ' || W_LN || ' added.');
 14   END;
 15   /

Procedure created.

SQL> _
```

FIGURE 8.48 TESTING THE PRC_INV_ADD AND PRC_LINE_ADD PROCEDURES

```
SQL Plus

SQL> EXEC PRC_INV_ADD(20012, '09-APR-2016');
Invoice added.

PL/SQL procedure successfully completed.

SQL> EXEC PRC_LINE_ADD(1, '13-Q2/P2', 1);
*** Balance updated for customer: 20012
Invoice line 1 added.

PL/SQL procedure successfully completed.

SQL> EXEC PRC_LINE_ADD(2, '23109-HB', 1);
*** Balance updated for customer: 20012
Invoice line 2 added.

PL/SQL procedure successfully completed.

SQL> SELECT * FROM INVOICE WHERE CUS_CODE = 20012;

INV_NUMBER   CUS_CODE INV_DATE
---------- ---------- ---------
      4011      20012 09-APR-16

SQL> SELECT * FROM LINE WHERE INV_NUMBER = (SELECT INV_NUMBER FROM INVOICE WHERE CUS_CODE = 20012);

INV_NUMBER LINE_NUMBER P_CODE    LINE_UNITS LINE_PRICE
---------- ----------- -------- ---------- ----------
      4011           1 13-Q2/P2          1      14.99
      4011           2 23109-HB          1       9.95

SQL> SELECT * FROM PRODUCT WHERE P_CODE IN ('13-Q2/P2', '23109-HB');

P_CODE   P_DESCRIPT                          P_INDATE  P_QOH P_MIN P_PRICE P_DISCOUNT V_CODE P_MIN_ORDER P_REORDER
-------- ----------------------------------- --------- ----- ----- ------- ---------- ------ ----------- ---------
13-Q2/P2 7.25-in. pwr. saw blade             13-DEC-15    31    15   14.99       0.15  21344          50         0
23109-HB Claw hammer                         20-JAN-16    22    10    9.95       0.20  21225          25         0

SQL> SELECT * FROM CUSTOMER WHERE CUS_CODE = 20012;

 CUS_CODE CUS_LNAME      CUS_FNAME        C CUS CUS_PHON CUS_BALANCE
--------- -------------- ---------------- - --- -------- -----------
    20012 Walker         James              615 84-HORSE       24.94

SQL> _
```

8-7c PL/SQL Processing with Cursors

Until now, all of the SQL statements you have used inside a PL/SQL block (trigger or stored procedure) have returned a single value. If the SQL statement returns more than one value, you will generate an error. If you want to use a SQL statement that returns more than one value inside your PL/SQL code, you need to use a cursor. A **cursor** is a special construct used in procedural SQL to hold the data rows returned by a SQL query. You can think of a cursor as a reserved area of memory in which the output of the query is stored, like an array holding columns and rows. Cursors are held in a reserved memory area in the DBMS server, not in the client computer.

There are two types of cursors: implicit and explicit. An **implicit cursor** is automatically created in procedural SQL when the SQL statement returns only one value. Up to this point, all of the examples created an implicit cursor. An **explicit cursor** is created to hold the output of a SQL statement that may return two or more rows (but could return zero rows or only one). To create an explicit cursor, you use the following syntax inside a PL/SQL DECLARE section:

CURSOR *cursor_name* IS *select-query*;

Once you have declared a cursor, you can use specific PL/SQL cursor processing commands (OPEN, FETCH, and CLOSE) anywhere between the BEGIN and END keywords of the PL/SQL block. Table 8.10 summarizes the main use of each command.

Cursor-style processing involves retrieving data from the cursor one row at a time. Once you open a cursor, it becomes an active data set. That data set contains a "current" row pointer. Therefore, after opening a cursor, the current row is the first row of the cursor.

When you fetch a row from the cursor, the data from the "current" row in the cursor is copied to the PL/SQL variables. After the fetch, the "current" row pointer moves to the next row in the set and continues until it reaches the end of the cursor.

How do you know what number of rows are in the cursor? Or how do you know when you have reached the end of the cursor data set? You know because cursors have special attributes that convey important information. Table 8.11 summarizes the cursor attributes.

cursor
A special construct used in procedural SQL to hold the data rows returned by a SQL query. A cursor may be considered a reserved area of memory in which query output is stored, like an array holding columns and rows. Cursors are held in a reserved memory area in the DBMS server, not in the client computer.

implicit cursor
A cursor that is automatically created in procedural SQL when the SQL statement returns only one row.

explicit cursor
In procedural SQL, a cursor created to hold the output of a SQL statement that may return two or more rows, but could return zero or only one row.

TABLE 8.10

CURSOR PROCESSING COMMANDS	
CURSOR COMMAND	**EXPLANATION**
OPEN	Opening the cursor executes the SQL command and populates the cursor with data, opening the cursor for processing. The cursor declaration command only reserves a named memory area for the cursor; it does not populate the cursor with the data. Before you can use a cursor, you need to open it. For example: OPEN *cursor_name*

FETCH	Once the cursor is opened, you can use the FETCH command to retrieve data from the cursor and copy it to the PL/SQL variables for processing. The syntax is:
	FETCH cursor_name INTO variable1 [, variable2, …]
	    The PL/SQL variables used to hold the data must be declared in the DECLARE section and must have data types compatible with the columns retrieved by the SQL command. If the cursor's SQL statement returns five columns, there must be five PL/SQL variables to receive the data from the cursor.
	    This type of processing resembles the one-record-at-a-time processing used in previous database models. The first time you fetch a row from the cursor, the first row of data from the cursor is copied to the PL/SQL variables; the second time you fetch a row from the cursor, the second row of data is placed in the PL/SQL variables; and so on.

CLOSE	The CLOSE command closes the cursor for processing.

TABLE 8.11

CURSOR ATTRIBUTES

ATTRIBUTE	DESCRIPTION
%ROWCOUNT	Returns the number of rows fetched so far. If the cursor is not OPEN, it returns an error. If no FETCH has been done but the cursor is OPEN, it returns 0.
%FOUND	Returns TRUE if the last FETCH returned a row, and FALSE if not. If the cursor is not OPEN, it returns an error. If no FETCH has been done, it contains NULL.
%NOTFOUND	Returns TRUE if the last FETCH did not return any row, and FALSE if it did. If the cursor is not OPEN, it returns an error. If no FETCH has been done, it contains NULL.
%ISOPEN	Returns TRUE if the cursor is open (ready for processing) or FALSE if the cursor is closed. Remember, before you can use a cursor, you must open it.

To illustrate the use of cursors, use a simple stored procedure example to list the products that have a greater quantity on hand than the average quantity on hand for all products. The code is shown in Figure 8.49.

FIGURE 8.49 A SIMPLE PRC_CURSOR_EXAMPLE

```
SQL> CREATE OR REPLACE PROCEDURE PRC_CURSOR_EXAMPLE IS
  2  W_P_CODE     PRODUCT.P_CODE%TYPE;
  3  W_P_DESCRIPT         PRODUCT.P_DESCRIPT%TYPE;
  4  W_TOT               NUMBER(3);
  5  CURSOR PROD_CURSOR IS
  6     SELECT P_CODE, P_DESCRIPT
  7        FROM PRODUCT
  8        WHERE P_QOH > (SELECT AVG(P_QOH) FROM PRODUCT);
  9  BEGIN
 10  DBMS_OUTPUT.PUT_LINE('PRODUCTS WITH P_QOH > AVG(P_QOH)');
 11  DBMS_OUTPUT.PUT_LINE('======================================');
 12  OPEN PROD_CURSOR;
 13  LOOP
 14     FETCH PROD_CURSOR INTO W_P_CODE, W_P_DESCRIPT;
 15     EXIT WHEN PROD_CURSOR%NOTFOUND;
 16     DBMS_OUTPUT.PUT_LINE(W_P_CODE ||' -> ' || W_P_DESCRIPT );
 17  END LOOP;
 18  DBMS_OUTPUT.PUT_LINE('======================================');
 19  DBMS_OUTPUT.PUT_LINE('TOTAL PRODUCT PROCESSED ' || PROD_CURSOR%ROWCOUNT);
 20  DBMS_OUTPUT.PUT_LINE('--- END OF REPORT ----');
 21  CLOSE PROD_CURSOR;
 22  END;
 23  /

Procedure created.

SQL> EXEC PRC_CURSOR_EXAMPLE;
PRODUCTS WITH P_QOH > AVG(P_QOH)
======================================
PVC23DRT -> PVC pipe, 3.5-in., 8-ft
SM-18277 -> 1.25-in. metal screw, 25
SW-23116 -> 2.5-in. wd. screw, 50
======================================
TOTAL PRODUCT PROCESSED 3
--- END OF REPORT ----

PL/SQL procedure successfully completed.

SQL>
```

As you examine the stored procedure code shown in Figure 8.49, note the following important characteristics:

- Lines 2 and 3 use the %TYPE data type in the variable definition section. As indicated in Table 8.9, the %TYPE data type indicates that the given variable inherits the data type from a previously declared variable or from an attribute of a database table. In this case, you are using the %TYPE to indicate that the W_P_CODE and W_P_DESCRIPT will have the same data type as the respective columns in the PRODUCT table. This way, you ensure that the PL/SQL variable will have a compatible data type.

- Line 5 declares the PROD_CURSOR cursor.

- Line 12 opens the PROD_CURSOR cursor and populates it.

- Line 13 uses the LOOP statement to loop through the data in the cursor, fetching one row at a time.

- Line 14 uses the FETCH command to retrieve a row from the cursor and place it in the respective PL/SQL variables.

- Line 15 uses the EXIT command to evaluate when there are no more rows in the cursor (using the %NOTFOUND cursor attribute) and to exit the loop.

- Line 19 uses the %ROWCOUNT cursor attribute to obtain the total number of rows processed.

- Line 21 issues the CLOSE PROD_CURSOR command to close the cursor.

The use of cursors, combined with standard SQL, makes working with relational databases very desirable because programmers can work in the best of both worlds: set-oriented processing and record-oriented processing. Any experienced programmer knows to use the tool that best fits the job. Sometimes you will be better off manipulating data in a set-oriented environment; at other times, it might be better to use a record-oriented environment. Procedural SQL lets you have your proverbial cake and eat it too. Procedural SQL provides functionality that enhances the capabilities of the DBMS while maintaining a high degree of manageability.

8-7d PL/SQL Stored Functions

Using programmable or procedural SQL, you can also create your own stored functions. Stored procedures and functions are very similar. A **stored function** is basically a named group of procedural and SQL statements that returns a value, as indicated by a RETURN statement in its program code. To create a function, you use the following syntax:

CREATE FUNCTION *function_name* (*argument* IN *data-type*, …) RETURN *data-type* [IS]
BEGIN
      PL/SQL statements;
      …
      RETURN (*value or expression*);
END;

Stored functions can be invoked only from within stored procedures or triggers, and cannot be invoked from SQL statements unless the function follows some very specific compliance rules. Remember not to confuse built-in SQL functions (such as MIN, MAX, and AVG) with stored functions.

stored function
A named group of procedural and SQL statements that returns a value, as indicated by a RETURN statement in its program code.

8-8 Embedded SQL

There is little doubt that SQL's popularity as a data manipulation language is due in part to its ease of use and its powerful data-retrieval capabilities. In the real world, however, database systems are related to other systems and programs, and you still need a conventional programming language such as Visual Basic .NET, C#, or COBOL to integrate database systems with other programs and systems. If you are developing web applications, you are most likely familiar with Visual Studio .NET, Java, ASP, or ColdFusion. Yet, almost regardless of the programming tools you use, if your web application or Windows-based GUI system requires access to a database such as MS Access, SQL Server, Oracle, or DB2, you will likely need to use SQL to manipulate the data in the database.

Embedded SQL is a term used to refer to SQL statements contained within an application programming language such as Visual Basic .NET, C#, COBOL, or Java. The program being developed might be a standard binary executable in Windows or Linux, or it might be a web application designed to run over the Internet. No matter what language you use, if it contains embedded SQL statements, it is called the **host language**. Embedded SQL is still the most common approach to maintaining procedural capabilities in DBMS-based applications. However, mixing SQL with procedural languages requires that you understand some key differences between the two.

- *Run-time mismatch.* Remember that SQL is a nonprocedural, interpreted language; that is, each instruction is parsed, its syntax is checked, and it is executed one instruction at a time. (The authors are particularly grateful for the thoughtful comments provided by Emil T. Cipolla.) All of the processing takes place at the server side. Meanwhile, the host language is generally a binary-executable program (also known as a *compiled program*). The host program typically runs at the client side in its own memory space, which is different from the DBMS environment.

- *Processing mismatch.* Conventional programming languages (COBOL, ADA, FORTRAN, Pascal, C++, and PL/I) process one data element at a time. Although you can use arrays to hold data, you still process the array elements one row at a time. This is especially true for file manipulation, where the host language typically manipulates data one record at a time. However, newer programming environments such as Visual Studio .NET have adopted several object-oriented extensions that help the programmer manipulate data sets in a cohesive manner.

- *Data type mismatch.* SQL provides several data types, but some of them might not match data types used in different host languages (for example, the DATE and VARCHAR2 data types).

To bridge the differences, the embedded SQL standard defines a framework to integrate SQL within several programming languages. The embedded SQL framework defines the following:

- A standard syntax to identify embedded SQL code within the host language (EXEC SQL/END-EXEC).

- A standard syntax to identify host variables, which are variables in the host language that receive data from the database (through the embedded SQL code) and process the data in the host language. All host variables are preceded by a colon (:).

- A communication area used to exchange status and error information between SQL and the host language. This communication area contains two variables—SQLCODE and SQLSTATE.

embedded SQL
SQL statements contained within application programming languages such as COBOL, C++, ASP, Java, and ColdFusion.

host language
Any language that contains embedded SQL statements.

Another way to interface host languages and SQL is through the use of a call-level interface (CLI), in which the programmer writes to an application programming interface (API). A common CLI in Windows is provided by the Open Database Connectivity (ODBC) interface.

Online Content

Additional coverage of CLIs and ODBC is available in Appendix F, Client/Server Systems, and Appendix J, Web Database Development with ColdFusion, at *www. cengagebrain.com*.

Before continuing, you should explore the process required to create and run an executable program with embedded SQL statements. If you have ever programmed in COBOL or C++, you are familiar with the multiple steps required to generate the final executable program. Although the specific details vary among language and DBMS vendors, the following general steps are standard:

1. The programmer writes embedded SQL code within the host language instructions. The code follows the standard syntax required for the host language and embedded SQL.

2. A preprocessor is used to transform the embedded SQL into specialized procedure calls that are DBMS- and language-specific. The preprocessor is provided by the DBMS vendor and is specific to the host language.

3. The program is compiled using the host language compiler. The compiler creates an object code module for the program containing the DBMS procedure calls.

4. The object code is linked to the respective library modules and generates the executable program. This process binds the DBMS procedure calls to the DBMS run-time libraries. Additionally, the binding process typically creates an "access plan" module that contains instructions to run the embedded code at run time.

5. The executable is run, and the embedded SQL statement retrieves data from the database.

Note that you can embed individual SQL statements or even an entire PL/SQL block. Up to this point in the book, you have used a DBMS-provided application (SQL*Plus) to write SQL statements and PL/SQL blocks in an interpretive mode to address one-time or ad hoc data requests. However, it is extremely difficult and awkward to use ad hoc queries to process transactions inside a host language. Programmers typically embed SQL statements within a host language that is compiled once and executed as often as needed. To embed SQL into a host language, follow this syntax:

```
EXEC SQL
        SQL statement;
END-EXEC.
```

The preceding syntax will work for SELECT, INSERT, UPDATE, and DELETE statements. For example, the following embedded SQL code will delete employee 109, George Smith, from the EMPLOYEE table:

```
EXEC SQL
        DELETE FROM EMPLOYEE WHERE EMP_NUM = 109;
END-EXEC.
```

Remember, the preceding embedded SQL statement is compiled to generate an executable statement. Therefore, the statement is fixed permanently and cannot change (unless, of course, the programmer changes it). Each time the program runs, it deletes the same row. In short, the preceding code is good only for the first run; all subsequent runs will likely generate an error. Clearly, this code would be more useful if you could specify a variable to indicate the employee number to be deleted.

In embedded SQL, all host variables are preceded by a colon (:). The host variables may be used to send data from the host language to the embedded SQL, or they may be used to receive the data from the embedded SQL. To use a host variable, you must first declare it in the host language. Common practice is to use similar host variable names as the SQL source attributes. For example, if you are using COBOL, you would define the host variables in the Working Storage section. Then you would refer to them in the embedded SQL section by preceding them with a colon. For example, to delete an employee whose employee number is represented by the host variable W_EMP_NUM, you would write the following code:

EXEC SQL

        DELETE FROM EMPLOYEE WHERE EMP_NUM = :W_EMP_NUM;
END-EXEC.

At run time, the host variable value will be used to execute the embedded SQL statement. What happens if the employee you are trying to delete does not exist in the database? How do you know that the statement has been completed without errors? As mentioned previously, the embedded SQL standard defines a SQL communication area to hold status and error information. In COBOL, such an area is known as the SQLCA area and is defined in the Data Division as follows:

EXEC SQL

        INCLUDE SQLCA
END-EXEC.

The SQLCA area contains two variables for status and error reporting. Table 8.12 shows some of the main values returned by the variables and their meaning.

TABLE 8.12

SQL STATUS AND ERROR REPORTING VARIABLES

VARIABLE NAME	VALUE	EXPLANATION
SQLCODE		Old-style error reporting supported for backward compatibility only; returns an integer value (positive or negative)
	0	Successful completion of command
	100	No data; the SQL statement did not return any rows and did not select, update, or delete any rows
	−999	Any negative value indicates that an error occurred
SQLSTATE		Added by SQL-92 standard to provide predefined error codes; defined as a character string (5 characters long)
	00000	Successful completion of command
		Multiple values in the format XXYYY where: XX-> represents the class code YYY-> represents the subclass code

The following embedded SQL code illustrates the use of the SQLCODE within a COBOL program.

```
EXEC SQL
        SELECT          EMP_LNAME, EMP_LNAME INTO :W_EMP_FNAME,
                        :W_EMP_LNAME WHERE EMP_NUM = :W_EMP_NUM;
END-EXEC.
IF SQLCODE = 0 THEN
        PERFORM DATA_ROUTINE
ELSE
        PERFORM ERROR_ROUTINE
END-IF.
```

In this example, the SQLCODE host variable is checked to determine whether the query completed successfully. If it did, the DATA_ROUTINE is performed; otherwise, the ERROR_ROUTINE is performed.

Just as with PL/SQL, embedded SQL requires the use of cursors to hold data from a query that returns more than one value. If COBOL is used, the cursor can be declared either in the Working Storage section or in the Procedure Division. The cursor must be declared and processed, as you learned earlier in Section 8-7c. To declare a cursor, you use the syntax shown in the following example:

```
EXEC SQL
        DECLARE PROD_CURSOR FOR
        SELECT          P_CODE, P_DESCRIPT FROM PRODUCT
        WHERE           P_QOH > (SELECT AVG(P_QOH) FROM PRODUCT);
END-EXEC.
```

Next, you must open the cursor to make it ready for processing:

```
EXEC SQL
        OPEN PROD_CURSOR;
END-EXEC.
```

To process the data rows in the cursor, you use the FETCH command to retrieve one row of data at a time and place the values in the host variables. The SQLCODE must be checked to ensure that the FETCH command completed successfully. This section of code typically constitutes part of a routine in the COBOL program. Such a routine is executed with the PERFORM command. For example:

```
EXEC SQL
        FETCH PROD_CURSOR INTO :W_P_CODE, :W_P_DESCRIPT;
END-EXEC.
IF SQLCODE = 0 THEN
        PERFORM DATA_ROUTINE
ELSE
        PERFORM ERROR_ROUTINE
END-IF.
```

When all rows have been processed, you close the cursor as follows:

EXEC SQL

      CLOSE PROD_CURSOR;

END-EXEC.

Thus far, you have seen examples of embedded SQL in which the programmer used predefined SQL statements and parameters. Therefore, the end users of the programs are limited to the actions that were specified in the application programs. That style of embedded SQL is known as **static SQL**, meaning that the SQL statements will not change while the application is running. For example, the SQL statement might read like this:

SELECT   P_CODE, P_DESCRIPT, P_QOH, P_PRICE

                FROM PRODUCT

WHERE   P_PRICE > 100;

Note that the attributes, tables, and conditions are known in the preceding SQL statement. Unfortunately, end users seldom work in a static environment. They are more likely to require the flexibility of defining their data access requirements on the fly. Therefore, the end user requires that SQL be as dynamic as the data access requirements.

Dynamic SQL is a term used to describe an environment in which the SQL statement is not known in advance; instead, the SQL statement is generated at run time. At run time in a dynamic SQL environment, a program can generate the SQL statements that are required to respond to ad hoc queries. In such an environment, neither the programmer nor the end user is likely to know precisely what kind of queries will be generated or how they will be structured. For example, a dynamic SQL equivalent of the preceding example could be:

SELECT   :W_ATTRIBUTE_LIST

FROM   :W_TABLE

WHERE   :W_CONDITION;

Note that the attribute list and the condition are not known until the end user specifies them. W_TABLE, W_ATTRIBUTE_LIST, and W_CONDITION are text variables that contain the end-user input values used in the query generation. Because the program uses the end-user input to build the text variables, the end user can run the same program multiple times to generate varying outputs. For example, in one instance, the end user might want to know which products cost less than $100; in another case, the end user might want to know how many units of a given product are available for sale at any given moment.

Although dynamic SQL is clearly flexible, such flexibility carries a price. Dynamic SQL tends to be much slower than static SQL. Dynamic SQL also requires more computer resources (overhead). Finally, you are more likely to find inconsistent levels of support and incompatibilities among DBMS vendors.

static SQL
A style of embedded SQL in which the SQL statements do not change while the application is running.

dynamic SQL
An environment in which the SQL statement is not known in advance, but instead is generated at run time. In a dynamic SQL environment, a program can generate the SQL statements that are required to respond to ad hoc queries.

Summary

- Operations that join tables can be classified as inner joins and outer joins. An inner join is the traditional join in which only rows that meet a given criterion are selected. An outer join returns the matching rows as well as the rows with unmatched attribute values for one table or both tables to be joined.

- A natural join returns all rows with matching values in the matching columns and eliminates duplicate columns. This style of query is used when the tables share a common attribute with a common name. One important difference between the syntax for a natural join and for the old-style join is that the natural join does not require the use of a table qualifier for the common attributes. In practice, natural joins are often discouraged because the common attribute is not specified within the command, making queries more difficult to understand and maintain.

- Joins may use keywords such as USING and ON. If the USING clause is used, the query will return only the rows with matching values in the column indicated in the USING clause; that column must exist in both tables. If the ON clause is used, the query will return only the rows that meet the specified join condition.

- Subqueries and correlated queries are used when it is necessary to process data based on *other* processed data. That is, the query uses results that were previously unknown and that are generated by another query. Subqueries may be used with the FROM, WHERE, IN, and HAVING clauses in a SELECT statement. A subquery may return a single row or multiple rows.

- Most subqueries are executed in a serial fashion. That is, the outer query initiates the data request, and then the inner subquery is executed. In contrast, a correlated subquery is a subquery that is executed once for each row in the outer query. That process is similar to the typical nested loop in a programming language. A correlated subquery is so named because the inner query is related to the outer query—the inner query references a column of the outer subquery.

- SQL functions are used to extract or transform data. The most frequently used functions are date and time functions. The results of the function output can be used to store values in a database table, to serve as the basis for the computation of derived variables, or to serve as a basis for data comparisons. Function formats can be vendor-specific. Aside from time and date functions, there are numeric and string functions as well as conversion functions that convert one data format to another.

- SQL provides relational set operators to combine the output of two queries to generate a new relation. The UNION and UNION ALL set operators combine the output of two or more queries and produce a new relation with all unique (UNION) or duplicate (UNION ALL) rows from both queries. The INTERSECT relational set operator selects only the common rows. The EXCEPT (MINUS) set operator selects only the rows that are different. UNION, INTERSECT, and EXCEPT require union-compatible relations.

- In Oracle and SQL Server, sequences may be used to generate values to be assigned to a record. For example, a sequence may be used to number invoices automatically. MS Access uses an AutoNumber data type to generate numeric sequences, and MySQL uses the AUTO_INCREMENT property during table creation. Oracle and SQL Server can use the Identity column property to designate the column that will have sequential numeric values automatically assigned to it. There can only be one Identity column per table.

- Procedural Language SQL (PL/SQL) can be used to create triggers, stored procedures, and PL/SQL functions. A trigger is procedural SQL code that is automatically invoked by the DBMS upon the occurrence of a specified data manipulation event (UPDATE, INSERT, or DELETE). Triggers are critical to proper database operation and management. They help automate various transaction and data management processes, and they can be used to enforce constraints that are not enforced at the DBMS design and implementation levels.

- A stored procedure is a named collection of SQL statements. Just like database triggers, stored procedures are stored in the database. One of the major advantages of stored procedures is that they can be used to encapsulate and represent complete business transactions. Use of stored procedures substantially reduces network traffic and increases system performance. Stored procedures also help reduce code duplication by creating unique PL/SQL modules that are called by the application programs, thereby minimizing the chance of errors and the cost of application development and maintenance.

- When SQL statements are designed to return more than one value inside the PL/SQL code, a cursor is needed. You can think of a cursor as a reserved area of memory in which the output of the query is stored, like an array holding columns and rows. Cursors are held in a reserved memory area in the DBMS server, rather than in the client computer. There are two types of cursors: implicit and explicit.

- Embedded SQL refers to the use of SQL statements within an application programming language such as Visual Basic .NET, C#, COBOL, or Java. The language in which the SQL statements are embedded is called the host language. Embedded SQL is still the most common approach to maintaining procedural capabilities in DBMS-based applications.

Key Terms

Online Content

Flashcards and crossword puzzles for key term practice are available at *www.cengagebrain.com*.

anonymous PL/SQL block	host language	set-oriented
base table	implicit cursor	statement-level trigger
batch update routine	inner join	static SQL
correlated subquery	outer join	stored function
CREATE VIEW	persistent stored module (PSM)	stored procedure
cross join		trigger
cursor	Procedural Language SQL (PL/SQL)	union-compatible
dynamic SQL	row-level trigger	updatable view
embedded SQL	sequence	view
explicit cursor		

Review Questions

1. What is a CROSS JOIN? Give an example of its syntax.

2. What three join types are included in the OUTER JOIN classification?

3. Using tables named T1 and T2, write a query example for each of the three join types you described in Question 2. Assume that T1 and T2 share a common column named C1.

4. What is a subquery, and what are its basic characteristics?

5. What are the three types of results that a subquery can return?

6. What is a correlated subquery? Give an example.

7. Explain the difference between a regular subquery and a correlated subquery.

8. What does it mean to say that SQL operators are set-oriented?

9. The relational set operators UNION, INTERSECT, and EXCEPT (MINUS) work properly only when the relations are union-compatible. What does union-compatible mean, and how would you check for this condition?

10. What is the difference between UNION and UNION ALL? Write the syntax for each.

11. Suppose you have two tables: EMPLOYEE and EMPLOYEE_1. The EMPLOYEE table contains the records for three employees: Alice Cordoza, John Cretchakov, and Anne McDonald. The EMPLOYEE_1 table contains the records for employees John Cretchakov and Mary Chen. Given that information, list the query output for the UNION query.

12. Given the employee information in Question 11, list the query output for the UNION ALL query.

13. Given the employee information in Question 11, list the query output for the INTERSECT query.

14. Given the employee information in Question 11, list the query output for the EXCEPT (MINUS) query of EMPLOYEE to EMPLOYEE_1.

15. Why does the order of the operands (tables) matter in an EXCEPT (MINUS) query but not in a UNION query?

16. What MS Access and SQL Server function should you use to calculate the number of days between your birth date and the current date?

17. What Oracle function should you use to calculate the number of days between your birth date and the current date?

18. Suppose a PRODUCT table contains two attributes, PROD_CODE and VEND_CODE. Those two attributes have values of ABC, 125, DEF, 124, GHI, 124, and JKL, 123, respectively. The VENDOR table contains a single attribute, VEND_CODE, with values 123, 124, 125, and 126, respectively. (The VEND_CODE attribute in the PRODUCT table is a foreign key to the VEND_CODE in the VENDOR table.) Given that information, what would be the query output for:

    a. A UNION query based on the two tables?

    b. A UNION ALL query based on the two tables?

    c. An INTERSECT query based on the two tables?

    d. An EXCEPT (MINUS) query based on the two tables?

19. What string function should you use to list the first three characters of a company's EMP_LNAME values? Give an example using a table named EMPLOYEE. Provide examples for Oracle and SQL Server.

20. What is a sequence? Write its syntax.

21. What is a trigger, and what is its purpose? Give an example.

22. What is a stored procedure, and why is it particularly useful? Give an example.

23. What is embedded SQL and how is it used?

24. What is dynamic SQL, and how does it differ from static SQL?

Problems

Use the database tables in Figure P8.1 as the basis for Problems 1–18.

FIGURE P8.1 CH08_SIMPLECO DATABASE TABLES

Database name: Ch08_SimpleCo

Table name: CUSTOMER

CUST_NUM	CUST_LNAME	CUST_FNAME	CUST_BALANCE
1000	Smith	Jeanne	1050.11
1001	Ortega	Juan	840.92

Table name: CUSTOMER_2

CUST_NUM	CUST_LNAME	CUST_FNAME
2000	McPherson	Anne
2001	Ortega	Juan
2002	Kowalski	Jan
2003	Chen	George

Table name: INVOICE

INV_NUM	CUST_NUM	INV_DATE	INV_AMOUNT
8000	1000	23-Mar-16	235.89
8001	1001	23-Mar-16	312.82
8002	1001	30-Mar-16	528.10
8003	1000	12-Apr-16	194.78
8004	1000	23-Apr-16	619.44

1. Create the tables. (Use the MS Access example shown in Figure P8.1 to see what table names and attributes to use.)

2. Insert the data into the tables you created in Problem 1.

3. Write the query that will generate a combined list of customers from the tables CUSTOMER and CUSTOMER_2 that do not include the duplicate customer records. Only the customer named Juan Ortega shows up in both customer tables. (Figure P8.3)

FIGURE P8.3 COMBINED LIST OF CUSTOMERS WITHOUT DUPLICATES

CUST_LNAME	CUST_FNAME
Chen	George
Kowalski	Jan
McPherson	Anne
Ortega	Juan
Smith	Jeanne

4. Write the query that will generate a combined list of customers to include the duplicate customer records. (Figure P8.4)

FIGURE P8.4   COMBINED LIST OF CUSTOMERS WITH DUPLICATES

CUST_LNAME	CUST_FNAME
Smith	Jeanne
Ortega	Juan
McPherson	Anne
Ortega	Juan
Kowalski	Jan
Chen	George

5. Write the query that will show only the duplicate customer records. (Figure P8.5)

FIGURE P8.5   DUPLICATE CUSTOMER RECORD

CUST_LNAME	CUST_FNAME
Ortega	Juan

6. Write the query that will generate only the records that are unique to the CUSTOMER_2 table. (Figure P8.6)

FIGURE P8.6   CUSTOMERS UNIQUE TO THE CUSTOMER_2 TABLE

CUST_LNAME	CUST_FNAME
McPherson	Anne
Kowalski	Jan
Chen	George

7. Write the query to show the invoice number, customer number, customer name, invoice date, and invoice amount for all customers in the CUSTOMER table with a balance of $1,000 or more. (Figure P8.7)

FIGURE P8.7   INVOICES OF CUSTOMERS WITH A BALANCE OVER $1000

INV_NUM	CUST_NUM	CUST_LNAME	CUST_FNAME	INV_DATE	INV_AMOUNT
8000	1000	Smith	Jeanne	23-Mar-16	235.89
8003	1000	Smith	Jeanne	12-Apr-16	194.78
8004	1000	Smith	Jeanne	23-Apr-16	619.44

8. Write the query for all the invoices that will show the invoice number, invoice amount, average invoice amount, and difference between the average invoice amount and the actual invoice amount. (Figure P8.8)

FIGURE P8.8   INVOICE AMOUNTS COMPARED TO THE AVERAGE INVOICE AMOUNT

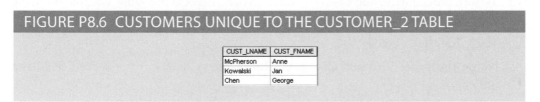

INV_NUM	INV_AMOUNT	AVG_INV	DIFF
8000	235.89	378.21	-142.32
8001	312.82	378.21	-65.39
8002	528.10	378.21	149.89
8003	194.78	378.21	-183.43
8004	619.44	378.21	241.23

9. Write the query that will write Oracle sequences to produce automatic customer number and invoice number values. Start the customer numbers at 1000 and the invoice numbers at 5000.

10. Modify the CUSTOMER table to include two new attributes: CUST_DOB and CUST_AGE. Customer 1000 was born on March 15, 1979, and customer 1001 was born on December 22, 1988.

11. Assuming that you completed Problem 10, write the query that will list the names and ages of your customers.

12. Assuming that the CUSTOMER table contains a CUST_AGE attribute, write the query to update the values in that attribute. (*Hint*: Use the results of the previous query.)

13. Write the query that lists the average age of your customers. (Assume that the CUSTOMER table has been modified to include the CUST_DOB and the derived CUST_AGE attribute.)

14. Write the trigger to update the CUST_BALANCE in the CUSTOMER table when a new invoice record is entered. (Assume that the sale is a credit sale.) Test the trigger using the following new INVOICE record:

    8005, 1001, '27-APR-16', 225.40

    Name the trigger **trg_updatecustbalance**.

15. Write a procedure to add a new customer to the CUSTOMER table. Use the following values in the new record:

    1002, 'Rauthor', 'Peter', 0.00

    Name the procedure **prc_cust_add**. Run a query to see if the record has been added.

16. Write a procedure to add a new invoice record to the INVOICE table. Use the following values in the new record:

    8006, 1000, '30-APR-16', 301.72

    Name the procedure **prc_invoice_add**. Run a query to see if the record has been added.

17. Write a trigger to update the customer balance when an invoice is deleted. Name the trigger **trg_updatecustbalance2**.

18. Write a procedure to delete an invoice, giving the invoice number as a parameter. Name the procedure **prc_inv_delete**. Test the procedure by deleting invoices 8005 and 8006.

Use the Ch08_LargeCo database shown in Figure P8.19 to work Problems 19–27. For problems with very large result sets, only the first several rows of output are shown in the following figures.

FIGURE P8.19 THE LARGECO ERD

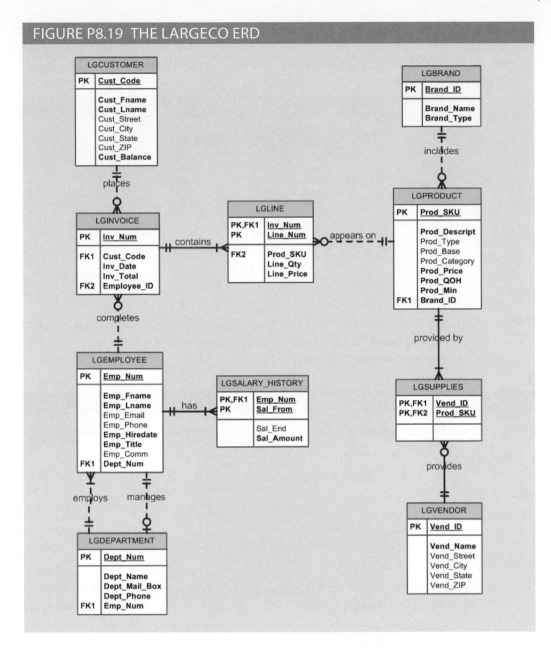

19. Write a query to display the products that have a price greater than $50.

20. Write a query to display the current salary for each employee in department 300. Assume that only current employees are kept in the system, and therefore the most current salary for each employee is the entry in the salary history with a NULL end date. Sort the output in descending order by salary amount. (Figure P8.20)

FIGURE P8.20 CURRENT SALARY FOR EMPLOYEES IN DEPARTMENT 300

Emp_Num	Emp_LName	Emp_FName	Sal_Amount
83746	RANKIN	SEAN	95550
84328	CARPENTER	FERN	94090
83716	RIVERA	HENRY	85920
84432	JAMISON	MERLE	85360
83902	VARGAS	ROCKY	79540
83695	MENDEZ	CARROLL	79200
84500	WESTON	CHRISTINE	78690
84594	TIDWELL	ODELL	77400
83910	AVERY	LAUREN	76110
83359	WATTS	MERLE	72240
83790	ACEVEDO	LAVINA	72000

21. Write a query to display the starting salary for each employee. The starting salary would be the entry in the salary history with the oldest salary start date for each employee. Sort the output by employee number. (Figure P8.21)

FIGURE P8.21 STARTING SALARY FOR EACH EMPLOYEE

Emp_Num	Emp_LName	Emp_FName	Sal_Amount
83304	MCDONALD	TAMARA	19770
83308	LOVE	CONNIE	11230
83312	BAKER	ROSALBA	39260
83314	DAVID	CHAROLETTE	15150
83318	PECK	DARCIE	22330
83321	FARMER	ANGELINA	18250
83332	LONG	WILLARD	23380
83341	CORTEZ	CHRISTINE	14510
83347	WINN	QUINTIN	17010
83349	SINGH	JENNIFFER	21220
83359	WATTS	MERLE	25370
83366	BLEDSOE	PHOEBE	23200

22. Write a query to display the invoice number, line numbers, product SKUs, product descriptions, and brand ID for sales of sealer and top coat products of the same brand on the same invoice. (Figure P8.22)

FIGURE P8.22 INVOICES FOR SEALER AND TOP COAT OF THE SAME BRAND

Inv_Num	l.Line_Num	p.Prod_Sku	p.Prod_Descript	l2.Line_Num	p2.Prod_Sku	p2.Prod_Descript	Brand_ID
115	2	5140-RTG	Fire Resistant Sealer, for Exterior Wood (ULC Approved)	1	1203-AIS	Fire Retardant Coating, Latex, Interior, Flat (ULC Approved)	35
118	2	5140-RTG	Fire Resistant Sealer, for Exterior Wood (ULC Approved)	5	5046-TTC	Aluminum Paint, Heat Resistant (Up to 427°C - 800°F)	35
135	5	3036-PCT	Sealer, for Knots	2	1074-VVJ	Light Industrial Coating, Exterior, Water Based ('eggshell-like' - MPI Gloss Level 3)	25
153	2	3701-YAW	Sealer, Solvent Based, for Concrete Floors	1	3955-NWD	Water Repellant, Clear (Not Paintable)	30
222	1	1336-FVM	Alkyd, Sanding Sealer, Clear	3	8199-YRF	Varnish, Exterior, Water Based, (Satin-Like) MPI Gloss Level 4	33
234	4	5728-ZPO	Shop Coat, Quick Dry, for Interior Steel	3	9272-LTP	Varnish, Marine Spar, Exterior, Gloss (MPI Gloss Level 6)	27
234	4	5728-ZPO	Shop Coat, Quick Dry, for Interior Steel	2	9126-PWF	Latex, Recycled (Consolidated), Interior (MPI Gloss Level 3)	27
243	1	4072-SWV	Sealer, Solvent Based, for Concrete Floors	3	5653-RTU	Aluminum Paint	23
287	1	8894-LUR	Lacquer, Sanding Sealer, Clear	5	9838-FUF	Fire Retardant Top-Coat, Clear, Alkyd, Interior (ULC Approved)	27
333	1	3701-YAW	Sealer, Solvent Based, for Concrete Floors	6	2584-CIJ	Stain, for Exterior Wood Decks	30
333	1	3701-YAW	Sealer, Solvent Based, for Concrete Floors	5	4784-SLU	Lacquer, Clear, Flat	30
369	2	1403-TUY	Sealer, Water Based, for Concrete Floors	1	8726-ZNM	Floor Paint, Alkyd, Low Gloss	29

23. The Binder Prime Company wants to recognize the employee who sold the most of its products during a specified period. Write a query to display the employee number, employee first name, employee last name, email address, and total units sold for the employee who sold the most Binder Prime brand products between November 1, 2015, and December 5, 2015. If there is a tie for most units sold, sort the output by employee last name. (Figure P8.23)

FIGURE P8.23 EMPLOYEES WITH MOST BINDER PRIME UNITS SOLD

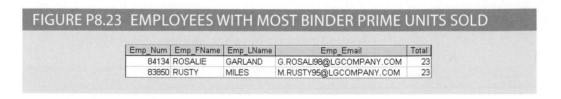

Emp_Num	Emp_FName	Emp_LName	Emp_Email	Total
84134	ROSALIE	GARLAND	G.ROSALI98@LGCOMPANY.COM	23
83850	RUSTY	MILES	M.RUSTY95@LGCOMPANY.COM	23

24. Write a query to display the customer code, first name, and last name of all customers who have had at least one invoice completed by employee 83649 and at least one invoice completed by employee 83677. Sort the output by customer last name and then first name. (Figure P8.24)

FIGURE P8.24 CUSTOMERS WITH INVOICES FILLED BY EMPLOYEES 83649 AND 83677

Cust_Code	Cust_FName	Cust_LName
684	WENDI	BEAN
340	MARCIA	BURRIS
211	GERALD	CAUDILL
292	VALARIE	DILLARD
293	CLAIR	ERICKSON
416	TATIANA	HOWE
996	EZRA	LYON
98	VALENTIN	MARINO
121	PETER	SMALL
1157	LUCIO	STALEY
617	CESAR	TALLEY
457	SHAUNA	WERNER
131	SAL	WHALEY

25. LargeCo is planning a new promotion in Alabama (AL) and wants to know about the largest purchases made by customers in that state. Write a query to display the customer code, customer first name, last name, full address, invoice date, and invoice total of the largest purchase made by each customer in Alabama. Be certain to include any customers in Alabama who have never made a purchase; their invoice dates should be NULL and the invoice totals should display as 0. (Figure P8.25)

FIGURE P8.25 LARGEST PURCHASES OF CUSTOMERS IN AL

Cust_Code	Cust_FName	Cust_LName	Cust_Street	Cust_City	Cust_State	Cust_ZIP	Inv_Date	Largest Invoice
903	ROBIN	ADDISON	323 LORETTA PLACE	Mobile	AL	36693	8/26/2015	230.63
643	NINA	ALLEN	680 RED TALON DRIVE	Robertsdale	AL	36574	6/21/2015	11.99
295	DORTHY	AUSTIN	829 BIG BEND LOOP	Diamond Shamrock	AL	36614	4/24/2015	589.75
393	FOSTER	BERNAL	1299 EAST 3RD AVENUE	Birmingham	AL	35280		0
853	GAYLORD	BOLTON	1069 LUGENE LANE	Montgomery	AL	36131	11/25/2015	372.68
925	ALANA	BOOKER	1874 I STREET	Mccullough	AL	36502	12/12/2015	208.85
1248	LISA	BRADY	491 LOWLAND AVENUE	Daphne	AL	36577	12/5/2015	414.47
538	CHIQUITA	CALDWELL	1501 BRIGGS COURT	Normal	AL	35762	5/26/2015	143.9
89	MONICA	CANTRELL	697 ADAK CIRCLE	Loachapoka	AL	36865	3/31/2015	516.58
1233	NATHALIE	CHURCH	1802 SNOWY OWL CIRCLE	Napier Field	AL	36303	11/24/2015	160.96
304	GERTRUDE	CONNORS	1042 PLEASANT DRIVE	Georgiana	AL	36033	12/29/2015	376.32
1131	CARMA	CORNETT	767 CHISANA WAY	Killen	AL	35645	10/25/2015	265.12
1407	FELICIA	CRUZ	643 TURNAGAIN PARKWAY	Coalburg	AL	35068	1/6/2016	387.93

26. One of the purchasing managers is interested in the impact of product prices on the sale of products of each brand. Write a query to display the brand name, brand type, average price of products of each brand, and total units sold of products of each brand. Even if a product has been sold more than once, its price should only be included once in the calculation of the average price. However, you must be careful because multiple products of the same brand can have the same price, and each of those products must be included in the calculation of the brand's average price. (Figure P8.26)

FIGURE P8.26 AVERAGE PRICE AND TOTAL UNITS SOLD OF PRODUCTS BY BRAND

Brand_Name	Brand_Type	Average Price	Units Sold
BINDER PRIME	PREMIUM	16.12	3753
BUSTERS	VALUE	22.59	3727
FORESTERS BEST	VALUE	20.94	2086
HOME COMFORT	CONTRACTOR	21.8	4842
LE MODE	PREMIUM	19.22	5284
LONG HAUL	CONTRACTOR	20.12	5728
OLDE TYME QUALITY	CONTRACTOR	18.33	3614
STUTTENFURST	CONTRACTOR	16.47	3671
VALU-MATTE	VALUE	16.84	2485

27. The purchasing manager is still concerned about the impact of price on sales. Write a query to display the brand name, brand type, product SKU, product description, and price of any products that are not a premium brand, but that cost more than the most expensive premium brand products. (Figure P8.27)

FIGURE P8.27 NON-PREMIUM PRODUCTS THAT ARE MORE EXPENSIVE THAN PREMIUM PRODUCTS

Brand_Name	Brand_Type	Prod_Sku	Prod_Descript	Prod_Price
LONG HAUL	CONTRACTOR	1964-OUT	Fire Resistant Top Coat, for Interior Wood	78.49

Use the Ch08_SaleCo2 database shown in Figure P8.28 to work Problems 28–31.

FIGURE P8.28 CH08_SALECO2 DATABASE TABLES

Database name: Ch08_SaleCo2

Table name: CUSTOMER

CUS_CODE	CUS_LNAME	CUS_FNAME	CUS_INITIAL	CUS_AREACODE	CUS_PHONE	CUS_BALANCE
10010	Ramas	Alfred	A	615	844-2573	0.00
10011	Dunne	Leona	K	713	894-1238	0.00
10012	Smith	Kathy	W	615	894-2285	345.86
10013	Olowski	Paul	F	615	894-2180	536.75
10014	Orlando	Myron		615	222-1672	0.00
10015	O'Brian	Amy	B	713	442-3361	0.00
10016	Brown	James	G	615	297-1228	221.19
10017	Williams	George		615	290-2556	768.93
10018	Farriss	Anne	G	713	382-7185	216.55
10019	Smith	Olette	K	615	297-3809	0.00

Table name: PRODUCT

P_CODE	P_DESCRIPT	P_INDATE	P_QOH	P_MIN	P_PRICE	P_DISCOUNT	V_CODE
11QER/31	Power painter, 15 psi., 3-nozzle	03-Nov-15	8	5	109.99	0.00	25595
13-Q2/P2	7.25-in. pwr. saw blade	13-Dec-15	32	15	14.99	0.05	21344
14-Q1/L3	9.00-in. pwr. saw blade	13-Nov-15	18	12	17.49	0.00	21344
1546-QQ2	Hrd. cloth, 1/4-in., 2x50	15-Jan-16	15	8	39.95	0.00	23119
1558-QW1	Hrd. cloth, 1/2-in., 3x50	15-Jan-16	23	5	43.99	0.00	23119
2232/QTY	B&D jigsaw, 12-in. blade	30-Dec-15	8	5	109.92	0.05	24288
2232/QWE	B&D jigsaw, 8-in. blade	24-Dec-15	6	5	99.87	0.05	24288
2238/QPD	B&D cordless drill, 1/2-in.	20-Jan-16	12	5	38.95	0.05	25595
23109-HB	Claw hammer	20-Jan-16	23	10	9.95	0.10	21225
23114-AA	Sledge hammer, 12 lb.	02-Jan-16	8	5	14.40	0.05	
54778-2T	Rat-tail file, 1/8-in. fine	15-Dec-15	43	20	4.99	0.00	21344
89-WRE-Q	Hicut chain saw, 16 in.	07-Feb-16	11	5	256.99	0.05	24288
PVC23DRT	PVC pipe, 3.5-in., 8-ft	20-Feb-16	188	75	5.87	0.00	
SM-18277	1.25-in. metal screw, 25	01-Mar-16	172	75	6.99	0.00	21225
SW-23116	2.5-in. wd. screw, 50	24-Feb-16	237	100	8.45	0.00	21231
WR3/TT3	Steel matting, 4'x8'x1/6", .5" mesh	17-Jan-16	18	5	119.95	0.10	25595

Table name: VENDOR

V_CODE	V_NAME	V_CONTACT	V_AREACODE	V_PHONE	V_STATE	V_ORDER
21225	Bryson, Inc.	Smithson	615	223-3234	TN	Y
21226	SuperLoo, Inc.	Flushing	904	215-8995	FL	N
21231	D&E Supply	Singh	615	228-3245	TN	Y
21344	Gomez Bros.	Ortega	615	889-2546	KY	N
22567	Dome Supply	Smith	901	678-1419	GA	N
23119	Randsets Ltd.	Anderson	901	678-3998	GA	Y
24004	Brackman Bros.	Browning	615	228-1410	TN	N
24288	ORDVA, Inc.	Hakford	615	898-1234	TN	Y
25443	B&K, Inc.	Smith	904	227-0093	FL	N
25501	Damal Supplies	Smythe	615	890-3529	TN	N
25595	Rubicon Systems	Orton	904	456-0092	FL	Y

Table name: INVOICE

INV_NUMBER	CUS_CODE	INV_DATE	INV_SUBTOTAL	INV_TAX	INV_TOTAL
1001	10014	16-Jan-16	24.90	1.99	26.89
1002	10011	16-Jan-16	9.98	0.80	10.78
1003	10012	16-Jan-16	153.85	12.31	166.16
1004	10011	17-Jan-16	34.97	2.80	37.77
1005	10018	17-Jan-16	70.44	5.64	76.08
1006	10014	17-Jan-16	397.83	31.83	429.66
1007	10015	17-Jan-16	34.97	2.80	37.77
1008	10011	17-Jan-16	399.15	31.93	431.08

Table name: LINE

INV_NUMBER	LINE_NUMBER	P_CODE	LINE_UNITS	LINE_PRICE	LINE_TOTAL
1001	1	13-Q2/P2	1	14.99	14.99
1001	2	23109-HB	1	9.95	9.95
1002	1	54778-2T	2	4.99	9.98
1003	1	2238/QPD	1	38.95	38.95
1003	2	1546-QQ2	1	39.95	39.95
1003	3	13-Q2/P2	5	14.99	74.95
1004	1	54778-2T	3	4.99	14.97
1004	2	23109-HB	2	9.95	19.90
1005	1	PVC23DRT	12	5.87	70.44
1006	1	SM-18277	3	6.99	20.97
1006	2	2232/QTY	1	109.92	109.92
1006	3	23109-HB	1	9.95	9.95
1006	4	89-WRE-Q	1	256.99	256.99
1007	1	13-Q2/P2	2	14.99	29.98
1007	2	54778-2T	1	4.99	4.99
1008	1	PVC23DRT	5	5.87	29.35
1008	2	WR3/TT3	3	119.95	359.85
1008	3	23109-HB	1	9.95	9.95

Online Content

The Ch08_SaleCo2 database used in Problems 28–31 is available at *www.cengagebrain.com*, as are the script files to duplicate this data set in Oracle, MySQL, and SQL Server.

28. Create a trigger named **trg_line_total** to write the LINE_TOTAL value in the LINE table every time you add a new LINE row. (The LINE_TOTAL value is the product of the LINE_UNITS and LINE_PRICE values.)

29. Create a trigger named **trg_line_prod** that automatically updates the quantity on hand for each product sold after a new LINE row is added.

30. Create a stored procedure named **prc_inv_amounts** to update the INV_SUBTOTAL, INV_TAX, and INV_TOTAL. The procedure takes the invoice number as a parameter. The INV_SUBTOTAL is the sum of the LINE_TOTAL amounts for the invoice, the INV_TAX is the product of the INV_SUBTOTAL and the tax rate (8 percent), and the INV_TOTAL is the sum of the INV_SUBTOTAL and the INV_TAX.

31. Create a procedure named **prc_cus_balance_update** that will take the invoice number as a parameter and update the customer balance. (*Hint*: You can use the DECLARE section to define a TOTINV numeric variable that holds the computed invoice total.)

Use the Ch08_AviaCo database shown in Figure P8.32 to work Problems 32–43.

FIGURE P8.32 CH08_AVIACO DATABASE TABLES

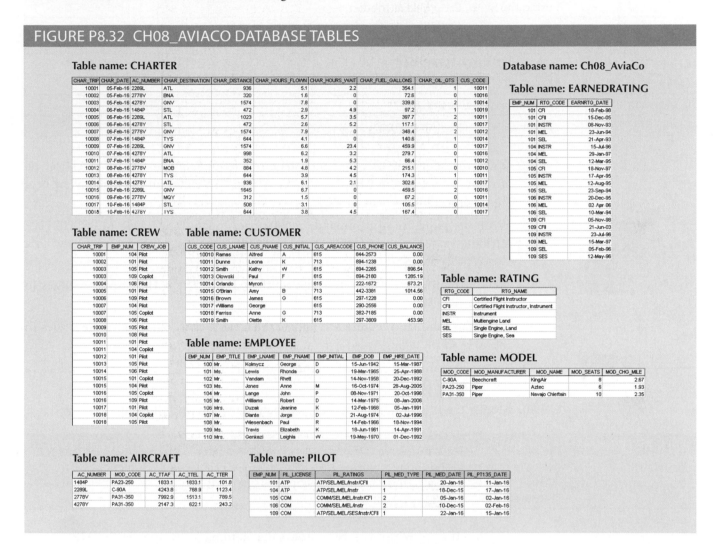

Table name: CHARTER

CHAR_TRIP	CHAR_DATE	AC_NUMBER	CHAR_DESTINATION	CHAR_DISTANCE	CHAR_HOURS_FLOWN	CHAR_HOURS_WAIT	CHAR_FUEL_GALLONS	CHAR_OIL_QTS	CUS_CODE
10001	05-Feb-16	2289L	ATL	936	5.1	2.2	354.1	1	10011
10002	05-Feb-16	2778V	BNA	320	1.6	0	72.6	0	10016
10003	05-Feb-16	4278Y	GNV	1574	7.8	0	339.8	2	10014
10004	06-Feb-16	1484P	STL	472	2.9	4.9	97.2	1	10019
10005	06-Feb-16	2289L	ATL	1023	5.7	3.5	397.7	2	10011
10006	06-Feb-16	4278Y	STL	472	2.6	5.2	117.1	0	10017
10007	06-Feb-16	2778V	GNV	1574	7.9	0	348.4	2	10012
10008	07-Feb-16	1484P	TYS	644	4.1	0	140.6	1	10014
10009	07-Feb-16	2289L	GNV	1574	6.6	23.4	459.9	0	10017
10010	07-Feb-16	4278Y	ATL	998	6.2	3.2	279.7	0	10016
10011	07-Feb-16	1484P	BNA	352	1.9	5.3	66.4	1	10012
10012	08-Feb-16	2778V	MOB	884	4.8	4.2	215.1	0	10010
10013	08-Feb-16	4278Y	TYS	644	3.9	4.5	174.3	1	10011
10014	09-Feb-16	4278Y	ATL	936	6.1	2.1	302.6	0	10017
10015	09-Feb-16	2289L	GNV	1645	6.7	0	459.5	2	10016
10016	09-Feb-16	2778V	MQY	312	1.5	0	67.2	0	10011
10017	10-Feb-16	1484P	STL	508	3.1	0	105.5	0	10014
10018	10-Feb-16	4278Y	TYS	644	3.8	4.5	167.4	0	10017

Database name: Ch08_AviaCo

Table name: EARNEDRATING

EMP_NUM	RTG_CODE	EARNRTG_DATE
101	CFI	18-Feb-98
101	CFII	15-Dec-05
101	INSTR	08-Nov-93
101	MEL	23-Jun-94
101	SEL	21-Apr-93
104	INSTR	15-Jul-96
104	MEL	29-Jan-97
104	SEL	12-Mar-95
105	CFI	18-Nov-97
105	INSTR	17-Apr-95
105	MEL	12-Aug-95
105	SEL	23-Sep-94
106	INSTR	20-Dec-95
106	MEL	02-Apr-06
106	SEL	10-Mar-94
109	CFI	05-Nov-98
109	CFII	21-Jun-03
109	INSTR	23-Jul-96
109	MEL	15-Mar-97
109	SEL	05-Feb-96
109	SES	12-May-96

Table name: CREW

CHAR_TRIP	EMP_NUM	CREW_JOB
10001	104	Pilot
10002	101	Pilot
10003	105	Pilot
10003	109	Copilot
10004	106	Pilot
10005	101	Pilot
10006	109	Pilot
10007	104	Pilot
10007	105	Copilot
10008	106	Pilot
10009	105	Pilot
10010	108	Pilot
10011	101	Pilot
10011	104	Copilot
10012	101	Pilot
10013	105	Pilot
10014	106	Pilot
10015	101	Copilot
10015	104	Pilot
10016	105	Copilot
10016	109	Pilot
10017	101	Pilot
10018	104	Copilot
10018	105	Pilot

Table name: CUSTOMER

CUS_CODE	CUS_LNAME	CUS_FNAME	CUS_INITIAL	CUS_AREACODE	CUS_PHONE	CUS_BALANCE
10010	Ramas	Alfred	A	615	844-2573	0.00
10011	Dunne	Leona	K	713	894-1238	0.00
10012	Smith	Kathy	W	615	894-2285	896.54
10013	Olowski	Paul	F	615	894-2180	1285.19
10014	Orlando	Myron		615	222-1672	673.21
10015	O'Brian	Amy	B	713	442-3381	1014.56
10016	Brown	James	G	615	297-1228	0.00
10017	Williams	George		615	290-2556	0.00
10018	Farriss	Anne	G	713	382-7185	0.00
10019	Smith	Olette	K	615	297-3809	453.98

Table name: RATING

RTG_CODE	RTG_NAME
CFI	Certified Flight Instructor
CFII	Certified Flight Instructor, Instrument
INSTR	Instrument
MEL	Multiengine Land
SEL	Single Engine, Land
SES	Single Engine, Sea

Table name: EMPLOYEE

EMP_NUM	EMP_TITLE	EMP_LNAME	EMP_FNAME	EMP_INITIAL	EMP_DOB	EMP_HIRE_DATE
100	Mr.	Kolmycz	George	D	15-Jun-1942	15-Mar-1987
101	Ms.	Lewis	Rhonda	G	19-Mar-1965	25-Apr-1988
102	Mr.	Vandam	Rhett		14-Nov-1958	20-Dec-1992
103	Ms.	Jones	Anne	M	16-Oct-1974	28-Aug-2005
104	Mr.	Lange	John	P	08-Nov-1971	20-Oct-1996
105	Mr.	Williams	Robert	D	14-Mar-1975	08-Jan-2006
106	Mrs.	Duzak	Jeanine	K	12-Feb-1968	05-Jan-1991
107	Mr.	Diante	Jorge	D	21-Aug-1974	02-Jul-1996
108	Mr.	Wiesenbach	Paul	R	14-Feb-1966	18-Nov-1994
109	Ms.	Travis	Elizabeth	K	18-Jun-1961	14-Apr-1991
110	Mrs.	Genkazi	Leighla	W	19-May-1970	01-Dec-1992

Table name: MODEL

MOD_CODE	MOD_MANUFACTURER	MOD_NAME	MOD_SEATS	MOD_CHG_MILE
C-90A	Beechcraft	KingAir	8	2.67
PA23-250	Piper	Aztec	6	1.93
PA31-350	Piper	Navajo Chieftain	10	2.35

Table name: AIRCRAFT

AC_NUMBER	MOD_CODE	AC_TTAF	AC_TTEL	AC_TTER
1484P	PA23-250	1833.1	1833.1	101.8
2289L	C-90A	4243.8	768.9	1123.4
2778V	PA31-350	7992.9	1513.1	789.5
4278Y	PA31-350	2147.3	622.1	243.2

Table name: PILOT

EMP_NUM	PIL_LICENSE	PIL_RATINGS	PIL_MED_TYPE	PIL_MED_DATE	PIL_PT135_DATE
101	ATP	ATP/SEL/MEL/Instr/CFII	1	20-Jan-16	11-Jan-16
104	ATP	ATP/SEL/MEL/Instr	1	18-Dec-15	17-Jan-16
105	COM	COMM/SEL/MEL/Instr/CFI	2	05-Jan-16	02-Jan-16
106	COM	COMM/SEL/MEL/Instr	2	10-Dec-15	02-Feb-16
109	COM	ATP/SEL/MEL/SES/Instr/CFII	1	22-Jan-16	15-Jan-16

32. Modify the MODEL table to add the attribute and insert the values shown in the following table.

ATTRIBUTE NAME	ATTRIBUTE DESCRIPTION	ATTRIBUTE TYPE	ATTRIBUTE VALUES
MOD_WAIT_CHG	Waiting charge per hour for each model	Numeric	$100 for C-90A $50 for PA23-250 $75 for PA31-350

33. Write the queries to update the MOD_WAIT_CHG attribute values based on Problem 32.

Online Content

The Ch08_AviaCo database used in Problems 32–43 is available at *www.cengagebrain.com*, as are the script files to duplicate this data set in Oracle, SQL Server, and MySQL.

34. Modify the CHARTER table to add the attributes shown in the following table.

ATTRIBUTE NAME	ATTRIBUTE DESCRIPTION	ATTRIBUTE TYPE
CHAR_WAIT_CHG	Waiting charge for each model (copied from the MODEL table)	Numeric
CHAR_FLT_CHG_HR	Flight charge per mile for each model (copied from the MODEL table using the MOD_CHG_MILE attribute)	Numeric
CHAR_FLT_CHG	Flight charge (calculated by CHAR_HOURS_FLOWN × CHAR_FLT_CHG_HR)	Numeric
CHAR_TAX_CHG	CHAR_FLT_CHG × tax rate (8%)	Numeric
CHAR_TOT_CHG	CHAR_FLT_CHG + CHAR_TAX_CHG	Numeric
CHAR_PYMT	Amount paid by customer	Numeric
CHAR_BALANCE	Balance remaining after payment	Numeric

35. Write the sequence of commands required to update the CHAR_WAIT_CHG attribute values in the CHARTER table. (*Hint*: Use either an updatable view or a stored procedure.)

36. Write the sequence of commands required to update the CHAR_FLT_CHG_HR attribute values in the CHARTER table. (*Hint*: Use either an updatable view or a stored procedure.)

37. Write the command required to update the CHAR_FLT_CHG attribute values in the CHARTER table.

38. Write the command required to update the CHAR_TAX_CHG attribute values in the CHARTER table.

39. Write the command required to update the CHAR_TOT_CHG attribute values in the CHARTER table.

40. Modify the PILOT table to add the attribute shown in the following table.

ATTRIBUTE NAME	ATTRIBUTE DESCRIPTION	ATTRIBUTE TYPE
PIL_PIC_HRS	Pilot in command (PIC) hours; updated by adding the CHARTER table's CHAR_HOURS_FLOWN to the PIL_PIC_HRS when the CREW table shows the CREW_JOB to be Pilot	Numeric

41. Create a trigger named **trg_char_hours** that automatically updates the AIRCRAFT table when a new CHARTER row is added. Use the CHARTER table's CHAR_HOURS_FLOWN to update the AIRCRAFT table's AC_TTAF, AC_TTEL, and AC_TTER values.

42. Create a trigger named **trg_pic_hours** that automatically updates the PILOT table when a new CREW row is added and the CREW table uses a Pilot CREW_JOB entry. Use the CHARTER table's CHAR_HOURS_FLOWN to update the PILOT table's PIL_PIC_HRS only when the CREW table uses a Pilot CREW_JOB entry.

43. Create a trigger named **trg_cust_balance** that automatically updates the CUSTOMER table's CUS_BALANCE when a new CHARTER row is added. Use the CHARTER table's CHAR_TOT_CHG as the update source. (Assume that all charter charges are charged to the customer balance.)

Problems 44–67 use the Ch08_Fact database shown in Figure P8.44. For problems with very large results sets, only the first several rows of output are shown in the following figures.

FIGURE P8.44 THE CH08_FACT ERD

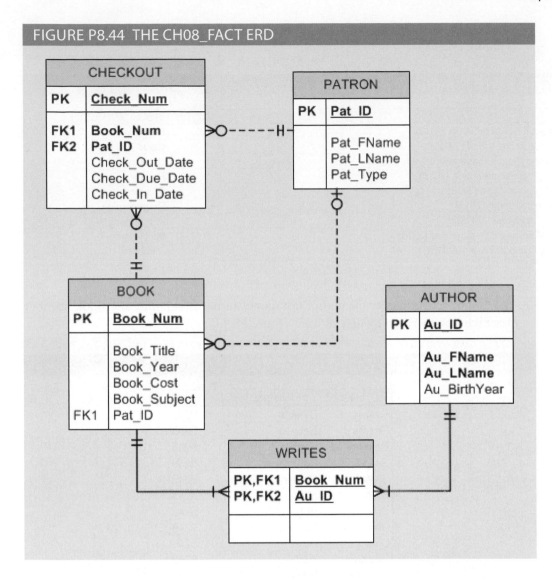

44. Write a query to display all rows in the PATRON table.

45. Write a query to display the patron ID, book number, and days kept for each check-out. "Days Kept" is the difference from the date on which the book is returned to the date it was checked out. (Figure P8.45)

46. Write a query to display the patron ID, patron full name, and patron type for each patron. (Figure P8.46)

FIGURE P8.45 DAYS KEPT

PATRON	BOOK	Days Kept
1165	5235	9
1209	5238	5
1160	5240	9
1160	5237	3
1202	5236	8
1203	5235	8
1174	5244	3
1181	5248	1
1170	5242	4
1161	5235	0

FIGURE P8.46 PATRON AND PATRON TYPE

PAT_ID	Patron Name	PAT_TYPE
1160	robert carter	Faculty
1161	Kelsey Koch	Faculty
1165	Cedric Baldwin	Faculty
1166	Vera Alvarado	Student
1167	Alan Martin	FACULTY
1170	Cory Barry	faculty
1171	Peggy Marsh	STUDENT

47. Write a query to display the book number, title with year, and subject for each book. (Figure P8.47)

FIGURE P8.47 BOOK TITLE WITH YEAR

BOOK_NUM	BOOK	BOOK_SUBJECT
5235	Beginner's Guide to JAVA (2012)	Programming
5236	Database in the Cloud (2012)	Cloud
5237	Mastering the database environment (2013)	Database
5238	Conceptual Programming (2013)	Programming
5239	J++ in Mobile Apps (2013)	Programming
5240	iOS Programming (2013)	Programming
5241	JAVA First Steps (2013)	Programming
5242	C# in Middleware Deployment (2013)	Middleware
5243	DATABASES in Theory (2013)	Database

48. Write a query to display the author last name, author first name, and book number for each book written by that author. (Figure P8.48)

FIGURE P8.48 BOOKS WRITTEN BY AUTHOR

AU_LNAME	AU_FNAME	BOOK_NUM
Reeves	Benson	5237
Reeves	Benson	5253
Beatney	Rachel	5240
Salvadore	Carmine	5239
Salvadore	Carmine	5248
Bruer	Hugo	5243
Bruer	Hugo	5246
Chiang	Xia	5244
Chiang	Xia	5249
Chiang	Xia	5252
Durante	Reba	5235
Tankersly	Trina	5244

49. Write a query to display the author ID, book number, title, and year for each book. (Figure P8.49)

FIGURE P8.49 AUTHORS OF BOOKS

AU_ID	BOOK_NUM	BOOK_TITLE	BOOK_YEAR
273	5235	Beginner's Guide to JAVA	2012
383	5236	Database in the Cloud	2012
185	5237	Mastering the database environment	2013
603	5238	Conceptual Programming	2013
229	5239	J++ in Mobile Apps	2013
460	5239	J++ in Mobile Apps	2013
592	5239	J++ in Mobile Apps	2013
218	5240	iOS Programming	2013
460	5241	JAVA First Steps	2013
559	5241	JAVA First Steps	2013
581	5242	C# in Middleware Deployment	2013
251	5243	DATABASES in Theory	2013

50. Write a query to display the author last name, first name, book title, and year for each book. (Figure P8.50)

FIGURE P8.50 AUTHOR NAME AND BOOK TITLE

AU_LNAME	AU_FNAME	BOOK_TITLE	BOOK_YEAR
Durante	Reba	Beginner's Guide to JAVA	2012
Walsh	Neal	Database in the Cloud	2012
Reeves	Benson	Mastering the database environment	2013
Palca	Julia	Conceptual Programming	2013
Salvadore	Carmine	J++ in Mobile Apps	2013
Paulsen	Connie	J++ in Mobile Apps	2013
Sheel	Lawrence	J++ in Mobile Apps	2013
Beatney	Rachel	iOS Programming	2013
Paulsen	Connie	JAVA First Steps	2013
McGill	Rachel	JAVA First Steps	2013

51. Write a query to display the patron ID, book number, patron first name and last name, and book title for all currently checked out books. (Remember to use the redundant relationship described in the assignment instructions for current check-outs.) Sort the output by patron last name and book title. (Figure P8.51)

FIGURE P8.51 CURRENTLY CHECKED OUT BOOKS

PAT_ID	BOOK_NUM	PAT_FNAME	PAT_LNAME	BOOK_TITLE
1229	5252	Gerald	Burke	Beyond the Database Veil
1229	5238	Gerald	Burke	Conceptual Programming
1228	5242	Homer	Goodman	C# in Middleware Deployment
1212	5240	Iva	McClain	iOS Programming
1172	5246	Tony	Miles	Capture the Cloud
1207	5249	Iva	Ramos	Starlight Applications

52. Write a query to display the patron ID, full name (first and last), and patron type for all patrons. Sort the results by patron type, then by last name and first name. Ensure that all sorting is case insensitive. (Figure P8.52)

FIGURE P8.52 SORTED PATRONS WITH FULL NAMES

PAT_ID	NAME	PAT_TYPE
1165	Cedric Baldwin	Faculty
1170	Cory Barry	faculty
1160	robert carter	Faculty
1183	Helena Hughes	Faculty
1161	Kelsey Koch	Faculty
1167	Alan Martin	FACULTY
1166	Vera Alvarado	Student
1202	Holly Anthony	Student
1180	Nadine Blair	STUDENT

53. Write a query to display the book number and the number of times each book has been checked out. Do not include books that have never been checked out. (Figure P8.53)

FIGURE P8.53 TIMES CHECKED OUT

BOOK_NUM	Times Checked Out
5236	12
5235	9
5240	7
5238	6
5237	5
5254	4
5252	4
5249	4
5246	4
5244	4
5242	4
5248	3
5243	2

54. Write a query to display the author ID, first and last name, book number, and book title of all books in the subject "Cloud". Sort the results by book title and then by author last name. (Figure P8.54)

FIGURE P8.54 BOOKS ON CLOUD COMPUTING

AU_ID	AU_FNAME	AU_LNAME	BOOK_NUM	BOOK_TITLE
251	Hugo	Bruer	5246	Capture the Cloud
262	Xia	Chiang	5244	Cloud-based Mobile Applications
284	Trina	Tankersly	5244	Cloud-based Mobile Applications
383	Neal	Walsh	5236	Database in the Cloud
262	Xia	Chiang	5249	Starlight Applications

55. Write a query to display the book number, title, author last name, author first name, patron ID, last name, and patron type for all books currently checked out to a patron. Sort the results by book title. (Figure P8.55)

FIGURE P8.55 CURRENTLY CHECKED OUT BOOKS WITH AUTHOR

BOOK_NUM	BOOK_TITLE	AU_LNAME	AU_FNAME	PAT_ID	PAT_LNAME	PAT_TYPE
5252	Beyond the Database Veil	Chiang	Xia	1229	Burke	Student
5242	C# in Middleware Deployment	Aggerwal	Manish	1228	Goodman	Student
5246	Capture the Cloud	Bruer	Hugo	1172	Miles	STUDENT
5238	Conceptual Programming	Palca	Julia	1229	Burke	Student
5240	iOS Programming	Beatney	Rachel	1212	McClain	Student
5249	Starlight Applications	Chiang	Xia	1207	Ramos	Student

56. Write a query to display the book number, title, and number of times each book has been checked out. Include books that have never been checked out. Sort the results in descending order by the number times checked out, then by title. (Figure P8.56)

FIGURE P8.56 NUMBER OF CHECKOUTS FOR EVERY BOOK

BOOK_NUM	BOOK_TITLE	Times Checked Out
5236	Database in the Cloud	12
5235	Beginner's Guide to JAVA	9
5240	iOS Programming	7
5238	Conceptual Programming	6
5237	Mastering the database environment	5
5252	Beyond the Database Veil	4
5242	C# in Middleware Deployment	4
5246	Capture the Cloud	4
5244	Cloud-based Mobile Applications	4
5254	Coding Style for Maintenance	4
5249	Starlight Applications	4
5248	What You Always Wanted to Know About Database, But Were Afraid to Ask	3
5243	DATABASES in Theory	2
5239	J++ in Mobile Apps	0
5241	JAVA First Steps	0
5250	Reengineering the Middle Tier	0
5247	Shining Through the Cloud: Sun Programming	0
5245	The Golden Road to Platform independence	0
5251	Thoughts on Revitalizing Ruby	0
5253	Virtual Programming for Virtual Environments	0

57. Write a query to display the book number, title, and number of times each book has been checked out. Limit the results to books that have been checked out more than five times. Sort the results in descending order by the number of times checked out, and then by title. (Figure P8.57)

FIGURE P8.57 BOOKS WITH MORE THAN 5 CHECKOUTS

BOOK_NUM	BOOK_TITLE	Times Checked Out
5236	Database in the Cloud	12
5235	Beginner's Guide to JAVA	9
5240	iOS Programming	7
5238	Conceptual Programming	6

58. Write a query to display the author ID, author last name, book title, checkout date, and patron last name for all the books written by authors with the last name "Bruer" that have ever been checked out by patrons with the last name "Miles". (Figure P8.58)

FIGURE P8.58 BOOKS BY AUTHOR FOR PATRON "MILES"

AU_ID	AU_LNAME	BOOK_TITLE	CHECK_OUT_DATE	PAT_LNAME
251	Bruer	Capture the Cloud	4/21/2015	Miles
251	Bruer	Capture the Cloud	5/15/2015	Miles

59. Write a query to display the patron ID, first and last name of all patrons that have never checked out any book. Sort the result by patron last name then first name. (Figure P8.59)

FIGURE P8.59 PATRONS THAT NEVER CHECKED OUT A BOOK

PAT_ID	PAT_FNAME	PAT_LNAME
1166	Vera	Alvarado
1180	Nadine	Blair
1238	Erika	Bowen
1208	Ollie	Cantrell
1227	Alicia	Dickson
1205	Claire	Gomez
1239	Elton	Irwin
1240	Jan	Joyce
1243	Roberto	Kennedy
1242	Mario	King
1237	Brandi	Larson
1167	Alan	Martin
1182	Jamal	Melendez
1201	Shelby	Noble
1244	Leon	Richmond
1200	Lorenzo	Torres
1241	Irene	West

60. Write a query to display the patron ID, last name, number of times that patron has ever checked out a book, and the number of different books the patron has ever checked out. For example, if a given patron has checked out the same book twice, that would count as two checkouts but only one book. Limit the results to only patrons that have made at least three checkouts. Sort the results in descending order by number of books, then in descending order by number of checkouts, then in ascending order by patron ID. (Figure P8.60)

FIGURE P8.60 CHECKOUTS AND BOOKS BY PATRON

PAT_ID	PAT_LNAME	NUM CHECKOUTS	NUM DIFFERENT BOOKS
1161	Koch	3	3
1165	Baldwin	3	3
1181	Horne	3	3
1185	Yang	3	3
1210	Cooley	3	3
1229	Burke	3	3
1160	carter	3	2
1171	Marsh	3	2
1172	Miles	3	2
1207	Ramos	3	2
1209	Mathis	3	2
1183	Hughes	3	1

61. Write a query to display the average number of days a book is kept during a check-out. (Figure P8.61)

FIGURE P8.61 AVERAGE DAYS KEPT

Average Days Kept
4.44

62. Write a query to display the patron ID and the average number of days that patron keeps books during a checkout. Limit the results to only patrons that have at least three checkouts. Sort the results in descending order by the average days the book is kept. (Figure P8.62)

FIGURE P8.62 AVERAGE DAYS KEPT BY PATRON

PAT_ID	Average Days Kept
1160	7
1185	6.67
1165	5.67
1207	5.5
1209	5.33
1172	4.5
1183	4.33
1181	3.67
1171	3.67
1161	3.33
1210	2.33
1229	2

63. Write a query to display the book number, title, and cost of books that have the lowest cost of any books in the system. Sort the results by book number. (Figure P8.63)

FIGURE P8.63 LEAST EXPENSIVE BOOKS

BOOK_NUM	BOOK_TITLE	BOOK_COST
5239	J++ in Mobile Apps	49.95
5241	JAVA First Steps	49.95
5248	What You Always Wanted to Know About Database, But Were Afraid to Ask	49.95
5254	Coding Style for Maintenance	49.95

64. Write a query to display the author ID, first and last name for all authors that have never written a book with the subject Programming. Sort the results by author last name. (Figure P8.64)

FIGURE P8.64 AUTHORS THAT HAVE NEVER WRITTEN ON PROGRAMMING

AU_ID	AU_FNAME	AU_LNAME
581	Manish	Aggerwal
251	Hugo	Bruer
262	Xia	Chiang
438	Perry	Pearson
284	Trina	Tankersly
383	Neal	Walsh

65. Write a query to display the book number, title, subject, average cost of books within that subject, and the difference between each book's cost and the average cost of books in that subject. Sort the results by book title. (Figure P8.65)

FIGURE P8.65 BOOKS WITH AVERAGE COST BY SUBJECT

BOOK_NUM	BOOK_TITLE	BOOK_SUBJECT	AVGCOST	DIFFERENCE
5235	Beginner's Guide to JAVA	Programming	66.62	-6.67
5236	Database in the Cloud	Cloud	72.45	7.5
5237	Mastering the database environment	Database	84.95	5
5238	Conceptual Programming	Programming	66.62	-6.67
5239	J++ in Mobile Apps	Programming	66.62	-16.67
5240	iOS Programming	Programming	66.62	13.33
5241	JAVA First Steps	Programming	66.62	-16.67
5242	C# in Middleware Deployment	Middleware	89.95	-30
5243	DATABASES in Theory	Database	84.95	45
5244	Cloud-based Mobile Applications	Cloud	72.45	-2.5
5245	The Golden Road to Platform independence	Middleware	89.95	30
5246	Capture the Cloud	Cloud	72.45	-2.5
5247	Shining Through the Cloud: Sun Programming	Programming	66.62	43.33
5248	What You Always Wanted to Know About Database, But Were Afraid to Ask	Database	84.95	-35
5249	Starlight Applications	Cloud	72.45	-2.5
5250	Reengineering the Middle Tier	Middleware	89.95	0
5251	Thoughts on Revitalizing Ruby	Programming	66.62	-6.67
5252	Beyond the Database Veil	Database	84.95	-15
5253	Virtual Programming for Virtual Environments	Programming	66.62	13.33
5254	Coding Style for Maintenance	Programming	66.62	-16.67

66. Write a query to display the book number, title, subject, author last name, and the number of books written by that author. Limit the results to books in the Cloud subject. Sort the results by book title and then author last name. (Figure P8.66)

FIGURE P8.66 NUMBER OF BOOKS BY CLOUD AUTHORS

BOOK_NUM	BOOK_TITLE	BOOK_SUBJECT	AU_LNAME	Num Books by Author
5246	Capture the Cloud	Cloud	Bruer	2
5244	Cloud-based Mobile Applications	Cloud	Chiang	3
5244	Cloud-based Mobile Applications	Cloud	Tankersly	1
5236	Database in the Cloud	Cloud	Walsh	2
5249	Starlight Applications	Cloud	Chiang	3

67. Write a query to display the lowest average cost of books within a subject and the highest average cost of books within a subject. (Figure P8.67)

FIGURE P8.67 LOWEST AND HIGHEST AVERAGE SUBJECT COSTS

Lowest Avg Cost	Highest Avg Cost
66.62	89.95

Cases

The following problems expand on the EliteVideo case from Chapter 7. To complete the following problems, you must have first completed the table creation and data-entry requirements specified in Problems 96 and 97 in Chapter 7.

68. Alter the DETAILRENTAL table to include a derived attribute named DETAIL_ DAYSLATE to store integers of up to three digits. The attribute should accept null values.

69. Update the DETAILRENTAL table to set the values in DETAIL_RETURNDATE to include a time component. Make each entry match the values shown in the following table.

RENT_NUM	VID_NUM	DETAIL_RETURNDATE
1001	34342	02-MAR-16 10:00am
1001	61353	03-MAR-16 11:30am
1002	59237	04-MAR-16 03:30pm
1003	54325	09-MAR-16 04:00pm
1003	61369	09-MAR-16 04:00pm
1003	61388	09-MAR-16 04:00pm
1004	44392	07-MAR-16 09:00am
1004	34367	07-MAR-16 09:00am
1004	34341	07-MAR-16 09:00am
1005	34342	05-MAR-16 12:30pm
1005	44397	05-MAR-16 12:30pm
1006	34366	04-MAR-16 10:15pm
1006	61367	
1007	34368	
1008	34369	05-MAR-16 09:30pm
1009	54324	
1001	34366	02-MAR-16 10:00am

70. Alter the VIDEO table to include an attribute named VID_STATUS to store character data up to four characters long. The attribute should not accept null values. The attribute should have a constraint to enforce the domain ("IN", "OUT", and "LOST") and have a default value of "IN".

71. Update the VID_STATUS attribute of the VIDEO table using a subquery to set the VID_STATUS to "OUT" for all videos that have a null value in the DETAIL_RETURNDATE attribute of the DETAILRENTAL table.

72. Alter the PRICE table to include an attribute named PRICE_RENTDAYS to store integers of up to two digits. The attribute should not accept null values, and it should have a default value of 3.

73. Update the PRICE table to place the values shown in the following table in the PRICE_RENTDAYS attribute.

PRICE_CODE	PRICE_RENTDAYS
1	5
2	3
3	5
4	7

74. Create a trigger named **trg_late_return** that will write the correct value to DETAIL_DAYSLATE in the DETAILRENTAL table whenever a video is returned. The trigger should execute as a BEFORE trigger when the DETAIL_RETURNDATE or DETAIL_DUEDATE attributes are updated. The trigger should satisfy the following conditions:

    a.   If the return date is null, then the days late should also be null.

    b.   If the return date is not null, then the days late should determine if the video is returned late.

    c.   If the return date is noon of the day after the due date or earlier, then the video is not considered late, and the days late should have a value of zero (0).

    d.   If the return date is past noon of the day after the due date, then the video is considered late, so the number of days late must be calculated and stored.

75. Create a trigger named **trg_mem_balance** that will maintain the correct value in the membership balance in the MEMBERSHIP table when videos are returned late. The trigger should execute as an AFTER trigger when the due date or return date attributes are updated in the DETAILRENTAL table. The trigger should satisfy the following conditions:

    a.   Calculate the value of the late fee prior to the update that triggered this execution of the trigger. The value of the late fee is the days late multiplied by the daily late fee. If the previous value of the late fee was null, then treat it as zero (0).

    b.   Calculate the value of the late fee after the update that triggered this execution of the trigger. If the value of the late fee is now null, then treat it as zero (0).

    c.   Subtract the prior value of the late fee from the current value of the late fee to determine the change in late fee for this video rental.

    d.   If the amount calculated in Part c is not zero (0), then update the membership balance by the amount calculated for the membership associated with this rental.

76. Create a sequence named **rent_num_seq** to start with 1100 and increment by 1. Do not cache any values.

77. Create a stored procedure named **prc_new_rental** to insert new rows in the RENTAL table. The procedure should satisfy the following conditions:

    a. The membership number will be provided as a parameter.

    b. Use a Count() function to verify that the membership number exists in the MEMBERSHIP table. If it does not exist, then a message should be displayed that the membership does not exist and no data should be written to the database.

    c. If the membership does exist, then retrieve the membership balance and display a message that the balance amount is the previous balance. (For example, if the membership has a balance of $5.00, then display "Previous balance: $5.00".)

    d. Insert a new row in the rental table using the sequence created in Case Question 76 to generate the value for RENT_NUM, the current system date for the RENT_DATE value, and the membership number provided as the value for MEM_NUM.

78. Create a stored procedure named **prc_new_detail** to insert new rows in the DETAILRENTAL table. The procedure should satisfy the following requirements:

    a. The video number will be provided as a parameter.

    b. Verify that the video number exists in the VIDEO table. If it does not exist, then display a message that the video does not exist, and do not write any data to the database.

    c. If the video number does exist, then verify that the VID_STATUS for the video is "IN". If the status is not "IN", then display a message that the video's return must be entered before it can be rented again, and do not write any data to the database.

    d. If the status is "IN", then retrieve the values of the video's PRICE_RENTFEE, PRICE_DAILYLATEFEE, and PRICE_RENTDAYS from the PRICE table.

    e. Calculate the due date for the video rental by adding the number of days in PRICE_RENTDAYS to 11:59:59PM (hours:minutes:seconds) in the current system date.

    f. Insert a new row in the DETAILRENTAL table using the previous value returned by RENT_NUM_SEQ as the RENT_NUM, the video number provided in the parameter as the VID_NUM, the PRICE_RENTFEE as the value for DETAIL_FEE, the due date calculated above for the DETAIL_DUEDATE, PRICE_DAILYLATEFEE as the value for DETAIL_DAILYLATEFEE, and null for the DETAIL_RETURNDATE.

79. Create a stored procedure named **prc_return_video** to enter data about the return of videos that have been rented. The procedure should satisfy the following requirements.

    a. The video number will be provided as a parameter.

    b. Verify that the video number exists in the VIDEO table. If it does not exist, display a message that the video number provided was not found and do not write any data to the database.

c. If the video number does exist, then use a Count() function to ensure that the video has only one record in DETAILRENTAL for which it does not have a return date. If more than one row in DETAILRENTAL indicates that the video is rented but not returned, display an error message that the video has multiple outstanding rentals and do not write any data to the database.

d. If the video does not have any outstanding rentals, then update the video status to "IN" for the video in the VIDEO table, and display a message that the video had no outstanding rentals but is now available for rental. If the video has only one outstanding rental, then update the return date to the current system date, and update the video status to "IN" for that video in the VIDEO table. Then display a message that the video was successfully returned.

Chapter 9

Database Design

In this chapter, you will learn:
- That a sound database design is the foundation for a successful information system, and that the database design must reflect the information system of which the database is a part
- That successful information systems are developed within a framework known as the Systems Development Life Cycle (SDLC)
- That within the information system, the most successful databases are subject to frequent evaluation and revision within a framework known as the Database Life Cycle (DBLC)
- How to conduct evaluation and revision within the SDLC and DBLC frameworks
- About database design strategies: top-down versus bottom-up design and centralized versus decentralized design

Preview

Databases are a part of a larger picture called an information system. Database designs that fail to recognize this fact are not likely to be successful. Database designers must recognize that the database is a critical means to an end rather than an end in itself. Managers want the database to serve their management needs, but too many databases seem to force managers to alter their routines to fit the database requirements.

Information systems don't just happen; they are the product of a carefully staged development process. Systems analysis is used to determine the need for an information system and to establish its limits. Within systems analysis, the actual information system is created through a process known as systems development.

The creation and evolution of information systems follows an iterative pattern called the Systems Development Life Cycle (SDLC), which is a continuous process of creation, maintenance, enhancement, and replacement of the information system. A similar cycle applies to databases: the database is created, maintained, enhanced, and eventually replaced. The Database Life Cycle (DBLC) is carefully traced in this chapter, and is shown in the context of the larger Systems Development Life Cycle.

At the end of the chapter, you will be introduced to some classical approaches to database design: top-down versus bottom-up and centralized versus decentralized.

Data Files Available on cengagebrain.com

Note

Because it is purely conceptual, this chapter does not reference any data files.

information system (IS)

A system that provides for data collection, storage, and retrieval; facilitates the transformation of data into information; and manages both data and information. An information system is composed of hardware, the DBMS and other software, database(s), people, and procedures.

9-1 The Information System

Basically, a database is a carefully designed and constructed repository of facts. The database is part of a larger whole known as an **information system**, which provides for data collection, storage, transformation, and retrieval. The information system also helps transform data into information, and it allows for the management of both data and information. Thus, a complete information system is composed of people, hardware, software, the database(s), application programs, and procedures. **Systems analysis** is the process that establishes the need for an information system and its extent. The process of creating an information system is known as **systems development**.

One key characteristic of current information systems is the strategic value of information in the age of global business. Therefore, information systems should always be aligned with strategic business mission and goals; the view of isolated and independent information systems is no longer valid. Current information systems should always be integrated with the company's enterprise-wide information systems architecture.

Note

This chapter does not mean to cover all aspects of systems analysis and development, which are usually covered in a separate course or book. However, this chapter should help you better understand the issues associated with database design, implementation, and management, all of which are affected by the information system in which the database is a critical component.

systems analysis

The process that establishes the need for an information system and its extent.

systems development

The process of creating an information system.

Within the framework of systems development, applications transform data into the information that forms the basis for decision making. Applications usually generate formal reports, tabulations, and graphic displays designed to produce insight from the information. Figure 9.1 illustrates that every application is composed of two parts: the data and the code (program instructions) by which the data is transformed into information. The data and code work together to represent real-world business functions and activities. At any given moment, physically stored data represents a snapshot of the business, but the picture is not complete without an understanding of the business activities represented by the code.

FIGURE 9.1 GENERATING INFORMATION FOR DECISION MAKING

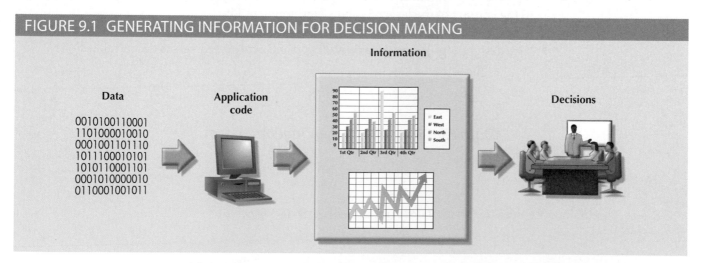

The performance of an information system depends on three factors:

- Database design and implementation
- Application design and implementation
- Administrative procedures

This book emphasizes the database design and implementation segment of the triad—arguably the most important of the three. However, failure to address the other two segments will likely yield a poorly functioning information system. Creating a sound information system is hard work: systems analysis and development require extensive planning to ensure that all of the activities will interface with each other, that they will complement each other, and that they will be completed on time.

In a broad sense, the term **database development** describes the process of database design and implementation. The primary objective in database design is to create complete, normalized, nonredundant (to the greatest extent possible), and fully integrated conceptual, logical, and physical database models. The implementation phase includes creating the database storage structure, loading data into the database, and providing for data management. Consideration should be taken to design and implement a database that is flexible and scalable over time. Although most designs typically focus on solving current problems, it is important to create a design that is flexible enough to adapt to future changes (such as performance, size, or reporting requirements).

To make the procedures discussed in this chapter broadly applicable, the chapter focuses on the elements that are common to all information systems. Most of the processes and procedures described in this chapter do not depend on the size, type, or complexity of the database being implemented. However, the procedures that would be used to design a small database, such as one for a neighborhood shoe store, do not precisely scale up to the procedures that would be needed to design a database for a large corporation or even a segment of such a corporation. To use an analogy, building a small house requires a blueprint, just as building the Golden Gate Bridge did, but the bridge required far more complex planning, analysis, and design.

The next sections will trace the overall Systems Development Life Cycle and the related Database Life Cycle. Once you are familiar with those processes and procedures, you will learn about various approaches to database design, such as top-down versus bottom-up and centralized versus decentralized design.

Note

The Systems Development Life Cycle is a general framework through which you can track and understand the activities required to develop and maintain information systems. Within that framework, there are several ways to complete various tasks specified in the SDLC. For example, this book focuses on ER modeling and on relational database design and implementation, and that focus is maintained in this chapter. However, there are alternative methodologies:

- Unified Modeling Language (UML) provides object-oriented tools to support the tasks associated with the development of information systems. UML is covered in Appendix H, Unified Modeling Language (UML), at *www.cengagebrain.com*.

- Rapid Application Development (RAD)[1] is an iterative software development methodology that uses prototypes, CASE tools, and flexible management to develop application systems. RAD started as an alternative to traditional structured development, which suffered from long deliverable times and unfulfilled requirements.

- Agile Software Development[2] is a framework for developing software applications that divides the work into smaller subprojects to obtain valuable deliverables in shorter times and with better cohesion. This method emphasizes close communication among all users and continuous evaluation with the purpose of increasing customer satisfaction.

  Although the development *methodologies* may change, the basic framework within which they are used does not change.

database development
The process of database design and implementation.

[1] *See Rapid Application Development*, James Martin, Prentice-Hall, Macmillan College Division, 1991.
[2] For more information about Agile Software Development, go to *www.agilealliance.org*.

9-2 The Systems Development Life Cycle

The **Systems Development Life Cycle (SDLC)** traces the history of an information system. Perhaps more important to the system designer, the SDLC provides the big picture within which the database design and application development can be mapped out and evaluated.

As illustrated in Figure 9.2, the traditional SDLC is divided into five phases: planning, analysis, detailed systems design, implementation, and maintenance. The SDLC is an iterative process rather than a sequential process. For example, the details of the feasibility study might help refine the initial assessment, and the details discovered during the user requirements portion of the SDLC might help refine the feasibility study.

FIGURE 9.2 THE SYSTEMS DEVELOPMENT LIFE CYCLE (SDLC)

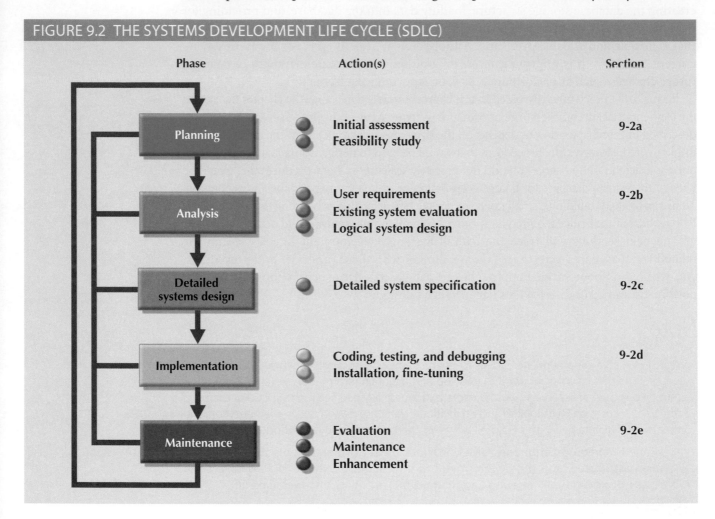

Phase	Action(s)	Section
Planning	Initial assessment Feasibility study	9-2a
Analysis	User requirements Existing system evaluation Logical system design	9-2b
Detailed systems design	Detailed system specification	9-2c
Implementation	Coding, testing, and debugging Installation, fine-tuning	9-2d
Maintenance	Evaluation Maintenance Enhancement	9-2e

Because the Database Life Cycle fits into and resembles the SDLC, a brief description of the SDLC is in order.

9-2a Planning

The SDLC planning phase yields a general overview of the company and its objectives. An initial assessment of the information flow-and-extent requirements must be made during this discovery portion of the SDLC. Such an assessment should answer some important questions:

- *Should the existing system be continued?* If the information generator does its job well, there is no point in modifying or replacing it. To quote an old saying, "If it ain't broke, don't fix it."

Systems Development Life Cycle (SDLC)
The cycle that traces the history of an information system. The SDLC provides the big picture within which database design and application development can be mapped out and evaluated.

- *Should the existing system be modified?* If the initial assessment indicates deficiencies in the extent and flow of the information, minor (or even major) modifications might be needed. When considering modifications, the participants in the initial assessment must remember the distinction between wants and needs.

- *Should the existing system be replaced?* The initial assessment might indicate that the current system's flaws are beyond fixing. Given the effort required to create a new system, a careful distinction between wants and needs is perhaps even more important in this case than it is when modifying the system.

Participants in the SDLC's initial assessment must begin to study and evaluate alternative solutions. If a new system is necessary, the next question is whether it is feasible. The feasibility study must address the following:

- *The technical aspects of hardware and software requirements.* The decisions might not yet be vendor-specific, but they must address the nature of the hardware requirements (desktop computer, multiprocessor computer, mainframe, or supercomputer) and the software requirements (single-user or multiuser operating systems, database type and software, programming languages to be used by the applications, and so on).

- *The system cost.* The admittedly mundane question "Can we afford it?" is crucial. The answer might force a careful review of the initial assessment. A million-dollar solution to a thousand-dollar problem is not defensible. At some point, the decision may be between building a system "in-house" or buying (and customizing) a third-party vendor system. In the long run, you need to find a cost-effective solution that best serves the needs (present and future) of the organization.

- *The operational cost.* Does the company possess the human, technical, and financial resources to keep the system operational? Should the feasibility study include the cost of management and end-user support needed to implement operational procedures to ensure the success of this system? What would be the impact of this new system in the company's culture? People's resistance to change should never be underestimated.[3]

Even if you choose to "buy" rather than to "build," the system implementation must be carefully planned for it to be successful. Whatever the chosen option (build or buy), an analysis must be done to deploy the solution across the organization in ways that minimize cost and culture changes, while maximizing value. The SDLC provides a framework for sound planning and implementation.

9-2b Analysis

Problems defined during the planning phase are examined in greater detail during the analysis phase. A macro analysis must be made both of individual needs and organizational needs, addressing questions such as:

- What are the requirements of the current system's end users?

- Do those requirements fit into the overall information requirements?

The analysis phase of the SDLC is, in effect, a thorough *audit* of user requirements.

The existing hardware and software systems are also studied during the analysis phase. The result of the analysis should be a better understanding of the system's functional areas, actual and potential problems, and opportunities.

---

[3] "At Zappos, 210 employees decide to leave rather than work with 'no bosses,'" Jena McGregor, Washington Post, May 8, 2015.

End users and the system designer(s) must work together to identify processes and uncover potential problem areas. Such cooperation is vital to defining the appropriate performance objectives by which the new system can be judged.

Along with a study of user requirements and the existing systems, the analysis phase also includes the creation of a logical systems design. The logical design must specify the appropriate conceptual data model, inputs, processes, and expected output requirements.

When creating a logical design, the designer might use tools such as data flow diagrams (DFDs), hierarchical input process output (HIPO) diagrams, entity relationship (ER) diagrams, and even some application prototypes. The database design's data-modeling activities take place at this point to discover and describe all entities and their attributes, and the relationships among the entities within the database.

Defining the logical system also yields functional descriptions of the system's components (modules) for each process within the database environment. All data transformations (processes) are described and documented, using systems analysis tools such as DFDs. The conceptual data model is validated against those processes.

9-2c Detailed Systems Design

In the detailed systems design phase, the designer completes the design of the system's processes. The design includes all the necessary technical specifications for the screens, menus, reports, and other devices that might help make the system a more efficient information generator. The steps are laid out for conversion from the old system to the new system. Training principles and methodologies are also planned and must be submitted for management's approval.

Note

Because this book has focused on the details of systems design, it has not explicitly recognized until now that management approval is needed at all stages of the process. Such approval is needed because a "go" decision requires funding. There are many "go" and "no go" decision points along the way to a completed systems design!

9-2d Implementation

During the implementation phase, the hardware, DBMS software, and application programs are installed, and the database design is implemented. During the initial stages of the implementation phase, the system enters into a cycle of coding, testing, and debugging until it is ready to be delivered. The actual database is created, and the system is customized by the creation of tables and views, user authorizations, and so on.

The database contents might be loaded interactively or in batch mode, using a variety of methods and devices:

- Customized user programs

- Database interface programs

- Conversion programs that import the data from a different file structure, using batch programs, a database utility, or both

The system is subjected to exhaustive testing until it is ready for use. Traditionally, the implementation and testing of a new system took 50 to 60 percent of the total development time. However, the advent of sophisticated application generators and debugging tools has substantially decreased coding and testing time. After testing is concluded, the final documentation is reviewed and printed and end users are trained. The system is in full operation at the end of this phase, but it will be continuously evaluated and fine-tuned.

9-2e Maintenance

Almost as soon as the system is operational, end users begin to request changes in it. Those changes generate system maintenance activities, which can be grouped into three types:

- *Corrective maintenance* in response to systems errors
- *Adaptive maintenance* due to changes in the business environment
- *Perfective maintenance* to enhance the system

Because every request for structural change requires retracing the SDLC steps, the system is, in a sense, always at some stage of the SDLC.

Each system has a predetermined operational life span, but its actual life span depends on its perceived utility. There are several reasons for reducing the operational life of certain systems. Rapid technological change is one reason, especially for systems based on processing speed and expandability. Another common reason is the cost of maintaining a system.

If the system's maintenance cost is high, its value becomes suspect. **Computer-aided software engineering (CASE)** tools, such as System Architect or Visio Professional, help produce better systems within a reasonable amount of time and at a reasonable cost. In addition, CASE-produced applications are more structured, better documented, and especially *standardized*, which tends to prolong the operational life of systems by making them easier and cheaper to update and maintain.

9-3 **The Database Life Cycle**

Within the larger information system, the database is subject to a life cycle as well. The **Database Life Cycle (DBLC)** contains six phases, as shown in Figure 9.3: database initial study, database design, implementation and loading, testing and evaluation, operation, and maintenance and evolution.

9-3a The Database Initial Study

If a designer has been called in, chances are that the current system has failed to perform functions deemed vital by the company. (You don't call the plumber unless the pipes leak.) Therefore, in addition to examining the current system's operation within the company, the designer must determine how and why the current system fails. That means spending a lot of time talking and listening to end users. Although database design is a technical business, it is also people-oriented. Database designers must be excellent communicators and must have finely tuned interpersonal skills.

Depending on the complexity and scope of the database environment, the database designer might be a lone operator or part of a systems development team composed of a project leader, one or more senior systems analysts, and one or more junior systems analysts. The word *designer* is used generically here to cover a wide range of design team compositions.

computer-aided systems engineering (CASE)
Tools used to automate part or all of the Systems Development Life Cycle.

Database Life Cycle (DBLC)
A cycle that traces the history of a database within an information system. The cycle is divided into six phases: initial study, design, implementation and loading, testing and evaluation, operation and maintenance, and evolution.

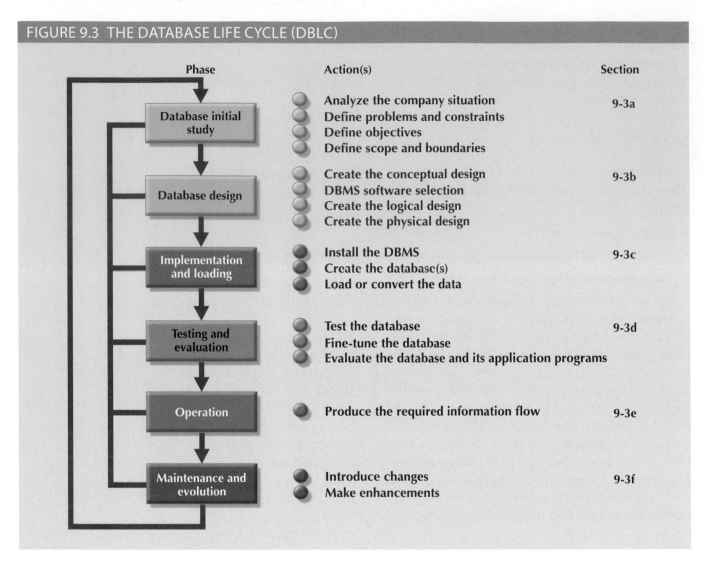

FIGURE 9.3 THE DATABASE LIFE CYCLE (DBLC)

The overall purpose of the database initial study is to:

- Analyze the company situation
- Define problems and constraints
- Define objectives
- Define scope and boundaries

Figure 9.4 depicts the interactive and iterative processes required to complete the first phase of the DBLC successfully. Note that the database initial study phase leads to the development of database system objectives. Using Figure 9.4 as a discussion template, examine each of its components in greater detail.

Analyze the Company Situation

The *company situation* describes the general conditions in which a company operates, its organizational structure, and its mission. To analyze the company situation, the database designer must learn the company's operational components, how they function, and how they interact.

The following issues must be resolved:

- *What is the organization's general operating environment, and what is its mission within that environment?* The design must satisfy the operational demands created by the

FIGURE 9.4 A SUMMARY OF ACTIVITIES IN THE DATABASE INITIAL STUDY

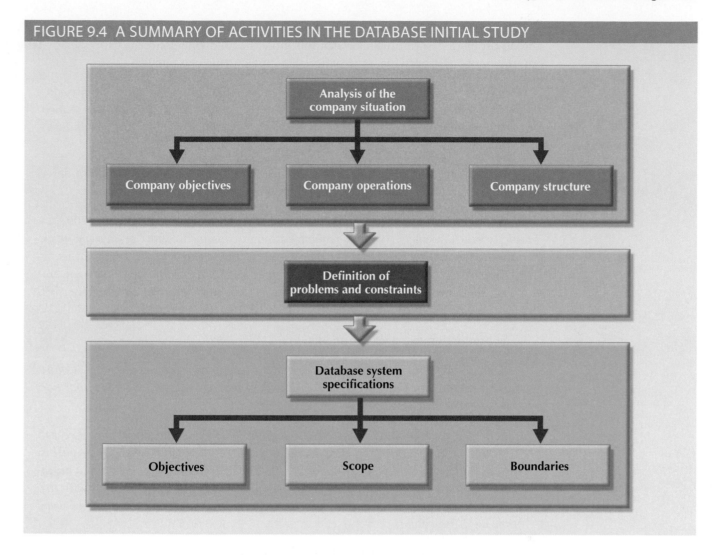

organization's mission. For example, a mail-order business probably has operational requirements for its database that are quite different from those of a manufacturing business.

- *What is the organization's structure?* Knowing who controls what and who reports to whom is quite useful when you need to define required information flows, specific report and query formats, and so on.

Define Problems and Constraints The designer has both formal and informal sources of information. If the company has existed for any length of time, it already has a system in place (either manual or computer-based). How does the existing system function? What input does the system require? What documents does the system generate? By whom and how is the system output used? Studying the paper trail can be very informative. In addition to the official version of the system's operation, there is also the more informal, perhaps more real version; the designer must be shrewd enough to see how these differ.

The process of defining problems might initially appear to be unstructured. Company end users often cannot precisely describe the larger scope of company operations or identify the real problems encountered during company operations. Often the managerial view of a company's operation and its problems is different from that of the end users, who perform the actual routine work.

During the initial problem definition process, the designer is likely to collect very broad problem descriptions. For example, note the following concerns expressed by the president of a fast-growing, transnational manufacturing company:

Although the rapid growth is gratifying, members of the management team are concerned that such growth is beginning to undermine the ability to maintain a high customer service standard, and perhaps worse, to diminish manufacturing standards control.

The problem definition process quickly leads to a host of general problem descriptions. For example, the marketing manager comments:

I'm working with an antiquated filing system. We manufacture more than 1,700 specialty machine parts. When a regular customer calls in, we can't get a very quick inventory scan. If a new customer calls in, we can't do a current parts search by using a simple description, so we often do a machine setup for a part that we have in inventory. That's wasteful. And of course, some new customers get irritated when we can't give a quick response.

The production manager comments:

At best, it takes hours to generate the reports I need for scheduling purposes. I don't have hours for quick turnarounds. It's difficult to manage what I don't have information about.

I don't get quick product request routing. Take machine setup. Right now I've got operators either waiting for the right stock or getting it themselves when a new part is scheduled for production. I can't afford to have an operator doing chores that a much lower-paid worker ought to be doing. There's just too much waiting around with the current scheduling. I'm losing too much time, and my schedules back up. Our overtime bill is ridiculous.

I sometimes produce parts that are already in inventory because we don't seem to be able to match what we've got in inventory with what we have scheduled. Shipping yells at me because I can't turn out the parts, and often they've got them in inventory one bay down. That's costing us big bucks sometimes.

New reports can take days or even weeks to get to my office. And I need a ton of reports to schedule personnel, downtime, training, etc. I can't get new reports that I need NOW. What I need is the ability to get quick updates on percent defectives, percent rework, the effectiveness of training, you name it. I need such reports by shift, by date, by any characteristic I can think of to help me manage scheduling, training, you name it.

A machine operator comments:

It takes a long time to set my stuff up. If I get my schedule banged up because John doesn't get the paperwork on time, I wind up looking for setup specs, startup material, bin assignments, and other stuff. Sometimes I spend two or three hours just setting up. Now you know why I can't meet schedules. I try to be productive, but I'm spending too much time getting ready to do my job.

After the initial declarations, the database designer must continue to probe carefully to generate additional information that will help define the problems within the larger framework of company operations. How does the problem of the marketing manager's customer fit within the broader set of marketing department activities? How does the solution to the customer's problem help meet the objectives of the marketing department and the rest of the company? How do the marketing department's activities relate to those of the other departments? That last question is especially important. Note that there are common threads in the problems described by the marketing and production department managers. If the inventory query process can be improved, both departments are likely to find simple solutions to at least some of their problems.

Finding precise answers is important, especially concerning the operational relationships among business units. If a proposed system will solve the marketing department's problems but exacerbate those of the production department, not much progress will have been made. Using an analogy, suppose that your home water bill is too high. You have determined the problem: the faucets leak. The solution? You step outside and cut off the water supply to the house. However, is that an adequate solution, or would the replacement of faucet washers do a better job of solving the problem? You might find this scenario simplistic, yet almost any experienced database designer can find similar instances of database problem solving, although they are admittedly more complicated.

Even the most complete and accurate problem definition does not always lead to the perfect solution. The real world usually intrudes to limit the design of even the most elegant database by imposing constraints such as time, budget, and personnel. If you must have a solution within a month and within a $12,000 budget, you cannot take two years to develop a database at a cost of $100,000. *The designer must learn to distinguish between what's perfect and what's possible.*

Define Objectives   A proposed database system must be designed to help solve at least the major problems identified during the problem discovery process. As the list of problems unfolds, several common sources are likely to be discovered. In the previous example, both the marketing manager and the production manager seem to be plagued by inventory inefficiencies. If the designer can create a database that sets the stage for more efficient parts management, both departments gain. The initial objective, therefore, might be to create an efficient inventory query and management system.

Note

When trying to develop solutions, the database designer must look for the source of the problems. Many database systems have failed to satisfy the end users because they were designed to treat the *symptoms* of the problems rather than their source.

Note that the initial study phase also yields proposed problem solutions. The designer's job is to make sure that his or her database system objectives correspond to those envisioned by the end user(s). In any case, the database designer must begin to address the following questions:

- What is the proposed system's initial objective?
- Will the system interface with other existing or future systems in the company?
- Will the system share the data with other systems or users?

Define Scope and Boundaries   The designer must recognize two sets of limits: scope and boundaries. The system's **scope** defines the extent of the design according to operational requirements. Will the database design encompass the entire organization, one or more departments within the organization, or one or more functions of a single

scope
The part of a system that defines the extent of the design, according to operational requirements.

department? The designer must know the "size of the ballpark." Knowing the scope helps define the required data structures, the type and number of entities, the physical size of the database, and so on.

The proposed system is also subject to limits known as **boundaries**, which are external to the system. Has any designer ever been told, "We have all the time in the world" or "Use an unlimited budget and as many people as needed to make the design come together"? Boundaries are also imposed by existing hardware and software. Ideally, the designer can choose the hardware and software that will best accomplish the system goals. In fact, software selection is an important aspect of the Systems Development Life Cycle. Unfortunately, in the real world, a system must often be designed around existing hardware. Thus, the scope and boundaries become the factors that force the design into a specific mold, and the designer's job is to design the best system possible within those constraints. (Note that problem definitions and the objectives must sometimes be reshaped to meet the system scope and boundaries.)

9-3b Database Design

The second phase of the DBLC focuses on the design of the database model that will support company operations and objectives. This is arguably the most critical DBLC phase: making sure that the final product meets user and system requirements. In the process of database design, you must concentrate on the data characteristics required to build the database model. At this point, there are two views of the data within the system: the business view of data as a source of information and the designer's view of the data structure, its access, and the activities required to transform the data into information. Figure 9.5 contrasts those views. Note that you can summarize the different views by looking at the terms *what* and *how*. Defining data is an integral part of the DBLC's second phase.

As you examine the procedures required to complete the design phase in the DBLC, remember these points:

- The process of database design is loosely related to the analysis and design of a larger system. The data component is only one element of a larger information system.

- The systems analysts or systems programmers are in charge of designing the other system components. Their activities create the procedures that will help transform the data within the database into useful information.

- The database design does not constitute a sequential process. Rather, it is an iterative process that provides continuous feedback designed to trace previous steps.

The database design process is depicted in Figure 9.6. The figure shows that there are three essential stages: conceptual, logical, and physical design, plus the DBMS selection decision, which is critical to determine the type of logical and physical designs to be created. The design process starts with conceptual design and moves to the logical and physical design stages. At each stage, more details about the data model design are determined and documented. You could think of the conceptual design as the overall data as seen by the end user, the logical design as the data as seen by the DBMS, and the physical design as the data as seen by the operating system's storage management devices.

It is important to note that the overwhelming majority of database designs and implementations are based on the relational model, and therefore use the relational model constructs and techniques. When you finish the design activities, you will have a complete database design ready to be implemented.

boundaries
The external limits to which any proposed system is subjected. These limits include budgets, personnel, and existing hardware and software.

FIGURE 9.5   TWO VIEWS OF DATA: BUSINESS MANAGER AND DESIGNER

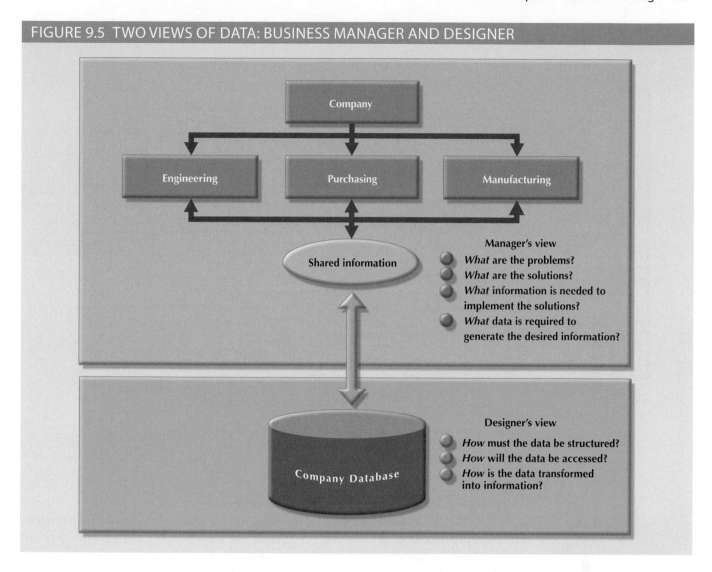

Database design activities are covered in detail in Sections 9-4 (Conceptual Design), 9-5 (DBMS Software Selection), 9-6 (Logical Design), and 9-7 (Physical Design).

9-3c Implementation and Loading

The output of the database design phase is a series of instructions detailing the creation of tables, attributes, domains, views, indexes, security constraints, and storage and performance guidelines. In this phase, you actually implement all these design specifications.

Install the DBMS This step is required only when a new dedicated instance of the DBMS is necessary for the system. In many cases, the organization will have made a particular DBMS the standard to leverage investments in the technology and the skills that employees have already developed. The DBMS may be installed on a new server or on existing servers. One current trend is called virtualization. **Virtualization** is a technique that creates logical representations of computing resources that are independent of the underlying physical computing resources. This technique is used in many areas of computing, such as the creation of virtual servers, virtual storage, and virtual private networks. In a database environment, database virtualization refers to

virtualization
A technique that creates logical representations of computing resources that are independent of the underlying physical computing resources.

FIGURE 9.6 DATABASE DESIGN PROCESS

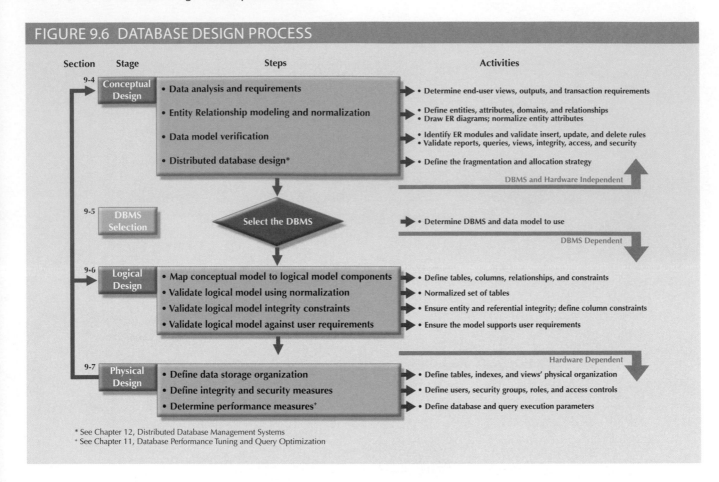

* See Chapter 12, Distributed Database Management Systems
+ See Chapter 11, Database Performance Tuning and Query Optimization

Online Content

Two appendixes at *www.cengagebrain.com* provide a concise example of simple real-world database development: Appendix B, The University Lab: Conceptual Design, and Appendix C, The University Lab: Conceptual Design Verification, Logical Design, and Implementation.

the installation of a new instance of the DBMS on a virtual server running on shared hardware. This is normally a task that involves system and network administrators to create appropriate user groups and services in the server configuration and network routing. Another common trend is the use of cloud database services such Microsoft SQL Database Service or Amazon Relational Database Services (RDS). This new generation of services allows user to create databases that could be easily managed, tested, and scaled up as needed.

Create the Database(s) In most modern relational DBMSs, a new database implementation requires the creation of special storage-related constructs to house the end-user tables. The constructs usually include the storage group (or file groups), the table spaces, and the tables. Figure 9.7 shows that a storage group can contain more than one table space and that a table space can contain more than one table.

For example, the implementation of the logical design in IBM's DB2 would require the following:

1. The system administrator (SYSADM) would create the database storage group. This step is mandatory for such mainframe databases as DB2. Other DBMS software may create equivalent storage groups automatically when a database is created. (See Step 2.) Consult your DBMS documentation to see if you must create a storage group, and if so, what the command syntax must be.

2. The SYSADM creates the database within the storage group.

3. The SYSADM assigns the rights to use the database to a database administrator (DBA).

4. The DBA creates the table space(s) within the database.

FIGURE 9.7 PHYSICAL ORGANIZATION OF A DB2 DATABASE ENVIRONMENT

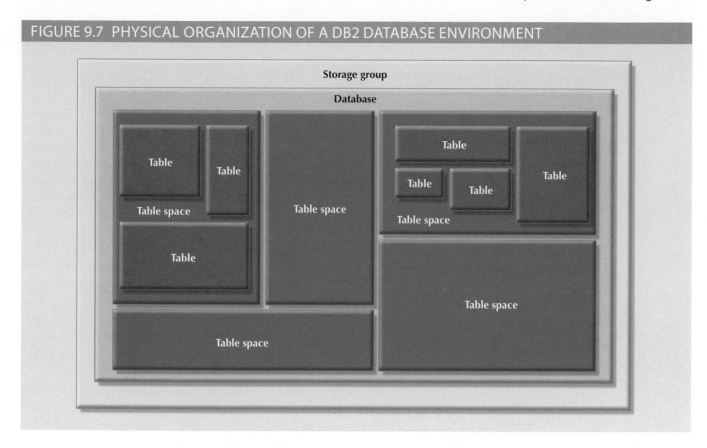

5. The DBA creates the table(s) within the table space(s).

6. The DBA assigns access rights to the table spaces and to the tables within specified table spaces. Access rights may be limited to views rather than to whole tables. The creation of views is not required for database access in the relational environment, but views are desirable from a security standpoint. For example, using the following command, access rights to a table named PROFESSOR may be granted to the user Lynn Eilers, whose identification code is LEILERS:

> GRANT SELECT ON PROFESSOR TO USER LEILERS;

Load or Convert the Data After the database has been created, the data must be loaded into the database tables. Typically, the data will have to be migrated from the prior version of the system. Often, data to be included in the system must be aggregated from multiple sources. In a best-case scenario, all of the data will be in a relational database so that it can be readily transferred to the new database. However, in some cases data may have to be imported from other relational databases, nonrelational databases, flat files, legacy systems, or even manual paper-and-pencil systems. If the data format does not support direct importing into the new database, conversion programs may have to be created to reformat the data for importing. In a worst-case scenario, much of the data may have to be manually entered into the database. Once the data has been loaded, the DBA works with the application developers to test and evaluate the database.

Loading existing data into a cloud-based database service sometimes can be expensive. The reason for this is that most cloud services are priced based not only on the volume of data to be stored but also on the amount of data that travels over the network. In such cases, loading a 1 TB database could be a very expensive proposition. Therefore, system administrators must be very careful in reading and negotiating the terms of cloud service contracts to ensure that there will be no "hidden" costs.

9-3d Testing and Evaluation

In the design phase, decisions were made to ensure integrity, security, performance, and recoverability of the database. During implementation and loading, these plans were put into place. In testing and evaluation, the DBA tests and fine-tunes the database to ensure that it performs as expected. This phase occurs in conjunction with application programming. Programmers use database tools to *prototype* the applications during coding of the programs. Tools such as report generators, screen painters, and menu generators are especially useful to application programmers.

Test the Database During this step, the DBA tests the database to ensure that it maintains the integrity and security of the data. Data integrity is enforced by the DBMS through the proper use of primary and foreign key rules. Many DBMSs also support the creation of domain constraints and database triggers. Testing will ensure that these constraints were properly designed and implemented. Data integrity is also the result of properly implemented data management policies, which are part of a comprehensive data administration framework. For a more detailed study of this topic, see The DBA's Managerial Role section in Chapter 16, Database Administration and Security.

Previously, users and roles were created to grant users access to the data. In this stage, not only must those privileges be tested, but the broader view of data privacy and security must be addressed. Data stored in the company database must be protected from access by unauthorized users. (It does not take much imagination to predict the likely results if students have access to a student database or if employees have access to payroll data!) Consequently, you must test for at least the following:

- *Physical security* allows only authorized personnel physical access to specific areas. Depending on the type of database implementation, however, establishing physical security might not always be practical. For example, a university student research database is not a likely candidate for physical security.

- *Password security* allows the assignment of access rights to specific authorized users. Password security is usually enforced at login time at the operating system level.

- *Access rights* can be established through the use of database software. The assignment of access rights may restrict operations (CREATE, UPDATE, DELETE, and so on) on predetermined objects such as databases, tables, views, queries, and reports.

- *Audit trails* are usually provided by the DBMS to check for access violations. Although the audit trail is an after-the-fact device, its mere existence can discourage unauthorized use.

- *Data encryption* can render data useless to unauthorized users who might have violated some of the database security layers.

- *Diskless workstations* allow end users to access the database without being able to download the information from their workstations.

For a more detailed discussion of security issues, refer to Chapter 16, Database Administration and Security.

Fine-Tune the Database Database performance can be difficult to evaluate because there are no standards for measuring it, but it is typically one of the most important factors in database implementation. Different systems will place different performance requirements on the database. Systems that support rapid transactions will require the database to be implemented so that they provide superior performance during high volumes of inserts, updates, and deletes. Other systems, like decision support systems, may require superior performance for complex data retrieval tasks. Many factors can affect the database's performance on various tasks, including the hardware and software environment in which the database exists. Naturally, the characteristics and volume of the data also affect database performance: a search of 10 tuples is faster than a search of 100,000 tuples. Other important factors in database performance include system and database configuration parameters such as data placement, access path definition, the use of indexes, and buffer size. For a more in-depth discussion of database performance issues, see Chapter 11, Database Performance Tuning and Query Optimization.

Evaluate the Database and Its Application Programs As the database and application programs are created and tested, the system must also be evaluated using a more holistic approach. Testing and evaluation of the individual components should culminate in a variety of broader system tests to ensure that all of the components interact properly to meet the needs of the users. At this stage, integration issues and deployment plans are refined, user training is conducted, and system documentation is finalized. Once the system receives final approval, it must be a sustainable resource for the organization. To ensure that the data contained in the database is protected against loss, backup and recovery plans are tested.

Timely data availability is crucial for almost every database. Unfortunately, the database can lose data through unintended deletions, power outages, and other causes. Data backup and recovery procedures create a safety valve, ensuring the availability of consistent data. Typically, database vendors encourage the use of fault-tolerant components such as uninterruptible power supply (UPS) units, RAID storage devices, clustered servers, and data replication technologies to ensure the continuous operation of the database in case of a hardware failure. Even with these components, backup and restore functions constitute a very important part of daily database operations. Some DBMSs provide functions that allow the database administrator to schedule automatic database backups to permanent storage devices such as disks, DVDs, tapes, and online storage. Database backups can be performed at different levels:

- A **full backup**, or *dump*, of the entire database. In this case, all database objects are backed up in their entirety.

- A **differential backup** of the database, in which only the objects that have been updated or modified since the last full backup are backed up.

- A **transaction log backup**, which backs up only the transaction log operations that are not reflected in a previous backup copy of the database. In this case, no other database objects are backed up. (For a complete explanation of the transaction log, see Chapter 10, Transaction Management and Concurrency Control.)

The database backup is stored in a secure place, usually in a different building from the database itself, and is protected against dangers such as fire, theft, flood, and other potential calamities. The main purpose of the backup is to guarantee database restoration following a hardware or software failure.

Failures that plague databases and systems are generally induced by software, hardware, programming exemptions, transactions, or external factors. Table 9.1 summarizes the most common sources of database failure.

Depending on the type and extent of the failure, the recovery process ranges from a minor short-term inconvenience to a major long-term rebuild. Regardless of the extent of the required recovery process, recovery is not possible without a usable backup.

full backup (database dump)
A complete copy of an entire database saved and periodically updated in a separate location. A full backup ensures a full recovery of all data after a physical disaster or database integrity failure.

differential backup
A level of database backup in which only the last modifications to the database are copied.

transaction log backup
A backup of only the transaction log operations that are not reflected in a previous backup copy of the database.

TABLE 9.1		
COMMON SOURCES OF DATABASE FAILURE		
SOURCE	**DESCRIPTION**	**EXAMPLE**
Software	Software-induced failures may be traceable to the operating system, the DBMS software, application programs, or viruses and other malware.	In January 2015, a security vulnerability was found for Oracle E-Business Suite that could cause serious data compromise.[4]
Hardware	Hardware-induced failures may include memory chip errors, disk crashes, bad disk sectors, and disk-full errors.	A bad memory module or a multiple hard disk failure in a database system can bring it to an abrupt stop.
Programming exemptions	Application programs or end users may roll back transactions when certain conditions are defined. Programming exemptions can also be caused by malicious or improperly tested code that can be exploited by hackers.	Hackers constantly search for ways to exploit unprotected web database systems. For example, in February 2015, Anthem, the second largest health insurer, announced that it was hacked and data for 80 million customers might have been exposed.[5]
Transactions	The system detects deadlocks and aborts one of the transactions. (See Chapter 10.)	Deadlock occurs when executing multiple simultaneous transactions.
External factors	Backups are especially important when a system suffers complete destruction from fire, earthquake, flood, or other natural disaster.	In 2012, Hurricane Sandy hit the northeastern United States, causing data and service losses worth billions of dollars across multiple states.

Database recovery generally follows a predictable scenario. First, the type and extent of the required recovery are determined. If the entire database needs to be recovered to a consistent state, the recovery uses the most recent backup copy of the database in a known consistent state. The backup copy is then rolled forward to restore all subsequent transactions by using the transaction log information. If the database needs to be recovered but the committed portion of the database is still usable, the recovery process uses the transaction log to "undo" all of the transactions that were not committed (see Chapter 10, Transaction Management and Concurrency Control).

At the end of this phase, the database completes an iterative process of testing, evaluation, and modification that continues until the system is certified as ready to enter the operational phase.

9-3e Operation

Once the database has passed the evaluation stage, it is considered operational. At that point, the database, its management, its users, and its application programs constitute a complete information system.

The beginning of the operational phase invariably starts the process of system evolution. As soon as all of the targeted end users have entered the operations phase, problems that could not have been foreseen during the testing phase begin to surface. Some of the problems are serious enough to warrant emergency "patchwork," while others are merely minor annoyances. For example, if the database design is implemented to interface with the web, the sheer volume of transactions might cause even a well-designed system to

---

[4]"Oracle Patches Backdoor Vulnerability, Recommends Disabling SSL," January 21, 2015. Url: https://threatpost.com/oracle-patches-backdoor-vulnerability-recommends-disabling-ssl/110555

[5]"Massive data hack of health insurer Anthem potentially exposes millions," Fred Barbash and Abby Phillip, February 5, 2015, Washington Post. http://www.washingtonpost.com/blogs/the-switch/wp/2015/03/20/2015-is-already-the-year-of-the-health-care-hack-and-its-only-going-to-get-worse/

bog down. In that case, the designers have to identify the source of the bottleneck and produce alternative solutions. Those solutions may include using load-balancing software to distribute the transactions among multiple computers, increasing the available cache for the DBMS, and so on. The demand for change is the designer's constant concern, which leads to phase 6, maintenance and evolution.

9-3f Maintenance and Evolution

The database administrator must be prepared to perform routine maintenance activities within the database. Some of the required periodic maintenance activities include:

- Preventive maintenance (backup)

- Corrective maintenance (recovery)

- Adaptive maintenance (enhancing performance, adding entities and attributes, and so on)

- Assignment of access permissions and their maintenance for new and old users

- Generation of database access statistics to improve the efficiency and usefulness of system audits and to monitor system performance

- Periodic security audits based on the system-generated statistics

- Monthly, quarterly, or yearly system usage summaries for internal billing or budgeting purposes

The likelihood of new information requirements and the demand for additional reports and new query formats require application changes and possible minor changes in the database components and contents. These changes can be easily implemented only when the database design is flexible and when all documentation is updated and online. Eventually, even the best-designed database environment will no longer be capable of incorporating such evolutionary changes, and then the whole DBLC process begins anew.

As you can see, many of the activities described in the DBLC are similar to those in the SDLC. This should not be surprising because the SDLC is the framework within which the DBLC activities take place. A summary of the parallel activities that occur within the SDLC and DBLC is shown in Figure 9.8.

9-4 Conceptual Design

Recall that the second phase of the DBLC is database design, which comprises three stages: conceptual design, logical design, and physical design, plus the critical decision of DBMS selection. **Conceptual design** is the first stage in the database design process. The goal at this stage is to design a database that is independent of database software and physical details. The output of this process is a conceptual data model that describes the main data entities, attributes, relationships, and constraints of a given problem domain. This design is descriptive and narrative in form. In other words, it is generally composed of a graphical representation as well as textual descriptions of the main data elements, relationships, and constraints.

In this stage, data modeling is used to create an abstract database structure that represents real-world objects in the most realistic way possible. The conceptual model must embody a clear understanding of the business and its functional areas. At this level of abstraction, the type of hardware and database model to be used might not have been identified yet. Therefore, the design must be software- and hardware-independent so that the system can be set up within any platform chosen later.

conceptual design
A process that uses data-modeling techniques to create a model of a database structure that represents real-world objects as realistically as possible. The design is both software- and hardware-independent.

FIGURE 9.8 PARALLEL ACTIVITIES IN THE DBLC AND THE SDLC

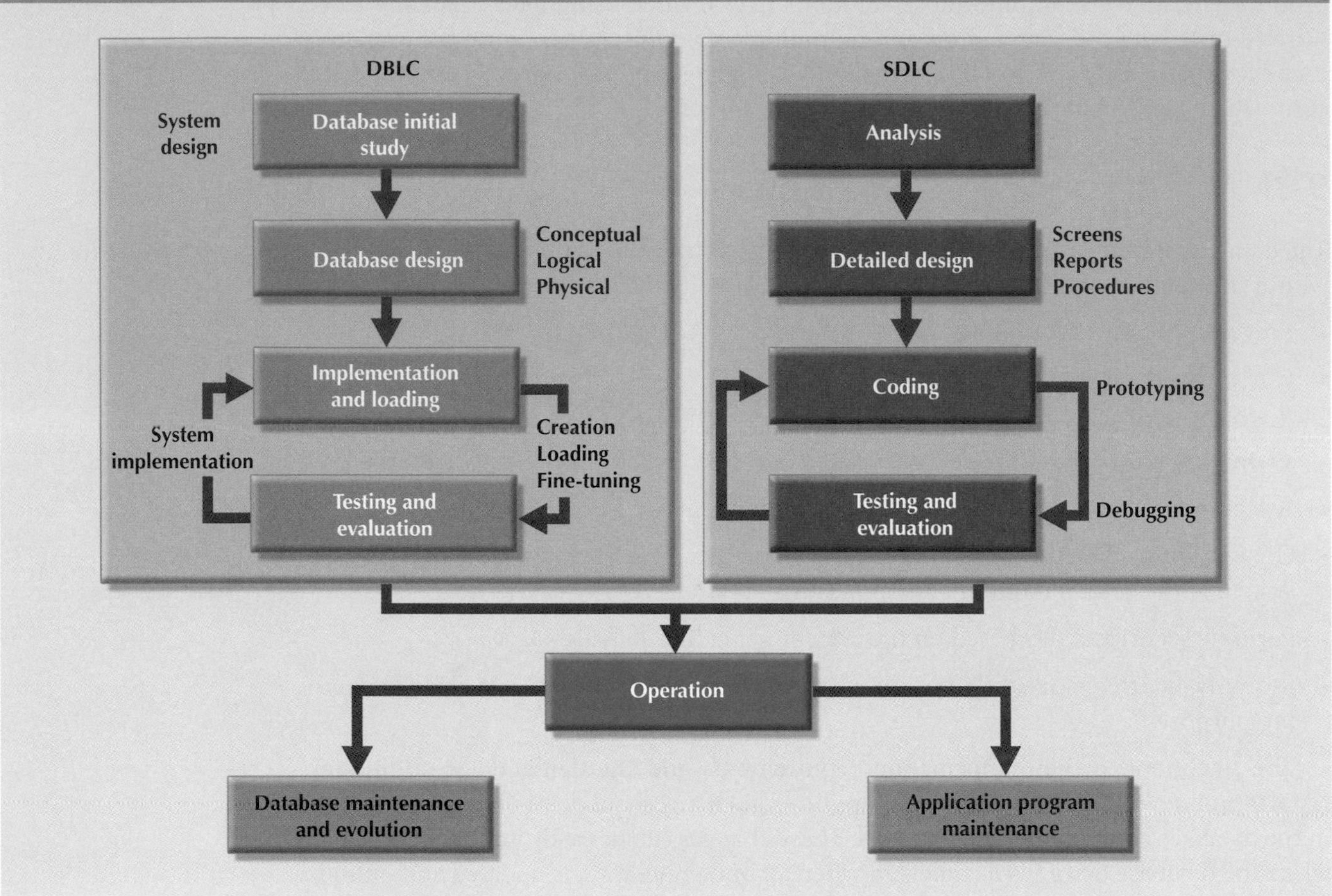

Keep in mind the following **minimal data rule**:

All that is needed is there, and all that is there is needed.

In other words, make sure that all data needed is in the model and that all data in the model is needed. All data elements required by the database transactions must be defined in the model, and all data elements defined in the model must be used by at least one database transaction.

However, as you apply the minimal data rule, avoid excessive short-term bias. Focus not only on the immediate data needs of the business but on future data needs. Thus, the database design must leave room for future modifications and additions, ensuring that the business's investment in information resources will endure.

The conceptual design has four steps, which are listed in Table 9.2.

minimal data rule
Defined as "All that is needed is there, and all that is there is needed." In other words, all data elements required by database transactions must be defined in the model, and all data elements defined in the model must be used by at least one database transaction.

TABLE 9.2

CONCEPTUAL DESIGN STEPS

STEP	ACTIVITY
1	Data analysis and requirements
2	Entity relationship modeling and normalization
3	Data model verification
4	Distributed database design

The following sections cover these steps in more detail.

9-4a Data Analysis and Requirements

The first step in conceptual design is to discover the characteristics of the data elements. An effective database is an information factory that produces key ingredients for successful decision making. Appropriate data element characteristics are those that can be transformed into appropriate information. Therefore, the designer's efforts are focused on:

- *Information needs.* What kind of information is needed? That is, what output (reports and queries) must be generated by the system, what information does the current system generate, and to what extent is that information adequate?

- *Information users.* Who will use the information? How is the information to be used? What are the various end-user data views?

- *Information sources.* Where is the information to be found? How is the information to be extracted once it is found?

- *Information constitution.* What data elements are needed to produce the information? What are the data attributes? What relationships exist in the data? What is the data volume? How frequently is the data used? What data transformations will be used to generate the required information?

The designer obtains the answers to those questions from a variety of sources to compile the necessary information:

- *Developing and gathering end-user data views.* The database designer and the end user(s) jointly develop a precise description of end-user data views, which in turn are used to help identify the database's main data elements.

- *Directly observing the current system: existing and desired output.* The end user usually has an existing system in place, whether it is manual or computer-based. The designer reviews the existing system to identify the data and its characteristics. The designer examines the input forms and files (tables) to discover the data type and volume. If the end user already has an automated system in place, the designer carefully examines the current and desired reports to describe the data required to support the reports.

- *Interfacing with the systems design group.* As noted earlier in this chapter, the database design process is part of the SDLC. In some cases, the systems analyst in charge of designing the new system will also develop the conceptual database model. (This is usually true in a decentralized environment.) In other cases, the database design is considered part of the DBA's job. The presence of a DBA usually implies the existence of a formal data-processing department. The DBA designs the database according to the specifications created by the systems analyst.

To develop an accurate data model, the designer must have a thorough understanding of the company's data types and their extent and uses. But data does not, by itself, yield the required understanding of the total business. From a database point of view, the collection of data becomes meaningful only when business rules are defined. Remember from Chapter 2, Data Models, that a *business rule* is a brief and precise description of a policy, procedure, or principle within a specific organization's environment. Business rules, derived from a detailed description of an organization's operations, help to create and enforce actions within that organization's environment. When business rules are written properly, they define entities, attributes, relationships, connectivities, cardinalities, and constraints.

To be effective, business rules must be easy to understand, and they must be widely disseminated to ensure that every person in the organization shares a common interpretation of the rules. Using simple language, business rules describe the main and distinguishing characteristics of the data as *viewed by the company*. Examples of business rules are as follows:

- A customer may make many payments on an account.

- Each payment on an account is credited to only one customer.

- A customer may generate many invoices.

- Each invoice is generated by only one customer.

Given their critical role in database design, business rules must not be established casually. Poorly defined or inaccurate business rules lead to database designs and implementations that fail to meet the needs of the organization's end users.

Ideally, business rules are derived from a formal **description of operations**, which is a document that provides a precise, up-to-date, and thoroughly reviewed description of the activities that define an organization's operating environment. (To the database designer, the operating environment is both the data sources and the data users.) Naturally, an organization's operating environment is dependent on the organization's mission. For example, the operating environment of a university would be quite different from that of a steel manufacturer, an airline, or a nursing home. Yet, no matter how different the organizations may be, the *data analysis and requirements* component of the database design is enhanced when the data environment and data use are described accurately and precisely within a description of operations.

In a business environment, the main sources of information for the description of operations—and therefore of business rules—are company managers, policymakers, department managers, and written documentation such as company procedures, standards, and operations manuals. A faster and more direct source of business rules is direct interviews with end users. Unfortunately, because perceptions differ, the end user can be a less reliable source when it comes to specifying business rules. For example, a maintenance department mechanic might believe that any mechanic can initiate a maintenance procedure, when actually only mechanics with inspection authorization should perform such a task. This distinction might seem trivial, but it has major legal consequences. Although end users are crucial contributors to the development of business rules, it pays to verify end-user perceptions. Often, interviews with several people who perform the same job yield very different perceptions of their job components. While such a discovery might point to "management problems," that general diagnosis does not help the database designer. Given the discovery of such problems, the database designer's job is to reconcile the differences and verify the results of the reconciliation to ensure that the business rules are appropriate and accurate.

Knowing the business rules enables the designer to fully understand how the business works and what role the data plays within company operations. Consequently, the designer must identify the company's business rules and analyze their impact on the nature, role, and scope of data.

Business rules yield several important benefits in the design of new systems:

- They help standardize the company's view of data.

- They constitute a communications tool between users and designers.

- They allow the designer to understand the nature, role, and scope of the data.

- They allow the designer to understand business processes.

- They allow the designer to develop appropriate relationship participation rules and foreign key constraints. See Chapter 4, Entity Relationship (ER) Modeling.

description of operations
A document that provides a precise, detailed, up-to-date, and thoroughly reviewed description of the activities that define an organization's operating environment.

The last point is especially noteworthy: whether a given relationship is mandatory or optional is usually a function of the applicable business rule.

9-4b Entity Relationship Modeling and Normalization

Before creating the ER model, the designer must communicate and enforce appropriate standards to be used in the documentation of the design. The standards include the use of diagrams and symbols, documentation writing style, layout, and any other conventions to be followed during documentation. Designers often overlook this very important requirement, especially when they are working as members of a design team. Failure to standardize documentation often means a failure to communicate later, and communications failures often lead to poor design work. In contrast, well-defined and enforced standards make design work easier and promise (but do not guarantee) a smooth integration of all system components.

Because the business rules usually define the nature of the relationship(s) among the entities, the designer must incorporate them into the conceptual model. The process of defining business rules and developing the conceptual model using ER diagrams can be described using the steps shown in Table 9.3.[6]

TABLE 9.3

DEVELOPING THE CONCEPTUAL MODEL USING ER DIAGRAMS

STEP	ACTIVITY
1	Identify, analyze, and refine the business rules.
2	Identify the main entities, using the results of Step 1.
3	Define the relationships among the entities, using the results of Steps 1 and 2.
4	Define the attributes, primary keys, and foreign keys for each of the entities.
5	Normalize the entities. (Remember that entities are implemented as tables in an RDBMS.)
6	Complete the initial ER diagram.
7	Validate the ER model against the end users' information and processing requirements.
8	Modify the ER model, using the results of Step 7.

Some of the steps listed in Table 9.3 take place concurrently, and some, such as the normalization process, can generate a demand for additional entities and/or attributes, thereby causing the designer to revise the ER model. For example, while identifying two main entities, the designer might also identify the composite bridge entity that represents the many-to-many relationship between the two main entities.

To review, suppose that you are creating a conceptual model for the JollyGood Movie Rental Corporation, whose end users want to track customers' DVD movie kiosk rentals. The simple ER diagram presented in Figure 9.9 shows a composite entity that helps track customers and their video rentals. Business rules define the optional nature of the relationships between the entities VIDEO and CUSTOMER. For example, customers are not required to check out a video. A video need not be checked out in order to exist in the kiosk. A customer may rent many videos, and a video may be rented by many customers. In particular, note the composite RENTAL entity that connects the two main entities.

[6]See "Linking Rules to Models," Alice Sandifer and Barbara von Halle, *Database Programming and Design*, 4(3), March 1991, pp. 13–16. Although the source seems dated, it remains the current standard. The technology has changed substantially, but the process has not.

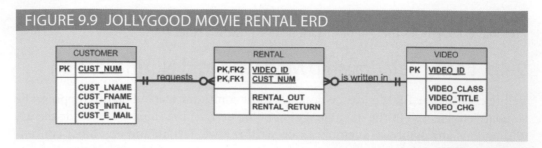

FIGURE 9.9 JOLLYGOOD MOVIE RENTAL ERD

As you will likely discover, the initial ER model may be subjected to several revisions before it meets the system's requirements. Such a revision process is quite natural. Remember that the ER model is a communications tool as well as a design blueprint. Therefore, when you meet with the proposed system users, the initial ER model should give rise to questions such as "Is this really what you meant?" For example, the ERD shown in Figure 9.9 is far from complete. Clearly, many more attributes must be defined and the dependencies must be checked before the design can be implemented. In addition, the design cannot yet support typical video rental transactions. For example, each video is likely to have many copies available for rental purposes. However, if the VIDEO entity shown in Figure 9.9 is used to store the titles as well as the copies, the design triggers the data redundancies shown in Table 9.4.

TABLE 9.4

DATA REDUNDANCIES IN THE VIDEO TABLE

VIDEO_ID	VIDEO_TITLE	VIDEO_COPY	VIDEO_CHG	VIDEO_DAYS
SF-12345FT-1	Adventures on Planet III	1	$1.09	1
SF-12345FT-2	Adventures on Planet III	2	$1.09	1
SF-12345FT-3	Adventures on Planet III	3	$1.09	1
WE-5432GR-1	TipToe Canoe and Tyler 2: A Journey	1	$1.09	2
WE-5432GR-2	TipToe Canoe and Tyler 2: A Journey	2	$1.09	2

The initial ERD shown in Figure 9.9 must be modified to reflect the answer to the question "Is more than one copy available for each title?" Also, payment transactions must be supported. (You will have an opportunity to modify this initial design in Problem 5 at the end of the chapter.)

From the preceding discussion, you might get the impression that ER modeling activities such as entity and attribute definition, normalization, and verification take place in a precise sequence. In fact, once you have completed the initial ER model, chances are that you will move back and forth among the activities until you are satisfied that the ER model accurately represents a database design that can meet the required system demands. The activities often take place in parallel, and the process is iterative. Figure 9.10 summarizes the ER modeling interactions. Figure 9.11 summarizes the array of design tools and information sources that the designer can use to produce the conceptual model.

All objects (entities, attributes, relations, views, and so on) are defined in a data dictionary, which is used in tandem with the normalization process to help eliminate data anomalies and redundancy problems. During this ER modeling process, the designer must:

- Define entities, attributes, primary keys, and foreign keys. (The foreign keys serve as the basis for the relationships among the entities.)

- Make decisions about adding new primary key attributes to satisfy end-user and processing requirements.

FIGURE 9.10   ER MODELING IS AN ITERATIVE PROCESS BASED ON MANY ACTIVITIES

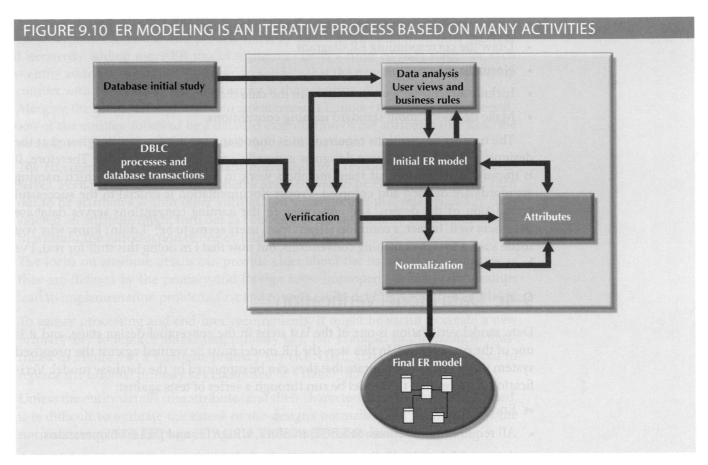

FIGURE 9.11   CONCEPTUAL DESIGN TOOLS AND INFORMATION SOURCES

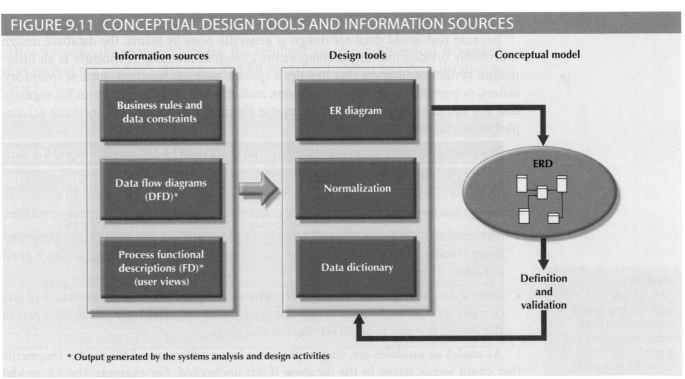

* Output generated by the systems analysis and design activities

- Make decisions about the treatment of composite and multivalued attributes.
- Make decisions about adding derived attributes to satisfy processing requirements.
- Make decisions about the placement of foreign keys in 1:1 relationships.

a primarily top-down approach may be easier. Most companies have standards for systems development and database design already in place.

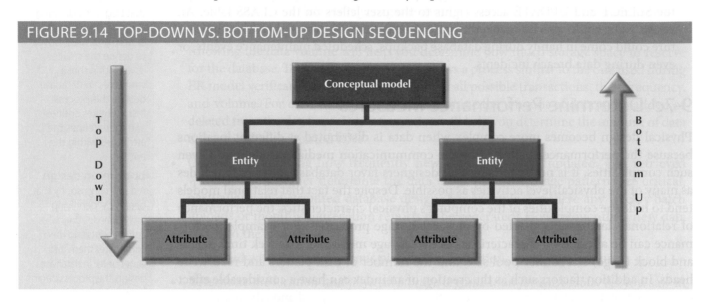

FIGURE 9.14 TOP-DOWN VS. BOTTOM-UP DESIGN SEQUENCING

Note

Even when a primarily top-down approach is selected, the normalization process that revises existing table structures is inevitably a bottom-up technique. ER models constitute a top-down process even when the selection of attributes and entities can be described as bottom-up. Because both the ER model and normalization techniques form the basis for most designs, the top-down versus bottom-up debate may be based on a theoretical distinction rather than an actual difference.

9-9 Centralized Versus Decentralized Design

centralized design
A process by which all database design decisions are carried out centrally by a small group of people. Suitable in a top-down design approach when the problem domain is relatively small, as in a single unit or department in an organization.

decentralized design
A process in which conceptual design models subsets of an organization's database requirements, which are then aggregated into a complete design. Such modular designs are typical of complex systems with a relatively large number of objects and procedures.

The two general approaches to database design (bottom-up and top-down) can be influenced by factors such as the scope and size of the system, the company's management style, and the company's structure (centralized or decentralized). Depending on these factors, the database design may be based on two very different design philosophies: centralized and decentralized.

Centralized design is productive when the data component has a relatively small number of objects and procedures. The design can be carried out and represented in a fairly simple database. Centralized design is typical of relatively simple, small databases and can be successfully done by a single database administrator or by a small, informal design team. The company operations and the scope of the problem are sufficiently limited to allow even a single designer to define the problem(s), create the conceptual design, verify the conceptual design with the user views, define system processes and data constraints to ensure the efficacy of the design, and ensure that the design will comply with all the requirements. (Although centralized design is typical for small companies, do not make the mistake of assuming that it is limited to small companies. Even large companies can operate within a relatively simple database environment.) Figure 9.15 summarizes the centralized design option. Note that a single conceptual design is completed and then validated in the centralized design approach.

Decentralized design might be used when the system's data component has a considerable number of entities and complex relations on which very complex operations are performed. Decentralized design is also often used when the problem itself is spread across several operational sites and each element is a subset of the entire data set. (See Figure 9.16.)

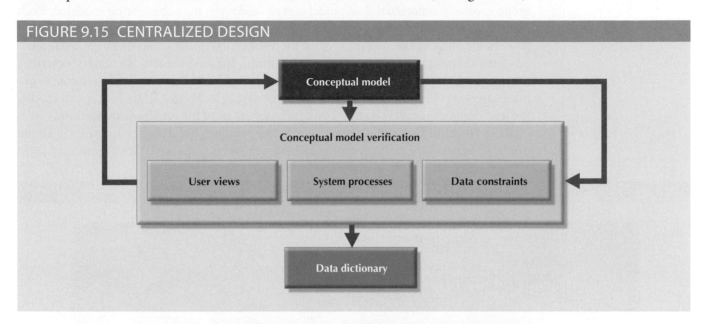

FIGURE 9.15   CENTRALIZED DESIGN

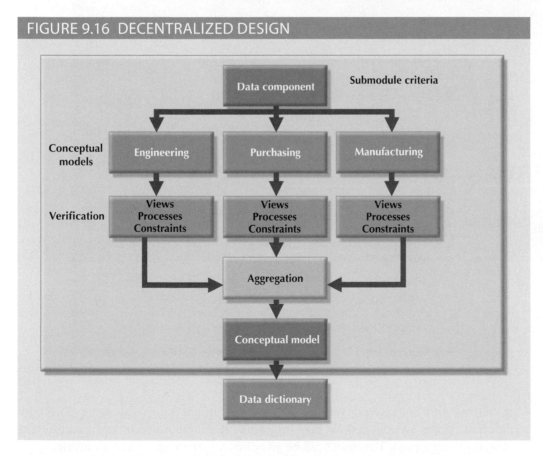

FIGURE 9.16   DECENTRALIZED DESIGN

In large and complex projects, the database typically cannot be designed by only one person. Instead, a carefully selected team of database designers tackles a complex database project. Within the decentralized design framework, the database design task is divided into several modules. Once the design criteria have been established, the lead designer assigns design subsets or modules to design groups within the team.

Because each design group focuses on modeling a subset of the system, the definition of boundaries and the interrelation among data subsets must be very precise. Each design group creates a conceptual data model corresponding to the subset being modeled. Each conceptual model is then verified individually against the user views, processes, and constraints for each of the modules. After the verification process has been completed, all modules are integrated into one conceptual model. Because the data dictionary describes the characteristics of all objects within the conceptual data model, it plays a vital role in the integration process. After the subsets have been aggregated into a larger conceptual model, the lead designer must verify that it still can support all of the required transactions.

Keep in mind that the aggregation process requires the designer to create a single model in which various aggregation problems must be addressed. (See Figure 9.17.)

FIGURE 9.17 SUMMARY OF AGGREGATION PROBLEMS

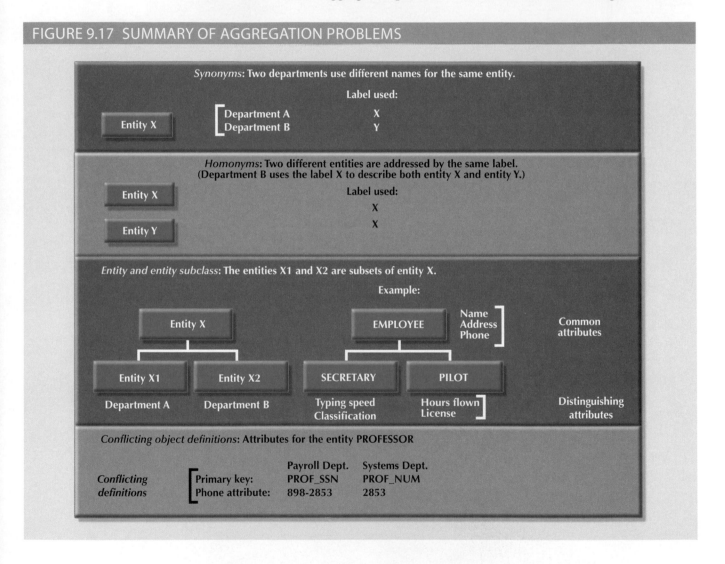

- *Synonyms and homonyms.* Various departments might know the same object by different names (synonyms), or they might use the same name to address different objects (homonyms). The object can be an entity, an attribute, or a relationship.

- *Entity and entity subtypes.* An entity subtype might be viewed as a separate entity by one or more departments. The designer must integrate such subtypes into a higher-level entity.

- *Conflicting object definitions.* Attributes can be recorded as different types (character, numeric), or different domains can be defined for the same attribute. Constraint definitions can vary as well. The designer must remove such conflicts from the model.

Summary

- An information system is designed to help transform data into information and to manage both data and information. Thus, the database is a very important part of the information system. Systems analysis is the process that establishes the need for an information system and its extent. Systems development is the process of creating an information system.

- The Systems Development Life Cycle (SDLC) traces the history of an application within the information system. The SDLC can be divided into five phases: planning, analysis, detailed systems design, implementation, and maintenance. The SDLC is an iterative process rather than a sequential process.

- The Database Life Cycle (DBLC) describes the history of the database within the information system. The DBLC is composed of six phases: database initial study, database design, implementation and loading, testing and evaluation, operation, and maintenance and evolution. Like the SDLC, the DBLC is iterative rather than sequential.

- The conceptual portion of the design may be subject to several variations based on two basic design philosophies: bottom-up versus top-down and centralized versus decentralized.

Key Terms

bottom-up design	Database Life Cycle (DBLC)	module coupling
boundaries	database role	physical design
centralized design	decentralized design	scope
clustered tables	description of operations	systems analysis
cohesivity	differential backup	systems development
computer-aided software engineering (CASE)	full backup	Systems Development Life Cycle (SDLC)
conceptual design	information system	top-down design
database development	logical design	transaction log backup
database fragment	minimal data rule	virtualization
	module	

Online Content

Flashcards and crossword puzzles for key term practice are available at *www.cengagebrain.com*.

Review Questions

1. What is an information system? What is its purpose?

2. How do systems analysis and systems development fit into a discussion about information systems?

3. What does the acronym SDLC mean, and what does an SDLC portray?

4. What does the acronym DBLC mean, and what does a DBLC portray?

5. Discuss the distinction between centralized and decentralized conceptual database design.

6. What is the minimal data rule in conceptual design? Why is it important?

7. Discuss the distinction between top-down and bottom-up approaches in database design.

8. What are business rules? Why are they important to a database designer?

9. What is the data dictionary's function in database design?

10. What steps are required in the development of an ER diagram? (*Hint*: See Table 9.3.)

11. List and briefly explain the activities involved in the verification of an ER model.

12. What factors are important in a DBMS software selection?

13. List and briefly explain the four steps performed during the logical design stage.

14. List and briefly explain the three steps performed during the physical design stage.

15. What three levels of backup may be used in database recovery management? Briefly describe what each backup level does.

Problems

1. The ABC Car Service & Repair Centers are owned by the Silent Car Dealership; ABC services and repairs only silent cars. Three ABC centers provide service and repair for the entire state.

   Each of the three centers is independently managed and operated by a shop manager, a receptionist, and at least eight mechanics. Each center maintains a fully stocked parts inventory.

   Each center also maintains a manual file system in which each car's maintenance history is kept: repairs made, parts used, costs, service dates, owner, and so on. Files are also kept to track inventory, purchasing, billing, employees' hours, and payroll.

   You have been contacted by one of the center's managers to design and implement a computerized database system. Given the preceding information, do the following:

   a. Indicate the most appropriate sequence of activities by labeling each of the following steps in the correct order. (For example, if you think that "Load the database" is the appropriate first step, label it "1.")

      _____ Normalize the conceptual model.

      _____ Obtain a general description of company operations.

      _____ Load the database.

      _____ Create a description of each system process.

      _____ Test the system.

      _____ Draw a data flow diagram and system flowcharts.

      _____ Create a conceptual model using ER diagrams.

      _____ Create the application programs.

      _____ Interview the mechanics.

      _____ Create the file (table) structures.

      _____ Interview the shop manager.

b. Describe the various modules that you believe the system should include.

c. How will a data dictionary help you develop the system? Give examples.

d. What general (system) recommendations might you make to the shop manager? For example, if the system will be integrated, what modules will be integrated? What benefits would be derived from such an integrated system? Include several general recommendations.

e. What is the best approach to conceptual database design? Why?

f. Name and describe at least four reports the system should have. Explain their use. Who will use those reports?

2. Suppose that you have been asked to create an information system for a manufacturing plant that produces nuts and bolts of many shapes, sizes, and functions. What questions would you ask, and how would the answers affect the database design?

a. What do you envision the SDLC to be?

b. What do you envision the DBLC to be?

3. Suppose that you perform the same functions noted in Problem 2 for a larger warehousing operation. How are the two sets of procedures similar? How and why are they different?

4. Using the same procedures and concepts employed in Problem 1, how would you create an information system for the Tiny College example in Chapter 4?

5. Write the proper sequence of activities for the design of a video rental database. (The initial ERD was shown in Figure 9.9.) The design must support all rental activities, customer payment tracking, and employee work schedules, as well as track which employees checked out the videos to the customers. After you finish writing the design activity sequence, complete the ERD to ensure that the database design can be successfully implemented. (Make sure that the design is normalized properly and that it can support the required transactions.)

6. In a construction company, a new system has been in place for a few months and now there is a list of possible changes/updates that need to be done. For each of the changes/updates, specify what type of maintenance needs to be done: (a) corrective, (b) adaptive, and (c) perfective.

a. An error in the size of one of the fields has been identified and it needs to be updated status field needs to be changed.

b. The company is expanding into a new type of service and this will require to enhancing the system with a new set of tables to support this new service and integrate it with the existing data.

c. The company has to comply with some government regulations. To do this, it will require adding a couple of fields to the existing system tables.

7. You have been assigned to design the database for a new soccer club. Indicate the most appropriate sequence of activities by labeling each of the following steps in the correct order. (For example, if you think that "Load the database" is the appropriate first step, label it "1.")

_____ Create the application programs.

_____ Create a description of each system process.

_____ Test the system.

_____ Load the database.

_____ Normalize the conceptual model.

_____ Interview the soccer club president.

_____ Create a conceptual model using ER diagrams.

_____ Interview the soccer club director of coaching.

_____ Create the file (table) structures.

_____ Obtain a general description of the soccer club operations.

_____ Draw a data flow diagram and system flowcharts.

PART 4

Advanced Database Concepts

Transaction Management and Concurrency Control

In this chapter, you will learn:

- About database transactions and their properties
- What concurrency control is and what role it plays in maintaining the database's integrity
- What locking methods are and how they work
- How stamping methods are used for concurrency control
- How optimistic methods are used for concurrency control
- How database recovery management is used to maintain database integrity

Preview

Database transactions reflect real-world transactions that are triggered by events such as buying a product, registering for a course, or making a deposit into a checking account. Transactions are likely to contain many parts, such as updating a customer's account, adjusting product inventory, and updating the seller's accounts receivable. All parts of a transaction must be successfully completed to prevent data integrity problems. Therefore, executing and managing transactions are important database system activities.

In this chapter you will learn about the main properties of database transactions (atomicity, consistency, isolation, and durability, plus serializability for concurrent transactions). After defining the transaction properties, the chapter shows how SQL can be used to represent transactions, and how transaction logs can ensure the DBMS's ability to recover transactions.

When many transactions take place at the same time, they are called *concurrent transactions*. Managing the execution of such transactions is called *concurrency control*. This chapter discusses some of the problems that can occur with concurrent transactions (lost updates, uncommitted data, and inconsistent retrievals) and the most common algorithms for concurrency control: locks, time stamping, and optimistic methods. Finally, you will see how database recovery management can ensure that a database's contents are restored to a valid consistent state in case of a hardware or software failure.

Data Files and Available Formats

	MS Access	Oracle	MS SQL	My SQL		MS Access	Oracle	MS SQL	My SQL
CH10_SaleCo	✓	✓	✓	✓	CH10_ABC_Markets	✓	✓	✓	✓

Data Files Available on cengagebrain.com

10-1 **What Is a Transaction?**

To illustrate what transactions are and how they work, use the Ch10_SaleCo database. The relational diagram for the database is shown in Figure 10.1.

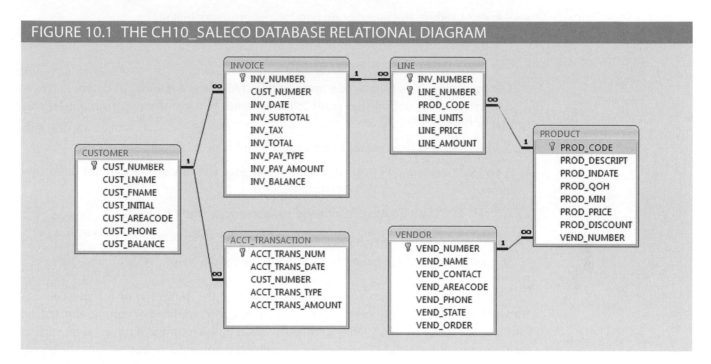

FIGURE 10.1 THE CH10_SALECO DATABASE RELATIONAL DIAGRAM

Note

Although SQL commands illustrate several transaction and concurrency control issues, you should be able to follow the discussions even if you have not studied Chapter 7, Introduction to Structured Query Language (SQL), and Chapter 8, Advanced SQL. If you don't know SQL, ignore the SQL commands and focus on the discussions. If you have a working knowledge of SQL, you can use the Ch10_SaleCo database to generate your own SELECT and UPDATE examples and to augment the material in Chapters 7 and 8 by writing your own triggers and stored procedures.

As you examine the relational diagram in Figure 10.1, note the following features:

- The design stores the customer balance (CUST_BALANCE) value in the CUSTOMER table to indicate the total amount owed by the customer. The CUST_BALANCE attribute is increased when the customer makes a purchase on credit, and it is decreased when the customer makes a payment. Including the current customer account balance in the CUSTOMER table makes it easy to write a query to determine the current balance for any customer and to generate important summaries such as total, average, minimum, and maximum balances.

- The ACCT_TRANSACTION table records all customer purchases and payments to track the details of customer account activity.

You could change the design of the Ch10_SaleCo database to reflect accounting practice more precisely, but the implementation provided here will enable you to track the transactions well enough to understand the chapter's discussions.

To understand the concept of a transaction, suppose that you sell a product to a customer. Furthermore, suppose that the customer may charge the purchase to his or her account. Given that scenario, your sales transaction consists of at least the following parts:

- You must write a new customer invoice.
- You must reduce the quantity on hand in the product's inventory.
- You must update the account transactions.
- You must update the customer balance.

The preceding sales transaction must be reflected in the database. In database terms, a transaction is any action that reads from or writes to a database. A transaction may consist of the following:

- A simple SELECT statement to generate a list of table contents.
- A series of related UPDATE statements to change the values of attributes in various tables.
- A series of INSERT statements to add rows to one or more tables.
- A combination of SELECT, UPDATE, and INSERT statements.

The sales transaction example includes a combination of INSERT and UPDATE statements.

Given the preceding discussion, you can augment the definition of a transaction. A **transaction** is a *logical* unit of work that must be entirely completed or entirely aborted; no intermediate states are acceptable. In other words, a multicomponent transaction, such as the previously mentioned sale, must not be partially completed. Updating only the inventory or only the accounts receivable is not acceptable. All of the SQL statements in the transaction must be completed successfully. If any of the SQL statements fail, the entire transaction is rolled back to the original database state that existed before the transaction started. A successful transaction changes the database from one consistent state to another. A **consistent database state** is one in which all data integrity constraints are satisfied.

To ensure consistency of the database, every transaction must begin with the database in a known consistent state. If the database is not in a consistent state, the transaction will yield an inconsistent database that violates its integrity and business rules. For that reason, subject to limitations discussed later, all transactions are controlled and executed by the DBMS to guarantee database integrity.

Most real-world database transactions are formed by two or more database requests. A **database request** is the equivalent of a single SQL statement in an application program or transaction. For example, if a transaction is composed of two UPDATE statements and one INSERT statement, the transaction uses three database requests. In turn, each database request generates several input/output (I/O) operations that read from or write to physical storage media.

10-1a Evaluating Transaction Results

Not all transactions update the database. Suppose that you want to examine the CUSTOMER table to determine the current balance for customer number 10016. Such a transaction can be completed by using the following SQL code:

```
SELECT          CUST_NUMBER, CUST_BALANCE
FROM            CUSTOMER
WHERE           CUST_NUMBER = 10016;
```

Although the query does not make any changes in the CUSTOMER table, the SQL code represents a transaction because it *accesses* the database. If the database existed in

transaction
A sequence of database requests that accesses the database. A transaction is a logical unit of work; that is, it must be *entirely* completed or aborted—no intermediate ending states are accepted. All transactions must have the properties of atomicity, consistency, isolation, and durability.

consistent database state
A database state in which all data integrity constraints are satisfied.

database request
The equivalent of a single SQL statement in an application program or a transaction.

a consistent state before the access, the database remains in a consistent state after the access because the transaction did not alter the database.

Remember that a transaction may consist of a single SQL statement or a collection of related SQL statements. Revisit the previous sales example to illustrate a more complex transaction, using the Ch10_SaleCo database. Suppose that on January 18, 2016, you register the credit sale of one unit of product 89-WRE-Q to customer 10016 for $277.55. The required transaction affects the INVOICE, LINE, PRODUCT, CUSTOMER, and ACCT_TRANSACTION tables. The SQL statements that represent this transaction are as follows:

```
INSERT INTO INVOICE
          VALUES (1009, 10016,'18-Jan-2016', 256.99, 20.56, 277.55, 'cred', 0.00, 277.55);
INSERT INTO LINE
          VALUES (1009, 1, '89-WRE-Q', 1, 256.99, 256.99);

UPDATE     PRODUCT
SET        PROD_QOH = PROD_QOH – 1
WHERE      PROD_CODE = '89-WRE-Q';

UPDATE     CUSTOMER
SET        CUST_BALANCE = CUST_BALANCE + 277.55
WHERE      CUST_NUMBER = 10016;

INSERT INTO ACCT_TRANSACTION
          VALUES (10007, '18-Jan-16', 10016, 'charge', 277.55);
COMMIT;
```

The results of the successfully completed transaction are shown in Figure 10.2. (All records involved in the transaction are outlined in red.)

FIGURE 10.2   TRACING THE TRANSACTION IN THE CH10_SALECO DATABASE

Database name: Ch10_SaleCo

Table name: INVOICE

INV_NUMBER	CUST_NUMBER	INV_DATE	INV_SUBTOTAL	INV_TAX	INV_TOTAL	INV_PAY_TYPE	INV_PAY_AMOUNT	INV_BALANCE
1001	10014	16-Jan-16	54.92	4.39	59.31	cc	59.31	0.00
1002	10011	16-Jan-16	9.98	0.80	10.78	cash	10.78	0.00
1003	10012	16-Jan-16	270.70	21.66	292.36	cc	292.36	0.00
1004	10011	17-Jan-16	34.87	2.79	37.66	cc	37.00	0.00
1005	10018	17-Jan-16	70.44	5.64	76.08	cc	76.08	0.00
1006	10014	17-Jan-16	397.83	31.83	429.66	cred	100.00	329.66
1007	10015	17-Jan-16	34.97	2.80	37.77	chk	37.77	0.00
1008	10011	17-Jan-16	1033.08	82.65	1115.73	cred	500.00	615.73
1009	10016	18-Jan-16	256.99	20.56	277.55	cred	0.00	277.55

Table name: LINE

INV_NUMBER	LINE_NUMBER	PROD_CODE	LINE_UNITS	LINE_PRICE	LINE_AMOUNT
1001	1	13-Q2/P2	3	14.99	44.97
1001	2	23109-HB	1	9.95	9.95
1002	1	54778-2T	2	4.99	9.90
1003	1	2238/QPD	4	38.95	155.80
1003	2	1546-QQ2	1	39.95	39.95
1003	3	13-Q2/P2	5	14.99	74.95
1004	1	54778-2T	3	4.99	14.97
1004	2	23109-HB	2	9.95	19.90
1005	1	PVC23DRT	12	5.87	70.44
1006	1	SM-18277	3	6.99	20.97
1006	2	2232/QTY	1	109.92	109.92
1006	3	23109-HB	1	9.95	9.95
1006	4	89-WRE-Q	1	256.99	256.99
1007	1	13-Q2/P2	2	14.99	29.98
1007	2	54778-2T	1	4.99	4.99
1008	1	PVC23DRT	5	5.87	29.35
1008	2	WR3/TT3	4	119.95	479.80
1008	3	23109-HB	1	9.95	9.95
1008	4	89-WRE-Q	2	256.99	513.98
1009	1	89-WRE-Q	1	256.99	256.99

Table name: PRODUCT

PROD_CODE	PROD_DESCRIPT	PROD_INDATE	PROD_QOH	PROD_MIN	PROD_PRICE	PROD_DISCOUNT	VEND_NUMBER
11QER/31	Power painter, 15 psi., 3-nozzle	03-Nov-15	8	5	109.99	0.00	25595
13-Q2/P2	7.25-in. pwr. saw blade	13-Dec-15	32	15	14.99	0.05	21344
14-Q1/L3	9.00-in. pwr. saw blade	13-Nov-15	18	12	17.49	0.00	21344
1546-QQ2	Hrd. cloth, 1/4-in., 2x50	15-Jan-16	15	8	39.95	0.00	23119
1558-QW1	Hrd. cloth, 1/2-in., 3x50	15-Jan-16	23	5	43.99	0.00	23119
2232/QTY	B&D jigsaw, 12-in. blade	30-Dec-15	8	5	109.92	0.05	24288
2232/QWE	B&D jigsaw, 8-in. blade	24-Dec-15	6	5	99.87	0.05	24288
2238/QPD	B&D cordless drill, 1/2-in.	20-Jan-16	12	5	38.95	0.05	25595
23109-HB	Claw hammer	20-Jan-16	23	10	9.95	0.10	21225
23114-AA	Sledge hammer, 12 lb.	02-Jan-16	8	5	14.40	0.05	
54778-2T	Rat-tail file, 1/8-in. fine	15-Dec-15	43	20	4.99	0.00	21344
89-WRE-Q	Hicut chain saw, 16 in.	07-Jan-16	11	5	256.99	0.05	24288
PVC23DRT	PVC pipe, 3.5-in., 8-ft	06-Jan-16	188	75	5.87	0.00	
SM-18277	1.25-in. metal screw, 25	01-Mar-16	172	75	6.99	0.00	21225
SW-23116	2.5-in. wd. screw, 50	24-Feb-16	237	100	8.45	0.00	21231
WR3/TT3	Steel matting, 4'x8'x1/6", .5" mesh	17-Jan-16	18	5	119.95	0.10	25595

Table name: CUSTOMER

CUST_NUMB	CUST_LNAME	CUST_FNAME	CUST_INITIAL	CUST_AREACODE	CUST_PHONE	CUST_BALANCE
10010	Ramas	Alfred	A	615	844-2573	0.00
10011	Dunne	Leona	K	713	894-1238	615.73
10012	Smith	Kathy	W	615	894-2285	0.00
10013	Olowski	Paul	F	615	894-2180	0.00
10014	Orlando	Myron		615	222-1672	0.00
10015	O'Brian	Amy	B	713	442-3381	0.00
10016	Brown	James	G	615	297-1228	277.55
10017	Williams	George		615	290-2556	0.00
10018	Farriss	Anne	G	713	382-7185	0.00
10019	Smith	Olotte	K	615	297-3809	0.00

Table name: ACCT_TRANSACTION

ACCT_TRANS_NUM	ACCT_TRANS_DATE	CUST_NUMBER	ACCT_TRANS_TYPE	ACCT_TRANS_AMOUNT
10003	17-Jan-16	10014	charge	329.66
10004	17-Jan-16	10011	charge	615.73
10006	29-Jan-16	10014	payment	329.66
10007	18-Jan-16	10016	charge	277.55

To better understand the transaction results, note the following:

- A new row 1009 was added to the INVOICE table. In this row, derived attribute values were stored for the invoice subtotal, the tax, the invoice total, and the invoice balance.

- The LINE row for invoice 1009 was added to reflect the purchase of one unit of product 89-WRE-Q with a price of $256.99. In this row, the derived attribute values for the line amount were stored.

- Product 89-WRE-Q's quantity on hand (PROD_QOH) in the PRODUCT table was reduced by one, from 12 to 11.

- The customer balance (CUST_BALANCE) for customer 10016 was updated by adding $277.55 to the existing balance (the initial value was $0.00).

- A new row was added to the ACCT_TRANSACTION table to reflect the new account transaction number 10007.

- The COMMIT statement was used to end a successful transaction. (See Section 10-1c.)

Now suppose that the DBMS completes the first three SQL statements. Furthermore, suppose that during the execution of the fourth statement (the UPDATE of the CUSTOMER table's CUST_BALANCE value for customer 10016), the computer system loses electrical power. If the computer does not have a backup power supply, the transaction cannot be completed. Therefore, the INVOICE and LINE rows were added, and the PRODUCT table was updated to represent the sale of product 89-WRE-Q, but customer 10016 was not charged, nor was the required record written in the ACCT_TRANSACTION table. The database is now in an inconsistent state, and it is not usable for subsequent transactions. Assuming that the DBMS supports transaction management, *the DBMS will roll back the database to a previous consistent state*.

Although the DBMS is designed to recover a database to a previous consistent

Note

By default, MS Access does not support transaction management as discussed here. More sophisticated DBMSs, such as Oracle, SQL Server, and DB2, support the transaction management components discussed in this chapter. MS Access supports transaction management though specialized application programing interfaces (API) such as the Workspace or the DBEngine objects of the Data Access Objects (DAO) database middleware (see Chapter 15, Database Connectivity and Web Technologies for more information.)

state when an interruption prevents the completion of a transaction, the transaction itself is defined by the end user or programmer and must be semantically correct. *The DBMS cannot guarantee that the semantic meaning of the transaction truly represents the real-world event*. For example, suppose that following the sale of 10 units of product 89-WRE-Q, the inventory UPDATE commands were written this way:

```
UPDATE      PRODUCT
SET         PROD_QOH = PROD_QOH + 10
WHERE       PROD_CODE = '89-WRE-Q';
```

The sale should have *decreased* the PROD_QOH value for product 89-WRE-Q by 10. Instead, the UPDATE *added* 10 to product 89-WRE-Q's PROD_QOH value.

Although the UPDATE command's syntax is correct, its use yields incorrect results, that is, a database inconsistent with the real-world event. Yet, the DBMS will execute the transaction anyway. The DBMS cannot evaluate whether the transaction represents the real-world event correctly; that is the end user's responsibility. End users and programmers are capable of introducing many errors in this fashion. Imagine the consequences of reducing the quantity on hand for product 1546-QQ2 instead of product 89-WRE-Q or of crediting the CUST_BALANCE value for customer 10012 rather than customer 10016.

Clearly, improper or incomplete transactions can have a devastating effect on database integrity. Some DBMSs—*especially* the relational variety—provide means by which the user can define enforceable constraints based on business rules. Other integrity rules, such as those governing referential and entity integrity, are enforced automatically by the DBMS when the table structures are properly defined, thereby letting the DBMS validate some transactions. For example, if a transaction inserts a new customer number into a customer table and the number already exists, the DBMS will end the transaction with an error code to indicate a violation of the primary key integrity rule.

10-1b Transaction Properties

Each individual transaction must display *atomicity*, *consistency*, *isolation*, and *durability*. These four properties are sometimes referred to as the ACID test. Let's look briefly at each of the properties.

- **Atomicity** requires that *all* operations (SQL requests) of a transaction be completed; if not, the transaction is aborted. If a transaction T1 has four SQL requests, all four requests must be successfully completed; otherwise, the entire transaction is aborted. In other words, a transaction is treated as a single, indivisible, logical unit of work.

- **Consistency** indicates the permanence of the database's consistent state. A transaction takes a database from one consistent state to another. When a transaction is completed, the database must be in a consistent state. If any of the transaction parts violates an integrity constraint, the entire transaction is aborted.

- **Isolation** means that the data used during the execution of a transaction cannot be used by a second transaction until the first one is completed. In other words, if transaction T1 is being executed and is using the data item X, that data item cannot be accessed by any other transaction (T2 … T*n*) until T1 ends. This property is particularly useful in multiuser database environments because several users can access and update the database at the same time.

- **Durability** ensures that once transaction changes are done and committed, they cannot be undone or lost, even in the event of a system failure.

In addition to the individual transaction properties indicated above, there is another important property that applies when executing multiple transactions concurrently. For example, let's assume that the DBMS has three transactions (T1, T2 and T3) executing at the same time. To properly carry out transactions, the DBMS must schedule the concurrent execution of the transaction's operations. In this case, each individual transaction must comply with the ACID properties and, at the same time, the schedule of such multiple transaction operations must exhibit the property of serializability. **Serializability** ensures that the schedule for the concurrent execution of the transactions yields consistent results. This property is important in multiuser and distributed databases in which multiple transactions are likely to be executed concurrently. Naturally, if only a single transaction is executed, serializability is not an issue.

atomicity
The transaction property that requires all parts of a transaction to be treated as a single, indivisible, logical unit of work. All parts of a transaction must be completed or the entire transaction is aborted.

consistency
A database condition in which all data integrity constraints are satisfied. To ensure consistency of a database, every transaction must begin with the database in a known consistent state. If not, the transaction will yield an inconsistent database that violates its integrity and business rules.

isolation
A database transaction property in which a data item used by one transaction is not available to other transactions until the first one ends.

durability
The transaction property that ensures that once transaction changes are done and committed, they cannot be undone or lost, even in the event of a system failure.

serializability
A property in which the selected order of concurrent transaction operations creates the same final database state that would have been produced if the transactions had been executed in a serial fashion.

A single-user database system automatically ensures serializability and isolation of the database because only one transaction is executed at a time. The atomicity, consistency, and durability of transactions must be guaranteed by single-user DBMSs. (Even a single-user DBMS must manage recovery from errors created by OS-induced interruptions, power interruptions, and abnormal application terminations or crashes.)

Multiuser databases are typically subject to multiple concurrent transactions. Therefore, the multiuser DBMS must implement controls to ensure serializability and isolation of transactions—in addition to atomicity and durability—to guard the database's consistency and integrity. For example, if several concurrent transactions are executed over the same data set and the second transaction updates the database before the first transaction is finished, the isolation property is violated and the database is no longer consistent. The DBMS must manage the transactions by using concurrency control techniques to avoid undesirable situations.

10-1c Transaction Management with SQL

The American National Standards Institute (ANSI) has defined standards that govern SQL database transactions. Transaction support is provided by two SQL statements: COMMIT and ROLLBACK. The ANSI standards require that when a transaction sequence is initiated by a user or an application program, the sequence must continue through all succeeding SQL statements until one of the following four events occurs:

- A COMMIT statement is reached, in which case all changes are permanently recorded within the database. The COMMIT statement automatically ends the SQL transaction.

- A ROLLBACK statement is reached, in which case all changes are aborted and the database is rolled back to its previous consistent state.

- The end of a program is successfully reached, in which case all changes are permanently recorded within the database. This action is equivalent to COMMIT.

- The program is abnormally terminated, in which case the database changes are aborted and the database is rolled back to its previous consistent state. This action is equivalent to ROLLBACK.

The use of COMMIT is illustrated in the following simplified sales example, which updates a product's quantity on hand (PROD_QOH) and the customer's balance when the customer buys two units of product 1558-QW1 priced at $43.99 per unit (for a total of $87.98) and charges the purchase to the customer's account:

```
UPDATE      PRODUCT
SET         PROD_QOH = PROD_QOH – 2
WHERE       PROD_CODE = '1558-QW1';
UPDATE      CUSTOMER
SET         CUST_BALANCE = CUST_BALANCE + 87.98
WHERE       CUST_NUMBER = '10011';
COMMIT;
```

(Note that the example is simplified to make it easy to trace the transaction. In the Ch10_SaleCo database, the transaction would involve several additional table updates.)

The COMMIT statement used in the preceding example is not necessary if the UPDATE statement is the application's last action and the application terminates normally. However, good programming practice dictates that you include the COMMIT statement at the end of a transaction declaration.

A transaction begins implicitly when the first SQL statement is encountered. Not all SQL implementations follow the ANSI standard; some (such as SQL Server) use transaction management statements such as the following to indicate the beginning of a new transaction:

BEGIN TRANSACTION;

Other SQL implementations allow you to assign characteristics for the transactions as parameters to the BEGIN statement. For example, the Oracle RDBMS uses the SET TRANSACTION statement to declare the start of a new transaction and its properties.

10-1d The Transaction Log

A DBMS uses a **transaction log** to keep track of all transactions that update the database. The DBMS uses the information stored in this log for a recovery requirement triggered by a ROLLBACK statement, a program's abnormal termination, or a system failure such as a network discrepancy or a disk crash. Some RDBMSs use the transaction log to recover a database *forward* to a currently consistent state. After a server failure, for example, Oracle automatically rolls back uncommitted transactions and rolls forward transactions that were committed but not yet written to the physical database. This behavior is required for transactional correctness and is typical of any transactional DBMS.

While the DBMS executes transactions that modify the database, it also automatically updates the transaction log. The transaction log stores the following:

- A record for the beginning of the transaction.

- For each transaction component (SQL statement):

  - The type of operation being performed (INSERT, UPDATE, DELETE).

  - The names of the objects affected by the transaction (the name of the table).

  - The "before" and "after" values for the fields being updated.

  - Pointers to the previous and next transaction log entries for the same transaction.

- The ending (COMMIT) of the transaction.

Although using a transaction log increases the processing overhead of a DBMS, the ability to restore a corrupted database is worth the price.

Table 10.1 illustrates a simplified transaction log that reflects a basic transaction composed of two SQL UPDATE statements. If a system failure occurs, the DBMS will examine the transaction log for all uncommitted or incomplete transactions and restore (ROLLBACK) the database to its previous state on the basis of that information. When the recovery process is completed, the DBMS will write in the log all committed transactions that were not physically written to the database before the failure occurred.

If a ROLLBACK is issued before the termination of a transaction, the DBMS will restore the database only for that particular transaction, rather than for all of them, to maintain the *durability* of the previous transactions. In other words, committed transactions are not rolled back.

The transaction log is a critical part of the database, and it is usually implemented as one or more files that are managed separately from the actual database files. The transaction log is subject to common dangers such as disk-full conditions and disk crashes. Because the transaction log contains some of the most critical data in a DBMS, some implementations support logs on several different disks to reduce the consequences of a system failure.

transaction log
A feature used by the DBMS to keep track of all transaction operations that update the database. The information stored in this log is used by the DBMS for recovery purposes.

TABLE 10.1

A TRANSACTION LOG

TRL_ID	TRX_NUM	PREV PTR	NEXT PTR	OPERATION	TABLE	ROW ID	ATTRIBUTE	BEFORE VALUE	AFTER VALUE
341	101	Null	352	START	****Start Transaction				
352	101	341	363	UPDATE	PRODUCT	1558-QW1	PROD_QOH	25	23
363	101	352	365	UPDATE	CUSTOMER	10011	CUST_BALANCE	525.75	615.73
365	101	363	Null	COMMIT	**** End of Transaction				

TRL_ID = Transaction log record ID
TRX_NUM = Transaction number
PTR = Pointer to a transaction log record ID
(*Note*: The transaction number is automatically assigned by the DBMS.)

10-2 Concurrency Control

Coordinating the simultaneous execution of transactions in a multiuser database system is known as **concurrency control**. The objective of concurrency control is to ensure the serializability of transactions in a multiuser database environment. To achieve this goal, most concurrency control techniques are oriented toward preserving the isolation property of concurrently executing transactions. Concurrency control is important because the simultaneous execution of transactions over a shared database can create several data integrity and consistency problems. The three main problems are lost updates, uncommitted data, and inconsistent retrievals.

10-2a Lost Updates

concurrency control
A DBMS feature that coordinates the simultaneous execution of transactions in a multiprocessing database system while preserving data integrity.

lost update
A concurrency control problem in which a data update is lost during the concurrent execution of transactions.

The **lost update** problem occurs when two concurrent transactions, T1 and T2, are updating the same data element and one of the updates is lost (overwritten by the other transaction). To see an illustration of lost updates, examine a simple PRODUCT table. One of the table's attributes is a product's quantity on hand (PROD_QOH). Assume that you have a product whose current PROD_QOH value is 35. Also assume that two concurrent transactions, T1 and T2, occur and update the PROD_QOH value for some item in the PRODUCT table. The transactions are shown in Table 10.2.

TABLE 10.2

TWO CONCURRENT TRANSACTIONS TO UPDATE QOH

TRANSACTION	COMPUTATION
T1: Purchase 100 units	PROD_QOH = PROD_QOH + 100
T2: Sell 30 units	PROD_QOH = PROD_QOH − 30

Table 10.3 shows the serial execution of the transactions under normal circumstances, yielding the correct answer PROD_QOH = 105.

TABLE 10.3

SERIAL EXECUTION OF TWO TRANSACTIONS

TIME	TRANSACTION	STEP	STORED VALUE
1	T1	Read PROD_QOH	35
2	T1	PROD_QOH = 35 + 100	
3	T1	Write PROD_QOH	135
4	T2	Read PROD_QOH	135
5	T2	PROD_QOH = 135 − 30	
6	T2	Write PROD_QOH	105

However, suppose that a transaction can read a product's PROD_QOH value from the table *before* a previous transaction has been committed, using the *same* product. The sequence depicted in Table 10.4 shows how the lost update problem can arise. Note that the first transaction (T1) has not yet been committed when the second transaction (T2) is executed. Therefore, T2 still operates on the value 35, and its subtraction yields 5 in memory. In the meantime, T1 writes the value 135 to disk, which is promptly overwritten by T2. In short, the addition of 100 units is "lost" during the process.

TABLE 10.4

LOST UPDATES

TIME	TRANSACTION	STEP	STORED VALUE
1	T1	Read PROD_QOH	35
2	T2	Read PROD_QOH	35
3	T1	PROD_QOH = 35 + 100	
4	T2	PROD_QOH = 35 − 30	
5	T1	Write PROD_QOH (lost update)	135
6	T2	Write PROD_QOH	5

10-2b Uncommitted Data

The phenomenon of **uncommitted data** occurs when two transactions, T1 and T2, are executed concurrently and the first transaction (T1) is rolled back after the second transaction (T2) has already accessed the uncommitted data—thus violating the isolation property of transactions. To illustrate that possibility, use the same transactions described during the lost updates discussion. T1 has two atomic parts, one of which is the update of the inventory; the other possible part is the update of the invoice total (not shown). T1 is forced to roll back due to an error during the updating of the invoice's total; it rolls back all the way, undoing the inventory update as well. This time the T1 transaction is rolled back to eliminate the addition of the 100 units. (See Table 10.5.) Because T2 subtracts 30 from the original 35 units, the correct answer should be 5.

> **uncommitted data**
> A concurrency control problem in which a transaction accesses uncommitted data from another transaction.

TABLE 10.5

TRANSACTIONS CREATING AN UNCOMMITTED DATA PROBLEM

TRANSACTION	COMPUTATION
T1: Purchase 100 units	PROD_QOH = PROD_QOH + 100 (Rolled back)
T2: Sell 30 units	PROD_QOH = PROD_QOH − 30

Table 10.6 shows how the serial execution of these transactions yields the correct answer under normal circumstances.

TABLE 10.6

CORRECT EXECUTION OF TWO TRANSACTIONS

TIME	TRANSACTION	STEP	STORED VALUE
1	T1	Read PROD_QOH	35
2	T1	PROD_QOH = 35 + 100	
3	T1	Write PROD_QOH	135
4	T1	*****ROLLBACK *****	35
5	T2	Read PROD_QOH	35
6	T2	PROD_QOH = 35 − 30	
7	T2	Write PROD_QOH	5

Table 10.7 shows how the uncommitted data problem can arise when the ROLLBACK is completed after T2 has begun its execution.

TABLE 10.7

AN UNCOMMITTED DATA PROBLEM

TIME	TRANSACTION	STEP	STORED VALUE
1	T1	Read PROD_QOH	35
2	T1	PROD_QOH = 35 + 100	
3	T1	Write PROD_QOH	135
4	T2	Read PROD_QOH (Read uncommitted data)	135
5	T2	PROD_QOH = 135 − 30	
6	T1	***** ROLLBACK *****	35
7	T2	Write PROD_QOH	105

10-2c Inconsistent Retrievals

Inconsistent retrievals occur when a transaction accesses data before and after one or more other transactions finish working with such data. For example, an inconsistent retrieval would occur if transaction T1 calculated some summary (aggregate) function over a set of data while another transaction (T2) was updating the same data. The problem is that the transaction might read some data before it is changed and other data *after* it is changed, thereby yielding inconsistent results.

To illustrate the problem, assume the following conditions:

1. T1 calculates the total quantity on hand of the products stored in the PRODUCT table.

2. At the same time, T2 updates the quantity on hand (PROD_QOH) for two of the PRODUCT table's products.

The two transactions are shown in Table 10.8.

inconsistent retrievals
A concurrency control problem that arises when a transaction-calculating summary (aggregate) functions over a set of data while other transactions are updating the data, yielding erroneous results.

TABLE 10.8

RETRIEVAL DURING UPDATE

TRANSACTION T1	TRANSACTION T2
SELECT SUM(PROD_QOH) FROM PRODUCT	UPDATE PRODUCT SET PROD_QOH = PROD_QOH + 10 WHERE PROD_CODE = 1546-QQ2
	UPDATE PRODUCT SET PROD_QOH = PROD_QOH − 10 WHERE PROD_CODE = 1558-QW1
	COMMIT;

While T1 calculates the total quantity on hand (PROD_QOH) for all items, T2 represents the correction of a typing error: the user added 10 units to product 1558-QW1's PROD_QOH but meant to add the 10 units to product 1546-QQ2's PROD_QOH. To correct the problem, the user adds 10 to product 1546-QQ2's PROD_QOH and subtracts 10 from product 1558-QW1's PROD_QOH. (See the two UPDATE statements in Table 10.8.) The initial and final PROD_QOH values are reflected in Table 10.9. (Only a few PROD_CODE values are shown for the PRODUCT table. To illustrate the point, the sum for the PROD_QOH values is shown for these few products.)

TABLE 10.9

TRANSACTION RESULTS: DATA ENTRY CORRECTION

PROD_CODE	BEFORE PROD_QOH	AFTER PROD_QOH
11QER/31	8	8
13-Q2/P2	32	32
1546-QQ2	15	(15 + 10) ⟶ 25
1558-QW1	23	(23 − 10) ⟶ 13
2232-QTY	8	8
2232-QWE	6	6
Total	**92**	**92**

Although the final results shown in Table 10.9 are correct after the adjustment, Table 10.10 demonstrates that inconsistent retrievals are possible during the transaction execution, making the result of T1's execution incorrect. The "After" summation shown in Table 10.10 reflects that the value of 25 for product 1546-QQ2 was read *after* the WRITE statement was completed. Therefore, the "After" total is 40 + 25 = 65. The "Before" total reflects that the value of 23 for product 1558-QW1 was read *before* the next WRITE statement was completed to reflect the corrected update of 13. Therefore, the "Before" total is 65 + 23 = 88.

The computed answer of 102 is obviously wrong because you know from Table 10.9 that the correct answer is 92. Unless the DBMS exercises concurrency control, a multiuser database environment can create havoc within the information system.

10-2d The Scheduler

You now know that severe problems can arise when two or more concurrent transactions are executed. You also know that a database transaction involves a series of database I/O operations that take the database from one consistent state to another. Finally, you know

TABLE 10.10

INCONSISTENT RETRIEVALS

TIME	TRANSACTION	ACTION	VALUE	TOTAL
1	T1	Read PROD_QOH for PROD_CODE = '11QER/31'	8	8
2	T1	Read PROD_QOH for PROD_CODE = '13-Q2/P2'	32	40
3	T2	Read PROD_QOH for PROD_CODE = '1546-QQ2'	15	
4	T2	PROD_QOH = 15 + 10		
5	T2	Write PROD_QOH for PROD_CODE = '1546-QQ2'	25	
6	T1	Read PROD_QOH for PROD_CODE = '1546-QQ2'	25	(After) 65
7	T1	Read PROD_QOH for PROD_CODE = '1558-QW1'	23	(Before) 88
8	T2	Read PROD_QOH for PROD_CODE = '1558-QW1'	23	
9	T2	PROD_QOH = 23 − 10		
10	T2	Write PROD_QOH for PROD_CODE = '1558-QW1'	13	
11	T2	***** COMMIT *****		
12	T1	Read PROD_QOH for PROD_CODE = '2232-QTY'	8	96
13	T1	Read PROD_QOH for PROD_CODE = '2232-QWE'	6	102

that database consistency can be ensured only before and after the execution of transactions. A database always moves through an unavoidable temporary state of inconsistency during a transaction's execution if such a transaction updates multiple tables and rows. (If the transaction contains only one update, then there is no temporary inconsistency.) The temporary inconsistency exists because a computer executes the operations serially, one after another. During this serial process, the isolation property of transactions prevents them from accessing the data not yet released by other transactions. This consideration is even more important today, with the use of multicore processors that can execute several instructions at the same time. What would happen if two transactions executed concurrently and they were accessing the same data?

In previous examples, the operations within a transaction were executed in an arbitrary order. As long as two transactions, T1 and T2, access *unrelated* data, there is no conflict and the order of execution is irrelevant to the final outcome. However, if the transactions operate on related data or the same data, conflict is possible among the transaction components and the selection of one execution order over another might have some undesirable consequences. So, how is the correct order determined, and who determines that order? Fortunately, the DBMS handles that tricky assignment by using a built-in scheduler.

The **scheduler** is a special DBMS process that establishes the order in which the operations are executed within concurrent transactions. The scheduler *interleaves* the execution of database operations to ensure serializability and isolation of transactions. To determine the appropriate order, the scheduler bases its actions on concurrency control algorithms, such as locking or time stamping methods, which are explained in the next sections. However, it is important to understand that not all transactions are serializable. The DBMS determines what transactions are serializable and proceeds to interleave the execution of the transaction's operations. Generally, transactions that are not serializable are executed on a first-come, first-served basis by the DBMS. The scheduler's main job is to create a **serializable schedule** of a transaction's operations, in which the interleaved execution of the transactions (T1, T2, T3, etc.) yields the same results as if the transactions were executed in serial order (one after another).

scheduler
The DBMS component that establishes the order in which concurrent transaction operations are executed. The scheduler *interleaves* the execution of database operations in a specific sequence to ensure *serializability*.

serializable schedule
In transaction management, a schedule of operations in which the interleaved execution of the transactions yields the same result as if they were executed in serial order.

The scheduler also makes sure that the computer's central processing unit (CPU) and storage systems are used efficiently. If there were no way to schedule the execution of transactions, all of them would be executed on a first-come, first-served basis. The problem with that approach is that processing time is wasted when the CPU waits for a READ or WRITE operation to finish, thereby losing several CPU cycles. In short, first-come, first-served scheduling tends to yield unacceptable response times within the multiuser DBMS environment. Therefore, some other scheduling method is needed to improve the efficiency of the overall system.

Additionally, the scheduler facilitates data isolation to ensure that two transactions do not update the same data element at the same time. Database operations might require READ and/or WRITE actions that produce conflicts. For example, Table 10.11 shows the possible conflict scenarios when two transactions, T1 and T2, are executed concurrently over the same data. Note that in Table 10.11, two operations are in conflict when they access the same data and at least one of them is a WRITE operation.

TABLE 10.11

READ/WRITE CONFLICT SCENARIOS: CONFLICTING DATABASE OPERATIONS MATRIX

	TRANSACTIONS		
	T1	T2	RESULT
Operations	Read	Read	No conflict
	Read	Write	Conflict
	Write	Read	Conflict
	Write	Write	Conflict

Several methods have been proposed to schedule the execution of conflicting operations in concurrent transactions. These methods are classified as locking, time stamping, and optimistic. Locking methods, discussed next, are used most frequently.

10-3 Concurrency Control with Locking Methods

Locking methods are one of the most common techniques used in concurrency control because they facilitate the isolation of data items used in concurrently executing transactions. A **lock** guarantees exclusive use of a data item to a current transaction. In other words, transaction T2 does not have access to a data item that is currently being used by transaction T1. A transaction acquires a lock prior to data access; the lock is released (unlocked) when the transaction is completed so that another transaction can lock the data item for its exclusive use. This series of locking actions assumes that concurrent transactions might attempt to manipulate the same data item at the same time. The use of locks based on the assumption that conflict between transactions is likely is usually referred to as **pessimistic locking**.

Recall from Sections 10-1a and 10-1b that data consistency cannot be guaranteed *during* a transaction; the database might be in a temporary inconsistent state when several updates are executed. Therefore, locks are required to prevent another transaction from reading inconsistent data.

Most multiuser DBMSs automatically initiate and enforce locking procedures. All lock information is handled by a **lock manager**, which is responsible for assigning and policing the locks used by the transactions.

lock
A device that guarantees unique use of a data item in a particular transaction operation. A transaction requires a lock prior to data access; the lock is released after the operation's execution to enable other transactions to lock the data item for their own use.

pessimistic locking
The use of locks based on the assumption that conflict between transactions is likely.

lock manager
A DBMS component that is responsible for assigning and releasing locks.

10-3a Lock Granularity

lock granularity
The level of lock use. Locking can take place at the following levels: database, table, page, row, and field (attribute).

Lock granularity indicates the level of lock use. Locking can take place at the following levels: database, table, page, row, or even field (attribute).

Database Level In a **database-level lock**, the entire database is locked, thus preventing the use of any tables in the database by transaction T2 while transaction T1 is being executed. This level of locking is good for batch processes, but it is unsuitable for multiuser DBMSs. You can imagine how s-l-o-w data access would be if thousands of transactions had to wait for the previous transaction to be completed before the next one could reserve the entire database. Figure 10.3 illustrates the database-level lock; because of it, transactions T1 and T2 cannot access the same database concurrently *even when they use different tables.*

database-level lock
A type of lock that restricts database access to the owner of the lock and allows only one user at a time to access the database. This lock works for batch processes but is unsuitable for online multiuser DBMSs.

FIGURE 10.3 DATABASE-LEVEL LOCKING SEQUENCE

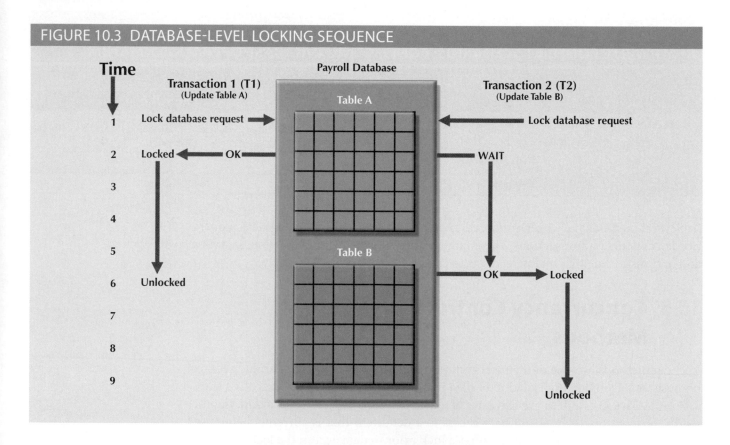

Table Level In a **table-level lock**, the entire table is locked, preventing access to any row by transaction T2 while transaction T1 is using the table. If a transaction requires access to several tables, each table may be locked. However, two transactions can access the same database as long as they access different tables.

Table-level locks, while less restrictive than database-level locks, cause traffic jams when many transactions are waiting to access the same table. Such a condition is especially irksome if the lock forces a delay when different transactions require access to different parts of the same table—that is, when the transactions would not interfere with each other. Consequently, table-level locks are not suitable for multiuser DBMSs. Figure 10.4 illustrates the effect of a table-level lock. Note that transactions T1 and T2 cannot access the same table even when they try to use different rows; T2 must wait until T1 unlocks the table.

table-level lock
A locking scheme that allows only one transaction at a time to access a table. A table-level lock locks an entire table, preventing access to any row by transaction T2 while transaction T1 is using the table.

FIGURE 10.4 AN EXAMPLE OF A TABLE-LEVEL LOCK

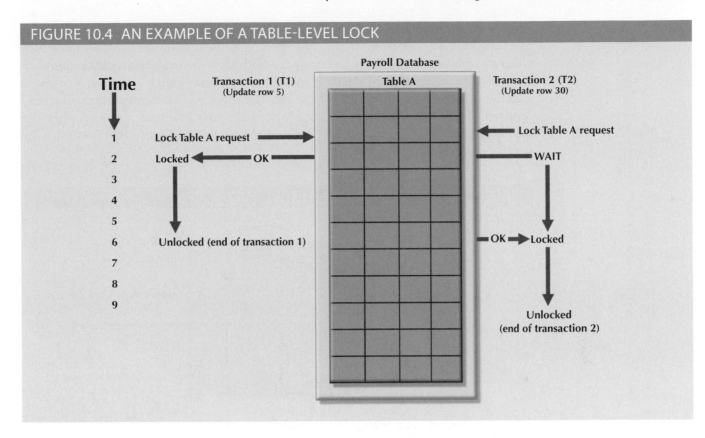

Page Level In a **page-level lock**, the DBMS locks an entire diskpage. A **diskpage**, or **page**, is the equivalent of a *diskblock*, which can be described as a directly addressable section of a disk. A page has a fixed size, such as 4K, 8K, or 16K. For example, if you want to write only 73 bytes to a 4K page, the entire 4K page must be read from disk, updated in memory, and written back to disk. A table can span several pages, and a page can contain several rows of one or more tables. Page-level locks are currently the most frequently used locking method for multiuser DBMSs. An example of a page-level lock is shown in Figure 10.5. Note that T1 and T2 access the same table while locking different diskpages. If T2 requires the use of a row located on a page that is locked by T1, T2 must wait until T1 unlocks the page.

FIGURE 10.5 AN EXAMPLE OF A PAGE-LEVEL LOCK

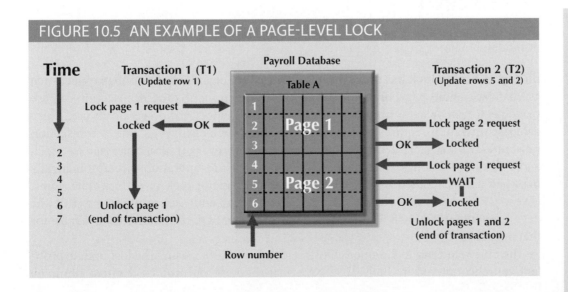

page-level lock
In this type of lock, the database management system locks an entire diskpage, or section of a disk. A diskpage can contain data for one or more rows and from one or more tables.

diskpage (page)
In permanent storage, the equivalent of a disk block, which can be described as a directly addressable section of a disk. A diskpage has a fixed size, such as 4K, 8K, or 16K.

Row Level A **row-level lock** is much less restrictive than the locks discussed earlier. The DBMS allows concurrent transactions to access different rows of the same table even when the rows are located on the same page. Although the row-level locking approach improves the availability of data, its management requires high overhead because a lock exists for each row in a table of the database involved in a conflicting transaction. Modern DBMSs automatically escalate a lock from a row level to a page level when the application session requests multiple locks on the same page. Figure 10.6 illustrates the use of a row-level lock.

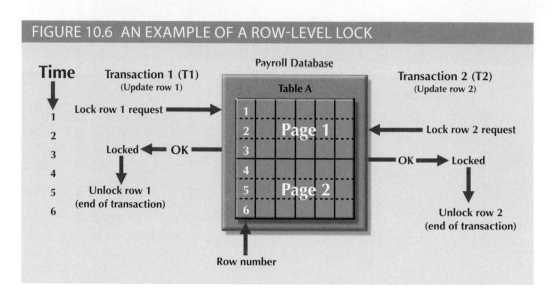

FIGURE 10.6 AN EXAMPLE OF A ROW-LEVEL LOCK

Note in Figure 10.6 that both transactions can execute concurrently, even when the requested rows are on the same page. T2 must wait only if it requests the same row as T1.

Field Level The **field-level lock** allows concurrent transactions to access the same row as long as they require the use of different fields (attributes) within that row. Although field-level locking clearly yields the most flexible multiuser data access, it is rarely implemented in a DBMS because it requires an extremely high level of computer overhead and because the row-level lock is much more useful in practice.

10-3b Lock Types

Regardless of the level of granularity of the lock, the DBMS may use different lock types or modes: binary or shared/exclusive.

Binary A **binary lock** has only two states: locked (1) or unlocked (0). If an object such as a database, table, page, or row is locked by a transaction, no other transaction can use that object. If an object is unlocked, any transaction can lock the object for its use. Every database operation requires that the affected object be locked. As a rule, a transaction must unlock the object after its termination. Therefore, every transaction requires a lock and unlock operation for each accessed data item. Such operations are automatically managed and scheduled by the DBMS; the user does not lock or unlock data items. (Every DBMS has a default-locking mechanism. If the end user wants to override the default settings, the LOCK TABLE command and other SQL commands are available for that purpose.)

The binary locking technique is illustrated in Table 10.12, using the lost update problem you encountered in Table 10.4. Note that the lock and unlock features eliminate

row-level lock
A less restrictive database lock in which the DBMS allows concurrent transactions to access different rows of the same table, even when the rows are on the same page.

field-level lock
A lock that allows concurrent transactions to access the same row as long as they require the use of different fields (attributes) within that row. This type of lock yields the most flexible multiuser data access but requires a high level of computer overhead.

binary lock
A lock that has only two states: *locked* (1) and *unlocked* (0). If a data item is locked by a transaction, no other transaction can use that data item.

TABLE 10.12

AN EXAMPLE OF A BINARY LOCK

TIME	TRANSACTION	STEP	STORED VALUE
1	T1	Lock PRODUCT	
2	T1	Read PROD_QOH	15
3	T1	PROD_QOH = 15 + 10	
4	T1	Write PROD_QOH	25
5	T1	Unlock PRODUCT	
6	T2	Lock PRODUCT	
7	T2	Read PROD_QOH	23
8	T2	PROD_QOH = 23 − 10	
9	T2	Write PROD_QOH	13
10	T2	Unlock PRODUCT	

the lost update problem because the lock is not released until the WRITE statement is completed. Therefore, a PROD_QOH value cannot be used until it has been properly updated. However, binary locks are now considered too restrictive to yield optimal concurrency conditions. For example, the DBMS will not allow two transactions to read the same database object even though neither transaction updates the database and therefore no concurrency problems can occur. Remember from Table 10.11 that concurrency conflicts occur only when two transactions execute concurrently and one of them updates the database.

Shared/Exclusive An **exclusive lock** exists when access is reserved specifically for the transaction that locked the object. The exclusive lock must be used when the potential for conflict exists (see Table 10.11). A **shared lock** exists when concurrent transactions are granted read access on the basis of a common lock. A shared lock produces no conflict as long as all the concurrent transactions are read-only.

A shared lock is issued when a transaction wants to read data from the database and no exclusive lock is held on that data item. An exclusive lock is issued when a transaction wants to update (write) a data item and no locks are currently held on that data item by any other transaction. Using the shared/exclusive locking concept, a lock can have three states: unlocked, shared (read), and exclusive (write).

As shown in Table 10.11, two transactions conflict only when at least one is a write transaction. Because the two read transactions can be safely executed at once, shared locks allow several read transactions to read the same data item concurrently. For example, if transaction T1 has a shared lock on data item X and transaction T2 wants to read data item X, T2 may also obtain a shared lock on data item X.

If transaction T2 updates data item X, an exclusive lock is required by T2 over data item X. *The exclusive lock is granted if and only if no other locks are held on the data item* (this condition is known as the **mutual exclusive rule**: only one transaction at a time can own an exclusive lock on an object.) Therefore, if a shared (or exclusive) lock is already held on data item X by transaction T1, an exclusive lock cannot be granted to transaction T2, and T2 must wait to begin until T1 commits. In other words, a shared lock will always block an exclusive (write) lock; hence, decreasing transaction concurrency.

exclusive lock
An exclusive lock is issued when a transaction requests permission to update a data item and no locks are held on that data item by any other transaction. An exclusive lock does not allow other transactions to access the database.

shared lock
A lock that is issued when a transaction requests permission to read data from a database and no exclusive locks are held on the data by another transaction. A shared lock allows other read-only transactions to access the database.

mutual exclusive rule
A condition in which only one transaction at a time can own an exclusive lock on the same object.

Although the use of shared locks renders data access more efficient, a shared/exclusive lock schema increases the lock manager's overhead for several reasons:

- The type of lock held must be known before a lock can be granted.
- Three lock operations exist: READ_LOCK to check the type of lock, WRITE_LOCK to issue the lock, and UNLOCK to release the lock.
- The schema has been enhanced to allow a lock upgrade from shared to exclusive and a lock downgrade from exclusive to shared.

Although locks prevent serious data inconsistencies, they can lead to two major problems:

- The resulting transaction schedule might not be serializable.
- The schedule might create deadlocks. A **deadlock** occurs when two transactions wait indefinitely for each other to unlock data. A database deadlock, which is similar to traffic gridlock in a big city, is caused when two or more transactions wait for each other to unlock data.

Fortunately, both problems can be managed: serializability is attained through a locking protocol known as two-phase locking, and deadlocks can be managed by using deadlock detection and prevention techniques. Those techniques are examined in the next two sections.

10-3c Two-Phase Locking to Ensure Serializability

Two-phase locking (2PL) defines how transactions acquire and relinquish locks. Two-phase locking guarantees serializability, but it does not prevent deadlocks. The two phases are:

1. A growing phase, in which a transaction acquires all required locks without unlocking any data. Once all locks have been acquired, the transaction is in its locked point.

2. A shrinking phase, in which a transaction releases all locks and cannot obtain a new lock.

The two-phase locking protocol is governed by the following rules:

- Two transactions cannot have conflicting locks.
- No unlock operation can precede a lock operation in the same transaction.
- No data is affected until all locks are obtained—that is, until the transaction is in its locked point.

Figure 10.7 depicts the two-phase locking protocol.

In this example, the transaction first acquires the two locks it needs. When it has the two locks, it reaches its locked point. Next, the data is modified to conform to the transaction's requirements. Finally, the transaction is completed as it releases all of the locks it acquired in the first phase. Two-phase locking increases the transaction processing cost and might cause additional undesirable effects, such as deadlocks.

10-3d Deadlocks

A deadlock occurs when two transactions wait indefinitely for each other to unlock data. For example, a deadlock occurs when two transactions, T1 and T2, exist in the following mode:

T1 = access data items X and Y

T2 = access data items Y and X

If T1 has not unlocked data item Y, T2 cannot begin; if T2 has not unlocked data item X, T1 cannot continue. Consequently, T1 and T2 each wait for the other to unlock the

deadlock
A condition in which two or more transactions wait indefinitely for the other to release the lock on a previously locked data item. Also called *deadly embrace*.

two-phase locking (2PL)
A set of rules that governs how transactions acquire and relinquish locks. Two-phase locking guarantees serializability, but it does not prevent deadlocks. The two-phase locking protocol is divided into two phases: (1) A *growing phase* occurs when the transaction acquires the locks it needs without unlocking any *existing* data locks. Once all locks have been acquired, the transaction is in its *locked* point. (2) A *shrinking phase* occurs when the transaction releases all locks and cannot obtain a new lock.

FIGURE 10.7 TWO-PHASE LOCKING PROTOCOL

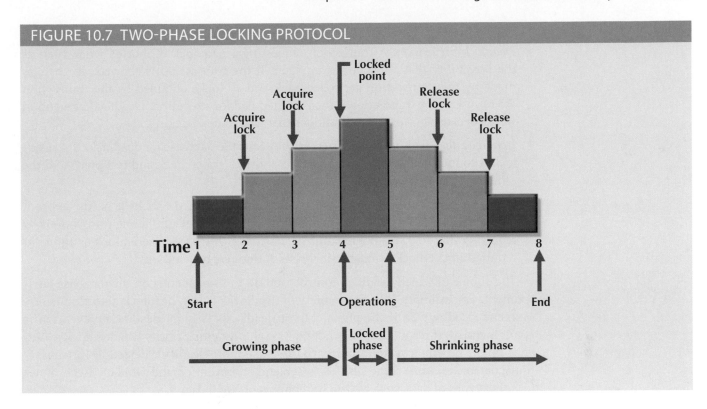

required data item. Such a deadlock is also known as a **deadly embrace**. Table 10.13 demonstrates how a deadlock condition is created.

The preceding example used only two concurrent transactions to demonstrate a deadlock condition. In a real-world DBMS, many more transactions can be executed simultaneously, thereby increasing the probability of generating deadlocks. Note that deadlocks are possible only when one of the transactions wants to obtain an exclusive lock on a data item; no deadlock condition can exist among *shared* locks.

deadly embrace
See deadlock.

TABLE 10.13

HOW A DEADLOCK CONDITION IS CREATED

TIME	TRANSACTION	REPLY	LOCK STATUS	
			DATA X	DATA Y
0			Unlocked	Unlocked
1	T1:LOCK(X)	OK	Locked	Unlocked
2	T2:LOCK(Y)	OK	Locked	Locked
3	T1:LOCK(Y)	WAIT	Locked	Locked
4	T2:LOCK(X)	WAIT	Locked	Locked
5	T1:LOCK(Y)	WAIT	Locked	Locked
6	T2:LOCK(X)	WAIT	Locked	Locked
7	T1:LOCK(Y)	WAIT	Locked	Locked
8	T2:LOCK(X)	WAIT	Locked	Locked
9	T1:LOCK(Y)	WAIT	Locked	Locked
...	....	....	....	....
...	....	....	....	....
...	....	....	....	....
...	....	....	....	....

Deadlock

The three basic techniques to control deadlocks are:

- *Deadlock prevention.* A transaction requesting a new lock is aborted when there is the possibility that a deadlock can occur. If the transaction is aborted, all changes made by this transaction are rolled back and all locks obtained by the transaction are released. The transaction is then rescheduled for execution. Deadlock prevention works because it avoids the conditions that lead to deadlocking.

- *Deadlock detection.* The DBMS periodically tests the database for deadlocks. If a deadlock is found, the "victim" transaction is aborted (rolled back and restarted) and the other transaction continues.

- *Deadlock avoidance.* The transaction must obtain all of the locks it needs before it can be executed. This technique avoids the rolling back of conflicting transactions by requiring that locks be obtained in succession. However, the serial lock assignment required in deadlock avoidance increases action response times.

The choice of which deadlock control method to use depends on the database environment. For example, if the probability of deadlocks is low, deadlock detection is recommended. However, if the probability of deadlocks is high, deadlock prevention is recommended. If response time is not high on the system's priority list, deadlock avoidance might be employed. All current DBMSs support deadlock detection in transactional databases, while some DBMSs use a blend of prevention and avoidance techniques for other types of data, such as data warehouses or XML data.

10-4 Concurrency Control with Time Stamping Methods

The **time stamping** approach to scheduling concurrent transactions assigns a global, unique time stamp to each transaction. The time stamp value produces an explicit order in which transactions are submitted to the DBMS. Time stamps must have two properties: uniqueness and monotonicity. **Uniqueness** ensures that no equal time stamp values can exist, and **monotonicity**[1] ensures that time stamp values always increase.

All database operations (read and write) within the same transaction must have the same time stamp. The DBMS executes conflicting operations in time stamp order, thereby ensuring serializability of the transactions. If two transactions conflict, one is stopped, rolled back, rescheduled, and assigned a new time stamp value.

The disadvantage of the time stamping approach is that each value stored in the database requires two additional time stamp fields: one for the last time the field was read and one for the last update. Time stamping thus increases memory needs and the database's processing overhead. Time stamping demands a lot of system resources because many transactions might have to be stopped, rescheduled, and restamped.

10-4a   Wait/Die and Wound/Wait Schemes

Time stamping methods are used to manage concurrent transaction execution. In this section, you will learn about two schemes used to decide which transaction is rolled back and which continues executing: the wait/die scheme and the wound/wait scheme.[2] An

---

time stamping
In transaction management, a technique used in scheduling concurrent transactions that assigns a global unique time stamp to each transaction.

uniqueness
In concurrency control, a property of time stamping that ensures no equal time stamp values can exist.

monotonicity
A quality that ensures that time stamp values always increase. (The time stamping approach to scheduling concurrent transactions assigns a global, unique time stamp to each transaction. The time stamp value produces an explicit order in which transactions are submitted to the DBMS.)

---

[1]The term *monotonicity* is part of the standard concurrency control vocabulary. The authors' first introduction to this term and its proper use was in an article written by W. H. Kohler, "A survey of techniques for synchronization and recovery in decentralized computer systems," *Computer Surveys* 3(2), June 1981, pp. 149–283.

[2]The procedure was first described by R. E. Stearnes and P. M. Lewis II in "System-level concurrency control for distributed database systems," *ACM Transactions on Database Systems*, No. 2, June 1978, pp. 178–198.

example illustrates the difference. Assume that you have two conflicting transactions: T1 and T2, each with a unique time stamp. Suppose that T1 has a time stamp of 11548789 and T2 has a time stamp of 19562545. You can deduce from the time stamps that T1 is the older transaction (the lower time stamp value), and T2 is the newer transaction. Given that scenario, the four possible outcomes are shown in Table 10.14.

TABLE 10.14

WAIT/DIE AND WOUND/WAIT CONCURRENCY CONTROL SCHEMES

TRANSACTION REQUESTING LOCK	TRANSACTION OWNING LOCK	WAIT/DIE SCHEME	WOUND/WAIT SCHEME
T1 (11548789)	T2 (19562545)	• T1 waits until T2 is completed and T2 releases its locks.	• T1 preempts (rolls back) T2. • T2 is rescheduled using the same time stamp.
T2 (19562545)	T1 (11548789)	• T2 dies (rolls back). • T2 is rescheduled using the same time stamp.	• T2 waits until T1 is completed and T1 releases its locks.

Using the wait/die scheme:

- If the transaction requesting the lock is the older of the two transactions, it will *wait* until the other transaction is completed and the locks are released.
- If the transaction requesting the lock is the younger of the two transactions, it will *die* (roll back) and is rescheduled using the same time stamp.

In short, in the **wait/die** scheme, the older transaction waits for the younger one to complete and release its locks.

In the wound/wait scheme:

- If the transaction requesting the lock is the older of the two transactions, it will preempt (*wound*) the younger transaction by rolling it back. T1 preempts T2 when T1 rolls back T2. The younger, preempted transaction is rescheduled using the same time stamp.
- If the transaction requesting the lock is the younger of the two transactions, it will wait until the other transaction is completed and the locks are released.

In short, in the **wound/wait** scheme, the older transaction rolls back the younger transaction and reschedules it.

In both schemes, one of the transactions waits for the other transaction to finish and release the locks. However, in many cases, a transaction requests multiple locks. How long does a transaction have to wait for each lock request? Obviously, that scenario can cause some transactions to wait indefinitely, causing a deadlock. To prevent a deadlock, each lock request has an associated time-out value. If the lock is not granted before the time-out expires, the transaction is rolled back.

10-5 Concurrency Control with Optimistic Methods

The **optimistic approach** is based on the assumption that the majority of database operations do not conflict. The optimistic approach requires neither locking nor time stamping techniques. Instead, a transaction is executed without restrictions until it is

wait/die
A concurrency control scheme in which an older transaction must wait for the younger transaction to complete and release the locks before requesting the locks itself. Otherwise, the newer transaction dies and is rescheduled.

wound/wait
A concurrency control scheme in which an older transaction can request the lock, preempt the younger transaction, and reschedule it. Otherwise, the newer transaction waits until the older transaction finishes.

optimistic approach
In transaction management, a concurrency control technique based on the assumption that most database operations do not conflict.

committed. Using an optimistic approach, each transaction moves through two or three phases, referred to as *read*, *validation*, and *write*.[3]

- During the *read phase*, the transaction reads the database, executes the needed computations, and makes the updates to a private copy of the database values. All update operations of the transaction are recorded in a temporary update file, which is not accessed by the remaining transactions.

- During the *validation phase*, the transaction is validated to ensure that the changes made will not affect the integrity and consistency of the database. If the validation test is positive, the transaction goes to the write phase. If the validation test is negative, the transaction is restarted and the changes are discarded.

- During the *write phase*, the changes are permanently applied to the database.

The optimistic approach is acceptable for most read or query database systems that require few update transactions. In a heavily used DBMS environment, the management of deadlocks—their prevention and detection—constitutes an important DBMS function. The DBMS will use one or more of the techniques discussed here, as well as variations on those techniques. To further understand how transaction management is implemented in a database, it is important that you learn about the transaction isolation levels as defined in ANSI SQL 1992 standard.

10-6 ANSI Levels of Transaction Isolation

The ANSI SQL standard (1992) defines transaction management based on transaction isolation levels. Transaction isolation levels refer to the degree to which transaction data is "protected or isolated" from other concurrent transactions. The isolation levels are described based on what data other transactions can see (read) during execution. More precisely, the transaction isolation levels are described by the type of "reads" that a transaction allows or not. The types of read operations are:

- **Dirty read**: a transaction can read data that is not yet committed.

- **Nonrepeatable read**: a transaction reads a given row at time t1, and then it reads the same row at time t2, yielding different results. The original row may have been updated or deleted.

- **Phantom read**: a transaction executes a query at time t1, and then it runs the same query at time t2, yielding additional rows that satisfy the query.

Based on the above operations, ANSI defined four levels of transaction isolation: Read Uncommitted, Read Committed, Repeatable Read, and Serializable. Table 10.15 shows the four ANSI transaction isolation levels. The table also shows an additional level of isolation provided by Oracle and MS SQL Server databases.

Read Uncommitted will read uncommitted data from other transactions. At this isolation level, the database does not place any locks on the data, which increases transaction performance but at the cost of data consistency. **Read Committed** forces transactions to read only committed data. This is the default mode of operation for most databases (including Oracle and SQL Server). At this level, the database will use exclusive locks on data, causing other transactions to wait until the original transaction commits. The **Repeatable Read** isolation level ensures that queries return consistent results. This type of isolation level uses shared locks to ensure other transactions do not update a row after

---

[3]The optimistic approach to concurrency control is described in an article by H. T. King and J. T. Robinson, "Optimistic methods for concurrency control," *ACM Transactions on Database Systems* 6(2), June 1981, pp. 213–226. Even the most current software is built on conceptual standards that were developed more than two decades ago.

---

dirty read
In transaction management, when a transaction reads data that is not yet committed.

nonrepeatable read
In transaction management, when a transaction reads a given row at time t1, then reads the same row at time t2, yielding different results because the original row may have been updated or deleted.

phantom read
In transaction management, when a transaction executes a query at time t1, then runs the same query at time t2, yielding additional rows that satisfy the query.

Read Uncommitted
An ANSI SQL transaction isolation level that allows transactions to read uncommitted data from other transactions, and which allows nonrepeatable reads and phantom reads. The least restrictive level defined by ANSI SQL.

Read Committed
An ANSI SQL transaction isolation level that allows transactions to read only committed data. This is the default mode of operations for most databases.

Repeatable Read
An ANSI SQL transaction isolation level that uses shared locks to ensure that other transactions do not update a row after the original query updates it. However, phantom reads are allowed.

TABLE 10.15

TRANSACTION ISOLATION LEVELS

	ISOLATION LEVEL	ALLOWED			COMMENT
		DIRTY READ	NONREPEATABLE READ	PHANTOM READ	
Less restrictive	Read Uncommitted	Y	Y	Y	The transaction reads uncommitted data, allows nonrepeatable reads, and phantom reads.
	Read Committed	N	Y	Y	Does not allow uncommitted data reads but allows nonrepeatable reads and phantom reads.
	Repeatable Read	N	N	Y	Only allows phantom reads.
More restrictive	Serializable	N	N	N	Does not allow dirty reads, nonrepeatable reads, or phantom reads.
Oracle / SQL Server Only	Read Only / Snapshot	N	N	N	Supported by Oracle and SQL Server. The transaction can only see the changes that were committed at the time the transaction started.

the original query reads it. However, new rows are read (phantom read) as these rows did not exist when the first query ran. The **Serializable** isolation level is the most restrictive level defined by the ANSI SQL standard. However, it is important to note that even with a Serializable isolation level, deadlocks are always possible. Most databases use a deadlock detection approach to transaction management, and, therefore, they will detect "deadlocks" during the transaction validation phase and reschedule the transaction.

The reason for the different levels of isolation is to increase transaction concurrency. The isolation levels go from the least restrictive (Read Uncommitted) to the more restrictive (Serializable). The higher the isolation level the more locks (shared and exclusive) are required to improve data consistency, at the expense of transaction concurrency performance. The isolation level of a transaction is defined in the transaction statement, for example using general ANSI SQL syntax:

```
BEGIN TRANSACTION ISOLATION LEVEL READ COMMITTED
… SQL STATEMENTS….
COMMIT TRANSACTION;
```

Oracle and MS SQL Server use the SET TRANSACTION ISOLATION LEVEL statement to define the level of isolation. SQL Server supports all four ANSI isolation levels. Oracle by default provides consistent statement-level reads to ensure Read Committed and Repeatable Read transactions. MySQL uses START TRANSACTION WITH CONSISTENT SNAPSHOT to provide transactions with consistent reads; that is, the transaction can only see the committed data at the time the transaction started.

As you can see from the previous discussion, transaction management is a complex subject and databases make use of various techniques to manage the concurrent execution of transactions. However, it may be necessary sometimes to employ database recovery techniques to restore the database to a consistent state.

Serializable
An ANSI SQL transaction isolation level that does not allow dirty reads, nonrepeatable reads, or phantom reads; the most restrictive level defined by the ANSI SQL standard.

10-7 Database Recovery Management

Database recovery restores a database from a given state (usually inconsistent) to a previously consistent state. Recovery techniques are based on the **atomic transaction property**: all portions of the transaction must be treated as a single, logical unit of work in which all operations are applied and completed to produce a consistent database. If a transaction operation cannot be completed for some reason, the transaction must be aborted and any changes to the database must be rolled back (undone). In short, transaction recovery reverses all of the changes that the transaction made to the database before the transaction was aborted.

Although this chapter has emphasized the recovery of *transactions*, recovery techniques also apply to the *database* and to the *system* after some type of critical error has occurred. Critical events can cause a database to stop working and compromise the integrity of the data. Examples of critical events are:

- *Hardware/software failures*. A failure of this type could be a hard disk media failure, a bad capacitor on a motherboard, or a failing memory bank. Other causes of errors under this category include application program or operating system errors that cause data to be overwritten, deleted, or lost. Some database administrators argue that this is one of the most common sources of database problems.

- *Human-caused incidents*. This type of event can be categorized as unintentional or intentional.

  – An unintentional failure is caused by a careless end user. Such errors include deleting the wrong rows from a table, pressing the wrong key on the keyboard, or shutting down the main database server by accident.

  – Intentional events are of a more severe nature and normally indicate that the company data is at serious risk. Under this category are security threats caused by hackers trying to gain unauthorized access to data resources and virus attacks caused by disgruntled employees trying to compromise the database operation and damage the company.

- *Natural disasters*. This category includes fires, earthquakes, floods, and power failures.

Whatever the cause, a critical error can render the database into an inconsistent state. The following section introduces the various techniques used to recover the database from an inconsistent state to a consistent state.

10-7a Transaction Recovery

In Section 10-1d, you learned about the transaction log and how it contains data for database recovery purposes. Database transaction recovery uses data in the transaction log to recover a database from an inconsistent state to a consistent state.

Before continuing, examine four important concepts that affect the recovery process:

- The **write-ahead-log protocol** ensures that transaction logs are always written *before* any database data is actually updated. This protocol ensures that, in case of a failure, the database can later be recovered to a consistent state using the data in the transaction log.

- **Redundant transaction logs** (several copies of the transaction log) ensure that a physical disk failure will not impair the DBMS's ability to recover data.

- Database **buffers** are temporary storage areas in primary memory used to speed up disk operations. To improve processing time, the DBMS software reads the data from the physical disk and stores a copy of it on a "buffer" in primary memory. When a transaction updates data, it actually updates the copy of the data in the buffer because that process is much faster than accessing the physical disk every time. Later, all buffers that contain updated data are written to a physical disk during a single operation, thereby saving significant processing time.

---

database recovery
The process of restoring a database to a previous consistent state.

atomic transaction property
A property that requires all parts of a transaction to be treated as a single, logical unit of work in which all operations must be completed (committed) to produce a consistent database.

write-ahead-log protocol
In concurrency control, a process that ensures transaction logs are written to permanent storage before any database data is actually updated. Also called a write-ahead protocol.

redundant transaction logs
Multiple copies of the transaction log kept by database management systems to ensure that the physical failure of a disk will not impair the DBMS's ability to recover data.

buffer
Temporary storage area in primary memory used to speed up disk operations.

- Database **checkpoints** are operations in which the DBMS writes all of its updated buffers in memory (also known as *dirty buffers*) to disk. While this is happening, the DBMS does not execute any other requests. A checkpoint operation is also registered in the transaction log. As a result of this operation, the physical database and the transaction log will be in sync. This synchronization is required because update operations update the copy of the data in the buffers and not in the physical database. Checkpoints are automatically and periodically executed by the DBMS according to certain operational parameters (such a high watermark for the transaction log size or volume of outstanding transactions) but can also be executed explicitly (as part of a database transaction statement) or implicitly (as part of a database backup operation). Of course, checkpoints that are too frequent would affect transaction performance; checkpoints that are too infrequent would affect database recovery performance. In any case, checkpoints serve a very practical function. As you will see next, checkpoints also play an important role in transaction recovery.

The database recovery process involves bringing the database to a consistent state after a failure. Transaction recovery procedures generally make use of deferred-write and write-through techniques.

When the recovery procedure uses a **deferred-write technique** (also called a **deferred update**), the transaction operations do not immediately update the physical database. Instead, only the transaction log is updated. The database is physically updated only with data from committed transactions, using information from the transaction log. If the transaction aborts before it reaches its commit point, no changes (no ROLLBACK or undo) need to be made to the database because it was never updated. The recovery process for all started and committed transactions (before the failure) follows these steps:

1. Identify the last checkpoint in the transaction log. This is the last time transaction data was physically saved to disk.

2. For a transaction that started and was committed before the last checkpoint, nothing needs to be done because the data is already saved.

3. For a transaction that performed a commit operation after the last checkpoint, the DBMS uses the transaction log records to redo the transaction and update the database, using the "after" values in the transaction log. The changes are made in ascending order, from oldest to newest.

4. For any transaction that had a ROLLBACK operation after the last checkpoint or that was left active (with neither a COMMIT nor a ROLLBACK) before the failure occurred, nothing needs to be done because the database was never updated.

When the recovery procedure uses a **write-through technique** (also called an **immediate update**), the database is immediately updated by transaction operations during the transaction's execution, even before the transaction reaches its commit point. If the transaction aborts before it reaches its commit point, a ROLLBACK or undo operation needs to be done to restore the database to a consistent state. In that case, the ROLLBACK operation will use the transaction log "before" values. The recovery process follows these steps:

1. Identify the last checkpoint in the transaction log. This is the last time transaction data was physically saved to disk.

2. For a transaction that started and was committed before the last checkpoint, nothing needs to be done because the data is already saved.

3. For a transaction that was committed after the last checkpoint, the DBMS re-does the transaction, using the "after" values of the transaction log. Changes are applied in ascending order, from oldest to newest.

checkpoint
In transaction management, an operation in which the database management system writes all of its updated buffers to disk.

deferred write technique
See *deferred update*.

deferred update
In transaction management, a condition in which transaction operations do not immediately update a physical database. Also called *deferred write technique*.

write-through technique
In concurrency control, a process that ensures a database is immediately updated by operations during the transaction's execution, even before the transaction reaches its commit point. Also called *immediate update*.

immediate update
See *write-through technique*.

4. For any transaction that had a ROLLBACK operation after the last checkpoint or that was left active (with neither a COMMIT nor a ROLLBACK) before the failure occurred, the DBMS uses the transaction log records to ROLLBACK or undo the operations, using the "before" values in the transaction log. Changes are applied in reverse order, from newest to oldest.

Use the transaction log in Table 10.16 to trace a simple database recovery process. To make sure you understand the recovery process, the simple transaction log includes three transactions and one checkpoint. This transaction log includes the transaction components used earlier in the chapter, so you should already be familiar with the basic process. Given the transaction, the transaction log has the following characteristics:

- Transaction 101 consists of two UPDATE statements that reduce the quantity on hand for product 54778-2T and increase the customer balance for customer 10011 for a credit sale of two units of product 54778-2T.

- Transaction 106 is the same credit sales event you saw in Section 10-1a. This transaction represents the credit sale of one unit of product 89-WRE-Q to customer 10016 for $277.55. This transaction consists of five SQL DML statements: three INSERT statements and two UPDATE statements.

- Transaction 155 represents a simple inventory update. This transaction consists of one UPDATE statement that increases the quantity on hand of product 2232/QWE from 6 units to 26 units.

- A database checkpoint writes all updated database buffers to disk. The checkpoint event writes only the changes for all previously committed transactions. In this case, the checkpoint applies all changes made by transaction 101 to the database data files.

Using Table 10.16, you can now trace the database recovery process for a DBMS using the deferred update method as follows:

1. Identify the last checkpoint—in this case, TRL ID 423. This was the last time database buffers were physically written to disk.

2. Note that transaction 101 started and finished before the last checkpoint. Therefore, all changes were already written to disk, and no additional action needs to be taken.

3. For each transaction committed after the last checkpoint (TRL ID 423), the DBMS will use the transaction log data to write the changes to disk, using the "after" values. For example, for transaction 106:

   a. Find COMMIT (TRL ID 457).

   b. Use the previous pointer values to locate the start of the transaction (TRL ID 397).

   c. Use the next pointer values to locate each DML statement, and apply the changes to disk using the "after" values. (Start with TRL ID 405, then 415, 419, 427, and 431.) Remember that TRL ID 457 was the COMMIT statement for this transaction.

   d. Repeat the process for transaction 155.

4. Any other transactions will be ignored. Therefore, for transactions that ended with ROLLBACK or that were left active (those that do not end with a COMMIT or ROLLBACK), nothing is done because no changes were written to disk.

TABLE 10.16

A TRANSACTION LOG FOR TRANSACTION RECOVERY EXAMPLES

TRL ID	TRX NUM	PREV PTR	NEXT PTR	OPERATION	TABLE	ROW ID	ATTRIBUTE	BEFORE VALUE	AFTER VALUE
341	101	Null	352	START	****Start Transaction				
352	101	341	363	UPDATE	PRODUCT	54778-2T	PROD_QOH	45	43
363	101	352	365	UPDATE	CUSTOMER	10011	CUST_BALANCE	615.73	675.62
365	101	363	Null	COMMIT	**** End of Transaction				
397	106	Null	405	START	****Start Transaction				
405	106	397	415	INSERT	INVOICE	1009			1009,10016, …
415	106	405	419	INSERT	LINE	1009,1			1009,1, 89-WRE-Q,1, …
419	106	415	427	UPDATE	PRODUCT	89-WRE-Q	PROD_QOH	12	11
423				CHECKPOINT					
427	106	419	431	UPDATE	CUSTOMER	10016	CUST_BALANCE	0.00	277.55
431	106	427	457	INSERT	ACCT_TRANSACTION	10007			1007,18-JAN-2016, …
457	106	431	Null	COMMIT	**** End of Transaction				
521	155	Null	525	START	****Start Transaction				
525	155	521	528	UPDATE	PRODUCT	2232/QWE	PROD_QOH	6	26
528	155	525	Null	COMMIT	**** End of Transaction				

***** C *R*A* S* H *****

Summary

- A transaction is a sequence of database operations that access the database. A transaction is a logical unit of work; that is, all parts are executed or the transaction is aborted. A transaction takes a database from one consistent state to another. A consistent database state is one in which all data integrity constraints are satisfied.

- Transactions have four main properties: atomicity, consistency, isolation, and durability. Atomicity means that all parts of the transaction must be executed; otherwise, the transaction is aborted. Consistency means that the database's consistent state is maintained. Isolation means that data used by one transaction cannot be accessed by another transaction until the first one is completed. Durability means that changes made by a transaction cannot be rolled back once the transaction is committed. In addition, transaction schedules have the property of serializability—the result of the concurrent execution of transactions is the same as that of the transactions being executed in serial order.

- SQL provides support for transactions through the use of two statements: COMMIT, which saves changes to disk, and ROLLBACK, which restores the previous database state. SQL transactions are formed by several SQL statements or database requests. Each database request originates several I/O database operations. The transaction log keeps track of all transactions that modify the database. The information stored in the transaction log is used for recovery (ROLLBACK) purposes.

- Concurrency control coordinates the simultaneous execution of transactions. The concurrent execution of transactions can result in three main problems: lost updates, uncommitted data, and inconsistent retrievals. The scheduler is responsible for establishing the order in which the concurrent transaction operations are executed. The transaction execution order is critical and ensures database integrity in multiuser database systems. The scheduler uses locking, time stamping, and optimistic methods to ensure the serializability of transactions.

- A lock guarantees unique access to a data item by a transaction. The lock prevents one transaction from using the data item while another transaction is using it. There are several levels of locks: database, table, page, row, and field. Two types of locks can be used in database systems: binary locks and shared/exclusive locks. A binary lock can have only two states: locked (1) or unlocked (0). A shared lock is used when a transaction wants to read data from a database and no other transaction is updating the same data. Several shared or "read" locks can exist for a particular item. An exclusive lock is issued when a transaction wants to update (write to) the database and no other locks (shared or exclusive) are held on the data.

- Serializability of schedules is guaranteed through the use of two-phase locking. The two-phase locking schema has a growing phase, in which the transaction acquires all of the locks that it needs without unlocking any data, and a shrinking phase, in which the transaction releases all of the locks without acquiring new locks. When two or more transactions wait indefinitely for each other to release a lock, they are in a deadlock, also called a deadly embrace. There are three deadlock control techniques: prevention, detection, and avoidance.

- Concurrency control with time stamping methods assigns a unique time stamp to each transaction and schedules the execution of conflicting transactions in time stamp order. Two schemes are used to decide which transaction is rolled back and which continues executing: the wait/die scheme and the wound/wait scheme.

- Concurrency control with optimistic methods assumes that the majority of database transactions do not conflict and that transactions are executed concurrently, using private, temporary copies of the data. At commit time, the private copies are updated to the database. The ANSI standard defines four transaction isolation levels: Read Uncommitted, Read Committed, Repeatable Read, and Serializable.

- Database recovery restores the database from a given state to a previous consistent state. Database recovery is triggered when a critical event occurs, such as a hardware error or application error.

Key Terms

atomicity	field-level lock	Repeatable Read
atomic transaction property	immediate update	row-level lock
binary lock	inconsistent retrieval	scheduler
buffer	isolation	serializability
checkpoint	lock	Serializable
concurrency control	lock granularity	serializable schedule
consistency	lock manager	shared lock
consistent database state	lost update	table-level lock
database-level lock	monotonicity	time stamping
database recovery	mutual exclusive rule	transaction
database request	nonrepeatable read	transaction log
deadlock	optimistic approach	two-phase locking (2PL)
deadly embrace	page	uncommitted data
deferred update	page-level lock	uniqueness
deferred-write technique	pessimistic locking	wait/die
dirty read	phantom read	wound/wait
diskpage	Read Committed	write-ahead-log protocol
durability	Read Uncommitted	write-through technique
exclusive lock	redundant transaction log	

Online Content

Flashcards and crossword puzzles for key term practice are available at *www.cengagebrain.com*.

Review Questions

1. Explain the following statement: A transaction is a logical unit of work.

2. What is a consistent database state, and how is it achieved?

3. The DBMS does not guarantee that the semantic meaning of the transaction truly represents the real-world event. What are the possible consequences of that limitation? Give an example.

4. List and discuss the four individual transaction properties.

5. What does serializability of transactions mean?

6. What is a transaction log, and what is its function?

7. What is a scheduler, what does it do, and why is its activity important to concurrency control?

8. What is a lock, and how does it work in general?

9. What are the different levels of lock granularity?

10. Why might a page-level lock be preferred over a field-level lock?

11. What is concurrency control, and what is its objective?

12. What is an exclusive lock, and under what circumstances is it granted?

13. What is a deadlock, and how can it be avoided? Discuss several strategies for dealing with deadlocks.

14. What are some disadvantages of time stamping methods for concurrency control?

15. Why might it take a long time to complete transactions when using an optimistic approach to concurrency control?

16. What are the three types of database-critical events that can trigger the database recovery process? Give some examples for each one.

17. What are the four ANSI transaction isolation levels? What type of reads does each level allow?

Problems

1. Suppose that you are a manufacturer of product ABC, which is composed of parts A, B, and C. Each time a new product ABC is created, it must be added to the product inventory, using the PROD_QOH in a table named PRODUCT. Also, each time the product is created, the parts inventory, using PART_QOH in a table named PART, must be reduced by one each of parts A, B, and C. The sample database contents are shown in Table P10.1.

TABLE P10.1

TABLE NAME: PRODUCT	
PROD_CODE	PROD_QOH
ABC	1,205

TABLE NAME: PART	
PART_CODE	PART_QOH
A	567
B	98
C	549

Given the preceding information, answer Questions a through e.

a. How many database requests can you identify for an inventory update for both PRODUCT and PART?

b. Using SQL, write each database request you identified in Step a.

c. Write the complete transaction(s).

d. Write the transaction log, using Table 10.1 as your template.

e. Using the transaction log you created in Step d, trace its use in database recovery.

2. Describe the three most common problems with concurrent transaction execution. Explain how concurrency control can be used to avoid those problems.

3. What DBMS component is responsible for concurrency control? How is this feature used to resolve conflicts?

4. Using a simple example, explain the use of binary and shared/exclusive locks in a DBMS.

5. Suppose that your database system has failed. Describe the database recovery process and the use of deferred-write and write-through techniques.

6. ABC Markets sell products to customers. The relational diagram shown in Figure P10.6 represents the main entities for ABC's database. Note the following important characteristics:

Online Content

The Ch10_ABC_Markets database is available at *www.cengagebrain.com*. Use this database to provide solutions for Problems 6–11.

- A customer may make many purchases, each one represented by an invoice.

   - The CUS_BALANCE is updated with each credit purchase or payment and represents the amount the customer owes.

   - The CUS_BALANCE is increased (+) with every credit purchase and decreased (–) with every customer payment.

   - The date of last purchase is updated with each new purchase made by the customer.

   - The date of last payment is updated with each new payment made by the customer.

- An invoice represents a product purchase by a customer.

   - An INVOICE can have many invoice LINEs, one for each product purchased.

   - The INV_TOTAL represents the total cost of the invoice, including taxes.

   - The INV_TERMS can be "30," "60," or "90" (representing the number of days of credit) or "CASH," "CHECK," or "CC."

   - The invoice status can be "OPEN," "PAID," or "CANCEL."

- A product's quantity on hand (P_QTYOH) is updated (decreased) with each product sale.

FIGURE P10.6 THE ABC MARKETS RELATIONAL DIAGRAM

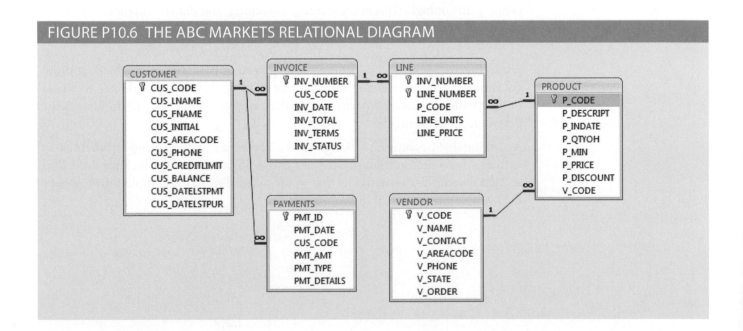

- A customer may make many payments. The payment type (PMT_TYPE) can be one of the following:
  - "CASH" for cash payments.
  - "CHECK" for check payments.
  - "CC" for credit card payments.
- The payment details (PMT_DETAILS) are used to record data about check or credit card payments:
  - The bank, account number, and check number for check payments.
  - The issuer, credit card number, and expiration date for credit card payments.

Note: Not all entities and attributes are represented in this example. Use only the attributes indicated.

Using this database, write the SQL code to represent each of the following transactions. Use BEGIN TRANSACTION and COMMIT to group the SQL statements in logical transactions.

  a. On May 11, 2016, customer 10010 makes a credit purchase (30 days) of one unit of product 11QER/31 with a unit price of $110.00; the tax rate is 8 percent. The invoice number is 10983, and this invoice has only one product line.

  b. On June 3, 2016, customer 10010 makes a payment of $100 in cash. The payment ID is 3428.

7. Create a simple transaction log (using the format shown in Table 10.14) to represent the actions of the transactions in Problems 6a and 6b.

8. Assuming that pessimistic locking is being used but the two-phase locking protocol is not, create a chronological list of the locking, unlocking, and data manipulation activities that would occur during the complete processing of the transaction described in Problem 6a.

9. Assuming that pessimistic locking is being used with the two-phase locking protocol, create a chronological list of the locking, unlocking, and data manipulation activities that would occur during the complete processing of the transaction described in Problem 6a.

10. Assuming that pessimistic locking is being used but the two-phase locking protocol is not, create a chronological list of the locking, unlocking, and data manipulation activities that would occur during the complete processing of the transaction described in Problem 6b.

11. Assuming that pessimistic locking with the two-phase locking protocol is being used with row-level lock granularity, create a chronological list of the locking, unlocking, and data manipulation activities that would occur during the complete processing of the transaction described in Problem 6b.

Chapter 11

Database Performance Tuning and Query Optimization

In this chapter, you will learn:

- Basic database performance-tuning concepts
- How a DBMS processes SQL queries
- About the importance of indexes in query processing
- About the types of decisions the query optimizer has to make
- Some common practices used to write efficient SQL code
- How to formulate queries and tune the DBMS for optimal performance

Preview

Database performance tuning is a critical topic, yet it usually receives minimal coverage in the database curriculum. Most databases used in classrooms have only a few records per table. As a result, the focus is often on making SQL queries perform an intended task, without considering the efficiency of the query process. In fact, even the most efficient query environment yields no visible performance improvements over the least efficient query environment when only 20 or 30 table rows (records) are queried. Unfortunately, that lack of attention to query efficiency can yield unacceptably slow results in the real world when queries are executed over tens of millions of records. In this chapter, you will learn what it takes to create a more efficient query environment.

Data Files and Available Formats

	MS Access	Oracle	MS SQL	My SQL		MS Access	Oracle	MS SQL	My SQL
CH11_SaleCo	✓	✓	✓	✓					

Data Files Available on cengagebrain.com

Note

Because this book focuses on databases, this chapter covers only the factors that directly affect *database* performance. Also, because performance-tuning techniques can be DBMS-specific, the material in this chapter might not be applicable under all circumstances, nor will it necessarily pertain to all DBMS types. This chapter is designed to build a foundation for the general understanding of database performance-tuning issues and to help you choose appropriate performance-tuning strategies. (For the most current information about tuning your database, consult the database vendor's documentation.)

11-1 Database Performance-Tuning Concepts

One of the main functions of a database system is to provide timely answers to end users. End users interact with the DBMS through the use of queries to generate information, using the following sequence:

1. The end-user (client-end) application generates a query.

2. The query is sent to the DBMS (server end).

3. The DBMS (server end) executes the query.

4. The DBMS sends the resulting data set to the end-user (client-end) application.

End users expect their queries to return results as quickly as possible. How do you know that the performance of a database is good? Good database performance is hard to evaluate. How do you know if a 1.06-second query response time is good enough? It is easier to identify bad database performance than good database performance—all it takes is end-user complaints about slow query results. Unfortunately, the same query might perform well one day and not so well two months later. Regardless of end-user perceptions, *the goal of database performance is to execute queries as fast as possible.* Therefore, database performance must be closely monitored and regularly tuned. **Database performance tuning** refers to a set of activities and procedures designed to reduce the response time of the database system—that is, to ensure that an end-user query is processed by the DBMS in the minimum amount of time.

The time required by a query to return a result set depends on many factors, which tend to be wide-ranging and to vary among environments and among vendors. In general, the performance of a typical DBMS is constrained by three main factors: CPU processing power, available primary memory (RAM), and input/output (hard disk and network) throughput. Table 11.1 lists some system components and summarizes general guidelines for achieving better query performance.

Naturally, the system will perform best when its hardware and software resources are optimized. However, in the real world, unlimited resources are not the norm; internal and external constraints always exist. Therefore, the system components should be optimized to obtain the best throughput possible with existing (and often limited) resources, which is why database performance tuning is important.

Fine-tuning the performance of a system requires a holistic approach. That is, *all* factors must be checked to ensure that each one operates at its optimum level and has sufficient resources to minimize the occurrence of bottlenecks. Because database design is such an important factor in determining the database system's performance efficiency, it is worth repeating this book's mantra:

Good database performance starts with good database design. *No amount of fine-tuning will make a poorly designed database perform as well as a well-designed database.*

database performance tuning
A set of activities and procedures designed to reduce the response time of a database system—that is, to ensure that an end-user query is processed by the DBMS in the minimum amount of time.

TABLE 11.1

GENERAL GUIDELINES FOR BETTER SYSTEM PERFORMANCE

	SYSTEM RESOURCES	CLIENT	SERVER
Hardware	CPU	The fastest possible Dual-core CPU or higher	The fastest possible Multiple processors (quad-core technology) Cluster of networked computers
	RAM	The maximum possible to avoid OS memory to disk swapping	The maximum possible to avoid OS memory to disk swapping
	Hard disk	Fast SATA/EIDE hard disk with sufficient free hard disk space Solid State Drives (SSD) for faster speed	Multiple high-speed, high-capacity disks Fast disk interface (SAS / SCSI / Firewire / Fibre Channel RAID configuration optimized for throughput Solid State Drives (SSD) for faster speed Separate disks for OS, DBMS, and data spaces
	Network	High-speed connection	High-speed connection
Software	Operating System (OS)	64-bit OS for larger address spaces Fine-tuned for best client application performance	64-bit OS for larger address spaces Fine-tuned for best server application performance
	Network	Fine-tuned for best throughput	Fine-tuned for best throughput
	Application	Optimize SQL in client application	Optimize DBMS server for best performance

This is particularly true when redesigning existing databases, where the end user expects unrealistic performance gains from older databases.

What constitutes a good, efficient database design? From the performance-tuning point of view, the database designer must ensure that the design makes use of features in the DBMS that guarantee the integrity and optimal performance of the database. This chapter provides fundamental knowledge that will help you optimize database performance by selecting the appropriate database server configuration, using indexes, understanding table storage organization and data locations, and implementing the most efficient SQL query syntax.

11-1a Performance Tuning: Client and Server

In general, database performance-tuning activities can be divided into those on the client side and those on the server side.

- On the client side, the objective is to generate a SQL query that returns the correct answer in the least amount of time, using the minimum amount of resources at the server end. The activities required to achieve that goal are commonly referred to as **SQL performance tuning**.

- On the server side, the DBMS environment must be properly configured to respond to clients' requests in the fastest way possible, while making optimum use of existing resources. The activities required to achieve that goal are commonly referred to as **DBMS performance tuning**.

SQL performance tuning
Activities to help generate a SQL query that returns the correct answer in the least amount of time, using the minimum amount of resources at the server end.

DBMS performance tuning
Activities to ensure that clients' requests are addressed as quickly as possible while making optimum use of existing resources.

Online Content

If you want to learn more about clients and servers, check Appendix F, Client/Server Systems, at *www.cengagebrain.com*.

Keep in mind that DBMS implementations are typically more complex than just a two-tier client/server configuration. The network component plays a critical role in delivering messages between clients and servers; this is especially important in distributed databases. In this chapter however, we assume a fully optimized network, and, therefore, our focus is on the database components. Even in multi-tier client/server environments that consist of a client front end, application middleware, and database server back end, performance-tuning activities are frequently divided into subtasks to ensure the fastest possible response time between any two component points. The database administrator must work closely with the network group to ensure that database traffic flows efficiently in the network infrastructure. This is even more important when you consider that most database systems service geographically dispersed users.

This chapter covers SQL performance-tuning practices on the client side and DBMS performance-tuning practices on the server side. However, before you start learning about the tuning processes, you must first learn more about the DBMS architectural components and processes, and how those processes interact to respond to end-users' requests.

data file
A named physical storage space that stores a database's data. It can reside in a different directory on a hard disk or on one or more hard disks. All data in a database is stored in data files. A typical enterprise database is normally composed of several data files. A data file can contain rows from one or more tables.

11-1b DBMS Architecture

The architecture of a DBMS is represented by the processes and structures (in memory and permanent storage) used to manage a database. Such processes collaborate with one another to perform specific functions. Figure 11.1 illustrates the basic DBMS architecture.

Note the following components and functions in Figure 11.1:

- All data in a database is stored in **data files**. A typical enterprise database is normally composed of several data files. A data file can contain rows from a single table, or it can contain rows from many different tables. A database administrator (DBA)

FIGURE 11.1 BASIC DBMS ARCHITECTURE

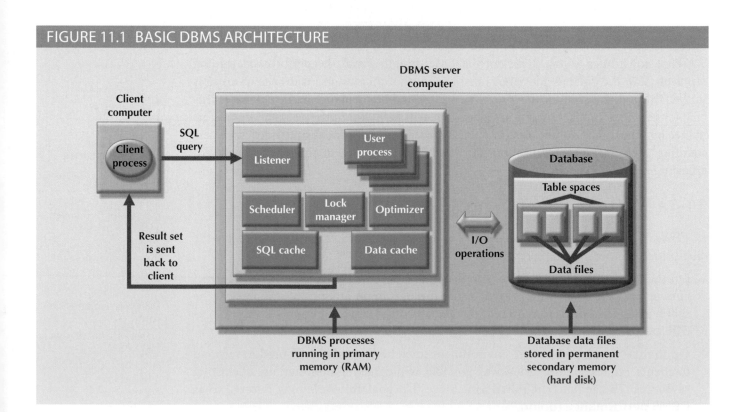

determines the initial size of the data files that make up the database; however, the data files can automatically expand as required in predefined increments known as **extents**. For example, if more space is required, the DBA can define that each new extent will be in 10 KB or 10 MB increments.

- Data files are generally grouped in file groups or table spaces. A **table space** or **file group** is a logical grouping of several data files that store data with similar characteristics. For example, you might have a *system* table space where the data dictionary table data is stored, a *user data* table space to store the user-created tables, an *index* table space to hold all indexes, and a *temporary* table space to do temporary sorts, grouping, and so on. Each time you create a new database, the DBMS automatically creates a minimum set of table spaces.

- The **data cache**, or **buffer cache**, is a shared, reserved memory area that stores the most recently accessed data blocks in RAM. The data read from the data files is stored in the data cache after the data has been read or before the data is written to the data files. The data cache also caches system catalog data and the contents of the indexes.

- The **SQL cache**, or **procedure cache**, is a shared, reserved memory area that stores the most recently executed SQL statements or PL/SQL procedures, including triggers and functions. (To learn more about PL/SQL procedures, triggers, and SQL functions, study Chapter 8, Advanced SQL.) The SQL cache does not store the SQL written by the end user. Rather, the SQL cache stores a "processed" version of the SQL that is ready for execution by the DBMS.

- To work with the data, the DBMS must retrieve the data from permanent storage and place it in RAM. In other words, the data is retrieved from the data files and placed in the data cache.

- To move data from permanent storage (data files) to RAM (data cache), the DBMS issues I/O requests and waits for the replies. An **input/output (I/O) request** is a low-level data access operation that reads or writes data to and from computer devices, such as memory, hard disks, video, and printers. Note that an I/O disk read operation retrieves an entire physical disk block, generally containing multiple rows, from permanent storage to the data cache, even if you will use only one attribute from only one row. The physical disk block size depends on the operating system and could be 4K, 8K, 16K, 32K, 64K, or even larger. Furthermore, depending on the circumstances, a DBMS might issue a single-block read request or a multiblock read request.

- Working with data in the data cache is many times faster than working with data in the data files because the DBMS does not have to wait for the hard disk to retrieve the data; no hard disk I/O operations are needed to work within the data cache.

- Most performance-tuning activities focus on minimizing the number of I/O operations because using I/O operations is many times slower than reading data from the data cache. For example, as of this writing, RAM access times range from 5 to 70 nanoseconds, while hard disk access times range from 5 to 15 milliseconds. This means that hard disks are about six orders of magnitude (a million times) slower than RAM.

Figure 11.1 also illustrates some typical DBMS processes. Although the number of processes and their names vary from vendor to vendor, the functionality is similar. The following processes are represented in Figure 11.1:

- *Listener*. The listener process listens for clients' requests and handles the processing of the SQL requests to other DBMS processes. Once a request is received, the listener passes the request to the appropriate user process.

extents
In a DBMS environment, refers to the ability of data files to expand in size automatically using predefined increments.

table space
In a DBMS, a logical storage space used to group related data. Also known as a *file group*.

file group
See *table space*.

data cache
A shared, reserved memory area that stores the most recently accessed data blocks in RAM. Also called *buffer cache*.

buffer cache
See *data cache*.

SQL cache
A shared, reserved memory area that stores the most recently executed SQL statements or PL/SQL procedures, including triggers and functions. Also called *procedure cache*.

procedure cache
See *SQL cache*.

input/output (I/O) request
A low-level data access operation that reads or writes data to and from computer devices.

- *User.* The DBMS creates a user process to manage each client session. Therefore, when you log on to the DBMS, you are assigned a user process. This process handles all requests you submit to the server. There are many user processes—at least one per logged-in client.

- *Scheduler.* The scheduler process organizes the concurrent execution of SQL requests. (See Chapter 10, Transaction Management and Concurrency Control.)

- *Lock manager.* This process manages all locks placed on database objects, including disk pages. (See Chapter 10.)

- *Optimizer.* The optimizer process analyzes SQL queries and finds the most efficient way to access the data. You will learn more about this process later in the chapter.

11-1c Database Query Optimization Modes

Most of the algorithms proposed for query optimization are based on two principles:

- The selection of the optimum execution order to achieve the fastest execution time

- The selection of sites to be accessed to minimize communication costs

Within those two principles, a query optimization algorithm can be evaluated on the basis of its *operation mode* or the *timing of its optimization.*

Operation modes can be classified as manual or automatic. **Automatic query optimization** means that the DBMS finds the most cost-effective access path without user intervention. **Manual query optimization** requires that the optimization be selected and scheduled by the end user or programmer. Automatic query optimization is clearly more desirable from the end user's point of view, but the cost of such convenience is the increased overhead that it imposes on the DBMS.

Query optimization algorithms can also be classified according to when the optimization is done. Within this timing classification, query optimization algorithms can be static or dynamic.

- **Static query optimization** takes place at compilation time. In other words, the best optimization strategy is selected when the query is compiled by the DBMS. This approach is common when SQL statements are embedded in procedural programming languages such as C# or Visual Basic .NET. When the program is submitted to the DBMS for compilation, it creates the plan necessary to access the database. When the program is executed, the DBMS uses that plan to access the database.

- **Dynamic query optimization** takes place at execution time. Database access strategy is defined when the program is executed. Therefore, access strategy is dynamically determined by the DBMS at run time, using the most up-to-date information about the database. Although dynamic query optimization is efficient, its cost is measured by run-time processing overhead. The best strategy is determined every time the query is executed; this could happen several times in the same program.

Finally, query optimization techniques can be classified according to the type of information that is used to optimize the query. For example, queries may be based on statistically based or rule-based algorithms.

- A **statistically based query optimization algorithm** uses statistical information about the database. The statistics provide information about database characteristics such as size, number of records, average access time, number of requests serviced, and number of users with access rights. These statistics are then used by the DBMS to determine the best access strategy. Within statistically based optimizers, some DBMSs allow setting a goal to specify that the optimizer should attempt to minimize

automatic query optimization
A method by which a DBMS finds the most efficient access path for the execution of a query.

manual query optimization
An operation mode that requires the end user or programmer to define the access path for the execution of a query.

static query optimization
A query optimization mode in which the access path to a database is predetermined at compilation time.

dynamic query optimization
The process of determining the SQL access strategy at run time, using the most up-to-date information about the database.

statistically based query optimization algorithm
A query optimization technique that uses statistical information about a database. The DBMS then uses these statistics to determine the best access strategy.

the time to retrieve the first row or the last row. Minimizing the time to retrieve the first row is often used in transaction systems and interactive client environments. In these cases, the goal is to present the first several rows to the user as quickly as possible. Then, while the DBMS waits for the user to scroll through the data, it can fetch the other rows for the query. Setting the optimizer goal to minimize retrieval of the last row is typically done in embedded SQL and inside stored procedures. In these cases, the control will not pass back to the calling application until all of the data has been retrieved; therefore, it is important to retrieve all of the data to the last row as quickly as possible so control can be returned.

- The statistical information is managed by the DBMS and is generated in one of two different modes: dynamic or manual. In the **dynamic statistical generation mode**, the DBMS automatically evaluates and updates the statistics after each data access operation. In the **manual statistical generation mode**, the statistics must be updated periodically through a user-selected utility such as IBM's RUNSTAT command, which is used by DB2 DBMSs.

- A **rule-based query optimization algorithm** is based on a set of user-defined rules to determine the best query access strategy. The rules are entered by the end user or database administrator, and they are typically general in nature.

Because database statistics play a crucial role in query optimization, this topic is explored in more detail in the next section.

11-1d Database Statistics

Another DBMS process that plays an important role in query optimization is gathering database statistics. The term **database statistics** refers to a number of measurements about database objects, such as number of processors used, processor speed, and temporary space available. Such statistics provide a snapshot of database characteristics.

As you will learn later in this chapter, the DBMS uses these statistics to make critical decisions about improving query processing efficiency. Database statistics can be gathered manually by the DBA or automatically by the DBMS. For example, many DBMS vendors support the ANALYZE command in SQL to gather statistics. In addition, many vendors have their own routines to gather statistics. For example, IBM's DB2 uses the RUNSTATS procedure, while Microsoft's SQL Server uses the UPDATE STATISTICS procedure and provides the Auto-Update and Auto-Create Statistics options in its initialization parameters. A sample of measurements that the DBMS may gather about various database objects is shown in Table 11.2.

dynamic statistical generation mode
In a DBMS, the capability to automatically evaluate and update the database access statistics after each data access operation.

manual statistical generation mode
A mode of generating statistical data access information for query optimization. In this mode, the DBA must periodically run a routine to generate the data access statistics—for example, running the RUNSTAT command in an IBM DB2 database.

rule-based query optimization algorithm
A query optimization technique that uses preset rules and points to determine the best approach to executing a query.

database statistics
In query optimization, measurements about database objects, such as the number of rows in a table, number of disk blocks used, maximum and average row length, number of columns in each row, and number of distinct values in each column. Such statistics provide a snapshot of database characteristics.

TABLE 11.2

SAMPLE DATABASE STATISTICS MEASUREMENTS

DATABASE OBJECT	SAMPLE MEASUREMENTS
Tables	Number of rows, number of disk blocks used, row length, number of columns in each row, number of distinct values in each column, maximum value in each column, minimum value in each column, and columns that have indexes
Indexes	Number and name of columns in the index key, number of key values in the index, number of distinct key values in the index key, histogram of key values in an index, and number of disk pages used by the index
Environment Resources	Logical and physical disk block size, location and size of data files, and number of extends per data file

If the object statistics exist, the DBMS will use them in query processing. Most newer DBMSs (such as Oracle, MySQL, SQL Server, and DB2) automatically gather statistics; others require the DBA to gather statistics manually. To generate the database object statistics manually, each DBMS has its own commands.

In Oracle, use ANALYZE <TABLE/INDEX> object_name COMPUTE STATISTICS;

In MySQL, use ANALYZE TABLE <table_name>;

In SQL Server, use UPDATE STATISTICS <object_name>, where object name refers to a table or a view.

For example, to generate statistics for the VENDOR table, you would use:

In Oracle: ANALYZE TABLE VENDOR COMPUTE STATISTICS;
In MySQL: ANALYZE TABLE VENDOR;
In SQL Server: UPDATE STATISTICS VENDOR;

When you generate statistics for a table, all related indexes are also analyzed. However, you could generate statistics for a single index by using the following command, where VEND_NDX is the name of the index:

ANALYZE INDEX VEND_NDX COMPUTE STATISTICS;

In SQL Server, use UPDATE STATISTICS <table_name> <index_name>. An example command would be UPDATE STATISTICS VENDOR VEND_NDX;.

Database statistics are stored in the system catalog in specially designated tables. It is common to periodically regenerate the statistics for database objects, especially database objects that are subject to frequent change. For example, if you have a video rental DBMS, your system will likely use a RENTAL table to store the daily video rentals. That RENTAL table and its associated indexes would be subject to constant inserts and updates as you record daily rentals and returns. Therefore, the RENTAL table statistics you generated last week do not accurately depict the table as it exists today. The more current the statistics are, the better the chances that the DBMS will properly select the fastest way to execute a given query.

Now that you know the basic architecture of DBMS processes and memory structures, and the importance and timing of the database statistics gathered by the DBMS, you are ready to learn how the DBMS processes a SQL query request.

11-2 Query Processing

What happens at the DBMS server end when the client's SQL statement is received? In simple terms, the DBMS processes a query in three phases:

1. *Parsing*. The DBMS parses the SQL query and chooses the most efficient access/execution plan.

2. *Execution*. The DBMS executes the SQL query using the chosen execution plan.

3. *Fetching*. The DBMS fetches the data and sends the result set back to the client.

The processing of SQL DDL statements (such as CREATE TABLE) is different from the processing required by DML statements. The difference is that a DDL statement actually updates the data dictionary tables or system catalog, while a DML statement (SELECT, INSERT, UPDATE, or DELETE) mostly manipulates end-user data. Figure 11.2 shows the general steps required for query processing. Each step will be discussed in the following sections.

FIGURE 11.2 QUERY PROCESSING

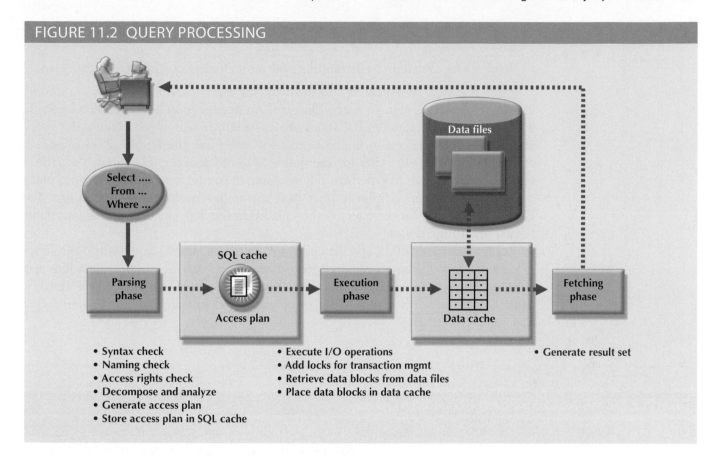

- Syntax check
- Naming check
- Access rights check
- Decompose and analyze
- Generate access plan
- Store access plan in SQL cache

- Execute I/O operations
- Add locks for transaction mgmt
- Retrieve data blocks from data files
- Place data blocks in data cache

- Generate result set

11-2a SQL Parsing Phase

The optimization process includes breaking down—parsing—the query into smaller units and transforming the original SQL query into a slightly different version of the original SQL code, but one that is fully equivalent and more efficient. *Fully equivalent* means that the optimized query results are always the same as the original query. *More efficient* means that the optimized query will almost always execute faster than the original query. (Note that it *almost* always executes faster because many factors affect the performance of a database, as explained earlier. Those factors include the network, the client computer's resources, and other queries running concurrently in the same database.) To determine the most efficient way to execute the query, the DBMS may use the database statistics you learned about earlier.

The SQL parsing activities are performed by the **query optimizer**, which analyzes the SQL query and finds the most efficient way to access the data. This process is the most time-consuming phase in query processing. Parsing a SQL query requires several steps, in which the SQL query is:

- Validated for syntax compliance

- Validated against the data dictionary to ensure that table names and column names are correct

- Validated against the data dictionary to ensure that the user has proper access rights

- Analyzed and decomposed into more atomic components

query optimizer
A DBMS process that analyzes SQL queries and finds the most efficient way to access the data. The query optimizer generates the access or execution plan for the query.

- Optimized through transformation into a fully equivalent but more efficient SQL query

- Prepared for execution by determining the most efficient execution or access plan

Once the SQL statement is transformed, the DBMS creates what is commonly known as an access plan or execution plan. An **access plan** is the result of parsing a SQL statement; it contains the series of steps a DBMS will use to execute the query and return the result set in the most efficient way. First, the DBMS checks to see if an access plan already exists for the query in the SQL cache. If it does, the DBMS reuses the access plan to save time. If it does not, the optimizer evaluates various plans and then decides which indexes to use and how to best perform join operations. The chosen access plan for the query is then placed in the SQL cache and made available for use and future reuse.

Access plans are DBMS-specific and translate the client's SQL query into the series of complex I/O operations required to read the data from the physical data files and generate the result set. Table 11.3 illustrates some I/O operations for an Oracle RDBMS. Most DBMSs perform similar types of operations when accessing and manipulating data sets.

TABLE 11.3

SAMPLE DBMS ACCESS PLAN I/O OPERATIONS

OPERATION	DESCRIPTION
Table scan (full)	Reads the entire table sequentially, from the first row to the last, one row at a time (slowest)
Table access (row ID)	Reads a table row directly, using the row ID value (fastest)
Index scan (range)	Reads the index first to obtain the row IDs and then accesses the table rows directly (faster than a full table scan)
Index access (unique)	Used when a table has a unique index in a column
Nested loop	Reads and compares a set of values to another set of values, using a nested loop style (slow)
Merge	Merges two data sets (slow)
Sort	Sorts a data set (slow)

In Table 11.3, note that a table access using a row ID is the fastest method. A row ID is a unique identification for every row saved in permanent storage; it can be used to access the row directly. Conceptually, a row ID is similar to a slip you get when you park your car in an airport parking lot. The parking slip contains the section number and lot number. Using that information, you can go directly to your car without searching every section and lot.

11-2b SQL Execution Phase

access plan
A set of instructions generated at application compilation time that is created and managed by a DBMS. The access plan predetermines how an application's query will access the database at run time.

In this phase, all I/O operations indicated in the access plan are executed. When the execution plan is run, the proper locks—if needed—are acquired for the data to be accessed, and the data is retrieved from the data files and placed in the DBMS's data cache. All transaction management commands are processed during the parsing and execution phases of query processing.

11-2c SQL Fetching Phase

After the parsing and execution phases are completed, all rows that match the specified condition(s) are retrieved, sorted, grouped, and aggregated (if required). During the fetching phase, the rows of the resulting query result set are returned to the client. The DBMS might use temporary table space to store temporary data. In this stage, the database server coordinates the movement of the result set rows from the server cache to the client cache. For example, a given query result set might contain 9,000 rows; the server would send the first 100 rows to the client and then wait for the client to request the next set of rows, until the entire result set is sent to the client.

11-2d Query Processing Bottlenecks

The main objective of query processing is to execute a given query in the fastest way possible with the least amount of resources. As you have seen, the execution of a query requires the DBMS to break down the query into a series of interdependent I/O operations to be executed in a collaborative manner. The more complex a query is, the more complex the operations are, which means that bottlenecks are more likely. A **query processing bottleneck** is a delay introduced in the processing of an I/O operation that causes the overall system to slow down. In the same way, the more components a system has, the more interfacing is required among the components, increasing the likelihood of bottlenecks. Within a DBMS, five components typically cause bottlenecks:

- *CPU.* The CPU processing power of the DBMS should match the system's expected work load. A high CPU utilization might indicate that the processor speed is too slow for the amount of work performed. However, heavy CPU utilization can be caused by other factors, such as a defective component, not enough RAM (the CPU spends too much time swapping memory blocks), a badly written device driver, or a rogue process. A CPU bottleneck will affect not only the DBMS but all processes running in the system.

- *RAM.* The DBMS allocates memory for specific usage, such as data cache and SQL cache. RAM must be shared among all running processes, including the operating system and DBMS. If there is not enough RAM available, moving data among components that are competing for scarce RAM can create a bottleneck.

- *Hard disk.* Other common causes of bottlenecks are hard disk speed and data transfer rates. Current hard disk storage technology allows for greater storage capacity than in the past; however, hard disk space is used for more than just storing end-user data. Current operating systems also use the hard disk for *virtual memory*, which refers to copying areas of RAM to the hard disk as needed to make room in RAM for more urgent tasks. Therefore, more hard disk storage space and faster data transfer rates reduce the likelihood of bottlenecks.

- *Network.* In a database environment, the database server and the clients are connected via a network. All networks have a limited amount of bandwidth that is shared among all clients. When many network nodes access the network at the same time, bottlenecks are likely.

- *Application code.* Not all bottlenecks are caused by limited hardware resources. Two of the most common sources of bottlenecks are inferior application code and poorly designed databases. Inferior code can be improved with code optimization techniques, as long as the underlying database design is sound. However, no amount of coding will make a poorly designed database perform better.

query processing bottleneck
In query optimization, a delay introduced in the processing of an I/O operation that causes the overall system to slow down.

Bottlenecks are the result of multiple database transactions competing for the use of database resources (CPU, RAM, hard disk, indexes, locks, buffers, etc.). As you learned earlier in this chapter, a DBMS uses many components and structures to perform its operations, such as processes, buffers, locks, table spaces, indexes, and log files. These resources are used by all transactions executing on the database, and, therefore, the transactions often compete for such resources. Because most (if not all) transactions work with data rows in tables, one of the most typical bottlenecks is caused by transactions competing for the same data rows. Another common source of contention is for shared memory resources, particularly shared buffers and locks. To speed up data update operations, the DMBS uses buffers to cache the data. At the same time, to manage access to data, the DBMS uses locks. Learning how to avoid these bottlenecks and optimize database performance is the main focus of this chapter.

11-3 Indexes and Query Optimization

Indexes are crucial in speeding up data access because they facilitate searching, sorting, and using aggregate functions and even join operations. The improvement in data access speed occurs because an index is an ordered set of values that contains the index key and pointers. The pointers are the row IDs for the actual table rows. Conceptually, a data index is similar to a book index. When you use a book index, you look up a word, which is similar to the index key. The word is accompanied by one or more page numbers where the word is used; these numbers are similar to pointers.

An index scan is more efficient than a full table scan because the index data is preordered and the amount of data is usually much smaller. Therefore, when performing searches, it is almost always better for the DBMS to use the index to access a table than to scan all rows in a table sequentially. For example, Figure 11.3 shows the index representation of a CUSTOMER table with 14,786 rows and the index STATE_NDX on the CUS_STATE attribute.

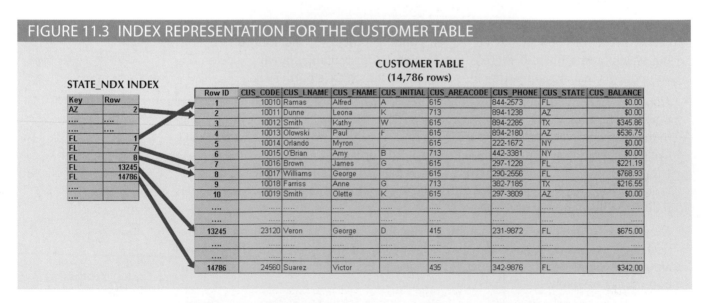

FIGURE 11.3 INDEX REPRESENTATION FOR THE CUSTOMER TABLE

Suppose you submit the following query:

```
SELECT      CUS_NAME, CUS_STATE
FROM        CUSTOMER
WHERE       CUS_STATE = 'FL';
```

If there is no index, the DBMS will perform a full-table scan and read all 14,786 customer rows. Assuming that the index STATE_NDX is created (and analyzed), the DBMS will automatically use the index to locate the first customer with a state equal to 'FL' and then proceed to read all subsequent CUSTOMER rows, using the row IDs in the index as a guide. Assuming that only five rows meet the condition CUS_STATE = 'FL' there are five accesses to the index and five accesses to the data, for a total of 10 I/O accesses. The DBMS would be saved from reading approximately 14,776 I/O requests for customer rows that do not meet the criteria. That is a lot of CPU cycles!

If indexes are so important, why not index every column in every table? The simple answer is that it is not practical to do so. Indexing every column in every table overtaxes the DBMS in terms of index-maintenance processing, especially if the table has many attributes and rows, or requires many inserts, updates, and deletes.

One measure that determines the need for an index is the data *sparsity* of the column you want to index. **Data sparsity** refers to the number of different values a column could have. For example, a STU_SEX column in a STUDENT table can have only two possible values, M or F; therefore, that column is said to have low sparsity. In contrast, the STU_DOB column that stores the student date of birth can have many different date values; therefore, that column is said to have high sparsity. Knowing the sparsity helps you decide whether the use of an index is appropriate. For example, when you perform a search in a column with low sparsity, you are likely to read a high percentage of the table rows anyway; therefore, index processing might be unnecessary work. In Section 11-5, you learn how to determine when an index is recommended.

Most DBMSs implement indexes using one of the following data structures:

- **Hash index**. A hash index is based on an ordered list of hash values. A hash algorithm is used to create a hash value from a key column. This value points to an entry in a hash table, which in turn points to the actual location of the data row. This type of index is good for simple and fast lookup operations based on equality conditions—for example, LNAME="Scott" and FNAME="Shannon".

- **B-tree index**. The B-tree index is an ordered data structure organized as an upside-down tree. (See Figure 11.4.) The index tree is stored separately from the data. The lower-level leaves of the B-tree index contain the pointers to the actual data rows. B-tree indexes are "self-balanced," which means that it takes approximately the same amount of time to access any given row in the index. This is the default and most common type of index used in databases. The B-tree index is used mainly in tables in which column values repeat a relatively small number of times.

- **Bitmap index**. A bitmap index uses a bit array (0s and 1s) to represent the existence of a value or condition. These indexes are used mostly in data warehouse applications in tables with a large number of rows in which a small number of column values repeat many times. (See Figure 11.4.) Bitmap indexes tend to use less space than B-tree indexes because they use bits instead of bytes to store their data.

Using the preceding index characteristics, a database designer can determine the best type of index to use. For example, assume that a CUSTOMER table has several thousand rows. The CUSTOMER table has two columns that are used extensively for query purposes: CUS_LNAME, which represents a customer's last name, and REGION_CODE, which can have one of four values (NE, NW, SW, and SE). Based on this information, you could conclude that:

- Because the CUS_LNAME column contains many different values that repeat a relatively small number of times compared to the total number of rows in the table, a B-tree index will be used.

data sparsity
A column distribution of values or the number of different values a column can have.

hash index
An index based on an ordered list of hash values.

B-tree index
An ordered data structure organized as an upside-down tree.

bitmap index
An index that uses a bit array (0s and 1s) to represent the existence of a value or condition.

FIGURE 11.4 B-TREE AND BITMAP INDEX REPRESENTATION

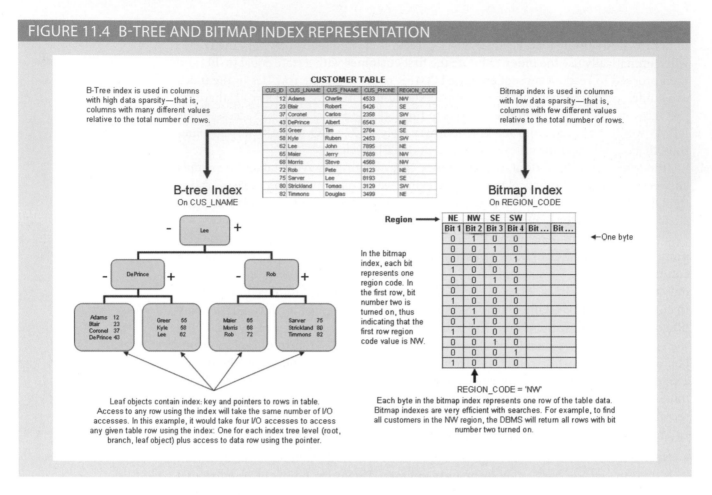

- Because the REGION_CODE column contains only a few different values that repeat a relatively large number of times compared to the total number of rows in the table, a bitmap index will be used. Figure 11.4 shows the B-tree and bitmap representations for the CUSTOMER table used in the previous discussion.

Current-generation DBMSs are intelligent enough to determine the best type of index to use under certain circumstances, provided that the DBMS has updated database statistics. Regardless of which index is chosen, the DBMS determines the best plan to execute a given query. The next section guides you through a simplified example of the type of choices the query optimizer must make.

11-4 **Optimizer Choices**

rule-based optimizer
A query optimization mode based on the rule-based query optimization algorithm.

cost-based optimizer
A query optimization mode that uses an algorithm based on statistics about the objects being accessed, including number of rows, indexes available, index sparsity, and so on.

Query optimization is the central activity during the parsing phase in query processing. In this phase, the DBMS must choose what indexes to use, how to perform join operations, which table to use first, and so on. Each DBMS has its own algorithms for determining the most efficient way to access the data. The query optimizer can operate in one of two modes:

- A **rule-based optimizer** uses preset rules and points to determine the best approach to execute a query. The rules assign a "fixed cost" to each SQL operation; the costs are then added to yield the cost of the execution plan. For example, a full table scan has a set cost of 10, while a table access by row ID has a set cost of 3.

- A **cost-based optimizer** uses sophisticated algorithms based on statistics about the objects being accessed to determine the best approach to execute a query. In this case, the optimizer process adds up the processing cost, the I/O costs, and the resource costs (RAM and temporary space) to determine the total cost of a given execution plan.

The optimizer's objective is to find alternative ways to execute a query—to evaluate the "cost" of each alternative and then to choose the one with the lowest cost. To understand the function of the query optimizer, consider a simple example. Assume that you want to list all products provided by a vendor based in Florida. To acquire that information, you could write the following query:

```
SELECT      P_CODE, P_DESCRIPT, P_PRICE, V_NAME, V_STATE
FROM        PRODUCT, VENDOR
WHERE       PRODUCT.V_CODE = VENDOR.V_CODE
            AND VENDOR.V_STATE = 'FL';
```

Furthermore, assume that the database statistics indicate the following:

- The PRODUCT table has 7,000 rows.
- The VENDOR table has 300 rows.
- Ten vendors are located in Florida.
- One thousand products come from vendors in Florida.

It is important to point out that only the first two items are available to the optimizer. The second two items are assumed to illustrate the choices that the optimizer must make. Armed with the information in the first two items, the optimizer would try to find the most efficient way to access the data. The primary factor in determining the most efficient access plan is the I/O cost. (Remember, the DBMS always tries to minimize I/O operations.) Table 11.4 shows two sample access plans for the previous query and their respective I/O costs.

To make the example easier to understand, the I/O Operations and I/O Cost columns in Table 11.4 estimate only the number of I/O disk reads the DBMS must perform.

TABLE 11.4

COMPARING ACCESS PLANS AND I/O COSTS

PLAN	STEP	OPERATION	I/O OPERATIONS	I/O COST	RESULTING SET ROWS	TOTAL I/O COST
A	A1	Cartesian product (PRODUCT, VENDOR)	7,000 + 300	7,300	2,100,000	7,300
	A2	Select rows in A1 with matching vendor codes	2,100,000	2,100,000	7,000	2,107,300
	A3	Select rows in A2 with V_STATE = 'FL'	7,000	7,000	1,000	**2,114,300**
B	B1	Select rows in VENDOR with V_STATE = 'FL'	300	300	10	300
	B2	Cartesian Product (PRODUCT, B1)	7,000 + 10	7,010	70,000	7,310
	B3	Select rows in B2 with matching vendor codes	70,000	70,000	1,000	**77,310**

For simplicity's sake, it is assumed that there are no indexes and that each row read has an I/O cost of 1. For example, in Step A1, the DBMS must calculate the Cartesian product of PRODUCT and VENDOR. To do that, the DBMS must read all rows from PRODUCT (7,000) and all rows from VENDOR (300), yielding a total of 7,300 I/O operations. The same computation is done in all steps. In Table 11.4, you can see how Plan A has a total I/O cost that is almost 30 times higher than Plan B. In this case, the optimizer will choose Plan B to execute the SQL.

Note

Not all DBMSs optimize SQL queries the same way. As a matter of fact, Oracle parses queries differently from the methods described in several sections in this chapter. Always read the documentation to examine the optimization requirements for your DBMS implementation.

Given the right conditions, some queries could be answered entirely by using only an index. For example, assume that you are using the PRODUCT table and the index P_QOH_NDX in the P_QOH attribute. Then a query such as SELECT MIN(P_QOH) FROM PRODUCT could be resolved by reading only the first entry in the P_QOH_NDX index, without the need to access any of the data blocks for the PRODUCT table. (Remember that the index defaults to ascending order.)

You learned in Section 11-3 that columns with low sparsity are not good candidates for index creation. However, in some cases an index in a low-sparsity column would be helpful. For example, assume that the EMPLOYEE table has 122,483 rows. If you want to find out how many female employees work at the company, you would write a query such as:

SELECT COUNT(EMP_SEX) FROM EMPLOYEE WHERE EMP_SEX = 'F';

If you do not have an index for the EMP_SEX column, the query would have to perform a full table scan to read all EMPLOYEE rows—and each full row includes attributes you do not need. However, if you have an index on EMP_SEX, the query can be answered by reading only the index data, without the need to access the employee data at all.

11-4a Using Hints to Affect Optimizer Choices

Although the optimizer generally performs very well under most circumstances, in some instances the optimizer might not choose the best execution plan. Remember, the optimizer makes decisions based on the existing statistics. If the statistics are old, the optimizer might not do a good job in selecting the best execution plan. Even with current statistics, the optimizer's choice might not be the most efficient one. Sometimes the end user would like to change the optimizer mode for the current SQL statement. To do that, you need to use hints. **Optimizer hints** are special instructions for the optimizer that are embedded inside the SQL command text. Table 11.5 summarizes a few of the most common optimizer hints used in standard SQL.

Now that you are familiar with the way the DBMS processes SQL queries, you can turn your attention to some general SQL coding recommendations to facilitate the work of the query optimizer.

optimizer hints
Special instructions for the query optimizer that are embedded inside the SQL command text.

TABLE 11.5

OPTIMIZER HINTS

HINT	USAGE
ALL_ROWS	Instructs the optimizer to minimize the overall execution time—that is, to minimize the time needed to return all rows in the query result set. This hint is generally used for batch mode processes. For example: SELECT /*+ ALL_ROWS */ * FROM PRODUCT WHERE P_QOH < 10;
FIRST_ROWS	Instructs the optimizer to minimize the time needed to process the first set of rows—that is, to minimize the time needed to return only the first set of rows in the query result set. This hint is generally used for interactive mode processes. For example: SELECT /*+ FIRST_ROWS */ * FROM PRODUCT WHERE P_QOH < 10;
INDEX(name)	Forces the optimizer to use the P_QOH_NDX index to process this query. For example: SELECT /*+ INDEX(P_QOH_NDX) */ * FROM PRODUCT WHERE P_QOH < 10

11-5 SQL Performance Tuning

SQL performance tuning is evaluated from the client perspective. Therefore, the goal is to illustrate some common practices used to write efficient SQL code. A few words of caution are appropriate:

- Most current-generation relational DBMSs perform automatic query optimization at the server end.

- Most SQL performance optimization techniques are DBMS-specific and, therefore, are rarely portable, even across different versions of the same DBMS. Part of the reason for this behavior is the constant advancement in database technologies.

Does this mean that you should not worry about how a SQL query is written because the DBMS will always optimize it? No, because there is considerable room for improvement. (The DBMS uses *general* optimization techniques rather than focusing on specific techniques dictated by the special circumstances of the query execution.) A poorly written SQL query can, *and usually will*, bring the database system to its knees from a performance point of view. The majority of current database performance problems are related to poorly written SQL code. Therefore, although a DBMS provides general optimizing services, a carefully written query almost always outperforms a poorly written one.

Although SQL data manipulation statements include many different commands such as INSERT, UPDATE, DELETE, and SELECT, most recommendations in this section are related to the use of the SELECT statement, and in particular, the use of indexes and how to write conditional expressions.

11-5a Index Selectivity

Indexes are the most important technique used in SQL performance optimization. The key is to know when an index is used. As a general rule, indexes are likely to be used:

- When an indexed column appears by itself in the search criteria of a WHERE or HAVING clause

- When an indexed column appears by itself in a GROUP BY or ORDER BY clause
- When a MAX or MIN function is applied to an indexed column
- When the data sparsity on the indexed column is high

Indexes are very useful when you want to select a small subset of rows from a large table based on a given condition. If an index exists for the column *used in the selection*, the DBMS may choose to use it. The objective is to create indexes with high selectivity. **Index selectivity** is a measure of the likelihood that an index will be used in query processing. Here are some general guidelines for creating and using indexes:

- *Create indexes for each single attribute used in a WHERE, HAVING, ORDER BY, or GROUP BY clause.* If you create indexes in all single attributes *used in search conditions*, the DBMS will access the table using an index scan instead of a full table scan. For example, if you have an index for P_PRICE, the condition P_PRICE > 10.00 can be solved by accessing the index instead of sequentially scanning all table rows and evaluating P_PRICE for each row. Indexes are also used in join expressions, such as in CUSTOMER.CUS_CODE = INVOICE.CUS_CODE.

- *Do not use indexes in small tables or tables with low sparsity.* Remember, small tables and low-sparsity tables are not the same thing. A search condition in a table with low sparsity may return a high percentage of table rows anyway, making the index operation too costly and making the full table scan a viable option. Using the same logic, do not create indexes for tables with few rows and few attributes—*unless you must ensure the existence of unique values in a column.*

- *Declare primary and foreign keys so the optimizer can use the indexes in join operations.* All natural joins and old-style joins will benefit if you declare primary keys and foreign keys because the optimizer will use the available indexes at join time. (The declaration of a PK or FK, primary key or foreign key, will automatically create an index for the declared column.) Also, for the same reason, it is better to write joins using the SQL JOIN syntax. (See Chapter 8, Advanced SQL.)

- *Declare indexes in join columns other than PK or FK.* If you perform join operations on columns other than the primary and foreign keys, you might be better off declaring indexes in those columns.

You cannot always use an index to improve performance. For example, using the data shown in Table 11.6 in the next section, the creation of an index for P_MIN will not help the search condition P_QOH > P_MIN * 1.10. The reason is that in some DBMSs, *indexes are ignored when you use functions in the table attributes.* However, major databases such as Oracle, SQL Server, and DB2 now support function-based indexes. A **function-based index** is an index based on a specific SQL function or expression. For example, you could create an index on YEAR(INV_DATE). Function-based indexes are especially useful when dealing with derived attributes. For example, you could create an index on EMP_SALARY + EMP_COMMISSION.

How many indexes should you create? It bears repeating that you should not create an index for every column in a table. Too many indexes will slow down INSERT, UPDATE, and DELETE operations, especially if the table contains many thousands of rows. Furthermore, some query optimizers will choose only one index to be the driving index for a query, even if your query uses conditions in many different indexed columns. Which index does the optimizer use? If you use the cost-based optimizer, the answer will change with time as new rows are added to or deleted from the tables. In any case, you should create indexes in all search columns and then let the optimizer choose. It is important to constantly evaluate the index usage—monitor, test, evaluate, and improve it if performance is not adequate.

index selectivity
A measure of how likely an index is to be used in query processing.

function-based index
A type of index based on a specific SQL function or expression.

11-5b Conditional Expressions

A conditional expression is normally placed within the WHERE or HAVING clauses of a SQL statement. Also known as conditional criteria, a conditional expression restricts the output of a query to only the rows that match the conditional criteria. Generally, the conditional criteria have the form shown in Table 11.6.

TABLE 11.6

CONDITIONAL CRITERIA

OPERAND1	CONDITIONAL OPERATOR	OPERAND2
P_PRICE	>	10.00
V_STATE	=	FL
V_CONTACT	LIKE	Smith%
P_QOH	>	P_MIN * 1.10

In Table 11.6, note that an operand can be:

- A simple column name such as P_PRICE or V_STATE
- A literal or a constant such as the value 10.00 or the text 'FL'
- An expression such as P_MIN * 1.10

Most of the query optimization techniques mentioned below are designed to make the optimizer's work easier. The following common practices are used to write efficient conditional expressions in SQL code.

- Use simple columns or literals as operands in a conditional expression—avoid the use of conditional expressions with functions whenever possible. Comparing the contents of a single column to a literal is faster than comparing to expressions. For example, P_PRICE > 10.00 is faster than P_QOH > P_MIN * 1.10 because the DBMS must evaluate the P_MIN * 1.10 expression first. The use of functions in expressions also adds to the total query execution time. For example, if your condition is UPPER (V_NAME) = 'JIM', try to use V_NAME = 'Jim' if all names in the V_NAME column are stored with proper capitalization.

- *Numeric field comparisons are faster than character, date, and NULL comparisons.* In search conditions, comparing a numeric attribute to a numeric literal is faster than comparing a character attribute to a character literal. In general, the CPU handles numeric comparisons (integer and decimal) faster than character and date comparisons. Because indexes do not store references to null values, NULL conditions involve additional processing, and therefore tend to be the slowest of all conditional operands.

- *Equality comparisons are generally faster than inequality comparisons.* For example, P_PRICE = 10.00 is processed faster because the DBMS can do a direct search using the index in the column. If there are no exact matches, the condition is evaluated as false. However, if you use an inequality symbol (>, >=, <, <=), the DBMS must perform additional processing to complete the request, because there will almost always be more "greater than" or "less than" values in the index than "equal" values. With the exception of NULL, the slowest of all comparison operators is LIKE with wildcard symbols, as in V_CONTACT LIKE "%glo%". Also, using the "not equal" symbol (< >) yields slower searches, especially when the sparsity of the data is high—that is, when there are many more different values than there are equal values.

- *Whenever possible, transform conditional expressions to use literals.* For example, if your condition is P_PRICE − 10 = 7, change it to read P_PRICE = 17. Also, if you have a composite condition such as:

P_QOH < P_MIN AND P_MIN = P_REORDER AND P_QOH = 10

change it to read:

P_QOH = 10 AND P_MIN = P_REORDER AND P_MIN > 10

- *When using multiple conditional expressions, write the equality conditions first.* Note that this was done in the previous example. Remember, equality conditions are faster to process than inequality conditions. Although most RDBMSs will automatically do this for you, paying attention to this detail lightens the load for the query optimizer. The optimizer will not have to do what you have already done.

- *If you use multiple AND conditions, write the condition most likely to be false first.* If you use this technique, the DBMS will stop evaluating the rest of the conditions as soon as it finds a conditional expression that is evaluated as false. Remember, for multiple AND conditions to be found true, all conditions must be evaluated as true. If one of the conditions evaluates to false, the whole set of conditions will be evaluated as false. If you use this technique, the DBMS will not waste time unnecessarily evaluating additional conditions. Naturally, the use of this technique implies knowledge of the sparsity of the data set. For example, look at the following condition list:

P_PRICE > 10 AND V_STATE = 'FL'

If you know that only a few vendors are located in Florida, you could rewrite this condition as:

V_STATE = 'FL' AND P_PRICE > 10

- *When using multiple OR conditions, put the condition most likely to be true first.* By doing this, the DBMS will stop evaluating the remaining conditions as soon as it finds a conditional expression that is evaluated as true. Remember, for multiple OR conditions to evaluate to true, only one of the conditions must be evaluated as true.

- *Whenever possible, try to avoid the use of the NOT logical operator.* It is best to transform a SQL expression that contains a NOT logical operator into an equivalent expression. For example:

NOT (P_PRICE > 10.00) can be written as P_PRICE <= 10.00.

Also, NOT (EMP_SEX = 'M') can be written as EMP_SEX = 'F'.

Note

Oracle does not evaluate queries as described here. Instead, Oracle evaluates conditions from last to first.

11-6 Query Formulation

Queries are usually written to answer questions. For example, if an end user gives you a sample output and tells you to match that output format, you must write the corresponding SQL. To get the job done, you must carefully evaluate what columns, tables, and computations are required to generate the desired output. To do that, you must have a good understanding of the database environment and the database that will be the focus of your SQL code.

This section focuses on SELECT queries because they are the queries you will find in most applications. To formulate a query, you would normally follow these steps:

1. *Identify what columns and computations are required.* The first step is needed to clearly determine what data values you want to return. Do you want to return just the names and addresses, or do you also want to include some computations? Remember that all columns in the SELECT statement should return single values.

   a. Do you need simple expressions? For example, do you need to multiply the price by the quantity on hand to generate the total inventory cost? You might need some single attribute functions such as DATE(), SYSDATE(), or ROUND().

   b. Do you need aggregate functions? If you need to compute the total sales by product, you should use a GROUP BY clause. In some cases, you might need to use a subquery.

   c. Determine the granularity of the raw data required for your output. Sometimes, you might need to summarize data that is not readily available in any table. In such cases, you might consider breaking the query into multiple subqueries and storing those subqueries as views. Then you could create a top-level query that joins those views and generates the final output.

2. *Identify the source tables.* Once you know what columns are required, you can determine the source tables used in the query. Some attributes appear in more than one table. In those cases, try to use the least number of tables in your query to minimize the number of join operations.

3. *Determine how to join the tables.* Once you know what tables you need in your query statement, you must properly identify how to join the tables. In most cases, you will use some type of natural join, but in some instances, you might need to use an outer join.

4. *Determine what selection criteria are needed.* Most queries involve some type of selection criteria. In this case, you must determine what operands and operators are needed in your criteria. Ensure that the data type and granularity of the data in the comparison criteria are correct.

   a. *Simple comparison.* In most cases, you will be comparing single values—for example, P_PRICE > 10.

   b. *Single value to multiple values.* If you are comparing a single value to multiple values, you might need to use an IN comparison operator—for example, V_STATE IN ('FL', 'TN', 'GA').

   c. *Nested comparisons.* In other cases, you might need to have some nested selection criteria involving subqueries—for example, P_PRICE >= (SELECT AVG (P_PRICE) FROM PRODUCT).

   d. *Grouped data selection.* On other occasions, the selection criteria might apply not to the raw data but to the aggregate data. In those cases, you need to use the HAVING clause.

5. *Determine the order in which to display the output.* Finally, the required output might be ordered by one or more columns. In those cases, you need to use the ORDER BY clause. Remember that the ORDER BY clause is one of the most resource-intensive operations for the DBMS.

11-7 DBMS Performance Tuning

DBMS performance tuning includes global tasks such as managing the DBMS processes in primary memory (allocating memory for caching purposes) and managing the structures in physical storage (allocating space for the data files).

Fine-tuning the performance of the DBMS also includes applying several practices examined in the previous section. For example, the DBA must work with developers to ensure that the queries perform as expected—creating the indexes to speed up query response time and generating the database statistics required by cost-based optimizers.

DBMS performance tuning at the server end focuses on setting the parameters used for:

- *Data cache.* The data cache size must be set large enough to permit as many data requests as possible to be serviced from the cache. Each DBMS has settings that control the size of the data cache; some DBMSs might require a restart. This cache is shared among all database users. The majority of primary memory resources will be allocated to the data cache.

- *SQL cache.* The SQL cache stores the most recently executed SQL statements (after the SQL statements have been parsed by the optimizer). Generally, if you have an application with multiple users accessing a database, the *same* query will likely be submitted by many different users. In those cases, the DBMS will parse the query only once and execute it many times, using the same access plan. In that way, the second and subsequent SQL requests for the same query are served from the SQL cache, skipping the parsing phase.

- *Sort cache.* The sort cache is used as a temporary storage area for ORDER BY or GROUP BY operations, as well as for index-creation functions.

- *Optimizer mode.* Most DBMSs operate in one of two optimization modes: cost-based or rule-based. Others automatically determine the optimization mode based on whether database statistics are available. For example, the DBA is responsible for generating the database statistics that are used by the cost-based optimizer. If the statistics are not available, the DBMS uses a rule-based optimizer.

From the performance point of view, it would be optimal to have the entire database stored in primary memory to minimize costly disk access. This is why several database vendors offer in-memory database options for their main products. **In-memory database** systems are optimized to store large portions (if not all) of the database in primary (RAM) storage rather than secondary (disk) storage. These systems are becoming popular because increasing performance demands of modern database applications (such as Business Analytics and Big Data), diminishing costs, and technology advances of components (such as flash-memory and solid state drives.) Even though these type of databases "eliminate" disk access bottlenecks, they are still subject to query optimization and performance tuning rules, especially when faced with poorly designed databases or poorly written SQL statements.

Although in-memory databases are carving a niche in selected markets, most database implementations still rely on data stored on disk drives. That is why managing the physical storage details of the data files plays an important role in DBMS performance tuning. Note the following general recommendations for physical storage of databases:

- Use **I/O accelerators**. This type of device uses flash solid-state drives (SSD) to store the database. A solid-state drive does not have any moving parts and, therefore performs I/O operations at a higher speed than traditional rotating disk drives. I/O accelerators deliver high transaction performance rates and reduce contention caused by typical storage drives.

in-memory database
A database optimized to store large portions (if not all) of the database in primary (RAM) storage rather than secondary (disk) storage.

I/O accelerator
A device used to improve throughput for input/output operations.

- Use **RAID** (Redundant Array of Independent Disks) to provide both performance improvement and fault tolerance, and a balance between them. Fault tolerance means that in case of failure, data can be reconstructed and retrieved. RAID systems use multiple disks to create virtual disks (storage volumes) formed by several individual disks. Table 11.7 describes the most common RAID configurations.

TABLE 11.7

COMMON RAID LEVELS

RAID LEVEL	DESCRIPTION
0	The data blocks are spread over separate drives. Also known as *striped array*. Provides increased performance but no fault tolerance. Requires a minimum of two drives.
1	The same data blocks are written (duplicated) to separate drives. Also referred to as *mirroring* or *duplexing*. Provides increased read performance and fault tolerance via data redundancy. Requires a minimum of two drives.
3	The data is striped across separate drives, and parity data is computed and stored in a dedicated drive. (Parity data is specially generated data that permits the reconstruction of corrupted or missing data.) Provides good read performance and fault tolerance via parity data. Requires a minimum of three drives.
5	The data and the parity data is striped across separate drives. Provides good read performance and fault tolerance via parity data. Requires a minimum of three drives.
1+0	The data blocks are spread over separate drives and mirrored (duplicated). This arrangement provides both speed and fault tolerance. This is the recommended RAID configuration for most database installations (if cost is not an issue).

- Minimize disk contention. Use multiple, independent storage volumes with independent spindles (rotating disks) to minimize hard disk cycles. Remember, a database is composed of many table spaces, each with a particular function. In turn, each table space is composed of several data files in which the data is actually stored. A database should have at least the following table spaces:

  - *System table space.* This is used to store the data dictionary tables. It is the most frequently accessed table space and should be stored in its own volume.

  - *User data table space.* This is used to store end-user data. You should create as many user data table spaces and data files as are required to balance performance and usability. For example, you can create and assign a different user data table space for each application and each distinct group of users, but this is not necessary for each user.

  - *Index table space.* This is used to store indexes. You can create and assign a different index table space for each application and each group of users. The index table space data files should be stored on a storage volume that is separate from user data files or system data files.

  - *Temporary table space.* This is used as a temporary storage area for merge, sort, or set aggregate operations. You can create and assign a different temporary table space for each application and each group of users.

  - *Rollback segment table space.* This is used for transaction-recovery purposes.

- Put high-usage tables in their own table spaces so the database minimizes conflict with other tables.

RAID
An acronym for Redundant Array of Independent Disks. RAID systems use multiple disks to create virtual disks (storage volumes) from several individual disks. RAID systems provide performance improvement, fault tolerance, and a balance between the two.

- Assign separate data files in separate storage volumes for the indexes, system, and high-usage tables. This ensures that index operations will not conflict with end-user data or data dictionary table access operations. Another advantage of this approach is that you can use different disk block sizes in different volumes. For example, the data volume can use a 16 K block size, while the index volume can use an 8 K block size. Remember that the index record size is generally smaller, and by changing the block size you will reduce contention and minimize I/O operations. This is very important; many database administrators overlook indexes as a source of contention. By using separate storage volumes and different block sizes, the I/O operations on data and indexes will happen asynchronously (at different times); more importantly, the likelihood of write operations blocking read operations is reduced, as page locks tend to lock fewer records.

- Take advantage of the various table storage organizations available in the database. For example, in Oracle consider the use of index-organized tables (IOT); in SQL Server, consider clustered index tables. An **index-organized table** (or **clustered index table**) is a table that stores the end-user data and the index data in consecutive locations on permanent storage. This type of storage organization provides a performance advantage to tables that are commonly accessed through a given index order, because the index contains the index key as well as the data rows. Therefore, the DBMS tends to perform fewer I/O operations.

- Partition tables based on usage. Some RDBMSs support the horizontal partitioning of tables based on attributes. (See Chapter 12, Distributed Database Management Systems.) By doing so, a single SQL request can be processed by multiple data processors. Put the table partitions closest to where they are used the most.

- Use denormalized tables where appropriate. In other words, you might be able to improve performance by taking a table from a higher normal form to a lower normal form—typically, from third to second normal form. This technique adds data duplication, but it minimizes join operations. (Denormalization was discussed in Chapter 6, Normalization of Database Tables.)

- Store computed and aggregate attributes in tables. In short, use derived attributes in your tables. For example, you might add the invoice subtotal, the amount of tax, and the total in the INVOICE table. Using derived attributes minimizes computations in queries and join operations, especially during the execution of aggregate queries.

11-8 Query Optimization Example

Now that you have learned the basis of query optimization, you are ready to test your new knowledge. A simple example illustrates how the query optimizer works and how you can help it work. The example is based on the QOVENDOR and QOPRODUCT tables, which are similar to tables you used in previous chapters. However, the QO prefix is used for the table name to ensure that you do not overwrite previous tables.

To perform this query optimization example, you will use the Oracle SQL*Plus interface. Some preliminary work must be done before you can start testing query optimization, as explained in the following steps:

1. Log in to Oracle SQL*Plus using the username and password provided by your instructor.

2. Create a fresh set of tables, using the QRYOPTDATA.SQL script file (available at *www.cengagebrain.com*). This step is necessary so that Oracle has a new set of tables and the new tables contain no statistics. At the SQL> prompt, type:

@path\QRYOPTDATA.SQL

where *path* is the location of the file in your computer.

index organized table
In a DBMS, a type of table storage organization that stores end-user data and index data in consecutive locations in permanent storage. Also known as *cluster-indexed table.*

clustered index table
See *index organized table.*

3. Create the PLAN_TABLE, which is a special table used by Oracle to store the access plan information for a given query. End users can then query the PLAN_TABLE to see how Oracle will execute the query. To create the PLAN_TABLE, run the UTLXPLAN.SQL script file in the RDBMS\ADMIN folder of your Oracle RDBMS installation. The UTLXPLAN.SQL script file is also available at *www.cengagebrain.com*. At the SQL prompt, type:

@path\UTLXPLAN.SQL

You use the EXPLAIN PLAN command to store the execution plan of a SQL query in the PLAN_TABLE. Then, you use the SELECT * FROM TABLE(DBMS_XPLAN.DISPLAY) command to display the access plan for a given SQL statement.

Note

Oracle 12c, MySQL, and SQL Server all default to cost-based optimization. In Oracle, if table statistics are not available, the DBMS will fall back to a rule-based optimizer.

To see the access plan used by the DBMS to execute your query, use the EXPLAIN PLAN and SELECT statements, as shown in Figure 11.5. Note that the first SQL statement generates the statistics for the QOVENDOR table. Also, the initial access plan in Figure 11.5 uses a full table scan on the QOVENDOR table, and the cost of the plan is 4.

FIGURE 11.5 INITIAL EXPLAIN PLAN

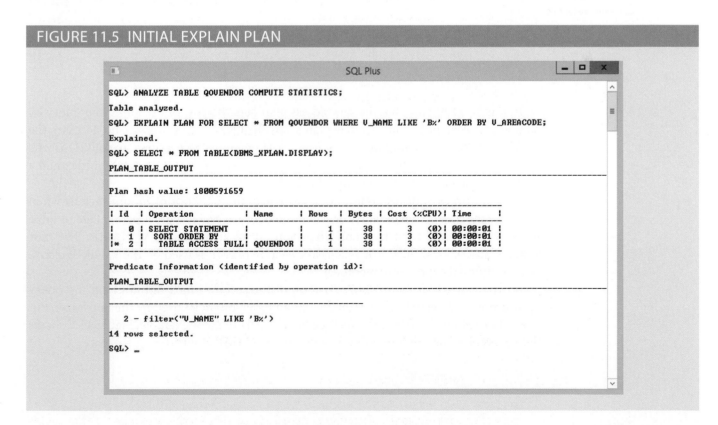

Now create an index on V_AREACODE (note that V_AREACODE is used in the ORDER BY clause) and see how it affects the access plan generated by the cost-based optimizer. The results are shown in Figure 11.6.

FIGURE 11.6   EXPLAIN PLAN AFTER INDEX ON V_AREACODE

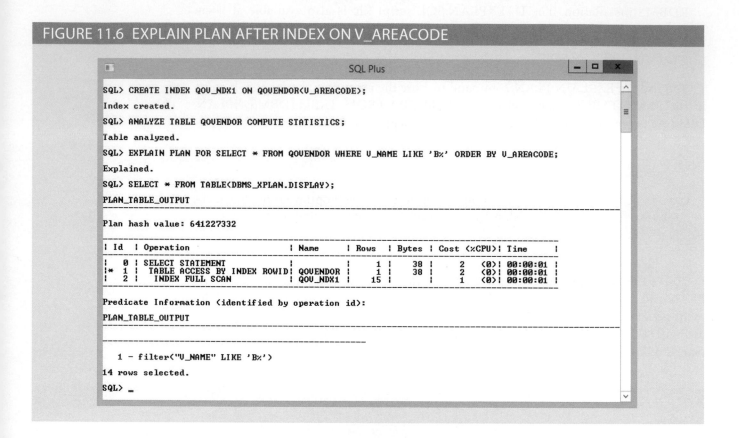

In Figure 11.6, note that the new access plan cuts the cost of executing the query by 30 percent! Also note that this new plan scans the QOV_NDX1 index and accesses the QOVENDOR rows, using the index row ID. (Remember that access by row ID is one of the fastest access methods.) In this case, the creation of the QOV_NDX1 index had a positive impact on overall query optimization results.

At other times, indexes do not necessarily help in query optimization, such as when you have indexes on small tables or when the query accesses a great percentage of table rows anyway. Note what happens when you create an index on V_NAME. The new access plan is shown in Figure 11.7. (Note that V_NAME is used on the WHERE clause as a conditional expression operand.)

As you can see in Figure 11.7, creation of the second index did not help the query optimization. However, on some occasions an index might be used by the optimizer, but it is not executed because of the way the query is written. For example, Figure 11.8 shows the access plan for a different query using the V_NAME column.

FIGURE 11.7 EXPLAIN PLAN AFTER INDEX ON V_NAME

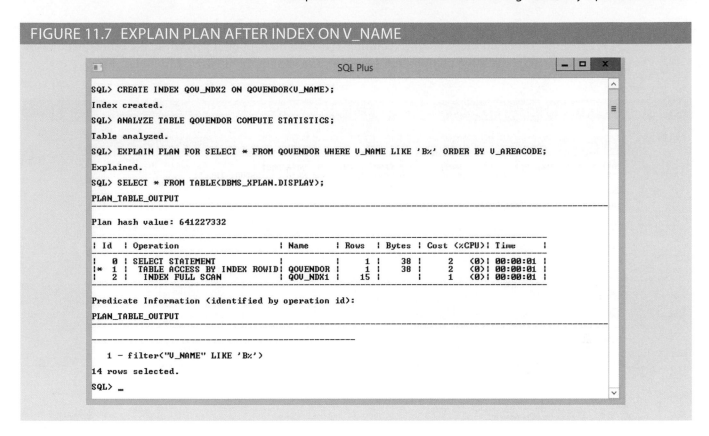

FIGURE 11.8 ACCESS PLAN USING INDEX ON V_NAME

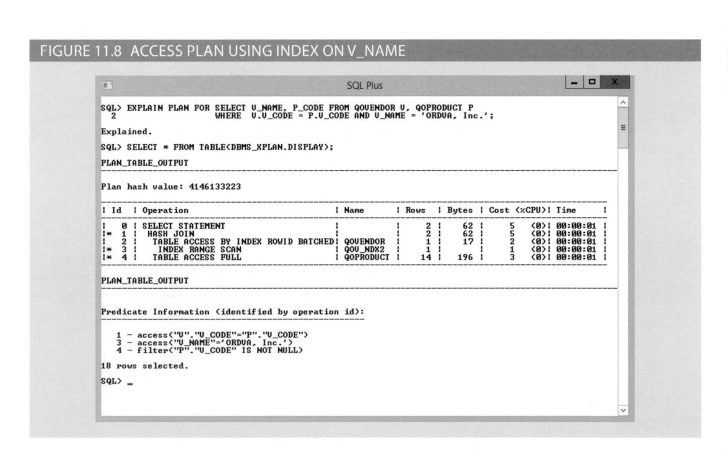

In Figure 11.8, note that the access plan for this new query uses the QOV_NDX2 index on the V_NAME column. What would happen if you wrote the same query, using the UPPER function on V_NAME? The results are illustrated in Figure 11.9.

FIGURE 11.9 ACCESS PLAN USING FUNCTIONS ON INDEXED COLUMNS

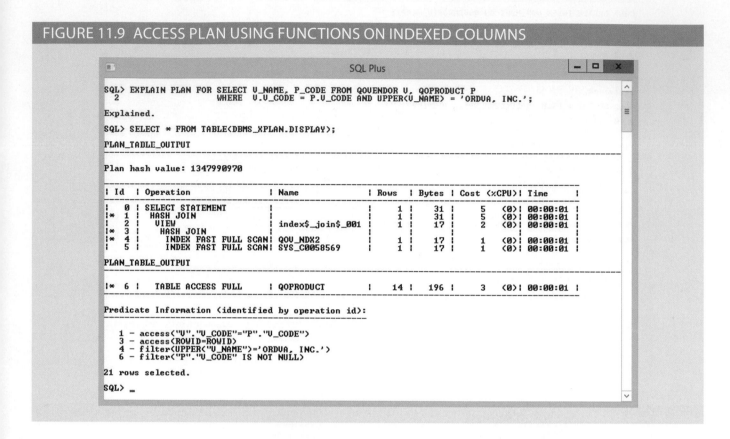

As Figure 11.9 shows, the use of a function on an indexed column caused the DBMS to perform additional operations that could potentially increase the cost of the query. The same query might produce different costs if your tables contain many more rows and if the index sparsity is different.

Now use the QOPRODUCT table to demonstrate how an index can help when aggregate function queries are being run. For example, Figure 11.10 shows the access plan for a SELECT statement using the MAX(P_PRICE) aggregate function. This plan uses a full table scan with a total cost of 3.

A cost of 3 is very low already, but you could improve the previous query performance by creating an index on P_PRICE. Figure 11.11 shows how the plan cost is reduced by two-thirds after the index is created and the QOPRODUCT table is analyzed. Also note that the second version of the access plan uses only the index QOP_NDX2 to answer the query; *the QOPRODUCT table is never accessed.*

FIGURE 11.10   FIRST EXPLAIN PLAN: AGGREGATE FUNCTION ON A NON-INDEXED COLUMN

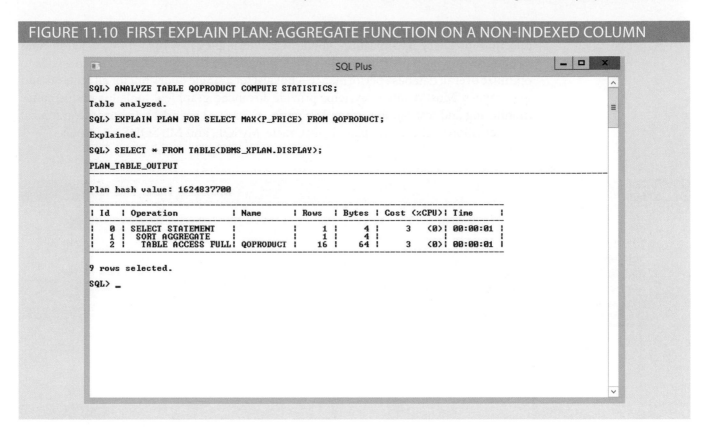

FIGURE 11.11   SECOND EXPLAIN PLAN: AGGREGATE FUNCTION ON AN INDEXED COLUMN

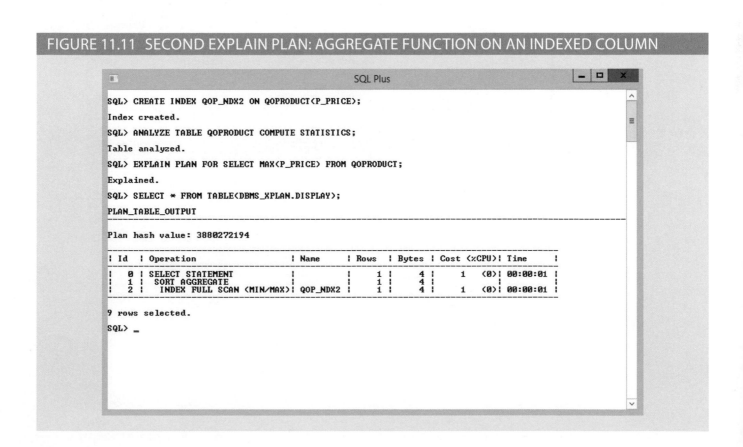

Although the few examples in this section show the importance of proper index selection for query optimization, you also saw examples in which index creation does not improve query performance. As a DBA, you should be aware that the main goal is to optimize overall database performance—not just for a single query but for all requests and query types. Most database systems provide advanced graphical tools for performance monitoring and testing. For example, Figures 11.12, 11.13, and 11.4 show the graphical representation of the access plan using Oracle, MySQL, and MS SQL Server tools.

FIGURE 11.12   ORACLE TOOLS FOR QUERY OPTIMIZATION

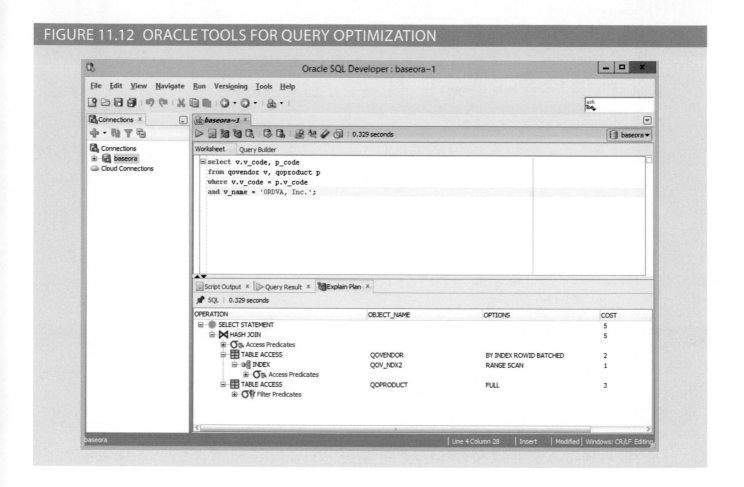

FIGURE 11.13 MYSQL TOOLS FOR QUERY OPTIMIZATION

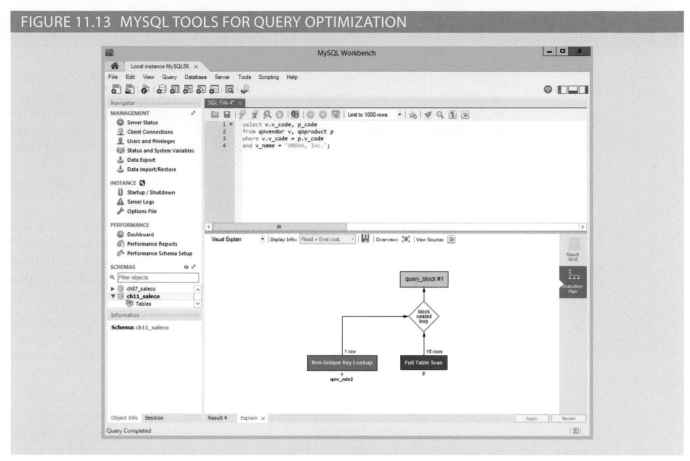

FIGURE 11.14 MICROSOFT SQL SERVER TOOLS FOR QUERY OPTIMIZATION

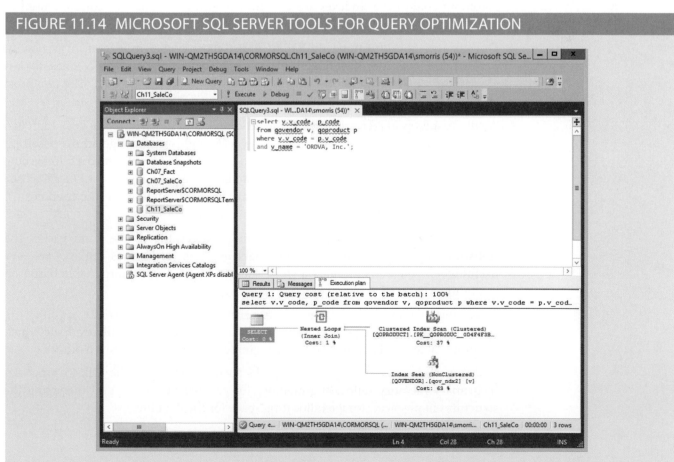

Summary

- Database performance tuning refers to a set of activities and procedures designed to ensure that an end-user query is processed by the DBMS in the least amount of time. SQL performance tuning refers to activities on the client side that are designed to generate SQL code that returns the correct answer in the least amount of time, using the minimum amount of resources at the server end. DBMS performance tuning refers to activities on the server side that are oriented so the DBMS is properly configured to respond to clients' requests in the fastest way possible while making optimum use of existing resources.

- Database statistics refer to a number of measurements gathered by the DBMS that describe a snapshot of the database objects' characteristics. The DBMS gathers statistics about objects such as tables, indexes, and available resources, which include the number of processors used, processor speed, and temporary space available. The DBMS uses the statistics to make critical decisions about improving query processing efficiency.

- DBMSs process queries in three phases. In the parsing phase, the DBMS parses the SQL query and chooses the most efficient access/execution plan. In the execution phase, the DBMS executes the SQL query using the chosen execution plan. In the fetching phase, the DBMS fetches the data and sends the result set back to the client.

- Indexes are crucial in the process that speeds up data access. Indexes facilitate searching, sorting, and using aggregate functions and join operations. The improvement in data access speed occurs because an index is an ordered set of values that contains the index key and pointers. Data sparsity refers to the number of different values a column could have. Indexes are recommended in high-sparsity columns used in search conditions.

- During query optimization, the DBMS must choose what indexes to use, how to perform join operations, which table to use first, and so on. Each DBMS has its own algorithms for determining the most efficient way to access the data. The two most common approaches are rule-based and cost-based optimization.

- A rule-based optimizer uses preset rules and points to determine the best approach to execute a query. A cost-based optimizer uses sophisticated algorithms based on statistics about the objects being accessed to determine the best approach to execute a query. In this case, the optimizer process adds up the processing cost, the I/O costs, and the resource costs (RAM and temporary space) to determine the total cost of a given execution plan.

- SQL performance tuning deals with writing queries that make good use of the statistics. In particular, queries should make good use of indexes. Indexes are very useful when you want to select a small subset of rows from a large table based on a condition.

- Query formulation deals with how to translate business questions into specific SQL code to generate the required results. To do this, you must carefully evaluate which columns, tables, and computations are required to generate the desired output.

- DBMS performance tuning includes tasks such as managing the DBMS processes in primary memory (allocating memory for caching purposes) and managing the structures in physical storage (allocating space for the data files).

Key Terms

access plan	dynamic query optimization	optimizer hints
automatic query optimization	dynamic statistical generation mode	procedure cache
bitmap index	extents	query optimizer
B-tree index	file group	query processing bottleneck
buffer cache	function-based index	RAID
clustered index table	hash index	rule-based optimizer
cost-based optimizer	in-memory database	rule-based query optimization algorithm
database performance tuning	index-organized table	static query optimization
database statistics	index selectivity	statistically based query optimization algorithm
data cache	input/output (I/O) request	
data files	I/O accelerator	SQL cache
data sparsity	manual query optimization	SQL performance tuning
DBMS performance tuning	manual statistical generation mode	table space

Online Content

Flashcards and crossword puzzles for key term practice are available at *www.cengagebrain.com*.

Review Questions

1. What is SQL performance tuning?

2. What is database performance tuning?

3. What is the focus of most performance-tuning activities, and why does that focus exist?

4. What are database statistics, and why are they important?

5. How are database statistics obtained?

6. What database statistics measurements are typical of tables, indexes, and resources?

7. How is the processing of SQL DDL statements (such as CREATE TABLE) different from the processing required by DML statements?

8. In simple terms, the DBMS processes a query in three phases. What are the phases, and what is accomplished in each phase?

9. If indexes are so important, why not index every column in every table? (Include a brief discussion of the role played by data sparsity.)

10. What is the difference between a rule-based optimizer and a cost-based optimizer?

11. What are optimizer hints, and how are they used?

12. What are some general guidelines for creating and using indexes?

13. Most query optimization techniques are designed to make the optimizer's work easier. What factors should you keep in mind if you intend to write conditional expressions in SQL code?

14. What recommendations would you make for managing the data files in a DBMS with many tables and indexes?

15. What does RAID stand for, and what are some commonly used RAID levels?

Problems

Problems 1 and 2 are based on the following query:

```
SELECT      EMP_LNAME, EMP_FNAME, EMP_AREACODE, EMP_SEX
FROM        EMPLOYEE
WHERE       EMP_SEX = 'F' AND EMP_AREACODE = '615'
ORDER BY    EMP_LNAME, EMP_FNAME;
```

1. What is the likely data sparsity of the EMP_SEX column?

2. What indexes should you create? Write the required SQL commands.

3. Using Table 11.4 as an example, create two alternative access plans. Use the following assumptions:

   a. There are 8,000 employees.

   b. There are 4,150 female employees.

   c. There are 370 employees in area code 615.

   d. There are 190 female employees in area code 615.

Problems 4–6 are based on the following query:

```
SELECT      EMP_LNAME, EMP_FNAME, EMP_DOB, YEAR(EMP_DOB) AS YEAR
FROM        EMPLOYEE
WHERE       YEAR(EMP_DOB) = 1976;
```

4. What is the likely data sparsity of the EMP_DOB column?

5. Should you create an index on EMP_DOB? Why or why not?

6. What type of database I/O operations will likely be used by the query? (See Table 11.3.)

Problems 7–10 are based on the ER model shown in Figure P11.7 and on the query shown after the figure.

```
SELECT      P_CODE, P_PRICE
FROM        PRODUCT
WHERE       P_PRICE >= (SELECT AVG(P_PRICE) FROM PRODUCT);
```

7. Assuming there are no table statistics, what type of optimization will the DBMS use?

8. What type of database I/O operations will likely be used by the query? (See Table 11.3.)

9. What is the likely data sparsity of the P_PRICE column?

10. Should you create an index? Why or why not?

FIGURE P11.7 THE CH11_SALECO ER MODEL

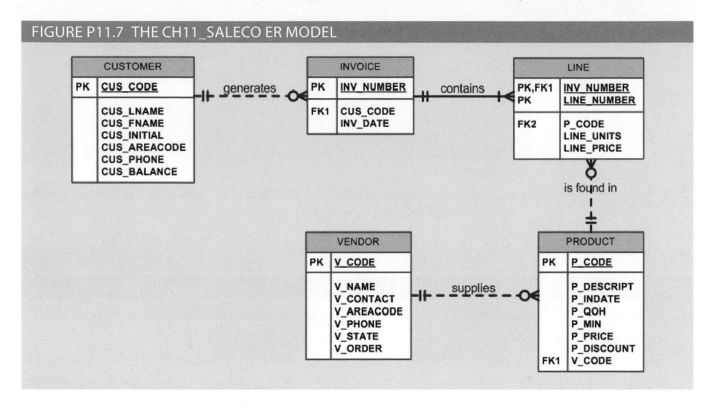

Problems 11–14 are based on the following query:

SELECT   P_CODE, SUM(LINE_UNITS)
FROM   LINE
GROUP BY   P_CODE
HAVING   SUM(LINE_UNITS) > (SELECT MAX(LINE_UNITS) FROM LINE);

11. What is the likely data sparsity of the LINE_UNITS column?

12. Should you create an index? If so, what would the index column(s) be, and why would you create the index? If not, explain your reasoning.

13. Should you create an index on P_CODE? If so, write the SQL command to create the index. If not, explain your reasoning.

14. Write the command to create statistics for this table.

Problems 15 and 16 are based on the following query:

SELECT   P_CODE, P_QOH*P_PRICE
FROM   PRODUCT
WHERE   P_QOH*P_PRICE > (SELECT AVG(P_QOH*P_PRICE)
                  FROM PRODUCT);

15. What is the likely data sparsity of the P_QOH and P_PRICE columns?

16. Should you create an index? If so, what would the index column(s) be, and why should you create the index?

Problems 17–21 are based on the following query:

```
SELECT          V_CODE, V_NAME, V_CONTACT, V_STATE
FROM            VENDOR
WHERE           V_STATE = 'TN'
ORDER BY        V_NAME;
```

17. What indexes should you create and why? Write the SQL command to create the indexes.

18. Assume that 10,000 vendors are distributed as shown in Table P11.18. What percentage of rows will be returned by the query?

TABLE P11.18

STATE	NUMBER OF VENDORS	STATE	NUMBER OF VENDORS
AK	15	MS	47
AL	55	NC	358
AZ	100	NH	25
CA	3244	NJ	645
CO	345	NV	16
FL	995	OH	821
GA	75	OK	62
HI	68	PA	425
IL	89	RI	12
IN	12	SC	65
KS	19	SD	74
KY	45	TN	113
LA	29	TX	589
MD	208	UT	36
MI	745	VA	375
MO	35	WA	258

19. What type of I/O database operations would most likely be used to execute the query?

20. Using Table 11.4 as an example, create two alternative access plans.

21. Assume that you have 10,000 different products stored in the PRODUCT table and that you are writing a web-based interface to list all products with a quantity on hand (P_QOH) that is less than or equal to the minimum quantity, P_MIN. What optimizer hint would you use to ensure that your query returns the result set to the web interface in the least time possible? Write the SQL code.

Problems 22–24 are based on the following query:

```
SELECT          P_CODE, P_DESCRIPT, P_PRICE, P.V_CODE, V_STATE
FROM            PRODUCT P, VENDOR V
WHERE           P.V_CODE = V.V_CODE
                AND V_STATE = 'NY'
                AND V_AREACODE = '212'
ORDER BY        P_PRICE;
```

22. What indexes would you recommend?

23. Write the commands required to create the indexes you recommended in Problem 22.

24. Write the command(s) used to generate the statistics for the PRODUCT and VENDOR tables.

Problems 25 and 26 are based on the following query:

```
SELECT          P_CODE, P_DESCRIPT, P_QOH, P_PRICE, V_CODE
FROM            PRODUCT
WHERE           V_CODE = '21344'
ORDER BY        P_CODE;
```

25. What index would you recommend, and what command would you use?

26. How should you rewrite the query to ensure that it uses the index you created in your solution to Problem 25?

Problems 27 and 28 are based on the following query:

```
SELECT          P_CODE, P_DESCRIPT, P_QOH, P_PRICE, V_CODE
FROM            PRODUCT
WHERE           P_QOH < P_MIN
                AND P_MIN = P_REORDER
                AND P_REORDER = 50
ORDER BY        P_QOH;
```

27. Use the recommendations given in Section 11-5b to rewrite the query and produce the required results more efficiently.

28. What indexes would you recommend? Write the commands to create those indexes.

Problems 29–32 are based on the following query:

```
SELECT          CUS_CODE, MAX(LINE_UNITS*LINE_PRICE)
FROM            CUSTOMER NATURAL JOIN INVOICE NATURAL JOIN LINE
WHERE           CUS_AREACODE = '615'
GROUP BY        CUS_CODE;
```

29. Assuming that you generate 15,000 invoices per month, what recommendation would you give the designer about the use of derived attributes?

30. Assuming that you follow the recommendations you gave in Problem 29, how would you rewrite the query?

31. What indexes would you recommend for the query you wrote in Problem 30, and what SQL commands would you use?

32. How would you rewrite the query to ensure that the index you created in Problem 31 is used?

Chapter 12

Distributed Database Management Systems

In this chapter, you will learn:
- About distributed database management systems (DDBMSs) and their components
- How database implementation is affected by different levels of data and process distribution
- How transactions are managed in a distributed database environment
- How distributed database design draws on data partitioning and replication to balance performance, scalability, and availability
- About the trade-offs of implementing a distributed data system

Preview

In this chapter, you will learn that a single database can be divided into several fragments stored on different computers within a geographically dispersed network. Processing also can be dispersed among several different network sites, or nodes.

The growth of distributed database systems has been fostered by the increased globalization of business operations, the accumulation of massive organizational data sets, and technological changes that have made distributed network-based services practical, more reliable, and cost-effective.

The distributed database management system (DDBMS) treats a distributed database as a single logical database; therefore, the basic design concepts you learned in earlier chapters apply. However, the distribution of data among different sites in a computer network adds to the system's complexity. For example, the design of a distributed database must consider the location of the data, partitioning the data into fragments, and replication of those fragments. Although a distributed database system requires a more sophisticated DBMS, the greater complexity of a distributed database system should be transparent to the end user.

In today's web-centric environment, any distributed data system must be highly scalable; in other words, it must grow dynamically as demand increases. To accommodate such dynamic growth, trade-offs must be made to achieve some desirable properties.

Data Files and Available Formats

	MS Access	Oracle	MS SQL	My SQL		MS Access	Oracle	MS SQL	My SQL
CH12_Text	✓	✓	✓	✓					

Data Files Available on cengagebrain.com

12-1 The Evolution of Distributed Database Management Systems

A **distributed database management system (DDBMS)** governs the storage and processing of logically related data over interconnected computer systems in which both data and processing are distributed among several sites. To understand how and why the DDBMS is different from the DBMS, it is useful to briefly examine the changes in the business environment that set the stage for the development of the DDBMS.

During the 1970s, corporations implemented centralized database management systems to meet their structured information needs. The use of a centralized database required that corporate data be stored in a single central site, usually a mainframe computer. Data access was provided through dumb terminals. The centralized approach, illustrated in Figure 12.1, worked well to fill the structured information needs of corporations, but it fell short when quickly moving events required faster response times and equally quick access to information. The slow progression from information request to approval to specialist to user simply did not serve decision makers well in a dynamic environment. What was needed was quick, unstructured access to databases, using ad hoc queries to generate on-the-spot information.

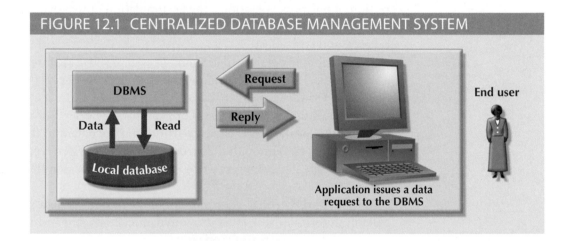

FIGURE 12.1 CENTRALIZED DATABASE MANAGEMENT SYSTEM

The last two decades gave birth to a series of crucial social and technological changes that affected the nature of the systems and the data they use:

- Business operations became global; with this change, competition expanded from the shop on the next corner to the web store in cyberspace.

- Customer demands and market needs favored an on-demand transaction style, mostly based on web-based services.

- Rapid social and technological changes fueled by low-cost, smart mobile devices increased the demand for complex and fast networks to interconnect them. As a consequence, corporations have increasingly adopted advanced network technologies as the platform for their computerized solutions. See Chapter 15, Database Connectivity and Web Technologies, for a discussion of cloud-based services.

- Data realms are converging in the digital world more frequently. As a result, applications must manage multiple types of data, such as voice, video, music, and images. Such data tends to be geographically distributed and remotely accessed from diverse locations via location-aware mobile devices.

distributed database management system (DDBMS)
A DBMS that supports a database distributed across several different sites; a DDBMS governs the storage and processing of logically related data over interconnected computer systems in which both data and processing functions are distributed among several sites.

- The advent of social media as a way to reach new customers and open new markets has fueled the need to store large amounts of digital data and created a revolution in the way data is managed and mined for knowledge. Businesses are looking for new ways to gain business intelligence through the analysis of vast stores of structured and unstructured data.

These factors created a dynamic business environment in which companies had to respond quickly to competitive and technological pressures. As large business units restructured to form leaner, quickly reacting, dispersed operations, two database requirements became obvious:

- *Rapid ad hoc data access* became crucial in the quick-response decision-making environment.

- *Distributed data access* was needed to support geographically dispersed business units.

During recent years, these factors became even more firmly entrenched. However, the way they were addressed was strongly influenced by the following factors:

- *The growing acceptance of the Internet as the platform for data access and distribution.* The web is effectively the repository for distributed data.

- *The mobile wireless revolution.* The widespread use of mobile wireless digital devices includes smartphones and tablets. These devices have created high demand for data access. They access data from geographically dispersed locations and require varied data exchanges in multiple formats, such as data, voice, video, music, and pictures. Although distributed data access does not necessarily imply distributed databases, performance and failure tolerance requirements often lead to the use of data replication techniques similar to those in distributed databases.

- *The accelerated growth of companies using "applications as a service."* This new type of service provides remote applications to companies that want to outsource their application development, maintenance, and operations. The company data is generally stored on central servers and is not necessarily distributed. Just as with mobile data access, this type of service may not require fully distributed data functionality; however, other factors such as performance and failure tolerance often require the use of data replication techniques similar to those in distributed databases.

- *The increased focus on mobile business intelligence.* More and more companies are embracing mobile technologies within their business plans. As companies use social networks to get closer to customers, the need for on-the-spot decision making increases. Although a data warehouse is not usually a distributed database, it does rely on techniques such as data replication and distributed queries that facilitate data extraction and integration. (You will learn more about this topic in Chapter 13, Business Intelligence and Data Warehouses.)

- *Emphasis on Big Data analytics.* The era of mobile communications unraveled an avalanche of data from many sources and of many types. Today's customers have significant influence on the spending habits of communities, and organizations are investing in ways to harvest such data to "discover" new ways to effectively and efficiently reach customers.

At this point, the long-term impact of the Internet and the mobile revolution on *distributed* database design and management is just starting to be felt. Perhaps the success of the Internet and mobile technologies will foster the use of distributed databases as bandwidth becomes a less troublesome bottleneck. Perhaps the resolution of bandwidth problems will simply confirm the centralized database standard. In any case, distributed

Online Content

To learn more about the Internet's impact on data access and distribution, see Appendix I, Databases in Electronic Commerce, at *www.cengagebrain.com*.

database concepts and components are likely to find a place in future database development, particularly for specialized mobile and location-aware applications.

The distributed database is especially desirable because centralized database management is subject to problems such as:

- *Performance degradation* because of a growing number of remote locations over greater distances.

- *High costs* associated with maintaining and operating large central (mainframe) database systems and physical infrastructure.

- *Reliability problems* created by dependence on a central site (single point of failure syndrome) and the need for data replication.

- *Scalability problems* associated with the physical limits imposed by a single location, such as physical space, temperature conditioning, and power consumption.

- *Organizational rigidity* imposed by the database, which means it might not support the flexibility and agility required by modern global organizations.

The dynamic business environment and the centralized database's shortcomings spawned a demand for applications based on accessing data from different sources at multiple locations. Such a multiple-source/multiple-location database environment is best managed by a DDBMS.

12-2 DDBMS Advantages and Disadvantages

Distributed database management systems deliver several advantages over traditional systems. At the same time, they are subject to some problems. Table 12.1 summarizes the advantages and disadvantages associated with a DDBMS.

Distributed databases are being used successfully in many web staples such as Google and Amazon, but they still have a long way to go before they yield the full flexibility and power they theoretically possess.

The remainder of this chapter explores the basic components and concepts of the distributed database. Because the distributed database is usually based on the relational database model, relational terminology is used to explain the basic concepts and components. Even though some of the most widely used distributed databases are part of the NoSQL movement (see Chapter 2, Data Models), the basic concepts and fundamentals of distributed data still apply to them.

12-3 Distributed Processing and Distributed Databases

In **distributed processing**, a database's logical processing is shared among two or more physically independent sites that are connected through a network. For example, the data input/output (I/O), data selection, and data validation might be performed on one computer, and a report based on that data might be created on another computer.

A basic distributed processing environment is illustrated in Figure 12.2, which shows that a distributed processing system shares the database processing chores among three sites connected through a communications network. Although the database resides at only one site (Miami), each site can access the data and update the database. The database is located on Computer A, a network computer known as the *database server*.

A **distributed database**, on the other hand, stores a logically related database over two or more physically independent sites. The sites are connected via a computer

distributed processing
Sharing the logical processing of a database over two or more sites connected by a network.

distributed database
A logically related database that is stored in two or more physically independent sites.

TABLE 12.1

DISTRIBUTED DBMS ADVANTAGES AND DISADVANTAGES

ADVANTAGES	DISADVANTAGES
Data is located near the site of greatest demand. The data in a distributed database system is dispersed to match business requirements.	*Complexity of management and control.* Applications must recognize data location, and they must be able to stitch together data from various sites. Database administrators must have the ability to coordinate database activities to prevent database degradation due to data anomalies.
Faster data access. End users often work with only the nearest stored subset of the data.	*Technological difficulty.* Data integrity, transaction management, concurrency control, security, backup, recovery, and query optimization must all be addressed and resolved.
Faster data processing. A distributed database system spreads out the system's workload by processing data at several sites.	*Security.* The probability of security lapses increases when data is located at multiple sites. The responsibility of data management will be shared by different people at several sites.
Growth facilitation. New sites can be added to the network without affecting the operations of other sites.	*Lack of standards.* There are no standard communication protocols at the database level. For example, different database vendors employ different and often incompatible techniques to manage the distribution of data and processing in a DDBMS environment.
Improved communications. Because local sites are smaller and located closer to customers, local sites foster better communication among departments and between customers and company staff.	*Increased storage and infrastructure requirements.* Multiple copies of data are required at different sites, thus requiring additional storage space.
Reduced operating costs. It is more cost-effective to add nodes to a network than to update a mainframe system. Development work is done more cheaply and quickly on low-cost PCs than on mainframes.	*Increased training cost.* Training costs are generally higher in a distributed model than they would be in a centralized model, sometimes even to the extent of offsetting operational and hardware savings.
User-friendly interface. PCs and workstations are usually equipped with an easy-to-use graphical user interface (GUI). The GUI simplifies training and use for end users.	*Costs.* Distributed databases require duplicated infrastructure to operate, such as physical location, environment, personnel, software, and licensing.
Less danger of a single-point failure. When one of the computers fails, the workload is picked up by other workstations. Data is also distributed at multiple sites.	
Processor independence. The end user can access any available copy of the data, and an end user's request is processed by any processor at the data location.	

network. In contrast, the distributed processing system uses only a single-site database but shares the processing chores among several sites. In a distributed database system, a database is composed of several parts known as **database fragments**. The database fragments are located at different sites and can be replicated among various sites. Each database fragment is, in turn, managed by its local database process. An example of a distributed database environment is shown in Figure 12.3.

The database in Figure 12.3 is divided into three database fragments (E1, E2, and E3) located at different sites. The computers are connected through a network system. In a fully distributed database, the users Alan, Betty, and Hernando do not need to know the name or location of each database fragment in order to access the database. Also, the

database fragment
A subset of a distributed database. Although the fragments may be stored at different sites within a computer network, the set of all fragments is treated as a single database. See also *horizontal fragmentation* and *vertical fragmentation*.

FIGURE 12.2 DISTRIBUTED PROCESSING ENVIRONMENT

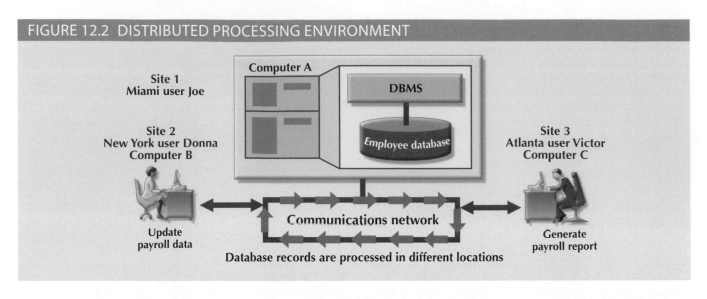

Site 1
Miami user Joe

Computer A

DBMS

Employee database

Site 2
New York user Donna
Computer B

Update
payroll data

Site 3
Atlanta user Victor
Computer C

Generate
payroll report

Communications network

Database records are processed in different locations

FIGURE 12.3 DISTRIBUTED DATABASE ENVIRONMENT

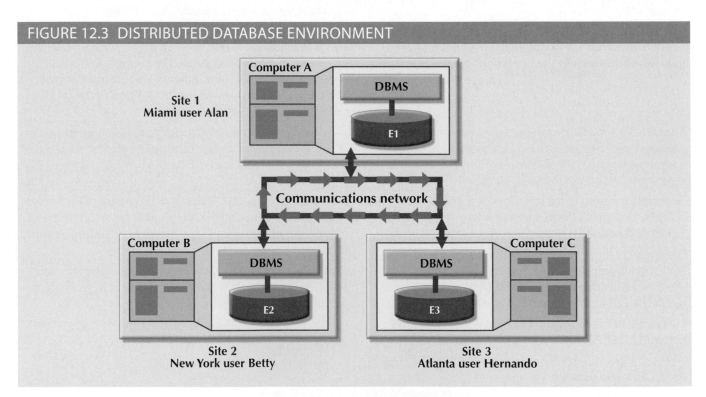

Site 1
Miami user Alan

Computer A

DBMS

E1

Communications network

Computer B

DBMS

E2

Computer C

DBMS

E3

Site 2
New York user Betty

Site 3
Atlanta user Hernando

users might be at sites other than Miami, New York, or Atlanta and still be able to access the database as a single logical unit.

As you examine Figures 12.2 and 12.3, keep the following points in mind:

- Distributed processing does not require a distributed database, but a distributed database requires distributed processing. (Each database fragment is managed by its own local database process.)

- Distributed processing may be based on a single database located on a single computer. For the management of distributed data to occur, copies or parts of the database processing functions must be distributed to all data storage sites.

- Both distributed processing and distributed databases require a network of interconnected components.

12-4 **Characteristics of Distributed Database Management Systems**

A DDBMS governs the storage and processing of logically related data over interconnected computer systems in which both data and processing functions are distributed among several sites. A DBMS must have at least the following functions to be classified as distributed:

- *Application interface* to interact with the end user, application programs, and other DBMSs within the distributed database
- *Validation* to analyze data requests for syntax correctness
- *Transformation* to decompose complex requests into atomic data request components
- *Query optimization* to find the best access strategy (which database fragments must be accessed by the query, and how must data updates, if any, be synchronized?)
- *Mapping* to determine the data location of local and remote fragments
- *I/O interface* to read or write data from or to permanent local storage
- *Formatting* to prepare the data for presentation to the end user or to an application program
- *Security* to provide data privacy at both local and remote databases
- *Backup and recovery* to ensure the availability and recoverability of the database in case of a failure
- *DB administration features* for the database administrator
- *Concurrency control* to manage simultaneous data access and to ensure data consistency across database fragments in the DDBMS
- *Transaction management* to ensure that the data moves from one consistent state to another; this activity includes the synchronization of local and remote transactions as well as transactions across multiple distributed segments

A fully distributed database management system must perform all of the functions of a centralized DBMS, as follows:

1. Receive the request of an application or end user.
2. Validate, analyze, and decompose the request. The request might include mathematical and logical operations such as the following: Select all customers with a balance greater than $1,000. The request might require data from only a single table, or it might require access to several tables.
3. Map the request's logical-to-physical data components.
4. Decompose the request into several disk I/O operations.
5. Search for, locate, read, and validate the data.
6. Ensure database consistency, security, and integrity.
7. Validate the data for the conditions, if any, specified by the request.
8. Present the selected data in the required format.

In addition, a distributed DBMS must handle all necessary functions imposed by the distribution of data and processing, and it must perform those additional functions *transparently* to the end user. The DDBMS's transparent data access features are illustrated in Figure 12.4.

The single logical database in Figure 12.4 consists of two database fragments, A1 and A2, located at Sites 1 and 2, respectively. Mary can query the database as if it were a local database; so can Tom. Both users "see" only one logical database and *do not need to know the names of the fragments*. In fact, the end users do not even need to know that the database is divided into fragments, *nor do they need to know where the fragments are located.*

To better understand the different types of distributed database scenarios, first consider the components of the distributed database system.

12-5 DDBMS Components

The DDBMS must include at least the following components:

- *Computer workstations or remote devices* (sites or nodes) that form the network system. The distributed database system must be independent of the computer system hardware.

- *Network hardware and software* components that reside in each workstation or device. The network components allow all sites to interact and exchange data. Because the components—computers, operating systems, network hardware, and so on—are likely to be supplied by different vendors, it is best to ensure that distributed database functions can be run on multiple platforms.

- *Communications media* that carry the data from one node to another. The DDBMS must be communications media-independent; that is, it must be able to support several types of communications media.

- The **transaction processor (TP)** is the software component found in each computer or device that requests data. The transaction processor receives and processes the application's remote and local data requests. The TP is also known as the **application processor (AP)** or the **transaction manager (TM)**.

- The **data processor (DP)** is the software component residing on each computer or device that stores and retrieves data located at the site. The DP is also known as the **data manager (DM)**. A data processor may even be a centralized DBMS.

---

transaction processor (TP)
In a DDBMS, the software component on each computer that requests data. The TP is responsible for the execution and coordination of all database requests issued by a local application that accesses data on any DP. Also called *transaction manager (TM)* or *application processor (AP)*.

application processor (AP)
See *transaction processor (TP)*.

transaction manager (TM)
See *transaction processor (TP)*.

data processor (DP)
The resident software component that stores and retrieves data through a DDBMS. The DP is responsible for managing the local data in the computer and coordinating access to that data. Also known as *data manager (DM)*.

data manager (DM)
See *data processor (DP)*.

Figure 12.5 illustrates the placement of the components and the interaction among them. The communication among TPs and DPs is made possible through a specific set of rules, or *protocols*, used by the DDBMS.

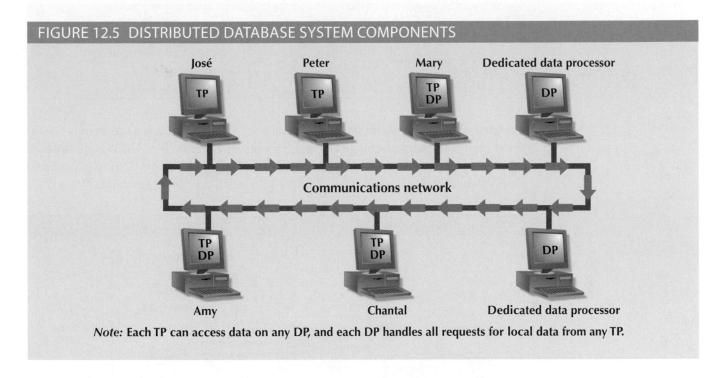

FIGURE 12.5 DISTRIBUTED DATABASE SYSTEM COMPONENTS

José · TP
Peter · TP
Mary · TP DP
Dedicated data processor · DP

Communications network

Amy · TP DP
Chantal · TP DP
Dedicated data processor · DP

Note: **Each TP can access data on any DP, and each DP handles all requests for local data from any TP.**

The protocols determine how the distributed database system will:

- Interface with the network to transport data and commands between DPs and TPs.
- Synchronize all data received from DPs (TP side) and route retrieved data to the appropriate TPs (DP side).
- Ensure common database functions in a distributed system. Such functions include data security, transaction management and concurrency control, data partitioning and synchronization, and data backup and recovery.

DPs and TPs should be added to the system transparently without affecting its operation. A TP and a DP can reside on the same computer, allowing the end user to access both local and remote data transparently. In theory, a DP can be an independent centralized DBMS with proper interfaces to support remote access from other independent DBMSs in the network.

12-6 Levels of Data and Process Distribution

Current database systems can be classified on the basis of how process distribution and data distribution are supported. For example, a DBMS may store data in a single site (using a centralized DB) or in multiple sites (using a distributed DB), and it may support data processing at one or more sites. Table 12.2 uses a simple matrix to classify database systems according to data and process distribution. These types of processes are discussed in the sections that follow.

12-6a Single-Site Processing, Single-Site Data

In the **single-site processing, single-site data (SPSD)** scenario, all processing is done on a single host computer, and all data is stored on the host computer's local disk system.

single-site processing, single-site data (SPSD)
A scenario in which all processing is done on a single host computer and all data is stored on the host computer's local disk.

TABLE 12.2

DATABASE SYSTEMS: LEVELS OF DATA AND PROCESS DISTRIBUTION

ADVANTAGES	SINGLE-SITE DATA	MULTIPLE-SITE DATA
Single-site process	Host DBMS	Not applicable (Requires multiple processes)
Multiple-site process	File server Client/server DBMS (LAN DBMS)	Fully distributed Client/server DDBMS

Processing cannot be done on the end user's side of the system. Such a scenario is typical of most mainframe and midrange UNIX/Linux server DBMSs. The DBMS is on the host computer, which is accessed by terminals connected to it (see Figure 12.6). This scenario is also typical of the first generation of single-user microcomputer databases.

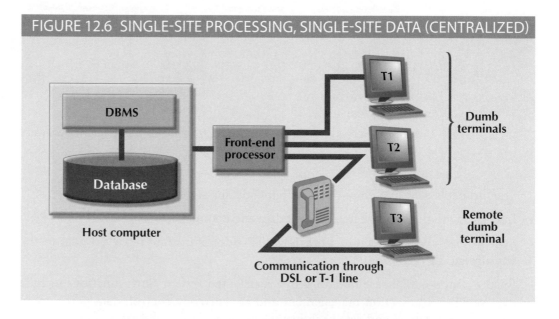

FIGURE 12.6 SINGLE-SITE PROCESSING, SINGLE-SITE DATA (CENTRALIZED)

Using Figure 12.6 as an example, you can see that the functions of the TP and DP are embedded within the DBMS on the host computer. The DBMS usually runs under a time-sharing, multitasking operating system, which allows several processes to run concurrently on a host computer accessing a single DP. All data storage and data processing are handled by a single host computer.

12-6b Multiple-Site Processing, Single-Site Data

Under the **multiple-site processing, single-site data (MPSD)** scenario, multiple processes run on different computers that share a single data repository. Typically, the MPSD scenario requires a network file server running conventional applications that are accessed through a network. Many multiuser accounting applications running under a personal computer network fit such a description (see Figure 12.7).

As you examine Figure 12.7, note that:

- The TP on each workstation acts only as a redirector to route all network data requests to the file server.

- The end user sees the file server as just another hard disk. Because only the data storage input/output (I/O) is handled by the file server's computer, the MPSD offers limited capabilities for distributed processing.

multiple-site processing, single-site data (MPSD) A scenario in which multiple processes run on different computers sharing a single data repository.

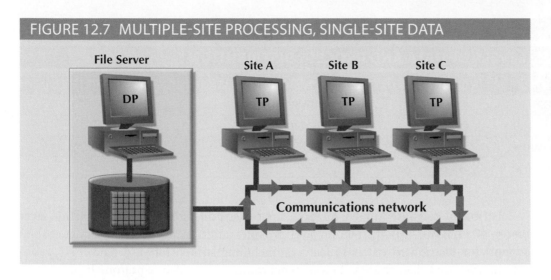

FIGURE 12.7   MULTIPLE-SITE PROCESSING, SINGLE-SITE DATA

Online Content

For more information about client/server architecture, see Appendix F, Client/Server Systems, available at *www.cengagebrain.com*.

- The end user must make a direct reference to the file server to access remote data. All record- and file-locking activities are performed at the end-user location.

- All data selection, search, and update functions take place at the workstation, thus requiring that entire files travel through the network for processing at the workstation. Such a requirement increases network traffic, slows response time, and increases communication costs.

The inefficiency of the last condition can be illustrated easily. For example, suppose that the file server computer stores a CUSTOMER table containing 100,000 data rows, 50 of which have balances greater than $1,000. Suppose that Site A issues the following SQL query:

```
SELECT          *
FROM            CUSTOMER
WHERE           CUS_BALANCE > 1000;
```

All 100,000 CUSTOMER rows must travel through the network to be evaluated at Site A. A variation of the multiple-site processing, single-site data approach is known as client/server architecture. **Client/server architecture** is similar to that of the network file server *except that all database processing is done at the server site, thus reducing network traffic*. Although both the network file server and the client/server systems perform multiple-site processing, the client/server system's processing is distributed. Note that the network file server approach requires the database to be located at a single site. In contrast, the client/server architecture is capable of supporting data at multiple sites.

12-6c Multiple-Site Processing, Multiple-Site Data

The **multiple-site processing, multiple-site data (MPMD)** scenario describes a fully distributed DBMS with support for multiple data processors and transaction processors at multiple sites. Depending on the level of support for various types of databases, DDBMSs are classified as either homogeneous or heterogeneous.

Homogeneous DDBMSs integrate multiple instances of the same DBMS over a network—for example, multiple instances of Oracle 11g running on different platforms. In contrast, **heterogeneous DDBMSs** integrate different types of DBMSs over a network, but all support the same data model. For example, Table 12.3 lists several relational database systems that could be integrated within a DDBMS. A **fully heterogeneous DDBMS** will support different DBMSs, each one supporting a different data model, running under different computer systems.

client/server architecture
A hardware and software system composed of clients, servers, and middleware. Features a user of resources (client) and a provider of resources (server).

multiple-site processing, multiple-site data (MPMD)
A scenario describing a fully distributed database management system with support for multiple data processors and transaction processors at multiple sites.

homogeneous DDBMS
A system that integrates only one type of centralized database management system over a network.

heterogeneous DDBMS
A system that integrates different types of centralized database management systems over a network.

TABLE 12.3

DATABASE SYSTEMS: LEVELS OF DATA AND PROCESS DISTRIBUTION			
PLATFORM	DBMS	OPERATING SYSTEM	NETWORK COMMUNICATIONS PROTOCOL
IBM 3090	DB2	MVS	APPC LU 6.2
IBM AS/400	SQL/400	OS/400	3270
RISC computer	Informix	UNIX	TCP/IP
Intel Xeon CPU	Oracle	Windows Server	TCP/IP

Distributed database implementations are better understood as an abstraction layer on top of a DBMS. This abstraction layer provides additional functionality that enables support for distributed database features, including straightforward data links, replication, advanced data fragmentation, synchronization, and integration. In fact, most database vendors provide for increasing levels of data fragmentation, replication, and integration. Therefore, the support for distributed databases can be better seen as a continuous spectrum that goes from homogeneous to fully heterogeneous distributed data management. Consequently, at any point on this spectrum, a DDBMS is subject to certain restrictions. For example:

- Remote access is provided on a read-only basis and does not support write privileges.

- Restrictions are placed on the number of remote tables that may be accessed in a single transaction.

- Restrictions are placed on the number of distinct databases that may be accessed.

- Restrictions are placed on the database model that may be accessed. Thus, access may be provided to relational databases but not to network or hierarchical databases.

The preceding list of restrictions is by no means exhaustive. The DDBMS technology continues to change rapidly, and new features are added frequently. Managing data at multiple sites leads to a number of issues that must be addressed and understood. The next section examines several key features of distributed database management systems.

12-7 Distributed Database Transparency Features

A distributed database system should provide some desirable transparency features that make all the system's complexities hidden to the end user. In other words, the end user should have the sense of working with a centralized DBMS. For this reason, the minimum desirable DDBMS transparency features are:

- **Distribution transparency** allows a distributed database to be treated as a single logical database. If a DDBMS exhibits distribution transparency, the user does not need to know:
  - The data is partitioned—meaning the table's rows and columns are split vertically or horizontally and stored among multiple sites.
  - The data is geographically dispersed among multiple sites.
  - The data is replicated among multiple sites.

fully heterogeneous distributed database system (fully heterogeneous DDBMS) A system that integrates different types of database management systems (hierarchical, network, and relational) over a network. It supports different database management systems that may even support different data models running under different computer systems. See also *heterogeneous DDBMS* and *homogeneous DDBMS*.

distribution transparency A DDBMS feature that allows a distributed database to look like a single logical database to an end user.

- **Transaction transparency** allows a transaction to update data at more than one network site. Transaction transparency ensures that the transaction will be either entirely completed or aborted, thus maintaining database integrity.

- **Failure transparency** ensures that the system will continue to operate in the event of a node or network failure. Functions that were lost because of the failure will be picked up by another network node. This is a very important feature, particularly in organizations that depend on web presence as the backbone for maintaining trust in their business.

- **Performance transparency** allows the system to perform as if it were a centralized DBMS. The system will not suffer any performance degradation due to its use on a network or because of the network's platform differences. Performance transparency also ensures that the system will find the most cost-effective path to access remote data. The system should be able to "scale out" in a transparent manner or increase performance capacity by adding more transaction or data-processing nodes, without affecting the overall performance of the system.

- **Heterogeneity transparency** allows the integration of several different local DBMSs (relational, network, and hierarchical) under a common, or global, schema. The DDBMS is responsible for translating the data requests from the global schema to the local DBMS schema.

The following sections discuss each of these transparency features in greater detail.

12-8 Distribution Transparency

Distribution transparency allows a physically dispersed database to be managed as though it were a centralized database. The level of transparency supported by the DDBMS varies from system to system. Three levels of distribution transparency are recognized:

- **Fragmentation transparency** is the highest level of distribution transparency. The end user or programmer does not need to know that a database is partitioned. Therefore, neither fragment names nor fragment locations are specified prior to data access.

- **Location transparency** exists when the end user or programmer must specify the database fragment names but does not need to specify where those fragments are located.

- **Local mapping transparency** exists when the end user or programmer must specify both the fragment names and their locations.

Transparency features are summarized in Table 12.4.

transaction transparency
A DDBMS property that ensures database transactions will maintain the distributed database's integrity and consistency, and that a transaction will be completed only when all database sites involved complete their part of the transaction.

failure transparency
A feature that allows continuous operation of a DDBMS, even if a network node fails.

performance transparency
A DDBMS feature that allows a system to perform as though it were a centralized DBMS.

heterogeneity transparency
A feature that allows a system to integrate several centralized DBMSs into one logical DDBMS.

fragmentation transparency
A DDBMS feature that allows a system to treat a distributed database as a single database even though it is divided into two or more fragments.

location transparency
A property of a DDBMS in which database access requires the user to know only the name of the database fragments. (Fragment locations need not be known.)

TABLE 12.4

SUMMARY OF TRANSPARENCY FEATURES

IF THE SQL STATEMENT REQUIRES:			
FRAGMENT NAME?	LOCATION NAME?	THEN THE DBMS SUPPORTS	LEVEL OF DISTRIBUTON TRANSPARENCY
Yes	Yes	Local mapping transparency	Low
Yes	No	Location transparency	Medium
No	No	Fragmentation transparency	High

12-9 **Transaction Transparency**

Transaction transparency is a DDBMS property that ensures database transactions will maintain the distributed database's integrity and consistency. Remember that a DDBMS database transaction can update data stored in many different computers connected in a network. Transaction transparency ensures that the transaction will be completed only when all database sites involved in the transaction complete their part of the transaction.

Distributed database systems require complex mechanisms to manage transactions and ensure the database's consistency and integrity. To understand how the transactions are managed, you should know the basic concepts governing remote requests, remote transactions, distributed transactions, and distributed requests.

12-9a Distributed Requests and Distributed Transactions[1]

Whether or not a transaction is distributed, it is formed by one or more database requests. The basic difference between a nondistributed transaction and a distributed transaction is that the distributed transaction can update or request data from several different remote sites on a network. To better understand distributed transactions, begin by learning the difference between remote and distributed transactions, using the BEGIN WORK and COMMIT WORK transaction format. Assume the existence of location transparency to avoid having to specify the data location.

A **remote request**, illustrated in Figure 12.9, lets a single SQL statement access the data that are to be processed by a single remote database processor. In other words, the SQL statement (or request) can reference data at only one remote site.

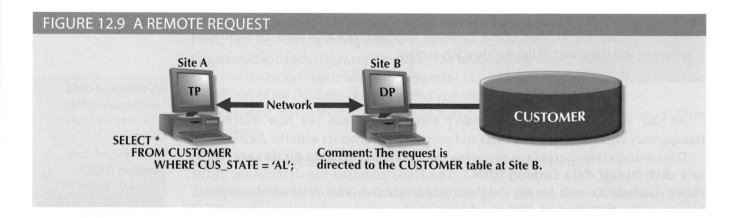

FIGURE 12.9 A REMOTE REQUEST

remote request
A DDBMS feature that allows a single SQL statement to access data in a single remote DP.

remote transaction
A DDBMS feature that allows a transaction (formed by several requests) to access data in a single remote DP.

Similarly, a **remote transaction**, composed of several requests, accesses data at a single remote site. A remote transaction is illustrated in Figure 12.10.

As you examine Figure 12.10, note the following remote transaction features:

- The transaction updates the PRODUCT and INVOICE tables (located at Site B).

- The remote transaction is sent to the remote Site B and executed there.

[1]The details of distributed requests and transactions were originally described by David McGoveran and Colin White, "Clarifying client/server," *DBMS* 3(12), November 1990, pp. 78–89.

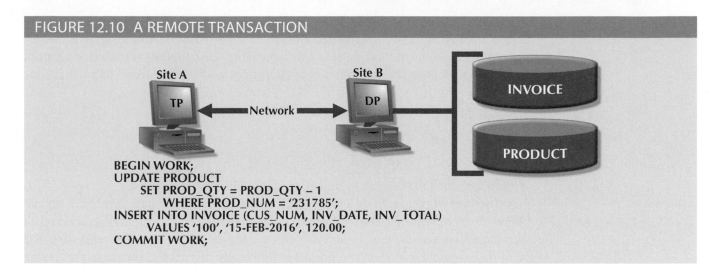

FIGURE 12.10   A REMOTE TRANSACTION

Site A

TP

Network

Site B

DP

INVOICE

PRODUCT

```
BEGIN WORK;
UPDATE PRODUCT
     SET PROD_QTY = PROD_QTY – 1
          WHERE PROD_NUM = '231785';
INSERT INTO INVOICE (CUS_NUM, INV_DATE, INV_TOTAL)
          VALUES '100', '15-FEB-2016', 120.00;
COMMIT WORK;
```

- The transaction can reference only one remote DP.

- Each SQL statement (or request) can reference only one (the same) remote DP at a time, and the entire transaction can reference and be executed at only one remote DP.

A **distributed transaction** can reference several different local or remote DP sites. Although each single request can reference only one local or remote DP site, the transaction as a whole can reference multiple DP sites because each request can reference a different site. The distributed transaction process is illustrated in Figure 12.11.

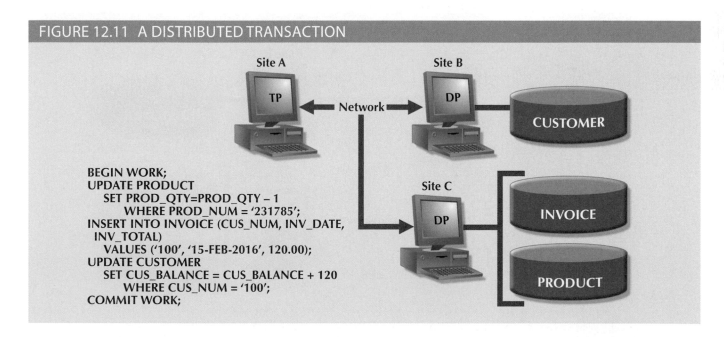

FIGURE 12.11   A DISTRIBUTED TRANSACTION

Site A

TP

Network

Site B

DP

CUSTOMER

Site C

DP

INVOICE

PRODUCT

```
BEGIN WORK;
UPDATE PRODUCT
     SET PROD_QTY=PROD_QTY – 1
          WHERE PROD_NUM = '231785';
INSERT INTO INVOICE (CUS_NUM, INV_DATE,
  INV_TOTAL)
     VALUES ('100', '15-FEB-2016', 120.00);
UPDATE CUSTOMER
     SET CUS_BALANCE = CUS_BALANCE + 120
          WHERE CUS_NUM = '100';
COMMIT WORK;
```

Note the following features in Figure 12.11:

- The transaction references two remote sites, B and C.

- The first two requests, UPDATE PRODUCT and INSERT INTO INVOICE, are processed by the DP at the remote Site C, and the last request (UPDATE CUSTOMER) is processed by the DP at the remote Site B.

- Each request can access only one remote site at a time.

distributed transaction
A database transaction that accesses data in several remote data processors (DPs) in a distributed database.

The third characteristic may create problems. For example, suppose the PRODUCT table is divided into two fragments, PROD1 and PROD2, located at Sites B and C, respectively. Given that scenario, the preceding distributed transaction cannot be executed because the following request cannot access data from more than one remote site:

```
SELECT      *
FROM        PRODUCT
WHERE       PROD_NUM = '231785';
```

Therefore, the DBMS must be able to support a distributed request.

A **distributed request** lets a single SQL statement reference data located at several different local or remote DP sites. Because each request (SQL statement) can access data from more than one local or remote DP site, a transaction can access several sites. The ability to execute a distributed request provides fully distributed database processing because you can:

- Partition a database table into several fragments.

- Reference one or more of those fragments with only one request. In other words, there is fragmentation transparency.

The location and partition of the data should be transparent to the end user. Figure 12.12 illustrates a distributed request. As you examine the figure, note that the transaction uses a single SELECT statement to reference two tables, CUSTOMER and INVOICE. The two tables are located at two different sites, B and C.

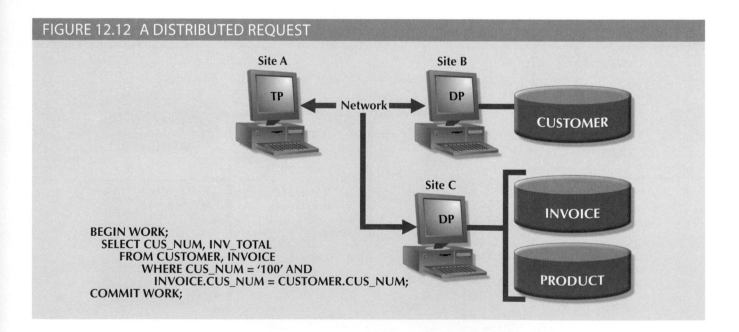

FIGURE 12.12 A DISTRIBUTED REQUEST

```
BEGIN WORK;
    SELECT CUS_NUM, INV_TOTAL
        FROM CUSTOMER, INVOICE
            WHERE CUS_NUM = '100' AND
                INVOICE.CUS_NUM = CUSTOMER.CUS_NUM;
COMMIT WORK;
```

distributed request
A database request that allows a single SQL statement to access data in several remote data processors (DPs) in a distributed database.

The distributed request feature also allows a single request to reference a physically partitioned table. For example, suppose that a CUSTOMER table is divided into two fragments, C1 and C2, located at Sites B and C, respectively. Further suppose that the end user wants to obtain a list of all customers whose balances exceed $250. The request is illustrated in Figure 12.13. Full-fragmentation transparency support is provided only by a DDBMS that supports distributed requests.

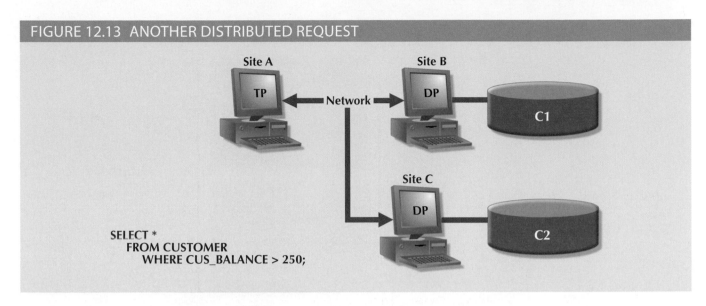

FIGURE 12.13   ANOTHER DISTRIBUTED REQUEST

Understanding the different types of database requests in distributed database systems helps you address the transaction transparency issue more effectively. Transaction transparency ensures that distributed transactions are treated as centralized transactions, ensuring their serializability. (Review Chapter 10, Transaction Management and Concurrency Control, if necessary.) That is, the execution of concurrent transactions, whether they are distributed or not, will take the database from one consistent state to another.

12-9b Distributed Concurrency Control

Concurrency control becomes especially important in distributed databases because multisite, multiple-process operations are more likely to create data inconsistencies and deadlocked transactions than single-site systems. For example, the TP component of a DDBMS must ensure that all parts of the transaction are completed at all sites before a final COMMIT is issued to record the transaction.

Suppose that a transaction updates data at three DP sites. The first two DP sites complete the transaction and commit the data at each local DP; however, the third DP site cannot commit the transaction. Such a scenario would yield an inconsistent database, with its inevitable integrity problems, because committed data cannot be uncommitted! This problem is illustrated in Figure 12.14.

The solution to this problem is a *two-phase commit protocol*, which you will explore next.

12-9c Two-Phase Commit Protocol

Centralized databases require only one DP. All database operations take place at only one site, and the consequences of database operations are immediately known to the DBMS. In contrast, distributed databases make it possible for a transaction to access data at several sites. A final COMMIT must not be issued until all sites have committed their parts of the transaction. The **two-phase commit protocol (2PC)** guarantees that if a portion of a transaction operation cannot be committed, all changes made at the other sites participating in the transaction will be undone to maintain a consistent database state.

two-phase commit protocol (2PC)
In a DDBMS, an algorithm used to ensure atomicity of transactions and database consistency as well as integrity in distributed transactions.

FIGURE 12.14   THE EFFECT OF A PREMATURE COMMIT

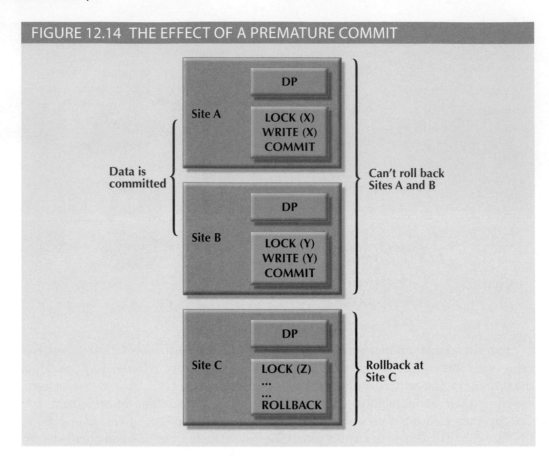

DO-UNDO-REDO protocol
A protocol used by a data processor (DP) to roll back or roll forward transactions with the help of a system's transaction log entries.

write-ahead protocol
A protocol that ensures transaction logs are written to permanent storage before any database data is actually updated.

coordinator
The transaction processor (TP) node that coordinates the execution of a two-phase COMMIT in a DDBMS.

subordinate
In a DDBMS, a data processor (DP) node that participates in a distributed transaction using the two-phase COMMIT protocol.

Each DP maintains its own transaction log. The two-phase commit protocol requires that the transaction log entry for each DP be written before the database fragment is actually updated (see Chapter 10). Therefore, the two-phase commit protocol requires a DO-UNDO-REDO protocol and a write-ahead protocol.

The **DO-UNDO-REDO protocol** is used by the DP to roll transactions back and forward with the help of the system's transaction log entries. The DO-UNDO-REDO protocol defines three types of operations:

- DO performs the operation and records the "before" and "after" values in the transaction log.

- UNDO reverses an operation, using the log entries written by the DO portion of the sequence.

- REDO redoes an operation, using the log entries written by the DO portion of the sequence.

To ensure that the DO, UNDO, and REDO operations can survive a system crash while they are being executed, a write-ahead protocol is used. The **write-ahead protocol** forces the log entry to be written to permanent storage before the actual operation takes place.

The two-phase commit protocol defines the operations between two types of nodes: the **coordinator** and one or more **subordinates**, or *cohorts*. The participating nodes agree on a coordinator. Generally, the coordinator role is assigned to the node that initiates the transaction. However, different systems implement various, more sophisticated election methods. The protocol is implemented in two phases, as illustrated in the following sections.

Phase 1: Preparation

The coordinator sends a PREPARE TO COMMIT message to all subordinates.

1. The subordinates receive the message, write the transaction log using the write-ahead protocol, and send an acknowledgment message (YES/PREPARED TO COMMIT or NO/NOT PREPARED) to the coordinator.

2. The coordinator makes sure that all nodes are ready to commit, or it aborts the action.

If all nodes are PREPARED TO COMMIT, the transaction goes to Phase 2. If one or more nodes reply NO or NOT PREPARED, the coordinator broadcasts an ABORT message to all subordinates.

Phase 2: The Final COMMIT

1. The coordinator broadcasts a COMMIT message to all subordinates and waits for the replies.

2. Each subordinate receives the COMMIT message and then updates the database using the DO protocol.

3. The subordinates reply with a COMMITTED or NOT COMMITTED message to the coordinator.

If one or more subordinates do not commit, the coordinator sends an ABORT message, thereby forcing them to UNDO all changes.

The objective of the two-phase commit is to ensure that each node commits its part of the transaction; otherwise, the transaction is aborted. If one of the nodes fails to commit, the information necessary to recover the database is in the transaction log, and the database can be recovered with the DO-UNDO-REDO protocol. (Remember that the log information was updated using the write-ahead protocol.)

12-10 Performance and Failure Transparency

One of the most important functions of a database is its ability to make data available. Web-based distributed data systems demand high availability, which means not only that data is accessible but that requests are processed in a timely manner. For example, the average Google search has a subsecond response time. When was the last time you entered a Google query and waited more than a couple of seconds for the results?

Performance transparency allows a DDBMS to perform as if it were a centralized database. In other words, no performance degradation should be incurred due to data distribution. Failure transparency ensures that the system will continue to operate in the case of a node or network failure. Although these are two separate issues, they are interrelated in that a failing node or congested network path could cause performance problems. Therefore, both issues are addressed in this section.

Note

Chapter 11, Database Performance Tuning and Query Optimization, provides additional details about query optimization.

The objective of query optimization is to minimize the total cost associated with the execution of a request. The costs associated with a request are a function of the following:

- Access time (I/O) cost involved in accessing the data from multiple remote sites

- Communication cost associated with data transmission among nodes in distributed database systems

- CPU time cost associated with the processing overhead of managing distributed transactions

Although costs are often classified either as communication or processing costs, it is difficult to separate the two. Not all query optimization algorithms use the same parameters, and not all algorithms assign the same weight to each parameter. For example, some algorithms minimize total time, others minimize the communication time, and still others do not factor in the CPU time, considering its cost insignificant relative to other costs.

As you learned in Chapter 11, a centralized database evaluates every data request to find the most-efficient way to access the data. This is a reasonable requirement, considering that all data is locally stored and all active transactions working on the data are known to the central DBMS. In contrast, in a DDBMS, transactions are distributed among multiple nodes; therefore, determining what data is being used becomes more complex. Hence, resolving data requests in a distributed data environment must take the following points into consideration:

- *Data distribution.* In a DDBMS, query translation is more complicated because the DDBMS must decide which fragment to access. (Distribution transparency was explained earlier in this chapter.) In this case, a TP executing a query must choose what fragments to access, create multiple data requests to the chosen remote DPs, combine the DP responses, and present the data to the application.

- *Data replication.* In addition, the data may also be replicated at several different sites. The data replication makes the access problem even more complex because the database must ensure that all copies of the data are consistent. Therefore, an important characteristic of query optimization in distributed database systems is that it must provide *replica* transparency. **Replica transparency** refers to the DDBMS's ability to hide multiple copies of data from the user. This ability is particularly important with data update operations. If a read-only request is being processed, it can be satisfied by accessing any available remote DP. However, processing a write request also involves "synchronizing" all existing fragments to maintain data consistency. The two-phase commit protocol you learned about in Section 12-9c ensures that the transaction will complete successfully. However, if data is replicated at other sites, the DDBMSs must also ensure the consistency of all the fragments—that is, all fragments should be mutually consistent. To accomplish this, a DP captures all changes and pushes them to each remote replica. This introduces delays in the system and basically means that not all data changes are immediately seen by all replicas. (The implications of this issue are explained in Section 12-12, The CAP Theorem.)

- *Network and node availability.* The response time associated with remote sites cannot be easily predetermined because some nodes finish their part of the query in less time than others and network path performance varies because of bandwidth and traffic loads. Hence, to achieve performance transparency, the DDBMS should consider issues such as **network latency**, the delay imposed by the amount of time required for a data packet to make a round trip from point A to point B, or **network partitioning**, the delay imposed when nodes become suddenly unavailable due to a network failure.

replica transparency
The DDBMS's ability to hide the existence of multiple copies of data from the user.

network latency
The delay imposed by the amount of time required for a data packet to make a round trip from point A to point B.

network partitioning
The delay that occurs when nodes become suddenly unavailable due to a network failure. In distributed databases, the system must account for the possibility of this condition.

Carefully planning how to partition a database and where to locate the database fragments can help ensure the performance and consistency of a distributed database. The following section discusses issues for distributed database design.

12-11 Distributed Database Design

Whether the database is centralized or distributed, the design principles and concepts described in Chapters 3, 4, and 6 are still applicable. However, the design of a distributed database introduces three new issues:

- How to partition the database into fragments
- Which fragments to replicate
- Where to locate those fragments and replicas

Data fragmentation and data replication deal with the first two issues, and data allocation deals with the third issue. Ideally, data in a distributed database should be evenly distributed to maximize performance, increase availability (reduce bottlenecks), and provide location awareness, which is an ever-increasing requirement for mobile applications.

12-11a Data Fragmentation

Data fragmentation allows you to break a single object into two or more segments, or fragments. The object might be a user's database, a system database, or a table. Each fragment can be stored at any site over a computer network. Information about data fragmentation is stored in the distributed data catalog (DDC), from which it is accessed by the TP to process user requests.

Data fragmentation strategies, as discussed here, are based at the table level and consist of dividing a table into logical fragments. You will explore three types of data fragmentation strategies: horizontal, vertical, and mixed. (Keep in mind that a fragmented table can always be re-created from its fragmented parts by a combination of unions and joins.)

- **Horizontal fragmentation** refers to the division of a relation into subsets (fragments) of tuples (rows). Each fragment is stored at a different node, and each fragment has unique rows. However, the unique rows all have the same attributes (columns). In short, each fragment represents the equivalent of a SELECT statement, with the WHERE clause on a single attribute.

- **Vertical fragmentation** refers to the division of a relation into attribute (column) subsets. Each subset (fragment) is stored at a different node, and each fragment has unique columns—with the exception of the key column, which is common to all fragments. This is the equivalent of the PROJECT statement in SQL.

- **Mixed fragmentation** refers to a combination of horizontal and vertical strategies. In other words, a table may be divided into several horizontal subsets (rows), each one having a subset of the attributes (columns).

To illustrate the fragmentation strategies, use the CUSTOMER table for the XYZ Company, as depicted in Figure 12.15. The table contains the attributes CUS_NUM, CUS_NAME, CUS_ADDRESS, CUS_STATE, CUS_LIMIT, CUS_BAL, CUS_RATING, and CUS_DUE.

data fragmentation
A characteristic of a DDBMS that allows a single object to be broken into two or more segments or fragments. The object might be a user's database, a system database, or a table. Each fragment can be stored at any site on a computer network.

horizontal fragmentation
The distributed database design process that breaks a table into subsets of unique rows.

vertical fragmentation
In distributed database design, the process that breaks a table into a subset of columns from the original table. Fragments must share a common primary key.

mixed fragmentation
A combination of horizontal and vertical strategies for data fragmentation, in which a table may be divided into several rows and each row has a subset of the attributes (columns).

FIGURE 12.15 A SAMPLE CUSTOMER TABLE

Table name: CUSTOMER   Database name: Ch12_Text

CUS_NUM	CUS_NAME	CUS_ADDRESS	CUS_STATE	CUS_LIMIT	CUS_BAL	CUS_RATING	CUS_DUE
10	Sinex, Inc.	12 Main St.	TN	3500.00	2700.00	3	1245.00
11	Martin Corp.	321 Sunset Blvd.	FL	6000.00	1200.00	1	0.00
12	Mynux Corp.	910 Eagle St.	TN	4000.00	3500.00	3	3400.00
13	BTBC, Inc.	Rue du Monde	FL	6000.00	5890.00	3	1090.00
14	Victory, Inc.	123 Maple St.	FL	1200.00	550.00	1	0.00
15	NBCC Corp.	909 High Ave.	GA	2000.00	350.00	2	50.00

Horizontal Fragmentation In this case, a table is divided into multiple subsets of rows. There are various ways to partition a table horizontally:

- *Round-robin partitioning.* Rows are assigned to a given fragment in a round-robin fashion (F1, F2, F3, ... , F*n*) to ensure an even distribution of rows among all fragments. However, this is not a good strategy if you require "location awareness"—the ability to determine which DP node will process a query based on the geospatial location of the requester. For example, you would want all queries from Florida customers to be resolved from a fragment that stores only Florida customers. Of course, you also would like this fragment to be located in a node close to Florida.

- *Range partitioning based on a partition key.* A **partition key** is one or more attributes in a table that determine the fragment in which a row will be stored. For example, if you want to provide location awareness, a good partition key would be the customer state field. This is the most common and useful data partitioning strategy.

Take a closer look at how to use a partition key to partition a table. Suppose that the XYZ Company's corporate management requires information about its customers in all three states, but company locations in each state (TN, FL, and GA) require data regarding local customers only. Based on such requirements, you decide to distribute the data by state. Therefore, you define the horizontal fragments to conform to the structure shown in Table 12.5.

TABLE 12.5

HORIZONTAL FRAGMENTATION OF THE CUSTOMER TABLE BY STATE

FRAGMENT NAME	LOCATION	CONDITION	NODE NAME	CUSTOMER NUMBERS	NUMBER OF ROWS
CUST_H1	Tennessee	CUS_STATE = 'TN'	NAS	10, 12	2
CUST_H2	Georgia	CUS_STATE = 'GA'	ATL	15	1
CUST_H3	Florida	CUS_STATE = 'FL'	TAM	11, 13, 14	3

The partition key will be the CUS_STATE field. Each horizontal fragment may have a different number of rows, but each fragment *must* have the same attributes. The resulting fragments yield the three tables depicted in Figure 12.16.

Vertical Fragmentation You may also divide the CUSTOMER relation into vertical fragments that are composed of a collection of attributes. For example, suppose that the company is divided into two departments: the service department and the collections department. Each department is located in a separate building, and each has an interest in only a few of the CUSTOMER table's attributes. In this case, the fragments are defined as shown in Table 12.6.

partition key
In partitioned databases, one or more attributes in a table that determine the fragment in which a row will be stored.

FIGURE 12.16 TABLE FRAGMENTS IN THREE LOCATIONS

Database name: Ch12_Text

Table name: CUST_H1 **Location: Tennessee** **Node: NAS**

CUS_NUM	CUS_NAME	CUS_ADDRESS	CUS_STATE	CUS_LIMIT	CUS_BAL	CUS_RATING	CUS_DUE
10	Sinex, Inc.	12 Main St.	TN	3500.00	2700.00	3	1245.00
12	Mynux Corp.	910 Eagle St.	TN	4000.00	3500.00	3	3400.00

Table name: CUST_H2 **Location: Georgia** **Node: ATL**

CUS_NUM	CUS_NAME	CUS_ADDRESS	CUS_STATE	CUS_LIMIT	CUS_BAL	CUS_RATING	CUS_DUE
15	NBCC Corp.	909 High Ave.	GA	2000.00	350.00	2	50.00

Table name: CUST_H3 **Location: Florida** **Node: TAM**

CUS_NUM	CUS_NAME	CUS_ADDRESS	CUS_STATE	CUS_LIMIT	CUS_BAL	CUS_RATING	CUS_DUE
11	Martin Corp.	321 Sunset Blvd.	FL	6000.00	1200.00	1	0.00
13	BTBC, Inc.	Rue du Monde	FL	6000.00	5890.00	3	1090.00
14	Victory, Inc.	123 Maple St.	FL	1200.00	550.00	1	0.00

TABLE 12.6

VERTICAL FRAGMENTATION OF THE CUSTOMER TABLE

FRAGMENT NAME	LOCATION	NODE NAME	ATTRIBUTE NAMES
CUST_V1	Service Bldg	SVC	CUS_NUM, CUS_NAME, CUS_ADDRESS, CUS_STATE
CUST_V2	Collection Bldg.	ARC	CUS_NUM, CUS_LIMIT, CUS_BAL, CUS_RATING, CUS_DUE

Each vertical fragment must have the same number of rows, but the inclusion of the different attributes depends on the key column. The vertical fragmentation results are displayed in Figure 12.17. Note that the key attribute (CUS_NUM) is common to both fragments CUST_V1 and CUST_V2.

FIGURE 12.17 VERTICALLY FRAGMENTED TABLE CONTENTS

Database name: Ch12_Text

Table name: CUST_V1 **Location: Service Building** **Node: SVC**

CUS_NUM	CUS_NAME	CUS_ADDRESS	CUS_STATE
10	Sinex, Inc.	12 Main St.	TN
11	Martin Corp.	321 Sunset Blvd.	FL
12	Mynux Corp.	910 Eagle St.	TN
13	BTBC, Inc.	Rue du Monde	FL
14	Victory, Inc.	123 Maple St.	FL
15	NBCC Corp.	909 High Ave.	GA

Table name: CUST_V2 **Location: Collection Building** **Node: ARC**

CUS_NUM	CUS_LIMIT	CUS_BAL	CUS_RATING	CUS_DUE
10	3500.00	2700.00	3	1245.00
11	6000.00	1200.00	1	0.00
12	4000.00	3500.00	3	3400.00
13	6000.00	5890.00	3	1090.00
14	1200.00	550.00	1	0.00
15	2000.00	350.00	2	50.00

Mixed Fragmentation The XYZ Company's structure requires that the CUSTOMER data be fragmented horizontally to accommodate the various company locations; within the locations, the data must be fragmented vertically to accommodate the two departments (service and collection). In short, the CUSTOMER table requires mixed fragmentation.

Mixed fragmentation requires a two-step procedure. First, horizontal fragmentation is introduced for each site based on the location within a state (CUS_STATE). The horizontal fragmentation yields the subsets of customer tuples (horizontal fragments) that are located at each site. Because the departments are located in different buildings, vertical fragmentation is used within each horizontal fragment to divide the attributes, thus meeting each department's information needs at each subsite. Mixed fragmentation yields the results displayed in Table 12.7.

TABLE 12.7

MIXED FRAGMENTATION OF THE CUSTOMER TABLE

FRAGMENT NAME	LOCATION	HORIZONTAL CRITERIA	NODE NAME	RESULTING ROWS AT SITE	VERTICAL CRITERIA ATTRIBUTES AT EACH FRAGMENT
CUST_M1	TN-Service	CUS_STATE = TN	NAS-S	10, 12	CUS_NUM, CUS_NAME CUS_ADDRESS, CUS_STATE
CUST_M2	TN-Collection	CUS_STATE = TN	NAS-C	10, 12	CUS_NUM, CUS_LIMIT, CUS_BAL, CUS_RATING, CUS_DUE
CUST_M3	GA-Service	CUS_STATE = GA	ATL-S	15	CUS_NUM, CUS_NAME CUS_ADDRESS, CUS_STATE
CUST_M4	GA-Collection	CUS_STATE = GA	ATL-C	15	CUS_NUM, CUS_LIMIT, CUS_BAL, CUS_RATING, CUS_DUE
CUST_M5	FL-Service	CUS_STATE = FL	TAM-S	11, 13, 14	CUS_NUM, CUS_NAME CUS_ADDRESS, CUS_STATE
CUST_M6	FL-Collection	CUS_STATE = FL	TAM-C	11, 13, 14	CUS_NUM, CUS_LIMIT, CUS_BAL, CUS_RATING, CUS_DUE

Each fragment displayed in Table 12.7 contains customer data by state and, within each state, by department location to fit each department's data requirements. The tables corresponding to the fragments listed in Table 12.7 are shown in Figure 12.18.

12-11b Data Replication

Data replication refers to the storage of data copies at multiple sites served by a computer network. Fragment copies can be stored at several sites to serve specific information requirements. Because the existence of fragment copies can enhance data availability and response time, data copies can help to reduce communication and total query costs.

Suppose database A is divided into two fragments, A1 and A2. Within a replicated distributed database, the scenario depicted in Figure 12.19 is possible: fragment A1 is stored at Sites S1 and S2, while fragment A2 is stored at Sites S2 and S3.

Replicated data is subject to the **mutual consistency rule**, which requires that all copies of data fragments be identical. Therefore, to maintain data consistency among the replicas, the DDBMS must ensure that a database update is performed at all sites where replicas exist.

There are basically two styles of replication:

- *Push replication*. After a data update, the originating DP node sends the changes to the replica nodes to ensure that data is immediately updated. This type of replication focuses on maintaining data consistency. However, it decreases data availability due to the latency involved in ensuring data consistency at all nodes.

data replication
The storage of duplicated database fragments at multiple sites on a DDBMS. Duplication of the fragments is transparent to the end user. Data replication provides fault tolerance and performance enhancements.

mutual consistency rule
A data replication rule that requires all copies of data fragments to be identical.

FIGURE 12.18 TABLE CONTENTS AFTER THE MIXED FRAGMENTATION PROCESS

Database name: Ch12_Text

Table name: CUST_M1   Location: TN-Service   Node: NAS-S

CUS_NUM	CUS_NAME	CUS_ADDRESS	CUS_STATE
10	Sinex, Inc.	12 Main St.	TN
12	Mynux Corp.	910 Eagle St.	TN

Table name: CUST_M2   Location: TN-Collection   Node: NAS-C

CUS_NUM	CUS_LIMIT	CUS_BAL	CUS_RATING	CUS_DUE
10	3500.00	2700.00	3	1245.00
12	4000.00	3500.00	3	3400.00

Table name: CUST_M3   Location: GA-Service   Node: ATL-S

CUS_NUM	CUS_NAME	CUS_ADDRESS	CUS_STATE
15	NBCC Corp.	909 High Ave.	GA

Table name: CUST_M4   Location: GA-Collection   Node: ATL-C

CUS_NUM	CUS_LIMIT	CUS_BAL	CUS_RATING	CUS_DUE
15	2000.00	350.00	2	50.00

Table name: CUST_M5   Location: FL-Service   Node: TAM-S

CUS_NUM	CUS_NAME	CUS_ADDRESS	CUS_STATE
11	Martin Corp.	321 Sunset Blvd.	FL
13	BTBC, Inc.	Rue du Monde	FL
14	Victory, Inc.	123 Maple St.	FL

Table name: CUST_M6   Location: FL-Collection   Node: TAM-C

CUS_NUM	CUS_LIMIT	CUS_BAL	CUS_RATING	CUS_DUE
11	6000.00	1200.00	1	0.00
13	6000.00	5890.00	3	1090.00
14	1200.00	550.00	1	0.00

FIGURE 12.19 DATA REPLICATION

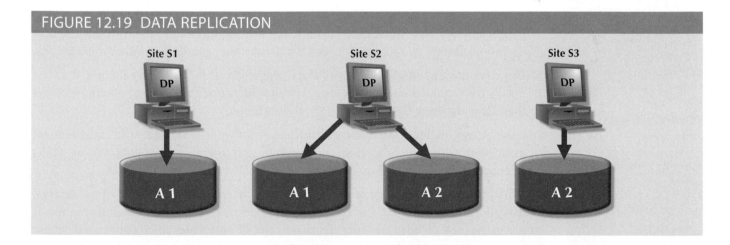

- *Pull replication.* After a data update, the originating DP node sends "messages" to the replica nodes to notify them of the update. The replica nodes decide when to apply the updates to their local fragment. In this type of replication, data updates propagate more slowly to the replicas. The focus is on maintaining data availability. However, this style of replication allows for temporary data inconsistencies.

Although replication has some benefits, such as improved data availability, better load distribution, improved data failure tolerance, and reduced query costs, it also imposes additional DDBMS processing overhead because each data copy must be maintained by the system. Furthermore, because the data is replicated at another site, there are associated storage costs and increased transaction times (as data must be updated at several sites concurrently to comply with the mutual consistency rule). To illustrate the replica overhead imposed on a DDBMS, consider the processes that the DDBMS must perform to use the database:

- If the database is fragmented, the DDBMS must decompose a query into subqueries to access the appropriate fragments.

- If the database is replicated, the DDBMS must decide which copy to access. A READ operation selects the *nearest copy* to satisfy the transaction. A WRITE operation requires that *all copies* be selected and updated to satisfy the mutual consistency rule.

- The TP sends a data request to each selected DP for execution.

- The DP receives and executes each request and sends the data back to the TP.

- The TP assembles the DP responses.

The problem becomes more complex when you consider additional factors such as network topology and communication throughputs.

Three replication scenarios exist: a database can be fully replicated, partially replicated, or unreplicated.

- A **fully replicated database** stores multiple copies of each database fragment at multiple sites. In this case, all database fragments are replicated. A fully replicated database can be impractical due to the amount of overhead it imposes on the system.

- A **partially replicated database** stores multiple copies of some database fragments at multiple sites. Most DDBMSs are able to handle the partially replicated database well.

- An **unreplicated database** stores each database fragment at a single site. Therefore, there are no duplicate database fragments.

Several factors influence the decision to use data replication:

- *Database size.* The amount of data replicated will have an impact on the storage requirements and the data transmission costs. Replicating large amounts of data requires a window of time and higher network bandwidth that could affect other applications.

- *Usage frequency.* The frequency of data usage determines how frequently the data needs to be updated. Frequently used data should be updated more often, for example, than large data sets that are used only every quarter.

- *Costs.* Costs include those for performance, software overhead, and management associated with synchronizing transactions and their components versus fault-tolerance benefits that are associated with replicated data.

When the usage frequency of remotely located data is high and the database is large, data replication can reduce the cost of data requests. Data replication information is stored in the DDC, whose contents are used by the TP to decide which copy of a database fragment to access. The data replication makes it possible to restore lost data.

12-11c Data Allocation

Data allocation describes the process of deciding where to locate data. Data allocation strategies are as follows:

- With **centralized data allocation**, the entire database is stored at one site.

- With **partitioned data allocation**, the database is divided into two or more disjoint parts (fragments) and stored at two or more sites.

- With **replicated data allocation**, copies of one or more database fragments are stored at several sites.

---

fully replicated database
In a DDBMS, the distributed database that stores multiple copies of each database fragment at multiple sites.

partially replicated database
A distributed database in which copies of only some database fragments are stored at multiple sites.

unreplicated database
A distributed database in which each database fragment is stored at a single site.

data allocation
In a distributed DBMS, the process of deciding where to locate data fragments.

centralized data allocation
A data allocation strategy in which the entire database is stored at one site. Also known as a *centralized database.*

partitioned data allocation
A data allocation strategy of dividing a database into two or more fragments that are stored at two or more sites.

replicated data allocation
A data allocation strategy in which copies of one or more database fragments are stored at several sites.

Data distribution over a computer network is achieved through data partitioning, through data replication, or through a combination of both. Data allocation is closely related to the way a database is divided or fragmented. Most data allocation studies focus on one issue: which data to locate where.

Data allocation algorithms consider a variety of factors, including:

- Performance and data availability goals

- Size, number of rows, and number of relations that an entity maintains with other entities

- Types of transactions to be applied to the database and the attributes accessed by each of those transactions

- Disconnected operation for mobile users

In some cases, the design might consider the use of loosely disconnected fragments for mobile users, particularly for read-only data that does not require frequent updates and for which the replica update windows may be longer. (A replica update window is the amount of time available to perform a data-processing task that cannot be executed concurrently with other tasks.)

Most algorithms include information such as network topology, network bandwidth and throughput, data size, and location. No optimal or universally accepted algorithm exists yet, and each database vendor implements its own version to showcase the strengths of its respective products.

12-12 The CAP Theorem

In a 2000 symposium on distributed computing, Dr. Eric Brewer stated in his presentation that "in any highly distributed data system there are three commonly desirable properties: consistency, availability, and partition tolerance. However, it is impossible for a system to provide all three properties at the same time."[2] The initials *CAP* stand for the three desirable properties. Consider these three properties in more detail:

- *Consistency*. In a distributed database, consistency takes a bigger role. All nodes should see the same data at the same time, which means that the replicas should be immediately updated. However, this involves dealing with latency and network partitioning delays, as you learned in Section 12-10.

- *Availability*. Simply speaking, a request is always fulfilled by the system. No received request is ever lost. If you are buying tickets online, you do not want the system to stop in the middle of the operation. This is a paramount requirement of all web-centric organizations.

- *Partition tolerance*. The system continues to operate even in the event of a node failure. This is the equivalent of failure transparency in distributed databases (see Section 12-7). The system will fail only if all nodes fail.

Do not mistake transaction management consistency (which you learned in Chapter 10) with CAP consistency. Transaction management consistency refers to the result when executing a transaction yields a database that complies with all integrity constraints. Consistency in CAP is based on the assumption that all transaction operations

---

[2] Eric A. Brewer, "Towards robust distributed systems," University of California at Berkeley and Inktomi Corporation, presentation at the Principles of Distributed Computing, ACM Symposium, July 2000. This theorem was later proven by Seth Gilbert and Nancy Lynch of MIT in their paper, "Brewer's conjecture and the feasibility of consistent, available, partition-tolerant web services," *ACM SIGACT News, 33*(2), 2002, pp. 51–59.

take place at the same time in all nodes, as if they were executing in a single-node database. ("All nodes see the same data at the same time.")

Although the CAP theorem focuses on highly distributed web-based systems, its implications are widespread for all distributed systems, including databases.

In Chapter 10, you learned that there are four database transaction properties: atomicity, consistency, isolation, and durability. The ACID properties ensure that all successful transactions result in a consistent database state—one in which all data operations always return the same results. For centralized and small distributed databases, latency is not an issue. As the business grows and the need for availability increases, database latency becomes a bigger problem. It is more difficult for a highly distributed database to ensure ACID transactions without paying a high price in network latency or data contention (delays imposed by concurrent data access).

For example, imagine that you are using Amazon.com to buy tickets for a Manchester United–Barcelona soccer game in Washington, D.C. You may spend a few minutes browsing through the available tickets and checking the stadium website to see which seats have the best view. At the same time, other users from all over the world may be doing exactly the same thing. By the time you click the checkout button, the tickets you selected may already have been purchased by someone else! In this case, you will start again and select other tickets until you get the ones you want. The website is designed to work this way because Amazon prefers the small probability of having a few customers restart their transactions to having to lock the database to ensure consistency and leaving thousands of customers waiting for their webpages to refresh. If you have noticed the small countdown clock when using Ticketmaster to buy concert tickets, you have seen the same principle at work.

As this example shows, when dealing with highly distributed systems, some companies tend to forfeit the consistency and isolation components of the ACID properties to achieve higher availability. This trade-off between consistency and availability has generated a new type of distributed data systems in which data is **basically available, soft state, eventually consistent (BASE)**. BASE refers to a data consistency model in which data changes are not immediate but propagate slowly through the system until all replicas are eventually consistent. For example, NoSQL databases provide a highly distributed database with eventual consistency (see Chapter 2, Data Models). In practice, the emergence of NoSQL distributed databases now provides a spectrum of consistency that ranges from the highly consistent (ACID) to the eventually consistent (BASE), as shown in Table 12.8.

basically available, soft state, eventually consistent (BASE)
A data consistency model in which data changes are not immediate but propagate slowly through the system until all replicas are eventually consistent.

TABLE 12.8

DISTRIBUTED DATABASE SPECTRUM

DBMS TYPE	CONSISTENCY	AVAILABILITY	PARTITION TOLERANCE	TRANSACTION MODEL	TRADE-OFF
Centralized DBMS	High	High	N/A	ACID	No distributed data processing
Relational DBMS	High	Relaxed	High	ACID (2PC)	Sacrifices availability to ensure consistency and isolation.
NoSQL DDBMS	Relaxed	High	High	BASE	Sacrifices consistency to ensure availability

12-13 C. J. Date's 12 Commandments for Distributed Databases

The notion of distributed databases has been around for over 20 years. With the rise of relational databases, most vendors implemented their own versions of distributed databases, generally highlighting their respective product's strengths. To make comparisons easier, C. J. Date formulated 12 "commandments" or basic principles of distributed databases.[3] Although no current DDBMS conforms to all of them, they constitute a useful target. The 12 rules are shown in Table 12.9.

TABLE 12.9

C. J. DATE'S 12 COMMANDMENTS FOR DISTRIBUTED DATABASES

RULE NUMBER	RULE NAME	RULE EXPLANATION
1	Local-site independence	Each local site can act as an independent, autonomous, centralized DBMS. Each site is responsible for security, concurrency control, backup, and recovery.
2	Central-site independence	No site in the network relies on a central site or any other site. All sites have the same capabilities.
3	Failure independence	The system is not affected by node failures. The system is in continuous operation even in the case of a node failure or an expansion of the network.
4	Location transparency	The user does not need to know the location of data to retrieve that data.
5	Fragmentation transparency	Data fragmentation is transparent to the user, who sees only one logical database. The user does not need to know the name of the database fragments to retrieve them.
6	Replication transparency	The user sees only one logical database. The DDBMS transparently selects the database fragment to access. To the user, the DDBMS manages all fragments transparently.
7	Distributed query processing	A distributed query may be executed at several different DP sites. Query optimization is performed transparently by the DDBMS.
8	Distributed transaction processing	A transaction may update data at several different sites, and the transaction is executed transparently.
9	Hardware independence	The system must run on any hardware platform.
10	Operating system independence	The system must run on any operating system platform.
11	Network independence	The system must run on any network platform.
12	Database independence	The system must support any vendor's database product.

[3]C. J. Date, "Twelve rules for a distributed database," *Computerworld* 2(23), June 8, 1987, pp. 77–81.

Summary

- A distributed database stores logically related data in two or more physically independent sites connected via a computer network. The database is divided into fragments, which can be a horizontal set of rows or a vertical set of attributes. Each fragment can be allocated to a different network node.

- Distributed processing is the division of logical database processing among two or more network nodes. Distributed databases require distributed processing. A distributed database management system (DDBMS) governs the processing and storage of logically related data through interconnected computer systems.

- The main components of a DDBMS are the transaction processor (TP) and the data processor (DP). The transaction processor component is the resident software on each computer node that requests data. The data processor component is the resident software on each computer that stores and retrieves data.

- Current database systems can be classified by the extent to which they support processing and data distribution. Three major categories are used to classify distributed database systems: single-site processing, single-site data (SPSD); multiple-site processing, single-site data (MPSD); and multiple-site processing, multiple-site data (MPMD).

- A homogeneous distributed database system integrates only one particular type of DBMS over a computer network. A heterogeneous distributed database system integrates several different types of DBMSs over a computer network.

- DDBMS characteristics are best described as a set of transparencies: distribution, transaction, performance, failure, and heterogeneity. All transparencies share the common objective of making the distributed database behave as though it were a centralized database system; that is, the end user sees the data as part of a single, logical centralized database and is unaware of the system's complexities.

- A transaction is formed by one or more database requests. An undistributed transaction updates or requests data from a single site. A distributed transaction can update or request data from multiple sites.

- Distributed concurrency control is required in a network of distributed databases. A two-phase COMMIT protocol is used to ensure that all parts of a transaction are completed.

- A distributed DBMS evaluates every data request to find the optimum access path in a distributed database. The DDBMS must optimize the query to reduce associated access costs, communication costs, and CPU costs.

- The design of a distributed database must consider the fragmentation and replication of data. The designer must also decide how to allocate each fragment or replica to obtain better overall response time and to ensure data availability to the end user. Ideally, a distributed database should evenly distribute data to maximize performance, availability, and location awareness.

- A database can be replicated over several different sites on a computer network. The replication of the database fragments has the objective of improving data availability, thus decreasing access time. A database can be partially, fully, or not replicated. Data allocation strategies are designed to determine the location of the database fragments or replicas.

- The CAP theorem states that a highly distributed data system has some desirable properties of consistency, availability, and partition tolerance. However, a system can only provide two of these properties at a time.

Key Terms

application processor (AP)

basically available, soft state, eventually consistent (BASE)

centralized data allocation

client/server architecture

coordinator

data allocation

data fragmentation

data manager (DM)

data processor (DP)

data replication

database fragments

distributed database

distributed database management system (DDBMS)

distributed data catalog (DDC)

distributed data dictionary (DDD)

distributed global schema

distributed processing

distributed request

distributed transaction

distribution transparency

DO-UNDO-REDO protocol

failure transparency

fragmentation transparency

fully heterogeneous DDBMS

fully replicated database

heterogeneity transparency

heterogeneous DDBMS

homogeneous DDBMS

horizontal fragmentation

local mapping transparency

location transparency

mixed fragmentation

multiple-site processing, multiple-site data (MPMD)

multiple-site processing, single-site data (MPSD)

mutual consistency rule

network latency

network partitioning

partially replicated database

partitioned data allocation

partition key

performance transparency

remote request

remote transaction

replica transparency

replicated data allocation

single-site processing, single-site data (SPSD)

subordinates

transaction manager (TM)

transaction processor (TP)

transaction transparency

two-phase commit protocol (2PC)

unique fragment

unreplicated database

vertical fragmentation

write-ahead protocol

Online Content

Flashcards and crossword puzzle for key term practice are available at *www. cengagebrain.com.*

Review Questions

1. Describe the evolution from centralized DBMSs to distributed DBMSs.

2. List and discuss some of the factors that influenced the evolution of the DDBMS.

3. What are the advantages of the DDBMS?

4. What are the disadvantages of the DDBMS?

5. Explain the difference between a distributed database and distributed processing.

6. What is a fully distributed database management system?

7. What are the components of a DDBMS?

8. List and explain the transparency features of a DDBMS.

9. Define and explain the different types of distribution transparency.

10. Describe the different types of database requests and transactions.

11. Explain the need for the two-phase commit protocol. Then describe the two phases.

12. What is the objective of query optimization functions?

13. To which transparency feature are the query optimization functions related?

14. What issues should be considered when resolving data requests in a distributed data environment?

15. Describe the three data fragmentation strategies. Give some examples of each.

16. What is data replication, and what are the three replication strategies?

17. What are the two basic styles of data replication?

18. What trade-offs are involved in building highly distributed data environments?

19. How does a BASE system differ from a traditional distributed database system?

Problems

Problem 1 is based on the DDBMS scenario in Figure P12.1.

FIGURE P12.1 THE DDBMS SCENARIO FOR PROBLEM 1

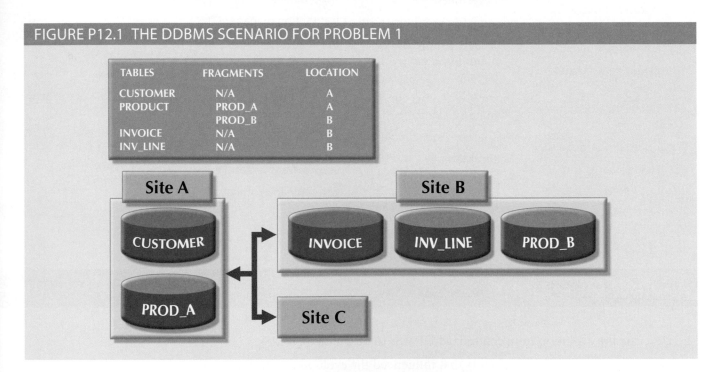

TABLES	FRAGMENTS	LOCATION
CUSTOMER	N/A	A
PRODUCT	PROD_A	A
	PROD_B	B
INVOICE	N/A	B
INV_LINE	N/A	B

1. Specify the minimum types of operations the database must support to perform the following operations. These operations include remote requests, remote transactions, distributed transactions, and distributed requests.

    At Site C

    a. SELECT   *
       FROM   CUSTOMER;

    b. SELECT   *
       FROM   INVOICE
       WHERE   INV_TOT < 1000;

    c. SELECT   *
       FROM   PRODUCT
       WHERE   PROD_ QOH < 10;

d. BEGIN WORK;

UPDATE	CUSTOMER
SET	CUS_BAL = CUS_BAL + 100
WHERE	CUS_NUM = '10936';
INSERT	INTO INVOICE(INV_NUM, CUS_NUM, INV_
	DATE, INV_TOTAL)
	VALUES ('986391', '10936', '15-FEB-2016', 100);
INSERT	INTO LINE(INV_NUM, PROD_NUM, LINE_PRICE)
	VALUES('986391', '1023', 100);
UPDATE	PRODUCT
SET	PROD_QOH = PROD_ QOH –1
WHERE	PROD_NUM = '1023';

COMMIT WORK;

e. BEGIN WORK;

INSERT	INTO CUSTOMER(CUS_NUM, CUS_NAME, CUS_
	ADDRESS, CUS_BAL)
	VALUES ('34210', 'Victor Ephanor', '123 Main St.', 0.00);
INSERT	INTO INVOICE(INV_NUM, CUS_NUM, INV_
	DATE, INV_TOTAL)
	VALUES ('986434', '34210', '10-AUG-2016', 2.00);

COMMIT WORK;

At Site A

f.
SELECT	CUS_NUM, CUS_NAME, INV_TOTAL
FROM	CUSTOMER, INVOICE
WHERE	CUSTOMER.CUS_NUM = INVOICE.CUS_NUM;

g.
SELECT	*
FROM	INVOICE
WHERE	INV_TOTAL > 1000;

h.
SELECT	*
FROM	PRODUCT
WHERE	PROD_QOH < 10;

At Site B

i.
SELECT	*
FROM	CUSTOMER;

j.
SELECT	CUS_NAME, INV_TOTAL
FROM	CUSTOMER, INVOICE
WHERE	INV_TOTAL > 1000 AND CUSTOMER.CUS_NUM =
	INVOICE.CUS_NUM;

k.
SELECT	*
FROM	PRODUCT
WHERE	PROD_QOH < 10;

2. The following data structure and constraints exist for a magazine publishing company:

a. The company publishes one regional magazine in each of four states: Florida (FL), South Carolina (SC), Georgia (GA), and Tennessee (TN).

b. The company has 300,000 customers (subscribers) distributed throughout the four states listed in Problem 2a.

c. On the first day of each month, an annual subscription INVOICE is printed and sent to each customer whose subscription is due for renewal. The INVOICE entity contains a REGION attribute to indicate the customer's state of residence (FL, SC, GA, TN):

CUSTOMER (CUS_NUM, CUS_NAME, CUS_ADDRESS, CUS_CITY, CUS_ZIP, CUS_SUBSDATE)

INVOICE (INV_NUM, INV_REGION, CUS_NUM, INV_DATE, INV_TOTAL)

The company is aware of the problems associated with centralized management and has decided to decentralize management of the subscriptions into the company's four regional subsidiaries. Each subscription site will handle its own customer and invoice data. The management at company headquarters, however, will have access to customer and invoice data to generate annual reports and to issue ad hoc queries such as:

- Listing all current customers by region
- Listing all new customers by region
- Reporting all invoices by customer and by region

Given these requirements, how must you partition the database?

3. Given the scenario and requirements in Problem 2, answer the following questions:

    a. What recommendations will you make regarding the type and characteristics of the required database system?

    b. What type of data fragmentation is needed for each table?

    c. What criteria must be used to partition each database?

    d. Design the database fragments. Show an example with node names, location, fragment names, attribute names, and demonstration data.

    e. What type of distributed database operations must be supported at each remote site?

    f. What type of distributed database operations must be supported at the headquarters site?

Business Intelligence and Data Warehouses

In this chapter, you will learn:
- How business intelligence provides a comprehensive business decision support framework
- About business intelligence architecture, its evolution, and reporting styles
- About the relationship and differences between operational data and decision support data
- What a data warehouse is and how to prepare data for one
- What star schemas are and how they are constructed
- About data analytics
- About online analytical processing (OLAP)
- How SQL extensions are used to support OLAP-type data manipulations

Preview

Business intelligence (BI) is the collection of best practices and software tools developed to support business decision making in this age of globalization, emerging markets, rapid change, and increasing regulation. The complexity and range of information required to support business decisions has increased, and operational database structures were unable to support all of these requirements. Therefore, a new data storage facility, called a *data warehouse*, developed. The data warehouse extracts its data from operational databases as well as from external sources, providing a more comprehensive data pool.

Additionally, new ways to analyze and present decision support data were developed. Online analytical processing (OLAP) provides advanced data analysis and visualization tools, including multidimensional data analysis. This chapter explores the main concepts and components of business intelligence and decision support systems that gather, generate, and present information for business decision makers, focusing especially on the use of data warehouses.

Data Files and Available Formats

	MS Access	Oracle	MS SQL	My SQL		MS Access	Oracle	MS SQL	My SQL
CH13_Text	✓	✓	✓	✓	CH13_PI	✓	✓	✓	✓
					CH13_P3	✓	✓	✓	✓
					CH13_P4	✓	✓	✓	✓
					CH13_SaleCo_DW	✓	✓	✓	✓

Data Files Available on cengagebrain.com

13-1 The Need for Data Analysis

Organizations tend to grow and prosper as they gain a better understanding of their environment. Most managers need to track daily transactions to evaluate how the business is performing. By tapping into the operational database, management can develop an understanding of how the company is performing and evaluate whether the current strategies meet organizational goals. In addition, analyzing the company data can provide insightful information about short-term tactical evaluations and strategic questions, such as: Are our sales promotions working? What market percentage are we controlling? Are we attracting new customers? Tactical and strategic decisions are also shaped by constant pressure from external and internal forces, including globalization, the cultural and legal environment, and technology.

Organizations are always looking for a competitive advantage through product development, market positioning, sales promotions, and customer service. Thanks to the Internet, customers are more informed than ever about the products they want and the prices they are willing to pay. Technology advances allow customers to place orders using their smart phones while they commute to work in the morning. Decision makers can no longer wait a couple of days for a report to be generated; they are compelled to make quick decisions if they want to remain competitive. Every day, TV ads offer low-price warranties, instant price matching, and so on. How can companies survive on lower margins and still make a profit? The key is in having the right data at the right time to support the decision-making process.

This process takes place at all levels of an organization. For example, transaction-processing systems, based on operational databases, are tailored to serve the information needs of people who deal with short-term inventory, accounts payable, and purchasing. Middle-level managers, general managers, vice presidents, and presidents focus on strategic and tactical decision making. Those managers require summarized information designed to help them make decisions in a complex business environment.

Companies and software vendors addressed these multilevel decision support needs by creating autonomous applications for particular groups of users, such as those in finance, customer management, human resources, and product support. Applications were also tailored to different industries such as education, retail, health care, and finance. This approach worked well for some time, but changes in the business world, such as globalization, expanding markets, mergers and acquisitions, increased regulation, and new technologies, called for new ways of integrating and managing decision support across levels, sectors, and geographic locations. This more comprehensive and integrated decision support framework within organizations became known as business intelligence.

13-2 Business Intelligence

business intelligence (BI)
A comprehensive, cohesive, and integrated set of tools and processes used to capture, collect, integrate, store, and analyze data with the purpose of generating and presenting information to support business decision making.

Business intelligence (BI)[1] is a term that describes a comprehensive, cohesive, and integrated set of tools and processes used to capture, collect, integrate, store, and analyze data with the purpose of generating and presenting information to support business decision making. This intelligence is based on learning and understanding the facts about the business environment. BI is a framework that allows a business to transform data into information, information into knowledge, and knowledge into wisdom. BI has the potential to positively affect a company's culture by creating continuous business performance improvement through active decision support at all levels in an organization.

[1] In 1989, while working at Gartner, Inc., Howard Dresner popularized BI as an umbrella term to describe a set of concepts and methods to improve business decision making by using fact-based support systems (*www.computerworld.com/s/article/266298/BI_at_age_17*).

This business insight empowers users to make sound decisions based on the accumulated knowledge of the business.

BI's initial adopters were high-volume industries such as financial services, insurance, and healthcare companies. As BI technology evolved, its usage spread to other industries such as telecommunications, retail/merchandising, manufacturing, media, government, and even education. Table 13.1 lists some companies that have implemented BI tools and shows how the tools benefited the companies. You will learn about these tools later in the chapter.

TABLE 13.1

SOLVING BUSINESS PROBLEMS AND ADDING VALUE WITH BI TOOLS

COMPANY	PROBLEM	BENEFIT
CiCi's Enterprises Eighth-largest pizza chain in the United States; operates 650 pizza restaurants in 30 states Source: Cognos Corp. *www.cognos.com*	• Information access was cumbersome and time-consuming • Needed to increase accuracy in the creation of marketing budgets • Needed an easy, reliable, and efficient way to access daily data	• Provided accurate, timely budgets in less time • Provided analysts with access to data for decision-making purposes • Received in-depth view of product performance by store to reduce waste and increase profits
NASDAQ Largest U.S. electronic stock market trading organization Source: Oracle Corp. *www.oracle.com*	• Inability to provide real-time, ad hoc query and standard reporting for executives, business analysts, and other users • Excessive storage costs for many terabytes of data	• Reduced storage costs by moving to a multitier storage solution • Implemented new data warehouse center with support for ad hoc query and reporting, and near real-time data access for end users
Pfizer Global pharmaceutical company Source: Oracle Corp. *www.oracle.com*	• Needed a way to control costs and adjust to tougher market conditions, international competition, and increasing government regulations • Needed better analytical capabilities and flexible decision-making framework	• Ability to get and integrate financial data from multiple sources in a reliable way • Streamlined, standards-based financial analysis to improve forecasting process • Faster and smarter decision making for business strategy formulation
Swisscom Switzerland's leading telecommunications provider Source: Microsoft Corp. *www.microsoft.com*	• Needed a tool to help employees monitor service-level compliance • Had a time-consuming process to generate performance reports • Needed a way to integrate data from 200 different systems	• Ability to monitor performance using dashboard technology • Quick and easy access to real-time performance data • Managers have closer and better control over costs

Implementing BI in an organization involves capturing not only internal and external business data, but also the metadata, or knowledge about the data. In practice, BI is a complex proposition that requires a deep understanding and alignment of the business processes, business data, and information needs of users at all levels in an organization. (See Appendix O, Data Warehouse Implementation Factors.)

BI is not a product by itself, but a framework of concepts, practices, tools, and technologies that help a business better understand its core capabilities, provide snapshots of the company situation, and identify key opportunities to create competitive advantage. In general, BI provides a framework for:

• Collecting and storing operational data

• Aggregating the operational data into decision support data

• Analyzing decision support data to generate information

• Presenting such information to the end user to support business decisions

- Making business decisions, which in turn generate more data that is collected, stored, and so on (restarting the process)

- Monitoring results to evaluate outcomes of the business decisions, which again provides more data to be collected, stored, and so on

- Predicting future behaviors and outcomes with a high degree of accuracy

The preceding points represent a system-wide view of the flow of data, processes, and outcomes within the BI framework. In practice, the first point, collecting and storing operational data, does not fall into the realm of a BI system per se; rather, it is the function of an operational system. However, the BI system will use the operational data as input material from which information will be derived. The rest of the processes and outcomes explained in the preceding points are oriented toward generating knowledge, and they are the focus of the BI system. In the following section, you will learn about the basic BI architecture.

13-2a Business Intelligence Architecture

BI covers a range of technologies and applications to manage the entire data life cycle from acquisition to storage, transformation, integration, presentation, analysis, monitoring, and archiving. BI functionality ranges from simple data gathering and transformation to very complex data analysis and presentation. BI architecture ranges from highly integrated single-vendor systems to loosely integrated, multivendor environments. However, some common functions are expected in most BI implementations.

Like any critical business IT infrastructure, the BI architecture is composed of many interconnected parts: people, processes, data, and technology working together to facilitate and enhance a business's management and governance. Figure 13.1 depicts how all these components fit together within the BI framework.

FIGURE 13.1 BUSINESS INTELLIGENCE FRAMEWORK

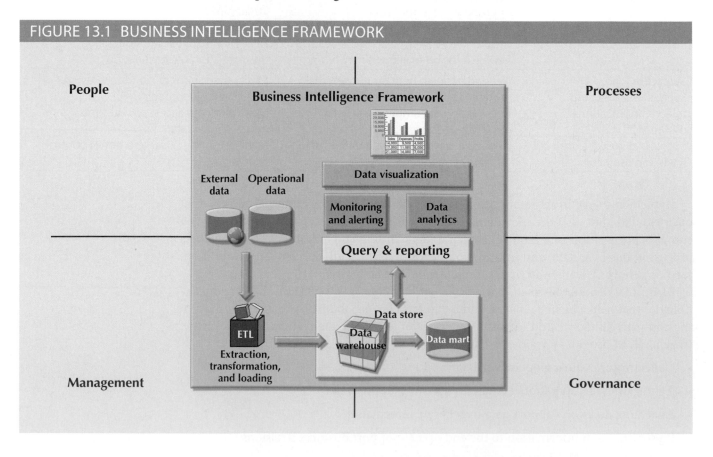

The general BI framework depicted in Figure 13.1 has six basic components that encompass the functionality required on most current-generation BI systems. You will learn more about these components later in this and future chapters. The components are briefly described in Table 13.2.

TABLE 13.2

BASIC BI ARCHITECTURAL COMPONENTS

COMPONENT	DESCRIPTION
ETL tools	Data **extraction, transformation, and loading (ETL)** tools collect, filter, integrate, and aggregate internal and external data to be saved into a data store optimized for decision support. Internal data is generated by the company during its day-to-day operations, such as product sales history, invoicing, and payments. The external data sources provide data that cannot be found within the company but is relevant to the business, such as stock prices, market indicators, marketing information (such as demographics), and competitors' data. Such data is generally located in external databases provided by industry groups or companies that market the data.
Data store	The data store is optimized for decision support and is generally represented by a *data warehouse* or a *data mart*. The data is stored in structures that are optimized for data analysis and query speed.
Query and reporting	This component performs data selection and retrieval, and it is used by the data analyst to create queries that access the database and create the required reports. Depending on the implementation, the query and reporting tool accesses the operational database, or more commonly, the data store.
Data visualization	This component presents data to the end user in a variety of meaningful and innovative ways. This tool helps the end user select the most appropriate presentation format, such as summary reports, maps, pie or bar graphs, mixed graphs, and static or interactive dashboards.
Data monitoring and alerting	This component allows real-time monitoring of business activities. The BI system will present the concise information in a single integrated view for the data analyst. This integrated view could include specific metrics about the system performance or activities, such as number of orders placed in the last four hours, number of customer complaints by product by month, and total revenue by region. Alerts can be placed on a given metric; once the value of a metric goes below or above a certain baseline, the system will perform a given action, such as emailing shop floor managers, presenting visual alerts, or starting an application.
Data analytics	This component performs data analysis and data-mining tasks using the data in the data store. This tool advises the user as to which data analysis tool to select and how to build a reliable business data model. Business models are generated by special algorithms that identify and enhance the understanding of business situations and problems. Data analysis can be either explanatory or predictive. Explanatory analysis uses the existing data in the data store to discover relationships and their types, and predictive analysis creates statistical models of the data that allow predictions of future values and events. Chapter 14, Big Data Analytics and NoSQL, covers these topics in more detail.

extraction, transformation, and loading (ETL) In a data warehousing environment, the integrated processes of getting data from original sources into the data warehouse. ETL includes retrieving data from original data sources (extraction), manipulating the data into an appropriate form (transformation), and storing the data in the data warehouse (loading).

Each BI component shown in Table 13.2 has generated a fast-growing market for specialized tools. Thanks to technological advancements, the components can interact with other components to form a truly open architecture. As a matter of fact, you can integrate multiple tools from different vendors into a single BI framework. Table 13.3 shows a sample of common BI tools and vendors.

TABLE 13.3

SAMPLE OF BUSINESS INTELLIGENCE TOOLS

TOOL	DESCRIPTION	SAMPLE VENDORS
Dashboards and business activity monitoring	**Dashboards** use web-based technologies to present key business performance indicators or information in a single integrated view, generally using graphics that are clear, concise, and easy to understand.	Salesforce IBM/Cognos BusinessObjects Information Builders iDashboards
Portals	**Portals** provide a unified, single point of entry for information distribution. Portals are a web-based technology that use a web browser to integrate data from multiple sources into a single webpage. Many different types of BI functionality can be accessed through a portal.	Oracle Portal Actuate Microsoft SAP
Data analysis and reporting tools	These advanced tools are used to query multiple and diverse data sources to create integrated reports.	Microsoft Reporting Services MicroStrategy SAS WebReportStudio
Data-mining tools	These tools provide advanced statistical analysis to uncover problems and opportunities hidden within business data. Chapter 14 covers data mining in more detail.	SAP Teradata MicroStrategy MS Analytics Services
Data warehouses (DW)	The data warehouse is the foundation of a BI infrastructure. Data is captured from the production system and placed in the DW on a near real-time basis. BI provides company-wide integration of data and the capability to respond to business issues in a timely manner.	Microsoft Oracle IBM/Cognos Teradata
OLAP tools	Online analytical processing provides multidimensional data analysis.	IBM/Cognos BusinessObjects Oracle Microsoft
Data visualization	These tools provide advanced visual analysis and techniques to enhance understanding and create additional insight of business data and its true meaning.	Dundas Tableau QlikView Actuate

dashboard
In business intelligence, a web-based system that presents key business performance indicators or information in a single, integrated view with clear and concise graphics.

portal
In terms of business intelligence, a unified, single point of entry for information distribution.

Note

You will learn about data warehouses and OLAP tools later in this chapter, and learn about data mining in Chapter 14.

As depicted in Figure 13.1, BI integrates people and processes using technology at all levels of the organization. A sound BI strategy adds value to an organization by providing the right data, in the right format, to the right people, at the right time. Such value is derived from how end users apply such information in their daily activities, and particularly in their daily business decision making.

The focus of traditional information systems was on operational automation and reporting; in contrast, BI tools focus on the strategic and tactical use of information. To achieve this goal, BI recognizes that technology alone is not enough. Therefore, BI uses an arrangement of best management practices to manage data as a corporate asset. One of the most recent developments in this area is the use of master data management techniques. **Master data management (MDM)** is a collection of concepts, techniques, and processes for the proper identification, definition, and management of data elements within an organization. MDM's main goal is to provide a comprehensive and consistent definition of all data within an organization. MDM ensures that all company resources (people, procedures, and IT systems) that work with data have uniform and consistent views of the company's data.

An added benefit of this meticulous approach to data management and decision making is that it provides a framework for business governance. **Governance** is a method or process of government. In this case, BI provides a method for controlling and monitoring business health and for consistent decision making. Furthermore, having such governance creates accountability for business decisions. In the present age of business flux, accountability is increasingly important. Had governance been as pivotal to business operations a few years back, crises precipitated by Enron, WorldCom, Arthur Andersen, and the 2008 financial meltdown might have been avoided.

Monitoring a business's health is crucial to understanding where the company is and where it is headed. To do this, BI makes extensive use of a special type of metrics known as key performance indicators. **Key performance indicators (KPIs)** are quantifiable numeric or scale-based measurements that assess the company's effectiveness or success in reaching its strategic and operational goals. Many different KPIs are used by different industries. Some examples of KPIs are:

- *General.* Year-to-year measurements of profit by line of business, same-store sales, product turnovers, product recalls, sales by promotion, and sales by employee
- *Finance.* Earnings per share, profit margin, revenue per employee, percentage of sales to account receivables, and assets to sales
- *Human resources.* Applicants to job openings, employee turnover, and employee longevity
- *Education.* Graduation rates, number of incoming freshmen, student retention rates, publication rates, and teaching evaluation scores

KPIs are determined after the main strategic, tactical, and operational goals are defined for a business. To tie the KPI to the strategic master plan of an organization, a KPI is compared to a desired goal within a specific time frame. For example, if you are in an academic environment, you might be interested in ways to measure student satisfaction or retention. In this case, a sample goal would be to increase the final exam grades of graduating high school seniors by Fall 2019. Another sample KPI would be to increase the returning student rate from freshman year to sophomore year from 60 percent to 75 percent by 2019. In this case, such performance indicators would be measured and monitored on a year-to-year basis, and plans to achieve such goals would be set in place.

Although BI has an unquestionably important role in modern business operations, the manager must initiate the decision support process by asking the appropriate questions.

master data management (MDM) In business intelligence, a collection of concepts, techniques, and processes for the proper identification, definition, and management of data elements within an organization.

governance In business intelligence, the methods for controlling and monitoring business health and promoting consistent decision making.

key performance indicators (KPIs) In business intelligence, quantifiable numeric or scale-based measurements that assess a company's effectiveness or success in reaching strategic and operational goals. Examples of KPIs are product turnovers, sales by promotion, sales by employee, and earnings per share.

The BI environment exists to support the manager; it does not replace the management function. If the manager fails to ask the appropriate questions, problems will not be identified and solved, and opportunities will be missed. In spite of the very powerful BI presence, the human component is still at the center of business technology.

Having a well-implemented BI environment (people, processes, technology, management, and governance) positions a company to react quickly to changes in the environment. Today's customers are more connected than ever with other customers (current or potential), companies, and organizations. In certain industries, social media plays a key role in marketing, brand recognition, and development. A simple tweet could generate millions of dollars in new sales or could cost a company millions of dollars in revenue. Companies monitor social media data to identify trends and quickly react to current or future threats or opportunities.

Data visualization is abstracting data to provide information in a visual format that enhances the user's ability to effectively comprehend the meaning of the data. The goal of data visualization is to allow the user to see the big picture in the most efficient way possible. Tables with hundreds, thousands, or millions of rows of data cannot be processed by the human mind. Providing summarized tabular data to managers does not give them the insight into the meaning of the data that they need to make informed decisions. Data visualization aggregates the data into a format that provides at-a-glance insight into overall trends and patterns.

Data visualization techniques can range from simple to very complex, and many are familiar. Techniques include: pie charts, line graphs, bar charts, scatter plots, Gantt charts, heat maps, and more.

An example of a heat map is shown in Figure 13.2. This heat map was created using Tableau (*www.tableau.com*), a data visualization tool, to analyze sales for a company.

data visualization
Abstracting data to provide information in a visual format that enhances the user's ability to effectively comprehend the meaning of the data.

FIGURE 13.2 VISUALIZING SALES TOTAL BY zip CODE

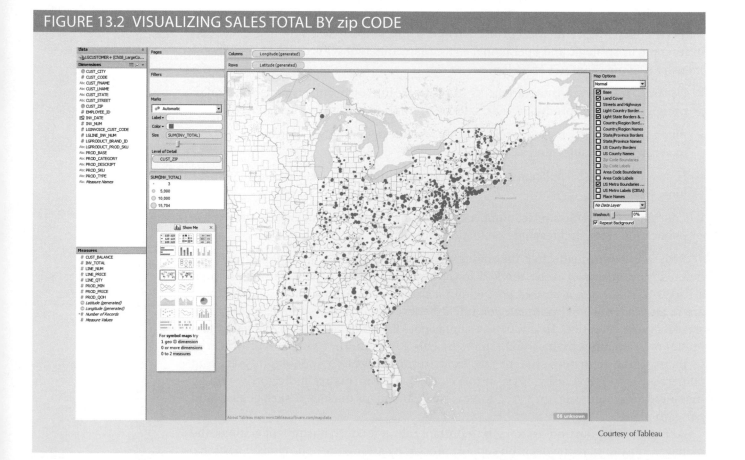

Courtesy of Tableau

The size of the circles is determined by the dollar value of the sales summed for all sales in each zip code, such that larger total sales produce a larger circle. The circles are then mapped against a geographical map of the United States based on the zip code. The figure makes it easy for a manager to quickly see the region of the northeastern United States that has the greatest sale penetration.

In addition to specialized data visualization software such as Tableau, R, and Gephi, common productivity tools such as Microsoft Excel can often provide surprisingly powerful data visualization. Excel has long provided basic charting abilities and PivotTable and PivotChart capabilities for visualizing spreadsheet data. More recently, the introduction of the PowerPivot add-in has eliminated row and column data limitations and allows for the integration of data from multiple sources. This puts powerful data visualization capabilities on the desktop of most business users.

Note

Data visualization plays an important role in discovering and understanding the meaning of data. New ways to present data are being constantly developed. Good data visualizations can be used in any discipline. For example, see the video from Dr. Hans Rosling, (*www.youtube.com/watch?v=jbkSRLYSojo*) in which he uses public health data to visualize the history of the world in the last 200 years.

The main BI architectural components were illustrated in Figure 13.1 and further explained in Tables 13.2 and 13.3. However, the heart of the BI system is its advanced information generation and decision support capabilities. A BI system's advanced decision support functions come to life via its intuitive and informational user interface, and particularly its reporting capabilities. A modern BI system provides three distinctive reporting styles:

- *Advanced reporting*. A BI system presents insightful information about the organization in a variety of presentation formats. Furthermore, the reports provide interactive features that allow the end user to study the data from multiple points of view—from highly summarized to very detailed data. The reports present key actionable information used to support decision making.

- *Monitoring and alerting*. After a decision has been made, the BI system offers ways to monitor the decision's outcome. The BI system provides the end user with ways to define metrics and other key performance indicators to evaluate different aspects of an organization. In addition, exceptions and alerts can be set to warn managers promptly about deviations or problem areas.

- *Advanced data analytics*. A BI system provides tools to help the end user discover relationships, patterns, and trends hidden within the organization's data. These tools are used to create two types of data analysis: explanatory and predictive. Explanatory analysis provides ways to discover relationships, trends, and patterns among data, while predictive analysis provides the end user with ways to create models that predict future outcomes.

Understanding the architectural components of a BI framework is the first step in properly implementing BI in an organization. A good BI infrastructure promises many benefits to an organization, as outlined in the next section.

13-2b Business Intelligence Benefits

As you have learned in previous sections, a properly implemented BI architecture could provide a framework for continuous performance improvements and business decision making. Improved decision making is the main goal of BI, but BI provides other benefits:

- *Integrating architecture.* Like any other IT project, BI has the potential of becoming the integrating umbrella for a disparate mix of IT systems within an organization. This architecture could support all types of company-generated data from operational to executive, as well as diverse hardware such as mainframes, servers, desktops for managers and executives, and mobile devices on the shop floor.

- *Common user interface for data reporting and analysis.* BI front ends can provide up-to-the-minute consolidated information using a common interface for all company users. IT departments no longer have to provide multiple training options for diverse interfaces. End users benefit from similar or common interfaces in different devices that use multiple clever and insightful presentation formats.

- *Common data repository fosters single version of company data.* In the past, multiple IT systems supported different aspects of an organization's operations. Such systems collected and stored data in separate data stores. Keeping the data synchronized and up to date has always been difficult. BI provides a framework to integrate such data under a common environment and present a single version of the data.

- *Improved organizational performance.* BI can provide competitive advantages in many different areas, from customer support to manufacturing processes. Such advantages can be reflected in added efficiency, reduced waste, increased sales, reduced employee and customer turnover, and most importantly, an increased bottom line for the business.

Achieving all these benefits takes a lot of human, financial, and technological resources, not to mention time. BI benefits are not achieved overnight, but are the result of a focused company-wide effort that could take a long time. As a matter of fact, as you will learn in the next section, the BI field has evolved over a long period of time itself.

13-2c Business Intelligence Evolution

Providing useful information to end users has been a priority of IT systems since mainframe computing became an integral part of corporations. Business decision support has evolved over many decades. Following computer technology advances, business intelligence started with centralized reporting systems and evolved into today's highly integrated BI environments. Table 13.4 summarizes the evolution of BI systems.

Using Table 13.4 as a guide, you can trace business intelligence from the mainframe environment to the desktop and then to the more current, cloud-based, mobile BI environments. (Chapter 15, Database Connectivity and Web Technologies, provides a detailed discussion of cloud-based systems.)

The precursor of the modern BI environment was the first-generation decision support system. A **decision support system (DSS)** is an arrangement of computerized tools used to assist managerial decision making. A DSS typically has a much narrower focus and reach than a BI solution. At first, decision support systems were the realm of a few selected managers in an organization. Over time, and with the introduction of the desktop computer, decision support systems migrated to more agile platforms, such as midrange computers, high-end servers, commodity servers, appliances, and cloud-based offerings. This evolution effectively changed the reach of decision support systems; BI is no longer limited to a small group of top-level managers with training in statistical

decision support system (DSS)
An arrangement of computerized tools used to assist managerial decision making within a business.

TABLE 13.4

BUSINESS INTELLIGENCE EVOLUTION

SYSTEM TYPE	DATA SOURCE	DATA EXTRACTION/ INTEGRATION PROCESS	DATA STORE	END-USER QUERY TOOL	END USER PRESENTATION TOOL
Traditional mainframe-based online transaction processing (OLTP)	Operational data	None Reports read and summarized data directly from operational data	None Temporary files used for reporting purposes	Very basic Predefined reporting formats Basic sorting, totaling, and averaging	Very basic Menu-driven, predefined reports, text and numbers only
Managerial information system (MIS)	Operational data	Basic extraction and aggregation Read, filter, and summarize operational data into intermediate data store	Lightly aggregated data in RDBMS	Same as above, in addition to some ad hoc reporting using SQL	Same as above, in addition to some ad hoc columnar report definitions
First-generation departmental decision support system (DSS)	Operational data External data	Data extraction and integration process populates DSS data store Run periodically	First DSS database generation Usually RDBMS	Query tool with some analytical capabilities and predefined reports	Spreadsheet style Advanced presentation tools with plotting and graphics capabilities
First-generation BI	Operational data External data	Advanced data extraction and integration Access diverse data sources, filters, aggregations, classifications, scheduling, and conflict resolution	Data warehouse RDBMS technology Optimized for query purposes Star schema model	Same as above	Same as above, in addition to multidimensional presentation tools with drill-down capabilities
Second-generation BI Online analytical processing (OLAP)	Same as above	Same as above	Data warehouse stores data in MDBMS Cubes with multiple dimensions	Adds support for end-user-based data analytics	Same as above, but uses cubes and multidimensional matrixes; limited by cube size Dashboards Scorecards Portals
Third-generation Mobile, cloud-based, and Big Data	Same as above Includes social media and machine-generated data	Same as above Cloud-based	Same as above Cloud-based Hadoop and NoSQL databases	Advanced analytics Limited ad hoc interactions	Mobile devices: smartphones and tablets

modeling. Instead, BI is now available to all users in an organization, from line managers to the shop floor to mobile agents in the field.

You can also use Table 13.4 to track the evolution of information dissemination styles used in business intelligence.

- Starting in the late 1970s, the need for information distribution was filled by centralized reports running on mainframes, minicomputers, or even central server environments. Such reports were predefined and took considerable time to process.

- With the introduction of desktop computers in the 1980s, a new style of information distribution, the spreadsheet, emerged as the dominant format for decision support systems. In this environment, managers downloaded information from centralized data stores and manipulated the data in desktop spreadsheets.

- As the use of spreadsheets multiplied, IT departments tried to manage the flow of data in a more formal way using enterprise reporting systems. These systems were developed in the early 1990s and basically integrated all data into an IT umbrella that started with the first-generation DSS. The systems still used spreadsheet-like features with which end users were familiar.

- Once DSSs were established, the evolution of business intelligence flourished with the introduction of the data warehouse and online analytical processing systems (OLAPs) in the mid-1990s.

- Rapid changes in information technology and the Internet revolution led to the introduction of advanced BI systems such as web-based dashboards in the early and mid 2000s and mobile BI later in the decade. With mobile BI, end users access BI reports via native applications that run on a mobile smart device, such as a smartphone or tablet.

- More recently, the social media revolution has generated large amounts of data. At the same time sensor-generated data is being collected and stored. Companies are using Big Data analytics tools to leverage such data and obtain critical information otherwise unavailable to them.

Figure 13.3 depicts the evolution of BI information dissemination.

FIGURE 13.3 EVOLUTION OF BI INFORMATION DISSEMINATION FORMATS

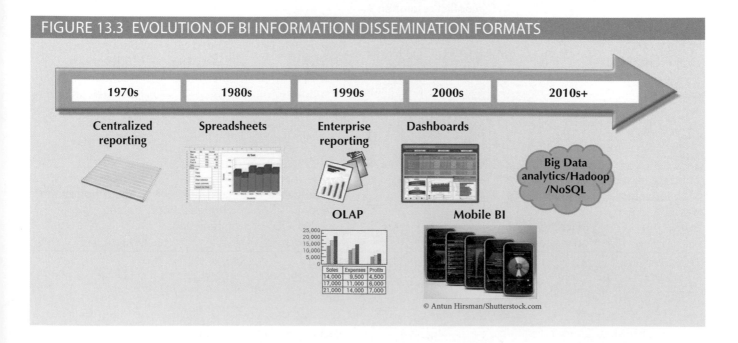

© Antun Hirsman/Shutterstock.com

Note

The OLAP environment is covered in Section 13-6 of this chapter.

Mobile BI technology is poised to have a significant impact on the way BI information is disseminated and processed. If the number of students using smartphones to communicate with friends, update their Facebook status, and send tweets on Twitter is any indicator, you can expect the next generation of consumers and workers to be highly mobile. Leading corporations are therefore starting to push decision making to agents in the field to facilitate customer relationships, sales and ordering, and product support. Such mobile technologies are so portable and interactive that some users call them "disruptive" technologies.

BI information technology has evolved from centralized reporting styles to the current, mobile BI and Big Data analytics style in the span of just a few years. The rate of technological change is not slowing down; to the contrary, technology advancements are accelerating the adoption of BI to new levels. The next section illustrates some BI technology trends.

13-2d Business Intelligence Technology Trends

Several technological advances are driving the growth of business intelligence technologies. These advances create new generations of more affordable products and services that are faster and easier to use. In turn, such products and services open new markets and work as driving forces in the increasing adoption of business intelligence technologies within organizations. Some of the more remarkable technological trends are:

- *Data storage improvements*. Newer data storage technologies, such as solid state drives (SSD) and Serial Advanced Technology Attachment (SATA) drives, offer increased performance and larger capacity that make data storage faster and more affordable. Currently you can buy single drives with a capacity approaching 4 terabytes.

- *Business intelligence appliances*. Vendors now offer plug-and-play appliances optimized for data warehouse and BI applications. These new appliances offer improved price-performance ratios, simplified administration, rapid installation, scalability, and fast integration. Some of these vendors include IBM, Netezza, Greenplum, and AsterData.

- *Business intelligence as a service*. Vendors now offer data warehouses and BI as a service. These cloud-based services allow any corporation to rapidly develop a data warehouse store without the need for hardware, software, or extra personnel. These prepackaged services offer "pay-as-you-go" models for specific industries and capacities, and they provide an opportunity for organizations to pilot-test a BI project without incurring large time or cost commitments. Such services are offered by Netezza, AppNexus, AsterData, MicroStrategy, and Kognitio.

- *Big Data analytics*. The Big Data phenomenon is creating a new market for data analytics. Organizations are turning to social media as the new source for information and knowledge to gain competitive advantages. Examples of Big Data analytics vendors include Vertica, AsterData, and Netezza. You'll learn more about Big Data analytics in Chapter 14.

- *Personal analytics*. OLAP brought data analytics to the desktop of every end user in an organization. Mobile BI is extending business decision making outside the walls of the organization. BI can now be deployed to mobile users who are closer to customers. The main requirement is for the BI end user to have a key understanding of the business. Some personal analytics vendors include MicroStrategy, QlikView, and Actuate. There is a growing trend toward self-service, personalized data analytics. It is not so far-fetched to imagine that in a few years, end users will have smart data analytics agents on their smartphones tailored to their personal interests. Such personal agents will provide users with up-to-the-minute "intelligent knowledge" about their personal interests.

One constant in this relentless technological evolution is the need for better decision support data and the importance of understanding the difference between decision support data and operational data.

13-3 Decision Support Data

Although BI is used at strategic and tactical managerial levels within organizations, *its effectiveness depends on the quality of data gathered at the operational level.* Yet, operational data is seldom well suited to the decision support tasks. The differences between operational data and decision support data are examined in the next section.

13-3a Operational Data Versus Decision Support Data

Operational data and decision support data serve different purposes. Therefore, it is not surprising to learn that their formats and structures differ. Most operational data is stored in a relational database in which the structures (tables) tend to be highly normalized. Operational data storage is optimized to support transactions that represent daily operations. For example, each time an item is sold, it must be accounted for. Customer data, inventory data, and other similar data need frequent updating. To provide effective update performance, operational systems store data in many tables, each with a minimum number of fields. Thus, a simple sales transaction might be represented by five or more different tables, such as INVOICE, INVOICE LINE, DISCOUNT, STORE, and DEPARTMENT. Although such an arrangement is excellent in an operational database, it is not efficient for query processing. For example, to extract a simple invoice, you would have to join several tables. Whereas operational data is useful for capturing daily business transactions, decision support data gives tactical and strategic business meaning to the operational data. From the data analyst's point of view, decision support data differs from operational data in three main areas: time span, granularity, and dimensionality.

- *Time span.* Operational data covers a short time frame. In contrast, decision support data tends to cover a longer time frame. Managers are seldom interested in a specific sales invoice to Customer X; rather, they tend to focus on sales generated during the last month, the last year, or the last five years.

- *Granularity (level of aggregation).* Decision support data must be presented at different levels of aggregation, from highly summarized to nearly atomic. For example, if managers analyze regional sales, they must be able to access data showing the sales by region, by city within the region, by store within the city within the region, and so on. In that case, summarized data to compare the regions is required, along with data in a structure that enables a manager to **drill down**, or decompose, the data into more atomic components—that is, finer-grained data at lower levels of aggregation. In contrast, when you **roll up** the data, you are aggregating the data to a higher level.

drill down
To decompose data into more atomic components—that is, data at lower levels of aggregation. This approach is used primarily in a decision support system to focus on specific geographic areas, business types, and so on.

roll up
(1) To aggregate data into summarized components, that is, higher levels of aggregation. (2) In SQL, an OLAP extension used with the GROUP BY clause to aggregate data by different dimensions. Rolling up the data is the exact opposite of drilling down the data.

- *Dimensionality.* Operational data focuses on representing individual transactions rather than the effects of the transactions over time. In contrast, data analysts tend to include many data dimensions and are interested in how the data relates over those dimensions. For example, an analyst might want to know how Product X fared relative to Product Z during the past six months by region, state, city, store, and customer. In that case, both place and time are part of the picture.

Figure 13.4 shows how decision support data can be examined from multiple dimensions such as product, region, and year, using a variety of filters to produce each dimension. The ability to analyze, extract, and present information in meaningful ways is one of the differences between decision support data and transaction-at-a-time operational data.

FIGURE 13.4 TRANSFORMING OPERATIONAL DATA INTO DECISION SUPPORT DATA

Operational Data

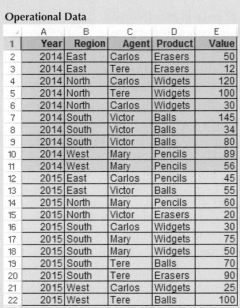

	A	B	C	D	E
1	Year	Region	Agent	Product	Value
2	2014	East	Carlos	Erasers	50
3	2014	East	Tere	Erasers	12
4	2014	North	Carlos	Widgets	120
5	2014	North	Tere	Widgets	100
6	2014	North	Carlos	Widgets	30
7	2014	South	Victor	Balls	145
8	2014	South	Victor	Balls	34
9	2014	South	Victor	Balls	80
10	2014	West	Mary	Pencils	89
11	2014	West	Mary	Pencils	56
12	2015	East	Carlos	Pencils	45
13	2015	East	Victor	Balls	55
14	2015	North	Mary	Pencils	60
15	2015	North	Victor	Erasers	20
16	2015	South	Carlos	Widgets	30
17	2015	South	Mary	Widgets	75
18	2015	South	Mary	Widgets	50
19	2015	South	Tere	Balls	70
20	2015	South	Tere	Erasers	90
21	2015	West	Carlos	Widgets	25
22	2015	West	Tere	Balls	100

Decision Support Data

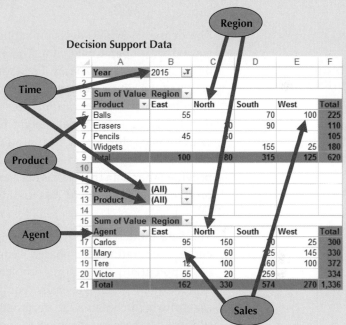

Operational data has a narrow time span, low granularity, and single focus. Such data is usually represented in tabular format, in which each row represents a single transaction. This format often makes it difficult to derive useful information.

Decision support system (DSS) data focuses on a broader time span, tends to have high levels of granularity, and can be examined in multiple dimensions. For example, note these possible aggregations:
- Sales by product, region, agent, and so on
- Sales for all years or only a few selected years
- Sales for all products or only a few selected products

Note

The decision support data in Figure 13.4 shows the output for the solution to Problem 2 at the end of this chapter.

From the designer's point of view, the differences between operational and decision support data are as follows:

- Operational data represents transactions as they happen in real time. Decision support data is a snapshot of the operational data at a given point in time. Therefore, decision support data is historic, representing a time slice of the operational data.

- Operational and decision support data are different in terms of transaction *type* and transaction *volume*. Whereas operational data is characterized by update transactions, decision support data is mainly characterized by read-only transactions. Decision support data also requires *periodic* updates to load new data that is summarized from the operational data. Finally, the concurrent transaction volume in operational data tends to be very high compared with the low to medium levels in decision support data.

- Operational data is commonly stored in many tables, and the stored data represents information about a given transaction only. Decision support data is generally stored in a few tables derived from the operational data. The decision support data does not include the details of each operational transaction. Instead, decision support data represents transaction *summaries*; therefore, the decision support database stores data that is integrated, aggregated, and summarized for decision support purposes.

- The degree to which decision support data is summarized is very high when contrasted with operational data. Therefore, you will see a great deal of derived data in decision support databases. For example, rather than storing all 10,000 sales transactions for a given store on a given day, the decision support database might simply store the total number of units sold and the total sales dollars generated during that day. Decision support data might be collected to monitor such aggregates as total sales for each store or for each product. The purpose of the summaries is simple: they are used to establish and evaluate sales trends and product sales comparisons and to provide other data that serves decision needs. (How well are items selling? Should this product be discontinued? Has the advertising been effective as measured by increased sales?)

- The data models that govern operational data and decision support data are different. The operational database's frequent and rapid data updates make data anomalies a potentially devastating problem. Therefore, the data in a relational transaction (operational) system generally requires normalized structures that yield many tables, each of which contains the minimum number of attributes. In contrast, the decision support database is not subject to such transaction updates, and the focus is on querying capability. Therefore, decision support databases tend to be non-normalized and include few tables, each of which contains a large number of attributes.

- The frequency and complexity of query activity in the operational database tends to be low to allow additional processing cycles for the more crucial update transactions. Therefore, queries against operational data typically are narrow in scope and low in complexity, and high speed is critical. In contrast, decision support data exists for the sole purpose of serving query requirements. Queries against decision support data typically are broad in scope and high in complexity, and less speed is needed.

- Finally, decision support data is characterized by very large amounts of data. The large data volume is the result of two factors. First, data is stored in non-normalized structures that are likely to display many data redundancies and duplications. Second, the same data can be categorized in many different ways to represent different snapshots. For example, sales data might be stored in relation to product, store, customer, region, and manager.

Table 13.5 summarizes the differences between operational and decision support data from the database designer's point of view.

TABLE 13.5

CONTRASTING OPERATIONAL AND DECISION SUPPORT DATA CHARACTERISTICS

CHARACTERISTIC	OPERATIONAL DATA	DECISION SUPPORT DATA
Data currency	Current operations Real-time data	Historic data Snapshot of company data Time component (week/month/year)
Granularity	Atomic-detailed data	Summarized data
Summarization level	Low; some aggregate yields	High; many aggregation levels
Data model	Highly normalized Mostly relational DBMSs	Non-normalized Complex structures Some relational, but mostly multidimensional DBMSs
Transaction type	Mostly updates	Mostly query
Transaction volumes	High-update volumes	Periodic loads and summary calculations
Transaction speed	Updates are critical	Retrievals are critical
Query activity	Low to medium	High
Query scope	Narrow range	Broad range
Query complexity	Simple to medium	Very complex
Data volumes	Hundreds of gigabytes	Terabytes to petabytes

The many differences between operational data and decision support data are good indicators of decision support database requirements, which are described in the next section.

13-3b Decision Support Database Requirements

A decision support database is a specialized DBMS tailored to provide fast answers to complex queries. There are three main requirements for a decision support database: the database schema, data extraction and filtering, and database size.

Database Schema The decision support database schema must support complex (non-normalized) data representations. As noted earlier, the decision support database must contain data that is aggregated and summarized. In addition to meeting those requirements, the queries must be able to extract multidimensional time slices. If you are using an RDBMS, the conditions suggest using non-normalized and even duplicated data. To see why this must be true, take a look at the 10-year sales history for a single store containing a single department. At this point, the data is fully normalized within the single table, as shown in Table 13.6.

This structure works well when you have only one store with only one department. However, it is very unlikely that such a simple environment has much need for a decision support database. A decision support database becomes a factor when you are dealing with more than one store, each of which has more than one department. To support all of the decision support requirements, the database must contain data for all of the stores and all of their departments—and the database must be able to support multi-dimensional queries that track sales by stores, by departments, and over time. For simplicity, suppose that there are only two stores (A and B) and two departments (1 and 2) within each store. Also, change the time dimension to include yearly data. Table 13.7 shows the sales figures under the specified conditions. Only 2006, 2012, and 2015 are shown; ellipses (…) are used to indicate that data values were omitted. You can see in

TABLE 13.6

TEN-YEAR SALES HISTORY FOR A SINGLE DEPARTMENT, IN MILLIONS OF DOLLARS

YEAR	SALES
2006	8,227
2007	9,109
2008	10,104
2009	11,553
2010	10,018
2011	11,875
2012	12,699
2013	14,875
2014	16,301
2015	19,986

TABLE 13.7

YEARLY SALES SUMMARIES, TWO STORES AND TWO DEPARTMENTS PER STORE, IN MILLIONS OF DOLLARS

YEAR	STORE	DEPARTMENT	SALES
2006	A	1	1,985
2006	A	2	2,401
2006	B	1	1,879
2006	B	2	1,962
...	...	...	...
2012	A	1	3,912
2012	A	2	4,158
2012	B	1	3,426
2012	B	2	1,203
...	...	...	...
2015	A	1	7,683
2015	A	2	6,912
2015	B	1	3,768
2015	B	2	1,623

Table 13.7 that the number of rows and attributes already multiplies quickly and that the table exhibits multiple redundancies.

Now suppose that the company has 10 departments per store and 20 stores nationwide, and suppose that you want to access yearly sales summaries. Now you are dealing with 200 rows and 12 monthly sales attributes per row. (Actually, there are 13 attributes per row if you add each store's sales total for each year.)

The decision support database schema must also be optimized for query (read-only) retrievals. To optimize query speed, the DBMS must support features such as bitmap indexes and data partitioning. In addition, the DBMS query optimizer must be enhanced to support the non-normalized and complex structures in decision support databases.

Data Extraction and Filtering The decision support database is created largely by extracting data from the operational database and by importing additional data from external sources. Thus, the DBMS must support advanced data extraction and data-filtering tools. To minimize the impact on the operational database, the data extraction capabilities should allow batch and scheduled data extraction, and should support different data sources: flat files and hierarchical, network, and relational databases, as well as multiple vendors. Data-filtering capabilities must include the ability to check for inconsistent data or data validation rules. Finally, to filter and integrate the operational data into the decision support database, the DBMS must support advanced data integration, aggregation, and classification.

Using data from multiple external sources also usually means having to solve data-formatting conflicts. For example, data such as Social Security numbers and dates can occur in different formats; measurements can be based on different scales, and the same data elements can have different names. In short, data must be filtered and purified to ensure that only the pertinent decision support data is stored in the database and that it is stored in a standard format.

Database Size Decision support databases tend to be very large; gigabyte and terabyte ranges are not unusual. For example, Walmart has more than 4 petabytes of data in its data warehouses. Therefore, the DBMS must be capable of supporting **very large databases (VLDBs)**. To support a VLDB adequately, the DBMS might be required to support advanced storage technologies, and even more importantly, to support multiple-processor technologies, such as a symmetric multiprocessor (SMP) or a massively parallel processor (MPP).

The complex information requirements and the ever-growing demand for sophisticated data analysis sparked the creation of a new type of data repository. This repository, called a data warehouse, contains data in formats that facilitate data extraction, data analysis, and decision making. It has become the foundation for a new generation of decision support systems.

13-4 The Data Warehouse

Bill Inmon, the acknowledged "father" of the **data warehouse**, defines the term as "an *integrated*, *subject-oriented*, *time-variant*, *nonvolatile* collection of data that provides support for decision making."[2] (Italics were added for emphasis.) To understand that definition, take a more detailed look at its components.

- *Integrated*. The data warehouse is a centralized, consolidated database that integrates data derived from the entire organization and from multiple sources with diverse formats. Data integration implies that all business entities, data elements, data characteristics, and business metrics are *described in the same way throughout the enterprise*. Although this requirement sounds logical, you would be amazed to discover how many different measurements for "sales performance" can exist within an organization; the same scenario can be true for any other business element. For instance, the status of an order might be indicated with text labels such as "open," "received," "canceled," and "closed" in one department and as "1," "2," "3," and "4" in another department. A student's status might be defined as "freshman," "sophomore," "junior," or "senior" in the accounting department and as "FR," "SO," "JR," or "SR" in the computer information systems department. To avoid the potential format tangle,

very large database (VLDB)
Database that contains huge amounts of data—gigabyte, terabyte, and petabyte ranges are not unusual.

data warehouse
An integrated, subject-oriented, time-variant, nonvolatile collection of data that provides support for decision making.

[2]Bill Inmon and Chuck Kelley, "The twelve rules of data warehouse for a client/server world," *Data Management Review* 4(5), May 1994, pp. 6–16.

the data in the data warehouse must conform to a common format that is acceptable throughout the organization. This integration can be time-consuming, but once accomplished, it enhances decision making and helps managers better understand the company's operations. This understanding can be translated into recognition of strategic business opportunities.

- *Subject-oriented*. Data warehouse data is arranged and optimized to provide answers to questions from diverse functional areas within a company. Data warehouse data is organized and summarized by topic, such as sales, marketing, finance, distribution, and transportation. For each topic, the data warehouse contains specific subjects of interest—products, customers, departments, regions, promotions, and so on. This form of data organization is quite different from the more functional or process-oriented organization of typical transaction systems. For example, an invoicing system designer concentrates on designing normalized data structures to support the business process by storing invoice components in two tables: INVOICE and INVLINE. In contrast, the data warehouse has a *subject* orientation. Data warehouse designers focus specifically on the data rather than on the processes that modify the data. (After all, data warehouse data is not subject to numerous real-time data updates!) Therefore, instead of storing an invoice, the data warehouse stores its "sales by product" and "sales by customer" components because decision support activities require the retrieval of sales summaries by product or customer.

- *Time-variant*. In contrast to operational data, which focuses on current transactions, warehouse data represents the flow of data through time. The data warehouse can even contain projected data generated through statistical and other models. It is also time-variant in the sense that when data is periodically uploaded to the data warehouse, all time-dependent aggregations are recomputed. For example, when data for previous weekly sales is uploaded to the data warehouse, the weekly, monthly, yearly, and other time-dependent aggregates for products, customers, stores, and other variables are also updated. Because data in a data warehouse constitutes a snapshot of the company history as measured by its variables, the time component is crucial. The data warehouse contains a time ID that is used to generate summaries and aggregations by week, month, quarter, year, and so on. Once the data enters the data warehouse, the time ID assigned to the data cannot be changed.

- *Nonvolatile*. Once data enters the data warehouse, it is never removed. Because the data in the warehouse represents the company's history, the operational data, which represents the near-term history, is always added to it. Because data is never deleted and new data is continually added, the data warehouse is always growing. Therefore, the DBMS must be able to support multiterabyte or greater databases operating on multiprocessor hardware.

Table 13.8 summarizes the differences between data warehouses and operational databases.

In summary, the data warehouse is a read-only database optimized for data analysis and query processing. Typically, data is extracted from various sources and are then transformed and integrated—in other words, passed through a data filter— before being loaded into the data warehouse. As mentioned, this process is known as ETL. Figure 13.5 illustrates the ETL process to create a data warehouse from operational data.

Although the centralized and integrated data warehouse can be an attractive proposition that yields many benefits, managers may be reluctant to embrace this strategy. Creating a data warehouse requires time, money, and considerable managerial effort.

TABLE 13.8

CHARACTERISTICS OF DATA WAREHOUSE DATA AND OPERATIONAL DATABASE DATA

CHARACTERISTIC	OPERATIONAL DATABASE DATA	DATA WAREHOUSE DATA
Integrated	Similar data can have different representations or meanings. For example, Social Security numbers may be stored as ###-##-#### or as #########, and a given condition may be labeled as T/F or 0/1 or Y/N. A sales value may be shown in thousands or in millions.	Provide a unified view of all data elements with a common definition and representation for all business units.
Subject-oriented	Data is stored with a functional, or process, orientation. For example, data may be stored for invoices, payments, and credit amounts.	Data is stored with a subject orientation that facilitates multiple views of the data and decision making. For example, sales may be recorded by product, division, manager, or region.
Time-variant	Data is recorded as current transactions. For example, the sales data may be the sale of a product on a given date, such as $342.78 on 12-MAY-2016.	Data is recorded with a historical perspective in mind. Therefore, a time dimension is added to facilitate data analysis and various time comparisons.
Nonvolatile	Data updates are frequent and common. For example, an inventory amount changes with each sale. Therefore, the data environment is fluid.	Data cannot be changed. Data is added only periodically from historical systems. Once the data is properly stored, no changes are allowed. Therefore, the data environment is relatively static.

FIGURE 13.5 THE ETL PROCESS

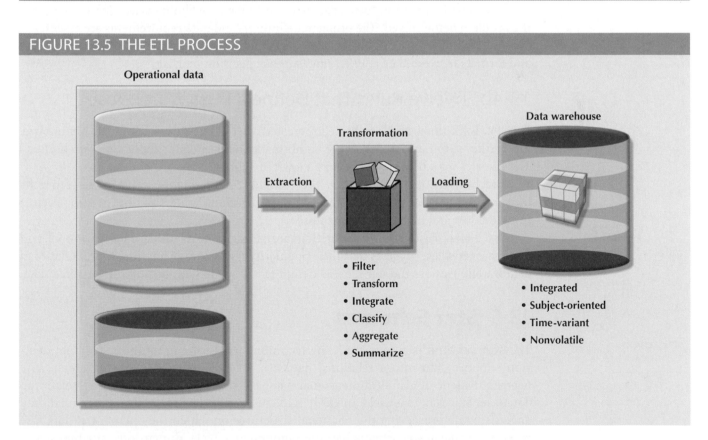

Therefore, it is not surprising that many companies begin their foray into data warehousing by focusing on more manageable data sets that are targeted to meet the special needs of small groups within the organization. These smaller data stores are called data marts.

13-4a Data Marts

A **data mart** is a small, single-subject data warehouse subset that provides decision support to a small group of people. In addition, a data mart could be created from data extracted from a larger data warehouse for the specific purpose of supporting faster data access to a target group or function. That is, data marts and data warehouses can coexist within a business intelligence environment.

Some organizations choose to implement data marts not only because of the lower cost and shorter implementation time but because of the technological advances and inevitable "people issues" that make data marts attractive. Powerful computers can provide a customized decision support system to small groups in ways that might not be possible with a centralized system. Also, a company's culture may predispose its employees to resist major changes, but they might quickly embrace relatively minor changes that lead to demonstrably improved decision support. In addition, people at different organizational levels are likely to require data with different summarization, aggregation, and presentation formats. Data marts can serve as a test vehicle for companies exploring the potential benefits of data warehouses. By gradually migrating from data marts to data warehouses, a specific department's decision support needs can be addressed within six months to one year, as opposed to the one- to three-year time frame usually required to implement a data warehouse. Information technology (IT) departments also benefit from this approach because their personnel can learn the issues and develop the skills required to create a data warehouse.

The only difference between a data mart and a data warehouse is the size and scope of the problem being solved. The problem definitions and data requirements are essentially the same for both. To be useful, the data warehouse must conform to uniform structures and formats to avoid data conflicts and support decision making.

13-4b Twelve Rules That Define a Data Warehouse

In 1994, Bill Inmon and Chuck Kelley created a set of rules to define a data warehouse. These rules summarize many of the points made in this chapter about data warehouses.[3] The 12 rules for a data warehouse are shown in Table 13.9.

Note how the 12 rules capture the complete data warehouse life cycle—from its introduction as an entity separate from the operational data store to its components, functionality, and management processes.

Most data warehouse implementations are based on the relational database model, and their market share suggests that their popularity will not fade anytime soon. Relational data warehouses use the star schema design technique to handle multidimensional data.

13-5 **Star Schemas**

The **star schema** is a data-modeling technique used to map multidimensional decision support data into a relational database. In effect, the star schema creates the near equivalent of a multidimensional database schema from the existing relational database. Star schemas yield an easily implemented model for multidimensional data analysis while preserving the relational structures on which the operational database is built. The basic star schema has four components: facts, dimensions, attributes, and attribute hierarchies.

data mart
A small, single-subject data warehouse subset that provides decision support to a small group of people.

star schema
A data modeling technique used to map multidimensional decision support data into a relational database. The star schema represents data using a central table known as a fact table in a 1:M relationship with one or more dimension tables.

---

[3]Bill Inmon, and Chuck Kelley, "The twelve rules of data warehouse for a client/server world," *Data Management Review* 4(5), May 1994, pp. 6–16.

TABLE 13.9

TWELVE RULES FOR A DATA WAREHOUSE

RULE NO.	DESCRIPTION
1	The data warehouse and operational environments are separated.
2	The data warehouse data is integrated.
3	The data warehouse contains historical data over a long time.
4	The data warehouse data is snapshot data captured at a given point in time.
5	The data warehouse data is subject oriented.
6	The data warehouse data is mainly read-only with periodic batch updates from operational data. No online updates are allowed.
7	The data warehouse development life cycle differs from classical systems development. Data warehouse development is data-driven; the classical approach is process-driven.
8	The data warehouse contains data with several levels of detail: current detail data, old detail data, lightly summarized data, and highly summarized data.
9	The data warehouse environment is characterized by read-only transactions to very large data sets. The operational environment is characterized by numerous update transactions to a few data entities at a time.
10	The data warehouse environment has a system that traces data sources, transformations, and storage.
11	The data warehouse's metadata is a critical component of this environment. The metadata identifies and defines all data elements. The metadata provides the source, transformation, integration, storage, usage, relationships, and history of each data element.
12	The data warehouse contains a chargeback mechanism for resource usage that enforces optimal use of the data by end users.

13-5a Facts

Facts are numeric measurements (values) that represent a specific business aspect or activity. For example, sales figures are numeric measurements that represent product and service sales. Facts commonly used in business data analysis are units, costs, prices, and revenues. Facts are normally stored in a fact table that is the center of the star schema. The **fact table** contains facts that are linked through their dimensions, which are explained in the next section.

Facts can also be computed or derived at run time. Such computed or derived facts are sometimes called **metrics** to differentiate them from stored facts. The fact table is updated periodically with data from operational databases.

13-5b Dimensions

Dimensions are qualifying characteristics that provide additional perspectives to a given fact. Recall that dimensions are of interest because *decision support data is almost always viewed in relation to other data*. For instance, sales might be compared by product from region to region and from one time period to the next. The kind of problem typically addressed by a BI system might be to compare the sales of unit X by region for the first quarters of 2006 through 2016. In that example, sales have product, location, and time dimensions. In effect, dimensions are the magnifying glass through which you study the facts. Such dimensions are normally stored in **dimension tables**. Figure 13.6 depicts a star schema for sales with product, location, and time dimensions.

facts
In a data warehouse, the measurements (values) that measure a specific business aspect or activity. For example, sales figures are numeric measurements that represent product or service sales. Facts commonly used in business data analysis include units, costs, prices, and revenues.

fact table
In a data warehouse, the star schema table that contains facts linked and classified through their common dimensions. A fact table is in a one-to-many relationship with each associated dimension table.

metrics
In a data warehouse, numeric facts that measure a business characteristic of interest to the end user.

dimensions
In a star schema design, qualifying characteristics that provide additional perspectives to a given fact.

dimension tables
In a data warehouse, tables used to search, filter, or classify facts within a star schema.

FIGURE 13.6 SIMPLE STAR SCHEMA

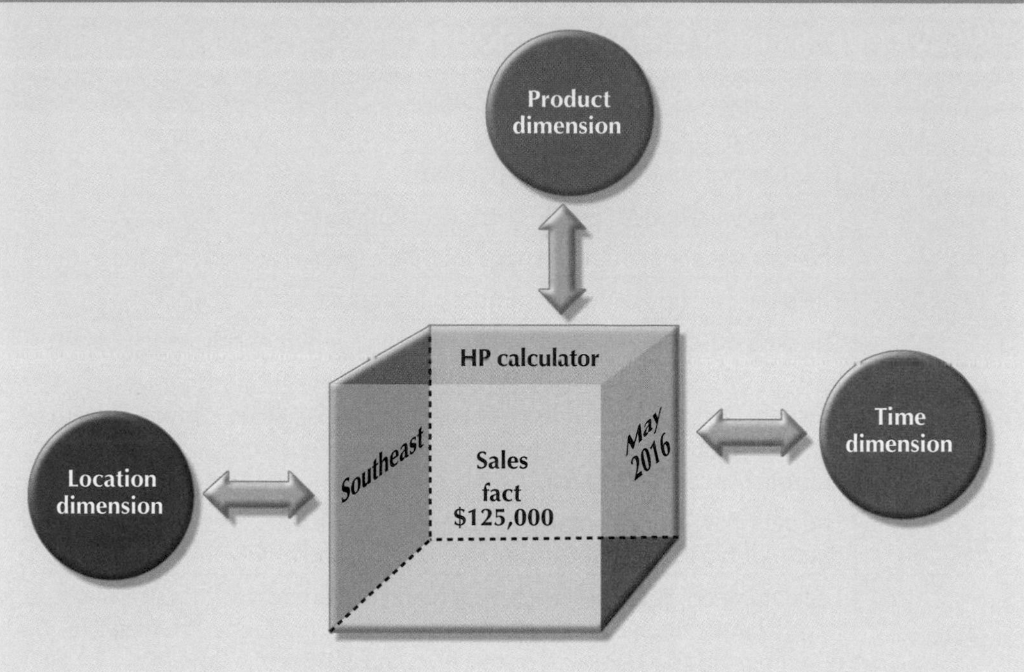

13-5c Attributes

Each dimension table contains attributes. Attributes are often used to search, filter, or classify facts. *Dimensions provide descriptive characteristics about the facts through their attributes.* Therefore, the data warehouse designer must define common business attributes that will be used by the data analyst to narrow a search, group information, or describe dimensions. Using a sales example, some possible attributes for each dimension are illustrated in Table 13.10.

TABLE 13.10

POSSIBLE ATTRIBUTES FOR SALES DIMENSIONS

DIMENSION NAME	DESCRIPTION	POSSIBLE ATTRIBUTES
Location	Anything that provides a description of the location—for example, Nashville, Store 101, South Region, and TN	Region, state, city, store, and so on
Product	Anything that provides a description of the product sold—for example, hair care product, shampoo, Natural Essence brand, 5.5-oz. bottle, and blue liquid	Product type, product ID, brand, package, presentation, color, size, and so on
Time	Anything that provides a time frame for the sales fact—for example, the year 2016, the month of July, the date 07/29/2016, and the time 4:46 p.m.	Year, quarter, month, week, day, time of day, and so on

These product, location, and time dimensions add a business perspective to the sales facts. The data analyst can now group the sales figures for a given product, in a given region, and at a given time. The star schema, through its facts and dimensions, can provide the data in a format suited for data analysis. Also, it can do so without imposing the burden of additional and unnecessary data, such as order number, purchase order number, and status that commonly exists in operational databases.

Conceptually, the sales example's multidimensional data model is best represented by a three-dimensional cube. Of course, this does not imply that there is a limit on the number of dimensions you can associate to a fact table. There is no mathematical limit to the number of dimensions used. However, using a three-dimensional model makes it easy to visualize the problem. The three-dimensional cube illustrated in Figure 13.7 represents a view of sales with product, location, and time dimensions.

FIGURE 13.7 THREE-DIMENSIONAL VIEW OF SALES

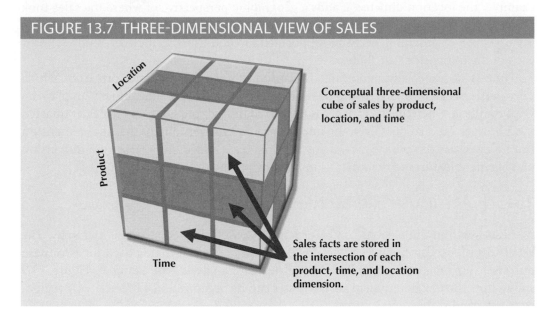

Conceptual three-dimensional cube of sales by product, location, and time

Sales facts are stored in the intersection of each product, time, and location dimension.

Keep in mind that this cube is only a *conceptual* representation of multidimensional data; it does not show how the data is physically stored in a data warehouse.

Whatever the underlying database technology, one of the main features of multidimensional analysis is its ability to focus on specific "slices" of the cube. For example, the product manager may be interested in examining the sales of a product while the store manager is interested in examining the sales made by a particular store. In multidimensional terms, the ability to focus on slices of the cube to perform a more detailed analysis is known as **slice and dice**. Figure 13.8 illustrates the slice-and-dice concept; note that each cut across the cube yields a slice. Intersecting slices produce small cubes that constitute the "dice" part of the slice-and-dice operation.

FIGURE 13.8 SLICE-AND-DICE VIEW OF SALES

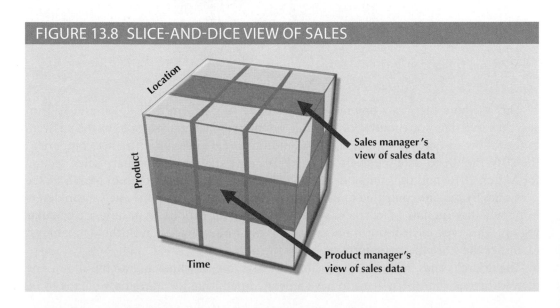

Sales manager's view of sales data

Product manager's view of sales data

To slice and dice, it must be possible to identify each slice of the cube. To do so, you use the values of each attribute in a given dimension. For example, to use the location dimension, you might need to define a STORE_ID attribute to focus on a particular store.

Given the requirement for attribute values in a slice-and-dice environment, re-examine Table 13.10. Note that each attribute adds perspective to the sales facts, thus setting the stage for finding new ways to search, classify, and possibly aggregate information. For example, the location dimension adds a geographic perspective of where the sales took place: in which region, state, city, store, and so on. All of the attributes are selected with the objective of providing decision support data to end users so they can study sales by each of the dimension's attributes.

Time is an especially important dimension; it provides a framework from which sales patterns can be analyzed and possibly predicted. Also, the time dimension plays an important role when the data analyst is interested in studying sales aggregates by quarter, month, week, and so on. Given the importance and universality of the time dimension from a data analysis perspective, many vendors have added automatic time dimension management features to their data-warehousing products.

13-5d Attribute Hierarchies

Attributes within dimensions can be ordered in a well-defined attribute hierarchy. The **attribute hierarchy** provides a top-down data organization that is used for two main purposes: aggregation and drill-down/roll-up data analysis. For example, Figure 13.9 shows how the location dimension attributes can be organized in a hierarchy by region, state, city, and store.

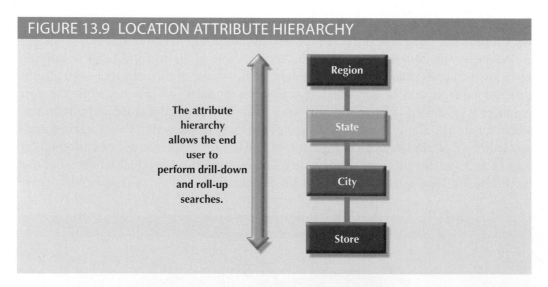

FIGURE 13.9 LOCATION ATTRIBUTE HIERARCHY

The attribute hierarchy allows the end user to perform drill-down and roll-up searches.

Region
State
City
Store

The attribute hierarchy provides the capability to perform drill-down and roll-up searches in a data warehouse. For example, suppose a data analyst looks at the answers to the following query: How does the 2015 month-to-date sales performance compare to the 2016 month-to-date sales performance? The data analyst spots a sharp sales decline for March 2016, and thus might decide to drill down inside the month of March to see how sales by regions compared to the previous year. By doing that, the analyst can determine whether the low March sales were reflected in all regions or in only a particular region. This type of drill-down operation can even be extended until the data analyst identifies the store that is performing below the norm.

The March sales scenario is possible because the attribute hierarchy allows the data warehouse and BI systems to have a defined path that identifies how data is to be

attribute hierarchy
A top-down data organization that is used for two main purposes: aggregation and drill-down/roll-up data analysis.

decomposed and aggregated for drill-down and roll-up operations. It is not necessary for all attributes to be part of an attribute hierarchy; some attributes exist merely to provide narrative descriptions of the dimensions. However, keep in mind that the attributes from different dimensions can be grouped to form a hierarchy. For example, after you drill down from city to store, you might want to drill down using the product dimension so the manager can identify slow-selling products in the store. The product dimension can be based on the product group (dairy, meat, and so on) or the product brand (Brand A, Brand B, and so on).

Figure 13.10 illustrates a scenario in which the data analyst studies sales facts using the product, time, and location dimensions. In this example, the product dimension is set to "All products," meaning that the data analyst will see all products on the *y*-axis. The time dimension (*x*-axis) is set to "Quarter," meaning that the data is aggregated by quarters—for example, total sales of products A, B, and C in Q1, Q2, Q3, and Q4. Finally, the location dimension is initially set to "Region," thus ensuring that each cell contains the total regional sales for a given product in a given quarter.

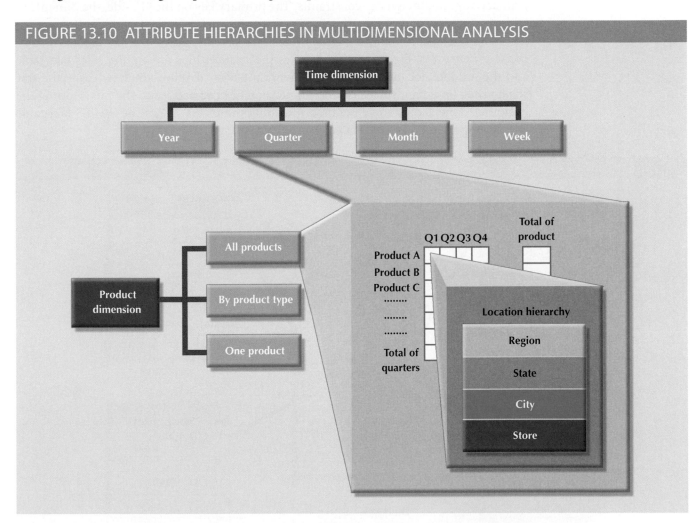

FIGURE 13.10 ATTRIBUTE HIERARCHIES IN MULTIDIMENSIONAL ANALYSIS

The simple scenario illustrated in Figure 13.10 provides the data analyst with three different information paths. On the product dimension (the *y*-axis), the data analyst can request to see all products, products grouped by type, or just one product. On the time dimension (the *x*-axis), the data analyst can request time–variant data at different levels of aggregation: year, quarter, month, or week. Each sales value initially shows the total sales, by region, of each product. When a GUI is used, clicking on the region cell enables the data analyst to drill down to see sales by states within the region. Clicking again on one of the state values yields the sales for each city in the state, and so forth.

As the preceding examples illustrate, attribute hierarchies determine how the data in the data warehouse is extracted and presented. The attribute hierarchy information is stored in the DBMS's data dictionary and is used by the BI tool to access the data warehouse properly. Once such access is ensured, query tools must be closely integrated with the data warehouse's metadata, and they must support powerful analytical capabilities.

13-5e Star Schema Representation

Facts and dimensions are normally represented by physical tables in the data warehouse database. The fact table is related to each dimension table in a many-to-one (M:1) relationship. In other words, many fact rows are related to each dimension row. Using the sales example, you can conclude that each product appears many times in the SALES fact table.

Fact and dimension tables are related by foreign keys and are subject to the familiar primary key and foreign key constraints. The primary key on the "1" side, the dimension table, is stored as part of the primary key on the "many" side, the fact table. *Because the fact table is related to many dimension tables, the primary key of the fact table is a composite primary key.* Figure 13.11 illustrates the relationships among the sales fact table and the product, location, and time dimension tables. To show you how easily the star schema can be expanded, a customer dimension has been added to the mix. Adding the customer dimension merely required including the CUST_ID in the SALES fact table and adding the CUSTOMER table to the database.

FIGURE 13.11 STAR SCHEMA FOR SALES

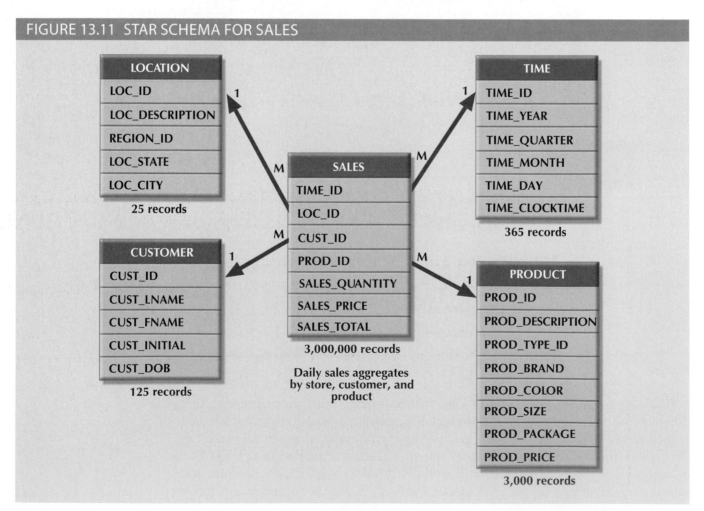

The composite primary key for the SALES fact table is composed of TIME_ID, LOC_ID, CUST_ID, and PROD_ID. Each record in the SALES fact table is uniquely identified by the combination of values for each of the fact table's foreign keys. *By default, the fact table's primary key is always formed by combining the foreign keys pointing to the dimension tables to which they are related.* In this case, each sales record represents each product sold to a specific customer, at a specific time, and in a specific location. In this schema, the TIME dimension table represents daily periods, so the SALES fact table represents daily sales aggregates by product and by customer. Because fact tables contain the actual values used in the decision support process, those values are repeated many times in the fact tables. Therefore, the fact tables are always the largest tables in the star schema. Because the dimension tables contain only nonrepetitive information, such as all unique salespersons and all unique products, the dimension tables are always smaller than the fact tables.

In a typical star schema, each dimension record is related to thousands of fact records. For example, "widget" appears only once in the product dimension, but it has thousands of corresponding records in the SALES fact table. This characteristic of the star schema facilitates data retrieval because the data analyst usually looks at the facts through the dimension's attributes. Therefore, a data warehouse DBMS that is optimized for decision support first searches the smaller dimension tables before accessing the larger fact tables.

Data warehouses usually have many fact tables. Each fact table is designed to answer specific decision support questions. For example, suppose that you develop a new interest in orders while maintaining your original interest in sales. In that scenario, you should maintain an ORDERS fact table and a SALES fact table in the same data warehouse. If orders are considered to be an organization's key interest, the ORDERS fact table should be the center of a star schema that might have vendor, product, and time dimensions. In that case, an interest in vendors yields a new vendor dimension, represented by a new VENDOR table in the database. The product dimension is represented by the same product table used in the initial sales star schema. However, given the interest in orders as well as sales, the time dimension now requires special attention. If the orders department uses the same time periods as the sales department, time can be represented by the same time table. If different time periods are used, you must create another table, perhaps named ORDER_TIME, to represent the time periods used by the orders department. In Figure 13.12, the ORDERS star schema shares the product, vendor, and time dimensions.

Multiple fact tables can also be created for performance and semantic reasons. The following section explains several performance-enhancing techniques that can be used within the star schema.

13-5f Performance-Improving Techniques for the Star Schema

Creating a database that provides fast and accurate answers to data analysis queries is the prime objective of data warehouse design. Therefore, performance enhancement might target query speed through the facilitation of SQL code and through better semantic representation of business dimensions. The following four techniques are often used to optimize data warehouse design:

- Normalizing dimensional tables
- Maintaining multiple fact tables to represent different aggregation levels
- Denormalizing fact tables
- Partitioning and replicating tables

FIGURE 13.12 ORDERS STAR SCHEMA

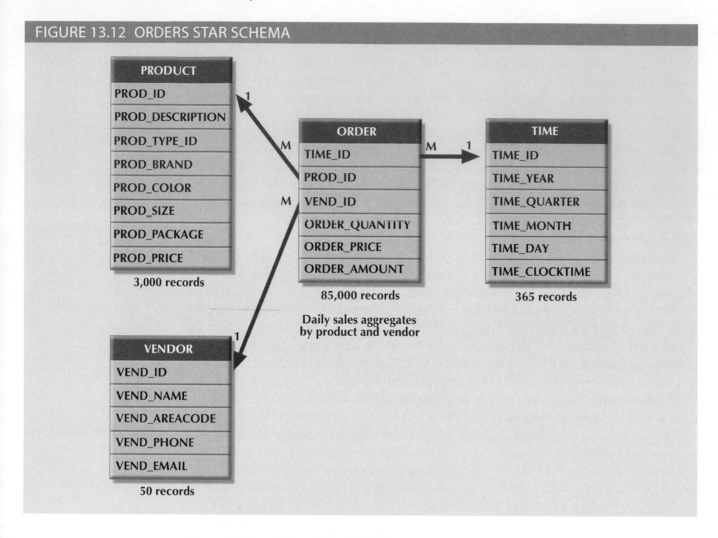

Daily sales aggregates
by product and vendor

Normalizing Dimensional Tables Dimensional tables are normalized to achieve semantic simplicity and facilitate end-user navigation through the dimensions. For example, if the location dimension table contains transitive dependencies among region, state, and city, you can revise those relationships to the 3NF (third normal form), as shown in Figure 13.13. (If necessary, review the normalization techniques in Chapter 6, Normalization of Database Tables.) The star schema shown in Figure 13.13 is known as a **snowflake schema**, which is a type of star schema in which the dimension tables can have their own dimension tables. The snowflake schema is usually the result of normalizing dimension tables.

By normalizing the dimension tables, you simplify the data-filtering operations related to the dimensions. In this example, the region, state, city, and location contain very few records compared to the SALES fact table. Only the location table is directly related to the SALES fact table.

snowflake schema
A type of star schema in which dimension tables can have their own dimension tables. The snowflake schema is usually the result of normalizing dimension tables.

Note

Although using the dimension tables shown in Figure 13.13 provides structural simplicity, there is a price to pay for that simplicity. For example, if you want to aggregate the data by region, you must use a four-table join, thus increasing the complexity of the SQL statements. The star schema in Figure 13.11 uses a LOCATION dimension table that greatly facilitates data retrieval by eliminating multiple join operations. This is yet another example of the trade-offs that designers must consider.

FIGURE 13.13 NORMALIZED DIMENSION TABLES

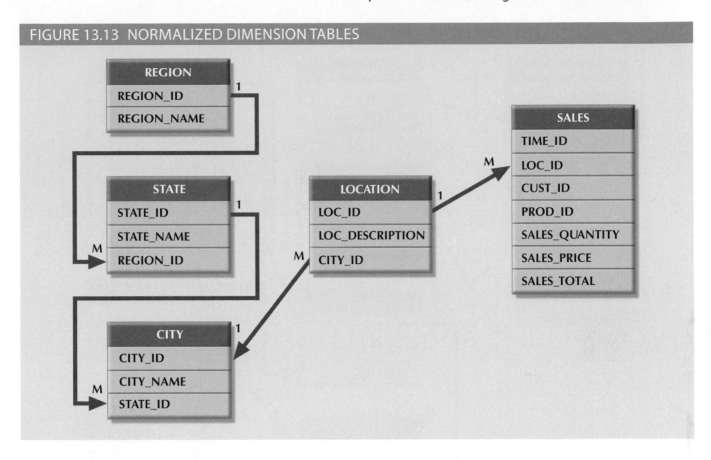

Maintaining Multiple Fact Tables that Represent Different Aggregation Levels

You can also speed up query operations by creating and maintaining multiple fact tables related to each level of aggregation (region, state, and city) in the location dimension. These aggregate tables are precomputed at the data-loading phase rather than at run time. The purpose of this technique is to save processor cycles at run time, thereby speeding up data analysis. An end-user query tool optimized for decision analysis then properly accesses the summarized fact tables instead of computing the values by accessing a fact table at a lower level of detail. This technique is illustrated in Figure 13.14, which adds aggregate fact tables for region, state, and city to the initial sales example.

The data warehouse designer must identify which levels of aggregation to precompute and store in the database. These multiple aggregate fact tables are updated during each load cycle in batch mode. Also, because the objective is to minimize access according to the expected frequency of use and to minimize the processing time required to calculate a given aggregation level at run time, the data warehouse designer must select which aggregation fact tables to create.

Denormalizing Fact Tables

Denormalizing fact tables improves data access performance and saves data storage space. The latter objective, however, is becoming less of an issue. Data storage costs decrease almost daily, and DBMS limitations on database and table size, record size, and the maximum number of records in a single table have far more negative effects than raw storage space costs.

Denormalization improves performance by using a single record to store data that normally takes many records. For example, to compute the total sales for all products in all regions, you might have to access the region sales aggregates and summarize all of the records in this table. If you have 300,000 product sales, you could be summarizing at least 300,000 rows. Although this might not be a taxing operation for a DBMS, a comparison

FIGURE 13.14 MULTIPLE FACT TABLES

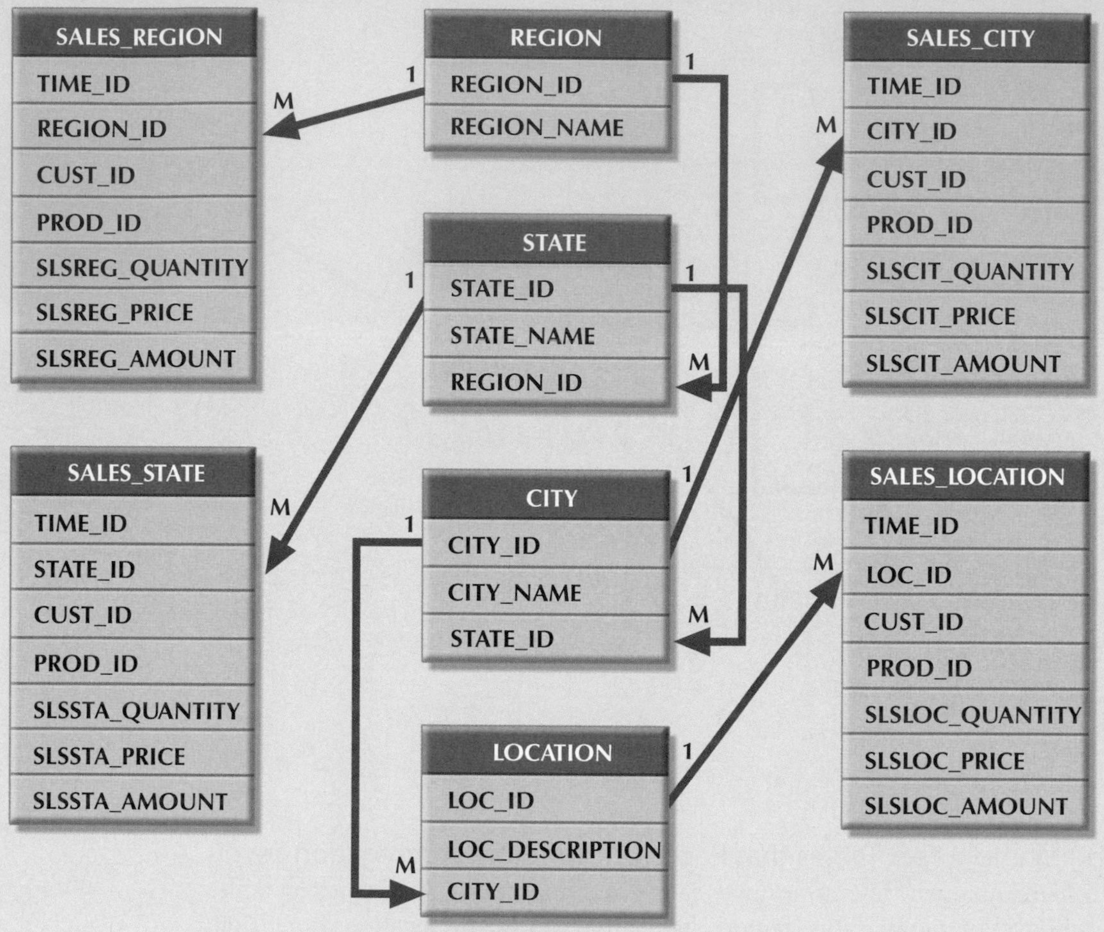

of 10 years' worth of previous sales begins to bog down the system. In such cases, it is useful to have special aggregate tables that are denormalized. For example, a YEAR_TOTALS table might contain the following fields: YEAR_ID, MONTH_1, MONTH_2 … MONTH_12, and each year's total. Such tables can easily be used to serve as a basis for year-to-year comparisons at the top month level, the quarter level, or the year level. Here again, design criteria such as frequency of use and performance requirements are evaluated against the possible overload placed on the DBMS to manage the denormalized relations.

Partitioning and Replicating Tables Because table partitioning and replication were covered in detail in Chapter 12, Distributed Database Management Systems, those techniques are discussed here only as they specifically relate to the data warehouse. Table partitioning and replication are particularly important when a BI system is implemented in dispersed geographic areas. **Partitioning** splits a table into subsets of rows or columns and places the subsets close to the client computer to improve data access time. **Replication** makes a copy of a table or partition and places it in a different location, also to improve access time.

No matter which performance-enhancement scheme is used, time is the most common dimension used in business data analysis. Therefore, it is very common to have one fact table for each level of aggregation defined within the time dimension. In the sales example, you might have five aggregate sales fact tables: daily, weekly, monthly,

partitioning
The process of splitting a table into subsets of rows or columns.

replication
The process of creating and managing duplicate versions of a database. Replication is used to place copies in different locations and to improve access time and fault tolerance.

quarterly, and yearly. These fact tables must have an implicit or explicit periodicity defined. **Periodicity**, which is usually expressed as current year only, previous years, or all years, provides information about the time span of the data stored in the table.

At the end of each year, daily sales for the current year are moved to another table that contains previous years' daily sales only. This table actually contains all sales records from the beginning of operations, with the exception of the current year. The data in the current year and previous years' tables thus represents the complete sales history of the company. The previous years' sales table can be replicated at several locations to avoid having to remotely access the historic sales data, which can cause a slow response time. The possible size of this table is enough to intimidate all but the bravest of query optimizers. Here is one case in which denormalization would be of value!

In this section you learned how the star schema design technique allows you to model data optimized for business decision making. A BI system uses all the previously mentioned components to provide decision support to all organizational users. In the next section you will learn about a widely used BI style known as *online analytical processing*.

13-6 **Online Analytical Processing**

Online analytical processing (OLAP) is a BI style whose systems share three main characteristics:

- Multidimensional data analysis techniques
- Advanced database support
- Easy-to-use end-user interfaces

This section examines each characteristic.

13-6a Multidimensional Data Analysis Techniques

The most distinctive characteristic of modern OLAP tools is their capacity for multidimensional analysis, in which data is processed and viewed as part of a multidimensional structure. This type of data analysis is particularly attractive to business decision makers because they tend to view business data as being related to other business data.

To better understand this view, you can examine how a business data analyst might investigate sales figures. In this case, the analyst is probably interested in the sales figures as they relate to other business variables such as customers and time. In other words, customers and time are viewed as different dimensions of sales. Figure 13.15 illustrates how the operational (one-dimensional) view differs from the multidimensional view of sales.

Note in Figure 13.15 that the operational (tabular) view of sales data is not well suited to decision support because the relationship between INVOICE and LINE does not provide a business perspective of the sales data. On the other hand, the end user's view of sales data *from a business perspective* is more closely represented by the multidimensional view of sales than by the tabular view of separate tables. Note also that the multidimensional view allows end users to consolidate or aggregate data at different levels: total sales figures by customers and by date. Finally, the multidimensional view of data allows a business data analyst to easily switch business perspectives (dimensions) from sales by customer to sales by division, by region, and so on.

Multidimensional data analysis techniques are augmented by the following functions:

- *Advanced data presentation functions*. These functions include 3D graphics, pivot tables, crosstabs, data rotation, and three-dimensional cubes. Such tools are compatible with desktop spreadsheets, statistical packages, and query and report packages.

periodicity
Information about the time span of data stored in a table, usually expressed as current year only, previous years, or all years.

online analytical processing (OLAP)
Decision support system (DSS) tools that use multidimensional data analysis techniques. OLAP creates an advanced data analysis environment that supports decision making, business modeling, and operations research.

FIGURE 13.15 OPERATIONAL VS. MULTIDIMENSIONAL VIEW OF SALES

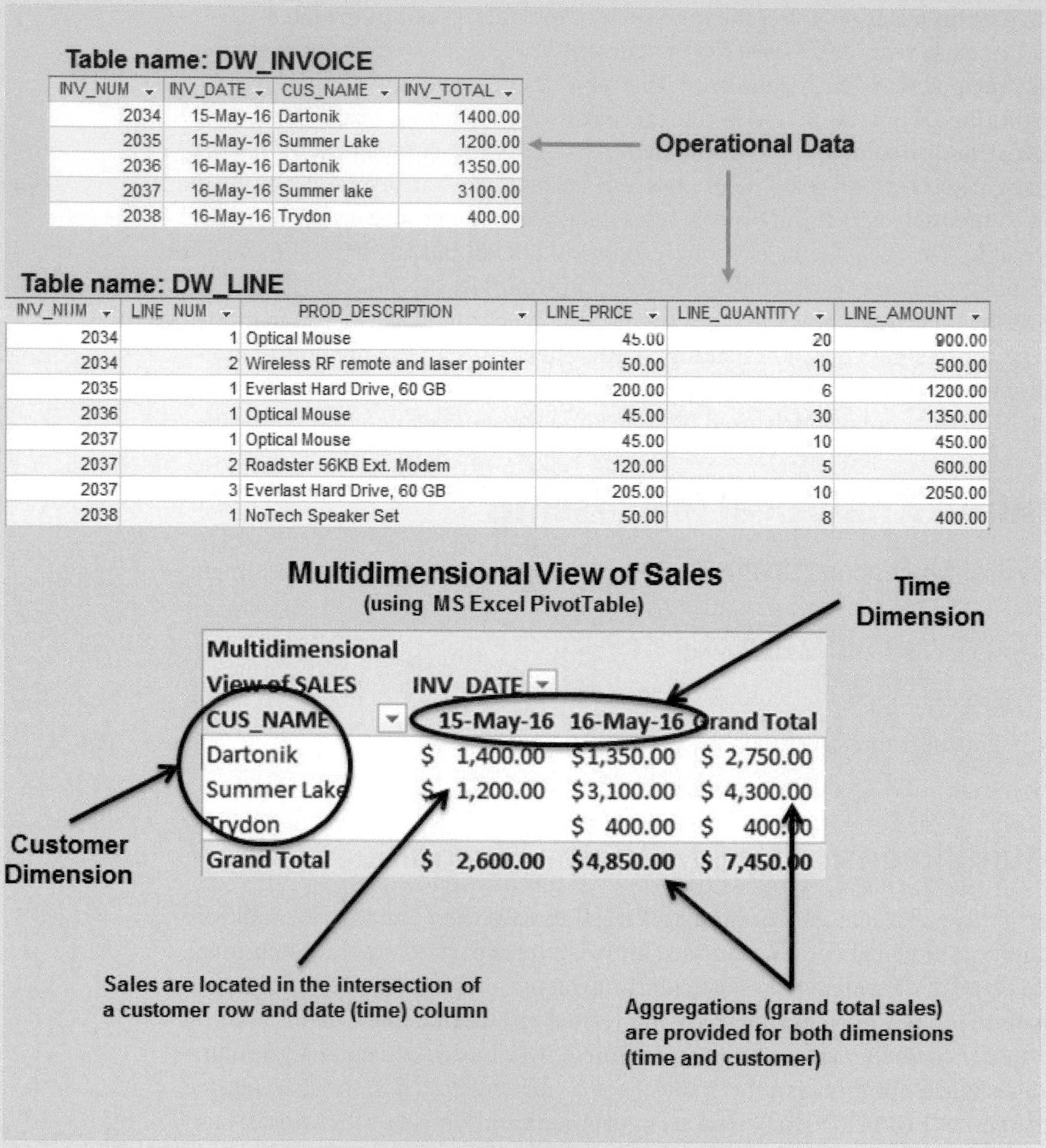

Table name: DW_INVOICE

INV_NUM	INV_DATE	CUS_NAME	INV_TOTAL
2034	15-May-16	Dartonik	1400.00
2035	15-May-16	Summer Lake	1200.00
2036	16-May-16	Dartonik	1350.00
2037	16-May-16	Summer lake	3100.00
2038	16-May-16	Trydon	400.00

→ **Operational Data**

Table name: DW_LINE

INV_NUM	LINE_NUM	PROD_DESCRIPTION	LINE_PRICE	LINE_QUANTITY	LINE_AMOUNT
2034	1	Optical Mouse	45.00	20	900.00
2034	2	Wireless RF remote and laser pointer	50.00	10	500.00
2035	1	Everlast Hard Drive, 60 GB	200.00	6	1200.00
2036	1	Optical Mouse	45.00	30	1350.00
2037	1	Optical Mouse	45.00	10	450.00
2037	2	Roadster 56KB Ext. Modem	120.00	5	600.00
2037	3	Everlast Hard Drive, 60 GB	205.00	10	2050.00
2038	1	NoTech Speaker Set	50.00	8	400.00

Multidimensional View of Sales
(using MS Excel PivotTable)

Time Dimension

Multidimensional View of SALES	INV DATE		
CUS_NAME	15-May-16	16-May-16	Grand Total
Dartonik	$ 1,400.00	$1,350.00	$ 2,750.00
Summer Lake	$ 1,200.00	$3,100.00	$ 4,300.00
Trydon		$ 400.00	$ 400.00
Grand Total	$ 2,600.00	$4,850.00	$ 7,450.00

Customer Dimension

Sales are located in the intersection of a customer row and date (time) column

Aggregations (grand total sales) are provided for both dimensions (time and customer)

- *Advanced data aggregation, consolidation, and classification functions.* These allow the data analyst to create multiple data aggregation levels, slice and dice data (see Section 13-5c), and drill down and roll up data across different dimensions and aggregation levels. For example, aggregating data by week, month, quarter, and year allows the data analyst to drill down and roll up across time dimensions.

- *Advanced computational functions.* These include business-oriented variables such as market share, period comparisons, sales margins, product margins, and percentage changes; financial and accounting ratios, including profitability, overhead, cost allocations, and returns; and statistical and forecasting functions. These functions are provided automatically, so the end user does not need to redefine the components each time they are accessed.

- *Advanced data-modeling functions.* These provide support for what-if scenarios, variable assessment, contributions to outcome, linear programming, and predictive modeling tools. Predictive modeling allows the system to build advanced statistical models to predict future values (business outcomes) with a high percentage of accuracy.

13-6b Advanced Database Support

To deliver efficient decision support, OLAP tools must have the following advanced data access features:

- Access to many different kinds of DBMSs, flat files, and internal and external data sources
- Access to aggregated data warehouse data as well as to the detail data found in operational databases
- Advanced data navigation features such as drill-down and roll-up
- Rapid and consistent query response times
- The ability to map end-user requests, expressed in either business or model terms, to the appropriate data source and then to the proper data access language (usually SQL). The query code must be optimized to match the data source, regardless of whether the source is operational or data warehouse data.
- Support for very large databases. As explained earlier, the data warehouse could easily and quickly grow to multiple terabytes in size.

To provide a seamless interface, OLAP tools map the data elements from the data warehouse and the operational database to their own data dictionaries. This metadata is used to translate end-user data analysis requests into the proper (optimized) query codes, which are then directed to the appropriate data sources.

13-6c Easy-to-Use End-User Interfaces

The end-user analytical interface is one of the most critical OLAP components. When properly implemented, an analytical interface permits the user to navigate the data in a way that simplifies and accelerates decision making or data analysis.

Advanced OLAP features become more useful when access to them is kept simple. OLAP tool vendors learned this lesson early and have equipped their sophisticated data extraction and analysis tools with easy-to-use graphical interfaces. Many of the interface features are "borrowed" from previous generations of data analysis tools that are already familiar to end users.

Because many analysis and presentation functions are common to desktop spreadsheet packages, most OLAP vendors have closely integrated their systems with spreadsheets such as Microsoft Excel. Using the features available in graphical end-user interfaces, OLAP simply becomes another option within the spreadsheet menu bar, as shown in Figure 13.16. This seamless integration is an advantage for OLAP systems and spreadsheet vendors because end users gain access to advanced data analysis features by using familiar programs and interfaces. Therefore, additional training and development costs are minimized.

13-6d OLAP Architecture

The OLAP architecture is designed to meet ease-of-use requirements while keeping the system flexible. An OLAP system has three main architectural components:

- Graphical user interface (GUI)
- Analytical processing logic
- Data-processing logic

FIGURE 13.16 INTEGRATION OF OLAP WITH A SPREADSHEET PROGRAM

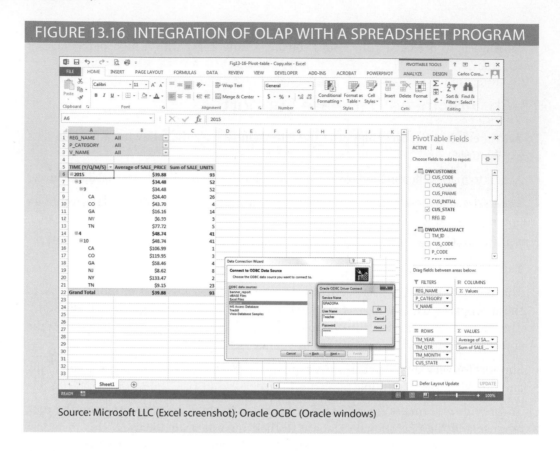

Source: Microsoft LLC (Excel screenshot); Oracle OCBC (Oracle windows)

These three components can exist on the same computer or be distributed among several computers. Figure 13.17 illustrates OLAP's architectural components.

FIGURE 13.17 OLAP ARCHITECTURE

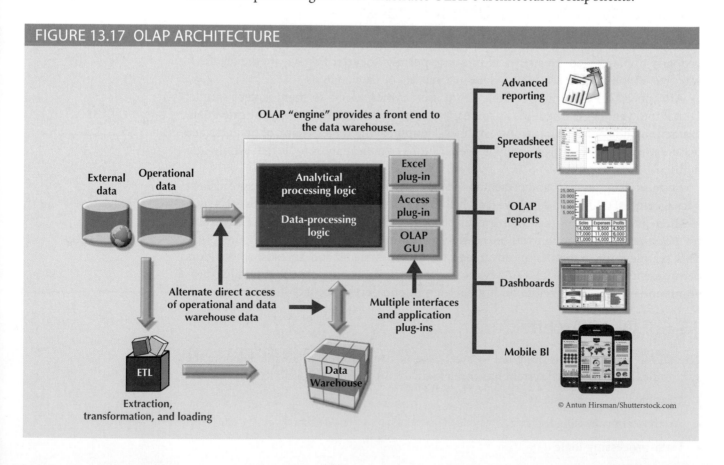

© Antun Hirsman/Shutterstock.com

As Figure 13.17 illustrates, OLAP systems are designed to use both operational and data warehouse data. The figure shows the OLAP system components on a single computer, but this single-user scenario is only one of many. In fact, one problem with the installation shown here is that each data analyst must have a powerful computer to store the OLAP system and perform all data processing locally.

A more common and practical architecture is one in which the OLAP GUI runs on client workstations while the OLAP data-processing logic (or OLAP "server") runs on a shared server computer. The OLAP analytical processing logic could be located on a client workstation, the OLAP server, or be split between the two sides. In any case, the OLAP server component acts as an intermediary between the OLAP GUI and the data warehouse. This middle layer accepts and handles the data-processing requests generated by the many end-user OLAP workstations. This flexible architecture allows for many different OLAP configurations. Figure 13.18 illustrates an OLAP server with local miniature data marts.

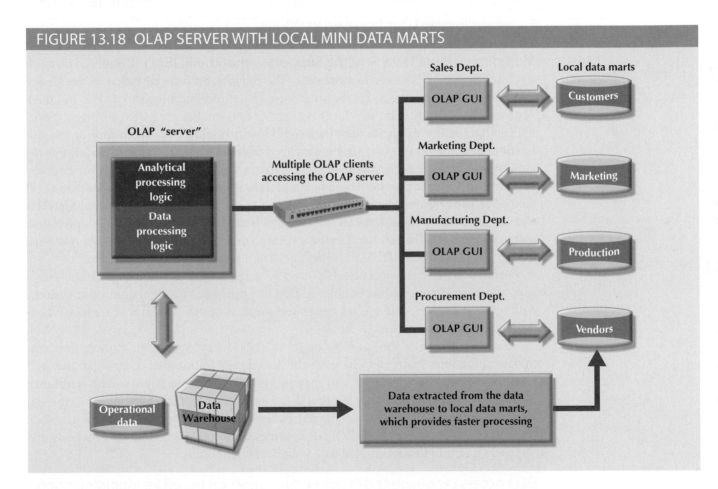

FIGURE 13.18 OLAP SERVER WITH LOCAL MINI DATA MARTS

As illustrated in Figure 13.18, the OLAP system could merge the data warehouse and data mart approaches by storing extracts of the data warehouse at end-user workstations. The objective is to increase the speed of data access and data visualization (the graphic representations of data trends and characteristics). The logic behind this approach is the assumption that most end users usually work with fairly small, stable data warehouse subsets. For example, a sales analyst is most likely to work with sales data, whereas a customer representative is likely to work with customer data.

Whatever the arrangement of the OLAP components, one thing is certain: multidimensional data must be used. But how is multidimensional data best stored and managed? OLAP proponents are sharply divided. Some favor the use of relational databases to store multidimensional data; others argue that specialized multidimensional databases are superior. The basic characteristics of each approach are examined next.

13-6e Relational OLAP

Relational online analytical processing (ROLAP) provides OLAP functionality by using relational databases and familiar relational query tools to store and analyze multidimensional data. This approach builds on existing relational technologies and represents a natural extension to companies that already use relational database management systems within their organizations. ROLAP adds the following extensions to traditional RDBMS technology:

- Multidimensional data schema support within the RDBMS
- Data access language and query performance optimized for multidimensional data
- Support for very large databases (VLDBs)

Multidimensional Data Schema Support within the RDBMS Relational technology uses normalized tables to store data. The reliance on normalization as the design methodology for relational databases is seen as a stumbling block to its use in OLAP systems. Normalization divides business entities into smaller pieces to produce the normalized tables. For example, sales data components might be stored in four or five different tables. The reason for using normalized tables is to reduce redundancies, thereby eliminating data anomalies, and to facilitate data updates. Unfortunately, for decision support purposes, it is easier to understand data when it is seen with respect to other data. (See the example in Figure 13.15.) Given that view of the data environment, this text has emphasized that decision support data tends to be non-normalized, duplicated, and preaggregated. Those characteristics seem to preclude the use of standard relational design techniques and RDBMSs as the foundation for multidimensional data.

Fortunately for companies heavily invested in relational technology, ROLAP uses a special design technique that enables RDBMS technology to support multidimensional data representations. This special design technique is known as a star schema, which is covered in detail in Section 13-5.

The star schema is designed to optimize data query operations rather than data update operations. Naturally, changing the data design foundation means that the tools used to access such data will have to change. End users who are familiar with traditional relational query tools will discover that those tools do not work efficiently with the star schema. However, ROLAP saves the day by adding support for the star schema when familiar query tools are used. ROLAP provides advanced data analysis functions and improves query optimization and data visualization methods.

Data Access Language and Query Performance Optimized for Multidimensional Data Another criticism of relational databases is that SQL is not suited for performing advanced data analysis. Most decision support data requests require the use of multiple-pass SQL queries or multiple nested SQL statements. To answer this criticism, ROLAP extends SQL so that it can differentiate between access requirements for data warehouse data (based on the star schema) and operational data (normalized tables). A ROLAP system therefore can generate the SQL code required to access the star schema data.

Query performance is also improved because the query optimizer is modified to identify the SQL code's intended query targets. For example, if the query target is the data warehouse, the optimizer passes the requests to the data warehouse. However, if the end user performs drill-down queries against operational data, the query optimizer identifies

relational online analytical processing (ROLAP)
Analytical processing functions that use relational databases and familiar relational query tools to store and analyze multidimensional data.

that operation and properly optimizes the SQL requests before passing them to the operational DBMS.

Another source of improved query performance is the use of advanced indexing techniques such as bitmapped indexes within relational databases. As the name suggests, a bitmapped index is based on 0 and 1 bits to represent a given condition. For example, if the REGION attribute in Figure 13.4 has only four outcomes—North, South, East, and West—those outcomes may be represented as shown in Table 13.11. Only the first 10 rows from Figure 13.4 are represented in the table. The "1" represents "bit on," and the "0" represents "bit off." For example, to represent a row with a REGION attribute = "East," only the "East" bit would be on. Note that each row must be represented in the index table.

Note that the index in Table 13.11 takes a minimal amount of space. Therefore, bitmapped indexes are more efficient at handling large amounts of data than the indexes typically found in many relational databases. However, keep in mind that bitmapped indexes are primarily used when the number of possible values for an attribute is fairly small. For example, REGION has only four outcomes in this example. Marital status—married, single, widowed, or divorced—would be another good bitmapped index candidate, as would gender—M or F.

TABLE 13.11

BITMAP REPRESENTATION OF REGION VALUES

NORTH	SOUTH	EAST	WEST
0	0	1	0
0	0	1	0
1	0	0	0
1	0	0	0
1	0	0	0
0	1	0	0
0	1	0	0
0	1	0	0
0	0	0	1
0	0	0	1

Support for Very Large Databases Recall that support for VLDBs is a requirement for decision support databases. Therefore, when the relational database is used in a decision support role, it also must be able to store very large amounts of data. Both the storage capability and the process of loading data into the database are crucial. Therefore, the RDBMS must have the proper tools to import, integrate, and populate the data warehouse with data. Decision support data is normally loaded in bulk (batch) mode from the operational data. However, batch operations require that both the source and the destination databases be reserved (locked). The speed of the data-loading operations is important, especially when you realize that most operational systems run 24 hours a day, 7 days a week. Therefore, the window of opportunity for maintenance and batch loading is open only briefly, typically during slack periods.

Clearly, ROLAP is a logical choice for companies that already use relational databases for their operational data. Given the size of the relational database market, it is hardly surprising that most current RDBMS vendors have extended their products to support data warehouses and OLAP capabilities.

13-6f Multidimensional OLAP

Multidimensional online analytical processing (MOLAP) extends OLAP functionality to **multidimensional database management systems (MDBMSs)**. An MDBMS uses proprietary techniques to store data in matrix-like n-dimensional arrays. MOLAP's premise is that multidimensional databases are best suited to manage, store, and analyze multidimensional data. Most of the proprietary techniques used in MDBMSs are derived from engineering fields such as computer-aided design/computer-aided manufacturing (CAD/CAM) and geographic information systems (GIS). MOLAP tools store data using multidimensional arrays, row stores, or column stores. (If necessary, review the NoSQL data model in Chapter 2, Data Models.)

Conceptually, MDBMS end users visualize the stored data as a three-dimensional cube known as a **data cube**. The location of each data value in the data cube is a function of the x-, y-, and z-axes in a three-dimensional space. The three axes represent the dimensions of the data value. The data cubes can grow to n number of dimensions, thus becoming *hypercubes*. Data cubes are created by extracting data from the operational databases or from the data warehouse. One important characteristic of data cubes is that they are static; that is, they are not subject to change and must be created before they can be used. Data cubes cannot be created by ad hoc queries. Instead, you query precreated cubes with defined axes; for example, a cube for sales will have the product, location, and time dimensions, and you can query only those dimensions. Therefore, the data cube creation process is critical and requires in-depth front-end design work. This design work may be well justified because MOLAP databases are known to be much faster than their ROLAP counterparts, especially when dealing with large data sets. To speed data access, data cubes are normally held in memory in the **cube cache**. (A data cube is only a window to a predefined subset of data in the database. A data cube and a database are not the same thing.) Because MOLAP also benefits from a client/server infrastructure, the cube cache can be located at the MOLAP server, the MOLAP client, or both.

Because the data cube is predefined with a set number of dimensions, the addition of a new dimension requires that the entire data cube be re-created, which is time-consuming. Therefore, when data cubes are created too often, the MDBMS loses some of its speed advantage over the relational database. In addition, the MDBMS uses proprietary data storage techniques that in turn require proprietary data access methods using a multidimensional query language.

Multidimensional data analysis is also affected by how the database system handles sparsity. **Sparsity** measures the density of the data held in the data cube; it is computed by dividing the total number of actual values in the cube by its total number of cells. Because the data cube's dimensions are predefined, not all cells are populated. In other words, some cells are empty. Returning to the sales example, many products might not be sold during a given time period in a given location. In fact, you will often find that less than 50 percent of the data cube's cells are populated. In any case, multidimensional databases must handle sparsity effectively to reduce processing overhead and resource requirements.

13-6g Relational versus Multidimensional OLAP

Table 13.12 summarizes some pros and cons of ROLAP and MOLAP. Keep in mind that the selection of one or the other often depends on the evaluator's vantage point. For example, a proper evaluation of OLAP must include price, supported hardware platforms, compatibility with the existing DBMS, programming requirements, performance, and availability of administrative tools. The summary in Table 13.12 provides a useful starting point for comparison.

multidimensional online analytical processing (MOLAP)
An extension of online analytical processing to multidimensional database management systems.

multidimensional database management system (MDBMS)
A database management system that uses proprietary techniques to store data in matrixlike arrays of n dimensions known as cubes.

data cube
The multidimensional data structure used to store and manipulate data in a multidimensional DBMS. The location of each data value in the data cube is based on its x-, y-, and z-axes. Data cubes are static, meaning they must be created before they are used, so they cannot be created by an ad hoc query.

cube cache
In multidimensional OLAP, the shared, reserved memory area where data cubes are held. Using the cube cache assists in speeding up data access.

sparsity
In multidimensional data analysis, a measurement of the data density held in the data cube.

TABLE 13.12

RELATIONAL VS. MULTIDIMENSIONAL OLAP

CHARACTERISTIC	ROLAP	MOLAP
Schema	Uses star schema Additional dimensions can be added dynamically	Uses data cubes Multidimensional arrays, row stores, column stores Additional dimensions require re-creation of the data cube
Database size	Medium to large	Large
Architecture	Client/server Standards-based	Client/server Open or proprietary, depending on vendor
Access	Supports ad hoc requests Unlimited dimensions	Limited to predefined dimensions Proprietary access languages
Speed	Good with small data sets; average for medium-sized to large data sets	Faster for large data sets with predefined dimensions

ROLAP and MOLAP vendors are working to integrate their respective solutions within a unified decision support framework. Many OLAP products can handle tabular and multidimensional data with the same ease. For example, if you use Excel OLAP functionality, as shown earlier in Figure 13.16, you can access relational OLAP data in a SQL server as well as cube (multidimensional) data in the local computer. In the meantime, relational databases have successfully extended SQL to support many OLAP tools.

13-7 SQL Extensions for OLAP

The proliferation of OLAP tools has fostered the development of SQL extensions to support multidimensional data analysis. Most SQL innovations are the result of vendor-centric product enhancements. However, many of the innovations have made their way into standard SQL. This section introduces some of the new SQL extensions that have been created to support OLAP-type data manipulations.

The SaleCo snowflake schema shown in Figure 13.19 demonstrates the use of the SQL extensions. Note that this snowflake schema has a central DWSALESFACT fact

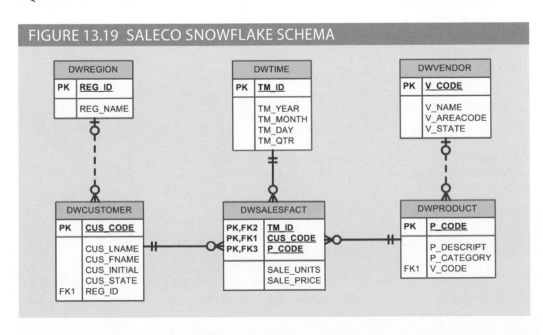

FIGURE 13.19 SALECO SNOWFLAKE SCHEMA

table and three dimension tables: DWCUSTOMER, DWPRODUCT, and DWTIME. The central fact table represents daily sales by product and customer. However, as you examine the schema shown in Figure 13.19, you will see that the DWCUSTOMER and DWPRODUCT dimension tables have their own dimension tables: DWREGION and DWVENDOR.

Keep in mind that a database is at the core of all data warehouses. Therefore, all SQL commands (such as CREATE, INSERT, UPDATE, DELETE, and SELECT) will work in the data warehouse as expected. However, most queries you run in a data warehouse tend to include a lot of data groupings and aggregations over multiple columns. Therefore, this section introduces two extensions to the GROUP BY clause that are particularly useful: ROLLUP and CUBE. In addition, you will learn about using materialized views to store preaggregated rows in the database.

Note

This section uses the Oracle RDBMS to demonstrate the use of SQL extensions to support OLAP functionality. If you use a different DBMS, consult the documentation to verify whether the vendor supports similar functionality and what the proper syntax is for your DBMS.

13-7a The ROLLUP Extension

The ROLLUP extension is used with the GROUP BY clause to generate aggregates by different dimensions. As you know, the GROUP BY clause will generate only one aggregate for each new value combination of attributes listed in the GROUP BY clause. The ROLLUP extension goes one step further; it enables you to get a subtotal for each column listed except for the last one, which gets a grand total instead. The syntax of the GROUP BY ROLLUP command sequence is as follows:

SELECT column1 [, column2, …], aggregate_function(expression)
FROM table1 [, table2, …]
[WHERE condition]
GROUP BY ROLLUP (column1 [, column2, …])
[HAVING condition]
[ORDER BY column1 [, column2, …]]

Note

MS SQL Server and MySQL both support ROLLUP functionality. Other than the GROUP BY clause, the same syntax used for working with aggregate functions in these DBMSs applies. The GROUP BY clause is written:

GROUP BY column1 [, column2, …] WITH ROLLUP

In MySQL, if the ROLLUP option is specified, then an ORDER BY clause is not allowed. Access does not support the ROLLUP extension.

The order of the column list within GROUP BY ROLLUP is very important. The last column in the list will generate a grand total, and all other columns will generate subtotals. For example, Figure 13.20 shows the use of the ROLLUP extension to generate subtotals by vendor and product.

FIGURE 13.20 ROLLUP EXTENSION

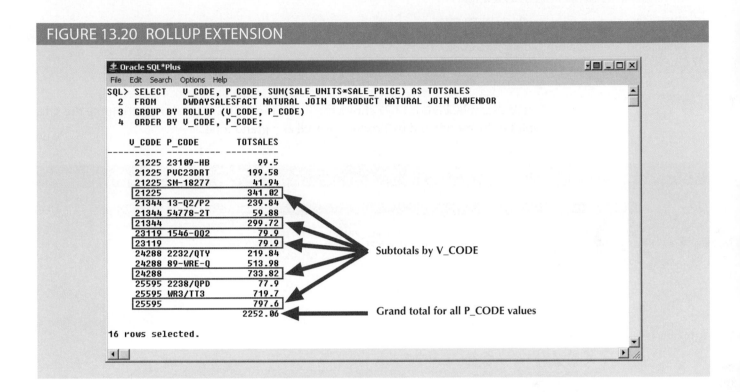

Figure 13.20 shows the subtotals by vendor code and a grand total for all product codes. Contrast that with the normal GROUP BY clause that generates only the subtotals for each vendor and product combination. The ROLLUP extension is particularly useful when you want to obtain multiple nested subtotals for a dimension hierarchy. For example, within a location hierarchy, you can use ROLLUP to generate subtotals by region, state, city, and store.

13-7b The CUBE Extension

The CUBE extension is also used with the GROUP BY clause to generate aggregates by the listed columns, including the last one. The CUBE extension enables you to get a subtotal for each column listed in the expression, in addition to a grand total for the last column listed. The syntax of the GROUP BY CUBE command sequence is as follows:

SELECT column1 [, column2, …], aggregate_function(expression)
FROM table1 [, table2, …]
[WHERE condition]
GROUP BY CUBE (column1 [, column2, …])
[HAVING condition]
[ORDER BY column1 [, column2, …]]

Note

MS SQL Server supports CUBE functionality, too. Other than the GROUP BY clause, the same syntax used for working with aggregate functions applies. The GROUP BY clause is written similarly to the ROLLUP extension:

GROUP BY column1 [, column2, …] WITH CUBE

MySQL and Access do not support the CUBE extension.

For example, Figure 13.21 shows the use of the CUBE extension to compute the sales subtotals by month and by product, as well as a grand total.

FIGURE 13.21 CUBE EXTENSION

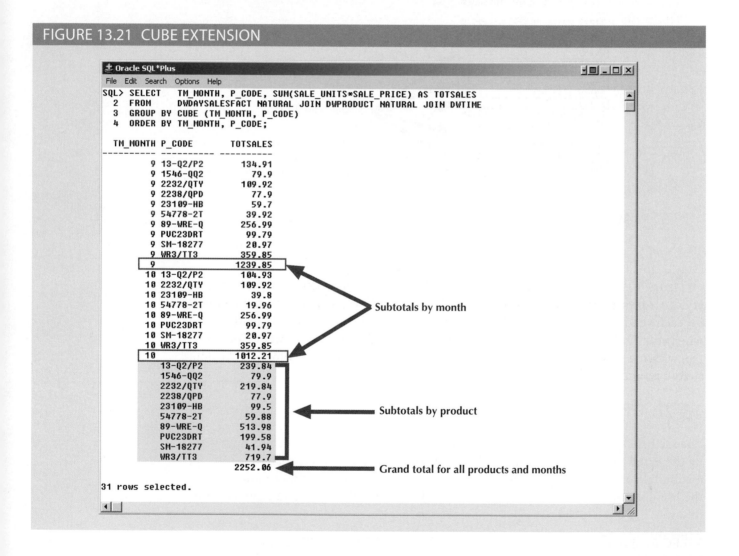

In Figure 13.21, the CUBE extension also generates subtotals for each combination of month and product. The CUBE extension is particularly useful when you want to compute all possible subtotals within groupings based on multiple dimensions. Cross-tabulations are especially good candidates for application of the CUBE extension.

13-7c Materialized Views

The data warehouse normally contains fact tables that store specific measurements of interest to an organization. Such measurements are organized by different dimensions. The vast majority of OLAP business analysis of everyday activity is based on data comparisons that are aggregated at different levels, such as totals by vendor, by product, and by store.

Because businesses normally use a predefined set of summaries for benchmarking, it is reasonable to predefine such summaries for future use by creating summary fact tables. (See Section 13-5f for a discussion of additional performance-improving techniques.) However, creating multiple summary fact tables that use GROUP BY queries with multiple table joins could become resource-intensive. In addition, data warehouses must be able to maintain up-to-date summarized data at all times. So what happens with the summary fact tables after new sales data has been added to the base fact tables? Under normal circumstances, the summary fact tables are re-created. This operation requires that the SQL code be run again to re-create all summary rows, even when only a few rows need updating. Clearly, this is a time-consuming process.

To save query processing time, most database vendors have implemented additional functions to manage aggregate summaries more efficiently. This new functionality resembles the standard SQL views for which the SQL code is predefined in the database. However, the added difference is that the views also store the preaggregated rows, something like a summary table. For example, Microsoft SQL Server provides indexed views, while Oracle provides materialized views. This section explains the use of materialized views.

A **materialized view** is a dynamic table that not only contains the SQL query command to generate the rows, it stores the actual rows. The materialized view is created the first time the query is run, and the summary rows are stored in the table. The materialized view rows are automatically updated when the base tables are updated. That way, the data warehouse administrator will create the view but will not have to worry about updating the view. The use of materialized views is totally transparent to the end user. The OLAP end user can create OLAP queries using the standard fact tables, and the DBMS query optimization feature will automatically use the materialized views if they provide better performance.

The basic syntax for the materialized view is:

```
CREATE MATERIALIZED VIEW view_name
BUILD {IMMEDIATE | DEFERRED}
REFRESH {[FAST | COMPLETE | FORCE]} ON COMMIT
[ENABLE QUERY REWRITE]
AS select_query;
```

The BUILD clause indicates when the materialized view rows are actually populated. IMMEDIATE indicates that the materialized view rows are populated right after the command is entered. DEFERRED indicates that the materialized view rows will be populated later. Until then, the materialized view is in an unusable state. The DBMS provides a special routine that an administrator runs to populate materialized views.

The REFRESH clause lets you indicate when and how to update the materialized view when new rows are added to the base tables. FAST indicates that whenever a change is made in the base tables, the materialized view updates only the affected rows. COMPLETE indicates that a complete update will be made for all rows in the materialized view when you rerun the SELECT query on which the view is based. FORCE indicates

materialized view
A dynamic table that not only contains the SQL query command to generate rows but stores the actual rows. The materialized view is created the first time the query is run and the summary rows are stored in the table. The materialized view rows are automatically updated when the base tables are updated.

that the DBMS will first try to do a FAST update; otherwise, it will do a COMPLETE update. The ON COMMIT clause indicates that the updates to the materialized view will take place as part of the commit process of the underlying DML statement—that is, as part of the commitment of the DML transaction that updated the base tables. The ENABLE QUERY REWRITE option allows the DBMS to use the materialized views in query optimization.

To create materialized views, you must have specified privileges and you must complete specified prerequisite steps. As always, you must consult the DBMS documentation for the latest updates. In the case of Oracle, you must create materialized view logs on the base tables of the materialized view. Figure 13.22 shows the steps required to create the SALES_MONTH_MV materialized view in the Oracle RDBMS.

FIGURE 13.22 CREATING A MATERIALIZED VIEW

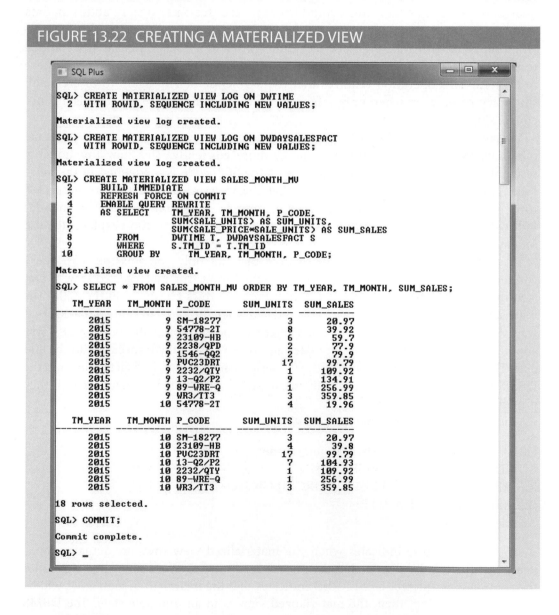

The materialized view in Figure 13.22 computes the monthly total units sold and the total sales aggregated by product. The SALES_MONTH_MV materialized view is configured to automatically update after each change in the base tables. The last row of SALES_MONTH_MV indicates that during October, three units of product "WR3/ TT3" were sold for a total of $359.85. Figure 13.23 shows the effects of updating the DWDAYSALESFACT base table.

FIGURE 13.23   REFRESHING A MATERIALIZED VIEW

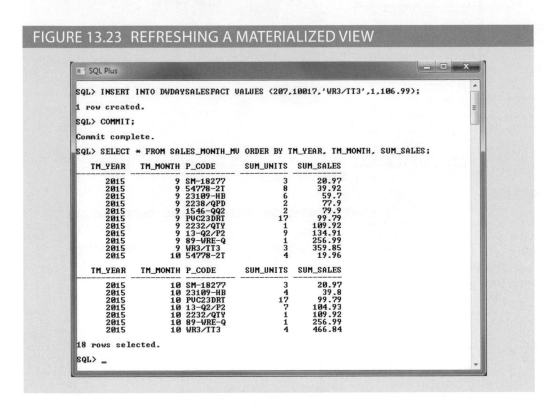

Figure 13.23 shows how the materialized view was automatically updated after the insertion of a new row in the DWDAYSALESFACT table. The last row of SALES_ MONTH_MV now shows that in October, four units of product "WR3/TT3" were sold for a total of $466.84.

Although all of the examples in this section focus on SQL extensions to support OLAP reporting in an Oracle DBMS, you have seen just a small fraction of the many business intelligence features currently provided by most DBMS vendors. For example, most vendors provide rich graphical user interfaces to manipulate, analyze, and present the data in multiple formats. Figure 13.24 shows two sample screens, one for Oracle and one for Microsoft SQL Server.

FIGURE 13.24 SAMPLE OLAP APPLICATIONS

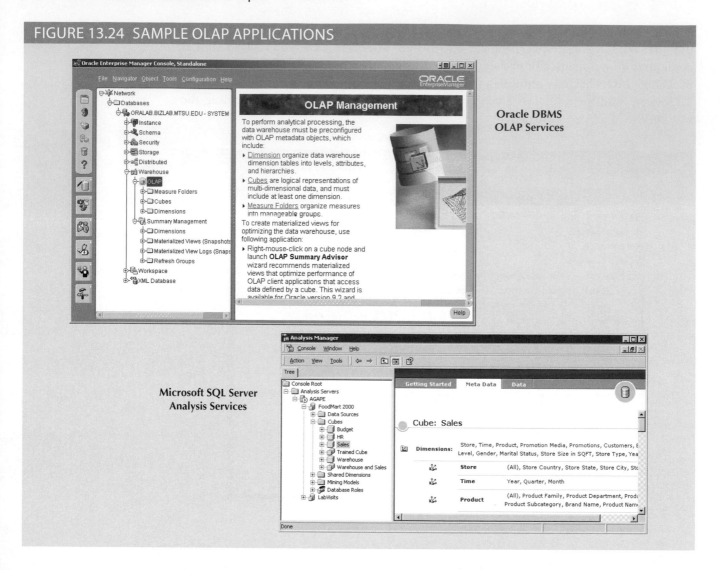

Oracle DBMS OLAP Services

Microsoft SQL Server Analysis Services

Summary

- Business intelligence (BI) is a term for a comprehensive, cohesive, and integrated set of applications used to capture, collect, integrate, store, and analyze data with the purpose of generating and presenting information to support business decision making.

- Decision support systems (DSSs) refer to an arrangement of computerized tools used to assist managerial decision making within a business. DSSs were the original precursor of current-generation BI systems.

- Operational data is not well suited for decision support. From the end user's point of view, decision support data differs from operational data in three main areas: time span, granularity, and dimensionality.

- The data warehouse is an integrated, subject-oriented, time-variant, nonvolatile collection of data that provides support for decision making. The data warehouse is usually a read-only database optimized for data analysis and query processing. A data mart is a small, single-subject data warehouse subset that provides decision support to a small group of people.

- The star schema is a data-modeling technique used to map multidimensional decision support data into a relational database for advanced data analysis. The basic star schema has four components: facts, dimensions, attributes, and attribute hierarchies. Facts are numeric measurements or values that represent a specific business aspect or activity. Dimensions are general qualifying categories that provide additional perspectives to facts. Conceptually, the multidimensional data model is best represented by a three-dimensional cube. Attributes can be ordered in well-defined hierarchies, which provide a top-down organization that is used for two main purposes: to permit aggregation and provide drill-down and roll-up data analysis.

- Online analytical processing (OLAP) refers to an advanced data analysis environment that supports decision making, business modeling, and operations research.

- SQL has been enhanced with extensions that support OLAP-type processing and data generation.

Key Terms

attribute hierarchy

business intelligence (BI)

cube cache

dashboard

data cube

data mart

data visualization

data warehouse

decision support system (DSS)

dimension tables

dimensions

drill down

extraction, transformation, and loading (ETL)

fact table

facts

governance

key performance indicator (KPI)

master data management (MDM)

materialized view

metrics

multidimensional database management system (MDBMS)

multidimensional online analytical processing (MOLAP)

online analytical processing (OLAP)

partitioning

periodicity

portal

relational online analytical processing (ROLAP)

replication

roll up

slice and dice

snowflake schema

sparsity

star schema

very large database (VLDB)

Online Content

Flashcards and crossword puzzles for key term practice are available at *www.cengagebrain.com*.

Review Questions

1. What is business intelligence? Give some recent examples of BI usage, using the Internet for assistance. What BI benefits have companies found?

2. Describe the BI framework. Illustrate the evolution of BI.

3. What are decision support systems, and what role do they play in the business environment?

4. Explain how the main components of the BI architecture interact to form a system. Describe the evolution of BI information dissemination formats.

5. What are the most relevant differences between operational data and decision support data?

6. What is a data warehouse, and what are its main characteristics? How does it differ from a data mart?

7. Give three examples of likely problems when operational data is integrated into the data warehouse.

    Use the following scenario to answer Questions 8–14.

    While working as a database analyst for a national sales organization, you are asked to be part of its data warehouse project team.

8. Prepare a high-level summary of the main requirements for evaluating DBMS products for data warehousing.

9. Your data warehousing project group is debating whether to create a prototype of a data warehouse before its implementation. The project group members are especially concerned about the need to acquire some data warehousing skills before implementing the enterprise-wide data warehouse. What would you recommend? Explain your recommendations.

10. Suppose that you are selling the data warehouse idea to your users. How would you define multidimensional data analysis for them? How would you explain its advantages to them?

11. The data warehousing project group has invited you to provide an OLAP overview. The group's members are particularly concerned about the OLAP client/server architecture requirements and how OLAP will fit the existing environment. Your job is to explain the main OLAP client/server components and architectures.

12. One of your vendors recommends using an MDBMS. How would you explain this recommendation to your project leader?

13. The project group is ready to make a final decision, choosing between ROLAP and MOLAP. What should be the basis for this decision? Why?

14. The data warehouse project is in the design phase. Explain to your fellow designers how you would use a star schema in the design.

15. Briefly discuss the OLAP architectural styles with and without data marts.

16. What is OLAP, and what are its main characteristics?

17. Explain ROLAP, and list the reasons you would recommend its use in the relational database environment.

18. Explain the use of facts, dimensions, and attributes in the star schema.

19. Explain multidimensional cubes, and describe how the slice-and-dice technique fits into this model.

20. In the star schema context, what are attribute hierarchies and aggregation levels, and what is their purpose?

21. Discuss the most common performance improvement techniques used in star schemas.

Problems

1. The university computer lab's director keeps track of lab usage, as measured by the number of students using the lab. This function is important for budgeting purposes. The computer lab director assigns you the task of developing a data warehouse to keep track of the lab usage statistics. The main requirements for this database are to:

   - Show the total number of users by different time periods.

   - Show usage numbers by time period, by major, and by student classification.

   - Compare usage for different majors and different semesters.

   Use the **Ch13_P1.mdb** database, which includes the following tables:

   - USELOG contains the student lab access data.

   - STUDENT is a dimension table that contains student data.

   Given the three preceding requirements, and using the **Ch13_P1.mdb** data, complete the following problems:

   a. Define the main facts to be analyzed. (*Hint*: These facts become the source for the design of the fact table.)

   b. Define and describe the appropriate dimensions. (*Hint*: These dimensions become the source for the design of the dimension tables.)

   c. Draw the lab usage star schema, using the fact and dimension structures you defined in Problems 1a and 1b.

   d. Define the attributes for each of the dimensions in Problem 1b.

   e. Recommend the appropriate attribute hierarchies.

   f. Implement your data warehouse design, using the star schema you created in Problem 1c and the attributes you defined in Problem 1d.

   g. Create the reports that will meet the requirements listed in this problem's introduction.

Online Content

The databases used for the following problems are available at *www.cengagebrain.com* (see the list of data files at the beginning of the chapter). The data for Problem 2 is stored in Microsoft Excel format at *www.cengagebrain.com*. The spreadsheet filename is Ch13_P2.xls.

2. Victoria Ephanor manages a small product distribution company. Because the business is growing fast, she recognizes that it is time to manage the vast information pool to help guide the accelerating growth. Ephanor, who is familiar with spreadsheet software, currently employs a sales force of four people. She asks you to develop a data warehouse application prototype that will enable her to study sales figures by year, region, salesperson, and product. (This prototype will be used as the basis for a future data warehouse database.)

   Using the data supplied in the **Ch13_P2.xls** file, complete the following seven problems:

   a. Identify the appropriate fact table components.

   b. Identify the appropriate dimension tables.

   c. Draw a star schema diagram for this data warehouse.

   d. Identify the attributes for the dimension tables that will be required to solve this problem.

   e. Using Microsoft Excel or any other spreadsheet program that can produce pivot tables, generate a pivot table to show the sales by product and by region. The end user must be able to specify the display of sales for any given year. The sample output is shown in the first pivot table in Figure P13.2E.

FIGURE P13.2E USING A PIVOT TABLE

	A	B	C	D	E	F
1	Year	2015 ▼				
2						
3	Sum of Value	Region ▼				
4	Product ▼	East	North	South	West	Total
5	Balls	55		70	100	225
6	Erasers		20	90		110
7	Pencils	45	60			105
8	Widgets			155	25	180
9	Total	100	80	315	125	620
10						
11						
12	Year	(All) ▼				
13	Product	(All) ▼				
14						
15	Sum of Value	Region ▼				
16	Agent ▼	East	North	South	West	Total
17	Carlos	95	150	30	25	300
18	Mary		60	125	145	330
19	Tere	12	100	160	100	372
20	Victor	55	20	259		334
21	Total	162	330	574	270	1,336

f. Using Problem 2e as your base, add a second pivot table (see Figure P13.2E) to show the sales by salesperson and by region. The end user must be able to specify sales for a given year or for all years, and for a given product or for all products.

g. Create a 3D bar graph to show sales by salesperson, by product, and by region. (See the sample output in Figure P13.2G.)

FIGURE P13.2G 3D BAR GRAPH SHOWING THE RELATIONSHIPS AMONG AGENT, PRODUCT, AND REGION

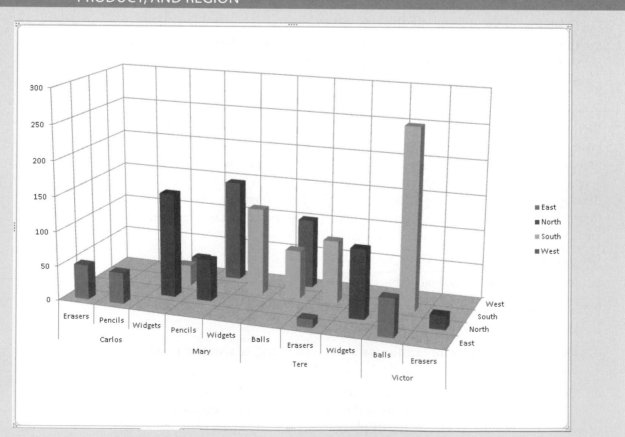

3. David Suker, the inventory manager for a marketing research company, wants to study the use of supplies within the different company departments. Suker has heard that his friend, Victoria Ephanor, has developed a spreadsheet-based data warehouse model that she uses to analyze sales data (see Problem 2). Suker is interested in developing a data warehouse model like Ephanor's so he can analyze orders by department and by product. He will use Microsoft Access as the data warehouse DBMS and Microsoft Excel as the analysis tool.

   a. Develop the order star schema.

   b. Identify the appropriate dimension attributes.

   c. Identify the attribute hierarchies required to support the model.

   d. Develop a crosstab report in Microsoft Access, using a 3D bar graph to show orders by product and by department. (The sample output is shown in Figure P13.3.)

FIGURE P13.3 CROSSTAB REPORT: ORDERS BY PRODUCT AND DEPARTMENT

Orders by Department and Product

Product	Accounting	Design	Marketing	Production	Row Total
Disks	$1.429.79	$17,268.80	$13,510.09	$6,312.24	$38,520.92
Envelopes	$329.16	$1,165.79	$17,074.33	$4,517.91	$23,087.19
Labels	$3.651.76	$1,514.15	$2,356.72	$8,464.79	$15,987.42
Paper	$1.761.90	$5,246.74	$14,222.35	$3,928.99	$25,159.98
Pencil	$741.83	$1,585.21	$2,014.56	$1,370.30	$5,711.90
Ribbons	$1.916.92	$525.00	$1,873.21	$3,203.82	$7,518.95
Toners	$110.47	$448.55	$358.25	$1,589.57	$2,506.84
	$9,941.83	$27,754.24	$51,409.51	$29,387.62	$118,493.20

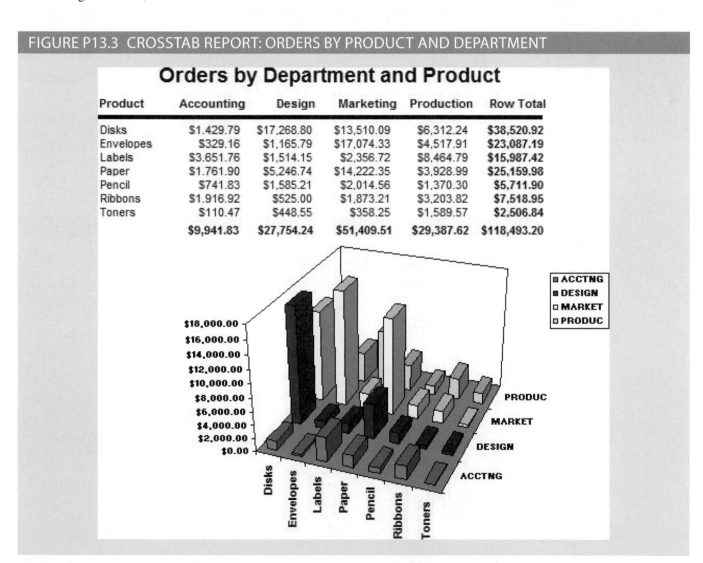

4. ROBCOR, whose sample data is contained in the database named **Ch13_P4.mdb**, provides "on-demand" aviation charters using a mix of different aircraft and aircraft types. Because ROBCOR has grown rapidly, its owner has hired you as its first database manager. The company's database, developed by an outside consulting team, is already in place to help manage all company operations. Your first critical assignment is to develop a decision support system to analyze the charter data. (Review the company's operations in Problems 24–31 of Chapter 3, The Relational Database Model.) The charter operations manager wants to be able to analyze charter data such as cost, hours flown, fuel used, and revenue. She also wants to be able to drill down by pilot, type of airplane, and time periods.

Given those requirements, complete the following:

a. Create a star schema for the charter data.

b. Define the dimensions and attributes for the charter operation's star schema.

c. Define the necessary attribute hierarchies.

d. Implement the data warehouse design using the design components you developed in Problems 4a–4c.

e. Generate the reports to illustrate that your data warehouse meets the specified information requirements.

Using the data provided in the **Ch13-SaleCo-DW** database, solve the following problems. (*Hint*: In Problems 5–11, use the ROLLUP command.)

5. What is the SQL command to list the total sales by customer and by product, with subtotals by customer and a grand total for all product sales? Figure P13.5 shows the abbreviated results of the query.

FIGURE P13.5 PROBLEM 5 ABBREVIATED RESULT

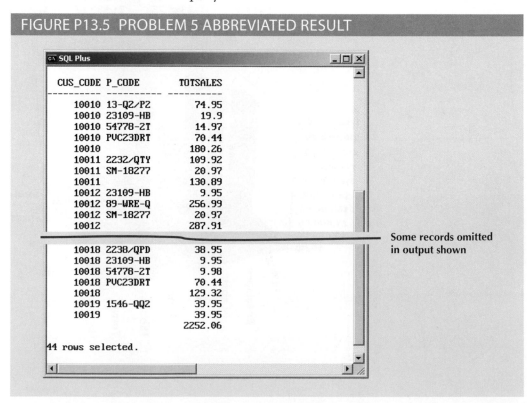

6. What is the SQL command to list the total sales by customer, month, and product, with subtotals by customer and by month and a grand total for all product sales? Figure P13.6 shows the abbreviated results of the query.

FIGURE P13.6   PROBLEM 6 ABBREVIATED RESULT

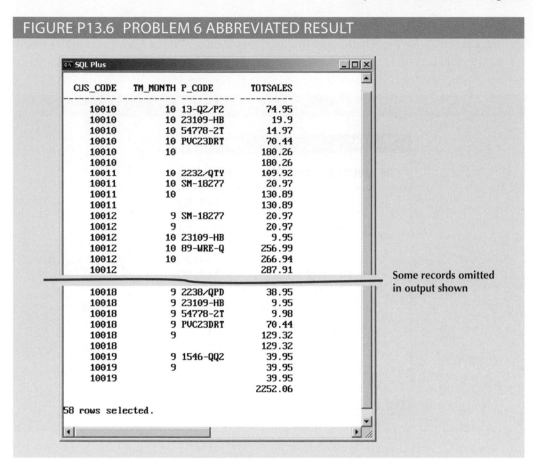

7. What is the SQL command to list the total sales by region and customer, with subtotals by region and a grand total for all sales? Figure P13.7 shows the result of the query.

FIGURE P13.7   PROBLEM 7 RESULT

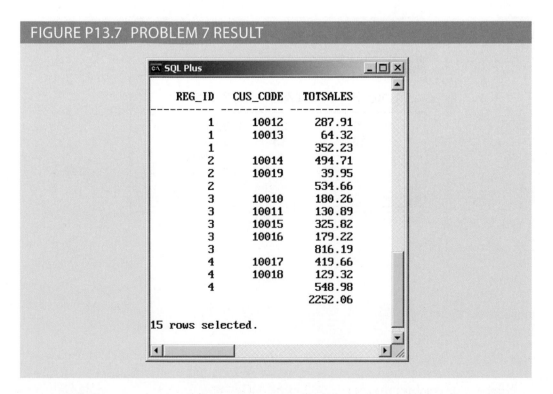

8. What is the SQL command to list the total sales by month and product category, with subtotals by month and a grand total for all sales? Figure P13.8 shows the result of the query.

FIGURE P13.8 PROBLEM 8 RESULT

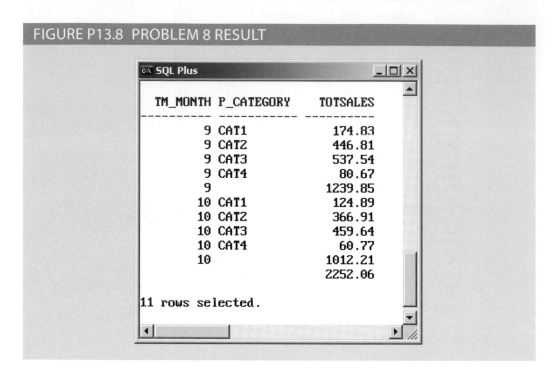

9. What is the SQL command to list the number of product sales (number of rows) and total sales by month, with subtotals by month and a grand total for all sales? Figure P13.9 shows the result of the query.

FIGURE P13.9 PROBLEM 9 RESULT

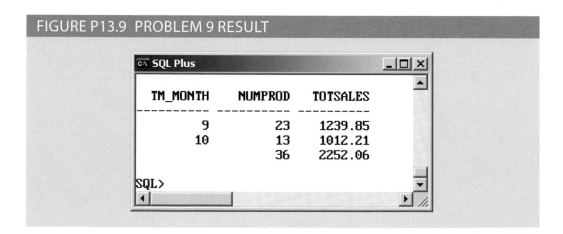

10. What is the SQL command to list the number of product sales (number of rows) and total sales by month and product category, with subtotals by month and product category and a grand total for all sales? Figure P13.10 shows the result of the query.

FIGURE P13.10 PROBLEM 10 RESULT

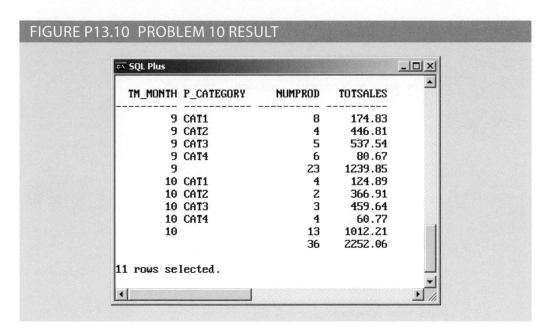

11. What is the SQL command to list the number of product sales (number of rows) and total sales by month, product category, and product, with subtotals by month and product category and a grand total for all sales? Figure P13.11 shows the result of the query.

FIGURE P13.11 PROBLEM 11 RESULT

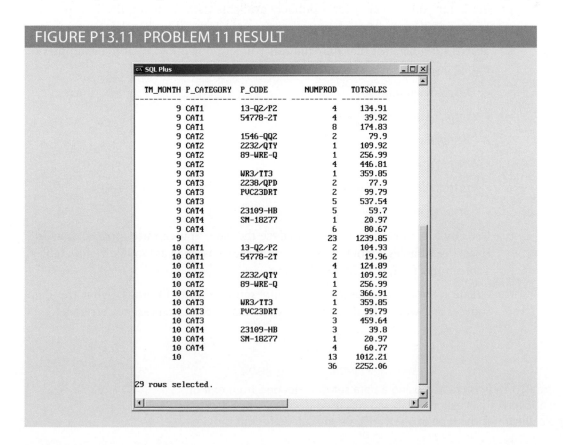

12. Using the answer to Problem 10 as your base, what command would you need to generate the same output but with subtotals in all columns? (*Hint*: Use the CUBE command.) Figure P13.12 shows the result of the query.

FIGURE P13.12 PROBLEM 12 RESULT

```
SQL Plus                                              _|□|×|

 TM_MONTH P_CATEGORY    NUMPROD    TOTSALES
 ---------- ----------- ---------- ----------
                              36     2252.06
            CAT1             12      299.72
            CAT2              6      813.72
            CAT3              8      997.18
            CAT4             10      141.44
         9                   23     1239.85
         9 CAT1              8      174.83
         9 CAT2              4      446.81
         9 CAT3              5      537.54
         9 CAT4              6       80.67
        10                   13     1012.21
        10 CAT1              4      124.89
        10 CAT2              2      366.91
        10 CAT3              3      459.64
        10 CAT4              4       60.77

15 rows selected.
```

13. Create your own data analysis and visualization presentation. The purpose of this project is for you to search for a publicly available data set using the Internet and create your own presentation using what you have learned in this chapter.

    a. Search for a data set that may interest you and download it. Some examples of public data sets sources are:

    - *http://www.data.gov*
    - *http://data.worldbank.org*
    - *http://aws.amazon.com/datasets*
    - *http://usgovxml.com/*
    - *https://data.medicare.gov/*
    - *http://www.faa.gov/data_research/*

    b. Use any tool available to you to analyze the data. You can use tools such as MS Excel Pivot Tables, Pivot Charts, or other free tools, such as Google Fusion tables, Tableau free trial, IBM Many Eyes, etc.

    c. Create a short presentation to explain some of your findings (what the data sources are, where the data comes from, what the data represents, etc.)

Note

The visualization in Figure P13.13 was created using a data set downloaded from one of the public sources listed above. A trial version of Tableau was used to create the visualizations. This simple example illustrates the type of quick analysis you can do for this project.

FIGURE P13.13 VISUALIZATION EXAMPLE USING TABLEAU

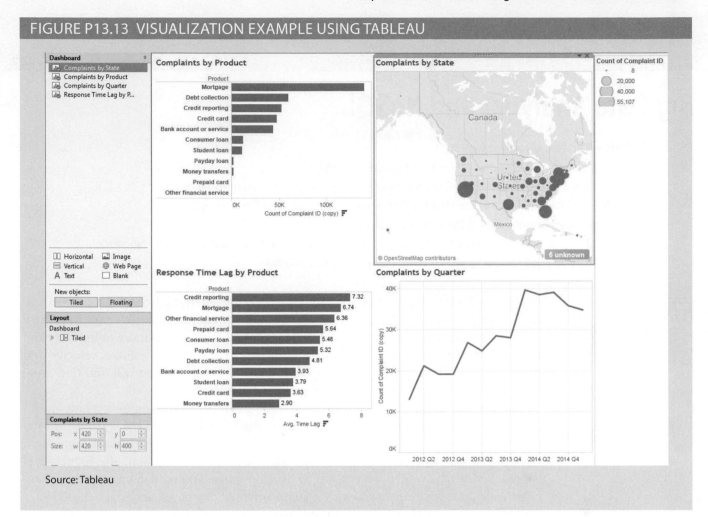

Source: Tableau

Big Data Analytics and NoSQL

In this chapter, you will learn:

- What Big Data is and why it is important in modern business
- The primary characteristics of Big Data and how these go beyond the traditional "3 Vs"
- How the core components of the Hadoop framework, HDFS and MapReduce, operate
- What the major components of the Hadoop ecosystem are
- The four major approaches of the NoSQL data model and how they differ from the relational model
- About data analytics, including data mining and predictive analytics

Preview

In Chapter 2, Data Models, you were introduced to the emerging NoSQL data model and the Big Data problem that has led to NoSQL's development. In this chapter, you learn about these issues in much greater detail. You will find that there is more to Big Data and the problem that it represents to modern businesses than just the volume, velocity, and variety ("3 V") characteristics introduced in Chapter 2. In fact, you will find that these characteristics themselves are more complex than previously discussed.

After learning about Big Data issues, you learn about the technologies that have developed, and continue to be developed, to address Big Data. First, you learn about the low-level technologies in the Hadoop framework. Hadoop has become a standard component in organizations' efforts to address Big Data. Next, you learn about the higher-level approaches of the NoSQL data model to develop nonrelational databases such as key-value databases, document databases, column-oriented databases, and graph databases.

Finally, you learn about the important area of data analytics and how statistical techniques are being used to help organizations turn the vast stores of data that are being collected into actionable information. Analytics are helping organizations understand not only what has happened in the business, but also to predict what is likely to happen.

Data Files Available on cengagebrain.com

Note

Because it is purely conceptual, this chapter does not reference any data files.

The relational database model has been dominant for decades, and during that time it has faced challenges such as object-oriented databases and the development of data warehouses. The relational model, and the tools based on it, have evolved to adapt to these challenges and remain dominant in the data management arena. In each case, the challenge arose because technological advances changed business's perceptions of what is possible and created new opportunities for organizations to create value from increased data leverage. The latest of these challenges is Big Data. Big Data is an ill-defined term that describes a new wave of data storage and manipulation possibilities and requirements. Organizations' efforts to store, manipulate, and analyze this new wave of data represent one of the most urgent emerging trends in the database field. The challenges of dealing with the wave of Big Data have led to the development of NoSQL databases that reject many of the underlying assumptions of the relational model. Although the term *Big Data* lacks a consistent definition, there is a set of characteristics generally associated with it.

14-1 **Big Data**

Big Data generally refers to a set of data that displays the characteristics of volume, velocity, and variety (the "3 Vs") to an extent that makes the data unsuitable for management by a relational database management system. These characteristics can be defined as follows:

- **Volume**—the quantity of data to be stored
- **Velocity**—the speed at which data is entering the system
- **Variety**—the variations in the structure of the data to be stored

Notice the lack of specific values associated with these characteristics. This lack of specificity is what leads to the ambiguity in defining Big Data. What was Big Data five years ago might not be considered Big Data now. Similarly, something considered Big Data now might not be considered Big Data five years from now. The key is that the characteristics are present to an extent that the current relational database technology struggles with managing the data.

Further adding to the problem of defining Big Data is that there is some disagreement among pundits about which of the 3 Vs must be present for a data set to be considered Big Data. Originally, Big Data was conceived as shown in Figure 14.1 as a combination of the 3 Vs. Web data, a combination of text, graphics, video, and audio sources combined

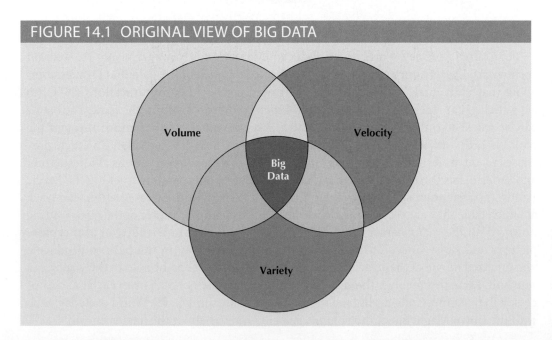

FIGURE 14.1 ORIGINAL VIEW OF BIG DATA

volume
A characteristic of Big Data that describes the quantity of data to be stored.

velocity
A characteristic of Big Data that describes the speed at which data enters the system and must be processed.

variety
A characteristic of Big Data that describes the variations in the structure of data to be stored.

into complex structures, is often cited as creating the new challenges for data management that involve all three characteristics. After the dot-com bubble burst in the 1990s, many startup web-based companies failed, but the companies that survived experienced significant growth as web commerce consolidated into a smaller set of businesses. As a result, companies like Google and Amazon experienced significant growth and were among the first to feel the pressure of managing Big Data. The success of social media giant Facebook quickly followed, and these companies became pioneers in creating new technologies to address Big Data problems. Google created the BigTable data store, Amazon created Dynamo, and Facebook created Cassandra to deal with the growing need to store and manage large sets of data that had the characteristics of the 3 Vs.

Although social media and web data have been at the forefront of perceptions of Big Data issues, other organizations have Big Data issues, too. More recently, changes in technology have increased the opportunities for businesses to generate and track data so that Big Data has been redefined as involving any, but not necessarily all, of the 3 Vs, as shown in Figure 14.2. Advances in technology have led to a vast array of user-generated data and machine-generated data that can spur growth in specific areas.

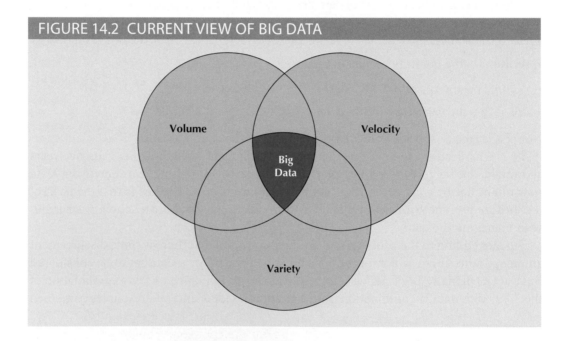

FIGURE 14.2 CURRENT VIEW OF BIG DATA

For example, Disney World has introduced "Magic Bands" for park visitors to wear on their wrists. Each visitor's Magic Band is connected to much of the data that Disney stores about that individual. These bands use RFID and near-field communications (NFC) to act as tickets for rides, hotel room keys, and even credit cards within the park. The bands can be tracked so the Disney systems can track individuals as they move through the park, record with which Disney characters (who are also tracked) they interact, purchases made, wait time in lines, and more. Visitors can make reservations at a restaurant and order meals through a Disney app on their smartphones, and by tracking the Magic Bands, the restaurant staff knows when the visitor arrives for their reservation, can track at which table they are seated, and deliver their meals within minutes of the guests sitting down. With the many cameras mounted throughout the park, Disney can also capture pictures and short videos of the visitor throughout their stay in the park to produce a personalized movie of their vacation experience, which can then be sold to the visitor as a souvenir. All of this involves the capture of a constant stream of data from each band, processed in real time. Considering the thousands of visitors in Disney World each day, each with their own Magic Band, the volume, velocity, and variety of the data is enormous.

14-1a Volume

Volume, the quantity of data to be stored, is a key characteristic of Big Data. The storage capacities associated with Big Data are extremely large. Table 14.1 provides definitions for units of data storage capacity.

TABLE 14.1

STORAGE CAPACITY UNITS

TERM	CAPACITY	ABBREVIATION
Bit	0 or 1 value	b
Byte	8 bits	B
Kilobyte	1024* bytes	KB
Megabyte	1024 KB	MB
Gigabyte	1024 MB	GB
Terabyte	1024 GB	TB
Petabyte	1024 TB	PB
Exabyte	1024 PB	EB
Zettabyte	1024 EB	ZB
Yottabyte	1024 ZB	YB

*Note that because bits are binary in nature and are the basis on which all other storage values are based, all values for data storage units are defined in terms of powers of 2. For example, the prefix *kilo* typically means 1000; however, in data storage, a kilobyte = 2^{10} = 1024 bytes.

Naturally, as the quantity of data needing to be stored increases, the need for larger storage devices increases as well. When this occurs, systems can either scale up or scale out. **Scaling up** is keeping the same number of systems, but migrating each system to a larger system: for example, changing from a server with 16 CPU cores and a 1 TB storage system to a server with 64 CPU cores and a 100 TB storage system. Scaling up involves moving to larger and faster systems. However, there are limits to how large and fast a single system can be. Further, the costs of these high-powered systems increase at a dramatic rate. On the other hand, **scaling out** means that when the workload exceeds the capacity of a server, the workload is spread out across a number of servers. This is also referred to as *clustering*—creating a cluster of low-cost servers to share a workload. This can help to reduce the overall cost of the computing resources since it is cheaper to buy ten 100 TB storage systems than it is to buy a single 1 PB storage system. Make no mistake, organizations need storage capacities in these extreme sizes. The eBay singularity system, which collects clickstream data among other things, is over 40 PB. This is in addition to the eBay enterprise data warehouse, which is over 14 PB and spread over hundreds of thousands of nodes.[1]

scaling up
A method for dealing with data growth that involves migrating the same structure to more powerful systems.

scaling out
A method for dealing with data growth that involves distributing data storage structures across a cluster of commodity servers.

[1] Cliff Saran, "Case study: How big data powers the eBay customer journey," *ComputerWeekly.com*, TechTarget, 2015, www.computerweekly.com/news/2240219736/Case-Study-How-big-data-powers-the-eBay-customer-journey, August 18, 2015.

Recall from Chapter 3 that one of the greatest advances represented by the relational model was the development of an RDBMS—a sophisticated database management system that could hide the complexity of the underlying data storage and manipulation from the user so that the data always appears to be in tables. To carry out these functions, the DBMS acts as the "brain" of the database system and must maintain control over all of the data within the database. As discussed in Chapter 12, it is possible to distribute a relational database over multiple servers using replication and fragmentation. However, because the DBMS must act as a single point of control for all of the data in the database, distributing the database across multiple systems requires a high degree of communication and coordination across the systems. There are significant limits associated with the ability to distribute the DBMS due to the increased performance costs of communication and coordination as the number of nodes grows. This limits the degree to which a relational database to be scaled out as data volume grows, and it makes RDBMSs ill-suited for clusters.

Note

Although some RDBMS products, such as SQL Server and Oracle Real-Application Clusters (RAC), legitimately claim to support clusters, these clusters are limited in scope and generally rely on a single, shared data storage subsystem, such as a storage area network (SAN).

14-1b Velocity

Velocity, another key characteristic of Big Data, refers to the rate at which new data enters the system as well as the rate at which the data must be processed. In many ways, the issues of velocity mirror those of volume. For example, consider a web retailer such as Amazon. In the past, a retail store might capture only the data about the final transaction of a customer making a purchase. A retailer like Amazon captures not only the final transaction, but every click of the mouse in the searching, browsing, comparing, and purchase process. Instead of capturing one event (the final sale) in a 20-minute shopping experience, it might capture data on 30 events during that 20-minute time frame—a 30× increase in the velocity of the data. Other advances in technology, such as RFID, GPS, and NFC, add new layers of data-gathering opportunities that often generate large amounts of data that must be stored in real-time. For example, RFID tags can be used to track items for inventory and warehouse management. The tags do not require line-of-sight between the tag and the reader, and the reader can read hundreds of tags simultaneously while the products are still in boxes. This means that instead of a single record for tracking a given quantity of a product being produced, each individual product is tracked, creating an increase of several orders of magnitude in the amount of data being delivered to the system at any one time.

In addition to the speed with which data is entering the system, for Big Data to be actionable, that data must be processed at a very rapid pace. The velocity of processing can be broken down into two categories.

- Stream processing

- Feedback loop processing

Stream processing focuses on input processing, and it requires analysis of the data stream as it enters the system. In some situations, large volumes of data can enter the system at such a rapid pace that it is not feasible to try to store all of the data. The data must be processed and filtered as it enters the system to determine which data to keep

stream processing
The processing of data inputs in order to make decisions about which data to keep and which data to discard before storage.

and which data to discard. For example, at the CERN Large Hadron Collider, the largest and most powerful particle accelerator in the world, experiments produce about 600 TB per second of raw data. Scientists have created **algorithms** to decide ahead of time which data will be kept. These algorithms are applied in a two-step process to filter the data down to only about 1 GB per second of data that will actually be stored.[2]

    Feedback loop processing refers to the analysis of the data to produce actionable results. While stream processing could be thought of as focused on inputs, feedback loop processing can be thought of as focused on outputs. The process of capturing the data, processing it into usable information, and then acting on that information is a feedback loop. Figure 14.3 shows a feedback loop for providing recommendations for book purchases. Feedback loop processing to provide immediate results requires analyzing large amounts of data within just a few seconds so that the results of the analysis can become a part of the product delivered to the user in real time. Not all feedback loops are used for inclusion of results within immediate data products. Feedback loop processing is also used to help organizations sift through terabytes and petabytes of data to inform decision makers to help them make faster strategic and tactical decisions, and it is a key component in data analytics.

FIGURE 14.3 FEEDBACK LOOP PROCESSING

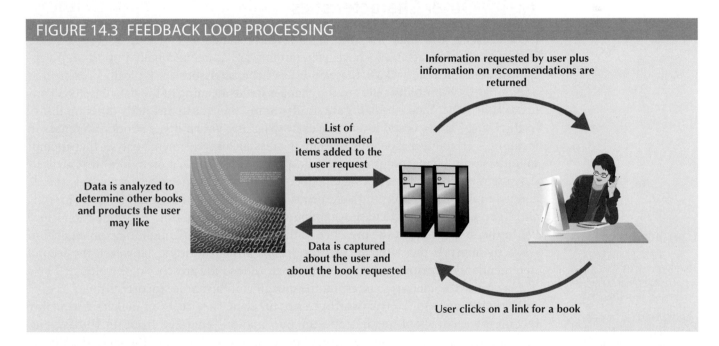

Information requested by user plus information on recommendations are returned

List of recommended items added to the user request

Data is analyzed to determine other books and products the user may like

Data is captured about the user and about the book requested

User clicks on a link for a book

14-1c Variety

In a Big Data context, variety refers to the vast array of formats and structures in which the data may be captured. Data can be considered to be structured, unstructured, or semistructured. **Structured data** is data that has been organized to fit a predefined data model. **Unstructured data** is data that is not organized to fit into a predefined data model. Semistructured data combines elements of both—some parts of the data fit a predefined model while other parts do not. Relational databases rely on structured data. A data model is created by the database designer based on the business rules, as discussed in Chapter 4. As data enters the database, the data is decomposed and routed for storage in the corresponding tables and columns as defined in the data model. Although much of the transactional data that organizations use works well in a structured environment,

algorithm
A process or set of operations in a calculation.

feedback loop processing
Analyzing stored data to produce actionable results.

structured data
Data that conforms to a predefined data model.

unstructured data
Data that does not conform to a predefined data model.

[2] CERN, "Processing: What to record?" *http://home.web.cern.ch/about/computing/processing-what-record*, August 20, 2015.

most of the data in the world is semistructured or unstructured. Unstructured data includes maps, satellite images, emails, texts, tweets, videos, transcripts, and a whole host of other data forms. Over the decades that the relational model has been dominant, relational databases have evolved to address some forms of unstructured data. For example, most large-scale RDBMSs support a binary large object (BLOB) data type that allows the storage of unstructured objects like audio, video, and graphic data as a single, atomic value. One problem with BLOB data is that the semantic value of the data, the meaning that the object conveys, is inaccessible and uninterpretable by data processing.

Big Data requires that the data be captured in whatever format it naturally exists, without any attempt to impose a data model or structure to the data. This is one of the key differences between processing data in a relational database and Big Data processing. Relational databases impose a structure on the data when the data is captured and stored. Big Data processing imposes a structure on the data as needed for applications as a part of retrieval and processing. One advantage to providing structure during retrieval and processing is the flexibility of being able to structure the data in different ways for different applications.

14-1d Other Characteristics

Characterizing Big Data with the 3 Vs is fairly standard. However, as the industry matures, other characteristics have been put forward as being equally important. Keeping with the spirit of the 3 Vs, these additional characteristics are typically presented as additional *Vs*. **Variability** refers to the changes in the meaning of the data based on context. While *variety* and *variability* are similar terms, they mean distinctly different things in Big Data. Variety is about differences in structure. Variability is about differences in meaning. Variability is especially relevant in areas such as sentiment analysis that attempt to understand the meanings of words. **Sentiment analysis** is a method of text analysis that attempts to determine if a statement conveys a positive, negative, or neutral attitude about a topic. For example, the statements, "I just bought a new smartphone—I love it!" and "The screen on my new smartphone shattered the first time I dropped it—I love it!" In the first statement the presence of the phrase "I love it" might help an algorithm correctly interpret the statement as expressing a positive attitude. However, the second statement uses sarcasm to express a negative attitude so the presence of the phrase "I love it" may cause the analysis to interpret the meaning of the phrase incorrectly.

Veracity refers to the trustworthiness of the data. Can decision makers reasonably rely on the accuracy of the data and the information generated from it? This is especially pertinent given the automation of data capture and some of the analysis. Uncertainty about the data can arise from several causes, such as having to capture only selected portions of the data due to high velocity. Also, in terms of sentiment analysis, customers' opinions and preferences can change over time, so comments at one point in time might not be suitable for action at another point in time.

Increasingly, value is being touted as an important characteristic for Big Data. **Value**, also called *viability*, refers to the degree to which the data can be analyzed to provide meaningful information that can add value to the organization. Just because a set of data *can* be captured does not mean that it *should* be captured. Only data that can form the basis for analysis that has the potential to impact organizational behavior should be included in a company's Big Data efforts.

The final characteristic of Big Data is visualization. **Visualization** is the ability to graphically present the data in such a way as to make it understandable. Volumes of data can leave decision makers awash in facts but with little understanding of what the facts mean. Visualization is a way of presenting the facts so that decision makers can comprehend the meaning of the information to gain insights.

variability
The characteristic of Big Data for the same data values to vary in meaning over time.

sentiment analysis
A method of text analysis that attempts to determine if a statement conveys a positive, negative, or neutral attitude.

veracity
The trustworthiness of a set of data.

value
The degree to which data can be analyzed to provide meaningful insights.

visualization
The ability to graphically present data in such a way as to make it understandable to users.

An argument could be made that these additional Vs are not necessarily characteristics of Big Data; or, perhaps more accurately, they are not characteristics of *only* Big Data. Veracity of data is an issue with even the smallest data store, which is why data management is so important in relational databases. Value of data also applies to traditional, structured data in a relational database. One of the keys to data modeling is that only the data that is of interest to the users should be included in the data model. Data that is not of value should not be recorded in any data store—Big Data or not. Visualization was discussed and illustrated at length in Chapter 13 as an important tool in working with data warehouses, which are often maintained as structured data stores in RDBMS products. The important thing to remember is that these characteristics that play an important part in working with data in the relational model are universal and also apply to Big Data.

Big Data represents a new wave in data management challenges, but it does not mean that relational database technology is going away. Structured data that depends on ACID transactions, as discussed in Chapter 10, will always be critical to business operations. Relational databases are still the best way for storing and managing this type of data. What has changed is that now, for the first time in decades, relational databases are not necessarily the best way for storing and managing *all* of an organization's data. Since the rise of the relational model, the decision for data managers when faced with new storage requirements was not whether to use a relational database, but rather which relational DBMS to use. Now, the decision of whether to use a relational database at all is a real question. This has led to **polyglot persistence**—the coexistence of a variety of data storage and management technologies within an organization's infrastructure. Scaling up, as discussed, is often considered a viable option as relational databases grow. However, it has practical limits and cost considerations that make it infeasible for many Big Data installations. Scaling out into clusters based on low-cost, commodity servers is the dominant approach that organizations are currently pursuing for Big Data management. As a result, new technologies not based on the relational model have been developed.

14-2 Hadoop

Big Data requires a different approach to distributed data storage that is designed for large-scale clusters. Although other implementation technologies are possible, Hadoop has become the de facto standard for most Big Data storage and processing. Hadoop is not a database. Hadoop is a Java-based framework for distributing and processing very large data sets across clusters of computers. While the Hadoop framework includes many parts, the two most important components are the Hadoop Distributed File System (HDFS) and MapReduce. HDFS is a low-level distributed file processing system, which means that it can be used directly for data storage. MapReduce is a programming model that supports processing large data sets in a highly parallel, distributed manner. While it is possible to use HDFS and MapReduce separately, the two technologies complement each other so that they work better together as a Hadoop system. Hadoop was engineered specifically to distribute and process enormous amounts of data across vast clusters of servers.

14-2a HDFS

The **Hadoop Distributed File System (HDFS)** approach to distributing data is based on several key assumptions:

- *High volume.* The volume of data in Big Data applications is expected to be in terabytes, petabytes, or larger. Hadoop assumes that files in the HDFS will be extremely large. Data in the HDFS is organized into physical blocks, just as in other file storage. For example, on a typical personal computer, file storage is organized into blocks that

polyglot persistence
The coexistence of a variety of data storage and data management technologies within an organization's infrastructure.

Hadoop Distributed File System (HDFS)
A highly distributed, fault-tolerant file storage system designed to manage large amounts of data at high speeds.

are often 512 bytes in size, depending on the hardware and operating system involved. Relational databases often aggregate these into database blocks. By default, Oracle organizes data into 8-KB physical blocks. Hadoop, on the other hand, has a default block size of 64 MB (8,000 times the size of an Oracle block!), and it can be configured to even larger values. As a result, the number of blocks per file is greatly reduced, simplifying the metadata overhead of tracking the blocks in each file.

- *Write-once, read-many.* Using a write-once, read-many model simplifies concurrency issues and improves overall data throughput. Using this model, a file is created, written to the file system, and then closed. Once the file is closed, changes cannot be made to its contents. This improves overall system performance and works well for the types of tasks performed by many Big Data applications. Although existing contents of the file cannot be changed, recent advancements in the HDFS allow for files to have new data appended to the end of the file. This is a key advancement for NoSQL databases because it allows for database logs to be updated.

- *Streaming access.* Unlike transaction processing systems where queries often retrieve small pieces of data from several different tables, Big Data applications typically process entire files. Instead of optimizing the file system to randomly access individual data elements, Hadoop is optimized for batch processing of entire files as a continuous stream of data.

- *Fault tolerance.* Hadoop is designed to be distributed across thousands of low-cost, commodity computers. It is assumed that with thousands of such devices, at any point in time, some will experience hardware errors. Therefore, the HDFS is designed to replicate data across many different devices so that when one device fails, the data is still available from another device. By default, Hadoop uses a replication factor of three, meaning that each block of data is stored on three different devices. Different replication factors can be specified for each file, if desired.

Hadoop uses several types of nodes. A *node* is just a computer that performs one or more types of tasks within the system. Within the HDFS, there are three types of nodes: the client node, the name node, and one or more data nodes, as depicted in Figure 14.4.

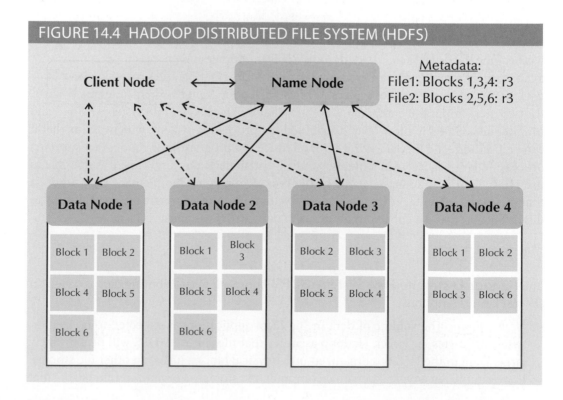

FIGURE 14.4 HADOOP DISTRIBUTED FILE SYSTEM (HDFS)

Data nodes store the actual file data within the HDFS. Recall that files in HDFS are broken into blocks and are replicated to ensure fault tolerance. As a result, each block is duplicated on more than one data node. Figure 14.4 shows the default replication factor of three, so each block appears on three data nodes.

The name node contains the metadata for the file system. There is typically only one name node within a HDFS cluster. The metadata is designed to be small, simple, and easily recoverable. Keeping the metadata small allows the name node to hold all of the metadata in memory to reduce disk accesses and improve system performance. This is important because there is only one name node so contention for the name node is minimized. The metadata is composed primarily of the name of each file, the block numbers that comprise each file, and the desired replication factor for each file. The client node makes requests to the file system, either to read files or to write new files, as needed to support the user application.

When a client node needs to create a new file, it communicates with the name node. The name node:

- Adds the new file name to the metadata.

- Determines a new block number for the file.

- Determines a list of which data nodes the block will be stored.

- Passes that information back to the client node.

The client node contacts the first data node specified by the name node and begins writing the file on that data node. At the same time, the client node sends the data node the list of other data nodes that will be replicating the block. As the data is received from the client node, the data node contacts the next data node in the list and begins sending the data to this node for replication. This second data node then contacts the next data node in the list and the process continues with the data being streamed across all of the data nodes that are storing the block. Once the first block is written, the client node can get another block number and list of data nodes from the name node for the next block. When the entire file has been written, the client node informs the name node that the file is closed. It is important to note that at no time was any of the data file actually transmitted to the name node. This helps to reduce the data flow to the name node to avoid congestion that could slow system performance.

Similarly, if a client node needs to read a file, it contacts the name node to request the list of blocks associated with that file and the data nodes that hold them. Given that each block may appear in many data nodes, for each block, the client attempts to retrieve the block from the data node that is closest to it on the network. Using this information, the client node reads the data directly from each of those nodes.

Periodically, each data node communicates with the name node. The data nodes send block reports and heartbeats. A **block report** is sent every 6 hours and informs the name node of which blocks are on that data node. Heartbeats are sent every 3 seconds. A **heartbeat** is used to let the name node know that the data node is still available. If a data node experiences a fault, due to hardware failure, power outage, etc., then the name node will not receive a heartbeat from that data node. As a result, the name node knows not to include that data node in lists to client nodes for reading or writing files. If the lack of a heartbeat from a data node causes a block to have fewer than the desired number of replicas, the name node can have a "live" data node initiate replicating the block on another data node.

Taken together, the components of the HDFS produce a powerful, yet highly specialized distributed file system that works well for the specialized processing requirements of Big Data applications. Next, we will consider how MapReduce provides data processing to complement data storage of HDFS.

block report
In the Hadoop Distributed File System (HDFS), a report sent every 6 hours by the data node to the name node informing the name node which blocks are on that data node.

heartbeat
In the Hadoop Distributed File System (HDFS), a signal sent every 3 seconds from the data node to the name node to notify the name node that the data node is still available.

14-2b MapReduce

MapReduce
An open-source application programming interface (API) that provides fast data analytics services; one of the main Big Data technologies that allows organizations to process massive data stores.

map
The function in a MapReduce job that sorts and filters data into a set of key-value pairs as a subtask within a larger job.

mapper
A program that performs a map function.

MapReduce is the computing framework used to process large data sets across clusters. Conceptually, MapReduce is easy to understand and follows the principle of *divide and conquer*. MapReduce takes a complex task, breaks it down into a collection of smaller subtasks, performs the subtasks all at the same time, and then combines the result of each subtask to produce a final result for the original task. As the name implies, it is a combination of a map function and a reduce function. A **map** function takes a collection of data and sorts and filters the data into a set of key-value pairs. The map function is performed by a program called a **mapper**. A **reduce** function takes a collection of key-value pairs, all with the same key value, and summarizes them into a single result. The reduce function is performed by a program called a **reducer**. Recall that Hadoop is a Java-based platform, therefore map and reduce functions are written as detailed, procedure-oriented Java programs.

Figure 14.5 provides a simple, conceptual illustration of MapReduce that determines the total number of units of each product that has been sold. The original data in Figure 14.5 is stored as key-value pairs, with the invoice number as the key and the remainder of the invoice data as a value. Remember, the data in Hadoop data storage is not a relational database so the data is not separated into tables and there is no form of normalization that ensures that each fact is stored only once. Therefore, there is a great deal of duplication of data in the original data store. Note that even in the very small subset of data that is shown in Figure 14.5, redundant data is kept for customer 10011, Leona Dunne. In the figure, map functions parse each invoice to find data about the products sold on that invoice. The result of the map function is a new list of key-value pairs in which the product code is the key and the line units are the value. The reduce function then takes that list of key-value pairs and combines them by summing the values associated with each key (product code) to produce the summary result.

FIGURE 14.5 MAPREDUCE

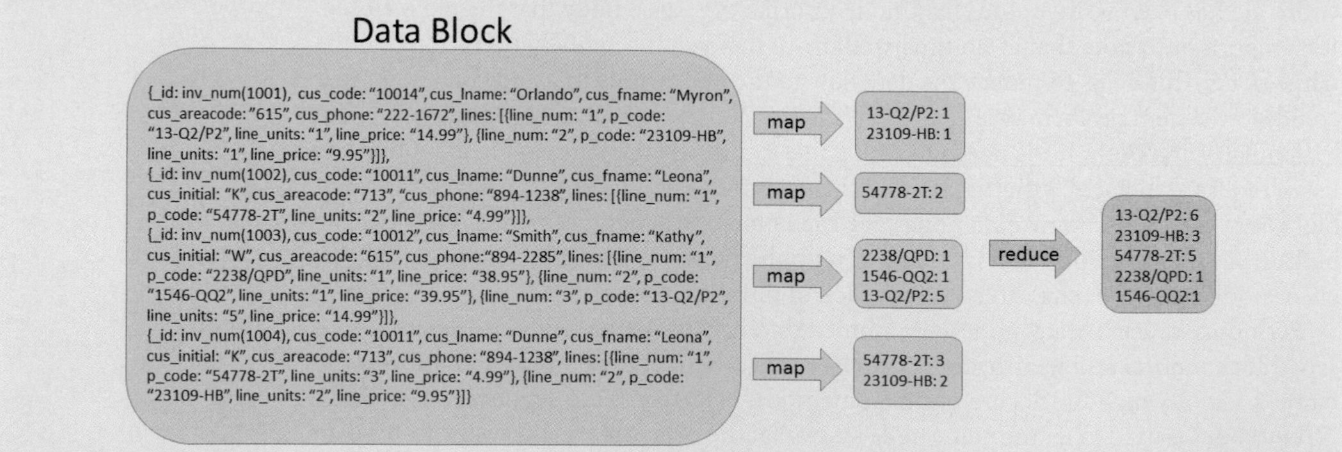

reduce
The function in a MapReduce job that collects and summarizes the results of map functions to produce a single result.

reducer
A program that performs a reduce function.

As previously stated, the data sets used in Big Data applications are extremely large. Transferring entire files from multiple nodes to a central node for processing would require a tremendous amount of network bandwidth, and place an incredible processing burden on the central node. Therefore, instead of the computational program retrieving the data for processing in a central location, copies of the program are "pushed" to the nodes containing the data to be processed. Each copy of the program produces results that are then aggregated across nodes and sent back to the client. This mirrors the distribution of data in the HDFS. Typically, the Hadoop framework will distribute a mapper for each block on each data node that must be processed. This can lead to a very large number of mappers. For example, if 1 TB of data is to be processed and the HDFS is using 64-MB blocks, that yields over 15,000 mapper programs. The number of reducers is configurable by the user, but best practices suggest about one reducer per data node.

Note

Best practices suggest that the number of mappers on a given node should be kept to 100 or less. However, there are cases of applications with simple map functions running as many as 300 mappers on a given node with satisfactory performance. Clearly, much depends on the computing resources available at each node.

The implementation of MapReduce complements the structure of the HDFS, which is an important reason why they work so well together. Just as the HDFS structure is composed of a name node and several data nodes, MapReduce uses a **job tracker** (the actual name of the program is JobTracker) and several **task trackers** (the programs are named TaskTrackers). The job tracker acts as a central control for MapReduce processing and it normally exists on the same server that is acting as the name node. Task tracker programs reside on the data nodes. One important feature of the MapReduce framework is that the user must write the Java code for the map and reduce functions, and must specify the input and output files to be read and written for the job that is being submitted. However, the job tracker will take care of locating the data, determining which nodes to use, dividing the job into tasks for the nodes, and managing failures of the nodes. All of this is done automatically without user intervention. When a user submits a MapReduce job for processing, the general process is as follows:

1. A client node (client application) submits a MapReduce job to the job tracker.

2. The job tracker communicates with the name node to determine which data nodes contain the blocks that should be processed for this job.

3. The job tracker determines which task trackers are available for work. Each task tracker can handle a set number of tasks. Remember, many MapReduce jobs from different users can be running on the Hadoop system simultaneously, so a data node may contain data that is being processed by multiple mappers from different jobs all at the same time. Therefore, the task tracker on that node might be busy running mappers for other jobs when this new request arrives. Because the data is replicated on multiple nodes, the job tracker may be able to select from multiple nodes for the same data.

4. The job tracker then contacts the task trackers on each of those nodes to begin mappers and reducers to complete that node's portion of the task.

5. The task tracker creates a new JVM (Java virtual machine) to run the map and reduce functions. This way, if a function fails or crashes, the entire task tracker is not halted.

6. The task tracker sends heartbeat messages to the job tracker to let the job tracker know that the task tracker is still working on the job (and about the nodes availability for more jobs).

7. The job tracker monitors the heartbeat messages to determine if a task manager has failed. If so, the job tracker can reassign that portion of the task to another node.

8. When the entire job is finished, the job tracker changes status to indicate that the job is completed.

9. The client node periodically queries the job tracker until the job status is completed.

job tracker
A central control program used to accept, distribute, monitor, and report on MapReduce processing jobs in a Hadoop environment.

task tracker
A program in the MapReduce framework responsible to running map and reduce tasks on a node.

The Hadoop system uses batch processing. **Batch processing** is when a program runs from beginning to end, either completing the task or halting with an error, without any interaction with the user. Batch processing is often used when the computing task requires an extended period of time or a large portion of the system's processing capacity. Businesses often use batch processing to run year-end financial reports in the evenings when systems are often idle, and universities might use batch processing for student fee payment processing. Batch processing is not bad, but it has limitations. As a result, a number of complementary programs have been developed to improve the integration of Hadoop within the larger IT infrastructure. The next section discusses some of these programs.

14-2c Hadoop Ecosystem

Hadoop is widely used by organizations tapping into the potential of analyzing extremely large data sets. Unfortunately, because Hadoop is a very low-level tool requiring considerable effort to create, manage, and use, it presents quite a few obstacles. As a result, a host of related applications have grown up around Hadoop to attempt to make it easier to use and more accessible to users who are not skilled at complex Java programming. Figure 14.6 shows examples of some of these types of applications. Most organizations that use Hadoop also use a set of other related products that interact and complement each other to produce an entire ecosystem of applications and tools. Like any ecosystem, the interconnected pieces are constantly evolving and their relationships are changing, so it is a rather fluid situation. The following are some of the more popular components in a Hadoop ecosystem and how they relate to each other.

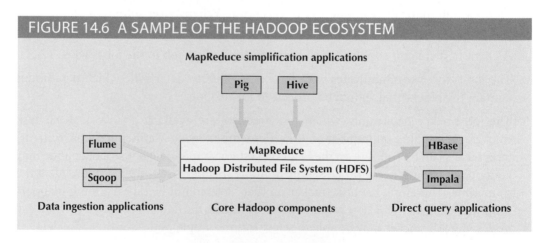

FIGURE 14.6 A SAMPLE OF THE HADOOP ECOSYSTEM

MapReduce Simplification Applications Creating MapReduce jobs requires significant programming skills. As the mapper and reducer programs become more complex, the skill requirements increase and the time to produce the programs becomes significant. These skills are beyond the capabilities of most data users. Therefore, applications to simplify the process of creating MapReduce jobs have been developed. Two of the most popular are Hive and Pig.

Hive is a data warehousing system that sits on top of HDFS. It is not a relational database, but it supports its own SQL-like language, called HiveQL, that mimics SQL commands to run ad hoc queries. HiveQL commands are processed by the Hive query engine into sets of MapReduce jobs. As a result, the underlying processing tends to be batch-oriented, producing jobs that are very scalable over extremely large sets of data. However, the batch nature of the jobs makes Hive a poor choice for jobs that only require a small subset of data to be returned very quickly.

batch processing
A data processing method that runs data processing tasks from beginning to end without any user interaction.

Pig is a tool for compiling a high-level scripting language, named Pig Latin, into MapReduce jobs for executing in Hadoop. In concept it is similar to Hive in that it provides a means of producing MapReduce jobs without the burden of low-level Java programming. The primary difference is that Pig Latin is a scripting language, which means it is procedural, while HiveQL, like SQL, is declarative. Declarative languages allow the user to specify what they want, not how to get it. This is very useful for query processing. Procedural languages require the user to specify how the data is to be manipulated. This is very useful for performing data transformations. As a result, Pig is often used for producing data pipeline tasks that transform data in a series of steps. This is often seen in ETL processes as described in Chapter 13.

Data Ingestion Applications   One challenge faced by organizations that are taking advantage of Hadoop's massive data storage and data processing capabilities is the issue of actually getting data from their existing systems into the Hadoop cluster. To simplify this task, applications have been developed to "ingest" or gather this data into Hadoop.

Flume is a component for ingesting data into Hadoop. It is designed primarily for harvesting large sets of data from server log files, like clickstream data from web server logs. It can be configured to import the data on a regular schedule or based on specified events. In addition to simply bringing the data into Hadoop, Flume contains a simple query processing component so the possibility exists of performing some transformations on the data as it is being harvested. Typically, Flume would move the data into the HDFS, but it can also be configured to input the data directly into another component of the Hadoop ecosystem named HBase.

Sqoop is a more recent addition to the Hadoop ecosystem. It is a tool for converting data back and forth between a relational database and the HDFS. The name Sqoop (pronounced, "scoop," as in a scoop of ice cream) is an amalgam of "SQL-to-Hadoop." In concept, Sqoop is similar to Flume in that it provides a way of bringing data into the HDFS. However, while Flume works primarily with log files, Sqoop works with relational databases such as Oracle, MySQL, and SQL Server. Further, while Flume operates in one direction only, Sqoop can transfer data in both directions—into and out of HDFS. When transferring data from a relational database into HDFS, the data is imported one table at a time with the process reading the table row-by-row. This is done in a highly parallelized manner using MapReduce, so the contents of the table will usually be distributed into several files with the rows stored in a delimited format. Once the data has been imported into HDFS, it can be processed by MapReduce jobs or using Hive. The resulting data can then be exported from HDFS back to the relational database, most often a traditional data warehouse.

Direct Query Applications   Direct query applications attempt to provide faster query access than is possible through MapReduce. These applications interact with HDFS directly, instead of going through the MapReduce processing layer.

HBase is a column-oriented NoSQL database designed to sit on top of the HDFS. One of HBase's primary characteristics is that it is highly distributed and designed to scale out easily. It does not support SQL or SQL-like languages, relying instead on lower-level languages such as Java for interaction. The system does not rely on MapReduce jobs, so it avoids the delays caused by batch processing, making it more suitable for fast processing involving smaller subsets of the data. HBase is very good at quickly processing sparse data sets. HBase is one of the more popular components of the Hadoop ecosystem, and is used by Facebook for its messaging system. Column-oriented databases will be discussed in more detail in the next section.

Impala was the first SQL-on-Hadoop application. It was produced by Cloudera as a query engine that supports SQL queries that pull data directly from HDFS. Prior to Impala, if an organization needed to make data from Hadoop available to analysts through an SQL interface, data would be extracted from HDFS and imported into a relational database. With Impala, analysts can write SQL queries directly against the data while it is still in HDFS. Impala makes heavy use of in-memory caching on data nodes. It is generally considered an appropriate tool for processing large amounts of data into a relatively small result set.

Note

Other than Impala, each of the components of the Hadoop ecosystem described in this section are all open-source, top-level projects of the Apache Software Foundation. More information on each of these projects and many others is available at *www.apache.org*.

14-3 NoSQL

NoSQL is the unfortunate name given to a broad array of nonrelational database technologies that have developed to address the challenges represented by Big Data. The name is unfortunate in that it does not describe what the NoSQL technologies are, but rather what they are not. In fact, the name also does a poor job of explaining what the technologies are not! The name was chosen as a Twitter hashtag to simplify coordinating a meeting of developers to discuss ideas about the nonrelational database technologies that were being developed by organizations like Google, Amazon, and Facebook to deal with the problems they were encountering as their data sets reached enormous sizes. The term "NoSQL" was never meant to imply that products in this category should never include support for SQL. In fact, many such products support query languages that mimic SQL in important ways. Although no one has yet produced a NoSQL system that implements standard SQL, given the large base of SQL users, the appeal of creating such a product is obvious. More recently, some industry observers have tried to interject that "NoSQL" could stand for "Not Only SQL." In fact, if the requirement to be considered a NoSQL product were simply that languages beyond SQL are supported, then all of the traditional RDBMS products such as Oracle, SQL Server, MySQL, and MS Access would all qualify. Regardless, you are better off focusing on understanding the array of technologies to which the term refers than worrying about the name itself.

There are literally hundreds of products that can be considered as being under the broadly defined term NoSQL. Most of these fit roughly into one of four categories: key-value data stores, document databases, column-oriented databases, and graph databases. Table 14.2 shows some popular NoSQL databases of each type. Although not all NoSQL databases have been produced as open-source software, most have been. As a result, NoSQL databases are generally perceived as a part of the open-source movement. Accordingly, they also tend to be associated with the Linux operating system. It makes sense from a cost standpoint that, if an organization is going to create a cluster containing tens of thousands of nodes, the organization does not want to purchase licenses for Windows or Mac OS for all of those nodes. The preference is to use a platform, like Linux, that is freely available and highly customizable. Therefore, most of the NoSQL products run only in a Linux or Unix environment. The following sections discuss each of the major NoSQL approaches.

NoSQL
A new generation of database management systems that is not based on the traditional relational database model.

TABLE 14.2

NoSQL DATABASES

NoSQL CATEGORY	EXAMPLE DATABASES
Key-value database	Dynamo Riak Redis Voldemort
Document databases	MongoDB CouchDB OrientDB RavenDB
Column-oriented databases	HBase Cassandra Hypertable
Graph databases	Neo4J ArangoDB GraphBase

14-3a Key-Value Databases

Key-value (KV) databases are conceptually the simplest of the NoSQL data models. A KV database is a NoSQL database that stores data as a collection of key-value pairs. The key acts as an identifier for the value. The value can be anything such as text, an XML document, or an image. The database does not attempt to understand the contents of the value component or its meaning—the database simply stores whatever value is provided for the key. It is the job of the applications that use the data to understand the meaning of the data in the value component. There are no foreign keys; in fact, relationships cannot be tracked among keys at all. This greatly simplifies the work that the DBMS must perform, making KV databases extremely fast and scalable for basic processing.

Key-value pairs are typically organized into "buckets." A **bucket** can roughly be thought of as the KV database equivalent of a table. A bucket is a logical grouping of keys. Key values must be unique within a bucket, but they can be duplicated across buckets. All data operations are based on the bucket plus the key. In other words, it is not possible to query the data based on anything in the value component of the key-value pair. All queries are performed by specifying the bucket and key. Operations on KV databases are rather simple—only *get*, *store*, and *delete* operations are used. *Get* or *fetch* is used to retrieve the value component of the pair. *Store* is used to place a value in a key. If the bucket + key combination does not exist, then it is added as a new key-value pair. If the bucket + key combination does exist, then the existing value component is replaced with the new value. *Delete* is used to remove a key-value pair. Figure 14.7 shows a customer bucket with three key-value pairs. Since the KV model does not allow queries based on data in the value component, it is not possible to query for a key-value pair based on customer last name, for example. In fact, the KV DBMS does not even know that there is such a thing as a customer last name because it does not understand the content of the value component. An application could issue a *get* command to have the KV DBMS return the key-value pair for bucket customer and key 10011, but it would be up to the application to know how to parse the value component to find the customer's last name, first name, and other characteristics. (One important note about Figure 14.7: Be aware that although key-value pairs appear in tabular form in the figure, the tabular format is just a convenience to help visually distinguish the components. Actual key-value pairs are not stored in a table-like structure.)

key-value (KV) database
A NoSQL database model that stores data as a collection of key-value pairs in which the value component is unintelligible to the DBMS.

bucket
In a key-value database, a logical collection of related key-value pairs.

FIGURE 14.7 KEY-VALUE DATABASE STORAGE

Bucket = Customer

Key	Value
10010	"LName Ramas FName Alfred Initial A Areacode 615 Phone 844-2573 Balance 0"
10011	"LName Dunne FName Leona Initial K Areacode 713 Phone 894-1238 Balance 0"
10014	"LName Orlando FName Myron Areacode 615 Phone 222-1672 Balance 0"

14-3b Document Databases

Document databases are conceptually similar to key-value databases, and they can almost be considered a subtype of KV databases. A document database is a NoSQL database that stores data in tagged documents in key-value pairs. Unlike a KV database where the value component can contain any type of data, a document database always stores a document in the value component. The document can be in any encoded format, such as XML, **JSON (JavaScript Object Notation)**, or **BSON (Binary JSON)**. Another important difference is that while KV databases do not attempt to understand the content of the value component, document databases do. Tags are named portions of a document. For example, a document may have tags to identify which text in the document represents the title, author, and body of the document. Within the body of the document, there may be additional tags to indicate chapters and sections. Despite the use of tags in documents, document databases are considered schema-less, that is, they do not impose a predefined structure on the data that is stored. For a document database, being schema-less means that although all documents have tags, not all documents are required to have the same tags, so each document can have its own structure. The tags in a document database are extremely important because they are the basis for most of the additional capabilities that document databases have over KV databases. Tags inside the document are accessible to the DBMS, which makes sophisticated querying possible.

Just as KV databases group key-value pairs into logical groups called *buckets*, document databases group documents into logical groups called *collections*. While a document may be retrieved by specifying the collection and key, it is also possible to query based on the contents of tags. For example, Figure 14.8 represents the same data from Figure 14.7, but in a tagged format for a document database. Because the DBMS is aware of the tags within the documents, it is possible to write queries that retrieve all of

document database
A NoSQL database model that stores data in key-value pairs in which the value component is composed of a tag-encoded document.

JSON (JavaScript Object Notation)
A human-readable text format for data interchange that defines attributes and values in a document.

BSON (Binary JSON)
A computer-readable format for data interchange that expands the JSON format to include additional data types including binary objects.

FIGURE 14.8 DOCUMENT DATABASE TAGGED FORMAT

Collection = Customer

Key	Document
10010	{LName: "Ramas", FName: "Alfred", Initial: "A", Areacode: "615", Phone: "844-2573", Balance: "0"}
10011	{LName: "Dunne", FName: "Leona", Initial: "K", Areacode: "713", Phone: "894-1238", Balance: "0"}
10014	{LName: "Orlando", FName: "Myron", Areacode: "615", Phone: "222-1672", Balance: "0"}

the documents where the Balance tag has the value 0. Document databases even support some aggregate functions such as summing or averaging balances in queries.

Document databases tend to operate on an implied assumption that a document is relatively self-contained, not a fragment of the data about a given topic. Relational databases decompose complex data in the business environment into a set of related tables. For example data about orders may be decomposed into customer, invoice, line, and product tables. A document database would expect all of the data related to an order to be in a single order document. Therefore, each order document in an *Orders* collection would contain data on the customer, the order itself, and the products purchased in that order all as a single self-contained document. Document databases do not store relationships as perceived in the relational model and generally have no support for join operations.

14-3c Column-Oriented Databases

The term *column-oriented database* can refer to two different sets of technologies that are often confused with each other. In one sense, column-oriented database or columnar database can refer to traditional, relational database technologies that use **column-centric storage** instead of **row-centric storage**. Relational databases present data in logical tables; however, the data is actually stored in data blocks containing rows of data. All of the data for a given row is stored together in sequence with many rows in the same data block. If a table has many rows of data, the rows will be spread across many data blocks. Figure 14.9 illustrates a relational table with 10 rows of data that is physically stored across five data blocks. Row-centric storage minimizes the number of disk reads necessary to retrieve a row of data. Retrieving one row of data requires accessing just

column-centric storage
A physical data storage technique in which data is stored in blocks, which hold data from a single column across many rows.

row-centric storage
A physical data storage technique in which data is stored in blocks, which hold data from all columns of a given set of rows.

FIGURE 14.9 COMPARISON OF ROW-CENTRIC AND COLUMN-CENTRIC STORAGE

CUSTOMER relational table

Cus_Code	Cus_LName	Cus_FName	Cus_City	Cus_State
10010	Ramas	Alfred	Nashville	TN
10011	Dunne	Leona	Miami	FL
10012	Smith	Kathy	Boston	MA
10013	Olowski	Paul	Nashville	TN
10014	Orlando	Myron		
10015	O'Brian	Amy	Miami	FL
10016	Brown	James		
10017	Williams	George	Mobile	AL
10018	Farriss	Anne	Opp	AL
10019	Smith	Olette	Nashville	TN

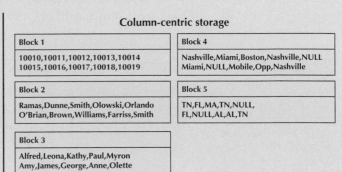

Row-centric storage

Block 1
10010,Ramas,Alfred,Nashville,TN
10011,Dunne,Leona,Miami,FL

Block 2
10012,Smith,Kathy,Boston,MA
10013,Olowski,Paul,Nashville,TN

Block 3
10014,Orlando,Myron,NULL,NULL
10015,O'Brian,Amy,Miami,FL

Block 4
10016,Brown,James,NULL,NULL
10017,Williams,George,Mobile,AL

Block 5
10018,Farriss,Anne,OPP,AL
10019,Smith,Olette,Nashville,TN

Column-centric storage

Block 1
10010,10011,10012,10013,10014
10015,10016,10017,10018,10019

Block 2
Ramas,Dunne,Smith,Olowski,Orlando
O'Brian,Brown,Williams,Farriss,Smith

Block 3
Alfred,Leona,Kathy,Paul,Myron
Amy,James,George,Anne,Olette

Block 4
Nashville,Miami,Boston,Nashville,NULL
Miami,NULL,Mobile,Opp,Nashville

Block 5
TN,FL,MA,TN,NULL,
FL,NULL,AL,AL,TN

one data block, as shown in Figure 14.9. Remember, in transactional systems, normalization is used to decompose complex data into related tables to reduce redundancy and to improve the speed of rapid manipulation of small sets of data. These manipulations tend to be row-oriented, so row-oriented storage works very well. However, in queries that retrieve a small set of columns across a large set of rows, a large number of disk accesses is required. For example, a query that wants to retrieve only the city and state of every customer will have to access every data block that contains a customer row to retrieve that data. In Figure 14.9, that would mean accessing five data blocks to get the city and state of every customer. A column-oriented or columnar database stores the data in blocks by column instead of by row. A single customer's data will be spread across several blocks, but all of the data from a single column will be in just a few blocks. In Figure 14.9, all of the city data for customers will be stored together, just as all of the state data will be stored together. In that case, retrieving the city and state for every customer might require accessing only two data blocks. This type of column-centric storage works very well for databases that are primarily used to run queries over few columns but many rows, as is done in many reporting systems and data warehouses. Though Figure 14.9 shows only a few rows and data blocks, it is easy to imagine that the gains would be significant if the table size grew to millions or billions of rows across hundreds of thousands of data blocks. At the same time, column-centric storage would be very inefficient for processing transactions since insert, update, and delete activities would be very disk intensive. It is worth noting that column-centric storage can be achieved within relational database technology, meaning that it still requires structured data and has the advantage of supporting SQL for queries.

The other use of the term *column-oriented database*, also called column family database, is to describe a type of NoSQL database that takes the concept of column-centric storage beyond the confines of the relational model. As NoSQL databases, these products do not require the data to conform to predefined structures nor do they support SQL for queries. This database model originated with Google's BigTable product. Other column-oriented database products include HBase, described earlier, and Cassandra. Cassandra began as a project at Facebook, but Facebook released it to the open-source community, which has continued to develop Cassandra into one of the most popular column-oriented databases. A **column family database** is a NoSQL database that organizes data in key-value pairs with keys mapped to a set of columns in the value component. While column family databases use many of the same terms as relational databases, the terms don't mean quite the same things. Fortunately, the column family databases are conceptually simple and are conceptually close enough to the relational model that your understanding of the relational model can help you understand the column family model. A column is a key-value pair that is similar to a cell of data in a relational database. The key is the name of the column, and the value component is the data that is stored in that column. Therefore, "cus_lname: Ramas" is a column; *cus_lname* is the name of the column, and *Ramas* is the data value in the column. Similarly, "cus_city: Nashville" is another column, with *cus_city* as the column name and *Nashville* as the data value.

column family database

A NoSQL database model that organizes data into key-value pairs, in which the value component is composed of a set of columns that vary by row.

Note

Even though column family databases do not (yet) support standard SQL, Cassandra developers have created a Cassandra query language (CQL). It is similar to SQL in many respects and is one of the more compelling reasons for adopting Cassandra.

As more columns are added, it becomes clear that some columns form natural groups, such as cus_fname, cus_lname, and cus_initial which would logically group together to form a customer's name. Similarly, cus_street, cus_city, cus_state, and cus_zip would logically group together to form a customer's address. These groupings are used to create super columns. A **super column** is a group of columns that are logically related. Recall the discussion in Chapter 4 about simple and composite attributes in the entity relationship model. In many cases, super columns can be thought of as the composite attribute and the columns that compose the super column as the simple attributes. Just as all simple attributes do not have to belong to a composite attribute, not all columns have to belong to a super column. Although this analogy is helpful in many contexts, it is not perfect. It is possible to group columns into a super column that logically belongs together for application processing reasons but does not conform to the relational idea of a composite attribute.

Row keys are created to identify objects in the environment. All of the columns or super columns that describe these objects are grouped together to create a **column family**; therefore, a column family is conceptually similar to a table in the relational model. While a column family is similar in concept to a relational table, Figure 14.10 shows that it is structurally very different. Notice in Figure 14.10 that each row key in the column family can have different columns.

Note

A column family can be composed of columns or super columns, but it cannot contain both.

FIGURE 14.10 COLUMN FAMILY DATABASE

Column Family Name	CUSTOMERS	
Key	Rowkey 1	
Columns	City	Nashville
	Fname	Alfred
	Lname	Ramas
	State	TN
Key	Rowkey 2	
Columns	Balance	345.86
	Fname	Kathy
	Lname	Smith
Key	Rowkey 3	
Columns	Company	Local Markets, Inc.
	Lname	Dunne

super column
In a column family database, a column that is composed of a group of other related columns.

column family
In a column family database, a collection of columns or super columns related to a collection of rows.

14-3d Graph Databases

A **graph database** is a NoSQL database based on graph theory to store data about relationship-rich environments. Graph theory is a mathematical and computer science field that models relationships, or edges, between objects called nodes. Modeling and storing data about relationships is the focus of graph databases. Graph theory is a well-established field of study going back hundreds of years. As a result, creating a database model based on graph theory immediately provides a rich source for algorithms and applications that have helped graph databases gain in sophistication very quickly. Since it also happens that much of the data explosion over the last decade has involved data that is relationship-rich, graph databases have been poised to experience significant interest in the business environment.

Interest in graph databases originated in the area of social networks. Social networks include a wide range of applications beyond the typical Facebook, Twitter, and Instagram that immediately come to mind. Dating websites, knowledge management, logistics and routing, master data management, and identity and access management are all areas that rely heavily on tracking complex relationships among objects. Of course, relational databases support relationships too. One of the great advances of the relational model was that relationships are easy to maintain. A relationship between a customer and an agent is as easy to implement in the relational model as adding a foreign key to create a common attribute, and the customer and agent rows are related by having the same value in the common attributes. If the customer changes to a different agent, then simply changing the value in the foreign key will change the relationship between the rows to maintain the integrity of the data. The relational model does all of these things very well. However, what if we want a "like" option so customers can "like" agents on our website? This would require a structural change to the database to add a new foreign key to support this second relationship. Next, what if the company wants to allow customers on its website to "friend" each other so a customer can see which agents their friends like, or the friends of their friends? In social networking data, there can be dozens of different relationships among individuals that need to be tracked, and often the relationships are tracked many layers deep (e.g., friends, friends of friends, friends of friends of friends, etc.). This results in a situation where the relationships become just as important as the data itself. This is the area where graph databases shine.

The primary components of graph databases are nodes, edges, and properties, as shown in Figure 14.11. A node corresponds to the idea of a relational entity instance. The **node** is a specific instance of something we want to keep data about. Each node (circle) in Figure 14.10 represents a single agent. Properties are like attributes; they are the data that we need to store about the node. All agent nodes might have properties like first name and last name, but all nodes are not required to have the same properties. An **edge** is a relationship between nodes. Edges (shown as arrows in Figure 14.10) can be in one direction, or they can be bidirectional. For example, in Figure 14.11, the *friends* relationships are bidirectional, but the *likes* relationships are not. Note that edges can also have **properties**. In Figure 14.11 the date on which customer Alfred Ramas *liked* agent Alex Alby is recorded in the graph database. A query in a graph database is called a **traversal**. Instead of *querying the database*, the correct terminology would be *traversing the graph*. Graph databases excel at traversals that focus on relationships between nodes, such as shortest path and degree of connectedness.

Graph database share some characteristics with other NoSQL databases in that graph databases do not force data to fit predefined structures, do not support SQL, and are optimized to provide velocity of processing, at least for relationship-intensive data. However, other key characteristics do not apply to graph databases. Graph databases do not scale

graph database
A NoSQL database model based on graph theory that stores data on relationship-rich data as a collection of nodes and edges.

node
In a graph database, the representation of a single entity instance.

edge
In a graph database, the representation of a relationship between nodes.

properties
In a graph database, the attributes or characteristics of a node or edge that are of interest to the users.

traversal
A query in a graph database.

FIGURE 14.11 GRAPH DATABASE REPRESENTATION

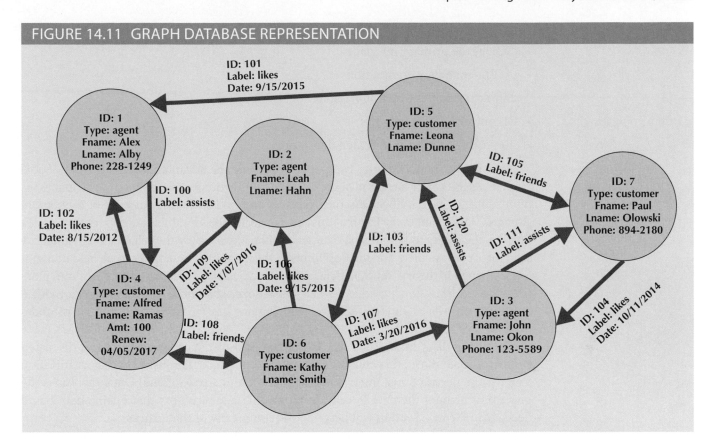

out very well to clusters. The other NoSQL database models achieve clustering efficiency by making each piece of data relatively independent. That allows a key-value pair to be stored on one node in the cluster without the DBMS needing to associate it with another key-value pair that may be on a different node on the cluster. The greater the number of nodes involved in a data operation, the greater the need for coordination and centralized control of resources. Separating independent pieces of data, often called *shards*, across nodes in the cluster is what allows NoSQL databases to scale out so effectively. Graph databases specialize in highly related data, not independent pieces of data. As a result, graph databases tend to perform best in centralized or lightly clustered environments, similar to relational databases.

14-3e NewSQL Databases

Relational databases are the mainstay of organizational data, and NoSQL databases do not attempt to replace them for supporting line-of-business transactions. These transactions that support the day-to-day operations of business rely on ACID-compliant transactions and concurrency control, as discussed in Chapter 10. NoSQL databases (except graph databases that focus on specific relationship-rich domains) are concerned with the distribution of user-generated and machine-generated data over massive clusters. NewSQL databases try to bridge the gap between RDBMS and NoSQL. **NewSQL** databases attempt to provide ACID-compliant transactions over a highly distributed infrastructure. NewSQL databases are the latest technologies to appear in the data management arena to address Big Data problems. As a new category of data management products, NewSQL databases have not yet developed a track record of success and have been adopted by relatively few organizations.

NewSQL products, such as ClusterixDB and NuoDB, are designed from scratch as hybrid products that incorporate features of relational databases and NoSQL databases.

NewSQL
A database model that attempts to provide ACID-compliant transactions across a highly distributed infrastructure.

Like RDBMS, NewSQL databases support:

- SQL as the primary interface
- ACID-compliant transactions

Similar to NoSQL, NewSQL databases also support:

- Highly distributed clusters
- Key-value or column-oriented data stores

As expected, no technology can perfectly provide the advantages of both RDBMS and NoSQL, so NewSQL has disadvantages (the CAP theorem still applies!). Principally, the disadvantages that have been discovered center around NewSQL's heavy use of in-memory storage. Critics point to the fact that this can jeopardize the "durability" component of ACID. Further, the ability to handle vast data sets can be impacted by the reliance on in-memory structures because there are practical limits to the amount of data that can be held in memory. Although in theory NewSQL databases should be able to scale out significantly, in practice little has been done to scale beyond a few dozen data nodes. While this is a marked improvement over traditional RDBMS distribution, it is far from the hundreds of nodes used by NoSQL databases.

Capturing data, in and of itself, is not the goal of data management. As discussed earlier, the data must add value to the organization. The data must help the organization to meet the needs of customers and provide value to shareholders. Data analysis is the process of turning the data into information that adds insights that enable data-based decisions. The next section will describe the complexity of that process.

14-4 Data Analytics

Data analytics is a subset of business intelligence (BI) functionality that encompasses a wide range of mathematical, statistical, and modeling techniques with the purpose of extracting knowledge from data. Data analytics is used at all levels within the BI framework, including queries and reporting, monitoring and alerting, and data visualization. Hence, data analytics is a "shared" service that is crucial to what BI adds to an organization. Data analytics represents what business managers really want from BI: the ability to extract actionable business insight from current events and foresee future problems or opportunities.

Data analytics discovers characteristics, relationships, dependencies, or trends in the organization's data, and then explains the discoveries and predicts future events based on the discoveries. In practice, data analytics is better understood as a continuous spectrum of knowledge acquisition that goes from *discovery* to *explanation* to *prediction*. The outcomes of data analytics then become part of the information framework on which decisions are built. Data analytics tools can be grouped into two separate (but closely related and often overlapping) areas:

- **Explanatory analytics** focuses on discovering and explaining data characteristics and relationships based on existing data. Explanatory analytics uses statistical tools to formulate hypotheses, test them, and answer the *how* and *why* of such relationships—for example, how do past sales relate to previous customer promotions?

- **Predictive analytics** focuses on *predicting future data outcomes* with a high degree of accuracy. Predictive analytics uses sophisticated statistical tools to help the end user create advanced models that answer questions about future data occurrences—for example, what would next month's sales be based on a given customer promotion?

data analytics
A subset of business intelligence functionality that encompasses a wide range of mathematical, statistical, and modeling techniques with the purpose of extracting knowledge from data.

explanatory analytics
Data analysis that provides ways to discover relationships, trends, and patterns among data.

predictive analytics
Data analytics that use advanced statistical and modeling techniques to predict future business outcomes with great accuracy.

You can think of explanatory analytics as explaining the past and present, while predictive analytics forecasts the future. However, you need to understand that both sciences work together; predictive analytics uses explanatory analytics as a stepping stone to create predictive models.

Data analytics has evolved over the years from simple statistical analysis of business data to dimensional analysis with OLAP tools, and then from data mining that discovers data patterns, relationships, and trends to its current status of predictive analytics. The next sections illustrate the basic characteristics of data mining and predictive analytics.

14-4a Data Mining

Data mining refers to analyzing massive amounts of data to uncover hidden trends, patterns, and relationships; to form computer models to simulate and explain the findings; and then to use such models to support business decision making. In other words, data mining focuses on the discovery and explanation stages of knowledge acquisition.

To put data mining in perspective, look at the pyramid in Figure 14.12, which represents how knowledge is extracted from data. *Data* forms the pyramid base and represents what most organizations collect in their operational databases. The second level contains *information* that represents the purified and processed data. Information forms the basis for decision making and business understanding. *Knowledge* is found at the pyramid's apex and represents highly distilled information that provides concise, actionable business insight.

FIGURE 14.12 EXTRACTING KNOWLEDGE FROM DATA

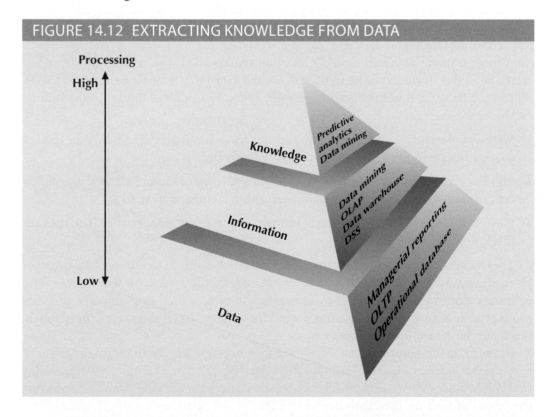

Current-generation data-mining tools contain many design and application variations to fit specific business requirements. Depending on the problem domain, data-mining tools focus on market niches such as banking, insurance, marketing, retailing, finance, and health care. Within a given niche, data-mining tools can use certain algorithms that are implemented in different ways and applied over different data. Despite the lack of precise standards, data mining consists of four general phases:

data mining
A process that employs automated tools to analyze data in a data warehouse and other sources and to proactively identify possible relationships and anomalies.

- Data preparation
- Data analysis and classification
- Knowledge acquisition
- Prognosis

In the *data preparation phase*, the main data sets to be used by the data-mining operation are identified and cleansed of any data impurities. Because the data in the data warehouse is already integrated and filtered, the data warehouse usually is the target set for data-mining operations.

The *data analysis and classification phase* studies the data to identify common data characteristics or patterns. During this phase, the data-mining tool applies specific algorithms to find:

- Data groupings, classifications, clusters, or sequences
- Data dependencies, links, or relationships
- Data patterns, trends, and deviations

The *knowledge acquisition phase* uses the results of the data analysis and classification phase. During the knowledge acquisition phase, the data-mining tool (with possible intervention by the end user) selects the appropriate modeling or knowledge acquisition algorithms. The most common algorithms used in data mining are based on neural networks, decision trees, rules induction, genetic algorithms, classification and regression trees, memory-based reasoning, and nearest neighbor. A data-mining tool may use many of these algorithms in any combination to generate a computer model that reflects the behavior of the target data set.

Although many data-mining tools focus on the knowledge–discovery phase, others continue to the *prognosis phase*. In that phase, the data-mining findings are used to predict future behavior and forecast business outcomes. Examples of data-mining findings can be:

- Sixty-five percent of customers who did not use a particular credit card in the last six months are 88 percent likely to cancel that account.

- Eighty-two percent of customers who bought a 42-inch or larger LCD TV are 90 percent likely to buy an entertainment center within the next four weeks.

- If age < 30, income <= 25,000, credit rating < 3, and credit amount > 25,000, then the minimum loan term is 10 years.

The complete set of findings can be represented in a decision tree, a neural network, a forecasting model, or a visual presentation interface that is used to project future events or results. For example, the prognosis phase might project the likely outcome of a new product rollout or a new marketing promotion. Figure 14.13 illustrates the different phases of the data-mining process.

Because of the nature of the data-mining process, some findings might fall outside the boundaries of what business managers expect. For example, a data-mining tool might find a close relationship between a customer's favorite brand of soda and the brand of tires on the customer's car. Clearly, that relationship might not be held in high regard among sales managers. (In regression analysis, those relationships are commonly described by the label "idiot correlation.") Fortunately, data mining usually yields more meaningful results. In fact, data mining has proven helpful in finding practical relationships among data that help define customer buying patterns, improve product development and acceptance, reduce health care fraud, analyze stock markets, and so on.

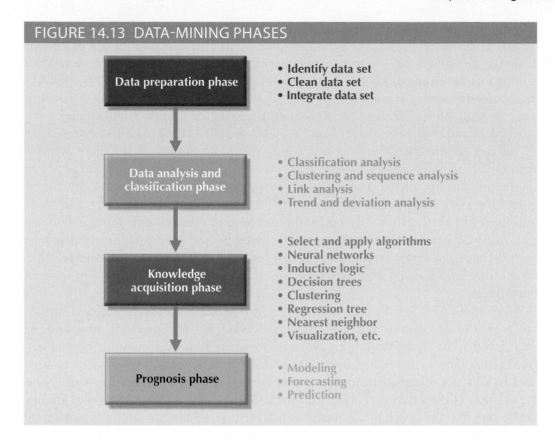

FIGURE 14.13 DATA-MINING PHASES

Data mining can be run in two modes:

- *Guided.* The end user guides the data-mining tool step by step to explore and explain known patterns or relationships. In this mode, the end user decides what techniques to apply to the data.

- *Automated.* In this mode, the end user sets up the data-mining tool to run automatically and uncover hidden patterns, trends, and relationships. The data-mining tool applies multiple techniques to find significant relationships.

As you learned in this section, data-mining methodologies focus on discovering and extracting information that describes and explains the data. For example, an explanatory model could create a customer profile that describes a given customer group. However, data mining can also be used as the basis to create advanced predictive data models. For example, a predictive model could be used to predict future customer behavior, such as a customer response to a target marketing campaign. The next section explains the use of predictive analytics in more detail.

14-4b Predictive Analytics

Although the term *predictive analytics* is used by many BI vendors to indicate many different levels of functionality, the promise of predictive analytics is very attractive for businesses looking for ways to improve their bottom line. Therefore, predictive analytics is receiving a lot of marketing buzz; vendors and businesses are dedicating extensive resources to this BI area. Predictive analytics refers to the use of advanced mathematical, statistical, and modeling tools to predict future business outcomes with high degrees of accuracy.

What is the difference between data mining and predictive analytics? As you learned earlier, data mining also has predictive capabilities. In fact, data mining and predictive analytics use similar and overlapping sets of tools, but with a slightly different focus. Data mining focuses on answering the "how" and "what" of *past* data, while predictive analytics focuses on creating actionable models to predict *future* behaviors and events. In some ways, you can think of predictive analytics as the next logical step after data mining; once you understand your data, you can use the data to predict future behaviors. In fact, most BI vendors are dropping the term *data mining* and replacing it with the more alluring term *predictive analytics*.

The origins of predictive analytics can be traced back to the banking and credit card industries. The need to profile customers and predict customer buying patterns in these industries was a critical driving force for the evolution of many modeling methodologies used in BI data analytics today. For example, based on your demographic information and purchasing history, a credit card company can use data-mining models to determine what credit limit to offer, what offers you are more likely to accept, and when to send those offers.

Predictive analytics received a big stimulus with the advent of social media. Companies turned to data mining and predictive analytics as a way to harvest the mountains of data stored on social media sites. Google was one of the first companies that offered targeted ads as a way to increase and personalize search experiences. Similar initiatives were used by all types of organizations to increase customer loyalty and drive up sales. Note the example of the airline and credit card industries and their frequent flyer and affinity card programs. Today, many organizations use predictive analytics to profile customers in an attempt to get and keep the right ones, which in turn will increase loyalty and sales.

Predictive analytics employs mathematical and statistical algorithms, neural networks, artificial intelligence, and other advanced modeling tools to create actionable predictive models based on available data. The algorithms used to build the predictive model are specific to certain types of problems and work with certain types of data. Therefore, it is important that the end user, who typically is trained in statistics and understands business, applies the proper algorithms to the problem in hand. However, thanks to constant technology advances, modern BI tools automatically apply multiple algorithms to find the optimum model.

Most predictive analytics models are used in areas such as customer relationships, customer service, customer retention, fraud detection, targeted marketing, and optimized pricing. Predictive analytics can add value to an organization in many different ways. For example, it can help optimize existing processes, identify hidden problems, and anticipate future problems or opportunities. However, predictive analytics is not the "secret sauce" to fix all business problems. Managers should carefully monitor and evaluate the value of predictive analytics models to determine their return on investment.

In Chapter 13, you learned about data warehouses and star schemas to model and store decision support data. In this chapter, you have added to that by exploring the vast stores of data that organizations are collecting in unstructured formats and the technologies that make that data available to users. Data analytics is used to extract knowledge from all of these sources of data—NoSQL databases, Hadoop data stores, and data warehouses—to provide decision support to all organizational users.

Summary

- Big Data is characterized by data of such volume, velocity, and/or variety that the relational model struggles to adapt to it. Volume refers to the quantity of data that must be stored. Velocity refers to both the speed at which data is entering storage as well as the speed with which it must be processed. Variety refers to the lack of uniformity in the structure of the data being stored. As a result of Big Data, organizations are having to employ a variety of data storage solutions that include technologies in addition to relational databases, a situation referred to as polyglot persistence.

- Volume, velocity, and variety are collectively referred to as the 3 Vs of Big Data. However, these are not the only characteristics of Big Data to which data administrators must be sensitive. Additional Vs that have been suggested by the data management industry include variability, veracity, value, and visualization. Variability is the variation in the meaning of data that can occur over time. Veracity is the trustworthiness of the data. Value is concerned with whether or not the data is useful. Finally, visualization is the requirement that the data must be able to be presented in a manner that makes it comprehendible to decision makers. Most of these additional Vs are not unique to Big Data. They are also concerns for data in relational databases as well.

- The Hadoop framework has quickly emerged as a standard for the physical storage of Big Data. The primary components of the framework include the Hadoop Distributed File System (HDFS) and MapReduce. HDFS is a coordinated technology for reliably distributing data over a very large cluster of commodity servers. MapReduce is a complementary process for distributing data processing across distributed data. One of the key concepts for MapReduce is to move the computations to the data instead of moving the data to the computations. MapReduce works by combining the functions of *map*, which distributes subtasks to the cluster servers that hold data to be processed, and *reduce*, which combines the map results into a single result set. The Hadoop framework also supports an entire ecosystem of additional tools and technologies, such as Hive, Pig, and Flume that work together to produce a complex system of Big Data processing.

- NoSQL is a broad term to refer to any of several nonrelational database approaches to data management. Most NoSQL databases fall into one of four categories: key-value databases, document databases, column-oriented databases, or graph databases. Due to the wide variability of products under the NoSQL umbrella, these categories are not necessarily all-encompassing, and many products can fit into multiple categories.

- Key-value databases store data in key-value pairs. In a key-value pair, the value of the key must be known to the DBMS, but the data in the value component can be of any type, and the DBMS makes no attempt to understand the meaning of the data in it. These types of databases are very fast when the data is completely independent, and the application programs can be relied on to understand the meaning of the data.

- Document databases also store data in key-value pairs, but the data in the value component is an encoded document. The document must be encoded using tags, such as in XML or JSON. The DBMS is aware of the tags in the documents, which makes querying on tags possible. Document databases expect documents to be self-contained and relatively independent of each other.

- Column-oriented databases, also called column family databases, organize data into key-value pairs in which the value component is composed of a series of columns, which are themselves key-value pairs. Columns can be grouped into super columns, similar to a composite attribute in the relational model being composed of simple attributes. All objects of a similar type are identified as rows, given a row key, and placed within a column family. Rows within a column family are not required to have the same structure, that is, they are not required to have the same columns.

- Graph databases are based on graph theory and represent data through nodes, edges, and properties. A node is similar to an instance of an entity in the relational model. Edges are the relationships between nodes. Both nodes and edges can have properties, which are attributes that describe the corresponding node or edge. Graph databases excel at tracking data that is highly interrelated, such as social media data. Due to the many relationships among the nodes, it is difficult to distribute a graph database across a cluster in a highly-distributed manner.

- NewSQL databases attempt to integrate features of both RDBMS (providing ACID-compliant transactions) and NoSQL databases (using a highly distributed infrastructure).

- Data analytics is a subset of BI functionality that provides advanced data analysis tools to extract knowledge from business data. Data analytics can be divided into explanatory and predictive analytics. Explanatory analytics focuses on discovering and explaining data characteristics and relationships. Predictive analytics focuses on creating models to predict future outcomes or events based on the existing data.

- Data mining automates the analysis of operational data to find previously unknown data characteristics, relationships, dependencies, and trends. The data-mining process has four phases: data preparation, data analysis and classification, knowledge acquisition, and prognosis.

- Predictive analytics uses the information generated in the data-mining phase to create advanced predictive models with high degrees of accuracy.

Key Terms

Online Content

Flashcards and crossword puzzles for key term practice are available at *www.cengagebrain.com*.

algorithm	heartbeat	scaling out
batch processing	job tracker	scaling up
block report	JSON (JavaScript Object Notation)	sentiment analysis
BSON (Binary JSON)	key-value (KV) database	stream processing
bucket	map	structured data
column family	MapReduce	super column
column family database	mapper	task tracker
column-centric storage	NewSQL	traversal
data analytics	node	unstructured data
data mining	NoSQL	value
document database	polyglot persistence	variability
edge	predictive analytics	variety
explanatory analytics	properties	velocity
feedback loop processing	reduce	veracity
graph database	reducer	visualization
Hadoop Distributed File System (HDFS)	row-centric storage	volume

Review Questions

1. What is Big Data? Give a brief definition.

2. What are the traditional 3 Vs of Big Data? Briefly, define each.

3. Explain why companies like Google and Amazon were among the first to address the Big Data problem.

4. Explain the difference between *scaling up* and *scaling out*.

5. What is stream processing, and why is it sometimes necessary?

6. How is stream processing different from feedback loop processing?

7. Explain why veracity, value, and visualization can also be said to apply to relational databases as well as Big Data.

8. What is polyglot persistence, and why is it considered a new approach?

9. What are the key assumptions made by the Hadoop Distributed File System approach?

10. What is the difference between a name node and a data node in HDFS?

11. Explain the basic steps in MapReduce processing.

12. Briefly explain how HDFS and MapReduce are complementary to each other.

13. What are the four basic categories of NoSQL databases?

14. How are the value components of a key-value database and a document database different?

15. Briefly explain the difference between row-centric and column-centric data storage.

16. What is the difference between a column and a super column in a column family database?

17. Explain why graph databases tend to struggle with scaling out.

18. What is data analytics? Briefly define explanatory and predictive analytics.

19. Describe and contrast the focus of data mining and predictive analytics. Give some examples.

20. How does data mining work? Discuss the different phases in the data mining process.

21. Describe the characteristics of predictive analytics. What is the impact of Big Data in predictive analytics?

PART 5
Databases and the Internet

Database Connectivity and Web Technologies

In this chapter, you will learn:

- About database connectivity fundamentals
- About various database connectivity technologies: ODBC, OLE, ADO.NET, JDBC
- How web-to-database middleware is used to integrate databases with the Internet
- What services are provided by web application servers
- What Extensible Markup Language (XML) is and why it is important for web database development
- About cloud computing and how it enables the database-as-a-service model

Preview

Databases are the central repository for critical data generated by business applications, including newer channels such as the web and mobile devices. For businesses to remain competitive, such data must be readily available, anywhere and anytime, to all business users and in all types of formats: a desktop spreadsheet, a Visual Basic application, a web front end, and using newer technologies such as smartphones and tablets. In this chapter, you will learn about various architectures used to connect applications to databases.

The Internet has changed how organizations of all types operate. Buying goods and services via the Internet has become commonplace. This chapter examines the fundamentals of web database technologies used to open databases to the Internet. In today's environment, interconnectivity occurs not only between an application and the database but between applications exchanging messages and data. Extensible Markup Language (XML) provides a standard way of exchanging unstructured and structured data between applications.

Companies that want to integrate database and web technologies within their applications portfolio can now choose from a range of Internet-based services. Therefore, you will learn how organizations can benefit from cloud computing by leveraging the database-as-a-service model within their IT environments. These cloud-based services offer a quick and cost-efficient way to provide new business services.

Data Files and Available Formats

	MS Access	Oracle	MS SQL	My SQL		MS Access	Oracle	MS SQL	My SQL
CH15_Orderdb	✓	✓	✓	✓					

Data Files Available on cengagebrain.com

15-1 **Database Connectivity**

Database connectivity refers to the mechanisms through which application programs connect and communicate with data repositories. Databases store data in persistent storage structures so it can be retrieved at a later time for processing. As you already learned, the database management system (DBMS) functions as an intermediary between the data (stored in the database) and the end-user's applications. Before learning about the various data connectivity options, it is important to review some important fundamentals you have learned in this book:

- DBMSs provide means to interact with the data in their databases. This could be in the form of administrative tools and data manipulation tools. DBMSs also provide a proprietary way for external application programs to connect to the database by the means of an application programing interface. See Chapter 1, Database Systems.

- Modern DBMSs have the option to store data locally or distributed in multiple locations. Locally stored data resides in the same processing host as the DBMS. A distributed database stores data in multiple geographically distributed nodes with data management capability. See Chapter 12, Distributed Database Management Systems.

- The database connectivity software we discuss in this chapter supports Structured Query Language (SQL) as the standard data manipulation language. However, depending on the type of database model, some database connectivity interfaces may support other proprietary data manipulation languages.

- Database connectivity software works in a client/server architecture, by which processing tasks are split among multiple software layers. In this model, the multiple layers exchange control messages and data. See Chapter 12 and Appendix F, Client/Server Systems, for more information on this topic.

To better understand database connectivity software, we use client/server concepts in which an application is broken down in interconnected functional layers. In the case of database connectivity software, you could break down its basic functionality into three broad layers:

1. A data layer where the data resides. You could think of this layer as the actual data repository interface. This layer resides closest to the database itself and normally is provided by the DBMS vendor.

2. A middle layer that manages multiple connectivity and data transformation issues. This layer is in charge of dealing with data logic issues, data transformations, ways to "talk" to the database below it, and so on. This would also include translating multiple data manipulation languages to the native language supported by the specific data repository.

3. A top layer that interfaces with the actual external application. This mostly comes in the form of an application programming interface that publishes specific protocols for the external programs to interact with the data.

From the previous discussion, you can understand why the database connectivity software is also known as **database middleware**—because it provides an interface between the application program and the database or data repository. The data repository, also known as the *data source*, represents the data management application, such as Oracle, SQL Server, IBM DB2, or NoSQL that will be used to store the data generated by the application program. Ideally, a data source or data repository could be located anywhere and hold any type of data. Furthermore, the same database connectivity middleware could support multiple data sources at the same time. For example, the data source could be a relational database, a NoSQL database, a spreadsheet, a MS Access database,

database middleware
Database connectivity software through which application programs connect and communicate with data repositories.

or a text data file. This multi-data-source type capability is based on the support of well-established data access standards.

The need for standard database connectivity interfaces cannot be overstated. Just as SQL has become the de facto data manipulation language, a standard database connectivity interface is necessary for enabling applications to connect to data repositories. Although there are many ways to achieve database connectivity, this section covers only the following interfaces:

- Native SQL connectivity (vendor provided)
- Microsoft's Open Database Connectivity (ODBC), Data Access Objects (DAO), and Remote Data Objects (RDO)
- Microsoft's Object Linking and Embedding for Database (OLE-DB)
- Microsoft's ActiveX Data Objects (ADO.NET)
- Oracle's Java Database Connectivity (JDBC)

The data connectivity interfaces illustrated here are dominant players in the market, and more importantly, they enjoy the support of most database vendors. In fact, ODBC, OLE-DB, and ADO.NET form the backbone of Microsoft's **Universal Data Access (UDA)** architecture, a collection of technologies used to access any type of data source and manage the data through a common interface. As you will see, Microsoft's database connectivity interfaces have evolved over time: each interface builds on top of the other, thus providing enhanced functionality, features, flexibility, and support.

15-1a Native SQL Connectivity

Most DBMS vendors provide their own methods for connecting to their databases. Native SQL connectivity refers to the connection interface that is provided by the database vendor and is unique to that vendor. The best example of this type of native interface is the Oracle RDBMS. To connect a client application to an Oracle database, you must install and configure Oracle's SQL*Net interface on the client computer. Figure 15.1 shows the configuration of the Oracle SQL*Net interface on the client computer.

Native database connectivity interfaces are optimized for "their" DBMS, and those interfaces support access to most or all of the database features. However, maintaining

Universal Data Access (UDA)
Within the Microsoft application framework, a collection of technologies used to access any type of data source and to manage the data through a common interface.

FIGURE 15.1 ORACLE NATIVE CONNECTIVITY

multiple native interfaces for different databases can become a burden for the programmer. Therefore, the need for universal database connectivity arises. Usually, the native database connectivity interface provided by the vendor is not the only way to connect to a database; most current DBMS products support other database connectivity standards, the most common being ODBC.

15-1b ODBC, DAO, and RDO

Developed in the early 1990s, **Open Database Connectivity (ODBC)** is Microsoft's implementation of a superset of the SQL Access Group **Call Level Interface (CLI)** standard for database access. ODBC is probably the most widely supported database connectivity interface. ODBC allows any Windows application to access relational data sources, using SQL via a standard **application programming interface (API)**. The Webopedia online dictionary (*www.webopedia.com*) defines an API as "a set of routines, protocols, and tools for building software applications." A good API makes it easy to develop a program by providing all of the building blocks; the programmer puts the blocks together. Most operating environments, such as Windows, provide an API so that programmers can write applications consistent with the operating environment. Although APIs are designed for programmers, they are ultimately good for users because they guarantee that all programs using a common API will have similar interfaces. That makes it easy for users to learn new programs.

ODBC was the first widely adopted database middleware standard, and it enjoyed rapid adoption in Windows applications. As programming languages evolved, ODBC did not provide significant functionality beyond the ability to execute SQL to manipulate relational-style data. Therefore, programmers needed a better way to access data. To answer that need, Microsoft developed two other data access interfaces:

- **Data Access Objects (DAO)** is an object-oriented API used to access desktop databases, such as MS Access and FileMaker Pro. DAO provides an optimized interface that exposes programmers to the functionality of the Jet data engine, on which MS Access is based. The DAO interface can also be used to access other relational-style data sources.

- **Remote Data Objects (RDO)** is a higher-level, object-oriented application interface used to access remote database servers. RDO uses the lower-level DAO and ODBC for direct access to databases. RDO is optimized to deal with server-based databases such as MS SQL Server, Oracle, and DB2.

Figure 15.2 illustrates how Windows applications can use ODBC, DAO, and RDO to access local and remote relational data sources.

The DAO and RDO object interfaces provide more functionality than ODBC. DAO and RDO make use of the underlying ODBC data services. ODBC, DAO, and RDO are implemented as shared code that is dynamically linked to the Windows operating environment through **dynamic-link libraries (DLLs)**, which are stored as files with a .dll extension. Running as a DLL, the code speeds up load and run times.

The basic ODBC architecture has three main components:

- A high-level *ODBC API* through which application programs access ODBC functionality

- A *driver manager* that is in charge of managing all database connections

- An *ODBC driver* that communicates directly to the DBMS

Defining a data source is the first step in using ODBC. To define a data source, you must create a **data source name (DSN)** for it. To create a DSN, you need to provide the following:

- *An ODBC driver*. You must identify the driver to use to connect to the data source. The ODBC driver is normally provided by the database vendor, although Microsoft provides several drivers that connect to most common databases. For example, if you

Open Database Connectivity (ODBC) Microsoft database middleware that provides a database access API to Windows applications.

Call Level Interface (CLI) A standard developed by the SQL Access Group for database access.

application programming interface (API) Software through which programmers interact with middleware. An API allows the use of generic SQL code, thereby allowing client processes to be database server-independent.

Data Access Objects (DAO) An object-oriented application programming interface used to access MS Access, FileMaker Pro, and other Jet-based databases.

Remote Data Objects (RDO) A higher-level, object-oriented application interface used to access remote database servers. RDO uses the lower-level DAO and ODBC for direct access to databases.

dynamic-link library (DLL) Shared code module that is treated as part of the operating system or server process so it can be dynamically invoked at run time.

data source name (DSN) A name that identifies and defines an ODBC data source.

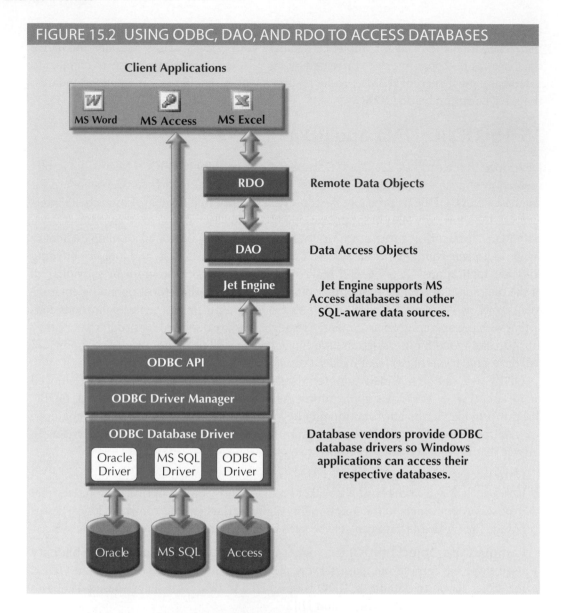

FIGURE 15.2 USING ODBC, DAO, AND RDO TO ACCESS DATABASES

are using an Oracle DBMS, you would select the Oracle ODBC driver provided by Oracle. Or, you could instead select the Microsoft-provided ODBC driver for Oracle.

- *A name.* This is a unique name by which the data source will be known to ODBC, and therefore to applications. ODBC offers two types of data sources: user and system. *User data sources* are available only to the user. *System data sources* are available to all users, including operating system services.

- *ODBC driver parameters.* Most ODBC drivers require specific parameters to establish a connection to the database. For example, if you are using an MS Access database, you must point to the location of the MS Access file and then provide a username and password if necessary. If you are using a DBMS server, you must provide the server name, the database name, the username, and the password needed to connect to the database. Figure 15.3 shows the ODBC screens required to create a system ODBC data source for an Oracle DBMS. Note that some ODBC drivers use the native driver provided by the DBMS vendor.

Once the ODBC data source is defined, application programmers can write to the ODBC API by issuing specific commands and providing the required parameters. The ODBC Driver Manager will properly route the calls to the appropriate data source.

FIGURE 15.3 CONFIGURING AN ORACLE ODBC DATA SOURCE

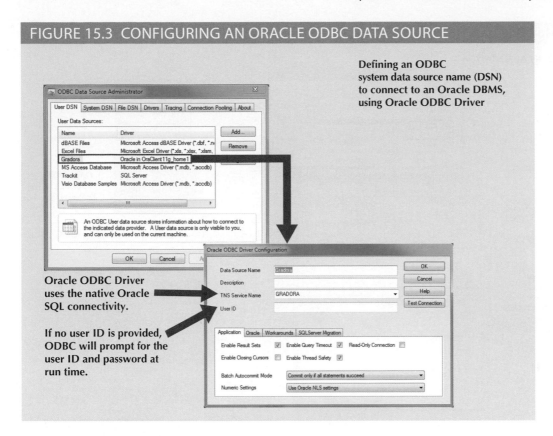

The ODBC API standard defines three levels of compliance: Core, Level-1, and Level-2, which provide increasing levels of functionality. For example, Level-1 might provide support for most SQL DDL and DML statements, including subqueries and aggregate functions, but not for procedural SQL or cursors. The database vendors can choose which level to support. However, to interact with ODBC, the database vendor must implement all of the features in the specified ODBC API support level.

Figure 15.4 shows how you could use MS Excel to retrieve data from an Oracle RDBMS using ODBC. Because much of the functionality provided by these interfaces is oriented toward accessing relational data sources, the use of the interfaces was limited with other data source types. With the advent of object-oriented programming languages, it has become more important to provide access to other nonrelational data sources.

15-1c OLE-DB

Although ODBC, DAO, and RDO are used, they do not provide support for nonrelational data. To answer that need and to simplify data connectivity, Microsoft developed **Object Linking and Embedding for Database (OLE-DB)**. Based on Microsoft's Component Object Model (COM), OLE-DB is database middleware that adds object-oriented functionality for access to relational and nonrelational data. OLE-DB was the first part of Microsoft's strategy to provide a unified object-oriented framework for the development of next-generation applications.

OLE-DB is composed of a series of COM objects that provide low-level database connectivity for applications. Because OLE-DB is based on COM, the objects contain data and methods, also known as the interface. The OLE-DB model is better understood when you divide its functionality into two types of objects:

- *Consumers* are objects (applications or processes) that request and use data. Consumers request data by invoking the methods exposed by the data provider objects (public interface) and passing the required parameters.

Object Linking and Embedding for Database (OLE-DB) Based on Microsoft's Component Object Model (COM), OLE-DB is database middleware that adds object-oriented functionality for accessing relational and nonrelational data.

FIGURE 15.4 MS EXCEL USES ODBC TO CONNECT TO AN ORACLE DATABASE

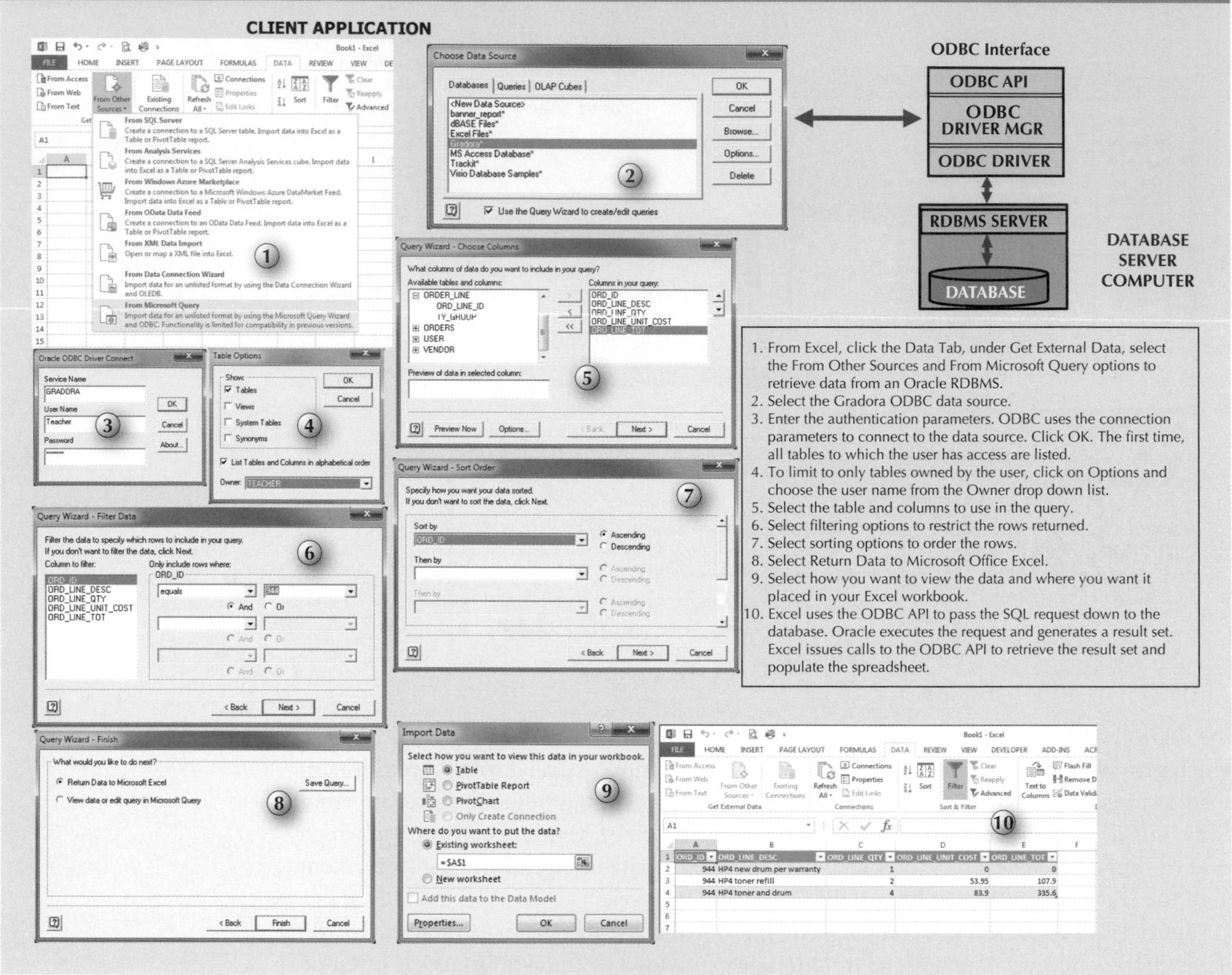
CLIENT APPLICATION
ODBC Interface
ODBC API
ODBC DRIVER MGR
ODBC DRIVER
RDBMS SERVER
DATABASE
DATABASE SERVER COMPUTER
1. From Excel, click the Data Tab, under Get External Data, select the From Other Sources and From Microsoft Query options to retrieve data from an Oracle RDBMS.
2. Select the Gradora ODBC data source.
3. Enter the authentication parameters. ODBC uses the connection parameters to connect to the data source. Click OK. The first time, all tables to which the user has access are listed.
4. To limit to only tables owned by the user, click on Options and choose the user name from the Owner drop down list.
5. Select the table and columns to use in the query.
6. Select filtering options to restrict the rows returned.
7. Select sorting options to order the rows.
8. Select Return Data to Microsoft Office Excel.
9. Select how you want to view the data and where you want it placed in your Excel workbook.
10. Excel uses the ODBC API to pass the SQL request down to the database. Oracle executes the request and generates a result set. Excel issues calls to the ODBC API to retrieve the result set and populate the spreadsheet.

- *Providers* are objects that manage the connection with a data source and provide data to the consumers. Providers are divided into two categories: data providers and service providers.
  - *Data providers* provide data to other processes. Database vendors create data provider objects that expose the functionality of the underlying data source (relational, object-oriented, text, and so on).
  - *Service providers* provide additional functionality to consumers. The service provider is located between the data provider and the consumer. The service provider requests data from the data provider, transforms the data, and then provides the transformed data to the data consumer. In other words, the service provider acts like a data consumer of the data provider and as a data provider for the data consumer (end-user application). For example, a service provider could offer cursor management services, transaction management services, query processing services, and indexing services.

As a common practice, many vendors provide OLE-DB objects to augment their ODBC support, effectively creating a shared object layer on top of their existing database connectivity (ODBC or native) through which applications can interact. The OLE-DB objects expose functionality about the database; for example, there are objects that deal with relational data, hierarchical data, and flat-file text data. Additionally, the objects implement specific tasks, such as establishing a connection, executing a query, invoking a stored procedure, defining a transaction, or invoking an OLAP function. By using OLE-DB objects, the database vendor can choose what functionality to implement in a modular way, instead of being forced to include all of the functionality all of the time. Table 15.1 shows a sample of the object-oriented classes used by OLE-DB and some of the methods (interfaces) exposed by the objects.

TABLE 15.1

SAMPLE OLE-DB CLASSES AND INTERFACES

OBJECT CLASS	USAGE	SAMPLE INTERFACES
Session	Used to create an OLE-DB session between a data consumer application and a data provider	IGetDataSource
Command	Used to process commands to manipulate a data provider's data; generally, the command object will create RowSet objects to hold the data returned by a data provider	ICommandPrepare
RowSet	Used to hold the result set returned by a relational-style database or a database that supports SQL; represents a collection of rows in a tabular format	IRowsetInfo IRowsetFind IRowsetScroll

OLE-DB provides additional capabilities for the applications accessing the data. However, it does not provide support for scripting languages, especially the ones used for web development, such as Active Server Pages (ASP) and ActiveX. (A **script** is written in a programming language that is not compiled but is interpreted and executed at run time.) To provide that support, Microsoft developed a new object framework called **ActiveX Data Objects (ADO)**, which provides a high-level, application-oriented interface to interact with OLE-DB, DAO, and RDO. ADO provides a unified interface to access data from any programming language that uses the underlying OLE-DB objects. Figure 15.5 illustrates the ADO/OLE-DB architecture and how it interacts with ODBC and native connectivity options.

ADO introduced a simpler object model that was composed of only a few interacting objects to provide the data manipulation services required by the applications. Sample objects in ADO are shown in Table 15.2.

Although the ADO model is a tremendous improvement over the OLE-DB model, Microsoft is actively encouraging programmers to use its newer data access framework, ADO.NET.

15-1d ADO.NET

Based on ADO, **ADO.NET** is the data access component of Microsoft's .NET application development framework. The **Microsoft .NET framework** is a component-based platform for developing distributed, heterogeneous, interoperable applications aimed at manipulating any type of data using any combination of network, operating system, and programming language. Comprehensive coverage of the .NET framework is beyond the scope of this book. Therefore, this section only introduces the basic data access component of the .NET architecture, ADO.NET.

It is important to understand that the .NET framework extends and enhances the functionality provided by the ADO/OLE-DB duo. ADO.NET introduced two new features that are critical for the development of distributed applications: DataSets and XML support.

script
A programming language that is not compiled, but is interpreted and executed at run time.

ActiveX Data Objects (ADO)
A Microsoft object framework that provides a high-level, application-oriented interface to OLE-DB, DAO, and RDO. ADO provides a unified interface to access data from any programming language that uses the underlying OLE-DB objects.

ADO.NET
The data access component of Microsoft's .NET application development framework.

Microsoft .NET framework
A component-based platform for the development of distributed, heterogeneous, interoperable applications aimed at manipulating any type of data over any network regardless of operating system and programming language.

FIGURE 15.5 OLE-DB ARCHITECTURE

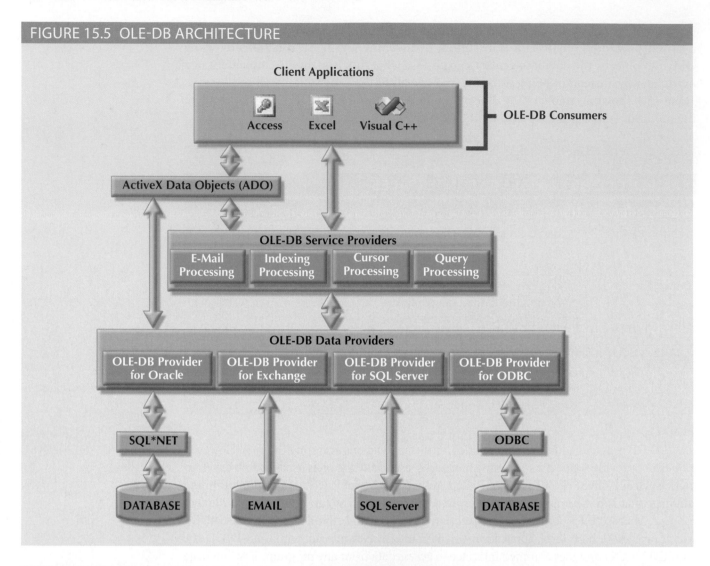

TABLE 15.2

SAMPLE ADO OBJECTS

OBJECT CLASS	USAGE
Connection	Used to set up and establish a connection with a data source. ADO will connect to any OLE-DB data source. The data source can be of any type.
Command	Used to execute commands against a specific connection (data source)
Recordset	Contains the data generated by the execution of a command. It will also contain any new data to be written to the data source. The Recordset can be disconnected from the data source.
Fields	Contains a collection of field descriptions for each column in the Recordset

DataSet
In ADO.NET, a disconnected, memory-resident representation of the database. The DataSet contains tables, columns, rows, relationships, and constraints.

To understand the importance of this new model, you should know that a **DataSet** is a disconnected, memory-resident representation of the database. That is, the DataSet contains tables, columns, rows, relationships, and constraints. Once the data is read from a data provider, it is placed in a memory-resident DataSet, which is then disconnected from the data provider. The data consumer application interacts with the data in the DataSet object to make inserts, updates, and deletes in the DataSet. Once the processing is done, the DataSet data is synchronized with the data source and the changes are made permanent.

The DataSet is internally stored in XML format, and the data in the DataSet can be made persistent as XML documents. This is critical in today's distributed environments. You can think of the DataSet as an XML-based, in-memory database that represents the persistent data stored in the data source. (You will learn about XML later in this chapter.)

Figure 15.6 illustrates the main components of the ADO.NET object model.

FIGURE 15.6 ADO.NET FRAMEWORK

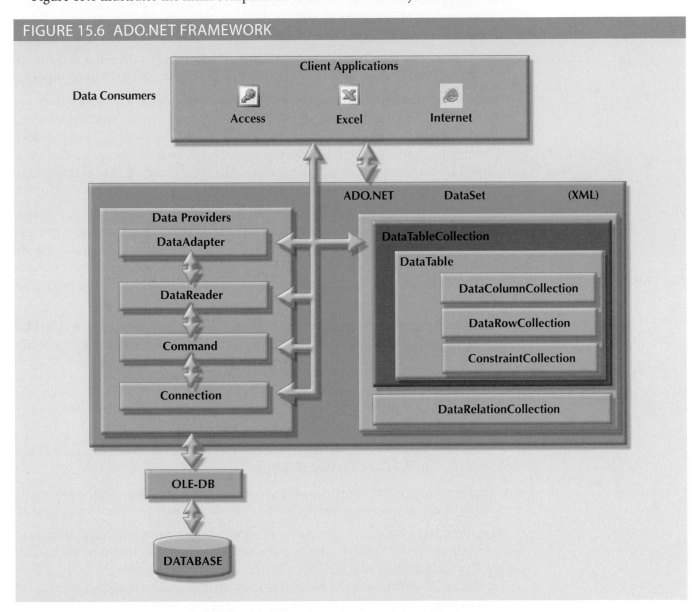

The ADO.NET framework consolidates all data access functionality under one integrated object model. In this object model, several objects interact with one another to perform specific data manipulations. These objects can be grouped as data providers and consumers.

Data provider objects are provided by the database vendors. However, ADO.NET comes with two standard data providers: one for OLE-DB data sources and one for SQL Server. That way, ADO.NET can work with any previously supported database, including an ODBC database with an OLE-DB data provider. At the same time, ADO.NET includes a highly optimized data provider for SQL Server.

Whatever the data provider is, it must support a set of specific objects to manipulate the data in the data source. Some of those objects are shown in Figure 15.6. A brief description of the objects follows.

- *Connection.* The Connection object defines the data source used, the name of the server, the database, and so on. This object enables the client application to open and close a connection to a database.

- *Command.* The Command object represents a database command to be executed within a specified database connection. This object contains the actual SQL code or a stored procedure call to be run by the database. When a SELECT statement is executed, the Command object returns a set of rows and columns.

- *DataReader.* The DataReader object is a specialized object that creates a read-only session with the database to retrieve data sequentially (forward only) and very quickly.

- *DataAdapter.* The DataAdapter object is in charge of managing a DataSet object, and it is the most specialized object in the ADO.NET framework. The DataAdapter object contains the following objects that aid in managing the data in the DataSet: Select-Command, InsertCommand, UpdateCommand, and DeleteCommand. The Data-Adapter object uses these objects to populate and synchronize the data in the DataSet with the permanent data source data.

- *DataSet.* The DataSet object is the in-memory representation of the data in the database. This object contains two main objects. The DataTableCollection object contains a collection of DataTable objects that make up the "in-memory" database, and the DataRelationCollection object contains a collection of objects that describe the data relationships and ways to associate one row in a table to the related row in another table.

- *DataTable.* The DataTable object represents the data in tabular format. This object has one very important property: PrimaryKey, which allows the enforcement of entity integrity. In turn, the DataTable object is composed of three main objects:

  - *DataColumnCollection* contains one or more column descriptions. Each column description has properties such as column name, data type, nulls allowed, maximum value, and minimum value.

  - *DataRowCollection* contains zero rows, one row, or more than one row with data as described in the DataColumnCollection.

  - *ConstraintCollection* contains the definition of the constraints for the table. Two types of constraints are supported: ForeignKeyConstraint and UniqueConstraint.

As you can see, a DataSet is a simple database with tables, rows, and constraints. Even more importantly, the DataSet does not require a permanent connection to the data source. The DataAdapter uses the SelectCommand object to populate the DataSet from a data source. However, once the DataSet is populated, it is completely independent of the data source, which is why it is called disconnected.

Additionally, DataTable objects in a DataSet can come from different data sources. This means that you could have an EMPLOYEE table in an Oracle database and a SALES table in a SQL Server database. You could then create a DataSet that relates both tables as though they were in the same database. In short, the DataSet object paves the way for truly heterogeneous, distributed database support within applications.

The ADO.NET framework is optimized to work in disconnected environments. In a disconnected environment, applications exchange messages in request/reply format. The most common example of a disconnected system is the Internet. Modern applications rely on the Internet as the network platform and on the web browser as the graphical user interface. In later sections, you will learn about how Internet databases work.

15-1e Java Database Connectivity (JDBC)

Java is an object-oriented programming language developed by Sun Microsystems (acquired by Oracle in 2010) that runs on top of web browser software. Java is one of the most-common programming languages for web development. Sun Microsystems created Java as a "write once, run anywhere" environment, which means that a programmer can write a Java application once and then run it in multiple environments without any modification. The cross-platform capabilities of Java are based on its portable architecture. Java code is normally stored in preprocessed "chunks" known as applets that run in a virtual machine environment in the host operating system. This environment has well-defined boundaries, and all interactivity with the host operating system is closely monitored. Java run-time environments are available for most operating systems, from computers to handheld mobile devices to TV set-top boxes. Another advantage of using Java is its "on-demand" architecture. When a Java application loads, it can dynamically download all its modules or required components via the Internet.

When Java applications need to access data outside the Java runtime environment, they use predefined application programming interfaces. **Java Database Connectivity (JDBC)** is an application programming interface that allows a Java program to interact with a wide range of data sources, including relational databases, tabular data sources, spreadsheets, and text files. JDBC allows a Java program to establish a connection with a data source, prepare and send the SQL code to the database server, and process the result set.

One main advantage of JDBC is that it allows a company to leverage its existing investment in technology and personnel training. JDBC allows programmers to use their SQL skills to manipulate the data in the company's databases. As a matter of fact, JDBC allows direct access to a database server or access via database middleware. Furthermore, JDBC provides a way to connect to databases through an ODBC driver. Figure 15.7 illustrates the basic JDBC architecture and the various database access styles.

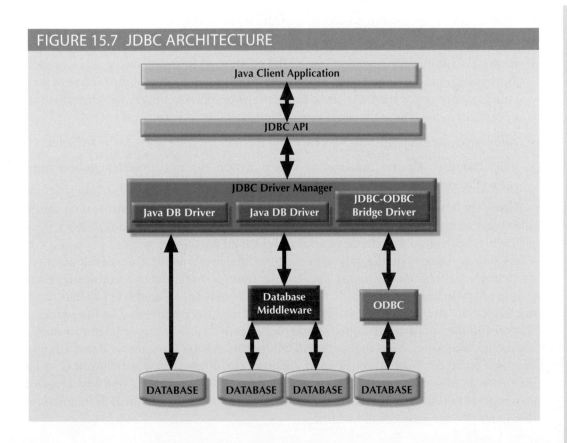

FIGURE 15.7 JDBC ARCHITECTURE

Java
An object-oriented programming language developed by Sun Microsystems that runs on top of the web browser software. Java applications are compiled and stored on the web server. Java's main advantage is its ability to let application developers create their applications once and then run them in many environments.

Java Database Connectivity (JDBC)
An application programming interface that allows a Java program to interact with a wide range of data sources, including relational databases, tabular data sources, spreadsheets, and text files.

As you see in Figure 15.7, the database access architecture in JDBC is very similar to the ODBC/OLE/ADO.NET architecture. All database access middleware shares similar components and functionality. One advantage of JDBC over other middleware is that it requires no configuration on the client side. The JDBC driver is automatically downloaded and installed as part of the Java applet download. Because Java is a web-based technology, applications can connect to a database directly using a simple URL. Once the URL is invoked, the Java architecture comes into play, the necessary applets are downloaded to the client (including the JDBC database driver and all configuration information), and then the applets are executed securely in the client's runtime environment. This framework is used successfully in many vertical database markets, in particular in the rapidly growing data analytics market, where open source players like Hadoop and MapReduce provide end-users with advanced application programming interfaces to high-performance data analytics functions using large-scale clusters of interconnected data stores.

Every day, more and more companies are investing resources to develop and expand their web presence and are finding ways to do more business on the Internet. Such business generates increasing amounts of data to be stored in databases. Java and the .NET framework are part of the trend toward increasing reliance on the Internet as a critical business resource. In fact, the Internet has become a major development platform for most businesses. In the next section, you will learn more about Internet databases and how they are used.

15-2 **Database Internet Connectivity**

Millions of people all over the world access the Internet and connect to databases via web browsers or data services. For example, they can use a smartphone app to get weather forecasts, stock prices, driving directions, concert tickets, or music downloads. Internet database connectivity opens the door to new, innovative services that do the following:

- Permit rapid responses to competitive pressures by bringing new services and products to market quickly.

- Increase customer satisfaction through the creation of innovative data services such as mapping data combined with GPS (Global Positioning System) information to provide location-aware services. These applications present end users with information or services located near the users' current location.

- Allow anywhere, anytime data access using mobile smart devices via the Internet.

- Yield fast and effective information dissemination through universal access from across the street or across the globe.

Given these advantages, many organizations rely on their IT departments to create universal data access architectures based on Internet standards. Table 15.3 shows a sample of Internet technology characteristics and the benefits they provide.

As you will learn in the following sections, database application development—particularly the creation and management of user interfaces and database connectivity—is profoundly affected by the web. However, having a web-based database interface does not negate the design and implementation issues that were addressed in the previous chapters. In the final analysis, whether you make a purchase by going online or by standing in line, the system-level transaction details are essentially the same, and they require the same basic database structures and relationships. If any immediate lesson is to be learned, it is this: *The effects of bad database design, implementation, and management are magnified in an environment in which transactions might be measured in hundreds of thousands per day rather than hundreds.*

TABLE 15.3

CHARACTERISTICS AND BENEFITS OF INTERNET TECHNOLOGIES

INTERNET CHARACTERISTIC	BENEFIT
Hardware and software independence	Savings in equipment and software acquisition Ability to run on most existing equipment Platform independence and portability No need for multiple platform development
Common and simple user interface	Reduced training time and cost Reduced end-user support cost No need for multiple platform development
Location independence	Global access through Internet infrastructure and mobile smart devices Creation of new location-aware services Reduced requirements (and costs!) for dedicated connections
Rapid development at manageable costs	Availability of multiple development tools Plug-and-play development tools (open standards) More interactive development Reduced development times Relatively inexpensive tools Free client access tools (web browsers) Low entry costs; frequent availability of free web servers Reduced costs of maintaining private networks Distributed processing and scalability using multiple servers

The simplicity of the web's interface and its cross-platform functionality are at the core of its success as a data access platform. In fact, the web has helped create a new information dissemination standard. The following sections examine how web-to-database middleware enables end users to interact with databases over the web.

15-2a Web-to-Database Middleware: Server-Side Extensions

In general, the web server is the main hub through which all Internet services are accessed. For example, when an end user uses a web browser to dynamically query a database, the client browser requests a webpage from the web server. When the web server receives the page request, it looks for the page on the hard disk; when it finds the page, the server sends it back to the client.

Dynamic webpages are at the heart of current websites. In this database query scenario, the web server generates the webpage contents before it sends the page to the client web browser. The only problem with the preceding query scenario is that the web server must include the database query result on the page *before* it sends that page back to the client. Unfortunately, neither the web browser nor the web server knows how to connect to and read data from the database. Therefore, to support this type of request, the web server's capability must be extended so it can understand and process database requests. This job is known as a server-side extension.

A **server-side extension** is a program that interacts directly with the web server to handle specific types of requests. In the preceding database query example, the server-side extension program retrieves the data from databases and passes the retrieved data to the web server, which in turn sends the data to the client's browser for display. The server-side extension makes it possible to retrieve and present the query results, but more importantly, *it provides its services to the web server in a way that is totally transparent to the client browser*. In short, the server-side extension adds significant functionality to the web server, and therefore to the Internet.

Online Content

Client/server systems are covered in detail in Appendix F, Client/Server Systems, at *www.cengagebrain.com*.

server-side extension
A program that interacts directly with the server process to handle specific types of requests. Server-side extensions add significant functionality to web servers and intranets.

A database server-side extension program is also known as **web-to-database middleware**. Figure 15.8 shows the interaction between the browser, the web server, and the web-to-database middleware.

FIGURE 15.8 WEB-TO-DATABASE MIDDLEWARE

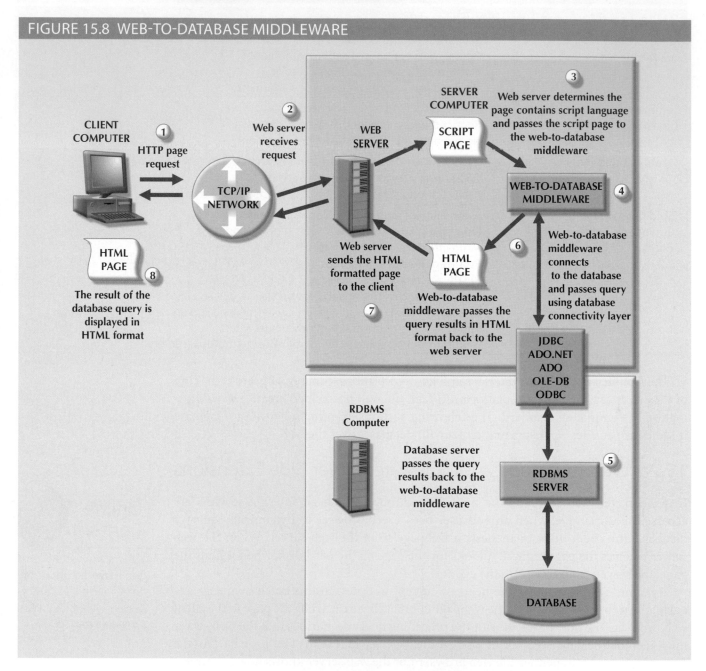

Trace the web-to-database middleware actions in Figure 15.8:

1. The client browser sends a page request to the web server.

2. The web server receives and passes the request to the web-to-database middleware for processing.

3. Generally, the requested page contains some type of scripting language to enable the database interaction. The web server passes the script to the web-to-database middleware.

4. The web-to-database middleware reads, validates, and executes the script. In this case, it connects to the database and passes the query using the database connectivity layer.

web-to-database middleware
A database server-side extension that retrieves data from databases and passes them to the web server, which in turn sends the data to the client's browser for display.

5. The database server executes the query and passes the result back to the web-to-database middleware.

6. The web-to-database middleware compiles the result set, dynamically generates an HTML-formatted page that includes the data retrieved from the database, and sends it to the web server.

7. The web server returns the just-created HTML page, which now includes the query result, to the client browser.

8. The client browser displays the page on the local computer.

The interaction between the web server and the web-to-database middleware is crucial to the development of a successful Internet database implementation. Therefore, the middleware must integrate closely via a well-defined web server interface.

15-2b Web Server Interfaces

Extending web server functionality implies that the web server and the web-to-database middleware will properly communicate with each other. (Database professionals often use the word *interoperate* to indicate that each party can respond to the communications of the other.) A web server interface defines a standard way to exchange messages with external programs. Currently, there are two well-defined web server interfaces:

- Common Gateway Interface (CGI)
- Application programming interface (API)

The **Common Gateway Interface (CGI)** uses script files that perform specific functions based on the client's parameters that are passed to the web server. The script file is a small program containing commands written in a programming language—usually Perl, C#, or Visual Basic. The script file's contents can be used to connect to the database and to retrieve data from it, using the parameters passed by the web server. Next, the script converts the retrieved data to HTML format and passes the data to the web server, which sends the HTML-formatted page to the client.

The main disadvantage of using CGI scripts is that the script file is an external program that executes separately for each user request and therefore causes a resource bottleneck. Performance also could be degraded by using an interpreted language or by writing the script inefficiently.

An application programming interface (API) is a newer web server interface standard that is more efficient and faster than a CGI script. APIs are more efficient because they are implemented as shared code or as dynamic-link libraries (DLLs). That means the API is treated as part of the web server program that is dynamically invoked when needed.

APIs are faster than CGI scripts because the code resides in memory, so there is no need to run an external program for each request. Instead, the same API serves all requests. Another advantage is that an API can use a shared connection to the database instead of creating a new one every time, as is the case with CGI scripts.

Although APIs are more efficient in handling requests, they have some disadvantages. Because the APIs share the same memory space as the web server, an API error can bring down the web server. Another disadvantage is that APIs are specific to the web server and to the operating system.

The web interface architecture is illustrated in Figure 15.9.

Common Gateway Interface (CGI)
A web server interface standard that uses script files to perform specific functions based on a client's parameters.

FIGURE 15.9 WEB SERVER CGI AND API INTERFACES

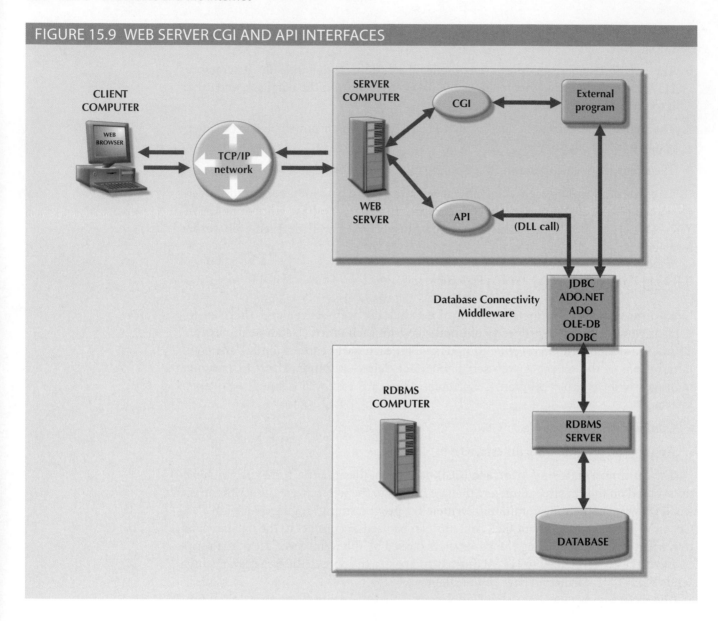

Regardless of the type of web server interface used, the web-to-database middle-ware program must be able to connect with the database. That connection can be accomplished in one of two ways:

- Use the native SQL access middleware provided by the vendor. For example, you can use SQL*Net if you are using Oracle.

- Use the services of general database connectivity standards such as ODBC, OLE-DB, ADO, ADO.NET, or JDBC.

15-2c The Web Browser

The web browser is software such as Microsoft Internet Explorer, Google Chrome, Apple Safari, or Mozilla Firefox that lets end users navigate the web from their client computer. Each time the end user clicks a hyperlink, the browser generates an HTTP GET page request that is sent to the designated web server using the TCP/IP Internet protocol.

The web browser's job is to *interpret* the HTML code that it receives from the web server and to present the various page components in a standard formatted way. Unfortunately, the browser's interpretation and presentation capabilities are not sufficient to develop web-based applications. The web is a **stateless system**—at any given time, a web server does not know the status of any of the clients communicating with it. That is, there is no open communication line between the server and each client accessing it, which of course is impractical in a *worldwide* web! Instead, client and server computers interact in very short "conversations" that follow the request-reply model. For example, the browser is concerned only with the *current* page, so there is no way for the second page to know what was done in the first page. The only time the client and server computers communicate is when the client requests a page—when the user clicks a link—and the server sends the requested page to the client. Once the client receives the page and its components, the client/server communication is ended. Therefore, although you may be browsing a page and *think* that the communication is open, you are actually just browsing the HTML document stored in the local cache (temporary directory) of your browser. The server does not have any idea what the end user is doing with the document, what data is entered in a form, what option is selected, and so on. On the web, if you want to act on a client's selection, you need to jump to a new page (go back to the web server), thus losing track of what was done before.

The web browser, through its use of HTML, does not have computational abilities beyond formatting output text and accepting form field inputs. Even when the browser accepts form field data, there is no way to perform immediate data entry validation. Therefore, to perform such crucial processing in the client, the web defers to other web programming languages such as Java, JavaScript, and VBScript. The browser resembles a dumb terminal that displays only data and can perform only rudimentary processing such as accepting form data inputs. To improve the capabilities of the web browser, you must use plug-ins and other client-side extensions. On the server side, web application servers provide the necessary processing power.

15-2d Client-Side Extensions

Client-side extensions add functionality to the web browser. Although client-side extensions are available in various forms, the most common are:

- Plug-ins
- Java and JavaScript
- ActiveX and VBScript

A **plug-in** is an external application that is automatically invoked by the browser when needed. The plug-in is associated with a data object—generally using the file extension—to allow the web server to properly handle data that is not originally supported. For example, if one of the page components is a PDF document, the web server will receive the data, recognize it as a Portable Document Format object, and launch Adobe Reader to present the document on the client computer.

JavaScript is a scripting language (one that enables the execution of a series of commands or macros) that allows web authors to design interactive sites. JavaScript code is embedded in the webpage and executed after a specific event, such as a mouse click on an object or a page being loaded from the server into memory.

ActiveX is Microsoft's alternative to Java. ActiveX is a specification for writing programs that run inside the Microsoft client browser, Internet Explorer. Because ActiveX is oriented toward Windows applications, it has low portability. ActiveX extends the web browser by adding controls to webpages, including drop-down lists, a slider, a calendar, and a calculator. Those controls are downloaded from the web server when needed so you can manipulate data inside the browser. ActiveX controls can be created in several

---

stateless system
A system in which a web server does not know the status of the clients communicating with it. The web does not reserve memory to maintain an open communications state between the client and the server.

client-side extension
Extension that adds functionality to a web browser. The most common extensions are plug-ins, Java, JavaScript, ActiveX, and VBScript.

plug-in
On the web, a client-side, external application that is automatically invoked by the browser when needed to manage specific types of data.

JavaScript
A scripting language that allows web authors to design interactive websites. JavaScript code is embedded in webpages, and then downloaded with the page and activated when a specific event takes place, such as a mouse click on an object.

ActiveX
Microsoft's alternative to Java. A specification for writing programs that will run inside the Microsoft client browser. Oriented mainly to Windows applications, it is not portable. It adds controls such as drop-down windows and calendars to webpages.

programming languages; C++ and Visual Basic are most commonly used. Microsoft's .NET framework allows for wider interoperability of ActiveX-based applications (such as ADO.NET) across multiple operating environments.

VBScript is another Microsoft product that is used to extend browser functionality. VBScript is derived from Microsoft Visual Basic. Like JavaScript, VBScript code is embedded inside an HTML page and is activated by triggering events such as clicking a link.

From the developer's point of view, using routines that permit data validation on the client side is an absolute necessity. For example, when data is entered in a web form and no data validation is done on the client side, the entire data set must be sent to the web server. That scenario requires the server to perform all data validation, thus wasting valuable CPU processing cycles. Therefore, client-side data input validation is one of the most basic requirements for web applications. Most of the data validation routines are done in Java, JavaScript, ActiveX, or VBScript.

15-2e Web Application Servers

A **web application server** is a middleware application that expands the functionality of web servers by linking them to a wide range of services, such as databases, directory systems, and search engines. The web application server also provides a consistent run-time environment for web applications. Web application servers can be used to perform the following:

- Connect to and query a database from a webpage.
- Present database data in a webpage using various formats.
- Create dynamic web search pages.
- Create webpages to insert, update, and delete database data.
- Enforce referential integrity in the application program logic.
- Use simple and nested queries and programming logic to represent business rules.

Web application servers provide features such as:

- An integrated development environment with session management and support for persistent application variables
- Security and authentication of users through user IDs and passwords
- Computational languages to represent and store business logic in the application server
- Automatic generation of HTML pages integrated with Java, JavaScript, VBScript, ASP, and so on
- Performance and fault-tolerant features
- Database access with transaction management capabilities
- Access to multiple services, such as file transfers (FTP), database connectivity, email, and directory services

Examples of web application servers include ColdFusion/JRun by Adobe, WebSphere Application Server by IBM, WebLogic Server by Oracle, Fusion by NetObjects, Visual Studio .NET by Microsoft, and WebObjects by Apple. All web application servers offer the ability to connect web servers to multiple data sources and other services. They vary in their range of available features, robustness, scalability, compatibility with other web and database tools, and extent of the development environment.

VBScript
A Microsoft client-side extension that extends a browser's functionality; VBScript is derived from Visual Basic.

web application server
A middleware application that expands the functionality of web servers by linking them to a wide range of services, such as databases, directory systems, and search engines.

15-2f Web Database Development

Web database development deals with the process of interfacing databases with the web browser—in short, how to create webpages that access data in a database. As you learned earlier in this chapter, multiple web environments can be used to develop web database applications. This section presents two simple code examples (ColdFusion and PHP). Because this is a database book, the examples focus only on the commands used to interface with the database rather than the specifics of HTML code.

A Microsoft Access database named **Ch15_Orderdb** is used to illustrate the web-to-database interface examples. The Ch15_Orderdb database, whose relational diagram is shown in Figure 15.10, was designed to track the purchase orders placed by users in a multidepartment company.

Online Content

To see and try a particular web-to-database interface in action, consult Appendix J, Web Database Development with ColdFusion, at *www.cengagebrain.com*. This appendix steps you through the process of creating and using a simple web-to-database interface, and provides more detailed information on developing web databases with Adobe ColdFusion middleware.

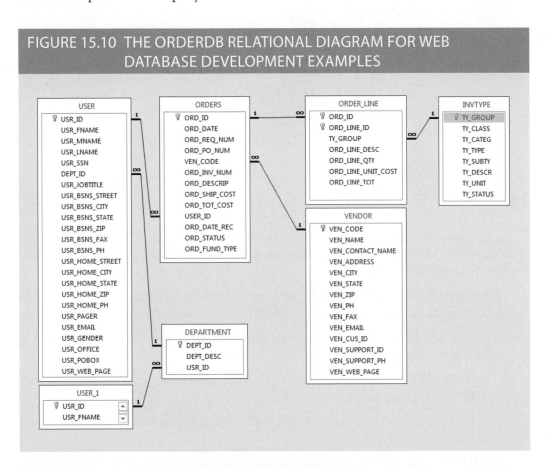

FIGURE 15.10 THE ORDERDB RELATIONAL DIAGRAM FOR WEB DATABASE DEVELOPMENT EXAMPLES

The following examples explain how to use ColdFusion and PHP to create a simple webpage to list the VENDOR rows. The scripts used in these examples perform two basic tasks:

1. Query the database using standard SQL to retrieve a data set that contains all records in the VENDOR table. The examples will use an ODBC data source named Rob-Cor. The ODBC data source was defined using the operating system tools shown in Section 15-1b.

2. Format the records generated in Step 1 in HTML so they are included in the webpage that is returned to the client browser.

Figure 15.11 shows the ColdFusion code to query the VENDOR table.

FIGURE 15.11 COLDFUSION CODE TO QUERY THE VENDOR TABLE

```
1   <HTML>
2   <HEAD>
3   <TITLE>Rob & Coronel -  ColdFusion Examples</TITLE>
4   <CFQUERY NAME="venlist" DATASOURCE="RobCor">
5       SELECT * FROM VENDOR ORDER BY VEN_CODE
6   </CFQUERY>
7   </HEAD>
8   <BODY BGCOLOR="LIGHTBLUE">
9   <H1>
10  <CENTER><B>Simple Query using CFQUERY and CFOUTPUT</B></CENTER>
11  <CENTER><B>(Vertical Output)</B></CENTER>
12  </H1>
13  <BR>
14  <HR>
15  <CFOUTPUT>
16      Your query returned #venlist.RecordCount# records
17  </CFOUTPUT>
18  <CFOUTPUT QUERY="venlist">
19  <PRE><B>
20  VENDOR CODE:        #VEN_CODE#
21  VENDOR NAME:        #VEN_NAME#
22  CONTACT PERSON:     #VEN_CONTACT_NAME#
23  ADDRESS:            #VEN_ADDRESS#
24  CITY:               #VEN_CITY#
25  STATE:              #VEN_STATE#
26  ZIP:                #VEN_ZIP#
27  PHONE:              #VEN_PH#
28  FAX:                #VEN_FAX#
29  E-MAIL:             #VEN_EMAIL#
30  CUSTOMER ID:        #VEN_CUS_ID#
31  SUPPORT ID:         #VEN_SUPPORT_ID#
32  SUPPORT PHONE:      #VEN_SUPPORT_PH#
33  VENDOR WEB PAGE:    #VEN_WEB_PAGE#
34  <HR></B></PRE>
35  </CFOUTPUT>
36  </FORM>
37  </BODY>
38  </HTML>
```

In the code in Figure 15.11, note that its ColdFusion tags are CFQUERY (to query a database) and CFOUTPUT (to display the data returned by the query). Take a closer look at these two CFML tags:

- <CFQUERY>tag (lines 4–6). This tag sets the stage for the database connection and the execution of the enclosed SQL statement. The CFQUERY tag uses the following parameters:

  – NAME = "*queryname*". This name uniquely identifies the record set returned by the database query.

  – DATASOURCE = "*datasourcename*". This parameter uses the previously defined ODBC data source name.

  – The SQL statement (line 5) is the SQL code used to retrieve the data rows from the VENDOR table.

- <CFOUTPUT>tag (lines 15–17 and 18–35). This tag is used to display the results from a CFQUERY or to call other ColdFusion variables or functions. Its parameters are as follows:

  – QUERY = "*queryname*". This is an optional parameter (see line 18). The tag works like a loop that is executed as many times as the number of rows in the named query set. You can include any valid HTML tags or text within the opening and closing CFOUTPUT tags.

  – ColdFusion uses pound signs (#) to reference query fields in the resulting query set or to call other ColdFusion variables. For example, #venlist.RecordCount# (line 16) displays the number of rows returned by the "venlist" query result set.

  – Lines 19–34 are repeated as a loop, one for each record returned in the named query.

Figure 15.12 shows the PHP code to query the VENDOR table.

FIGURE 15.12   PHP CODE TO QUERY THE VENDOR TABLE

```
1   <HTML>
2   <HEAD>
3   <TITLE>Rob & Coronel -  PHP Example</TITLE>
4   </HEAD>
5   <BODY BGCOLOR="LIGHTBLUE">
6   <H1><CENTER><B>Simple Query using PHP and ODBC functions</B></CENTER>
7   <CENTER><B>(Vertical Output)</B></CENTER></H1>
8   <BR>
9   <HR>
10  <?php
11  $dbc = odbc_connect("RobCor","","");
12  $sql = "SELECT * FROM VENDOR ORDER BY VEN_CODE";
13  $rs  = odbc_exec( $dbc, $sql );
14
15  while (odbc_fetch_row( $rs ))
16  {
17  $VEN_CODE          = odbc_result($rs,"VEN_CODE");
18  $VEN_NAME          = odbc_result($rs,"VEN_NAME");
19  $VEN_CONTACT_NAME  = odbc_result($rs,"VEN_CONTACT_NAME");
20  $VEN_ADDRESS       = odbc_result($rs,"VEN_ADDRESS");
21  $VEN_CITY          = odbc_result($rs,"VEN_CITY");
22  $VEN_STATE         = odbc_result($rs,"VEN_STATE");
23  $VEN_ZIP           = odbc_result($rs,"VEN_ZIP");
24  $VEN_PH            = odbc_result($rs,"VEN_PH");
25  $VEN_FAX           = odbc_result($rs,"VEN_FAX");
26  $VEN_EMAIL         = odbc_result($rs,"VEN_EMAIL");
27  $VEN_CUS_ID        = odbc_result($rs,"VEN_CUS_ID");
28  $VEN_SUPPORT_ID    = odbc_result($rs,"VEN_SUPPORT_ID");
29  $VEN_SUPPORT_PH    = odbc_result($rs,"VEN_SUPPORT_PH");
30  $VEN_WEB_PAGE      = odbc_result($rs,"VEN_WEB_PAGE");
31
32  echo "<BR>";
33  echo "VENDOR CODE:    ". $VEN_CODE . "<BR>";
34  echo "VENDOR NAME:    ". $VEN_NAME . "<BR>";
35  echo "CONTACT PERSON:". $VEN_CONTACT_NAME . "<BR>";
36  echo "ADDRESS:        ". $VEN_ADDRESS . "<BR>";
37  echo "CITY:           ". $VEN_CITY . "<BR>";
38  echo "STATE:          ". $VEN_STATE . "<BR>";
39  echo "ZIP:            ". $VEN_ZIP . "<BR>";
40  echo "PHONE:          ". $VEN_PH . "<BR>";
41  echo "FAX:            ". $VEN_FAX . "<BR>";
42  echo "E-MAIL:         ". $VEN_EMAIL . "<BR>";
43  echo "CUSTOMER ID:    ". $VEN_CUS_ID . "<BR>";
44  echo "SUPPORT ID:     ". $VEN_SUPPORT_ID . "<BR>";
45  echo "SUPPORT PHONE:  ". $VEN_SUPPORT_PH . "<BR>";
46  echo "VENDOR WEB PAGE:". $VEN_WEB_PAGE . "<BR>";
47  echo "<HR>";
48  }
49  odbc_close($dbc);
50  ?>
51  </BODY>
52  </HTML>
```

In the figure, note that PHP uses multiple tags to query and display the data returned by the query. Take a closer look at the PHP functions:

- The **odbc_connect** function (line 11) opens a connection to the ODBC data source. A handle to this database is set in the $dbc variable.

- The **odbc_exec** function (line 13) executes the SQL query stored in the $sql variable against the $dbc database connection. The query's result set is stored in the $rs variable.

- The **while** function (line 15) loops through the result set ($rs) and uses the ODBC_FETCH_ROW function to get one row at a time from the result set. Notice that PHP variables start with the dollar sign ($).

- The **odbc_result** function (lines 17–30) gets a column value from a row in the result set and stores it in a variable. This function extracts the different values for each field to be displayed and stores them in variables.

- The **echo** function (lines 32–47) outputs text to the webpage using the variables defined in the previous lines. You can also combine text (HTML code) and PHP variables (lines 33–46) using the "." delimiter.

- The **odbc_close** function closes the database connection.

The previous examples are just two of the many ways you can interface webpages and databases to web applications. These examples only scratch the surface of the multiple features that web application servers provide.

Current-generation systems involve more than just the development of web-enabled database applications. They also require applications that can communicate with each other and with other systems not based on the web. Clearly, systems must be able to exchange data in a standard-based format. That is the role of XML.

15-3 Extensible Markup Language (XML)

Companies use the Internet to generate business transactions and integrate data to increase efficiency and reduce costs. These transactions are known as electronic commerce (e-commerce); it enables all types of organizations to sell products and services to a global market. E-commerce transactions—the sale of products or services—can take place between businesses (business-to-business, or B2B) or between a business and a consumer (business-to-consumer, or B2C).

Most e-commerce transactions take place between businesses. Because B2B e-commerce integrates business processes among companies, it requires the transfer of business information among different business entities. However, the way in which businesses represent, identify, and use data tends to differ substantially from company to company. As a simple example, some companies use the term *product code*, while others use *item ID*.

Until recently, a purchase order traveling over the web was expected to be in the form of an HTML document. The HTML webpage displayed on the web browser would include formatting as well as the order details. HTML **tags** describe how something *looks* on the webpage, such as typefaces and heading styles, and they often come in pairs to start and end formatting features. For example, the following tags in angle brackets would display FOR SALE in bold Arial font:

<strong><font face=Arial>FOR SALE</font><strong>

If an application needs to get the order data from the webpage, there is no easy way to extract details such as the order number, date, customer number, product code, quantity, or price from an HTML document. The HTML document can only describe how to display the order in a web browser; it does not permit the manipulation of the order's data elements. To solve that problem, a new markup language known as Extensible Markup Language was developed.

Extensible Markup Language (XML) is a meta-language used to represent and manipulate data elements. XML is designed to facilitate the exchange of structured documents, such as orders and invoices, over the Internet. The World Wide Web Consortium (W3C) published the first XML 1.0 standard definition in 1998, setting the stage for giving XML the real-world appeal of being a true vendor-independent platform. It is not surprising that XML has rapidly become the data exchange standard for e-commerce applications.

The XML meta-language allows the definition of new tags, such as <ProdPrice>, to describe the data elements used in an XML document. This ability to *extend* the language explains the X in XML; the language is said to be *extensible*. XML is derived from the Standard Generalized Markup Language (SGML), an international standard for the publication and distribution of highly complex technical documents. For example, documents used by the aviation industry and the military services are too complex and unwieldy for the web. Just like HTML, which was also derived from

Online Content

To learn more about e-commerce, consult Appendix I, Databases in Electronic Commerce, at *www.cengagebrain.com*.

tag
In markup languages such as HTML and XML, a command inserted in a document to specify how the document should be formatted. Tags are used in server-side markup languages and interpreted by a web browser for presenting data.

Extensible Markup Language (XML)
A meta-language used to represent and manipulate data elements. Unlike other markup languages, XML permits the manipulation of a document's data elements. XML facilitates the exchange of structured documents such as orders and invoices over the Internet.

SGML, an XML document is a text file. However, it has a few important additional characteristics:

- XML allows the definition of new tags to describe data elements.

- XML is case sensitive: <ProductID>is not the same as <Productid>.

- XML must be well formed; that is, tags must be properly formatted. Most openings also have a corresponding closing. For example, a product's identification would require the format <ProductId>2345-AA</ProductId>.

- XML must be properly nested. For example, properly nested XML might look like this: <Product><ProductId>2345-AA</ProductId></Product>.

- You can use the <-- and --> symbols to enter comments in the XML document.

- 'lhe *XML* and *xml* prefixes are reserved for XML only.

XML is *not* a new version or replacement for HTML. XML is concerned with the description and representation of the data, rather than the way the data is displayed. XML provides the semantics that facilitate the sharing, exchange, and manipulation of structured documents over organizational boundaries. XML and HTML perform complementary functions rather than overlapping functions. Extensible Hypertext Markup Language (XHTML) is the next generation of HTML based on the XML framework. The XHTML specification expands the HTML standard to include XML features. Although it is more powerful than HTML, XHTML requires strict adherence to syntax requirements.

To illustrate the use of XML for data exchange purposes, consider a B2B example in which Company A uses XML to exchange product data with Company B over the Internet. Figure 15.13 shows the contents of the productlist.xml document.

FIGURE 15.13 CONTENTS OF THE PRODUCTLIST.XML DOCUMENT

```
productlist.xml - Notepad
File  Edit  Format  View  Help
<?xml version ="1.0"?>
<ProductList>
        <Product>
                <P_CODE>23109-HB</P_CODE>
                <P_DESCRIPT>Claw hammer</P_DESCRIPT>
                <P_INDATE>08/19/2016</P_INDATE>
                <P_QOH>23</P_QOH>
                <P_MIN>10</P_MIN>
                <P_PRICE>5.95</P_PRICE>
        </Product>
        <Product>
                <P_CODE>23114-AA</P_CODE>
                <P_DESCRIPT>Sledge Hammer, 12 lb.</P_DESCRIPT>
                <P_INDATE>09/01/2016</P_INDATE>
                <P_QOH>8</P_QOH>
                <P_MIN>5</P_MIN>
                <P_PRICE>14.40</P_PRICE>
        </Product>
</ProductList>
```

The preceding example illustrates several important XML features:

- The first line represents the XML document declaration, and it is mandatory.

- Every XML document has a *root element*. In the example, the second line declares the ProductList root element.

- The root element contains *child elements* or subelements. In the example, line 3 declares Product as a child element of ProductList.

- Each element can contain *subelements*. For example, each Product element is composed of several child elements, represented by P_CODE, P_DESCRIPT, P_INDATE, P_QOH, P_MIN, and P_PRICE.

Once Company B receives productlist.xml, it can process the document, assuming that it understands the tags created by Company A. The meaning of the XML in Figure 15.13 is fairly self-evident, but there is no easy way to validate the data or to check whether the data is complete. For example, you could encounter a P_INDATE value of "25/14/2016," but is that value correct? What happens if Company B expects a Vendor element as well? How can companies share data descriptions about their business data elements? The next section shows how document type definitions and XML schemas are used to address such concerns.

15-3a Document Type Definitions (DTD) and XML Schemas

Companies that use B2B transactions must have a way to understand and validate each other's tags. One way to accomplish that task is through the use of document type definitions. A **document type definition (DTD)** is a file with a .dtd extension that describes XML elements—in effect, a DTD file provides the composition of the database's logical model and defines the syntax rules or valid elements for each type of XML document. (The DTD component is similar to having a public data dictionary for business data.) Companies that intend to engage in e-commerce transactions must develop and share DTDs. Figure 15.14 shows the productlist.dtd document for the productlist.xml document shown earlier in Figure 15.13.

FIGURE 15.14 CONTENTS OF THE PRODUCTLIST.DTD DOCUMENT

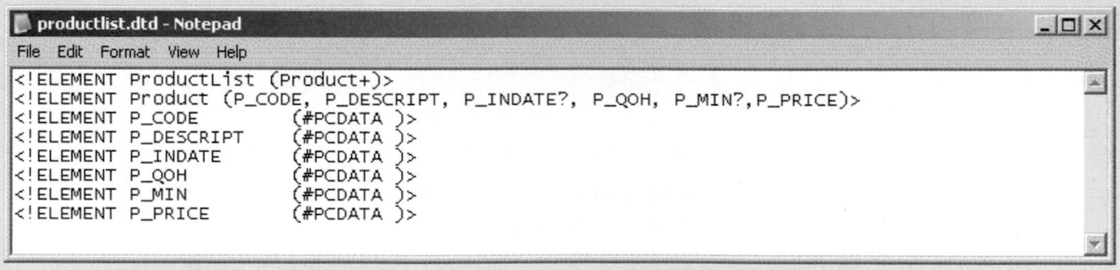

```
productlist.dtd - Notepad
File  Edit  Format  View  Help
<!ELEMENT ProductList (Product+)>
<!ELEMENT Product (P_CODE, P_DESCRIPT, P_INDATE?, P_QOH, P_MIN?,P_PRICE)>
<!ELEMENT P_CODE         (#PCDATA )>
<!ELEMENT P_DESCRIPT     (#PCDATA )>
<!ELEMENT P_INDATE       (#PCDATA )>
<!ELEMENT P_QOH          (#PCDATA )>
<!ELEMENT P_MIN          (#PCDATA )>
<!ELEMENT P_PRICE        (#PCDATA )>
```

In Figure 15.14, the productlist.dtd file provides definitions of the elements in the productlist.xml document. In particular, note the following:

- The first line declares the ProductList root element.

- The ProductList root element has one child, the Product element. The second line describes the Product element.

- The plus symbol (+) indicates that Product occurs one or more times within ProductList.

- An asterisk (*) would mean that the child element occurs zero or more times.

- The question mark (?) after P_INDATE and P_MIN indicates that they are optional child elements.

- The third through eighth lines show that the Product element has six child elements.

- The #PCDATA keyword represents the actual text data.

document type definition (DTD)
A file with a .dtd extension that describes XML elements; in effect, a DTD file describes a document's composition and defines the syntax rules or valid tags for each type of XML document.

To be able to use a DTD file to define elements within an XML document, the DTD must be referenced within that XML document. Figure 15.15 shows the productlistv2.xml document that includes the reference to productlist.dtd in the second line.

FIGURE 15.15 CONTENTS OF THE PRODUCTLISTV2.XML DOCUMENT

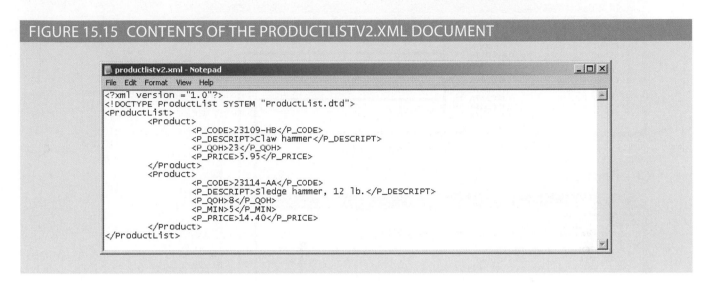

In Figure 15.15, note that P_INDATE and P_MIN do not appear in all Product definitions because they were declared to be optional elements. The DTD can be referenced by many XML documents of the same type. For example, if Company A routinely exchanges product data with Company B, it will need to create the DTD only once. All subsequent XML documents will refer to the DTD, and Company B will be able to verify the data being received.

To further demonstrate the use of XML and DTD for e-commerce data exchanges, consider the case of two companies exchanging order data. Figure 15.16 shows the DTD and XML documents for that scenario.

Although the use of DTDs is a great improvement for data sharing over the web, a DTD only provides descriptive information for understanding how the elements— root, parent, child, mandatory, or optional—relate to one another. A DTD provides limited additional semantic value, such as data type support or data validation rules. That information is very important for database administrators who are in charge of large e-commerce databases. To solve the DTD problem, the W3C published an XML schema standard that better describes XML data.

The **XML schema** is an advanced data definition language that is used to describe the structure of XML data documents. This structure includes elements, data types, relationship types, ranges, and default values. One of the main advantages of an XML schema is that it more closely maps to database terminology and features. For example, an XML schema can define common database types such as date, integer, or decimal; minimum and maximum values; a list of valid values; and required elements. Using the XML schema, a company would be able to validate data for values that may be out of range, have incorrect dates, contain invalid values, and so on. For example, a university application must be able to specify that a GPA value is between 0 and 4.0, and it must be able to detect an invalid birth date such as "14/13/2016." (There is no 14th month.) Many vendors are adopting this new standard and are supplying tools to translate DTD documents into XML schema definition documents. It is widely expected that XML schemas will replace DTD as the method to describe XML data.

XML schema
An advanced data definition language used to describe the elements, data types, relationship types, ranges, and default values of XML data documents. One of the main advantages of an XML schema is that it more closely maps to database terminology and features.

FIGURE 15.16 DTD AND XML DOCUMENTS FOR ORDER DATA

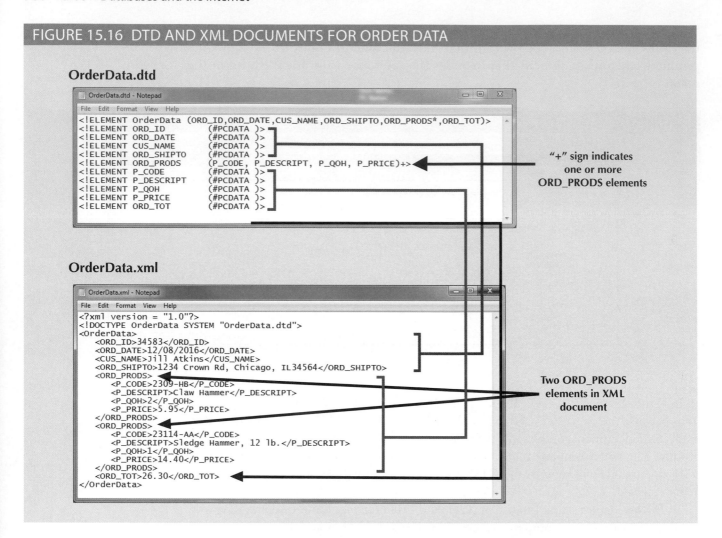

OrderData.dtd

```
OrderData.dtd - Notepad
File  Edit  Format  View  Help
<!ELEMENT OrderData (ORD_ID,ORD_DATE,CUS_NAME,ORD_SHIPTO,ORD_PRODS*,ORD_TOT)>
<!ELEMENT ORD_ID        (#PCDATA )>
<!ELEMENT ORD_DATE      (#PCDATA )>
<!ELEMENT CUS_NAME      (#PCDATA )>
<!ELEMENT ORD_SHIPTO    (#PCDATA )>
<!ELEMENT ORD_PRODS     (P_CODE, P_DESCRIPT, P_QOH, P_PRICE)+>
<!ELEMENT P_CODE        (#PCDATA )>
<!ELEMENT P_DESCRIPT    (#PCDATA )>
<!ELEMENT P_QOH         (#PCDATA )>
<!ELEMENT P_PRICE       (#PCDATA )>
<!ELEMENT ORD_TOT       (#PCDATA )>
```

"+" sign indicates one or more ORD_PRODS elements

OrderData.xml

```
OrderData.xml - Notepad
File  Edit  Format  View  Help
<?xml version = "1.0"?>
<!DOCTYPE OrderData SYSTEM "OrderData.dtd">
<OrderData>
    <ORD_ID>34583</ORD_ID>
    <ORD_DATE>12/08/2016</ORD_DATE>
    <CUS_NAME>Jill Atkins</CUS_NAME>
    <ORD_SHIPTO>1234 Crown Rd, Chicago, IL34564</ORD_SHIPTO>
    <ORD_PRODS>
        <P_CODE>2309-HB</P_CODE>
        <P_DESCRIPT>Claw Hammer</P_DESCRIPT>
        <P_QOH>2</P_QOH>
        <P_PRICE>5.95</P_PRICE>
    </ORD_PRODS>
    <ORD_PRODS>
        <P_CODE>23114-AA</P_CODE>
        <P_DESCRIPT>Sledge Hammer, 12 lb.</P_DESCRIPT>
        <P_QOH>1</P_QOH>
        <P_PRICE>14.40</P_PRICE>
    </ORD_PRODS>
    <ORD_TOT>26.30</ORD_TOT>
</OrderData>
```

Two ORD_PRODS elements in XML document

Unlike a DTD document, which uses a unique syntax, an **XML schema definition (XSD)** file uses a syntax that resembles an XML document. Figure 15.17 shows the XSD document for the OrderData XML document.

The code shown in Figure 15.17 is a simplified version of the XML schema document. As you can see, the XML schema syntax is similar to the XML document syntax. However, the XML schema introduces additional semantic information for the OrderData XML document, such as string, date, and decimal data types; required elements; and minimum and maximum cardinalities for the data elements.

15-3b XML Presentation

One of the main benefits of XML is that it separates data structure from its presentation and processing. By separating the two, you can present the same data in different ways—which is similar to having views in SQL. The Extensible Style Language (XSL) specification provides the mechanism to display XML data. *XSL* is used to define the rules by which XML data is formatted and displayed. The XSL specification is divided into two parts: Extensible Style Language Transformations (XSLT) and XSL style sheets.

- *Extensible Style Language Transformations (XSLT)* describes the general mechanism that is used to extract and process data from one XML document and enable its transformation within another document. Using XSLT, you can extract data from an XML document and convert it into a text file, an HTML webpage, or a webpage that is formatted

XML schema definition (XSD)
A file that contains the description of an XML document.

FIGURE 15.17 THE XML SCHEMA DOCUMENT FOR THE ORDER DATA

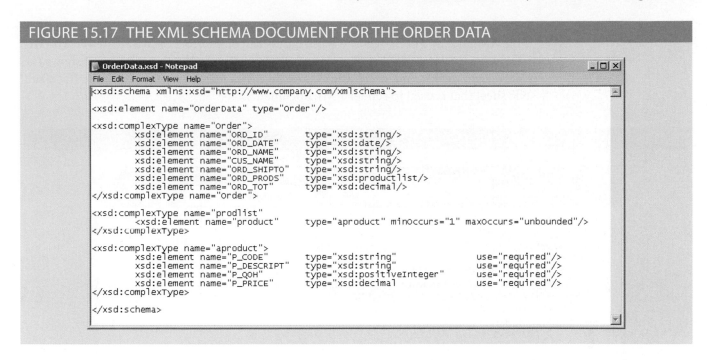

```
OrderData.xsd - Notepad
File  Edit  Format  View  Help
<xsd:schema xmlns:xsd="http://www.company.com/xmlschema">

<xsd:element name="OrderData" type="Order"/>

<xsd:complexType name="Order">
        xsd:element name="ORD_ID"        type="xsd:string/>
        xsd:element name="ORD_DATE"      type="xsd:date/>
        xsd:element name="ORD_NAME"      type="xsd:string/>
        xsd:element name="CUS_NAME"      type="xsd:string/>
        xsd:element name="ORD_SHIPTO"    type="xsd:string/>
        xsd:element name="ORD_PRODS"     type="xsd:productlist/>
        xsd:element name="ORD_TOT"       type="xsd:decimal/>
</xsd:complexType name="Order">

<xsd:complexType name="prodlist"
        <xsd:element name="product"      type="aproduct" minoccurs="1" maxoccurs="unbounded"/>
</xsd:complexType>

<xsd:complexType name="aproduct">
        xsd:element name="P_CODE"        type="xsd:string"           use="required"/>
        xsd:element name="P_DESCRIPT"    type="xsd:string"           use="required"/>
        xsd:element name="P_QOH"         type="xsd:positiveInteger"  use="required"/>
        xsd:element name="P_PRICE"       type="xsd:decimal           use="required"/>
</xsd:complexType>

</xsd:schema>
```

for a mobile device. What the user sees in those cases is actually a view (or HTML representation) of the XML data. XSLT can also be used to extract certain elements from an XML document, such as product codes and product prices, to create a product catalog. XSLT can even be used to transform one XML document into another.

- *XSL style sheets* define the presentation rules applied to XML elements—somewhat like presentation templates. The XSL style sheet describes the formatting options to apply to XML elements when they are displayed on a browser, smartphone, tablet screen, and so on.

Figure 15.18 illustrates the framework used by the various components to translate XML documents into viewable webpages, an XML document, or some other document.

FIGURE 15.18 FRAMEWORK FOR XML TRANSFORMATIONS

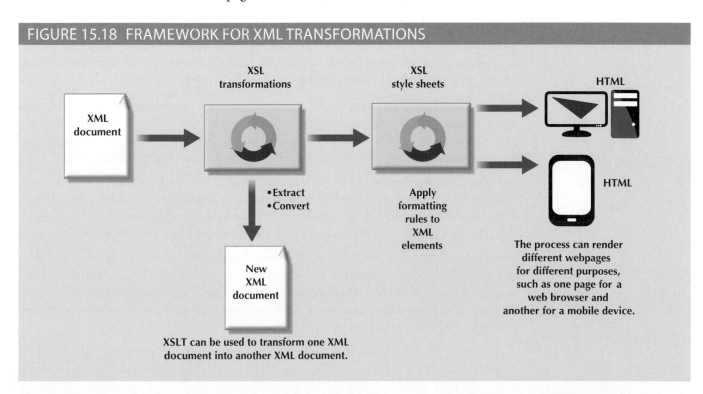

To display the XML document with Windows Internet Explorer (IE), enter the URL of the XML document in the browser's address bar. Figure 15.19 is based on the product-list.xml document created earlier. As you examine Figure 15.19, note that IE shows the XML data in a color-coded, collapsible, tree-like structure. (Actually, this is the IE default style sheet that is used to render XML documents.)

FIGURE 15.19 DISPLAYING XML DOCUMENTS

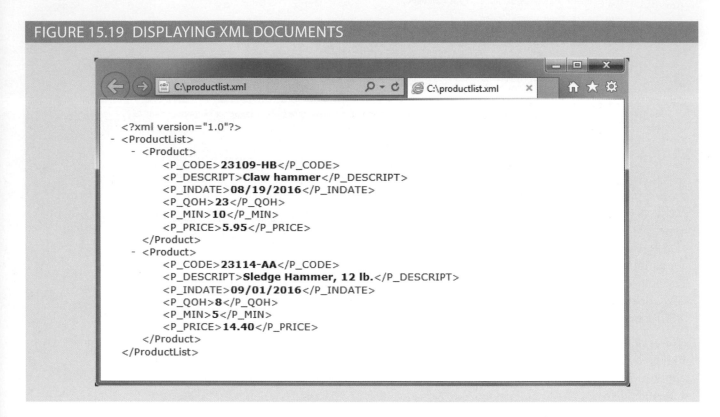

15-3c XML Applications

Now that you have some idea what XML is, how can you use it? What kinds of applications lend themselves particularly well to XML? This section lists some of the uses of XML. Keep in mind that the future use of XML is limited only by the imagination and creativity of developers, designers, and programmers.

- *B2B exchanges.* XML enables the exchange of B2B data, providing the standard for all organizations that need to exchange data with partners, competitors, the government, or customers. In particular, XML is positioned to replace EDI as the standard for automation of the supply chain because it is less expensive and more flexible.

- *Legacy systems integration.* XML provides the "glue" to integrate legacy system data with modern e-commerce web systems. Web and XML technologies could be used to inject some new life into old but trusted legacy applications. Another example is the use of XML to import transaction data from multiple databases to a data warehouse database.

- *Webpage development.* XML provides several features that make it a good fit for certain web development scenarios. For example, web portals with large amounts of personalized data can use XML to pull data from multiple external sources (such as news, weather, and stock sites) and apply different presentation rules to format pages on desktop computers as well as mobile devices.

- *Database support.* A DBMS that supports XML exchanges can integrate with external systems such as the web, mobile data, and legacy systems, thus enabling the creation of new types of systems. These databases can import or export data in XML format or generate XML documents from SQL queries while still storing the data using their native data model format. An example is the use of the FOR XML clause in the SQL SELECT statement in SQL Server. Alternatively, a DBMS can also support an XML data type to store XML data in its native format—enabling support to store tree-like hierarchical structures inside a relational structure.

- *Database metadictionaries.* XML is also used to create metadictionaries, or vocabularies, for entire industries. Examples of metadictionaries include HR-XML for the human resources industry, the metadata encoding and transmission standard (METS) from the Library of Congress, the clinical accounting information (CLAIM) data exchange standard for patient data exchange in electronic medical record systems, and the extensible business reporting language (XBRL) standard for exchanging business and financial information.

- *XML databases.*[1] Most databases on the market support XML to manage data in some shape or form. The approaches range from simple middleware XML software to object databases with XML interfaces to full XML database engines and servers. XML databases provide for the storage of data in complex relationships. For example, an XML database would be well suited to store the contents of a book. The book's structure would dictate its database structure: a book typically consists of chapters, sections, paragraphs, figures, charts, footnotes, endnotes, and so on. Examples of databases with XML data type support are Oracle, IBM DB2, and MS SQL Server. Fully XML databases examples are Tamino from Software AG *(www.softwareag.com)* and the open source dbXML from *http://sourceforge.net/projects/dbxml-core.*

- *XML services.* Many companies are already working to develop a new breed of services based on XML and web technologies. These services break down the interoperability barriers among systems and companies alike. XML provides the infrastructure that helps heterogeneous systems to work together across the desk, the street, and the world. Services would use XML and other Internet technologies to publish their interfaces. Other services that want to interact with existing services would locate them and learn their vocabulary (service request and replies) to establish a "conversation."

One area in which Internet, web, virtualization, and XML technologies work together in innovative ways to leverage IT services is cloud computing.

15-4 **Cloud Computing Services**

You have almost certainly heard about the "cloud" from the thousands of publications and TV ads that have used the term over the years, although it has represented different concepts. In the late 1980s, the term *cloud* was used by telecommunication companies to describe their data networks. In the late 1990s, during the peak of Internet growth, the term depicted the Internet itself. Then, in 2006, Google and Amazon began using the term *cloud computing* to describe a new set of innovative web-based services. Google, Yahoo, eBay, and Amazon were early adopters of this new computing paradigm.

But what exactly is cloud computing? According to the National Institute of Standards and Technology (NIST),[2] **cloud computing** is "a computing model for enabling ubiquitous, convenient, on-demand network access to a shared pool of configurable computer resources

cloud computing
A computing model that provides ubiquitous, on-demand access to a shared pool of configurable resources that can be rapidly provisioned.

[1] For a comprehensive analysis of XML database products, see "XML Database Products" by Ronald Bourret at *www.rpbourret.com.*
[2] *Recommendations of the National Institute of Standards and Technology*, Peter Mell and Timothy Grance, Special Publication 800–145 (Draft), January 2011.

(e.g., networks, servers, storage, applications and services) that can be rapidly provisioned and released with minimal management effort or service provider interaction." The term **cloud services** is used in this book to refer to the services provided by cloud computing. Cloud services allow any organization to quickly and economically add information technology services such as applications, storage, servers, processing power, databases, and infrastructure to its IT portfolio. Figure 15.20 shows a representation of cloud computing services on the Internet.

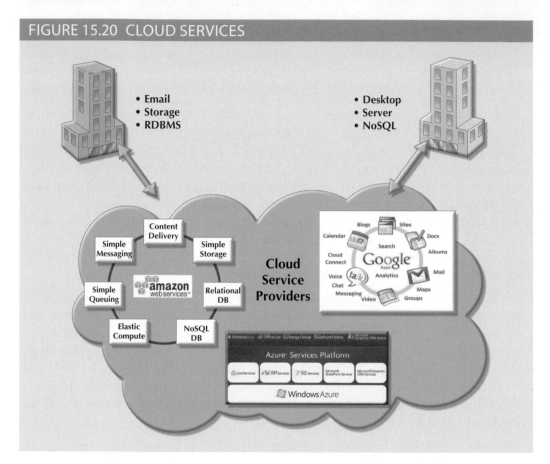

FIGURE 15.20 CLOUD SERVICES

Cloud computing allows IT-savvy organizations such as Amazon, Google, and Microsoft to build high-performance, fault-tolerant, flexible, and scalable IT services. These services include applications, storage, servers, processing power, databases, and email, which are delivered via the Internet to individuals and organizations using a pay-as-you-go price model.

For example, imagine that the chief technology officer of a nonprofit organization wants to add email services to the IT portfolio. A few years ago, this proposition would have implied building the email system's infrastructure from the ground up, including hardware, software, setup, configuration, operation, and maintenance. However, in today's cloud computing era, you can use Google Apps for Business or Microsoft Exchange Online and get a scalable, flexible, and more reliable email solution for a fraction of the cost. The best part is that you do not have to worry about the daily chores of managing and maintaining the IT infrastructure, such as OS updates, patches, security, fault tolerance, and recovery. What used to take months or years to implement can now be done in a matter of minutes. If you need more space, you just add another storage unit to your storage cloud. If you need more processing power to handle last-minute orders during the busy holiday season, you simply add more processing units to your cloud servers. Even more importantly, you can scale down as easily as you scaled up. Once your need for additional processing or storage subsides, you can go back to your previous

cloud services
The services provided by cloud computing. Cloud services allow any organization to quickly and economically add information technology services such as applications, storage, servers, processing power, databases, and infrastructure.

levels of usage and pay only for what you use. The beauty of cloud services is that you can scale down automatically, without an administrator's intervention.

Cloud computing is important for database technologies because it has the potential to become a "game changer." Cloud computing eliminates financial and technological barriers so organizations can leverage database technologies in their business processes with minimal effort and cost. In fact, cloud services have the potential to turn basic IT services into "commodity" services such as electricity, gas, and water, and to enable a revolution that could change not only the way that companies do business, but the IT business itself. As Nicholas Carr put it so vividly: "Cloud computing is for IT what the invention of the power grid was for electricity."[3]

The technologies that make cloud computing work have been around for a few years now; these technologies include the web, messaging, virtualization, remote desktop protocols, VPN, and XML. However, cloud computing itself is still in the early years and needs to mature further before it can be widely adopted. Despite this, more and more organizations are tapping into cloud services to secure advanced database services (relational or NoSQL) for their organizations. Currently, you can log in to Amazon Web Services (AWS) or Microsoft Azure and have a relational database ready for use in a matter of minutes. Instead of spending large amounts of cash buying hardware and software, organizations can employ a pay-per-use model for their IT services. Figure 15.21 depicts the cost of provisioning a relational database instance in Microsoft Azure and Amazon RDS services, respectively.

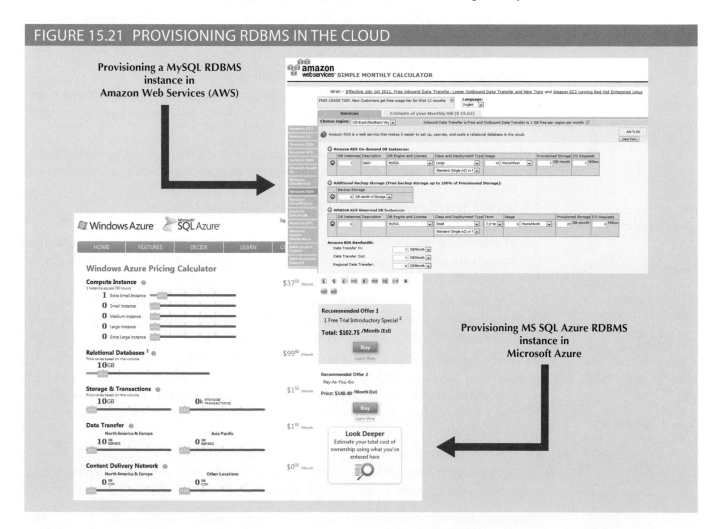

FIGURE 15.21 PROVISIONING RDBMS IN THE CLOUD

[3] Nicholas Carr, *The Big Switch: Rewiring the World, from Edison to Google.* W.W. Norton & Co., 2009.

Although Figure 15.21 shows a cloud that requires some degree of customization on the customer's part, other cloud computing services are more transparent to the user and require less customization. For example, Dropbox is a simple cloud service that lets you synchronize your documents, photos, music, and other files transparently over the Internet across many devices. Apple provides a similar service, known as iCloud, to seamlessly exchange data among all its mobile and nonmobile devices. Both services work transparently behind the scenes with minimal end-user intervention. As you can see, cloud computing implementations vary; the next section explains the basic types.

15-4a Cloud Implementation Types

Cloud computing has different types of implementations based on who the target customers are:

- **Public cloud.** This type of cloud infrastructure is built by a third-party organization to sell cloud services to the general public. The public cloud is the most common type of cloud implementation; examples include Amazon Web Services (AWS), Google Application Engine, and Microsoft Azure. In this model, cloud consumers share resources with other consumers transparently. The public cloud infrastructure is managed exclusively by the third-party provider.

- **Private cloud.** This type of internal cloud is built by an organization for the sole purpose of servicing its own needs. Private clouds are often used by large, geographically dispersed organizations to add agility and flexibility to internal IT services. The cloud infrastructure could be managed by internal IT staff or an external third party.

- **Community cloud.** This type of cloud is built by and for a specific group of organizations that share a common trade, such as agencies of the federal government, the military, or higher education. The cloud infrastructure could be managed by internal IT staff or an external third party.

Regardless of the implementation an organization uses, most cloud services share a common set of core characteristics. These characteristics are explored in the next section.

15-4b Characteristics of Cloud Services

Cloud computing services share a set of guiding principles. The characteristics listed in this section are shared by prominent public cloud providers such as Amazon, Google, Salesforce, SAP, and Microsoft. The prevalent characteristics are:

- *Ubiquitous access via Internet technologies.* All cloud services use Internet and web technologies to provision, deliver, and manage the services they provide. The basic requirement is that the device has access to the Internet.

- *Shared infrastructure.* The cloud service infrastructure is shared by multiple users. Sharing is made possible by web and virtualization technologies. Cloud services effectively provide an organization with a virtual IT infrastructure, which is locally managed by the consumer's organization as if it were the only user of the infrastructure.

- *Lower costs and variable pricing.* The initial costs of using cloud services tend to be significantly lower than building on-premise IT infrastructures. According to some studies,[4] the savings could range from 35 percent to 55 percent depending on company

public cloud
A form of computing in which the cloud infrastructure is built by a third-party organization to sell cloud services to the general public.

private cloud
A form of cloud computing in which an internal cloud is built by an organization to serve its own needs.

community cloud
A type of cloud built by and for a specific group of organizations that share a common trade, such as agencies of the federal government, the military, or higher education.

[4]"The Compelling TCO Case for Cloud Computing in SMB and Mid-Market Enterprises: A 4-year total cost of ownership (TCO) perspective comparing cloud and on-premise business application development," Sanjeev Aggarwal, Partner; Laurie McCabe, Partner: Hurwitz & Associates, 2009.

size, although more research is needed in this area. Because the web service's usage is metered per volume and time utilization, consumers benefit from lower and flexible pricing options. These options range from pay-as-you-go to fixed pricing based on minimum levels of service.

- *Flexible and scalable services.* The cloud services are built on an infrastructure that is highly scalable, fault tolerant, and very reliable. The services can scale up and down on demand according to resource demands.

- *Dynamic provisioning.* The consumer can quickly provision any needed resources, including servers, processing power, storage, and email, by accessing the web management dashboard and then adding and removing services on demand. This process also could be automated via other services.

- *Service orientation.* Cloud computing focuses on providing consumers with specific, well-defined services that use well-known interfaces. These interfaces hide the complexity from the end user, and can be delivered anytime and anywhere.

- *Managed operations.* Cloud computing minimizes the need for extensive and expensive in-house IT staff. The system infrastructure is managed by the cloud provider. The consumer organization's IT staff is free from routine management and maintenance tasks so they can focus on other tasks within the organization. Managed operations apply to organizations that use public clouds and that outsource cloud management to an external third party.

The preceding list is not exhaustive, but it is a starting point to understand most cloud computing offerings. Although most companies move to cloud services because of cost savings, some companies move to them because they are the best way to gain access to specific IT resources that would otherwise be unavailable. Not all cloud services are the same; in fact, there are several different types, as explained in the next section.

15-4c Types of Cloud Services

Cloud services come in different shapes and forms; no single type of service works for all consumers. In fact, cloud services often follow an à la carte model; consumers can choose multiple service options according to their individual needs. These services can build on top of each other to provide sophisticated solutions. Based on the types of services provided, cloud services can be classified by the following categories:

- **Software as a Service (SaaS).** The cloud service provider offers turnkey applications that run in the cloud. Consumers can run the provider's applications internally in their organizations via the web or any mobile device. The consumer can customize certain aspects of the application but cannot make changes to the application itself. The application is actually shared among users from multiple organizations. Examples of SaaS include MS Office 365, Google Docs, Intuit's TurboTax Online, and SCALA digital signage.

- **Platform as a Service (PaaS).** The cloud service provider offers the capability to build and deploy consumer-created applications using the provider's cloud infrastructure. In this scenario, the consumer can build, deploy, and manage applications using the provider's cloud tools, languages, and interfaces. However, the consumer does not manage the underlying cloud infrastructure. Examples of PaaS include the Microsoft Azure platform with .NET and the Java development environment, and Google Application Engine with Python or Java.

- **Infrastructure as a Service (IaaS).** In this case, the cloud service provider offers consumers the ability to provision their own resources on demand; these resources include

Software as a Service (SaaS)
A model in which the cloud service provider offers turnkey applications that run in the cloud.

Platform as a Service (PaaS)
A model in which the cloud service provider can build and deploy consumer-created applications using the provider's cloud infrastructure.

Infrastructure as a Service (IaaS)
A model in which the cloud service provider offers consumers the ability to provision their own resources on demand; these resources include storage, servers, databases, processing units, and even a complete virtualized desktop.

storage, servers, databases, processing units, and even a complete virtualized desktop. The consumer then can add or remove the resources as needed. For example, a consumer can use Amazon Web Services (AWS) and provision a server computer that runs Linux and Apache Web server using 16 GB of RAM and 160 GB of storage.

Figure 15.22 illustrates a sample of the different types of cloud services; these services can be accessed from any computing device.

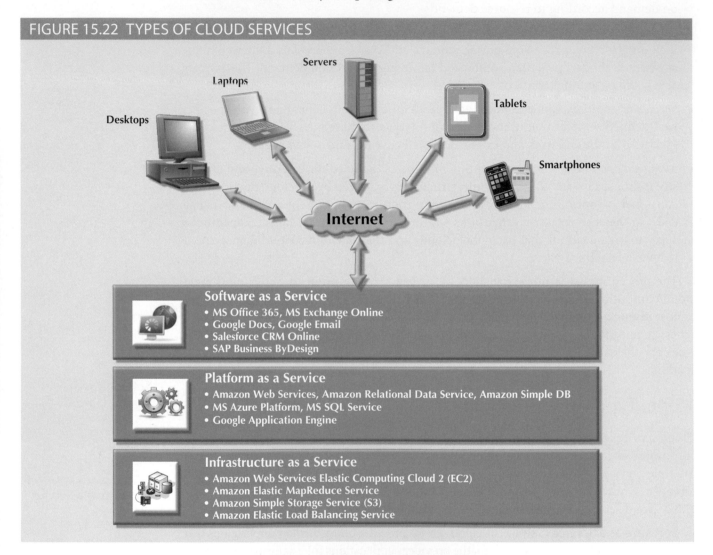

FIGURE 15.22 TYPES OF CLOUD SERVICES

Desktops
Laptops
Servers
Tablets
Smartphones

Internet

Software as a Service
- MS Office 365, MS Exchange Online
- Google Docs, Google Email
- Salesforce CRM Online
- SAP Business ByDesign

Platform as a Service
- Amazon Web Services, Amazon Relational Data Service, Amazon Simple DB
- MS Azure Platform, MS SQL Service
- Google Application Engine

Infrastructure as a Service
- Amazon Web Services Elastic Computing Cloud 2 (EC2)
- Amazon Elastic MapReduce Service
- Amazon Simple Storage Service (S3)
- Amazon Elastic Load Balancing Service

Cloud computing services have evolved in their sophistication and flexibility. The merging of new technologies has enabled the creation of new options such as "desktop as a service," which effectively creates a virtual computer on the cloud that can be accessed from any device over the Internet. For example, you can use a service such as Desktone (*http://www.vmwhorizonair.com/*) and get a Windows desktop running over the web for your personal use in a matter of minutes. Moreover, you can access your virtual desktop via the web browser or using any Remote Desktop Protocol (RDP) application.

15-4d Cloud Services: Advantages and Disadvantages

Cloud computing has grown remarkably in the past few years. Companies of all sizes are enjoying the advantages of cloud computing, but its widespread adoption is still limited by several factors. Table 15.4 summarizes the main advantages and disadvantages of cloud computing.

TABLE 15.4

ADVANTAGES AND DISADVANTAGES OF CLOUD COMPUTING

ADVANTAGE	DISADVANTAGE
Low initial cost of entry. Cloud computing has lower costs of entry when compared with the alternative of building in house.	*Issues of security, privacy, and compliance.* Trusting sensitive company data to external entities is difficult for most data-cautious organizations.
Scalability/elasticity. It is easy to add and remove resources on demand.	*Hidden costs of implementation and operation.* It is hard to estimate bandwidth and data migration costs.
Support for mobile computing. Cloud computing providers support multiple types of mobile computing devices.	*Data migration is a difficult and lengthy process.* Migrating large amounts of data to and from the cloud infrastructure can be difficult and time-consuming.
Ubiquitous access. Consumers can access the cloud resources from anywhere at any time, as long as they have Internet access.	*Complex licensing schemes.* Organizations that implement cloud services are faced with complex licensing schemes and complicated service-level agreements.
High reliability and performance. Cloud providers build solid infrastructures that otherwise are difficult for the average organization to leverage.	*Loss of ownership and control.* Companies that use cloud services are no longer in complete control of their data. What is the responsibility of the cloud provider if data are breached? Can the vendor use your data without your consent?
Fast provisioning. Resources can be provisioned on demand in a matter of minutes with minimal effort.	*Organization culture.* End users tend to be resistant to change. Do the savings justify being dependent on a single provider? Will the cloud provider be around in 10 years?
Managed infrastructure. Most cloud implementations are managed by dedicated internal or external staff. This allows the organization's IT staff to focus on other areas.	*Difficult integration with internal IT system.* Configuring the cloud services to integrate transparently with internal authentication and other internal services could be a daunting task.

As the table shows, the top-perceived benefit of cloud computing is the lower cost of entry. At the same time, the chief concern of cloud computing is data security and privacy, particularly in companies that deal with sensitive data and are subject to high levels of regulation and compliance. This concern leads to the perception that cloud services are mainly implemented in small to medium-sized companies where the risk of service loss is minimal. In fact, some companies that are subject to strict data security regulations tend to favor private clouds rather than public ones.

One of the biggest growth segments in cloud services is mobile computing. For example, Netflix, the video-on-demand trailblazer, moved significant parts of its IT infrastructure to AWS. Netflix decided to move to the cloud because of the challenges of building IT infrastructure fast enough to keep up with its relentless growth.

Note

Cloud Reality Check: Is the Cloud Enterprise-Ready?

Cloud service outages and security breach incidents are reported every year. Such incidents affect all types and sizes of organizations from data breaches in large universities to service interruptions in cloud infrastructure providers. Some are very public, such as the iCloud security breach that allowed hackers to steal thousands of private pictures from well-known celebrities. Other incidents could affect millions of people all over the world, such as recent interruptions in social media services (Instagram, Vines, and Twitter.) These incidents can cause service interruption, data loss, performance degradation, or cost millions of dollars in lost business.

Regardless of a company's size, databases remain at the center of all system development. Cloud computing brings a new dimension to data management that is within reach of any type of organization.

15-4e SQL Data Services

As you have seen in this chapter, data access technologies have evolved from simple ODBC data retrieval to advanced remote data processing using ADO.NET and XML. At the same time, companies are looking for ways to better manage ever-growing amounts of data while controlling costs without sacrificing data management features. Cloud computing provides a relatively stable and reliable platform for developing and deploying business services; cloud vendors have expanded their business to offer SQL data services. **SQL data services (SDS)** refers to a cloud computing-based data management service that provides relational data storage, access, and management to companies of all sizes without the typically high costs of in-house hardware, software, infrastructure, and personnel. This type of service provides some unique benefits:

- *Hosted data management.* SDS typically uses a cluster of database servers that provide a large subset of database functionality over the Internet to database administrators and users. Typically, features such as SQL queries, indexing, stored procedures, triggers, reporting, and analytical functions are available to end users. Other features such as data synchronization, data backup and restore, and data importing and exporting are available for administrative purposes.

- *Standard protocols.* SDS uses standard data communication and relational data access protocols. Typically, these services encapsulate SQL networking protocols, such as SQL-Net for Oracle databases and Tabular Data Services (TDS) for Microsoft SQL Server databases, inside the TCP/IP networking protocol.

- *A common programming interface.* SDS is transparent to application developers. Programmers continue to use familiar programming interfaces such as ADO.NET and Visual Studio .NET to manipulate the data. Programmers write embedded SQL code in their applications and connect to the database as if the data was stored locally instead of in a remote location on the Internet. One potential disadvantage, however, is that some specialized data types may not be supported by SDS.

SQL data services offer the following advantages when compared with in-house systems:

- Highly reliable and scalable relational database for a fraction of the cost
- High level of failure tolerance because data is normally distributed and replicated among multiple servers
- Dynamic and automatic load balancing
- Automated data backup and disaster recovery included with the service
- Dynamic creation and allocation of database processes and storage

Cloud providers such as Amazon and Microsoft allow you to get your own database server running in a matter of minutes. Even better, you do not have to worry about backups, fault tolerance, scalability, and routine maintenance tasks. The use of SQL data services enables rapid application development for businesses with limited information technology resources, and allows them to rapidly deploy business solutions. A consumer of cloud services is free to use the database to create the best solution for the problem at hand. However, having access to relational database technology via a SQL data service is just the start—you still need to be knowledgeable in database design and SQL to develop high-quality applications.

SQL data services (SDS)
Data management services that provide relational data storage, access, and management over the Internet.

Summary

- Database connectivity refers to the mechanisms through which application programs connect and communicate with data repositories. Database connectivity software is also known as *database middleware.*

- Microsoft database connectivity interfaces are dominant players in the market and enjoy the support of most database vendors. ODBC, OLE-DB, and ADO.NET form the backbone of Microsoft's Universal Data Access (UDA) architecture.

- Native database connectivity refers to the connection interface that is provided by the database vendor and is unique to that vendor. ODBC is probably the most widely supported database connectivity interface. ODBC allows any Windows application to access relational data sources using standard SQL. Data Access Objects (DAO) is an older, object-oriented application interface. Remote Data Objects (RDO) is a higher-level, object-oriented application interface used to access remote database servers. RDO was optimized to deal with server-based databases such as MS SQL Server and Oracle.

- Object Linking and Embedding for Database (OLE-DB) is database middleware developed with the goal of adding object-oriented functionality for access to relational and nonrelational data. ActiveX Data Objects (ADO) provides a high-level, application-oriented interface to interact with OLE-DB, DAO, and RDO. Based on ADO, ADO.NET is the data access component of Microsoft's .NET application development framework. Java Database Connectivity (JDBC) is the standard way to interface Java applications with data sources.

- Database access through the web is achieved through middleware. To improve the capabilities on the client side of the web browser, you must use plug-ins and other client-side extensions such as Java and JavaScript, or ActiveX and VBScript. On the server side, web application servers are middleware that expand the functionality of web servers by linking them to a wide range of services, such as databases, directory systems, and search engines.

- Extensible Markup Language (XML) facilitates the exchange of B2B and other data over the Internet. XML provides the semantics that facilitate the exchange, sharing, and manipulation of structured documents across organizational boundaries. XML produces the description and the representation of data, thus setting the stage for data manipulation in ways that were not possible before. XML documents can be validated through the use of document type definition (DTD) documents and XML schema definition (XSD) documents.

- Cloud computing is a computing model that provides ubiquitous, on-demand access to a shared pool of configurable resources that can be rapidly provisioned.

- SQL data services (SDS) refers to a cloud computing-based data management service that provides relational data storage, ubiquitous access, and local management to companies of all sizes. This service enables rapid application development for businesses with limited information technology resources. SDS allows rapid deployment of business solutions using standard protocols and common programming interfaces.

Key Terms

ActiveX

ActiveX Data Objects (ADO)

ADO.NET

application programming interface (API)

Call Level Interface (CLI)

client-side extension

cloud computing

cloud services

Common Gateway Interface (CGI)

community cloud

Data Access Objects (DAO)

data source name (DSN)

database middleware

DataSet

document type definition (DTD)

dynamic-link library (DLL)

Extensible Markup Language (XML)

Infrastructure as a Service (IaaS)

Java

Java Database Connectivity (JDBC)

JavaScript

Microsoft .NET framework

Object Linking and Embedding for Database (OLE-DB)

Open Database Connectivity (ODBC)

Platform as a Service (PaaS)

plug-in

private cloud

public cloud

Remote Data Objects (RDO)

script

server-side extension

Software as a Service (SaaS)

SQL data services (SDS)

stateless system

tags

Universal Data Access (UDA)

VBScript

web application server

web-to-database middleware

XML schema

XML schema definition (XSD)

Review Questions

1. Give some examples of database connectivity options and what they are used for.

2. What are ODBC, DAO, and RDO? How are they related?

3. What is the difference between DAO and RDO?

4. What are the three basic components of the ODBC architecture?

5. What steps are required to create an ODBC data source name?

6. What is OLE-DB used for, and how does it differ from ODBC?

7. Explain the OLE-DB model based on its two types of objects.

8. How does ADO complement OLE-DB?

9. What is ADO.NET, and what two new features make it important for application development?

10. What is a DataSet, and why is it considered to be disconnected?

11. What are web server interfaces used for? Give some examples.

12. Search the Internet for web application servers. Choose one and prepare a short presentation for your class.

13. What does this statement mean: "The web is a stateless system." What implications does a stateless system have for database application developers?

14. What is a web application server, and how does it work from a database perspective?

15. What are scripts, and what is their function? (Think in terms of database application development.)

16. What is XML, and why is it important?

17. What are document type definition (DTD) documents, and what do they do?

18. What are XML schema definition (XSD) documents, and what do they do?

19. What is JDBC, and what is it used for?

20. What is cloud computing, and why is it a "game changer"?

21. Name and contrast the types of cloud computing implementation.

22. Name and describe the most prevalent characteristics of cloud computing services.

23. Using the Internet, search for providers of cloud services. Then, classify the types of services they provide (SaaS, PaaS, and IaaS).

24. Summarize the main advantages and disadvantages of cloud computing services.

25. Define SQL data services and list their advantages.

Online Content

The Ch02 databases used in the Problems for this chapter are available at *www.cengagebrain.com*.

Problems

In the following exercises, you will set up database connectivity using MS Excel.

1. Use MS Excel to connect to the Ch02_InsureCo MS Access database using ODBC, and retrieve all of the AGENTs.

2. Use MS Excel to connect to the Ch02_InsureCo MS Access database using ODBC, and retrieve all of the CUSTOMERs.

3. Use MS Excel to connect to the Ch02_InsureCo MS Access database using ODBC, and retrieve the customers whose AGENT_CODE is equal to 503.

4. Create a System DSN ODBC connection called Ch02_SaleCo using the Administrative Tools section of the Windows Control Panel.

5. Use MS Excel to list all of the invoice lines for Invoice 103 using the Ch02_SaleCo System DSN.

6. Create a System DSN ODBC connection called Ch02_Tinycollege using the Administrative Tools section of the Windows Control Panel.

7. Use MS Excel to list all classes taught in room KLR200 using the Ch02_TinyCollege System DSN.

To answer Problems 8–11, use Section 15-3a as your guide.

8. Create a sample XML document and DTD for the exchange of customer data.

9. Create a sample XML document and DTD for the exchange of product and pricing data.

10. Create a sample XML document and DTD for the exchange of order data.

11. Create a sample XML document and DTD for the exchange of student transcript data. Use your college transcript as a sample.

PART 6
Database Administration

16 Database Administration and Security

Database Administration and Security

In this chapter, you will learn:

- That data is a valuable business asset requiring careful management
- How a database plays a critical role in an organization
- That the introduction of a DBMS has important technological, managerial, and cultural consequences for an organization
- About the database administrator's managerial and technical roles
- About data security, database security, and the information security framework
- About several database administration tools and strategies
- How cloud-based data services impact the DBA's role
- How various technical tasks of database administration are performed with Oracle

Preview

This chapter shows you the basis for a successful database administration strategy. Such a strategy requires that data be treated and managed as a valuable corporate asset.

In this chapter, you will learn about important data management issues by looking at the managerial and technical roles of the database administrator (DBA). This chapter also explores database security issues, such as the confidentiality, integrity, and availability of data. In our information-based society, a key aspect of data management is ensuring that data is protected against intentional or unintentional access by unauthorized personnel. It is also essential to ensure that data is available as needed, even in the face of natural disaster or hardware failure, and to maintain the integrity of the data in the database.

The chapter includes a discussion of database administration tools and the corporate-wide data architectural framework. You will also learn how database administration management fits within classical organizational structures. Furthermore, you will learn about several considerations when evaluating cloud-based data services. Even though many new types of databases have emerged, recent studies[1] show that relational databases still dominate the market share of the enterprise. Therefore, with the preponderance of relational databases in the market, it is important that you learn about some basic database administration tasks in Oracle RDBMS. Similar tasks can be performed in all major databases, such as Microsoft SQL Server, IBM DB2, Oracle MySQL, and so on.

Data Files Available on cengagebrain.com

Note

Because it is purely conceptual, this chapter does not reference any data files

---

[1] Emison, Joe Masters, "2014 State of Database Tech: Think Retro," *InformationWeek.com*, 3/10/2014.

16-1 **Data as a Corporate Asset**

In Chapter 1, Database Systems, you learned that data is the raw material from which information is produced. Therefore, in today's information-driven environment, data is a valuable asset that requires careful management.

To assess data's monetary value, consider what is stored in a company database: data about customers, suppliers, inventory, operations, and so on. How many opportunities are lost if the data is lost? What is the actual cost of data loss? For example, an accounting firm that lost its entire database would incur significant direct and indirect costs. The firm's problems would be magnified if the data loss occurred during tax season. Data loss puts any company in a difficult position. The company might be unable to handle daily operations effectively, it might lose customers who require quick and efficient service, and it might lose the opportunity to gain new customers.

Data is a valuable *resource* that can translate into *information*. If the information is accurate and timely, it can enhance the company's competitive position and generate wealth. In effect, an organization is subject to a *data-information-decision cycle*; that is, the data *user* applies intelligence to data to produce information that is the basis of knowledge used in *decision making*. This cycle is illustrated in Figure 16.1.

FIGURE 16.1 THE DATA-INFORMATION-DECISION-MAKING CYCLE

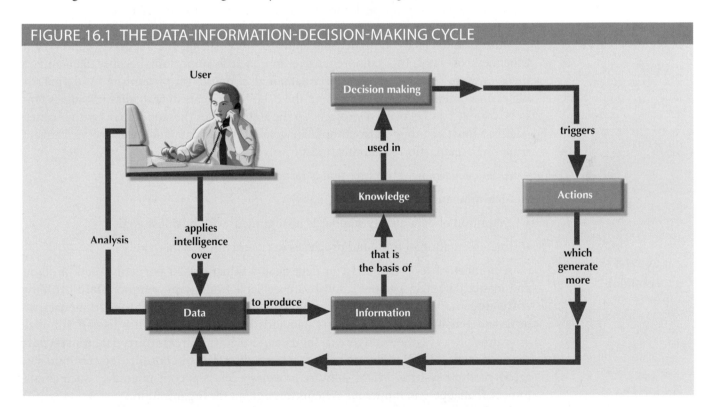

Note in Figure 16.1 that decisions made by high-level managers trigger actions within the organization's lower levels. Such actions produce additional data to be used for monitoring company performance. In turn, the additional data must be recycled within the data-information-decision framework. Thus, data forms the basis for decision making, strategic planning, control, and operations monitoring.

Efficient asset management is critical to the success of an organization. To manage data as a corporate asset, managers must understand the value of information. For some companies, such as credit reporting agencies, their only product is information, and their success is solely a function of information management.

Most organizations continually seek new ways to leverage their data resources to get greater returns. This leverage can take many forms, from data warehouses that support improved customer relationships to tighter integration with customers and suppliers in support of the electronic supply chain. As organizations become more dependent on information, that information's accuracy becomes more critical. **Dirty data**, or data that suffers from inaccuracies and inconsistencies, becomes an even greater threat. Data can become dirty for many reasons:

- Lack of enforcement of integrity constraints, such as not null, uniqueness, and referential integrity

- Data-entry errors and typographical errors

- Use of synonyms and homonyms across systems

- Nonstandard use of abbreviations in character data

- Different decompositions of composite attributes into simple attributes across systems

Some causes of dirty data, such as improper implementation of constraints, can be addressed within an individual database. However, addressing other causes is more complicated. Some dirty data comes from the movement of data across systems, as in the creation of a data warehouse. Efforts to control dirty data are generally referred to as data quality initiatives.

Data quality is a comprehensive approach to ensuring the accuracy, validity, and timeliness of data. This comprehensive approach is important because data quality involves more than just clening dirty data; it also focuses on preventing future inaccuracies and building user confidence in the data. Large-scale data quality initiatives tend to be complex and expensive projects, so the alignment of these initiatives with business goals is a must, as is buy-in from top management. While data quality efforts vary greatly from one organization to another, most involve the following:

- A data governance structure that is responsible for data quality

- Measurements of current data quality

- Definition of data quality standards in alignment with business goals

- Implementation of tools and processes to ensure future data quality

A number of tools can assist in data quality initiatives. In particular, data-profiling and master data management software are available from many vendors. **Data-profiling software** gathers statistics, analyzes existing data sources and metadata to determine data patterns, and compares the patterns against standards that the organization has defined. This analysis can help to assess the quality of existing data and identify sources of dirty data. **Master data management (MDM) software** helps to prevent dirty data by coordinating common data across multiple systems. MDM software provides a "master" copy of entities, such as customers, that appear in numerous systems throughout the organization.

While these technological approaches provide an important part of data quality, the overall solution to high-quality data within an organization still relies heavily on data administration and management.

16-2 The Need for a Database and its Role in an Organization

Data is used by different people in different departments for various reasons. Therefore, data management must address the concept of shared data. Chapter 1 showed how

---

dirty data
Data that contain inaccuracies and/or inconsistencies.

data quality
A comprehensive approach to ensuring the accuracy, validity, and timeliness of data.

data profiling software
Programs that analyze data and metadata to determine patterns that can help assess data quality.

master data management (MDM) software
Software the provides a "master copy" of entities such as customers, that appear in numerous systems throughout the organization. This software helps prevent dirty data by coordinating common data across multiple systems.

the need for data sharing made the DBMS almost inevitable. Used properly, the DBMS facilitates:

- *Interpretation* and *presentation* of data in useful formats by transforming raw data into information

- *Distribution* of data and information to the right people at the right time

- Data *preservation* and *monitoring* data usage for adequate periods of time

- *Control* over data duplication and use, both internally and externally

Regardless of the organization, the database's predominant role is *to support managerial decision making at all levels in the organization while preserving data privacy and security.*

An organization's managerial structure might be divided into three levels: top-level management makes strategic decisions, middle management makes tactical decisions, and operational management makes daily working decisions. *Operational* decisions are short term; for example, a manager might change the price of a product to clear it from inventory. *Tactical* decisions involve a longer time frame and affect larger-scale operations—for example, changing the price of a product in response to competitive pressures. *Strategic* decisions affect the long-term well-being of the company or even its survival—for example, changing the pricing strategy across product lines to capture market share.

The DBMS must give each level of management a useful view of the data and support the required level of decision making. The following activities are typical of each management level.

At the *top management* level, the database must be able to:

- Provide the information necessary for strategic decision making, strategic planning, policy formulation, and goals definition.

- Provide access to external and internal data to identify growth opportunities and to chart the direction of such growth. (Direction refers to the nature of the operations: will a company become a service organization, a manufacturing organization, or some combination of the two?)

- Provide a framework for defining and enforcing organizational policies that are translated into business rules at lower levels in the organization.

- Improve the likelihood of a positive return on investment by searching for new ways to reduce costs and boost productivity in the company.

- Provide feedback to monitor whether the company is achieving its goals.

At the *middle management* level, the database must be able to:

- Deliver the data necessary for tactical decisions and planning.

- Monitor and control the allocation and use of company resources and evaluate the performance of various departments.

- Provide a framework for enforcing and ensuring the security and privacy of the data in the database. **Security** means protecting the data against accidental or intentional use by unauthorized users. In the context of database administration, **privacy** is the extent to which individuals and organizations have the right to determine the details of data usage (who, what, when, where, and how).

At the *operational management* level, the database must be able to:

- Represent and support company operations as closely as possible. The data model must be flexible enough to incorporate all current and future data.

security
Activities and measures to ensure the confidentiality, integrity, and availability of an information system and its main asset, data.

privacy
The rights of individuals and organizations to determine access to data about themselves.

- Produce query results within specified performance levels. Keep in mind that the performance requirements increase for lower levels of management and operations. Thus, the database must support fast responses to a greater number of transactions at the operational management level.

- Enhance the company's short-term operations by providing timely information for customer support and for application development and computer operations.

A general objective for any database is to provide a seamless flow of information throughout the company.

The company's database is also known as the corporate or enterprise database. The **enterprise database** might be defined as the company's data representation that provides support for all present and expected future operations. Most of today's successful organizations depend on the enterprise database to provide support for all of their operations—from design to implementation, from sales to services, and from daily decision making to strategic planning.

16-3 Introduction of a Database: Special Considerations

Having a computerized database management system does not guarantee that the data will be properly used to provide the best solutions required by managers. A DBMS is a tool for managing data; like any tool, it must be used effectively to produce the desired results. In the hands of a carpenter, a hammer can help produce furniture, but in the hands of a child it might do damage. The solution to company problems is not the mere existence of a computer system or its database, but its effective management and use.

The introduction of a DBMS represents a big change and challenge. Throughout the organization, the DBMS is likely to have a profound impact, which might be positive or negative depending on how it is administered. For example, one key consideration is to adapt the DBMS to the organization rather than forcing the organization to adapt to the DBMS. The main issue should be the organization's needs rather than the DBMS's technical capabilities. However, the introduction of a DBMS (internally hosted or outsourced to a cloud service) cannot be accomplished without affecting the organization. The flood of new information has a profound effect on the way the organization functions and therefore on its corporate culture.

The introduction of a DBMS has been described as a process that includes three important aspects:[2]

- *Technological*—DBMS software and hardware
- *Managerial*—Administrative functions
- *Cultural*—Corporate resistance to change

The *technological* aspect includes selecting, installing, configuring, and monitoring the DBMS to make sure that it efficiently handles data storage, access, and security. The personnel in charge of installing the DBMS must have the technical skills to provide or secure adequate support for various users of the system: programmers, managers, and end users. Therefore, database administration staffing is a key technological consideration. The selected personnel must have the right mix of technical and managerial skills to provide a smooth transition to the new shared-data environment. In today's IT world, the technological aspects would apply to both internally hosted DBMS as well as cloud-based data environments.

enterprise database
The overall company data representation, which provides support for present and expected future needs.

[2] Murray, John P., "The Managerial and Cultural Issues of a DBMS," *370/390 Database Management* 1(8), September 1991, pp. 32–33.

The *managerial* aspect of the DBMS introduction should not be taken lightly. A high-quality DBMS does not guarantee a high-quality information system, just as having the best race car does not guarantee winning a race. Such managerial aspects would also include the management of the services and the relationship with the cloud-based data services provider.

The introduction of a DBMS requires careful planning to create an appropriate organizational structure and accommodate the personnel responsible for administering the system. This structure must also be subject to well-developed monitoring and controls. The administrative personnel must have excellent interpersonal and communications skills combined with broad organizational and business understanding. Top management must be committed to the new system and must define and support data administration functions, goals, and roles within the organization.

The *cultural* impact of the new database system must be assessed carefully. The DBMS is likely to have an effect on people, functions, and interactions. For example, additional personnel might be hired, new roles might be allocated to existing personnel, and employee performance might be evaluated using new standards.

A cultural impact is likely because the database approach creates a more controlled and structured information flow. Department managers who are accustomed to handling their own data must surrender ownership and share their data with the rest of the company. Application programmers must learn and follow new design and development standards. Managers might perceive an information overload and require time to adjust to the new environment.

When the new database comes online, people might be reluctant to use its information and might question its value or accuracy. Many might be disappointed that the information does not fit their preconceived notions and strongly held beliefs. Database administrators must be prepared to open their doors to end users, listen to their concerns, act on those concerns when possible, and explain the system's uses and benefits.

16-4 The Evolution of Database Administration

Data administration has its roots in the old, decentralized world of the file system. The cost of data and managerial duplication in these systems gave rise to centralized data administration known as the electronic data processing (EDP) or data processing (DP) department. The DP department's task was to pool all computer resources to support all departments *at the operational level*. DP administrators were given the authority to manage all company file systems as well as resolve data and managerial conflicts created by the duplication and misuse of data.

The advent of the DBMS and its shared view of data produced a new level of data management sophistication and led the DP department to evolve into an **information systems (IS) department**. The responsibilities of the IS department were broadened to include the following:

- A *service* function to provide end users with data management support

- A *production* function to provide end users with solutions for their information needs through integrated application or management information systems

The function of the IS department was reflected in its internal structure; a typical structure is shown in Figure 16.2.

As demand grew, the IS application development segment was subdivided by the type of system it supported: accounting, inventory, marketing, and so on. However, this development meant that database administration responsibilities were divided. The application development segment was in charge of gathering database requirements and logical

information systems (IS) department
A department responsible for all information technology services and production functions in an organization.

FIGURE 16.2 THE IS DEPARTMENT'S INTERNAL ORGANIZATION

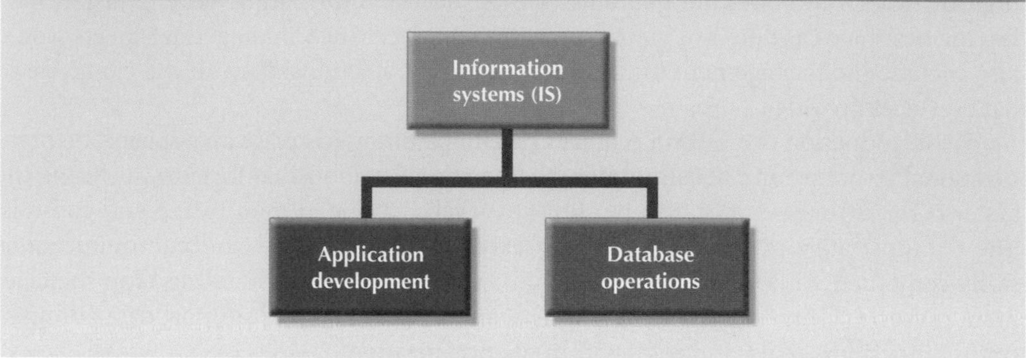

database design, whereas the database operations segment took charge of implementing, monitoring, and controlling DBMS operations.

As the number of database applications grew, data management became increasingly complex, thus leading to the development of database administration. The person responsible for control of the centralized and shared database became known as the **database administrator (DBA)**.

The size and role of the DBA function varies from company to company, as does its placement within the organizational structure. On the organizational chart, the DBA function might be defined as either a staff or line position. In a staff position, the DBA often takes on a consulting role; the DBA can devise the data administration strategy but does not have the authority to enforce it or resolve possible conflicts.[3] In a line position, the DBA has both the responsibility and authority to plan, define, implement, and enforce the policies, standards, and procedures used in data administration. The two possible DBA positions are illustrated in Figure 16.3.

There is no standard for how the DBA function fits in an organization's structure, partly because the function itself is probably the most dynamic of any in an organization. In fact, the fast-paced changes in DBMS technology dictate changing organizational styles. For example:

- The development of distributed databases can force an organization to decentralize data administration further. The distributed database requires the system DBA to define and delegate the responsibilities of each local DBA, thus imposing new and more complex *coordinating* activities on the system DBA.

- The growing use of Internet-accessible data and the growing number of data warehousing applications are likely to expand the DBA's data-modeling and design activities.

- The increasing sophistication and power of personal-computer-based DBMS packages provide an easy platform for developing user-friendly, cost-effective, and efficient solutions. However, such an environment also invites data duplication, not to mention the problems created by people who lack the technical qualifications to produce good database designs. In short, the new computing environment requires the DBA to develop a new set of technical and managerial skills.

- The increasing use of cloud data services is pushing many database platforms and infrastructures into the cloud. This can free the DBA from many lower-level

database administrator (DBA)
The person responsible for planning, organizing, controlling, and monitoring the centralized and shared corporate database. The DBA is the general manager of the database administration department.

---

[3]For a historical perspective on the development of the DBA function, refer to Jay-Louise Weldon's classic *Data Base Administration* (New York, Plenum Press, 1981). Although you might think that the book's publication date renders it obsolete, a surprising number of its topics are relevant to current databases.

FIGURE 16.3 THE PLACEMENT OF THE DBA FUNCTION

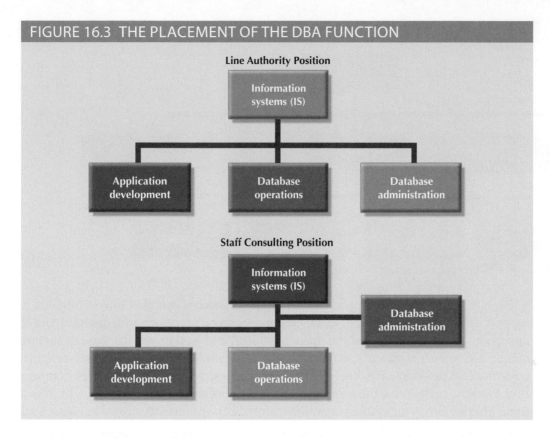

technology-oriented tasks, allowing DBAs to focus on higher-value strategic issues. In such environments, the DBA becomes a data use service provider and advisor for the organization.

- Conversely, the growing use of Big Data in organizations can force the DBA to become more technology-oriented. Ongoing efforts to integrate Hadoop storage systems with both NoSQL and relational databases require DBAs to be familiar with the lower-level storage and access issues that are still dominant in those emerging disciplines.

DBA operations are commonly defined and divided according to the phases of the Database Life Cycle (DBLC). If that approach is used, the DBA function requires personnel to cover the following activities:

- Database planning, including the definition of standards, procedures, and enforcement
- Database requirements gathering and conceptual design
- Database logical and transaction design
- Database physical design and implementation
- Database testing and debugging
- Database operations and maintenance, including installation, conversion, and migration
- Database training and support
- Data quality monitoring and management

Figure 16.4 represents a DBA functional organization according to the preceding model.

FIGURE 16.4 A DBA FUNCTIONAL ORGANIZATION

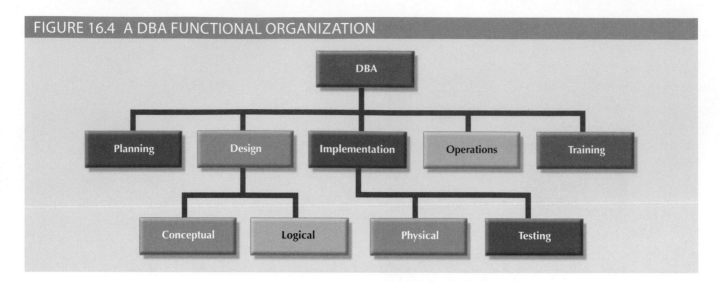

Keep in mind that a company might have several incompatible DBMSs installed to support different operations. For example, some corporations have a hierarchical DBMS to support daily transactions at the operational level and a relational database to support middle and top management's ad hoc information needs. A variety of personal computer DBMSs might be installed in different departments. In such an environment, the company might have one DBA assigned for each DBMS. The general coordinator of all DBAs is sometimes known as the **systems administrator**; that position is illustrated in Figure 16.5.

systems administrator
The person responsible for coordinating and performing day-to-day data-processing activities.

FIGURE 16.5 MULTIPLE DATABASE ADMINISTRATORS IN AN ORGANIZATION

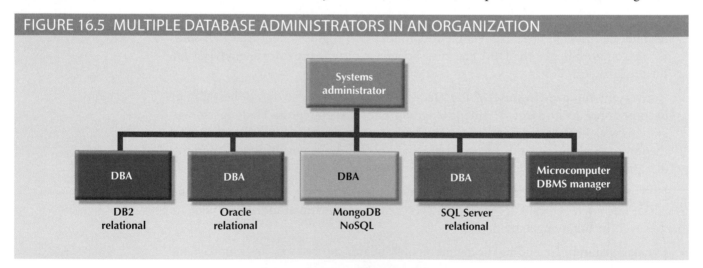

data administrator (DA)
The person responsible for managing the entire data resource, whether it is computerized or not. The DA has broader authority and responsibility than the database administrator (DBA). Also known as an *information resource manager (IRM)*.

information resource manager (IRM)
See *data administrator (DA)*.

There is a growing trend toward specialization in data management. For example, the organizational charts used by some larger corporations make a distinction between a DBA and the **data administrator (DA)**. The DA, also known as the **information resource manager (IRM)**, usually reports directly to top management and is given a higher degree of responsibility and authority than the DBA, although the two roles can overlap.

The DA is responsible for controlling the overall corporate data resources, both computerized and manual. Thus, the DA's job covers more operations than the DBA's because the DA controls data outside the scope of the DBMS in addition to computerized data. Depending on an organization's structure, the DBA might report to the DA, the IRM, the IS manager, or directly to the company's CEO.

16-5 The Database Environment's Human Component

A substantial portion of this book is devoted to relational database design and implementation, and to DBMS features and characteristics. Thus far, the book has focused on very important technical aspects of the database. However, even the most carefully crafted database system cannot operate without human assistance. In this section, you will explore how people perform the data administration activities that make a good database design useful.

Effective data administration requires both technical and managerial skills. For example, the DA's job typically has a strong managerial orientation with company-wide scope, along with a technical orientation that has a narrower, DBMS-specific scope. However, the DBA also must have considerable people skills. For example, both the DA and DBA direct and control personnel staffing and training within their respective departments.

Table 16.1 contrasts the general characteristics of both positions by summarizing typical DA and DBA activities. All of these activities are assigned to the DBA if the organization does not employ both a DA and a DBA.

TABLE 16.1

CONTRASTING DA AND DBA ACTIVITIES AND CHARACTERISTICS

DATA ADMINISTRATOR (DA)	DATABASE ADMINISTRATOR (DBA)
Performs strategic planning	Controls and supervises
Sets long-term goals	Executes plans to reach goals
Sets policies and standards	Enforces policies and procedures Enforces programming standards
Job is broad in scope	Job is narrow in scope
Focuses on the long term	Focuses on the short term (daily operations)
Has a managerial orientation	Has a technical orientation
Is DBMS-independent	Is DBMS-specific

Note that the DA provides a global and comprehensive administrative strategy for the organization's data. In other words, the DA's plans must consider the entire data spectrum. Thus, the DA is responsible for the consolidation and consistency of both manual and computerized data.

The DA must also set data administration goals. Those goals are defined by issues such as:

- Data "sharability" and time availability
- Data consistency and integrity
- Data security and privacy
- Data quality standards
- Extent and type of data use

Naturally, the list can be expanded to fit an organization's specific data needs. Regardless of how data management is conducted—and despite the fact that great authority is invested in the DA or DBA to define and control the way company data is used—the DA and DBA do not own the data. Instead, their functions are defined to emphasize that data is a shared company asset.

The preceding discussion should not lead you to believe that there are universally accepted DA and DBA administrative standards. The style, duties, organizational placement, and internal structure of both functions vary from company to company. For example, many companies distribute DA duties between the DBA and the manager of information systems. For simplicity and to avoid confusion, the label *DBA* is used here as a general title that encompasses all appropriate data administration.

The arbitration of interactions between the two most important assets of any organization, people and data, places the DBA in the dynamic environment portrayed in Figure 16.6.

FIGURE 16.6 A SUMMARY OF DBA ACTIVITIES

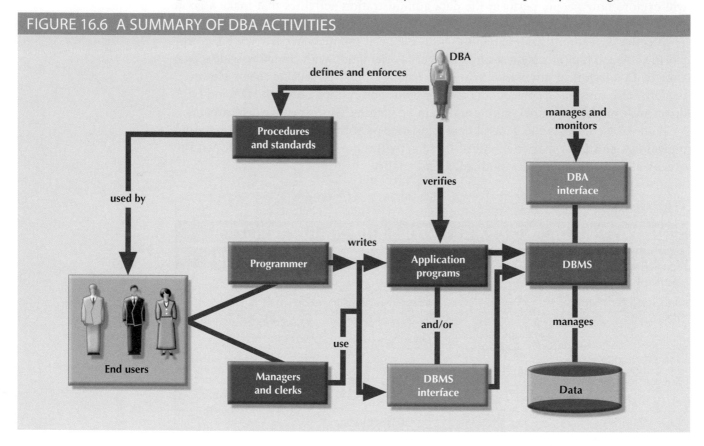

As you examine Figure 16.6, note that the DBA is the focal point for data and user interaction. The DBA defines and enforces the procedures and standards to be used by programmers and end users during their work with the DBMS. The DBA also verifies that programmer and end-user access meets the required quality and security standards.

Database users might be classified by the following criteria:

- Type of decision-making support required (operational, tactical, or strategic)
- Degree of computer knowledge (novice, proficient, or expert)
- Frequency of access (casual, periodic, or frequent)

These classifications are not exclusive and usually overlap. For example, an operational user can be an expert with casual database access, or a top-level manager might be a strategic novice user with periodic database access. On the other hand, a database application programmer is an operational expert and frequent database user. Thus, each organization employs people whose levels of database expertise span an entire spectrum. The DBA must be able to interact with all of them, understand their different needs, answer questions at all levels of expertise, and communicate effectively.

The DBA activities portrayed in Figure 16.6 suggest the need for a diverse mix of skills. In large companies, such skills are likely to be distributed among several DBAs.

In small companies, the skills might be the domain of just one DBA. The skills can be divided into two categories—managerial and technical—as summarized in Table 16.2.

TABLE 16.2

DESIRED DBA SKILLS

MANAGERIAL	TECHNICAL
Broad business understanding	Broad data-processing background and up-to-date knowledge of database technologies
Coordination skills	Understanding of Systems Development Life Cycle
Analytical skills	Structured methodologies • Data flow diagrams • Structure charts • Programming languages
Conflict resolution skills	Knowledge of Database Life Cycle
Communication skills (oral and written)	Database modeling and design skills • Conceptual • Logical • Physical
Negotiation skills	Operational skills: Database implementation, data dictionary management, security, and so on
Experience: 10 years in a large DP department	

As you examine Table 16.2, keep in mind that the DBA must perform two distinct roles. The DBA's managerial role is focused on personnel management and on interactions with end users. The DBA's technical role involves the use of the DBMS—database design, development, and implementation—as well as the production, development, and use of application programs. Both roles are examined in greater detail in the following sections.

16-5a The DBA's Managerial Role

As a manager, the DBA must concentrate on the control and planning of database administration. Therefore, the DBA is responsible for the following:

• Coordinating, monitoring, and allocating database administration resources: people and data

• Defining goals and formulating strategic plans for database administration

More specifically, the DBA's responsibilities are shown in Table 16.3.

TABLE 16.3

DBA ACTIVITIES AND SERVICES

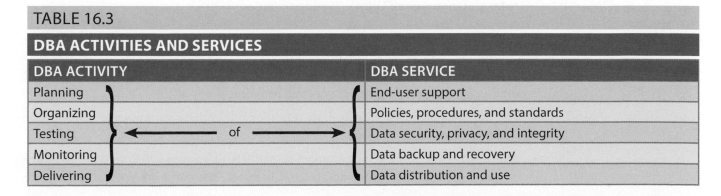

DBA ACTIVITY		DBA SERVICE
Planning		End-user support
Organizing		Policies, procedures, and standards
Testing	of	Data security, privacy, and integrity
Monitoring		Data backup and recovery
Delivering		Data distribution and use

734 Part 6 Database Administration

Table 16.3 illustrates that the DBA is generally responsible for planning, organizing, testing, monitoring, and delivering quite a few services. Those services might be performed by the DBA, although they are more likely to be performed by the DBA's personnel. The following sections examine the services in greater detail.

End-User Support The DBA interacts with end users by providing data and information support to their departments. Because end users usually have dissimilar computer backgrounds, support services include the following:

- *Gathering user requirements.* The DBA must work with end users to help gather the data required to identify and describe their present and future information needs. The DBA's communication skills are important in working closely with people who have varying computer backgrounds and communication styles.

- *Building end-user confidence.* Finding adequate solutions to end users' problems increases their trust and confidence in the DBA. The DBA also should educate end users about the services provided and how they enhance data stewardship and data security.

- *Resolving conflicts and problems.* Finding solutions to end users' problems in one department might trigger conflicts with other departments. End users are typically concerned with their own data needs rather than those of others, and they might not consider how their data might affect other departments within the organization. When conflicts arise, the DBA must have the authority and responsibility to resolve them.

- *Finding solutions to information needs.* The ability and authority to resolve data conflicts enables the DBA to develop solutions that will properly fit within the data management framework and address end users' information needs. Given the growing importance of the Internet, those solutions are likely to require the development and management of web servers to interface with the databases. In fact, the explosive growth of e-commerce requires the use of *dynamic* interfaces to facilitate interactive product queries and product sales.

- *Ensuring quality and integrity of data and applications.* Once the right solution has been found, it must be properly implemented and used. The DBA must work with application programmers and end users to teach them the database standards and procedures required for data quality, access, and manipulation. The DBA must also make sure that the database transactions do not adversely affect data quality. Likewise, certifying the quality of application programs that access the database is a crucial DBA function. Special attention must be given to DBMS Internet interfaces because they are prone to security issues, particularly when using cloud data services.

- *Managing the training and support of DBMS users.* One of the most time-consuming DBA activities is teaching end users how to use the database. The DBA must ensure that all users understand the basic functions of the DBMS software. The DBA coordinates and monitors all DBMS training activities.

Policies, Procedures, and Standards A successful data administration strategy requires the continuous enforcement of policies, procedures, and standards for correct data creation, usage, and distribution within the database. The DBA must define, document, and communicate the following before they can be enforced:

- **Policies** are general statements of direction or action that communicate and support DBA goals.

- **Standards** describe the minimum requirements of a given DBA activity; they are more detailed and specific than policies. In effect, standards are rules that evaluate

policy
General statement of direction that is used to manage company operations through the communication and support of the organization's objectives.

standard
A detailed and specific set of instructions that describes the minimum requirements for a given activity. Standards are used to evaluate the quality of the output.

the quality of the activity. For example, standards define the structure of application programs and the naming conventions programmers must use.

- **Procedures** are written instructions that describe a series of steps to be followed during the performance of a given activity. Procedures must be developed within existing working conditions, and they must support and enhance the work environment.

To illustrate the distinctions among policies, standards, and procedures, look at the following examples:

Policies

- All users must have passwords.
- Passwords must be changed every six months.

Standards

- A password must have a minimum of 5 characters.
- A password must have a maximum of 12 characters.
- Social Security numbers, names, and birth dates cannot be used as passwords.

Procedures

To create a password, (1) the end user sends the DBA a written request for the creation of an account; (2) the DBA approves the request and forwards it to the computer operator; (3) the computer operator creates the account, assigns a temporary password, and sends the account information to the end user; (4) a copy of the account information is sent to the DBA; and (5) the user changes the temporary password to a permanent one.

Standards and procedures defined by the DBA apply to all end users who want to benefit from the database. Standards and procedures must complement each other and must constitute an extension of data administration policies. Procedures must facilitate the work of end users and the DBA. The DBA must define, communicate, and enforce procedures that cover areas such as:

- *End-user database requirements gathering.* What documentation is required? What forms must be used?
- *Database design and modeling.* What database design methodology will be used (normalization or object-oriented)? What tools will be used (CASE tools, data dictionaries, UML or ER diagrams)?
- *Documentation and naming conventions.* What documentation must be used in the definition of all data elements, sets, and programs that access the database?
- *Design, coding, and testing of database application programs.* The DBA must define the standards for application program coding, documentation, and testing. The DBA standards and procedures are given to the application programmers, and the DBA must enforce those standards.
- *Database software selection.* The selected DBMS must properly interface with existing software, have the features needed by the organization, and provide a positive return on investment. In today's Internet environment, the DBA must also work with web and network administrators to implement efficient and secure web and cloud database connectivity.
- *Database security and integrity.* The DBA must define policies that govern security and integrity. Database security is especially crucial. Security standards must be

procedure
Series of steps to be followed during the performance of an activity or process.

clearly defined and strictly enforced. Security procedures must handle a multitude of scenarios to ensure that problems are minimized. Although no system can ever be completely secure, procedures must meet critical standards. The growing use of Internet interfaces to databases opens the door to new security threats that are far more complex and difficult to manage than those in traditional interfaces—this is particularly important when working with cloud data services. Therefore, the DBA must work closely with Internet security specialists to ensure that the databases are properly protected from attacks.

- *Database backup and recovery.* Database backup and recovery procedures must include information that guarantees proper execution and management of the backups. The DBA must work closely with any cloud-based data services provider to ensure the proper procedures are in place to manage data backups and restores and to ensure ownership and security of the data.

- *Database maintenance and operation.* The DBMS's daily operations must be clearly documented. Operators must keep job logs and must write operator instructions and notes. Such notes help pinpoint the causes and solutions of problems. Operational procedures must also include precise instructions for backup and recovery procedures.

- *End-user training.* A full-featured training program must be established within the organization, and training procedures must be clearly specified. Each end user must be aware of available training.

Procedures and standards must be revised at least annually to keep them up to date and to ensure that the organization can adapt quickly to changes in the work environment. Naturally, the introduction of new DBMS software, the discovery of security or integrity violations, company reorganizations, and similar changes require revision of procedures and standards.

Data Security, Privacy, and Integrity Data security, privacy, and integrity are of great concern to DBAs who manage DBMS installations. Technology has pointed the way to greater productivity through information management, and it has enabled the distribution of data across multiple sites, making it more difficult to maintain data control, security, and integrity. Thus, the DBA must use the security and integrity mechanisms provided by the DBMS to enforce the database administration policies defined in the previous section. In addition, DBAs must team up with Internet security experts to build security mechanisms that safeguard data from possible attacks or unauthorized access. Section 16-6 covers security issues in more detail.

Data Backup and Recovery When data is not readily available, companies face potentially ruinous losses. Therefore, data backup and recovery procedures are critical in all database installations. The DBA must also ensure that data can be fully recovered in case of data loss or loss of database integrity. These losses can be partial or total; therefore, backup and recovery procedures are the cheapest database insurance you can buy.

The management of database security, integrity, backup, and recovery is so critical that many DBA departments have created a position called the **database security officer (DSO)**. The DSO's sole job is to ensure database security and integrity. In large organizations, the DSO's activities are often classified as *disaster management*.

Disaster management includes all of the DBA activities designed to secure data availability following a physical disaster or a database integrity failure. Disaster management includes all planning, organizing, and testing of database contingency plans

---

database security officer (DSO)
The person responsible for the security, integrity, backup, and recovery of the database.

disaster management
The set of DBA activities dedicated to securing data availability following a physical disaster or a database integrity failure.

and recovery procedures. The backup and recovery measures must include at least the following:

- *Periodic data and application backups.* Some DBMSs include tools to ensure automatic backup and recovery of the database. Products such as IBM's DB2 allow different types of backups: full, incremental, and concurrent. A **full backup**, also known as a **database dump**, produces a complete copy of the entire database. An **incremental backup** produces a backup of all data since the last backup date. A **concurrent backup** takes place while the user is working on the database.

- *Proper backup identification.* Backups must be clearly identified through detailed descriptions and date information, thus enabling the DBA to ensure that the correct backups are used to recover the database. The most common backup medium has traditionally been tape; computer operators must diligently store and label the tapes, and the DBA must keep track of the current tape's location. However, organizations that are large enough to hire a DBA do not typically use tapes for enterprise backup. Other solutions include optical and disk-based backup devices. Such backup solutions include online storage based on network-attached storage (NAS), storage area networks (SAN), and cloud-based data storage. Enterprise backup solutions use a layered approach in which the data is first backed up to fast disk media for intermediate storage and fast restoration. Later, the data is transferred to tape for archival storage.

- *Convenient and safe backup storage.* Multiple backups of the same data are required, and each backup copy must be stored in a different location. The storage locations must include sites inside and outside the organization. (Keeping different backups in the same place defeats the purpose of having multiple backups.) The storage locations must be properly prepared, and they may include fire-safe and quakeproof vaults as well as humidity and temperature controls. The DBA must establish a policy to respond to two questions: (1) Where are the backups to be stored? (2) How long are backups to be stored?

- *Physical protection of both hardware and software.* Protection might include the use of closed installations with restricted access, as well as preparation of the computer sites to provide air conditioning, backup power, and fire protection. Physical protection also includes a backup computer and DBMS to be used in case of emergency. For example, when Hurricane Sandy hit the east coast of North America in 2012, the U.S. Northeast suffered widespread destruction of its communications infrastructure. The storm served as a wake-up call for many organizations and educational institutions that did not have adequate disaster recovery plans for such an extreme level of service interruption.

- *Personal access control to the software of a database installation.* Multilevel passwords and privileges as well as hardware and software challenge/response tokens can be used to identify authorized users of resources.

- *Insurance coverage for the data in the database.* The DBA or security officer must buy an insurance policy to provide financial protection in the event of a database failure. The insurance might be expensive, but it is less expensive than the disaster created by massive data loss.

Two additional points are worth making are:

- Data recovery and contingency plans must be thoroughly tested and evaluated, and they must be practiced frequently. So-called fire drills should not be disparaged, and they require top-level management's support and enforcement.

- A backup and recovery program is not likely to cover all components of an information system. Therefore, it is appropriate to establish priorities for the nature and extent of data recovery.

full backup (database dump)
A complete copy of an entire database saved and periodically updated in a separate memory location. A full backup ensures a full recovery of all data after a physical disaster or database integrity failure.

incremental backup
A process that only backs up data that has changed in the database since the last incremental or full backup.

concurrent backup
A backup that takes place while one or more users are working on a database.

Data Distribution and Use Data is useful only when it reaches the right users in a timely fashion. The DBA is responsible for ensuring that data is distributed to the right people, at the right time, and in the right format. These tasks can become very time-consuming, especially when data delivery capacity is based on a typical applications programming environment, where users depend on programmers to deliver the programs that access the database. Although the Internet and its intranet and extranet extensions have opened databases to corporate users, they have also created a new set of challenges for the DBA.

Current data distribution philosophy makes it easy for *authorized* end users to access the database. One way to accomplish this task is to facilitate the use of new, more sophisticated query tools and new web front ends. They enable the DBA to educate end users to produce required information without being dependent on applications programmers. Naturally, the DBA must ensure that users adhere to appropriate standards and procedures.

This data-sharing philosophy is common today, and it probably will become more common as database technology marches on. Such an environment is more flexible for end users; by becoming more self-sufficient in the acquisition and use of data, they can make better decisions. Yet, this "data democracy" can also produce some troublesome side effects. Letting end users micromanage their data subsets could inadvertently sever the connection between those users and data administrators. The DBA's job could become more complicated, and the efficiency of data administration could be compromised. Data duplication might flourish again without checks at the organizational level to ensure the uniqueness of data elements. Thus, end users who do not completely understand the nature and sources of data might use the data elements improperly.

16-5b The DBA's Technical Role

The DBA's technical role requires a broad understanding of DBMS functions, configuration, programming languages, and data-modeling and design methodologies. For example, the DBA's technical activities include the selection, installation, operation, maintenance, and upgrading of the DBMS and utility software, as well as the design, development, implementation, and maintenance of application programs that interact with the database.

Many of the DBA's technical activities are a logical extension of the DBA's managerial activities. For example, the DBA deals with database security and integrity, backup and recovery, and training and support. The technical aspects of the DBA's job are rooted in the following areas of operation:

- Evaluating, selecting, and installing the DBMS and related utilities
- Designing and implementing databases and applications
- Testing and evaluating databases and applications
- Operating the DBMS, utilities, and applications
- Training and supporting users
- Maintaining the DBMS, utilities, and applications

The following sections explore the details of each area.

Evaluating, Selecting, and Installing the DBMS and Utilities One of the DBA's first and most important technical responsibilities is selecting the database management system, utility software, and supporting hardware to be used in the organization. The DBMS selection might also include the consideration of cloud-based data services. This task requires extensive planning, which must be based on the organization's needs rather than specific software and hardware features. The DBA must recognize that the objective is solving problems rather than buying a computer or DBMS software. Put simply, a DBMS is a management tool and not a technological toy.

The first and most important step of the plan is to determine company needs. The DBA must make sure that all end users, including top-level and midlevel managers, are involved in the process. Once the needs are identified, the objectives of data administration can be clearly established and the DBMS features and selection criteria can be defined.

To match DBMS capability to the organization's needs, the DBA would be wise to develop a checklist of desired DBMS features that addresses at least the following issues:

- *DBMS model.* Are the company's needs better served by a relational, object-oriented, object/relational, or a NoSQL DBMS? If a data warehouse application is required, should a relational or multidimensional DBMS be used? Does the DBMS support star schemas? To determine which model is best, you need to identify the main goal of the application: is it high availability, high performance, transaction accuracy (ACID enforcement), or being able to manage a variety of types of data and complex relationships?

- *DBMS storage capacity.* What maximum disk and database sizes are required? How many disk packages must be supported? What is the minimum number of independent disk spindles required for the "recommended" installation? What are other storage needs? If using a cloud storage service, in addition to initial data size, special attention should be given to expected data growth rates because of contracted incremental data storage costs. Cloud storage introduces issues such as location, security, replication, redundancy, and data synchronization.

- *Application development support.* Which programming languages are supported? What application development tools are available? (Options include database schema design, a data dictionary, performance monitoring, and screen and menu painters.) Are end-user query tools provided? Does the DBMS provide web front-end access?

- *Security and integrity.* Does the DBMS support referential and entity integrity rules, access rights, and so on? Does the DBMS support the use of audit trails to spot errors and security violations? Can the audit trail's size be modified? If the data is stored in a public cloud, how secure is the data?

- *Backup and recovery.* Does the DBMS provide automated backup and recovery tools? Does the DBMS support tape, optical disc, or network-based backups? Does the DBMS automatically back up the transaction logs?

- *Concurrency control.* Does the DBMS support multiple users? What levels of isolation (table, page, row) does the DBMS offer? How much manual coding is needed in the application programs?

- *Performance.* How many transactions per second does the DBMS support? Are additional transaction processors needed? Is an in-memory database required to ensure top performance?

- *Database administration tools.* Does the DBMS offer some type of DBA management interface? What type of information does the DBA interface provide? Does the DBMS provide alerts to the DBA when errors or security violations occur?

- *Interoperability and data distribution.* Can the DBMS work with other DBMS types in the same environment? What coexistence or interoperability level is achieved? Does the DBMS support read and write operations to and from other DBMS packages? Does the DBMS support a client/server architecture? Would a cloud-based data service be a better choice for the given system?

- *Portability and standards.* Can the DBMS run on different operating systems and platforms? Can the DBMS run on mainframes, midrange computers, and personal computers? Can the DBMS applications run without modification on all platforms? What national and industry standards does the DBMS follow?

- *Hardware.* What hardware does the DBMS require? Can the DBMS run in a virtual machine? Does the DBMS implementation require the use of hardware clusters or a distributed environment?

- *Data dictionary.* Does the DBMS have an "accessible" data dictionary? Does the DBMS interface with any data dictionary tool? Does the DBMS support any open management tools?

- *Vendor training and support.* Does the vendor offer in-house training? What type and level of support does the vendor provide? Is the DBMS documentation easy to read and helpful? What is the vendor's upgrade policy?

- *Available third-party tools.* What additional tools are offered by third-party vendors? Do they include query tools, a data dictionary, access management and control, and storage allocation management tools?

- *Costs.* What costs are involved in the acquisition of the software and hardware? How many additional personnel are required, and what level of expertise is required of them? What are the recurring costs? What is the expected payback period?

If cloud data services are being considered, there are additional issues that need to be addressed with any potential cloud provider. Recall that the use of cloud databases frees the client organization from costs of acquiring and implementing the infrastructure as well as daily costs of maintenance. However, these services come with a loss of control over the data and the infrastructure. Any potential cloud-based vendors need to be evaluated based on several factors, including:

- *Downtime history.* Historically, how often are the cloud provider's services unavailable, and what provisions will they make to ensure that your data is always accessible?

- *Security.* How does the provider secure your data using firewalls, authentication, security audits, and encryption? Who at the cloud company will have access to your data files?

- *Support.* What customer support options are available if the client has issues or concerns with the data services provided?

- *Data loss contingencies.* The expectation is that the cloud provider will keep the data safe. However, what happens if they lose the client's data? What type of compensation or insurance against data loss is provided? What types of redundancies and backups are used to ensure that data loss will not happen? Where are the backups and redundancies kept to ensure that a natural disaster in one geographic area cannot cause the loss of all copies of the data?

Pros and cons of several alternative solutions must be evaluated during the selection process. Available alternatives are often restricted because software must be compatible with the organization's existing computer system. Remember that a DBMS is just part of a solution; it requires support from collateral hardware, application software, and utility programs. For example, the DBMS's use is likely to be constrained by the available CPU(s), front-end processor(s), auxiliary storage devices, data communication devices, the operating system, a transaction processor system, and so on. The costs associated with the hardware and software components must be included in the estimations.

The selection process must also consider the site's preparation costs. For example, the DBA must include both one-time and recurring expenditures for preparing and maintaining the computer room installations.

The DBA must supervise the installation of all software and hardware that supports the data administration strategy, and must thoroughly understand the components being installed, including their installation, configuration, and startup procedures. The installation procedures include the location of backup and transaction log files, network configuration information, and physical storage details.

Keep in mind that installation and configuration details are DBMS-dependent. Therefore, such details cannot be addressed in this book. Consult the installation and configuration sections of your system's DBMS administration guide for details.

Designing and Implementing Databases and Applications The DBA also provides data-modeling and design services to end users. Such services are often coordinated with an application development group within the data-processing department. Therefore, one of the primary activities of a DBA is to determine and enforce standards and procedures to be used. Once a framework of appropriate standards and procedures are in place, the DBA must ensure that the database-modeling and design activities are performed within the framework. The DBA then provides necessary assistance and support during the design of the database at the conceptual, logical, and physical levels. (Remember that the conceptual design is both DBMS- and hardware-independent, the logical design is DBMS-dependent and hardware-independent, and the physical design is both DBMS- and hardware-dependent.)

The DBA function usually requires that several people be dedicated to database modeling and design activities. Those people might be grouped according to the organizational areas covered by the application. For example, database modeling and design personnel may be assigned to production systems, financial and managerial systems, or executive and decision support systems. The DBA schedules the design jobs to coordinate the data design and modeling activities. That coordination may require reassignment of available resources based on externally determined priorities.

The DBA also works with application programmers to ensure the quality and integrity of database design and transactions. Such support services include reviewing the database application design to ensure that transactions are:

- *Correct.* The transactions mirror real-world events.

- *Efficient.* The transactions do not overload the DBMS.

- *Compliant.* Transactions comply with integrity rules and standards.

These activities require personnel with broad database design and programming skills.

The implementation of the applications requires the implementation of the physical database. Therefore, the DBA must provide assistance and oversight during the physical design, including determination and creation of storage space, data loading, conversion, and database migration services. The DBA's implementation tasks also include the generation, compilation, and storage of the application's access plan. An **access plan** is a set of instructions generated when the application is compiled that predetermines how the application will access the database at run time. To be able to create and validate the access plan, the user must have the required rights to access the database (see Chapter 11, Database Performance Tuning and Query Optimization).

Before an application comes online, the DBA must develop, test, and implement the operational procedures required by the new system. Such procedures include training, security, and backup and recovery plans, as well as assigning responsibility for database

access plan
A set of instructions generated at application compilation time that is created and managed by a DBMS. The access plan predetermines how an application's query will access the database at run time.

control and maintenance. Finally, the DBA must authorize application users to access the database from which the applications draw the required data.

The addition of a new database might require fine-tuning or reconfiguring of the DBMS. Remember that the DBMS assists all applications by managing the shared corporate data repository. Therefore, when data structures are added or modified, the DBMS might require the assignment of additional resources to serve new and original users with equal efficiency (see Chapter 11).

Testing and Evaluating Databases and Applications The DBA must also provide testing and evaluation services for all database and end-user applications. These services are the logical extension of the design, development, and implementation services described in the preceding section. Testing procedures and standards must already be in place before any application program can be approved for use in the company.

Although testing and evaluation services are closely related to database design and implementation services, they usually are maintained independently. The reason for the separation is that application programmers and designers are often too close to the problem being studied to detect errors and omissions.

Testing usually starts with the loading of the "test bed" database, which contains test data for the applications. Its purpose is to check the data definition and integrity rules of the database and application programs.

The testing and evaluation of a database application cover all aspects of the system, from the simple collection and creation of data to its use and retirement. The evaluation process covers the following:

- Technical aspects of both the applications and the database; backup and recovery, security and integrity, use of SQL, and application performance must be evaluated

- Evaluation of the written documentation and procedures to ensure that they are accurate and easy to follow

- Observance of standards for naming, documenting, and coding

- Checking for data duplication conflicts with existing data

- The enforcement of all data validation rules

Following the thorough testing of all applications, the database, and the procedures, the system is declared operational and can be made available to end users.

Operating the DBMS, Utilities, and Applications DBMS operations can be divided into four main areas:

- System support

- Performance monitoring and tuning

- Backup and recovery

- Security auditing and monitoring

System support activities cover all tasks directly related to the day-to-day operations of the DBMS and its applications. These activities include filling out job logs, changing tape, and verifying the status of computer hardware, disk packages, and emergency power sources. System-related activities include periodic tasks such as running special programs and resource configurations for new and upgraded versions of database applications.

Performance monitoring and tuning require much of the DBA's attention and time. These activities are designed to ensure that the DBMS, utilities, and applications maintain satisfactory performance levels. To carry out performance monitoring and tuning tasks, the DBA must:

- Establish DBMS performance goals.

- Monitor the DBMS to evaluate whether the performance objectives are being met.

- Isolate the problem and find solutions if performance objectives are not met.

- Implement the selected performance solutions.

DBMSs often include performance-monitoring tools that allow the DBA to query database usage information. Performance-monitoring tools are available from many different sources: DBMS utilities are provided by third-party vendors, or they might be included in operating system utilities or transaction processor facilities. Most of the performance-monitoring tools allow the DBA to focus on selected system bottlenecks. The most common bottlenecks in DBMS performance tuning are related to the use of indexes, query optimization algorithms, and management of storage resources.

Because improper index selection can have a deleterious effect on system performance, most DBMS installations adhere to a carefully defined index creation and usage plan. Such a plan is especially important in a relational database environment.

To produce satisfactory performance, the DBA might train programmers and end users in the proper use of SQL statements. Typically, DBMS programming manuals and administration manuals contain useful performance guidelines and examples that demonstrate the proper use of SQL statements, both at the command line and within application programs. Because relational systems do not give the user an index choice within a query, the DBMS makes the index selection for the user. Therefore, the DBA should create indexes that can be used to improve system performance. (For examples of database performance tuning, see Chapter 11.)

Query optimization routines are usually integrated into the DBMS package, allowing few tuning options. Query optimization routines are oriented toward improving concurrent access to the database. Several database packages let the DBA specify parameters for determining the desired level of concurrency. Concurrency is also affected by the types of locks used by the DBMS and requested by the applications. Because concurrency is important to the efficient operation of the system, the DBA must be familiar with the factors that influence concurrency. (See Chapter 10, Transaction Management and Concurrency Control, for more information.)

During DBMS performance tuning, the DBA must also consider available storage resources in terms of both primary and secondary memory. The allocation of storage resources is determined when the DBMS is configured. Storage configuration parameters can be used to determine:

- The number of databases that may be opened concurrently

- The number of application programs or users supported concurrently

- The amount of primary memory (buffer pool size) assigned to each database and each database process

- The size and location of the log file (remember that these files are used to recover the database; the log files can be located in a separate volume to reduce the disk's head movement and to increase performance)

Performance-monitoring issues are DBMS-specific. Therefore, the DBA must become familiar with the DBMS manuals to learn the technical details involved in performance monitoring (see Chapter 11).

Because data loss could be devastating to the organization, *backup and recovery activities* are of primary concern during the DBMS operation. The DBA must establish a schedule for backing up database and log files at appropriate intervals. Backup frequency is dependent on the application type and on the relative importance of the data. All critical system components—the database, the database applications, and the transaction logs—must be backed up periodically.

Most DBMS packages include utilities that schedule automated database backups, either full or incremental. Although incremental backups are faster than full backups, an incremental backup requires the existence of a periodic full backup to be useful for recovery purposes.

Database recovery after a media or systems failure requires application of the transaction log to the correct database copy. The DBA must plan, implement, test, and enforce a "bulletproof" backup and recovery procedure.

Security auditing and monitoring assumes the appropriate assignment of access rights and the proper use of access privileges by programmers and end users. The technical aspects of security auditing and monitoring involve creating users, assigning access rights, and using SQL commands to grant and revoke access rights to users and database objects. The DBA also must periodically generate an audit trail report to find actual or attempted security violations. If any are found, the DBA must ascertain where the violations occurred, and if possible, who committed them. For a comprehensive discussion of database security, see Section 16-6.

Training and Supporting Users

Training people to use the DBMS and its tools is part of the DBA's technical activities. In addition, the DBA provides or secures technical training for applications programmers in the use of the DBMS and its utilities. Applications programmer training covers the use of the DBMS tools as well as the procedures and standards required for database programming.

Unscheduled, on-demand technical support for end users and programmers is also part of the DBA's activities. A technical troubleshooting procedure can be developed to facilitate such support. The procedure might include the development of a technical database to find solutions to common technical problems.

Part of the support is provided by interaction with DBMS vendors. Establishing good relationships with software suppliers is one way to ensure that the company has a good external support source. Vendors are the source for up-to-date information concerning new products and personnel retraining. Good vendor-company relations also are likely to give organizations an edge in determining the future direction of database development.

Maintaining the DBMS, Utilities, and Applications

The maintenance activities of the DBA are an extension of the operational activities. Maintenance activities are dedicated to the preservation of the DBMS environment.

Periodic DBMS maintenance includes management of the physical or secondary storage devices. One of the most common maintenance activities is reorganizing the physical location of data in the database. (This is usually done as part of the DBMS fine-tuning activities.) The reorganization of a database might be designed to allocate contiguous disk-page locations to the DBMS to increase performance. The reorganization process also might free the space allocated to deleted data, thus providing more disk space for new data.

Maintenance activities also include upgrading the DBMS and utility software. The upgrade might require installing a new version of the DBMS software or an Internet front-end tool. Or, it might create an additional DBMS gateway to allow access to a host DBMS running on a different host computer. DBMS gateway services are very common in distributed DBMS applications running in a client/server environment. Also, new-generation databases include features such as spatial data support, data warehousing and star query support, and support for Java programming interfaces for Internet access (see Chapter 15, Database Connectivity and Web Technologies).

Quite often companies are faced with the need to exchange data in dissimilar formats or between databases. The maintenance efforts of the DBA include migration and conversion services for data in incompatible formats or for different DBMS software. Such conditions are common when the system is upgraded from one version to another or when the existing DBMS is replaced by an entirely new DBMS. Database conversion services also include downloading data from the host DBMS (mainframe-based) to an end user's personal computer to allow the user to perform a variety of activities—spreadsheet analysis, charting, statistical modeling, and so on. Migration and conversion services can be done at the logical level (DBMS-specific or software-specific) or at the physical level (storage media or operating system-specific). Current-generation DBMSs support XML as a standard format for data exchange among database systems and applications (see Chapter 15).

16-6 Security

Information system security refers to activities and measures that ensure the confidentiality, integrity, and availability of an information system and its main asset, data.[4] Securing data requires a comprehensive, company-wide approach. That is, you cannot secure data if you do not secure all the processes and systems around it, including hardware systems, software applications, the network and its devices, internal and external users, procedures, and the data itself. To understand the scope of data security, consider each of the three security goals in more detail:

- **Confidentiality** deals with ensuring that data is protected against unauthorized access, and if the data is accessed by an authorized user, that it is used only for an authorized purpose. In other words, confidentiality entails safeguarding data against disclosure of any information that would violate the privacy rights of a person or organization. Data must be evaluated and classified according to the level of confidentiality: highly restricted (very few people have access), confidential (only certain groups have access), and unrestricted (can be accessed by all users). The data security officer spends a great amount of time ensuring that the organization is in compliance with desired levels of confidentiality. **Compliance** refers to activities that meet data privacy and security reporting guidelines. These guidelines are either part of internal procedures or are imposed by external regulatory agencies such as the federal government. Examples of U.S. legislation enacted to ensure data privacy and confidentiality include the Health Insurance Portability and Accountability Act (HIPAA), Gramm-Leach-Bliley Act (GLBA), and Sarbanes-Oxley Act (SOX).

- **Integrity**, within the data security framework, is concerned with keeping data consistent and free of errors or anomalies. (See Chapter 1 to review the concepts of data inconsistencies and data anomalies.) The DBMS plays a pivotal role in ensuring the integrity of the data in the database. However, from the security point of view, the organizational processes, users, and usage patterns also must maintain integrity. For example, a work-at-home employee using the Internet to access product costing could be considered an acceptable use; however, security standards might require the employee to use a secure connection and follow strict procedures to manage the data at home, such as shredding printed reports and using encryption to copy data to the local hard drive. Maintaining data integrity is a process that starts with data collection

confidentiality
In the context of data security, ensuring that data is protected against unauthorized access, and if the data is accessed by an authorized user, that the data is used only for an authorized purpose.

compliance
Activities that meet data privacy and security reporting guidelines or requirements.

integrity
In a data security framework, refers to keeping data consistent and free of errors or anomalies. See also *data integrity*.

[4]Krause, M. and Tipton, H., *Handbook of Information Security Management*, CRC Press LLC, 1999.

and continues with data storage, processing, usage, and archiving (see Chapter 13, Business Intelligence and Data Warehouses). The rationale behind integrity is to treat data as the most-valuable asset in the organization and to ensure that rigorous data validation is carried out at all levels within the organization.

- **Availability** refers to the accessibility of data whenever required by authorized users and for authorized purposes. To ensure data availability, the entire system must be protected from service degradation or interruption caused by any internal or external source. Service interruptions could be very costly for companies and users alike. System availability is an important goal of security.

16-6a Security Policies

Normally, the tasks of securing the system and its main asset, the data, are performed by the database security officer and the database administrator(s), who work together to establish a cohesive data security strategy. Such a strategy begins with defining a sound and comprehensive security policy. A **security policy** is a collection of standards, policies, and procedures created to guarantee the security of a system and ensure auditing and compliance. The security audit process starts by identifying security vulnerabilities in the organization's information system infrastructure and identifying measures to protect the system and data against those vulnerabilities.

16-6b Security Vulnerabilities

A **security vulnerability** is a weakness in a system component that could be exploited to allow unauthorized access or cause service disruptions. Such vulnerabilities could fall under one of the following categories:

- *Technical.* An example would be a flaw in the operating system or web browser.
- *Managerial.* For example, an organization might not educate users about critical security issues.
- *Cultural.* Users might hide passwords under their keyboards or forget to shred confidential reports.
- *Procedural.* Company procedures might not require complex passwords or the checking of user IDs.

When a security vulnerability is left unchecked, it could become a security threat. A **security threat** is an imminent security violation.

A **security breach** occurs when a security threat is exploited to endanger the integrity, confidentiality, or availability of the system. Security breaches can lead to a database whose integrity is either preserved or corrupted:

- *Preserved.* In these cases, action is required to avoid the recurrence of similar security problems, but data recovery may not be necessary. As a matter of fact, most security violations are produced by unauthorized and unnoticed access for information purposes, but such snooping does not disrupt the database.
- *Corrupted.* Action is required to avoid the recurrence of similar security problems, and the database must be recovered to a consistent state. Corrupting security breaches include database access by computer viruses and by hackers who intend to destroy or alter data.

Table 16.4 illustrates some security vulnerabilities of system components and typical protective measures against them.

availability In the context of data security, it refers to the accessibility of data whenever required by authorized users and for authorized purposes.

security policy A collection of standards, policies, and procedures created to guarantee the security of a system and ensure auditing and compliance.

security vulnerability A weakness in a system component that could be exploited to allow unauthorized access or cause service disruptions.

security threat An imminent security violation that could occur due to unchecked security vulnerabilities.

security breach An event in which a security threat is exploited to endanger the integrity, confidentiality, or availability of the system.

TABLE 16.4

SAMPLE SECURITY VULNERABILITIES AND RELATED PROTECTIVE MEASURES

SYSTEM COMPONENT	SECURITY VULNERABILITY	SECURITY MEASURES
People	• The user sets a blank password. • The password is short or includes a birth date. • The user leaves the office door open all the time. • The user leaves payroll information on the screen for long periods of time.	• Enforce complex password policies. • Use multilevel authentication. • Use security screens and screen savers. • Educate users about sensitive data. • Install security cameras. • Use automatic door locks.
Workstation and servers	• The user copies data to a flash drive. • The workstation is used by multiple users. • A power failure crashes the computer. • Unauthorized personnel can use the computer. • Sensitive data is stored on a laptop computer. • Data is lost due to a stolen hard disk or laptop. • A natural disaster occurs.	• Use group policies to restrict the use of flash drives. • Assign user access rights to workstations. • Install uninterrupted power supplies (UPSs). • Add security locks to computers. • Implement a kill switch for stolen laptops. • Create and test data backup and recovery plans. • Protect the system against natural disasters—use co-location strategies.
Operating system	• Buffer overflow attacks • Virus attacks • Root kits and worm attacks • Denial-of-service attacks • Trojan horses • Spyware applications • Password crackers	• Apply OS security patches and updates. • Apply application server patches. • Install antivirus and antispyware software. • Enforce audit trails on the computers. • Perform periodic system backups. • Install only authorized applications. • Use group policies to prevent unauthorized installations.
Applications	• Application bugs—buffer overflow • SQL injection, session hijacking, etc. • Application vulnerabilities—cross-site scripting, nonvalidated inputs • Email attacks—spamming, phishing, etc. • Social engineering emails	• Test application programs extensively. • Build safeguards into code. • Do extensive vulnerability testing in applications. • Install spam filters and antivirus software for email systems. • Use secure coding techniques (see *www.owasp.org*). • Educate users about social engineering attacks.
Network	• IP spoofing • Packet sniffers • Hacker attacks • Clear passwords on network	• Install firewalls. • Use virtual private networks (VPNs). • Use intrusion detection systems (IDSs). • Use network access control (NAC). • Use network activity monitoring.
Data	• Data shares are open to all users. • Data can be accessed remotely. • Data can be deleted from a shared resource.	• Implement file system security. • Implement share access security. • Use access permission. • Encrypt data at the file system or database level.

16-6c Database Security

Database security refers to DBMS features and other related measures that comply with the organization's security requirements. From the DBA's point of view, security measures should be implemented to protect the DBMS against service degradation and to protect the database against loss, corruption, or mishandling. In short, the DBA should secure the DBMS from the point of installation through operation and maintenance.

Note

James Martin's excellent description of the desirable attributes of a database security strategy remains relevant today (*Managing the Database Environment*, Prentice-Hall, 1977). Martin's security strategy is based on the seven essentials of database security, and may be summarized as one in which data is protected, reconstructable, auditable, and tamper-proof, and users are identifiable, authorized, and monitored.

To protect the DBMS against service degradation, some security safeguards are recommended. For example:

- Change default system passwords.
- Change default installation paths.
- Apply the latest patches.
- Secure installation folders with proper access rights.
- Make sure that only required services are running.
- Set up auditing logs.
- Set up session logging.
- Require session encryption.

Furthermore, the DBA should work closely with the network administrator to implement network security that protects the DBMS and all services running on the network. In modern organizations, one of the most critical components in the information architecture is the network.

Protecting the data in the database is a function of authorization management. **Authorization management** defines procedures to protect and guarantee database security and integrity. Those procedures include the following:

- *User access management.* This function is designed to limit access to the database; it likely includes at least the following procedures:

  - *Define each user to the database.* The DBA performs this function at the operating system level and the DBMS level. At the operating system level, the DBA can request the creation of a unique user ID for each end user who logs on to the computer system. At the DBMS level, the DBA can either create a different user ID or employ the same one to authorize the end user to access the DBMS.

  - *Assign passwords to each user.* The DBA also performs this function at both the operating system and DBMS levels. The database passwords can be assigned with predetermined expiration dates, which enable the DBA to screen end users periodically and remind them to change their passwords, thus making unauthorized access less likely.

database security
The use of DBMS features and other related measures to comply with the security requirements of an organization.

authorization management
Procedures that protect and guarantee database security and integrity. Such procedures include user access management, view definition, DBMS access control, and DBMS usage monitoring.

- *Define user groups.* Classifying users into groups according to common access needs can help the DBA control and manage the access privileges of individual users. Also, the DBA can use database roles and resource limits to minimize the impact of rogue users in the system. (See Section 16-10d for more information about these topics.)

- *Assign access privileges.* The DBA assigns access privileges to specific users to access certain databases. Access rights may be limited to read-only, or the authorized access might include read, write, and delete privileges. Access privileges in relational databases are assigned through SQL GRANT and REVOKE commands.

- *Control physical access.* Physical security can prevent unauthorized users from directly accessing the DBMS installation and facilities. Common physical security for large database installations includes secured entrances, password-protected workstations, electronic personnel badges, closed-circuit video, voice recognition, and biometric technology.

- *View definition.* The DBA must define data views to protect and control the scope of the data that are accessible to an authorized user. The DBMS must provide tools that allow the definition of views composed of one or more tables, and must assign access rights to users. The SQL CREATE VIEW command is used in relational databases to define views. Oracle DBMS offers Virtual Private Database (VPD), which allows the DBA to create customized views of the data for different users. With this feature, the DBA could restrict regular users who query a payroll database to see only the necessary rows and columns, while department managers would see only the rows and columns pertinent to their departments.

- *DBMS access control.* Database access can be controlled by placing limits on the use of DBMS query and reporting tools. The DBA must make sure the tools are used properly and only by authorized personnel.

- *DBMS usage monitoring.* The DBA must also audit the use of data in the database. Several DBMS packages contain features that allow the creation of an **audit log**, which automatically records a brief description of database operations performed by all users. Such audit trails enable the DBA to pinpoint access violations. The audit trails can be tailored to record all database accesses or just failed ones.

The integrity of a database could be lost because of external factors beyond the DBA's control. For example, the database might be damaged or destroyed by an explosion, a fire, or an earthquake. Whatever the reason, the specter of database corruption or destruction makes backup and recovery procedures crucial to any DBA.

16-7 **Database Administration Tools**

The extraordinary growth of data management activities within organizations created the need for better management standards, processes, and tools. Over the years, a new industry arose dedicated exclusively to data administration tools. These tools cover the entire spectrum of data administration tasks, from selection to inception, deployment, migration, and day-to-day operations. For example, you can find sophisticated data administration tools for:

- Database monitoring
- Database load testing
- Database performance tuning

audit log
A security feature of a database management system that automatically records a brief description of the database operations performed by all users.

- SQL code optimization
- Database bottleneck identification and remediation
- Database modeling and design
- Database data extraction, transformation, and loading

All the above-mentioned tools have something in common. They all expand the database's metadata or data dictionary. The importance of the data dictionary as a DBA tool cannot be overstated. This section examines the data dictionary as a data administration tool, as well as the DBA's use of computer-aided systems engineering (CASE) tools to support database analysis and design.

16-7a The Data Dictionary

In Chapter 1, a *data dictionary* was defined as "a DBMS component that stores the definition of data characteristics and relationships." You may recall that such "data about data" are called *metadata*. The DBMS data dictionary provides the DBMS with its self-describing characteristic. In effect, the data dictionary resembles an x-ray of the company's entire data set, and it is a crucial element in data administration.

Two main types of data dictionaries exist: *integrated* and *standalone*. An integrated data dictionary is included with the DBMS. For example, all relational DBMSs include a built-in data dictionary or system catalog that is frequently accessed and updated by the RDBMS. Other DBMSs, especially older types, do not have a built-in data dictionary; instead, the DBA may use third-party *standalone* systems.

Data dictionaries can also be classified as *active* or *passive*. An **active data dictionary** is automatically updated by the DBMS with every database access to keep its access information up to date. A **passive data dictionary** is not updated automatically and usually requires running a batch process. Data dictionary access information is normally used by the DBMS for query optimization.

The data dictionary's main function is to store the description of all objects that interact with the database. Integrated data dictionaries tend to limit their metadata to the data managed by the DBMS. Standalone data dictionary systems are usually more flexible and allow the DBA to describe and manage all of the organization's data, whether they are computerized or not. Whatever the data dictionary's format, it provides database designers and end users with a much-improved ability to communicate. In addition, the data dictionary is the tool that helps the DBA resolve data conflicts.

Although there is no standard format for the information stored in the data dictionary, several features are common. For example, the data dictionary typically stores descriptions of the following:

- *Data elements that are defined in all tables of all databases.* Specifically, the data dictionary stores element names, data types, display format, internal storage format, and validation rules. The data dictionary explains where an element is used, who used it, and so on.

- *Tables defined in all databases.* For example, the data dictionary is likely to store the name of the table creator, the date of creation, access authorizations, and the number of columns.

- *Indexes defined for each database table.* For each index, the DBMS stores at least the index name, the attributes used, the location, specific index characteristics, and the creation date.

- *Defined databases.* This information includes who created each database, when the database was created, where the database is located, the DBA's name, and so on.

active data dictionary
A data dictionary that is automatically updated by the database management system every time the database is accessed, thereby keeping its information current.

passive data dictionary
A DBMS data dictionary that requires a command initiated by an end user to update its data access statistics.

- *End users and administrators of the database.* This information defines the users of the database.

- *Programs that access the database.* This information includes screen formats, report formats, application programs, and SQL queries.

- *Access authorizations for all users of all databases.* This information defines who can manipulate which objects and what types of operations can be performed.

- *Relationships among data elements.* This information includes which elements are involved, whether the relationships are mandatory or optional, and connectivity and cardinality requirements.

If the data dictionary can be organized to include data external to the DBMS itself, it becomes an especially flexible tool for more general corporate resource management. Such an extensive data dictionary thus makes it possible to manage the use and allocation of all of the organization's information, regardless of whether it has its roots in the database data. For this reason, some managers consider the data dictionary to be a key element of information resource management, which is why the data dictionary can be described as the **information resource dictionary**.

The metadata stored in the data dictionary is often the basis for monitoring database use and for assigning access rights to database users. The information stored in the data dictionary is usually based on a relational table format, thus enabling the DBA to query the database with SQL commands. For example, SQL commands can be used to extract information about the users of a specific table or the access rights of a particular user. In the following section, the IBM DB2 system catalog tables are the basis for several examples of how a data dictionary is used to derive information:

- SYSTABLES stores one row for each table or view.

- SYSCOLUMNS stores one row for each column of each table or view.

- SYSTABAUTH stores one row for each authorization given to a user for a table or view in a database.

Examples of Data Dictionary Usage

Example 1

List the names and creation dates of all tables created by the user JONESVI in the current database.

SELECT NAME, CTIME
FROM SYSTABLES
WHERE CREATOR = 'JONESVI';

Example 2

List the names of the columns for all tables created by JONESVI in the current database.

SELECT NAME
FROM SYSCOLUMNS
WHERE TBCREATOR = 'JONESVI';

Example 3

List the names of all tables for which the user JONESVI has DELETE authorization.

information resource dictionary
Another name for *data dictionary.*

```
SELECT TTNAME
FROM SYSTABAUTH
WHERE GRANTEE = 'JONESVI' AND DELETEAUTH = 'Y';
```

Example 4

List the names of all users who have some type of authority over the INVENTORY table.

```
SELECT DISTINCT GRANTEE
FROM SYSTABAUTH
WHERE TTNAME = 'INVENTORY';
```

Example 5

List the user and table names for all users who can alter the database structure for any table in the database.

```
SELECT GRANTEE, TTNAME
FROM SYSTABAUTH
WHERE ALTERAUTH = 'Y'
ORDER BY GRANTEE, TTNAME;
```

As you can see in the preceding examples, the data dictionary can be a tool for monitoring database security by checking the assignment of data access privileges. Although the preceding examples targeted database tables and users, information about the application programs that access the database can also be drawn from the data dictionary.

The DBA can use the data dictionary to support data analysis and design. For example, the DBA can create a report that lists all data elements to be used in a particular application; a list of all users who access a particular program; a report that checks for data redundancies, duplications, and the use of homonyms and synonyms; and a number of other reports that describe data users, data access, and data structure. The data dictionary can also be used to ensure that application programmers have met the naming standards for data elements in the database, and that the data validation rules are correct. Thus, the data dictionary can be used to support a wide range of data administration activities and facilitate the design and implementation of information systems. Integrated data dictionaries are also essential to the use of computer-aided systems engineering tools.

16-7b Case Tools

computer-aided systems engineering (CASE)
Tools used to automate part or all of the Systems Development Life Cycle.

front-end CASE tool
A computer-aided software tool that provides support for the planning, analysis, and design phases of the SDLC.

back-end CASE tool
A computer-aided software tool that provides support for the coding and implementation phases of the SDLC.

CASE is the acronym for **computer-aided systems engineering**. A CASE tool provides an automated framework for the Systems Development Life Cycle (SDLC). CASE uses structured methodologies and powerful graphical interfaces. Because they automate many tedious system design and implementation activities, CASE tools play an increasingly important role in information systems development.

CASE tools are usually classified according to the extent of support they provide for the SDLC. For example, **front-end CASE tools** provide support for the planning, analysis, and design phases; **back-end CASE tools** provide support for the coding and implementation phases. The benefits associated with CASE tools include:

- A reduction in development time and costs

- Automation of the SDLC

- Standardization of systems development methodologies

- Easier maintenance of application systems developed with CASE tools

One of the CASE tools' most important components is an extensive data dictionary, which keeps track of all objects created by the systems designer. For example, the CASE data dictionary stores data flow diagrams, structure charts, descriptions of all external and internal entities, data stores, data items, report formats, and screen formats. A CASE data dictionary also describes the relationships among system components.

Several CASE tools provide interfaces that work with the DBMS and allow the CASE tool to store its data dictionary information using the DBMS. Such interaction demonstrates the interdependence that exists between systems development and database development, and it helps create a fully integrated development environment.

In a CASE development environment, database and application designers use the CASE tool to store the description of the database schema, data elements, application processes, screens, reports, and other data relevant to development. The CASE tool integrates all systems development information in a common repository, which the DBA can check for consistency and accuracy.

As an additional benefit, a CASE environment tends to improve the extent and quality of communication among the DBA, application designers, and end users. The DBA can use the CASE tool to check the definition of the application's data schema, the observance of naming conventions, the duplication of data elements, validation rules for the data elements, and a host of other developmental and managerial variables. When the CASE tool finds conflicts, rules violations, and inconsistencies, it facilitates making corrections. Better yet, the CASE tool can make a correction and then cascade its effects throughout the applications environment, which greatly simplifies the job of the DBA and the application designer.

A typical CASE tool provides five components:

- Graphics designed to produce structured diagrams such as data flow diagrams, ER diagrams, class diagrams, and object diagrams

- Screen painters and report generators to produce the information system's input and output formats (for example, the end-user interface)

- An integrated repository for storing and cross-referencing the system design data; this repository includes a comprehensive data dictionary

- An analysis segment to provide a fully automated check on system consistency, syntax, and completeness

- A program documentation generator

Figure 16.7 illustrates how Microsoft Visio Professional can be used to produce an ER diagram.

Most CASE tools, produce fully documented ER diagrams that can be displayed at different abstraction levels. For example, ERwin Data Modeler by Computer Associates can produce detailed relational designs. The user specifies the attributes and primary keys for each entity and describes the relations. Current generation data modeling tools assign foreign keys based on the specified relationships among the entities. Changes in primary keys are always updated automatically throughout the system. Table 16.5 shows a short list of the many available CASE Data Modeling tool vendors.

FIGURE 16.7 AN EXAMPLE OF A CASE TOOL: MICROSOFT VISIO PROFESSIONAL

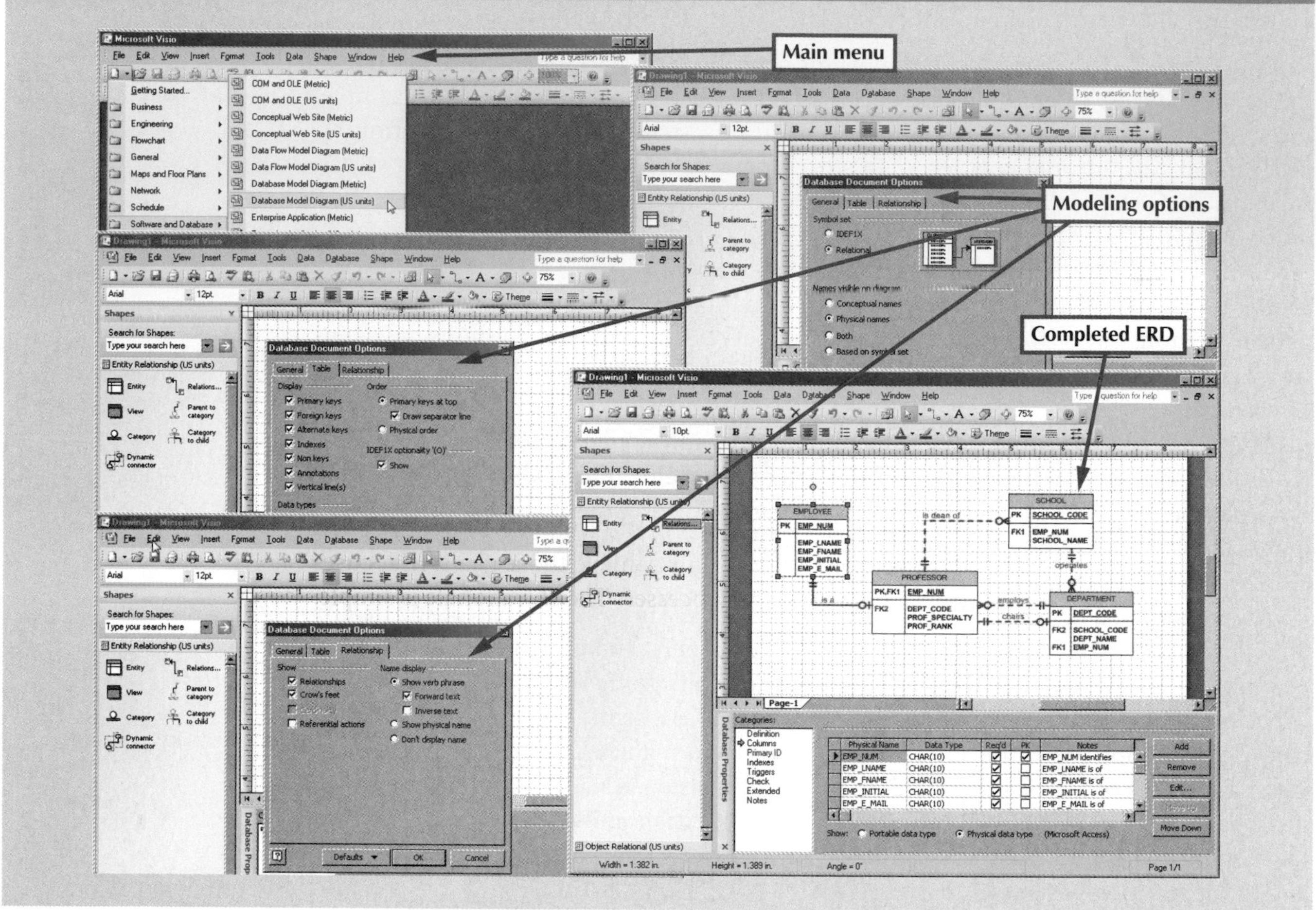

TABLE 16.5

CASE DATA MODELING TOOLS

COMPANY	PRODUCT	WEBSITE
Casewise	Corporate Modeler Suite	*www.casewise.com*
Computer Associates	ERwin	*www.erwin.com*
Embarcadero Technologies	ER/Studio	*www.embarcadero.com/products/er-studio-data-architect*
Microsoft	Visio	*office.microsoft.com/en-us/visio*
Oracle	SQL Developer Data Modeler	*www.oracle.com/technetwork/developer-tools/datamodeler/overview/index.html*
IBM	Rational Software Architect	*www-01.ibm.com/software/rational/products/swarchitect/*
SAP	Power Designer	*http://www.sap.com/pc/tech/database/software/model-driven-architecture/index.html*
Visible	Visible Analyst	*www.visible.com/Products/Analyst*

Major relational DBMS vendors, such as Oracle, now provide fully integrated CASE tools for their own DBMS software as well as for RDBMSs supplied by other vendors. For example, Oracle's CASE tools can be used with IBM's DB2, and Microsoft's SQL Server to produce fully documented database designs. Some vendors even take nonrelational DBMSs, develop their schemas, and produce the equivalent relational designs automatically.

There is no doubt that CASE tools have enhanced the efficiency of database designers and application programmers. However, no matter how sophisticated the CASE tool, its users must be well versed in conceptual design. In the hands of database novices, CASE tools produce impressive-looking but bad designs.

16-8 Developing a Data Administration Strategy

For a company to succeed, its activities must be committed to its main objectives or mission. Therefore, regardless of its size, a critical step for any organization is to ensure that its information system supports its strategic plans for each business area.

The database administration strategy must not conflict with the information systems plans. After all, these plans are derived from a detailed analysis of the company's goals, its condition or situation, and its business needs. Several methodologies are available to ensure the compatibility of data administration and information systems plans and to guide strategic plan development. The most commonly used methodology is known as information engineering.

Information engineering (IE) allows for translation of the company's strategic goals into the data and applications that will help the company achieve those goals. IE focuses on the description of corporate data instead of the processes. The IE rationale is simple: business data types tend to remain fairly stable, but processes change often and thus require frequent modification of existing systems. By placing the emphasis on data, IE helps decrease the impact on systems when processes change.

The output of the IE process is an **information systems architecture (ISA)** that serves as the basis for planning, development, and control of future information systems. Figure 16.8 shows the forces that affect ISA development.

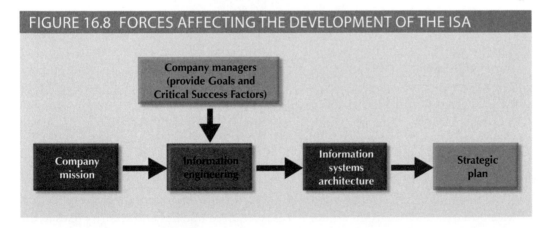

FIGURE 16.8 FORCES AFFECTING THE DEVELOPMENT OF THE ISA

Implementing IE in an organization is a costly process that involves planning, a commitment of resources, management liability, well-defined objectives, identification of critical factors, and control. An ISA provides a framework that includes computerized, automated, and integrated tools such as a DBMS and CASE tools.

The success of the overall information systems strategy and data administration strategy depends on several critical success factors that the DBA needs to understand. Critical success factors include the following managerial, technological, and corporate culture issues:

- *Management commitment.* The commitment of top-level management is necessary to enforce the use of standards, procedures, planning, and controls. The example must be set at the top.

- *Thorough analysis of the company situation.* The current state of the corporate data administration must be analyzed to understand the company's position and to have

information engineering (IE)
A methodology that translates a company's strategic goals into helpful data and applications. IE focuses on the description of corporate data instead of the processes.

information systems architecture (ISA)
The output of the information engineering (IE) process that serves as the basis for planning, developing, and controlling future information systems.

a clear vision of what must be done. For example, how are database analysis, design, documentation, implementation, standards, codification, and other issues handled? Needs and problems should be identified first and then prioritized.

- *End-user involvement.* What degree of organizational change is involved? Successful change requires that people be able to adapt to it. Users should have an open communication channel to upper management to ensure success of the implementation. Good communication is key to the overall process.

- *Defined standards.* Analysts and programmers must be familiar with appropriate methodologies, procedures, and standards. If not, they might need training.

- *Training.* The vendor must train DBA personnel in the use of the DBMS and other tools. End users must be trained to use the tools, standards, and procedures. Key personnel should be trained first so they can train others.

- *A small pilot project.* A small project is recommended to ensure that the DBMS will work in the company, that it produces expected output, and that the personnel have been trained properly.

This list of factors is not comprehensive, but it does provide the framework for developing a successful strategy. Remember that no matter how comprehensive you make the list, it must be based on developing and implementing a data administration strategy that is tightly integrated with the organization's overall information systems planning.

Developing a comprehensive data administration strategy within an organization is a large undertaking encompassing technical, operational, and managerial roles. Enterprises today also have the option of moving entire computing functions (such as servers, storage, backup, and even the database) outside the walls of the enterprise and into the cloud.

16-9 The DBA's Role in the Cloud

The use of cloud-based data services does not signal the end of DBAs, but it does have a significant impact on their role. As discussed in previous chapters, services such as Microsoft Azure and Amazon Web Services (AWS) allow outsourcing database technology as a highly scalable, capability-on-demand service. In this new world, some of the tasks that once resided in a single "in-house" DBA function are now split between the internal DBA and the cloud service provider. As a result, the use of cloud-based data services alters and expands the typical DBA's role in both technical and managerial dimensions. In general, the cloud services partner company provides:

- *DBMS installation and updates.* The DBMS is installed on a virtual server by the service provider. As the DBMS vendor releases required updates and security fixes to the DBMS software, the service provider manages the application of the updates within a specified maintenance window. The DBA's role now has to carefully coordinate such updates with the external cloud-based data service provider.

- *Server/network management.* The service provider configures and manages the server where the DBMS resides, including scaling the database across multiple servers as needed. If the database is distributed across multiple servers, the service provider can supply load balancing to ensure a high level of performance. However, the DBA must work with his or her /her company's network department to ensure that the network is properly configured for security, performance, availability and management.

- *Backup and recovery operations.* The service provider performs regular backups and stores backups in secure facilities. The DBA must ensure that internal data privacy and retention policies are enforced and maintained.

Although these services are valuable and free the DBA from these tasks, the primary benefit of cloud-based data services is their ability to provide and manage computing hardware and software configuration at a low cost. The preceding tasks are only a small part of the DBA's responsibilities; the DBA's managerial role is largely unchanged and sometimes is even augmented with the new cloud data services dimension. User requirements must still be gathered; data solutions must still be designed; end users need training; and policies, standards, and procedures must be developed and enforced.

Even the technical role of the DBA still exists with the use of cloud data services. There are many cloud data service providers, and some offer a variety of DBMS products, including proprietary systems. Only some versions of these DBMSs are available, including multiple versions of the same DBMS. For example, a given service provider may support both MySQL 5.1 and MySQL 5.5. In this environment, the DBA evaluates different DBMSs to determine which software product to use, and evaluates from which provider to purchase the DBMS. In addition, the DBA must work with the cloud data services provider to reconcile the required database technical features with the ones supported by the cloud data service provider and ensure data availability, security, and integrity within the expanded boundaries of the company network.

A variety of pricing schemes are offered by cloud data service providers. Pricing is typically based on factors such as storage space, computing resources (CPU cycles and memory), and data transfer sizes. Service users are billed monthly for the amount of resources used. Service providers have a vested interest in their clients' databases being as large as possible; it is also in their interest for database designs to be inefficient in processing queries because clients will have to buy more memory and CPU capacity. Service providers benefit if your database is filled with poorly designed tables that contain lots of unnecessarily redundant data, with every attribute in every table indexed, and queries that take a long time to run or return thousands of rows of data that must be transferred to a front-end application for additional processing. Therefore, the DBA can save the organization significant time and money by ensuring that databases are properly designed with minimal redundancy and that database coding is efficient. Clearly, the DBA's technical role is still critical to organizations that use cloud-based data services. The DBA's efforts in efficient and effective database design, coding, monitoring database performance, and database tuning still affect the organization's ability to use data and information as a resource, and they have an immediate visible impact on the monthly data service bill.

Regardless of whether the database is stored in the enterprise's server or in the cloud, the DBA must ensure the data's availability, security, and integrity.

16-10 The DBA at Work: Using Oracle for Database Administration

Thus far, you have learned about the DBA's work environment and responsibilities in general terms. This section provides a more detailed look at how a DBA might handle the following technical tasks in a specific DBMS:

- Creating and expanding database storage structures

- Managing the end-user database environment, including the type and extent of database access

- Customizing database initialization parameters

Many of these tasks require the DBA to use software tools and utilities that are commonly provided by the database vendor. In fact, all DBMS vendors provide a set of programs to interface with the database and to perform a wide range of database administrative tasks.

Oracle 12c for Windows is used to illustrate selected DBA tasks in this section because Oracle is typically used in organizations that are large and complex enough to employ a DBA. Also, Oracle has good market presence and is often used in small colleges and universities.

Note

Although Microsoft Access is a superb DBMS, it is typically used in smaller organizations or in organizations and departments with relatively simple data environments. Access has a superior database prototyping environment, and its easy-to-use GUI tools enable rapid front-end application development. Also, Access is a component in the MS Office suite, which makes applications integration relatively simple and seamless for end users. Finally, while Access does provide some important database administration tools, an Access-based database environment does not typically require a DBA, so MS Access is not a good fit for this section.

Most of the tasks described in this section are not particular to any DBMS or operating system. However, the *execution* of those tasks tends to be specific to the DBMS and operating system. Therefore, if you use IBM DB2 Universal Database or Microsoft SQL Server, you must adapt the procedures shown here to your DBMS. Also, these examples run under the Windows operating system, so you must adapt the procedures shown in this section if you use a different OS.

This section is not a database administration manual; it offers a brief introduction to performing typical DBA tasks in Oracle. Before learning these tasks, you should become familiar with Oracle's database administration tools and its procedures for logging on. These tools and procedures are discussed in the next two sections.

Note

Although the format of creating a database tends to be generic, its execution tends to be DBMS-specific. For a step-by-step procedure of creating a database using the Oracle Database Configuration Assistant, see Appendix N, Creating a New Database Using Oracle 12c.

16-10a Oracle Database Administration Tools

All database vendors supply a set of database administration tools. In Oracle, you perform most DBA tasks via the Oracle Enterprise Manager interface. (See Figure 16.9.)

Note that the interface shows the status of the current database. (This section uses the BASEORA database.) In the following sections, you examine the tasks most commonly encountered by a DBA.

16-10b Ensuring that the RDBMS Starts Automatically

One of a DBA's basic tasks is to ensure that database access starts automatically when you turn on the computer. Startup procedures are different for each operating system. Oracle is used for this section's examples; if you use a different system, you need to identify the required services to ensure automatic database startup. A *service* is the Windows name

FIGURE 16.9 THE ORACLE ENTERPRISE MANAGER EXPRESS INTERFACE

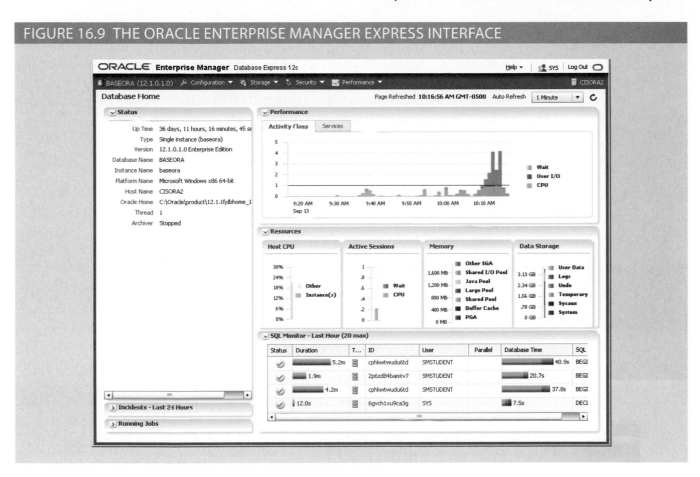

for a special program that runs automatically as part of the operating system. This program ensures the availability of required services to the system and to end users on the local computer or the network. Figure 16.10 shows the required Oracle services that are started automatically when Windows starts.

FIGURE 16.10 ORACLE RDBMS SERVICES

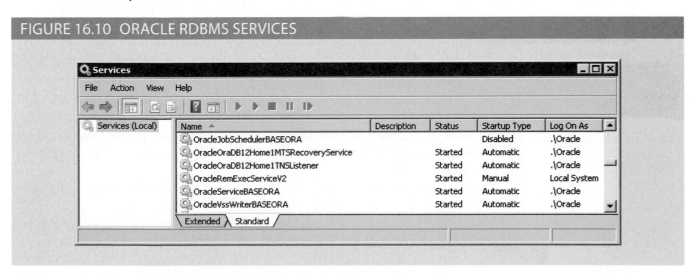

As you examine Figure 16.10, note the following Oracle services:

- *OracleOraDB12Home1TNSListener* is the process that "listens to" and processes end-user connection requests over the network. For example, when a SQL connection request such as "connect userid/password@BASEORA" is sent over the network, the listener service will validate the request and establish the connection.

- *OracleServiceBASEORA* refers to the Oracle processes running in memory that are associated with the BASEORA database instance. You can think of a **database instance** as a separate location in memory that is reserved to run your database. Because you can have several databases (and therefore several instances) running in memory at the same time, you need to identify each database instance uniquely, using a different suffix for each one.

16-10c Creating Tablespaces and Datafiles

Each DBMS manages data storage differently. In this example, the Oracle RDBMS is used to illustrate how the database manages data storage at the logical and physical levels. In Oracle,

- A database is *logically* composed of one or more tablespaces. A **tablespace** is a logical storage space. Tablespaces are used primarily to group related data logically.

- The tablespace data is *physically* stored in one or more datafiles. A **datafile** physically stores the database's data. Each datafile is associated with only one tablespace, but each datafile can reside in a different directory on the physical storage devices. For example, in Figure 16.11, the USERS tablespace data is physically stored in the datafile *users01.dbf*.

FIGURE 16.11 ORACLE STORAGE MANAGEMENT

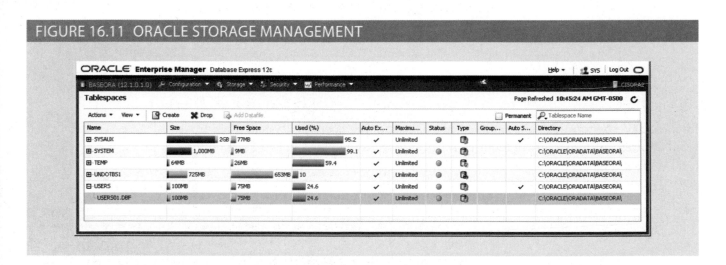

Given the preceding descriptions, you can conclude that a database has a one-to-many relationship with tablespaces and that a tablespace has a one-to-many relationship with datafiles. This set of 1:M hierarchical relationships isolates the end user from any physical details of data storage. However, *the DBA must be aware of these details to properly manage the database.*

To manage database storage, such as creating and managing tablespaces and datafiles, the DBA uses the Enterprise Manager → Server → Tablespaces option.

When the DBA creates a database, Oracle automatically creates the tablespaces and datafiles shown in Figure 16.11. A few of them are described as follows:

- The *SYSTEM* tablespace is used to store the data dictionary data.

- The *USERS* tablespace stores the table data created by the end users.

- The *TEMP* tablespace stores the temporary tables and indexes created during the execution of SQL statements. For example, temporary tables are created when your SQL statement contains an ORDER BY, GROUP BY, or HAVING clause.

database instance
In an Oracle DBMS, the collection of processes and data structures used to manage a specific database.

tablespace
In a DBMS, a logical storage space used to group related data. Also known as a *file group*.

datafile
A file on the hard drive or storage system where the data in a tablespace is physically stored.

- The *UNDOTBS1* tablespace stores database transaction recovery information. If a transaction must be rolled back (usually to preserve database integrity), the UNDOTBS1 tablespace stores the undo information.

Using the Enterprise Manager, the DBA can:

- Create additional tablespaces to organize the data in the database. Therefore, if you have a database with several hundred users, you can create several user tablespaces to segment data storage for different types of users. For example, you might create a teacher tablespace and a student tablespace.

- Create additional tablespaces to organize the various subsystems within the database. For example, you might create different tablespaces for human resources data, payroll data, accounting data, and manufacturing data. Figure 16.12 shows the wizard used to create a tablespace called CORMOR that holds the tables used in this book. This tablespace is stored in the datafile named CORMOR01.DBF, and its initial size is 100 megabytes. Note that the tablespace is available to users for data storage purposes. Also, you can click the Show SQL button at the top of the page to see the SQL code generated by Oracle to create the tablespace. (All DBA tasks can be accomplished through the direct use of SQL commands. In fact, some die-hard DBAs prefer writing their own SQL code rather than using the GUI.)

- Expand the tablespace storage capacity by creating additional datafiles. Remember that the datafiles can be stored in the same directory or on different disks to increase access performance. For example, you could increase storage and access performance to the USERS tablespace by creating a new datafile on a different drive.

FIGURE 16.12 CREATING A NEW ORACLE TABLESPACE

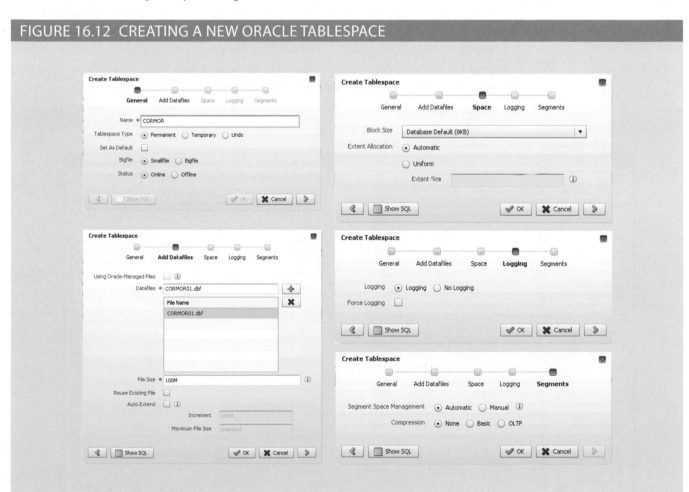

16-10d Managing Users and Establishing Security

One of the most common database administration activities is creating and managing database users. The creation of user IDs is the first component of any well-planned database security function.

The Security section of the Oracle Enterprise Manager enables the DBA to create users, roles, and profiles.

- A **user** is a uniquely identifiable object that allows a given person to log on to the database. The DBA assigns privileges for accessing the objects in the database. Within the privilege assignment, the DBA may specify a set of limits that define how many database resources the user can use.

- A **role** is a named collection of database access privileges that authorize a user to connect to the database and use its system resources. Examples of roles are as follows:

  - *CONNECT* allows a user to connect to the database and then create and modify tables, views, and other data-related objects.

  - *RESOURCE* allows a user to create triggers, procedures, and other data management objects.

  - *DBA* gives the user database administration privileges.

- A **profile** is a named collection of settings that control how much of the database resource a given user can access. For example, a runaway query could cause the database to lock up or stop responding to the user's commands, so it is important to limit access to the database resource. By specifying profiles, the DBA can limit how much storage space a user can have, how long a user can be connected, how much idle time may be used before the user is disconnected, and so on. In an ideal world, all users would have unlimited access to all resources at all times, but realistically, such access is neither possible nor desirable.

Figure 16.13 shows the Oracle Enterprise Manager Users page. From here, the DBA can manage the database and create security objects such as users, roles, and profiles.

user
In a system, a uniquely identifiable object that allows a given person or process to log on to the database.

role
In Oracle, a named collection of database access privileges that authorize a user to connect to a database and use its system resources.

profile
In Oracle, a named collection of settings that controls how much of the database resource a given user can use.

FIGURE 16.13 THE ORACLE ENTERPRISE MANAGER USERS PAGE

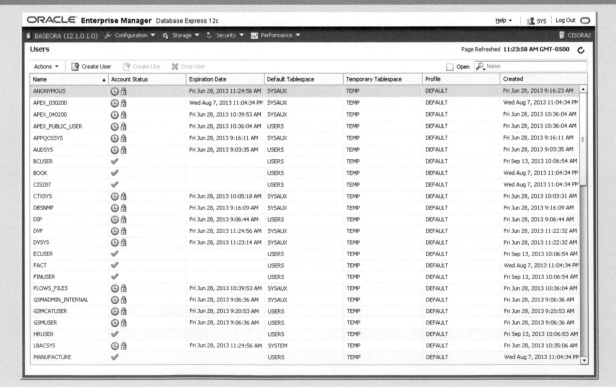

To create a new user, the DBA clicks the Create User button to start the wizard shown in Figure 16.14.

FIGURE 16.14 THE CREATE USER WIZARD

The User page buttons and menu items support many actions. For example, from this page the DBA can:

- Drop the user from the database.
- Alter the user's default and temporary tablespaces.
- Alter the privileges and roles assigned to the user.
- View the user details to adjust object privileges and quotas. Quotas allow the DBA to specify the maximum amount of storage that the user can have in each tablespace. For example, Figure 16.15 shows a user being assigned a maximum storage allocation of 20 megabytes on the CORMOR tablespace.

16-10e Customizing the Database Initialization Parameters

Fine-tuning a database is another important DBA task that usually requires the modification of database configuration parameters, some of which can be changed in real time using SQL commands. Changes to other parameters require the database to be shut down and restarted. Also, some parameters may affect only the database instance, while others affect the entire RDBMS and all instances that are running. So, it is very important

FIGURE 16.15 ASSIGNING A USER QUOTA

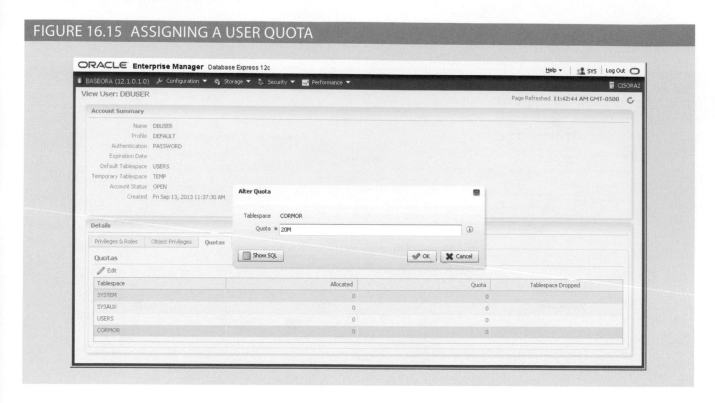

that the DBA become familiar with database configuration parameters, especially those that affect performance.

Each database has an associated initialization file that stores its run-time configuration parameters. The initialization file is read at instance startup and is used to set the working environment for the database. Oracle's Enterprise Manager allows the DBA to start, shut down, view, and edit the database configuration parameters of a database instance; these parameters are stored in the initialization file. The Oracle Enterprise Manager provides a GUI to modify the file, as shown in Figure 16.16.

FIGURE 16.16 ORACLE ENTERPRISE MANAGER INITIALIZATION PARAMETERS

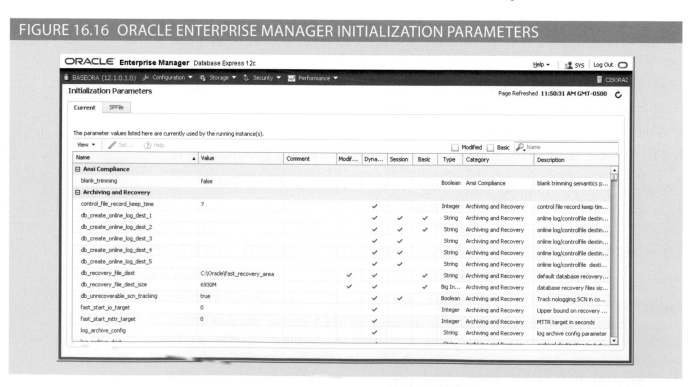

One of the important functions of the initialization parameters is to reserve the resources that the database uses at run time. One of those resources is the primary memory reserved for database caching. Such caching is used to fine-tune database performance. For example, the "db_cache_size" parameter sets the amount of memory reserved for database caching. This parameter should be set to a value that is large enough to support all concurrent transactions. Once you modify the initialization parameters, you may be required to restart the database.

As you have seen in this brief section, the DBA is responsible for a wide range of tasks. The quality and completeness of administration tools go a long way toward making the DBA's job easier. Even so, the DBA must become familiar with the tools and technical details of the RDBMS to perform tasks properly and efficiently.

Summary

- Data management is a critical activity for any organization, so data must be treated as a corporate asset. The value of a data set is measured by the utility of the information derived from it. Good data management is likely to produce good information, which is the basis for better decision making.

- Data quality is a comprehensive approach to ensure the accuracy, validity, and timeliness of data. Data quality focuses on correcting dirty data, preventing future inaccuracies in the data, and building user confidence in the data.

- The DBMS is the most commonly used tool for corporate data management. The DBMS supports strategic, tactical, and operational decision making at all levels of the organization. The introduction of a DBMS into an organization is a delicate job; the impact of the DBMS on the organization's managerial and cultural framework must be carefully examined.

- The database administrator (DBA) is responsible for managing the corporate database. The internal organization of database administration varies from company to company. Although no standard exists, it is common practice to divide DBA operations according to phases of the Database Life Cycle. Some companies have created a position with a broader mandate to manage computerized data and other data; this activity is handled by the data administrator (DA).

- The DA and DBA functions tend to overlap. Generally speaking, the DA has more managerial tasks than the more technically oriented DBA. Compared to the DBA function, the DA function is DBMS-independent, with a broader and longer-term focus. However, when the organization does not include a DA position, the DBA executes all of the DA's functions. In this combined role, the DBA must have a diverse mix of technical and managerial skills.

- A DBA's managerial services include supporting end users; defining and enforcing policies, procedures, and standards for the database; ensuring data security, privacy, and integrity; providing data backup and recovery services; and monitoring distribution and use of the data in the database.

- The DBA's technical role requires involvement in at least the following activities: evaluating, selecting, and installing the DBMS; designing and implementing databases and applications; testing and evaluating databases and applications; operating and maintaining the DBMS, utilities, and applications; and training and supporting users.

- Security refers to activities and measures that ensure the confidentiality, integrity, and availability of an information system and its main asset, data. A security policy is a collection of standards, policies, and practices that guarantee the security of a system and ensure auditing and compliance.

- A security vulnerability is a weakness in a system component that could be exploited to allow unauthorized access or service disruption. A security threat is an imminent security violation caused by an unchecked vulnerability. Security vulnerabilities exist in all components of an information system: people, hardware, software, network, procedures, and data. Therefore, it is critical to have robust database security. Database security refers to DBMS features and related measures that comply with the organization's security requirements.

- The development of a data administration strategy is closely related to the company's mission and objectives. Therefore, the strategic plan requires a detailed analysis of company goals, its situation, and its business needs. To guide the development of this data administration plan, an integrating methodology is required. The most commonly used integrating methodology is known as information engineering (IE).

- To help translate strategic plans into operational plans, the DBA has access to an arsenal of database administration tools, including a data dictionary and computer-aided systems engineering (CASE) tools.

- With the introduction of reliable cloud-based data services, the role of the DBA has expanded beyond corporate walls.

Key Terms

Online Content

Flashcards and crossword puzzles for key term practice are available at *www.cengagebrain.com*.

access plan
active data dictionary
audit log
authorization management
availability
back-end CASE tool
computer-aided systems engineering (CASE)
compliance
concurrent backup
confidentiality
data administrator (DA)
data quality
database administrator (DBA)
database dump
database instance (Oracle)
database security
database security officer (DSO)

datafile (Oracle)
data-profiling software
dirty data
disaster management
enterprise database
front-end CASE tool
full backup
incremental backup
information engineering (IE)
information resource dictionary
information resource manager (IRM)
information systems (IS) department
information systems architecture (ISA)
integrity
master data management (MDM) software

passive data dictionary
policies
privacy
procedures
profile (Oracle)
role (Oracle)
security
security breach
security policy
security threat
security vulnerability
standards
systems administrator
tablespace (Oracle)
user (Oracle)

Review Questions

1. Explain the difference between data and information. Give some examples of raw data and information.

2. Define dirty data, and identify some of its sources.

3. What is data quality, and why is it important?

4. Explain the interactions among end users, data, information, and decision making. Draw a diagram and explain the interactions.

5. Suppose that you are a DBA. What data dimensions would you describe to top-level managers to obtain their support for data administration?

6. How and why did database management systems become the data management standard in organizations? Discuss some advantages of the database approach over the file-system approach.

7. Using a single sentence, explain the role of databases in organizations. Then explain your answer in more detail.

8. Define *security* and *privacy*. How are the two concepts related?

9. Describe and contrast information needs at the strategic, tactical, and operational levels of an organization. Use examples to explain your answer.

10. What special considerations must you take into account when introducing a DBMS into an organization?

11. Describe the DBA's responsibilities.

12. How can the DBA function be placed within the organization chart? What effects will that placement have on the DBA function?

13. Why and how are new technological advances in computers and databases changing the DBA's role?

14. Explain the DBA department's internal organization based on the DBLC approach.

15. Explain and contrast differences and similarities between the DBA and DA.

16. Explain how the DBA plays an arbitration role between an organization's two main assets. Draw a diagram to illustrate your explanation.

17. Describe and characterize the skills desired for a DBA.

18. What are the DBA's managerial roles? Describe the managerial activities and services provided by the DBA.

19. What DBA activities support end users?

20. Explain the DBA's managerial role in the definition and enforcement of policies, procedures, and standards.

21. Protecting data security, privacy, and integrity are important database functions. What activities are required in the DBA's managerial role of enforcing those functions?

22. Discuss the importance and characteristics of database backup and recovery procedures. Then describe the actions that must be detailed in backup and recovery plans.

23. Assume that your company has assigned you the responsibility of selecting the corporate DBMS. Develop a checklist of the technical issues and other aspects involved in the selection process.

24. Describe the activities that are typically associated with the design and implementation services of the DBA's technical function. What technical skills are desirable in a DBA?

25. Why are testing and evaluation of the database and applications not done by the same people who are responsible for design and implementation? What minimum standards must be met during testing and evaluation?

26. Identify some bottlenecks in DBMS performance, and then propose some solutions used in DBMS performance tuning.

27. What are typical activities in the maintenance of the DBMS and its utilities and applications? Would you consider application performance tuning to be part of the maintenance activities? Explain your answer.

28. How do you normally define security? How is your definition similar to or different from the definition of database security in this chapter?

29. What are the levels of data confidentiality?

30. What are security vulnerabilities? What is a security threat? Give some examples of security vulnerabilities in different IS components.

31. Define the concept of a data dictionary, and discuss the different types of data dictionaries. If you managed an organization's entire data set, what characteristics would you want for the data dictionary?

32. Using SQL statements, give some examples of how you would use the data dictionary to monitor database security.

33. What characteristics do a CASE tool and a DBMS have in common? How can those characteristics be used to enhance data administration?

34. Briefly explain the concepts of information engineering (IE) and information systems architecture (ISA). How do those concepts affect the data administration strategy?

35. Identify and explain some critical success factors in the development and implementation of a good data administration strategy.

36. How have cloud-based data services affected the DBA's role?

37. What tool is used in Oracle to create users?

38. In Oracle, what is a tablespace?

39. In Oracle, what is a database role?

40. In Oracle, what is a datafile? How does it differ from a file systems file?

41. In Oracle, what is a database profile?

GLOSSARY

A

abstract data type (ADT)
Data type that describes a set of similar objects with shared and encapsulated data representation and methods. An abstract data type is generally used to describe complex objects. See also *class*.

access plan
A set of instructions generated at application compilation time that is created and managed by a DBMS. The access plan predetermines how an application's query will access the database at run time.

access point
In the case of wireless networks, this allows you to connect wireless devices to a wired or wireless network.

active data dictionary
A data dictionary that is automatically updated by the database management system every time the database is accessed, thereby keeping its information current.

ActiveX
Microsoft's alternative to Java. A specification for writing programs that will run inside the Microsoft client browser. Oriented mainly to Windows applications, it is not portable. It adds controls such as drop-down windows and calendars to webpages.

ActiveX Data Objects (ADO)
A Microsoft object framework that provides a high-level, application-oriented interface to OLE-DB, DAO, and RDO. ADO provides a unified interface to access data from any programming language that uses the underlying OLE-DB objects.

ad hoc query
A "spur-of-the-moment" question.

ADO.NET
The data access component of Microsoft's .NET application development framework.

algorithms
A process or set of operations in a calculation.

alias
An alternative name for a column or table in a SQL statement.

ALTER TABLE
The SQL command used to make changes to table structure. When the command is followed by a keyword (ADD or MODIFY), it adds a column or changes column characteristics.

American National Standards Institute (ANSI)
The group that accepted the DBTG recommendations and augmented database standards in 1975 through its SPARC committee.

analytical database
A database focused primarily on storing historical data and business metrics used for tactical or strategic decision making.

AND
The SQL logical operator used to link multiple conditional expressions in a WHERE or HAVING clause. It requires that all conditional expressions evaluate to true.

anonymous PL/SQL block
A PL/SQL block that has not been given a specific name.

application processor (AP)
See *transaction processor (TP)*.

Application Program-to-Program Communications (APPC)
A communications protocol used in IBM mainframe systems network architecture (SNA). Allows for communications between personal computers and IBM mainframe applications.

application programming interface (API)
Software through which programmers interact with middleware. An API allows the use of generic SQL code, thereby allowing client processes to be database server-independent.

AREA
In DB2, a named section of permanent storage space that is reserved to store the database.

associative entity
See *composite entity*.

associative object
In Object-oriented modeling, an object used to represent a relationship between two or more objects.

asymmetric encryption
A form of encryption that uses two numeric keys—the public key and the private key. Both keys are able to encrypt and decrypt each other's messages. See also *public-key encryption*.

atomic attribute
An attribute that cannot be further subdivided to produce meaningful components. For example, a person's last name attribute cannot be meaningfully subdivided.

atomic transaction property
A property that requires all parts of a transaction to be treated as a single, logical unit of work in which all operations must be completed (committed) to produce a consistent database.

atomicity
The transaction property that requires all parts of a transaction to be treated as a single, indivisible, logical unit of work. All parts of a transaction must be completed or the entire transaction is aborted.

attribute
A characteristic of an entity or object. An attribute has a name and a data type.

attribute hierarchy
A top-down data organization that is used for two main purposes: aggregation and drill-down/roll-up data analysis.

audit log
A security feature of a database management system that automatically records a brief description of the database operations performed by all users.

authentication
The process through which a DBMS verifies that only registered users can access the database.

authorization management
Procedures that protect and guarantee database security and integrity. Such procedures include user access management, view definition, DBMS access control, and DBMS usage monitoring.

automatic query optimization
A method by which a DBMS finds the most efficient access path for the execution of a query.

availability
In the context of data security, it refers to the accessibility of data whenever required by authorized users and for authorized purposes.

AVG
A SQL aggregate function that outputs the mean average for a specified column or expression.

B

B-tree index
An ordered data structure organized as an upside-down tree.

back-end application
The process that provides service to clients.

back-end CASE tool
A computer-aided software tool that provides support for the coding and implementation phases of the SDLC. In comparison, front-end CASE tools provide support for the planning, analysis, and design phases.

balancing
Ensuring that the processing load is distributed evenly among multiple servers.

base data types
A term used to describe the data types frequently used in traditional programming languages. Base data types include *real*, *integer*, and *string*.

base tables
The table on which a view is based.

basically available, soft state, eventually consistent (BASE)
A data consistency model in which data changes are not immediate but propagate slowly through the system until all replicas are eventually consistent.

batch processing
A data processing method that runs data processing tasks from beginning to end without any user interaction.

batch update routine
A routine that pools transactions into a single group to update a master table in a single operation.

BETWEEN
In SQL, a special comparison operator used to check whether a value is within a range of specified values.

bidirectional physically paired logical relationships
In the hierarchical model, a relationship that links a logical child with its logical parent in two directions.

bidirectional virtually paired logical relationships
In the hierarchical model, a relationship created when a logical child segment is linked to its logical parent in two directions. The virtually paired relationship is different from the physically paired relationship in that no duplicates are created.

Big Data
A movement to find new and better ways to manage large amounts of web-generated data and derive business insight from it, while simultaneously providing high performance and scalability at a reasonable cost.

binary lock
A lock that has only two states: *locked* (1) and *unlocked* (0). If a data item is locked by a transaction, no other transaction can use that data item.

binary relationship
An ER term for an association (relationship) between two entities. For example, PROFESSOR teaches CLASS.

bitmap index
An index that uses a bit array (0s and 1s) to represent the existence of a value or condition.

block report
In the Hadoop Distributed File System (HDFS), a report sent every 6 hours by the data node to the name node informing the name node which blocks are on that data node.

Boolean algebra
A branch of mathematics that uses the logical operators OR, AND, and NOT.

bottom-up design
A design philosophy that begins by identifying individual design components and then aggregates them into larger units. In database design, the process begins by defining attributes and then groups them into entities.

boundaries
The external limits to which any proposed system is subjected. These limits include budgets, personnel, and existing hardware and software.

Boyce-Codd normal form (BCNF)
A special type of third normal form (3NF) in which every determinant is a candidate key. A table in BCNF must be in 3NF. See also *determinant*.

bridge
A device that connects similar networks. Allows computers in one network to communicate with computers in another network.

bridge entity
See *composite entity*.

BSON (Binary JSON)
A computer-readable format for data interchange that expands the JSON format to include additional data types including binary objects.

bucket
In a key-value database, a logical collection of related key-value pairs.

buffer cache
See *data cache*.

buffers
Temporary storage area in primary memory used to speed up disk operations.

bus topology
Network topology requiring that all computers be connected to a main network cable. It bears the disadvantage that a single lost computer can result in network segment breakdown.

Business intelligence (BI)
A comprehensive, cohesive, and integrated set of tools and processes used to capture, collect, integrate, store, and analyze data with the purpose of generating and presenting information to support business decision making.

business rule
A description of a policy, procedure, or principle within an organization. For example, a pilot cannot be on duty for more than 10 hours during a 24-hour period, or a professor may teach up to four classes during a semester.

Business to business (B2B)
Electronic commerce between businesses.

Business to consumer (B2C)
Electronic commerce between a business and consumers.

C

Call Level Interface (CLI)
A standard developed by the SQL Access Group for database access.

campuswide network (CWN)
A typical college or university network in which buildings containing LANs are connected through a network backbone.

candidate key
A minimal superkey; that is, a key that does not contain a subset of attributes that is itself a superkey. See *key*.

cardinality
A property that assigns a specific value to connectivity and expresses the range of allowed entity occurrences associated with a single occurrence of the related entity.

cascading order sequence
A nested ordering sequence for a set of rows, such as a list in which all last names are alphabetically ordered and, within the last names, all first names are ordered.

CASE
See *Computer-aided software engineering*.

centralized data allocation
A data allocation strategy in which the entire database is stored at one site. Also known as a *centralized database*.

centralized database
A database located at a single site.

centralized design
A process by which all database design decisions are carried out centrally by a small group of people. Suitable in a top-down design approach when the problem domain is relatively small, as in a single unit or department in an organization.

certification authority (CA)
In the context of Internet security, a private entity or company that certifies the user or vendor is who (s) he claims to be.

checkpoints
In transaction management, an operation in which the database management system writes all of its updated buffers to disk.

Chen notation
See *entity relationship (ER) model*.

class
A collection of similar objects with shared structure (attributes) and behavior (methods). A class encapsulates an object's data representation and a method's implementation. Classes are organized in a class hierarchy.

class diagram notation
The set of symbols used in the creation of class diagrams in UML object modeling.

class diagrams
A diagram used to represent data and their relationships in UML object notation.

class hierarchy
The organization of classes in a hierarchical tree in which each parent class is a *superclass* and each child class is a *subclass*. See also *inheritance*.

class instance
Each individual object stored in a class. Each class instance must share the same structure and respond to the same messages if they are located in the same class. Also known as *object instance*.

class lattice
The class hierarchy is known as a class lattice if its classes can have multiple parent classes.

client
Any process that requests specific services from server processes in a client/server environment.

client node
One of three types of nodes used in the Hadoop Distributed File System (HDFS). The client node acts as the interface between the user application and the HDFS. See also *name node* and *data node*.

client-side extensions
Extension that adds functionality to a web browser. The most common extensions are plug-ins, Java, JavaScript, ActiveX, and VBScript.

client/server architecture
A hardware and software system composed of clients, servers, and middleware. Features a user of resources (client) and a provider of resources (server).

closure
A property of relational operators that permits the use of relational algebra operators on existing tables (relations) to produce new relations.

cloud computing
A computing model that provides ubiquitous, on-demand access to a shared pool of configurable resources that can be rapidly provisioned.

cloud database
A database that is created and maintained using cloud services, such as Microsoft Azure or Amazon AWS.

cloud services
The services provided by cloud computing. Cloud services allow any organization to quickly and economically add information technology services such as applications, storage, servers, processing power, databases, and infrastructure.

cluster tables
A data storage structure that physically stores related rows from different tables together to improve the speed at which related data can be accessed.

clustered index table
See *index organized table*.

coaxial cable
Copper cables enclosed in two layers of insulation or shielding. Often referred to as "coax." Very similar to cable used for home cable TV.

cohesivity
The strength of the relationships between a module's components. Module cohesivity must be high.

ColdFusion Markup Language (CFML)
A server-side markup language (HTML extensions or tags) that is used to create ColdFusion application pages known as *scripts*.

collection object
An object that contains one or more objects.

column family
In a column family database, a collection of columns or super columns related to a collection of rows.

column family database
A NoSQL database model that organizes data into key-value pairs, in which the value component is composed of a set of columns that vary by row.

column-centric storage
A physical data storage technique in which data is stored in blocks, which hold data from a single column across many rows.

COMMIT
The SQL command that permanently writes data changes to a database.

Common Gateway Interface (CGI)
A web server interface standard that uses script files to perform specific functions based on a client's parameters.

community cloud
A type of cloud built by and for a specific group of organizations that share a common trade, such as agencies of the federal government, the military, or higher education.

completeness constraint
A constraint that specifies whether each entity supertype occurrence must also be a member of at least one subtype. The completeness constraint can be partial or total.

complex object
An object formed by several different objects in complex relationships. See also *abstract data types*.

compliance
In the context of data security, activities that meet data privacy and security reporting guidelines or requirements.

composite attribute
An attribute that can be further subdivided to yield additional attributes. For example, a phone number such as 615-898-2368 may be divided into an area code (615), an exchange number (898), and a four-digit code (2368). Compare to *simple attribute*.

composite entity
An entity designed to transform an M:N relationship into two 1:M relationships. The composite entity's primary key comprises at least the primary keys of the entities that it connects. Also known as a *bridge entity or associative entity*. See also *linking table*.

composite identifier
In ER modeling, a key composed of more than one attribute.

composite key
A multiple-attribute key.

composite object
An object that contains at least one multivalued attribute and has no attributes that refer to another object.

compound object
An object that contains at least one attribute that references another object.

computer-aided software engineering (CASE)
Tools used to automate part or all of the Systems Development Life Cycle. Also known as computer-aided systems engineering.

computer-aided systems engineering
See *computer-aided software engineering*.

concentrator
A device that takes multiple wires and combines them into a single method of transfer to allow multiple users to access the line simultaneously. It resembles a network wiring closet.

conceptual design
A process that uses data-modeling techniques to create a model of a database structure that represents real-world objects as realistically as possible. The The design is both software- and hardware-independent.

conceptual model
The output of the conceptual design process. The conceptual model provides a global view of an entire database and describes the main data objects, avoiding details.

conceptual schema
A representation of the conceptual model, usually expressed graphically. See also *conceptual model*.

concurrency control
A DBMS feature that coordinates the simultaneous execution of transactions in a multiprocessing database system while preserving data integrity.

concurrent backup
A backup that takes place while one or more users are working on a database.

confidentiality
In the context of data security, ensuring that data is protected against unauthorized access, and if the data is accessed by an authorized user, that the data is used only for an authorized purpose.

connectivity
The type of relationship between entities. Classifications include 1:1, 1:M, and M:N.

consistency
A database condition in which all data integrity constraints are satisfied. To ensure consistency of a database, every transaction must begin with the database in a known consistent state. If not, the transaction will yield an inconsistent database that violates its integrity and business rules.

consistent database state
A database state in which all data integrity constraints are satisfied.

constraint
A restriction placed on data, usually expressed in the form of rules. For example, "A student's GPA must be between 0.00 and 4.00." Constraints are important because they help to ensure data integrity.

content management
Automation of the creation and management of a Web site's contents.

conventional data types
See *base data types*.

coordinator
The transaction processor (TP) node that coordinates the execution of a two-phase COMMIT in a DDBMS.

correlated subquery
A subquery that executes once for each row in the outer query.

cost-based optimizer
A query optimization mode that uses an algorithm based on statistics about the objects being accessed, including number of rows, indexes available, index sparsity, and so on.

COUNT
A SQL aggregate function that outputs the number of rows containing not null values for a given column or expression, sometimes used in conjunction with the DISTINCT clause.

CREATE INDEX
A SQL command that creates indexes on the basis of a selected attribute or attributes.

CREATE TABLE
A SQL command that creates a table's structures using the characteristics and attributes given.

CREATE VIEW
A SQL command that creates a logical, "virtual" table. The view can be treated as a real table.

cross join
A join that performs a relational product (or Cartesian product) of two tables.

Crow's Foot notation
A representation of the entity relationship diagram that uses a three-pronged symbol to represent the "many" sides of the relationship.

cube cache
In multidimensional OLAP, the shared, reserved memory area where data cubes are held. Using the cube cache assists in speeding up data access.

currency
In the Network Data Model, this term indicates the position of the record pointer within the database and *refers to the most recently accessed record*.

cursor
A special construct used in procedural SQL to hold the data rows returned by a SQL query. A cursor may be considered a reserved area of memory in which query output is stored, like an array holding columns and rows. Cursors are held in a reserved memory area in the DBMS server, not in the client computer.

D

dashboards
In business intelligence, a web-based system that presents key business performance indicators or information in a single, integrated view with clear and concise graphics.

data
Raw facts, or facts that have not yet been processed to reveal their meaning to the end user.

Data Access Objects (DAO)
An object-oriented application programming interface used to access MS Access, FileMaker Pro, and other Jet-based databases.

data administrator (DA)
The person responsible for managing the entire data resource, whether it is computerized or not. The DA has broader authority and responsibility than the database administrator (DBA). Also known as an *information resource manager (IRM)*.

data allocation
In a distributed DBMS, the process of deciding where to locate data fragments.

data analytics
A subset of business intelligence functionality that encompasses a wide range of mathematical, statistical, and modeling techniques with the purpose of extracting knowledge from data.

data anomaly
A data abnormality in which inconsistent changes have been made to a database. For example, an employee moves, but the address change is not corrected in all files in the database.

data cache
A shared, reserved memory area that stores the most recently accessed data blocks in RAM. Also called *buffer cache*.

data cube
The multidimensional data structure used to store and manipulate data in a multidimensional DBMS. The location of each data value in the data cube is based on its x-, y-, and z-axes. Data cubes are static, meaning they must be created before they are used, so they cannot be created by an ad hoc query.

data definition language (DDL)
The language that allows a database administrator to define the database structure, schema, and subschema.

data dependence
A data condition in which data representation and manipulation are dependent on the physical data storage characteristics.

data dictionary
A DBMS component that stores metadata—data about data. Thus, the data dictionary contains the data definition as well as their characteristics and relationships. A data dictionary may also include data that are external to the DBMS. Also known as an *information resource dictionary*. See also

active data dictionary, *metadata*, and *passive data dictionary*.

Data Encryption Standard (DES)
The most widely used standard for private-key encryption. DES is used by the U.S. government.

data files
A named physical storage space that stores a database's data. It can reside in a different directory on a hard disk or on one or more hard disks. All data in a database is stored in data files. A typical enterprise database is normally composed of several data files. A data file can contain rows from one or more tables.

data fragmentation
A characteristic of a DDBMS that allows a single object to be broken into two or more segments or fragments. The object might be a user's database, a system database, or a table. Each fragment can be stored at any site on a computer network.

data inconsistency
A condition in which different versions of the same data yield different (inconsistent) results.

data independence
A condition in which data access is unaffected by changes in the physical data storage characteristics.

data integrity
In a relational database, a condition in which the data in the database complies with all entity and referential integrity constraints.

data management
A process that focuses on data collection, storage, and retrieval. Common data management functions include addition, deletion, modification, and listing.

data manager (DM)
See *data processor (DP)*.

data manipulation language
The set of commands that allows an end user to manipulate the data in the database, such as SELECT, INSERT, UPDATE, DELETE, COMMIT, and ROLLBACK.

data mart
A small, single-subject data warehouse subset that provides decision support to a small group of people.

data mining
A process that employs automated tools to analyze data in a data warehouse and other sources and to proactively identify possible relationships and anomalies.

data model
A representation, usually graphic, of a complex "real-world" data structure. Data models are used in the database design phase of the Database Life Cycle.

data modeling
The process of creating a specific data model for a determined problem domain.

data node
One of three types of nodes used in the Hadoop Distributed File System (HDFS). The data node stores fixed-size data blocks (that could be replicated to other data nodes). See also *client node* and *name node*.

data processing (DP) specialist
The person responsible for developing and managing a computerized file processing system.

data processor (DP)
The resident software component that stores and retrieves data through a DDBMS. The DP is responsible for managing the local data in the computer and coordinating access to that data. Also known as *data manager (DM)*.

data quality
A comprehensive approach to ensuring the accuracy, validity, and timeliness of data.

data redundancy
Exists when the same data is stored unnecessarily at different places.

data replication
The storage of duplicated database fragments at multiple sites on a DDBMS. Duplication of the fragments is transparent to the end user. Data replication provides fault tolerance and performance enhancements.

data source name (DSN)
With ODBC, a name that identifies and defines an ODBC data source.

data sparsity
A column distribution of values or the number of different values a column can have.

data visualization
Abstracting data to provide information in a visual format that enhances the user's ability to effectively comprehend the meaning of the data.

data warehouse
An integrated, subject-oriented, time-variant, nonvolatile collection of data in a specialized database that stores historical and aggregated data in a format that provides support for decision making.

data-profiling software
Programs that analyze data and metadata to determine patterns that can help assess data quality.

database
A shared, integrated computer structure that houses a collection of related data. A database contains two types of data: end-user data (raw facts) and metadata.

database administrator (DBA)
The person responsible for planning, organizing, controlling, and monitoring the centralized and shared corporate database. The DBA is the general manager of the database administration department.

Database Administrator Control System (DBACS)
In the network model, the database definition processor that reads the database definition and validates the schema (The DBACS works like a compiler.).

database description (DBD) statement
In the hierarchical model, the series of commands that define the hierarchical tree structure and how the segments are stored in the database.

database design
The process that yields the description of the database structure and determines the database components. The second phase of the Database Life Cycle.

database development
The process of database design and implementation.

database dump
A complete copy of an entire database saved and periodically updated in a separate memory location. A full backup ensures a full recovery of all data after a physical disaster or database integrity failure.

database fragment
A subset of a distributed database. Although the fragments may be stored at different sites within a computer network, the set of all fragments is treated as a single database. See also *horizontal fragmentation* and *vertical fragmentation*.

database instance
In an Oracle DBMS, the collection of processes and data structures used to manage a specific database.

database integration
In the context of e-commerce databases, a business enabling service where a company's database is used in their Web page development.

Database Life Cycle (DBLC)
A cycle that traces the history of a database within an information system. The cycle is divided into six phases: initial study, design, implementation and loading, testing and evaluation, operation and maintenance, and evolution.

database management system (DBMS)
The collection of programs that manages the database structure and controls access to the data stored in the database.

database middleware
Database connectivity software through which application programs connect and communicate with data repositories.

database performance tuning
A set of activities and procedures designed to reduce the response time of a database system—that is, to ensure that an end-user query is processed by the DBMS in the minimum amount of time.

database recovery
The process of restoring a database to a previous consistent state.

database request
The equivalent of a single SQL statement in an application program or a transaction.

database role
A set of database privileges that could be assigned as a unit to a user or group.

database security
The use of DBMS features and other related measures to comply with the security requirements of an organization.

database security officer (DSO)
The person responsible for the security, integrity, backup, and recovery of the database.

database statistics
In query optimization, measurements about database objects, such as the number of rows in a table, number of disk blocks used, maximum and average row length, number of columns in each row, and number of distinct values in each column. Such statistics provide a snapshot of database characteristics.

database system
An organization of components that defines and regulates the collection, storage, management, and use of data in a database environment.

database translator
A middleware component that translates generic SQL calls into specific database server syntax to create database server independence.

database-level lock
A type of lock that restricts database access to the owner of the lock and allows only one user at a time to access the database. This lock works for batch processes but is unsuitable for online multiuser DBMSs.

datafile
A file on the hard drive or storage system where the data in a tablespace is physically stored.

DataSet
In ADO.NET, a disconnected, memory-resident representation of the database. The DataSet contains tables, columns, rows, relationships, and constraints.

DBGEN
In the hierarchical model, the component that generates the physical database with all its necessary structures.

DBMS performance tuning
Activities to ensure that clients' requests are addressed as quickly as possible while making optimum use of existing resources.

deadlock
A condition in which two or more transactions wait indefinitely for the other to release the lock on a previously locked data item. Also called *deadly embrace*.

deadly embrace
See *deadlock*.

decentralized design
A process in which conceptual design models subsets of an organization's database requirements, which are then aggregated into a complete design. Such modular designs are typical of complex systems with a relatively large number of objects and procedures.

decision support system (DSS)
An arrangement of computerized tools used to assist managerial decision making within a business.

deferred update
In transaction management, a condition in which transaction operations do not immediately update a physical database. Also called *deferred write technique*.

deferred-write technique
See *deferred update*.

DELETE
A SQL command that allows data rows to be deleted from a table.

denial-of-service
One of the most common hacker activities. This attack overloads Web servers and routers with millions of requests for service, rendering the services unavailable to legitimate users.

denormalization
A process by which a table is changed from a higher-level normal form to a lower-level normal form, usually to increase processing speed. Denormalization potentially yields data anomalies.

dependency diagram
A representation of all data dependencies (primary key, partial, or transitive) within a table.

dependent
An attribute whose value is determined by another attribute.

derived attribute
An attribute that does not physically exist within the entity and is derived via an algorithm. For example, the Age attribute might be derived by subtracting the birth date from the current date.

description of operations
A document that provides a precise, detailed, up-to-date, and thoroughly reviewed description of the activities that define an organization's operating environment.

design trap
A problem that occurs when a relationship is improperly or incompletely identified and therefore is represented in a way that is not consistent with the real world. The most common design trap is known as a *fan trap*.

desktop database
A single-user database that runs on a personal computer.

determinant
Any attribute in a specific row whose value directly determines other values in that row. See also *Boyce-Codd normal form (BCNF)*.

determination
The role of a key. In the context of a database table, the statement "A determines B" indicates that knowing the value of attribute A means that the value of attribute B can be looked up.

DIFFERENCE
In relational algebra, an operator used to yield all rows from one table that are not found in another union-compatible table.

differential backup
A level of database backup in which only the last modifications to the database are copied.

digital cash
In an e-commerce environment, the digital equivalent of hard currency (coins or bills of a given denomination).

digital certificate
A unique identifier given to an entity. The certificate holder may be an end user, a Web site, a computer, a Web page, or even a program. Digital certificates are used in combination with encryption to provide security and authentication.

digital signature
An encrypted attachment added to an electronic message to verify the sender's identity.

dimension tables
In a data warehouse, tables used to search, filter, or classify facts within a star schema.

dimensions
In a star schema design, qualifying characteristics that provide additional perspectives to a given fact.

dirty data
Data that contain inaccuracies and/or inconsistencies.

dirty read
In transaction management, when a transaction reads data that is not yet committed.

disaster management
The set of DBA activities dedicated to securing data availability following a physical disaster or a database integrity failure.

discipline-specific databases
A database that contains data focused on specific subject areas.

disjoint subtypes
In a specialization hierarchy, a unique and nonoverlapping subtype entity set.

diskpage
In permanent storage, the equivalent of a disk block, which can be described as a directly addressable section of a disk. A diskpage has a fixed size, such as 4K, 8K, or 16K.

DISTINCT
A SQL clause that produces only a list of values that are different from one another.

distributed data catalog (DDC)
A data dictionary that contains the description (fragment names and locations) of a distributed database.

distributed data dictionary (DDD)
See *distributed data catalog*.

distributed database
A logically related database that is stored in two or more physically independent sites.

distributed database management system (DDBMS)
A DBMS that supports a database distributed across several different sites; a DDBMS governs the storage and processing of logically related data over interconnected computer systems in which both data and processing functions are distributed among several sites.

distributed global schema
The database schema description of a distributed database as seen by the database administrator.

distributed processing
Sharing the logical processing of a database over two or more sites connected by a network.

distributed request
A database request that allows a single SQL statement to access data in several remote data processors (DPs) in a distributed database.

distributed transaction
A database transaction that accesses data in several remote data processors (DPs) in a distributed database.

distribution transparency
A DDBMS feature that allows a distributed database to look like a single logical database to an end user.

DIVIDE
In relational algebra, an operator that answers queries about one set of data being associated with all values of data in another set of data.

DO-UNDO-REDO protocol
A protocol used by a data processor (DP) to roll back or roll forward transactions with the help of a system's transaction log entries.

document databases
A NoSQL database model that stores data in key-value pairs in which the value component is composed of a tag-encoded document.

document type definition (DTD)
A file with a .dtd extension that describes XML elements; in effect, a DTD file describes a document's composition and defines the syntax rules or valid tags for each type of XML document.

domain
In data modeling, the construct used to organize and describe an attribute's set of possible values.

drill down
To decompose data into more atomic components—that is, data at lower levels of aggregation. This approach is used primarily in a decision support system to focus on specific geographic areas, business types, and so on.

DROP INDEX
A SQL command used to delete database objects such as tables, views, indexes, and users.

DROP TABLE
A SQL command used to delete database objects such as tables, views, indexes, and users.

durability
The transaction property that ensures that once transaction changes are done and committed, they cannot be undone or lost, even in the event of a system failure.

dynamic query optimization
The process of determining the SQL access strategy at run time, using the most up-to-date information about the database.

dynamic SQL
An environment in which the SQL statement is not known in advance, but instead is generated at run time. In a dynamic SQL environment, a program can generate the SQL statements that are required to respond to ad hoc queries.

dynamic statistical generation mode
In a DBMS, the capability to automatically evaluate and update the database access statistics after each data access operation.

dynamic webpage
Page with contents that change over time and cannot be anticipated; for example, an online ordering system. The content of dynamic Web pages is not predetermined, but is generated at run-time.

dynamic-link libraries (DLLs)
Shared code module that is treated as part of the operating system or server process so it can be dynamically invoked at run time.

E

early binding
A property by which the data type of an object's attribute must be known at definition time, bonding the data type to the object's attribute. Characteristic of an object oriented data model. See also *late binding*.

edge
In a graph database, the representation of a relationship between nodes.

EER diagram (EERD)
The entity relationship diagram resulting from the application of extended entity relationship concepts that provide additional semantic content in the ER model.

electronic commerce (e-commerce)
The use of electronic computer-based technology to bring products, services, or ideas to market and to support or enhance business operations.

electronic data interchange (EDI)
A communications protocol that enabled companies to exchange business documents over private phone networks.

electronic mail (email)
The graphics and text messages sent to other, specific computer end users connected to the same network. Widely used for its low cost and ability to instantly send and receive information to one or more people. Also called e-mail.

electronic wallet
Software that securely stores digital cash to facilitate online transactions.

embedded SQL
SQL statements contained within application programming languages such as COBOL, C++, ASP, Java, and ColdFusion.

encapsulation
A feature by which the object can hide the internal data representation and method's implementation from external objects. Characteristic of an object oriented data model.

encryption
A process of inputting data in "plain text" to yield an output "encoded" version of the data, making the data unintelligible to unauthorized users.

encryption key
Used by encryption algorithms to encode data. The encryption key is a very large number used to encrypt and decrypt data.

enterprise database
The overall company data representation, which provides support for present and expected future needs.

entity
A person, place, thing, concept, or event for which data can be stored. See also *attribute*.

entity cluster
A "virtual" entity type used to represent multiple entities and relationships in the ERD. An entity cluster is formed by combining multiple interrelated entities into a single abstract entity object. An entity cluster is considered "virtual" or "abstract" because it is not actually an entity in the final ERD.

entity instance
A row in a relational table. Also known as *entity occurrence*.

entity integrity
The property of a relational table that guarantees each entity has a unique value in a primary key and that the key has no null values.

entity occurrence
A row in a relational table. Also known as *entity instance*.

entity relationship (ER) model
A data model that describes relationships (1:1, 1:M, and M:N) among entities at the conceptual level with the help of ER diagrams. The model was developed by Peter Chen.

entity relationship diagram (ERD)
A diagram that depicts an entity relationship model's entities, attributes, and relations.

entity set
A collection of like entities.

entity subtypes
In a generalization/specialization hierarchy, a subset of an entity supertype. The entity supertype contains the common characteristics and the subtypes contain the unique characteristics of each entity.

entity supertype
In a generalization/specialization hierarchy, a generic entity type that contains the common characteristics of entity subtypes.

equijoin
A join operator that links tables based on an equality condition that compares specified columns of the tables.

ERM
A data model that describes relationships (1:1, 1:M, and M:N) among entities at the conceptual level with the help of ER diagrams. The model was developed by Peter Chen.

Ethernet
The dominant LAN standard used to interconnect computer systems. Ethernet is based on a bus or star topology that can use coaxial, twisted-pair, or fiber-optic cabling.

eventual consistency
A model for database consistency in which updates to the database will propagate through the system so that all data copies will be consistent eventually.

exclusive lock
An exclusive lock is issued when a transaction requests permission to update a data item and no locks are held on that data item by any other transaction. An exclusive lock does not allow other transactions to access the database.

existence-dependent
A property of an entity whose existence depends on one or more other entities. In such an environment, the existence-independent table must be created and loaded first because the existence-dependent key cannot reference a table that does not yet exist.

existence-independent
A property of an entity that can exist apart from one or more related entities. Such a table must be created first when referencing an existencedependent table.

EXISTS
In SQL, a comparison operator that checks whether a subquery returns any rows.

explanatory analytics
Data analysis that provides ways to discover relationships, trends, and patterns among data.

explicit cursor
In procedural SQL, a cursor created to hold the output of a SQL statement that may return two or more rows, but could return zero or only one row.

extended entity relationship model (EERM)
Sometimes referred to as the enhanced entity relationship model; the result of adding more semantic constructs, such as entity supertypes, entity subtypes, and entity clustering, to the original entity relationship (ER) model.

extended relational data model
A model that includes the object-oriented model's best features in an inherently simpler relational database structural environment. See *extended entity relationship model (EERM)*.

extensible
Capable of being extended by adding new data types and the operations to be performed on them.

Extensible Markup Language (XML)
A meta-language used to represent and manipulate data elements. Unlike other markup languages, XML permits the manipulation of a document's data elements. XML facilitates the exchange of structured documents such as orders and invoices over the Internet.

extents
In a DBMS environment, refers to the ability of data files to expand in size automatically using predefined increments.

external model
The application programmer's view of the data environment. Given its business focus, an external model works with a data subset of the global database schema.

external schema
The specific representation of an external view; the end user's view of the data environment.

extraction, transformation, and loading (ETL)
In a data warehousing environment, the integrated processes of getting data from original sources into the data warehouse. ETL includes retrieving data from original data sources (extraction), manipulating the data into an appropriate form (transformation), and storing the data in the data warehouse (loading).

F

fact table
In a data warehouse, the star schema table that contains facts linked and classified through their common dimensions. A fact table is in a one-to-many relationship with each associated dimension table.

facts
In a data warehouse, the measurements (values) that measure a specific business aspect or activity. For example, sales figures are numeric measurements that represent product or service sales. Facts commonly used in business data analysis include units, costs, prices, and revenues.

failure transparency
A feature that allows continuous operation of a DDBMS, even if a network node fails.

fan trap
A design trap that occurs when one entity is in two 1:M relationships with other entities, thus producing an association among the other entities that is not expressed in the model.

fat client
A client that carries a relatively larger proportion of the processing load than compared to the server. Fat clients are always paired with thin servers.

fat server
A server that carries a relatively larger proportion of the processing load than compared to the client. Fat servers are always paired with thin clients.

feedback loop processing
Analyzing stored data to produce actionable results.

fiber-optic cable
Data transmission medium for computer networks. It used light pulses to transmit the data from node to node and allows for the highest speed of information transfer available.

field
An alphabetic or numeric character or group of characters that defines a characteristic of a person, place, or thing. For example, a person's Social Security number, address, phone number, and bank balance all constitute fields.

field-level lock
A lock that allows concurrent transactions to access the same row as long as they require the use of different fields (attributes) within that row. This type of lock yields the most flexible multiuser data access but requires a high level of computer overhead.

file
A named collection of related records.

file group
See *table space*.

File Transfer Protocol (FTP)
Used to provide file transfer capabilities among computers on the Internet/intranet using a public or known name for classifying and grouping the messages.

firewall
Used to protect a network from unauthorized access from the outside world (public Internet). Specifically, a firewall is a hardware and/or software component that is used to limit and control Internet traffic going into a company's network infrastructure and data that are allowed to be moved outside a company's network.

first normal form (1NF)
The first stage in the normalization process. It describes a relation depicted in tabular format, with no repeating groups and a primary key identified. All nonkey attributes in the relation are dependent on the primary key.

flags
Special codes implemented by designers to trigger a required response, alert end users to specified conditions, or encode values. Flags may be used to prevent nulls by bringing attention to the absence of a value in a table.

foreign key (FK)
An attribute or attributes in one table whose values must match the primary key in another table or whose values must be null. See *key*.

fourth normal form (4NF)
A table is in 4NF if it is in 3NF and contains no multiple independent sets of multivalued dependencies.

fragmentation transparency
A DDBMS feature that allows a system to treat a distributed database as a single database even though it is divided into two or more fragments.

frame
Create in the data-link layer to add control to the information by specifying the network and physical media being used. The information is added at the beginning (header) and at the end (trailer) of a network packet to enclose (frame) the packet data.

FROM
A SQL clause that specifies the table or tables from which data is to be retrieved.

front-end application
Any process that the end user interacts with to request services from a server process.

front-end CASE tool
A computer-aided software tool that provides support for the planning, analysis, and design phases of the SDLC.

full backup
A complete copy of an entire database saved and periodically updated in a separate memory location. A full backup ensures a full recovery of all data after a physical disaster or database integrity failure.

full functional dependence
A condition in which an attribute is functionally dependent on a composite key but not on any subset of the key.

fully heterogeneous DDBMS
A system that integrates different types of database management systems (hierarchical, network, and relational) over a network. It supports different database management systems that may even support different data models running under different computer systems. See also *heterogeneous DDBMS* and *homogeneous DDBMS*.

fully replicated database
In a DDBMS, the distributed database that stores multiple copies of each database fragment at multiple sites.

function-based index
A type of index based on a specific SQL function or expression.

functional dependence
Within a relation R, an attribute B is functionally dependent on an attribute A if and only if a given value of attribute A determines exactly one value of attribute B. The relationship "B is dependent on A" is equivalent to "A determines B," and is written as A B.

G

gateway
A type of middleware software that is used to translate client requests into the appropriate protocols needed to access specific services.

gateway server firewall
A type of firewall that operates at the application level.

general-purpose databases
A database that contains a wide variety of data used in multiple disciplines.

generalization
In a specialization hierarchy, the grouping of common attributes into a supertype entity.

Get Hold (GH)
In an IMS hierarchical DBMS, this statement is used to hold a segment for delete or replace operations. There are three different Get Hold statements: Get Hold Next (GHN), Get Hold Next within Parent (GHNP), and Get Hold Unique (GHU).

Get Next (GN)
In hierarchical databases, a statement to retrieve sequential segments.

Get Next Within Parent (GNP)
In hierarchical databases, a statement to return all segments within the current parent.

Get Unique (GU)
In an IMS hierarchical DBMS, a statement that is used to retrieve a database segment into the application program input area or record area.

governance
In business intelligence, the methods for controlling and monitoring business health and promoting consistent decision making.

government to business (G2B)
Special case of the Business to Business and Business to Commerce e-commerce styles. See also *government to consumer (G2C)*.

government to consumer (G2C)
Special case of the Business to Business and Business to Commerce e-commerce styles. See also *government to business (G2B)*.

granularity
The level of detail represented by the values stored in a table's row. Data stored at its lowest level of granularity is said to be *atomic data*.

graph database
A NoSQL database model based on graph theory that stores data on relationship-rich data as a collection of nodes and edges.

GROUP BY
A SQL clause used to create frequency distributions when combined with any of the aggregate functions in a SELECT statement.

H

hacker
A person who maliciously and illegally accesses a Web site with the intention of stealing data, changing Web pages, or impairing Web site operations.

Hadoop
A Java based, open source, high speed, fault-tolerant distributed storage and computational framework. Hadoop uses low-cost hardware to create clusters of thousands of computer nodes to store and process data.

Hadoop Distributed File System (HDFS)
A highly distributed, fault-tolerant file storage system designed to manage large amounts of data at high speeds.

hardware independence
A condition in which a model does not depend on the hardware used in the model's implementation. Therefore, changes in the hardware will have no effect on the database design at the conceptual level.

hash index
An index based on an ordered list of hash values.

HAVING
A clause applied to the output of a GROUP BY operation to restrict selected rows.

heartbeat
In the Hadoop Distributed File System (HDFS), a signal sent every 3 seconds from the data node to the name node to notify the name node that the data node is still available.

heterogeneity transparency
A feature that allows a system to integrate several centralized DBMSs into one logical DDBMS.

heterogeneous DDBMSs
A system that integrates different types of centralized database management systems over a network.

hierarchical model
An early database model whose basic concepts and characteristics formed the basis for subsequent database development. This model is based on an upside-down tree structure in which each record is called a segment. The top record is the root segment. Each segment has a 1:M relationship to the segment directly below it.

homogeneous DDBMSs
A system that integrates only one type of centralized database management system over a network.

homonyms
The use of the same name to label different attributes. Homonyms generally should be avoided. Some relational software automatically checks for homonyms and either alerts the user to their existence or automatically makes the appropriate adjustments. See also *synonym*.

horizontal fragmentation
The distributed database design process that breaks a table into subsets of unique rows.

host language
Any language that contains embedded SQL statements.

hub
A warehouse of data packets housed in a central location on a local area network. It contains multiple ports that copy the data in the data packets to make it accessible to selected or all segments of the network.

hybrid object
The type of object classification that contains and represents a repeating group of attributes. One or more of the attributes reference another object that usually summarizes the contents of the hybrid object.

hyperlink
Link between Web pages in hypertext or other electronic document types.

Hypertext Markup Language (HTML)
Standard document-formatting language for Web pages.

Hypertext Transfer Protocol (HTTP)
Standard protocol used by Web browser and Web server to communicate.

I

I/O accelerators
A device used to improve throughput for input/output operations.

identifiers
One or more attributes that uniquely identify each entity instance.

imaging server
A process that runs on a computer and provides image management services to client computers.

immediate update
See *write-through technique*.

implicit cursor
A cursor that is automatically created in procedural SQL when the SQL statement returns only one row.

IN
In SQL, a comparison operator used to check whether a value is among a list of specified values.

in-memory database
A database optimized to store large portions (if not all) of the database in primary (RAM) storage rather than secondary (disk) storage.

inconsistent retrievals
A concurrency control problem that arises when a transaction-calculating summary (aggregate) functions over a set of data while other transactions are updating the data, yielding erroneous results.

incremental backup
A process that only backs up data that has changed in the database since the last incremental or full backup.

index
An ordered array of index key values and row ID values (pointers). Indexes are generally used to speed up and facilitate data retrieval. Also known as an *index key*.

index key
See *index*.

index selectivity
A measure of how likely an index is to be used in query processing.

index-organized table
In a DBMS, type of table storage organization that stores end-user data and index data in consecutive locations in permanent storage. Also known as *cluster-indexed table*.

information
The result of processing raw data to reveal its meaning. Information consists of transformed data and facilitates decision making.

Information Engineering (IE)
A methodology that translates a company's strategic goals into helpful data and applications. IE focuses on the description of corporate data instead of the processes.

information resource dictionary
Another name for *data dictionary*.

information resource manager (IRM)
See *data administrator (DA)*.

information system
A system that provides for data collection, storage, and retrieval; facilitates the transformation of data into information; and manages both data and information. An information system is composed of hardware, the DBMS and other software, database(s), people, and procedures.

information systems (IS) department
A department responsible for all information technology services and production functions in an organization.

information systems architecture (ISA)
The output of the information engineering (IE) process that serves as the basis for planning, developing, and controlling future information systems.

Infrastructure as a Service (IaaS)
A model in which the cloud service provider offers consumers the ability to provision their own resources on demand; these resources include storage, servers, databases, processing units, and even a complete virtualized desktop.

inheritance
In the object-oriented data model, the ability of an object to inherit the data structure and methods of the classes above it in the class hierarchy. See also *class hierarchy*.

inner join
A join operation in which only rows that meet a given criterion are selected. The join criterion can be an equality condition (natural join or equijoin) or an inequality condition (theta join). The inner join is the most commonly used type of join. Contrast with *outer join*.

inner query
A query that is embedded or nested inside another query. Also known as a *nested query* or a *subquery*.

input/output (I/O) request
A low-level data access operation that reads or writes data to and from computer devices.

Insert (ISRT)
In hierarchical databases, a statement used to add a segment to the database.

INSERT
A SQL command that allows the insertion of one or more data rows into a table.

instance variables
In the object-oriented model, another term for an attribute. See *attribute*.

Institute of Electrical and Electronics Engineers (IEEE)
An organization that develops standards to provide uniformity among the technical details that define network topology and data transmission across shared media for networks.

Integrity
In a data security framework, refers to keeping data consistent and free of errors or anomalies. See also *data integrity*.

intelligent terminals
A device that provides enhanced I/O functions to a mainframe system such as a PC connected to a mainframe computer.

Interactive Database Processor (IDP)
In an IDS/II network DBMS, a processor that allows users to manipulate databases. The IDP front end is intended for users who have some programming knowledge and is not well-suited for most end users.

internal model
In database modeling, a level of data abstraction that adapts the conceptual model to a specific DBMS model for implementation. The internal model is the representation of a database as "seen" by the DBMS. In other words, the internal model requires a designer to match the conceptual model's characteristics and constraints to those of the selected implementation model.

internal schema
A representation of an internal model using the database constructs supported by the chosen database.

International Organization for Standardization (ISO)
An organization formed to develop standards for diverse network systems.

Internet
A global network of computers connected together through a standard network protocol known as Transmission Control Protocol/Internet Protocol (TCP/IP). You can think of the Internet as the "highway" on which the data travel. The terms *Internet* and *World Wide Web* are often used interchangeably, but they are not synonyms.

Internetwork Packet Exchange/ Sequenced Packet Exchange (IPX/SPX)
A data communications protocol that determines how messages between computers are sent interpreted and processed.

interobject relationship
An attribute-class relationship created when an object's attribute references another object of the same or a different class.

interprocess communication (IPC)
A capability supported by various operating systems to allow two processes to communicate with each other so that applications can share data without interfering with each other.

interrogate
To ask for the interrogated object's instance variable value or values. An object may send messages to interrogate another object's state.

INTERSECT
In relational algebra, an operator used to yield only the rows that are common to two union-compatible tables.

intrabusiness
A style of e-commerce that involves interactions internal to a company.

intranets
Company-owned and -operated computer networks that are restricted to the company's internal use. Such systems can only be accessed by the computers inside the company's computer network. The purpose of such a system is to enhance company operations through improved data access management and communication.

IS NULL
In SQL, a comparison operator used to check whether an attribute has a value.

islands of information
In the old file system environment, pools of independent, often duplicated, and inconsistent data created and managed by different departments.

isolation
A database transaction property in which a data item used by one transaction is not available to other transactions until the first one ends.

iterative process
A process based on repetition of steps and procedures.

J

Java
An object-oriented programming language developed by Sun Microsystems that runs on top of the web browser software. Java applications are compiled and stored on the web server. Java's main advantage is its ability to let application developers create their applications once and then run them in many environments.

Java Database Connectivity (JDBC)
An application programming interface that allows a Java program to interact with a wide range of data sources, including relational databases, tabular data sources, spreadsheets, and text files.

JavaScript
A scripting language that allows web authors to design interactive websites. JavaScript code is embedded in webpages, and then downloaded with the page and activated when a specific event takes place, such as a mouse click on an object.

job tracker
A central control program used to accept, distribute, monitor, and report on MapReduce processing jobs in a Hadoop environment.

JOIN
In relational algebra, a type of operator used to yield rows from two tables based on criteria. There are many types of joins, such as natural join, theta join, equijoin, and outer join.

join columns
Columns that are used in the criteria of join operations. The join columns generally share similar values (have a compatible domain).

JSON (JavaScript Object Notation)
A human-readable text format for data interchange that defines attributes and values in a document.

K

key
One or more attributes that determine other attributes. See also *superkey, candidate key, primary key (PK), secondary key,* and *foreign key*.

key attribute
The attributes that form a primary key. See also *prime attribute*.

key performance indicators (KPIs)
In business intelligence, quantifiable numeric or scale-based measurements that assess a company's effectiveness or success in reaching strategic and operational goals. Examples of KPIs are product turnovers, sales by promotion, sales by employee, and earnings per share.

key-value (KV) databases
A NoSQL database model that stores data as a collection of key-value pairs in which the value component is unintelligible to the DBMS.

key-value
A data model based on a structure composed of two data elements: a key and a value, in which every key has a corresponding value or set of values. The keyvalue data model is also called the associative or attribute-value data model.

knowledge
The body of information and facts about a specific subject. Knowledge implies familiarity, awareness, and understanding of information as it applies to an environment. A key characteristic is that new knowledge can be derived from old knowledge.

L

late binding
A characteristic in which the data type of an attribute is not known until execution time or run-time.

left outer join
In a pair of tables to be joined, a join that yields all the rows in the left table, including those that have no matching values in the other table. For example, a left outer join of CUSTOMER with AGENT will yield all of the CUSTOMER rows, including the ones that do not have a matching AGENT row. See also *outer join* and *right outer join*.

LIKE
In SQL, a comparison operator used to check whether an attribute's text value matches a specified string pattern.

linking table
In the relational model, a table that implements an M:M relationship. See also *composite entity*.

load testing
Services to ensure that an application will support the load imposed by having thousands of users access it.

local area network
A network of computers that spans a small area, such as a single building.

local mapping transparency
A property of a DDBMS in which database access requires the user to know both the name and location of the fragments.

LOCATION MODE clause
In an IDS/II network DBMS, this clause determines where a record will be (physically) stored in the database and how the record will be retrieved.

location transparency
A property of a DDBMS in which database access requires the user to know only the name of the database fragments. (Fragment locations need not be known.)

lock
A device that guarantees unique use of a data item in a particular transaction operation. A transaction requires a lock prior to data access; the lock is released after the operation's execution to enable other transactions to lock the data item for their own use.

lock granularity
The level of lock use. Locking can take place at the following levels: database, table, page, row, and field (attribute).

lock manager
A DBMS component that is responsible for assigning and releasing locks.

logical data format
The way a person views data within the context of a problem domain.

logical design
A stage in the design phase that matches the conceptual design to the requirements of the selected DBMS and is therefore software dependent. Logical design is used to translate the conceptual design into the internal model for a selected database management system, such as DB2, SQL Server, Oracle, IMS, Informix, Access, or Ingress.

logical independence
A condition in which the internal model can be changed without affecting the conceptual model. (The internal model is hardware independent because it is unaffected by the computer on which the software is installed. Therefore, a change in storage devices or operating systems will not affect the internal model.)

lost update
A concurrency control problem in which a data update is lost during the concurrent execution of transactions.

M

mandatory participation
A relationship in which one entity occurrence must have a corresponding occurrence in another entity. For example, an EMPLOYEE works in a DIVISION. (A person cannot be an employee without being assigned to a company's division.)

manual query optimization
An operation mode that requires the end user or programmer to define the access path for the execution of a query.

manual statistical generation mode
A mode of generating statistical data access information for query optimization. In this mode, the DBA must periodically run a routine to generate the data access statistics—for example, running the RUNSTAT command in an IBM DB2 database.

many-to-many (M:N or *..*) relationship
Association among two or more entities in which one occurrence of an entity is associated with many occurrences of a related entity and one occurrence of the related entity is associated with many occurrences of the first entity.

map
The function in a MapReduce job that sorts and filters data into a set of key-value pairs as a subtask within a larger job.

mapper
A program that performs a map function.

MapReduce
An open-source application programming interface (API) that provides fast data analytics services; one of the main Big Data technologies that allows organizations to process massive data stores.

master data management (MDM)
In business intelligence, a collection of concepts, techniques, and processes for the proper identification, definition, and management of data elements within an organization.

master data management (MDM) software
Software the provides a "master copy" of entities such as customers, that appear in numerous systems throughout the organization. This software helps prevent dirty data by coordinating common data across multiple systems.

materialized view
A dynamic table that not only contains the SQL query command to generate rows but stores the actual rows. The materialized view is created the first time the query is run and the summary rows are stored in the table. The materialized view rows are automatically updated when the base tables are updated.

MAX
A SQL aggregate function that yields the maximum attribute value in a given column.

message
In the OO data model, the name of a method sent to an object in order to perform an action. A message triggers the object's behavior. See *method*.

messaging
A service to ensure the proper routing and delivery of application-oriented data among multiple services.

metadata
Data about data; that is, data about data characteristics and relationships. See also *data dictionary*.

method
In the object-oriented data model, a named set of instructions to perform an action. Methods represent real-world actions, and are invoked through messages.

metrics
In a data warehouse, numeric facts that measure a business characteristic of interest to the end user.

metropolitan area network (MAN)
Network type used to connect computers across a city or metropolitan area.

Microsoft .NET framework
A component-based platform for the development of distributed, heterogeneous, interoperable applications aimed at manipulating any type of data over any network regardless of operating system and programming language.

middleware
The computer software that allows clients and servers to communicate within the client/server architecture. It is used to insulate client processes from the network protocols and the details of the server process protocols.

MIN
A SQL aggregate function that yields the minimum attribute value in a given column.

minimal data rule
Defined as "All that is needed is there, and all that is there is needed." In other words, all data elements required by database transactions must be defined in the model, and all data elements defined in the model must be used by at least one database transaction.

mixed fragmentation
A combination of horizontal and vertical strategies for data fragmentation, in which a table may be divided into several rows and each row has a subset of the attributes (columns).

master data management (MDM) software
Software the provides a "master copy" of entities such as customers, that appear in numerous systems throughout the organization. This software helps prevent dirty data by coordinating common data across multiple systems.

module
(1) A design segment that can be implemented as an autonomous unit, and is sometimes linked to produce a system. (2) An information system component that handles a specific function, such as inventory, orders, or payroll.

module coupling
The extent to which modules are independent of one another.

monotonicity
A quality that ensures that time stamp values always increase. (The time stamping approach to scheduling concurrent transactions assigns a global, unique time stamp to each transaction. The time stamp value produces an explicit order in which transactions are submitted to the DBMS.)

multidimensional database management systems (MDBMSs)
A database management system that uses proprietary techniques to store data in matrixlike arrays of *n* dimensions known as cubes.

multidimensional online analytical processing (MOLAP)
An extension of online analytical processing to multidimensional database management systems.

multiple access unit (MAU)
A wiring concentrator through which a token ring's computers are connected physically.

multiple inheritance
Exists when a class can have more than one immediate (parent) superclass above it in an object-oriented database environment.

multiple-site processing, multiple-site data (MPMD)
A scenario describing a fully distributed database management system with support for multiple data processors and transaction processors at multiple sites.

multiple-site processing, single-site data (MPSD)
A scenario in which multiple processes run on different computers sharing a single data repository.

multitenant database
A database environment in which a container database can hold other databases.

multiuser database
A database that supports multiple concurrent users.

multivalued attributes
An attribute that can have many values for a single entity occurrence. For example, an EMP_ DEGREE attribute might store the string "BBA, MBA, PHD" to indicate three different degrees held.

mutual consistency rule
A data replication rule that requires all copies of data fragments to be identical.

mutual exclusive rule
A condition in which only one transaction at a time can own an exclusive lock on the same object.

N

name node
One of three types of nodes used in the Hadoop Distributed File System (HDFS). The name node stores all the metadata about the file system. See also *client node* and *data node*.

natural identifier
A generally accepted identifier for real-world objects. As its name implies, a natural key is familiar to end users and forms part of their day-to-day business vocabulary.

natural join
A relational operation that yields a new table composed of only the rows with common values in their common attribute(s).

natural key
See *natural identifier*.

nested query
In SQL, a query that is embedded in another query. See *subquery*.

network backbone
The main network cabling system for one or more local area networks.

Network Basic Input/Output System (NetBIOS)
A network protocol originally developed by IBM and SYTEK Corporation in 1984.

Network interface cards (NICs)
Electronic circuit board that allows computers to communicate within a network.

network latency
The delay imposed by the amount of time required for a data packet to make a round trip from point A to point B.

network model
An early data model that represented data as a collection of record types in 1:M relationships.

network operating system (NOS)
A computer operating system oriented toward providing server side services to clients (such as file and printer sharing and security management).

network partitioning
The delay that occurs when nodes become suddenly unavailable due to a network failure. In distributed databases, the system must account for the possibility of this condition.

network protocol
A set of rules (at the physical level) that determines how messages between computers are sent, processed, and interpreted.

network segment
A single section of cable that connects several computers.

network translator
A middleware component that manages the network communications protocols.

news and discussion group services
Specialized services that allow the creation of "virtual communities" in which users exchange messages regarding specific topics.

NewSQL
A database model that attempts to provide ACID-compliant transactions across a highly distributed infrastructure.

node
In a graph database, the representation of a single entity instance.

non-identifying relationship
A relationship in which the primary key of the related entity does not contain a primary key component of the parent entity.

nonkey attribute
See *nonprime attribute*.

nonoverlapping subtypes
See *disjoint subtype*.

nonprime attribute
An attribute that is not part of a key.

nonrepeatable read
In transaction management, when a transaction reads a given row at time t1, then reads the same row at time t2, yielding different results because the original row may have been updated or deleted.

nonserialized items
Items for which the attributes describe a generalized view of the that kind of item, without identifying each individual instance of the item.

normalization
A process that assigns attributes to entities so that data redundancies are reduced or eliminated.

NoSQL
A new generation of database management systems that is not based on the traditional relational database model.

NOT
A SQL logical operator that negates a given predicate.

null
The absence of an attribute value. Note that a null is not a blank.

O

object
An abstract representation of a real world entity that has a unique identity, embedded properties, and the ability to interact with other objects and itself.

object ID (OID)
In an object-oriented database environment, a system-generated object identifier that is independent of the object state and any physical address in memory.

object instance
Each particular object belonging to a class.

Object Linking and Embedding for Database (OLE-DB)
Based on Microsoft's Component Object Model (COM), OLE-DB is database middleware that adds object-oriented functionality for accessing relational and nonrelational data.

object orientation
A set of modeling and development principles focused on an autonomous entity with embedded intelligence to interact with other objects and itself.

object query language (OQL)
The database query language used by an object oriented database management system.

object schema
See Object space.

object space
The equivalent of the database schema, as seen by the designer in an object oriented database.

object state
The set of values that the object's attributes have at a given time.

object table
The equivalent of a relational table composed of many rows, where each row is an object of the same type. Each row object has a unique system generated object ID (OID) or object identifier.

object-oriented data model (OODM)
A data model whose basic modeling structure is an object.

object-oriented database management system (OODBMS)
Data management software used to manage data in an object-oriented database model.

object-oriented programming (OOP)
An alternative to conventional programming methods based on object oriented concepts. It reduces programming time and lines of code, and increases programmers' productivity.

object-oriented programming languages (OOPLs)
A programming language based on object oriented concepts.

object/relational database management system (O/R DBMS)
A DBMS based on the extended relational model (ERDM). The ERDM, championed by many relational database researchers, constitutes the relational model's response to the OODM. This model includes many of the object-oriented model's best features within an inherently simpler relational database structure.

one-to-many (1:M or 1..*) relationship
Associations among two or more entities that are used by data models. In a 1:M relationship, one entity instance is associated with many instances of the related entity.

one-to-one (1:1 or 1..1) relationship
Associations among two or more entities that are used by data models. In a 1:1 relationship, one entity instance is associated with only one instance of the related entity.

online analytical processing (OLAP)
Decision support system (DSS) tools that use multidimensional data analysis techniques. OLAP creates an advanced data analysis environment that supports decision making, business modeling, and operations research.

online transaction processing (OLTP) database
See *operational database*.

Open Database Connectivity (ODBC)
Microsoft database middleware that provides a database access API to Windows applications.

Open Systems Interconnection (OSI)
A seven-layer reference model developed by the International Organization for Standardization (ISO) to help standardize diverse network systems.

operational database
A database designed primarily to support a company's day-to-day operations. Also known as a *transactional database*, *OLTP database*, or *production database*.

optimistic approach
In transaction management, a concurrency control technique based on the assumption that most database operations do not conflict.

optimizer hints
Special instructions for the query optimizer that are embedded inside the SQL command text.

optional attribute
In ER modeling, an attribute that does not require a value; therefore, it can be left empty.

optional participation
In ER modeling, a condition In which one entity occurrence does not require a corresponding entity occurrence in a particular relationship.

OR
The SQL logical operator used to link multiple conditional expressions in a WHERE or HAVING clause. It requires only one of the conditional expressions to be true.

ORDER BY
A SQL clause that is useful for ordering the output of a SELECT query (for example, in ascending or descending order).

outer join
A relational algebra join operation that produces a table in which all unmatched pairs are retained; unmatched values in the related table are left null. Contrast with *inner join*. See also *left outer join* and *right outer join*.

overlapping subtypes
In a specialization hierarchy, a condition in which each entity instance (row) of the supertype can appear in more than one subtype.

P

packet filter firewall
A type of firewall that works at the TCP/IP packet level.

page
In permanent storage, the equivalent of a disk block, which can be described as a directly addressable section of a disk. A diskpage has a fixed size, such as 4K, 8K, or 16K.

page-level lock
In this type of lock, the database management system locks an entire diskpage, or section of a disk. A diskpage can contain data for one or more rows and from one or more tables.

partial completeness
In a generalization/specialization hierarchy, a condition in which some supertype occurrences might not be members of any subtype.

partial dependency
A condition in which an attribute is dependent on only a portion (subset) of the primary key.

partially replicated database
A distributed database in which copies of only some database fragments are stored at multiple sites.

participants
An ER term for entities that participate in a relationship. For example, in the relationship "PROFESSOR

teaches CLASS," the *teaches* relationship is based on the participants PROFESSOR and CLASS.

partition key
In partitioned databases, one or more attributes in a table that determine the fragment in which a row will be stored.

partitioned data allocation
A data allocation strategy of dividing a database into two or more fragments that are stored at two or more sites.

partitioning
The process of splitting a table into subsets of rows or columns.

passive data dictionary
A DBMS data dictionary that requires a command initiated by an end user to update its data access statistics.

performance transparency
A DDBMS feature that allows a system to perform as though it were a centralized DBMS.

performance tuning
Activities that make a database perform more efficiently in terms of storage and access speed.

periodicity
Information about the time span of data stored in a table, usually expressed as current year only, previous years, or all years.

persistent stored module (PSM)
A block of code with standard SQL statements and procedural extensions that is stored and executed at the DBMS server.

personalization
In an e-commerce environment, customization of a Web page for individual users.

pessimistic locking
The use of locks based on the assumption that conflict between transactions is likely.

phantom read
In transaction management, when a transaction executes a query at time t1, then runs the same query at time t2, yielding additional rows that satisfy the query.

physical data format
The way a computer "sees" (stores) data.

physical design
A stage of database design that maps the data storage and access characteristics of a database. Because these characteristics are a function of the types of devices supported by the hardware, the data access methods supported by the system physical design are both hardware- and software-dependent. See also *physical model*.

physical independence
A condition in which the physical model can be changed without affecting the internal model.

physical model
A model in which physical characteristics such as location, path, and format are described for the data. The physical model is both hardware- and software dependent. See also *physical design*.

Platform as a Service (PaaS)
A model in which the cloud service provider can build and deploy consumer-created applications using the provider's cloud infrastructure.

plug-in
On the web, a client-side, external application that is automatically invoked by the browser when needed to manage specific types of data.

pluggable database
In a multitenant database environment, a database that can be contained within a container database.

policies
General statement of direction that is used to manage company operations through the communication and support of the organization's objectives.

polyglot persistence
The coexistence of a variety of data storage and data management technologies within an organization's infrastructure.

polymorphism
An object oriented data model characteristic by which different objects can respond to the same message in different ways.

portals
In terms of business intelligence, a unified, single point of entry for information distribution.

predicate logic
Used extensively in mathematics to provide a framework in which an assertion (statement of fact) can be verified as either true or false.

predictive analytics
Data analytics that use advanced statistical and modeling techniques to predict future business outcomes with great accuracy.

Pretty Good Privacy (PGP)
An example of public-key encryption by Pretty Good Privacy Inc. PGP is a fairly popular and inexpensive method for encrypting e-mail messages on the Internet.

primary key (PK)
In the relational model, an identifier composed of one or more attributes that uniquely identifies a row. Also, a candidate key selected as a unique entity identifier. See also *key*.

prime attribute
A key attribute; that is, an attribute that is part of a key or is the whole key. See also *key attributes*.

privacy
The rights of individuals and organizations to determine access to data about themselves.

private cloud
A form of cloud computing in which an internal cloud is built by an organization to serve its own needs.

private key
A key that is known only to the owner of the key.

private-key encryption
Encryption that uses a single numeric key to encode and decode data. Both sender and receiver must know the encryption key. See also *symmetric encryption*.

Procedural Language SQL (PL/SQL)
An Oracle-specific programming language based on SQL with procedural extensions designed to run inside the Oracle database.

procedure cache
See *SQL cache*.

procedures
Series of steps to be followed during the performance of an activity or process.

processing option (PROCOPT)
A type of access granted to a program.

PRODUCT
In relational algebra, an operator used to yield all possible pairs of rows from two tables. Also known as the Cartesian product.

production database
See *operational database*.

profile
In Oracle, a named collection of settings that controls how much of the database resource a given user can use.

program communication block (PCB)
In a hierarchical database, after the physical database has been defined through the DBD, a way through which application programs are given a subset of the physical database.

program specification block (PSB)
In a hierarchical database, this represents a logical view of a selected portion of the database and also defines the database(s), segments, and types of operations that can be performed by the application. Using PSBs yields better data security as well as improved program efficiency by allowing access to only the portion of the database that is required to perform a given function.

PROJECT
In relational algebra, an operator used to select a subset of columns.

properties
In a graph database, the attributes or characteristics of a node or edge that are of interest to the users.

protocol
A specific set of rules to accomplish a specific function. In the object oriented data model, protocol refers to a collection of messages to which an object responds.

proxy server firewall
A firewall that operates as an intermediary between client computers inside a private network and the Internet.

public cloud
A form of computing in which the cloud infrastructure is built by a third-party organization to sell cloud services to the general public.

public key
A key that is available to anyone wanting to communicate securely with the key's owner.

public-key encryption
A form of encryption that uses two numeric keys—the public key and the private key. Both keys are able to encrypt and decrypt each other's messages. See also *asymmetric encryption*.

Q

query
A question or task asked by an end user of a database in the form of SQL code. A specific request for data manipulation issued by the end user or the application to the DBMS.

query language
A nonprocedural language that is used by a DBMS to manipulate its data. An example of a query language is SQL.

query optimizer
A DBMS process that analyzes SQL queries and finds the most efficient way to access the data. The query optimizer generates the access or execution plan for the query.

query processing bottleneck
In query optimization, a delay introduced in the processing of an I/O operation that causes the overall system to slow down.

query result set
The collection of data rows returned by a query.

R

RAID
An acronym for Redundant Array of Independent Disks. RAID systems use multiple disks to create virtual disks (storage volumes) from several individual disks. RAID systems provide performance improvement, fault tolerance, and a balance between the two.

read committed
An ANSI SQL transaction isolation level that allows transactions to read only committed data. This is the default mode of operations for most databases.

read uncommitted
An ANSI SQL transaction isolation level that allows transactions to read uncommitted data from other transactions, and which allows nonrepeatable reads and phantom reads. The least restrictive level defined by ANSI SQL.

record
A collection of related (logically connected) fields.

record at a time
This term indicates that the database commands affect a single record at a time.

RECORD NAME clause
In the IDS/II network DBMS, this clause initiates the record's definition by assigning it a unique name. An IDS/II network database must contain at least one record type.

recursive query
A nested query that joins a table to itself.

recursive relationship
A relationship found within a single entity type. For example, an EMPLOYEE is married to an EMPLOYEE or a PART is a component of another PART.

reduce
The function in a MapReduce job that collects and summarizes the results of map functions to produce a single result.

reducer
A program that performs a reduce function.

redundant transaction logs
Multiple copies of the transaction log kept by database management systems to ensure that the physical failure of a disk will not impair the DBMS's ability to recover data.

referential integrity
A condition by which a dependent table's foreign key must have either a null entry or a matching entry in the related table.

referential object sharing
When an object instance is referenced by other objects. That is, two or more different objects point to the same object instance, a change in the referenced object instance values is automatically reflected in all other referring objects.

regular entity
See *strong entity*.

relation
A logical construct perceived to be a two dimensional structure composed of intersecting rows (entities) and columns (attributes) that represents an entity set in the relational model.

relational algebra
A set of mathematical principles that form the basis for manipulating relational table contents; the eight main functions are SELECT, PROJECT, JOIN, INTERSECT, UNION, DIFFERENCE, PRODUCT, and DIVIDE.

relational database management system (RDBMS)
A collection of programs that manages a relational database. The RDBMS software translates a user's logical requests (queries) into commands that physically locate and retrieve the requested data.

relational diagram
A graphical representation of a relational database's entities, the attributes within those entities, and the relationships among the entities.

relational model
Developed by E. F. Codd of IBM in 1970, the relational model is based on mathematical set theory and represents data as independent relations. Each relation (table) is conceptually represented as a two dimensional structure of intersecting rows and columns. The relations are related to each other through the sharing of common entity characteristics (values in columns).

relational online analytical processing (ROLAP)
Analytical processing functions that use relational databases and familiar relational query tools to store and analyze multidimensional data.

relational schema
The organization of a relational database as described by the database administrator.

relationship
An association between entities.

relationship degree
The number of entities or participants associated with a relationship. A relationship degree can be unary, binary, ternary, or higher.

relvar
Short for relation variable, a variable that holds a relation. A relvar is a container (variable) for holding relation data, not the relation itself.

Remote Data Objects (RDO)
A higher-level, object-oriented application interface used to access remote database servers. RDO uses the lower-level DAO and ODBC for direct access to databases.

remote request
A DDBMS feature that allows a single SQL statement to access data in a single remote DP.

remote transaction
A DDBMS feature that allows a transaction (formed by several requests) to access data in a single remote DP.

repeatable read
An ANSI SQL transaction isolation level that uses shared locks to ensure that other transactions do not update a row after the original query updates it. However, phantom reads are allowed.

repeater
A device used in Ethernet networks to add network segments to the network and to extend its signal reach.

repeating group
In a relation, a characteristic describing a group of multiple entries of the same type for a single key attribute occurrence. For example, a car can have multiple colors for its top, interior, bottom, trim, and so on.

replica transparency
The DDBMS's ability to hide the existence of multiple copies of data from the user.

replicated data allocation
A data allocation strategy in which copies of one or more database fragments are stored at several sites.

replication
The process of creating and managing duplicate versions of a database. Replication is used to place copies in different locations and to improve access time and fault tolerance.

required attribute
In ER modeling, an attribute that must have a value. In other words, it cannot be left empty.

reserved words
Words used by a system that cannot be used for any other purpose. For example, in Oracle SQL, the word INITIAL cannot be used to name tables or columns.

resource security
The protection of the resource(s) from external and internal threats. Specifically, resource security means protecting the resource from viruses, unauthorized access by hackers, or denial of service attacks.

RESTRICT
See *SELECT*.

right outer join
In a pair of tables to be joined, a join that yields all of the rows in the right table, including the ones with no matching values in the other table. For example, a right outer join of CUSTOMER with AGENT will yield all of the AGENT rows, including the ones that do not have a matching CUSTOMER row. See also *left outer join* and *outer join*.

ring topology
Network topology in which computers are connected to one another via a cabling setup that, as the name implies, resembles a ring; it is more flexible than a bus topology, because addition or loss of a computer does not have a negative impact on other network activities.

role
In Oracle, a named collection of database access privileges that authorize a user to connect to a database and use its system resources.

roll up
(1) To aggregate data into summarized components, that is, higher levels of aggregation. (2) In SQL, an OLAP extension used with the GROUP BY clause to aggregate data by different dimensions. Rolling up the data is the exact opposite of drilling down the data.

ROLLBACK
A SQL command that restores the database table contents to the condition that existed after the last COMMIT statement.

router
(1) An intelligent device used to connect dissimilar networks. (2) Hardware/software equipment that connects multiple and diverse networks.

row-centric storage
A physical data storage technique in which data is stored in blocks, which hold data from all columns of a given set of rows.

row-level lock
A less restrictive database lock in which the DBMS allows concurrent transactions to access different rows of the same table, even when the rows are on the same page.

row-level trigger
A trigger that is executed once for each row affected by the triggering SQL statement. A row-level trigger requires the use of the FOR EACH ROW keywords in the trigger declaration.

rule-based optimizer
A query optimization mode based on the rule-based query optimization algorithm.

rule-based query optimization algorithm
A query optimization technique that uses preset rules and points to determine the best approach to executing a query.

rules of precedence
Basic algebraic rules that specify the order in which operations are performed. For example, operations within parentheses are executed first, so in the equation $2 + (3 \times 5)$, the multiplication portion is calculated first, making the correct answer 17.

S

scaling out
A method for dealing with data growth that involves distributing data storage structures across a cluster of commodity servers.

scaling up
A method for dealing with data growth that involves migrating the same structure to more powerful systems.

scheduler
The DBMS component that establishes the order in which concurrent transaction operations are executed. The scheduler *interleaves* the execution of database operations in a specific sequence to ensure *serializability*.

schema
A logical grouping of database objects, such as tables, indexes, views, and queries, that are related to each other. Usually, a schema belongs to a single user or application.

scope
The part of a system that defines the extent of the design, according to operational requirements.

script
A programming language that is not compiled, but is interpreted and executed at run time.

search services
A business enabling web service that allows Web sites to perform searches on their contents.

second normal form (2NF)
The second stage in the normalization process, in which a relation is in 1NF and there are no partial dependencies (dependencies in only part of the primary key).

secondary key
A key used strictly for data retrieval purposes. For example, customers are not likely to know their customer number (primary key), but the combination of last name, first name, middle initial, and telephone number will probably match the appropriate table row. See also *key*.

Secure Electronic Transaction (SET)
Initiative to provide a standard for secure credit card transactions over the Internet.

Secure Hypertext Transfer Protocol (S-HTTP)
Protocol used to securely transfer Web documents over the Internet. S-HTTP supports use of private and public keys for authentication and encryption. S-HTTP has not been widely used, because it only supports encrypted HTTP data and does not support other Internet protocols as does SSL.

Secure Sockets Layer (SSL)
A protocol used to implement secure communication channels between client and server computers on the Internet.

security
Activities and measures to ensure the confidentiality, integrity, and availability of an information system and its main asset, data.

security breach
An event in which a security threat is exploited to endanger the integrity, confidentiality, or availability of the system.

security policy
A collection of standards, policies, and procedures created to guarantee the security of a system and ensure auditing and compliance.

security threat
An imminent security violation that could occur due to unchecked security vulnerabilities.

security vulnerability
A weakness in a system component that could be exploited to allow unauthorized access or cause service disruptions.

segment (SEGM)
In the hierarchical data model, the equivalent of a file system's record type.

SELECT
(1) A SQL command that yields the values of all rows or a subset of rows in a table. The SELECT statement is used to retrieve data from tables.

(2) In relational algebra, an operator to select a subset of rows. Also known as *RESTRICT.*

semantic data model
The first of a series of data models that more closely represented the real world, modeling both data and their relationships in a single structure known as an object. The SDM, published in 1981, was developed by M. Hammer and D. McLeod.

semistructured data
Data that has already been processed to some extent.

SENSEG (SENsitive SEGment)
In the IMS hierarchical database, this keyword declares the segments that will be available, starting with the root segment.

sentiment analysis
A method of text analysis that attempts to determine if a statement conveys a positive, negative, or neutral attitude.

sequence
(1) A nested ordering sequence for a set of rows, such as a list in which all last names are alphabetically ordered and, within the last names, all first names are ordered. (2) An object for generating unique sequential values for a sequence field.

sequence field
An attribute that contains values that are unique and sequential (ascending or descending). Some DBMS allows the explicit definition of sequence or autonumber attributes that are generally used to uniquely identify each row.

serializability
A property in which the selected order of concurrent transaction operations creates the same final database state that would have been produced if the transactions had been executed in a serial fashion.

serializable
An ANSI SQL transaction isolation level that does not allow dirty reads, nonrepeatable reads, or phantom reads; the most restrictive level defined by the ANSI SQL standard.

serializable schedule
In transaction management, a schedule of operations in which the interleaved execution of the transactions yields the same result as if they were executed in serial order.

serialized items
Items for which each instance of the item must be tracked as an individually identifiable item.

server
Any process that provides requested services to clients. See *client/server architecture.*

server-side extension
A program that interacts directly with the server process to handle specific types of requests. Server-side extensions add significant functionality to web servers and intranets.

set theory
A part of mathematical science that deals with sets, or groups of things, and is used as the basis for data manipulation in the relational model.

set-oriented
Dealing with or related to sets, or groups of things. In the relational model, SQL operators are set-oriented because they operate over entire sets of rows and columns at once.

shared lock
A lock that is issued when a transaction requests permission to read data from a database and no exclusive locks are held on the data by another transaction. A shared lock allows other read-only transactions to access the database.

simple attribute
An attribute that cannot be subdivided into meaningful components. Compare to *composite attribute.*

simple object
An object that contains only single-valued attributes and has no attributes that refer to another object.

single inheritance
In the object oriented data model, the property of an object that allows it to have only one parent superclass from which it inherits its data structure and methods. See also *inheritance, multiple inheritance.*

single-site processing, single-site data (SPSD)
A scenario in which all processing is done on a single host computer and all data is stored on the host computer's local disk.

single-user database
A database that supports only one user at a time.

single-valued attribute
An attribute that can have only one value.

site monitoring and data analysis
In an e-commerce environment, services to ensure that a Web site performs at an optimal level.

slice and dice
The ability to focus on slices of a data cube (drill down or roll up) to perform a more detailed analysis.

sneakernet
One of the original ways to share data. When users needed to share data, they would simply copy the data to a disk and walk to the coworker's office, disk in hand.

snowflake schema
A type of star schema in which dimension tables can have their own dimension tables. The snowflake schema is usually the result of normalizing dimension tables.

social media
Web and mobile technologies that enable "anywhere, anytime, always on" human interactions.

Software as a Service (SaaS)
A model in which the cloud service provider offers turnkey applications that run in the cloud.

software independence
A property of any model or application that does not depend on the software used to implement it.

sparse data
A case in which the number of table attributes is very large but the number of actual data instances is low.

sparsity
In multidimensional data analysis, a measurement of the data density held in the data cube.

specialization
In a specialization hierarchy, the grouping of unique attributes into a subtype entity.

specialization hierarchy
A hierarchy based on the top-down process of identifying lower-level, more specific entity subtypes from a higher-level entity supertype. Specialization is based on grouping unique characteristics and relationships of the subtypes.

SQL cache
A shared, reserved memory area that stores the most recently executed SQL statements or PL/SQL procedures, including triggers and functions. Also called *procedure cache.*

SQL data services (SDS)
Data management services that provide relational data storage, access, and management over the Internet.

SQL performance tuning
Activities to help generate a SQL query that returns the correct answer in the least amount of time, using the minimum amount of resources at the server end.

standards
A detailed and specific set of instructions that describes the minimum requirements for a given activity. Standards are used to evaluate the quality of the output.

star schema
A data modeling technique used to map multidimensional decision support data into a relational database. The star schema represents data using a central table known as a fact table in a 1:M relationship with one or more dimension tables.

star topology
Network topology with all computers connected to one another in a star configuration through a central computer or network hub. Like a ring topology, allows for computers to be added to or released from the network without having an impact on other computers.

state inspection firewall
A type of firewall that compares parts of incoming packets and related outgoing packets.

stateless system
A system in which a web server does not know the status of the clients communicating with it. The web does not reserve memory to maintain an open communications state between the client and the server.

statement-level trigger
A SQL trigger that is assumed if the FOR EACH ROW keywords are omitted. This type of trigger is executed once, before or after the triggering statement completes, and is the default case.

static query optimization
A query optimization mode in which the access path to a database is predetermined at compilation time.

static SQL
A style of embedded SQL in which the SQL statements do not change while the application is running.

static webpage
Web page used to display information that does not change much over time or is not time-critical.

statistically based query optimization algorithm
A query optimization technique that uses statistical information about a database. The DBMS then uses these statistics to determine the best access strategy.

stored function
A named group of procedural and SQL statements that returns a value, as indicated by a RETURN statement in its program code.

stored procedure
(1) A named collection of procedural and SQL statements. (2) Business logic stored on a server in the form of SQL code or another DBMS-specific procedural language.

stream processing
The processing of data inputs in order to make decisions about which data to keep and which data to discard before storage.

strong (identifying) relationship
A relationship that occurs when two entities are existence-dependent; from a database design perspective, this relationship exists whenever the primary key of the related entity contains the primary key of the parent entity.

strong entity
An entity that is existence-independent, that is, it can exist apart from all of its related entities. Also called a *regular entity.*

structural dependence
A data characteristic in which a change in the database schema affects data access, thus requiring changes in all access programs.

structural independence
A data characteristic in which changes in the database schema do not affect data access.

structured data
Data that has been formatted to facilitate storage, use, and information generation.

Structured Query Language (SQL)
A powerful and flexible relational database language composed of commands that enable users to create database and table structures, perform various types of data manipulation and data administration, and query the database to extract useful information.

subclasses
See *class hierarchy.*

subordinates
In a DDBMS, a data processor (DP) node that participates in a distributed transaction using the two-phase COMMIT protocol.

subquery
A query that is embedded (or nested) inside another query. Also known as a *nested query* or an *inner query.*

subschema
The portion of the database that interacts with application programs.

subtype discriminator
The attribute in the supertype entity that determines to which entity subtype each supertype occurrence is related.

SUM
A SQL aggregate function that yields the sum of all values for a given column or expression.

super column
In a column family database, a column that is composed of a group of other related columns.

superclass
In a class hierarchy, the superclass is the more general classification from which the subclasses inherit data structures and behaviors.

superkey
An attribute or attributes that uniquely identify each entity in a table. See *key*.

surrogate key
A system-assigned primary key, generally numeric and autoincremented.

switch
An intelligent device that connects computers. Unlike a hub, a switch allows multiple simultaneous transmissions between two ports (computers). Therefore, switches have greater throughput and speed than regular hubs.

symmetric encryption
Encryption that uses a single numeric key to encode and decode data. Both sender and receiver must know the encryption key. See also *private-key encryption*.

synonym
The use of different names to identify the same object, such as an entity, an attribute, or a relationship; synonyms should generally be avoided. See also *homonym*.

system catalog
A detailed system data dictionary that describes all objects in a database.

systems administrator
The person responsible for coordinating and performing day-to-day data-processing activities.

systems analysis
The process that establishes the need for an information system and its extent.

systems development
The process of creating an information system.

Systems Development Life Cycle (SDLC)
The cycle that traces the history of an information system. The SDLC provides the big picture within which database design and application development can be mapped out and evaluated.

Systems Network Architecture (SNA)
A network environment used by IBM mainframe computers

T

table
A logical construct perceived to be a two dimensional structure composed of intersecting rows (entities) and columns (attributes) that represents an entity set in the relational model.

table space
In a DBMS, a logical storage space used to group related data. Also known as a *file group*.

table-level lock
A locking scheme that allows only one transaction at a time to access a table. A table-level lock locks an entire table, preventing access to any row by transaction T2 while transaction T1 is using the table.

tablespace
In a DBMS, a logical storage space used to group related data. Also known as a *file group*.

tags
In markup languages such as HTML and XML, a command inserted in a document to specify how the document should be formatted. Tags are used in server-side markup languages and interpreted by a web browser for presenting data.

task trackers
A program in the MapReduce framework responsible to running map and reduce tasks on a node.

ternary relationship
An ER term used to describe an association (relationship) between three entities. For example, a DOCTOR prescribes a DRUG for a PATIENT.

theta join
A join operator that links tables using an inequality comparison operator ($<, >, <=, >=$) in the join condition.

thin client
A client that carries a relatively smaller proportion of the processing load than compared to the server, thin clients are always paired with fat servers.

thin server
A server that carries a lesser processing load than the client processes, thin servers are always paired with fat clients.

third normal form (3NF)
A table is in 3NF when it is in 2NF and no nonkey attribute is functionally dependent on another nonkey attribute; that is, it cannot include transitive dependencies.

three-tier client/server system
A system design where the client's requests are handled by intermediate servers, which coordinate the execution of the client requests with subordinate servers. See also *client/server*.

3 Vs
Three basic characteristics of Big Data databases: volume, velocity, and variety.

time stamping
In transaction management, a technique used in scheduling concurrent transactions that assigns a global unique time stamp to each transaction.

time-variant data
Data whose values are a function of time. For example, timevariant data can be seen at work when a company's history of all administrative appointments is tracked.

token
In a ring topology network, the marker that passes from computer to computer, similar to a baton in a relay race. Only the computer with the token can transmit at a given time.

token ring networks
Networks that use a ring topology and token passing access control.

top-down design
A design philosophy that begins by defining the main structures of a system and then moves to define the smaller units within those structures. In database design, this process first identifies entities and then defines the attributes within the entities.

total completeness
In a generalization/specialization hierarchy, a condition in which every supertype occurrence must be a member of at least one subtype.

transaction
A sequence of database requests that accesses the database. A transaction is a logical unit of work; that is, it must be *entirely* completed or aborted—no intermediate ending states are accepted. All transactions must have the properties of atomicity, consistency, isolation, and durability.

transaction log
A feature used by the DBMS to keep track of all transaction operations that update the database. The information stored in this log is used by the DBMS for recovery purposes.

transaction log backup
A backup of only the transaction log operations that are not reflected in a previous backup copy of the database.

transaction manager (TM)
See *transaction processor (TP)*.

transaction processing
In electronic-commerce, this is the series of actions, changes, and/or functions among thousands of connected customers.

transaction processor (TP)
In a DDBMS, the software component on each computer that requests data. The TP is responsible for the execution and coordination of all database requests issued by a local application that accesses data on any DP. Also called *transaction manager (TM)* or *application processor (AP)*.

transaction transparency
A DDBMS property that ensures database transactions will maintain the distributed database's integrity and consistency, and that a transaction will be completed only when all database sites involved complete their part of the transaction.

transactional database
See *operational database*.

transitive dependency
A condition in which an attribute is dependent on another attribute that is not part of the primary key.

Transmission Control Protocol/ Internet Protocol (TCP/IP)
The official communications protocol of the Internet, a worldwide network of heterogeneous computer systems.

transparent
Indicating that the user is unaware of the system's operations.

Transport Layer Security (TLS)
An updated version of Secure Sockets Layer (SSL) that supports more secure encrypted communication between a client and server on the Internet.

traversal
A query in a graph database.

trigger
A procedural SQL code that is automatically invoked by the relational database management system when a data manipulation event occurs.

tuple
In the relational model, a table row.

twisted pair cable
Network cable formed by pairs of wires that are twisted inside a protective insulating cover; choice cabling for most network installations because it is easy to install and carries a low price tag. It resembles typical telephone cable.

two-phase commit protocol (2PC)
In a DDBMS, an algorithm used to ensure atomicity of transactions and database consistency as well as integrity in distributed transactions.

two-phase locking (2PL)
A set of rules that governs how transactions acquire and relinquish locks. Two-phase locking guarantees serializability, but it does not prevent deadlocks. The two-phase locking protocol is divided into two phases: (1) A *growing phase* occurs when the transaction acquires the locks it needs without unlocking any *existing* data locks. Once all locks have been acquired, the transaction is in its *locked* point. (2) A *shrinking phase* occurs when the transaction releases all locks and cannot obtain a new lock.

two-tier client/server system
A system within which a client requests services directly from the server. See also *client/server*.

U

unary relationship
An ER term used to describe an association *within* an entity. For example, an EMPLOYEE might manage another EMPLOYEE.

uncommitted data
A concurrency control problem in which a transaction accesses uncommitted data from another transaction.

unidirectional logical relationships
In a hierarchical database, relationships that are established by linking a logical child with a logical parent in a one-way arrangement.

Unified Modeling Language (UML)
A language based on object-oriented concepts that provides tools such as diagrams and symbols to graphically model a system.

Uniform Resource Locator (URL) or web address
Identifier for a resource on the Internet. Also called Web address.

UNION
In relational algebra, an operator used to merge (append) two tables into a new table, dropping the duplicate rows. The tables must be *union-compatible*.

union-compatible
Two or more tables that have the same number of columns and the corresponding columns have compatible domains.

unique fragment
In a DDBMS, a condition in which each row is unique, regardless of which fragment it is located in.

unique index
An index in which the index key can have only one associated pointer value (row).

uniqueness
In concurrency control, a property of time stamping that ensures no equal time stamp values can exist.

Universal Data Access (UDA)
Within the Microsoft application framework, a collection of technologies used to access any type of data source and to manage the data through a common interface.

unreplicated database
A distributed database in which each database fragment is stored at a single site.

unstructured data
Data that exists in its original, raw state; that is, in the format in which it was collected.

updatable view
A view that can update attributes in base tables that are used in the view.

UPDATE
A SQL command that allows attribute values to be changed in one or more rows of a table.

usability testing
Testing performed to evaluate the degree to which a user interface is considered friendly and easy to use.

user
In a system, a uniquely identifiable object that allows a given person or process to log on to the database.

UWA
In the Network Database Model, a specific area of memory that contains several fields used to access and inform regarding the status of the database. The UWA also contains space for each record type defined in the subschema.

V

value
The degree to which data can be analyzed to provide meaningful insights.

value chain
All activities required to design, plan, manufacture, market, sell, and support a product or service.

variability
The characteristic of Big Data for the same data values to vary in meaning over time.

variety
A characteristic of Big Data that describes the variations in the structure of data to be stored.

VBScript
A Microsoft client-side extension that extends a browser's functionality; VBScript is derived from Visual Basic.

velocity
A characteristic of Big Data that describes the speed at which data enters the system and must be processed.

veracity
The trustworthiness of a set of data.

verification
The process of refining a conceptual data model into a detailed design that is capable of supporting all required database transactions, and input and output requirements.

versioning
A property of an OODBMS that allows the database to keep track of the different transformations performed on an object.

vertical fragmentation
In distributed database design, the process that breaks a table into a subset of columns from the original table. Fragments must share a common primary key.

very large databases (VLDBs)
Database that contains huge amounts of data— gigabyte, terabyte, and petabyte ranges are not unusual.

view
A virtual table based on a SELECT query that is saved as an object in the database.

virtualization
A technique that creates logical representations of computing resources that are independent of the underlying physical computing resources.

virus
A malicious program that affects the normal operation of a computer system.

visualization
The ability to graphically present data in such a way as to make it understandable to users.

volume
A characteristic of Big Data that describes the quantity of data to be stored.

W

wait/die
A concurrency control scheme in which an older transaction must wait for the younger transaction to complete and release the locks before requesting the locks itself. Otherwise, the newer transaction dies and is rescheduled.

weak entity
An entity that displays existence dependence and inherits the primary key of its parent entity. For example, a DEPENDENT requires the existence of an EMPLOYEE.

weak relationship
A relationship in which the primary key of the related entity does not contain a primary key component of the parent entity.

web application server
A middleware application that expands the functionality of web servers by linking them to a wide range of services, such as databases, directory systems, and search engines.

web browser
The end-user application used to navigate the Internet; runs on a client computer and requests services from a Web server.

web caching
Performance enhancing technique in which a temporary storage area is created to provide Web pages at an optimal speed.

web development
The process of adding "business logic" to Web pages, thereby making them business-enabled.

web server
A specialized application whose only function is to "listen" for client requests, process them, and send the requested Web resource back to the client browser.

web-to-database middleware
A database server-side extension that retrieves data from databases and passes them to the web server, which in turn sends the data to the client's browser for display.

webpage
A text document on the World Wide Web containing text and special commands written in Hypertext Markup Language.

website
Refers to the Web server and the collection of Web pages stored on the local hard disk of the server computer.

WHERE
A SQL clause that adds conditional restrictions to a SELECT statement that limit the rows returned by the query.

wide area network (WAN)
Network type used to connect computer users across distant geographical areas; generally makes use of telephone or specialized communications companies.

wildcard character
A symbol that can be used as a general substitute for: (1) all columns in a table (*) when used in an attribute list of a SELECT statement or, (2) zero or more characters in a SQL LIKE clause condition (% and _).

wireless adapter
In the case of wireless networks, this adapter, sometimes called a wireless NIC, allows a computer to communicate using a wireless network.

wireless device support
Services that facilitate data integration with mobile communication devices, such as smart phones; allow users to conduct business, make hotel reservations, purchase groceries, and pay bills from virtually anywhere.

wireless LANs (WLANs)
Local area networks that are connected by wireless technology rather than wires.

workgroup database
A multiuser database that usually supports fewer than 50 users or is used for a specific department in an organization.

World Wide Web (WWW or the web)
Worldwide network collection of specially formatted and interconnected documents known as Web pages. Also called the Web.

wound/wait
A concurrency control scheme in which an older transaction can request the lock, preempt the younger transaction, and reschedule it. Otherwise, the newer transaction waits until the older transaction finishes.

write-ahead protocol
A protocol that ensures transaction logs are written to permanent storage before any database data is actually updated.

write-ahead-log protocol
In concurrency control, a process that ensures transaction logs are written to permanent storage before any database data is actually updated. Also called a write-ahead protocol.

write-through technique
In concurrency control, a process that ensures a database is immediately updated by operations during the transaction's execution, even before the transaction reaches its commit point. Also called *immediate update*.

X

XML database
A database system that stores and manages semistructured XML data.

XML schema
An advanced data definition language used to describe the elements, data types, relationship types, ranges, and default values of XML data documents. One of the main advantages of an XML schema is that it more closely maps to database terminology and features.

XML schema definition (XSD)
A file that contains the description of an XML document.

INDEX